A Textbook of Neuroanatomy

A Textbook of Neuroanatomy

Third Edition

Maria A. Patestas, PhD

Professor Emerita of Anatomy (retired)
Department of Anatomy
Des Moines University Medicine & Health Sciences
West Des Moines, Iowa

Amanda J. Meyer, PhD, FHEA

Senior Lecturer
Discipline of Anatomy and Pathology
College of Medicine and Dentistry
James Cook University
Townsville QLD
Australia

Leslie P. Gartner, PhD

Professor of Anatomy (retired)
Department of Biomedical Sciences
Baltimore College of Dental Surgery
Dental School
University of Maryland
Baltimore, Maryland

Copyright © 2025 by John Wiley & Sons, Inc. All rights reserved, including rights for text and data mining and training of artificial technologies or similar technologies.

Published by John Wiley & Sons, Inc., Hoboken, New Jersey.
Published simultaneously in Canada.

No part of this publication may be reproduced, stored in a retrieval system, or transmitted in any form or by any means, electronic, mechanical, photocopying, recording, scanning, or otherwise, except as permitted under Section 107 or 108 of the 1976 United States Copyright Act, without either the prior written permission of the Publisher, or authorization through payment of the appropriate per-copy fee to the Copyright Clearance Center, Inc., 222 Rosewood Drive, Danvers, MA 01923, (978) 750-8400, fax (978) 750-4470, or on the web at www.copyright.com. Requests to the Publisher for permission should be addressed to the Permissions Department, John Wiley & Sons, Inc., 111 River Street, Hoboken, NJ 07030, (201) 748-6011, fax (201) 748-6008, or online at http://www.wiley.com/go/permission.

The manufacturer's authorized representative according to the EU General Product Safety Regulation is Wiley-VCH GmbH, Boschstr. 12, 69469 Weinheim, Germany, e-mail: Product_Safety@wiley.com.

Trademarks: Wiley and the Wiley logo are trademarks or registered trademarks of John Wiley & Sons, Inc. and/or its affiliates in the United States and other countries and may not be used without written permission. All other trademarks are the property of their respective owners. John Wiley & Sons, Inc. is not associated with any product or vendor mentioned in this book.

Limit of Liability/Disclaimer of Warranty: While the publisher and author have used their best efforts in preparing this book, they make no representations or warranties with respect to the accuracy or completeness of the contents of this book and specifically disclaim any implied warranties of merchantability or fitness for a particular purpose. No warranty may be created or extended by sales representatives or written sales materials. The advice and strategies contained herein may not be suitable for your situation. You should consult with a professional where appropriate. Further, readers should be aware that websites listed in this work may have changed or disappeared between when this work was written and when it is read. Neither the publisher nor authors shall be liable for any loss of profit or any other commercial damages, including but not limited to special, incidental, consequential, or other damages.

For general information on our other products and services or for technical support, please contact our Customer Care Department within the United States at (800) 762-2974, outside the United States at (317) 572-3993 or fax (317) 572-4002.

Wiley also publishes its books in a variety of electronic formats. Some content that appears in print may not be available in electronic formats. For more information about Wiley products, visit our web site at www.wiley.com.

Library of Congress Cataloging-in-Publication Data:

Names: Patestas, Maria Antoniou, author. | Meyer, Amanda J., author. |
  Gartner, Leslie P., 1943- author.
Title: A textbook of neuroanatomy / Maria A. Patestas, Amanda J. Meyer,
  Leslie P. Gartner.
Description: Third edition. | Hoboken, N.J. : Wiley, [2025] | Includes
  bibliographical references and index.
Identifiers: LCCN 2024047936 (print) | LCCN 2024047937 (ebook) | ISBN
  9781394237067 (paperback) | ISBN 9781394237074 (adobe pdf) | ISBN
  9781394237081 (epub)
Subjects: MESH: Nervous System—anatomy & histology
Classification: LCC QP360 (print) | LCC QP360 (ebook) | NLM WL 101 | DDC
  612.8—dc23/eng/20250108
LC record available at https://lccn.loc.gov/2024047936
LC ebook record available at https://lccn.loc.gov/2024047937

Cover Design: Wiley
Cover Image: © Alones/Shutterstock

Set in 9/11.5pt Palatino LT Std by Straive, Chennai, India

SKY10100491_031825

Dedication

In loving memory of my father, Antonios;
and to my mother, Garifalia, and
my sister, Oursikía
MAP

To my husband, Nigel,
my mother, Pamela, and
my sons Ayden and Oliver
AJM

To my wife, Roseann,
my daughter, Jen, and
in loving memory of my parents
LPG

XXV
The brain within its groove
Runs evenly and true;
But let a splinter swerve,
'T were easier for you
To put the water back
When floods have slit the hills,
And scooped a turnpike for themselves,
And blotted out the mills!

Emily Dickinson

Contents

Preface	xi
Acknowledgments	xiii
About the Companion Website	xv

Part I General Principles of the Nervous System   1

Chapter 1 Introduction to the Nervous System   3
Cells of the Central Nervous System	5
Central Nervous System	6
Peripheral Nervous System	9
Questions to Ponder	10

Chapter 2 Development of the Nervous System   11
Clinical Case	11
Early Development	12
Neurulation	14
Early Development of the Spinal Cord and Brain	19
Development of the Spinal Cord	21
Development of the Brain	23
Clinical Considerations	30
Synonyms and Eponyms of the Nervous System	31
Follow-up to Clinical Case	31
Questions to Ponder	31

Chapter 3 Histophysiology of the Nervous System   32
Clinical Case	32
Neurons	33
Neuroglia	40
Generation and Conduction of Nerve Impulses	44
Clinical Considerations	47
Synonyms and Eponyms of Nervous System Histophysiology	48
Follow-up to Clinical Case	48
Questions to Ponder	48

Chapter 4 Neurotransmitter Substances   49
Clinical Case	49
Classification of Neurotransmitter Substances	52
Clinical Considerations	58
Follow-up to Clinical Case	59
Questions to Ponder	59

Chapter 5 Spinal Cord   60
Clinical Case	60
Morphology of the Spinal Cord	61
Internal Morphology of the Spinal Cord	67
Vascular Supply of the Spinal Cord	70
Clinical Considerations	73

Synonyms and Eponyms of the Spinal Cord	74
Follow-up to Clinical Case	74
Questions to Ponder	74

Chapter 6 Gross Anatomy of the Brain   75
Clinical Case	75
Cerebrum	76
Diencephalon	84
Cerebellum	85
Brainstem	87
Clinical Considerations	88
Synonyms and Eponyms of the Brain	90
Follow-up to Clinical Case	90
Questions to Ponder	90

Chapter 7 Brainstem   91
Clinical Case	91
Internal Organization of the Brainstem	91
Medulla Oblongata	96
Pons	102
Midbrain	107
Clinical Considerations	111
Synonyms and Eponyms of the Brainstem	116
Follow-up to Clinical Case	116
Questions to Ponder	116

Chapter 8 Meninges and Cerebrospinal Fluid   117
Clinical Case	117
Cranial Meninges	118
Spinal Meninges	126
Venous Sinuses of the Cranial Dura Mater	127
Cerebrospinal Fluid	130
Ventricles of the Brain	130
Clinical Considerations	131
Synonyms and Eponyms of the Cranial Meninges	132
Follow-up to Clinical Case	132
Questions to Ponder	132

Chapter 9 Vascular Supply of the Central Nervous System   133
Clinical Case	133
Vascular Supply of the Spinal Cord	134
Arterial Supply of the Brain	135
Venous Drainage of the Brain	149
Clinical Considerations	152
Synonyms and Eponyms of the Vascular Supply of the Central Nervous System	153
Follow-up to Clinical Case	153
Questions to Ponder	153

VIII ● ● ● **CONTENTS**

Chapter 10 Autonomic Nervous System 154
Clinical Case 154
Sympathetic Nervous System 157
Parasympathetic Nervous System 164
Enteric Nervous System 167
Neurotransmitters and Receptors of the Autonomic
  Nervous System 168
Pelvic Autonomic Functions 169
Clinical Considerations 171
Synonyms and Eponyms of the Autonomic
  Nervous System 172
Follow-up to Clinical Case 172
Questions to Ponder 172

Chapter 11 Spinal Reflexes 173
Clinical Case 173
Components of Reflexes 173
Lower Motor Neurons 174
Skeletal Muscle Innervation 175
Skeletal Muscle Receptors 176
Muscle Stretch Reflex 176
Reciprocal Inhibition 177
Autogenic Inhibition (Inverse Myotatic Reflex) 178
Flexor Reflex (Withdrawal Reflex, Nociceptive
  Reflex) 178
Crossed Extension Reflex 179
Maintenance of Muscle tone Via the Gamma Loop 180
Alpha–Gamma Coactivation 180
Synonyms and Eponyms of the Spinal Reflexes 182
Follow-up to Clinical Case 182
Questions to Ponder 182

**Part II Integrative Components of the
Nervous System** 183

Chapter 12 Ascending Sensory Pathways 185
Clinical Case 185
Sensory Receptors 187
Anterolateral System 197
Tactile Sensation and Proprioception 207
Sensory Pathways to the Cerebellum 214
Clinical Considerations 218
Modulation of Nociception 226
Neuroplasticity 229
Synonyms and Eponyms of the Ascending Sensory
  Pathways 229
Follow-up to Clinical Case 230
Questions to Ponder 230

**Chapter 13 Motor Cortex and Descending Motor
  Pathways** 231
Clinical Case 231
Cortical Areas Controlling Motor Activity 232
Descending Motor Pathways 235
Clinical Considerations 249
Synonyms and Eponyms of the Motor Cortex
  and Descending Motor Pathways 253
Follow-up to Clinical Case 254

Questions to Ponder 254

Chapter 14 Basal Nuclei 255
Clinical Case 255
Components of the Basal Nuclei 256
Nuclei Associated with the Basal Nuclei 260
Input, Intrinsic, and Output Nuclei of the Basal Nuclei 261
Connections of the Basal Nuclei 263
Circuits Connecting the Basal Nuclei, Thalamus,
  and Cerebral Cortex 270
Other Circuits of the Basal Nuclei 273
Neurotransmitters of the Basal Nuclei 274
"Direct," "Indirect," and "Hyperdirect" Loops
  (Pathways) of the Basal Nuclei 275
Circuits that Modulate Activity of the Basal Nuclei 277
Clinical Considerations 278
Synonyms and Eponyms of the Basal Nuclei 283
Follow-up to Clinical Case 284
Questions to Ponder 284

Chapter 15 Cerebellum 285
Clinical Case 285
Morphology of the Cerebellum 287
Cerebellar Peduncles 296
Deep Cerebellar Nuclei 298
Afferents (Input) to the Cerebellum 301
Efferents (Output) from the Cerebellum 304
Functional Organization of the Cerebellum:
  Intrinsic Circuitry 305
Clinical Considerations 307
Synonyms and Eponyms of the Cerebellum 309
Follow-up to Clinical Case 309
Questions to Ponder 309

Chapter 16 Reticular Formation 310
Clinical Case 310
Morphology of the Reticular Formation 311
Zones of the Reticular Formation 312
Nuclei Associated with the Reticular Formation 314
Input to and Output from the Reticular Formation 314
Functions of the Reticular Formation 315
Clinical Considerations 321
Synonyms and Eponyms of the Reticular Formation 322
Follow-up to Clinical Case 322
Questions to Ponder 322

Chapter 17 Cranial Nerves 323
Clinical Case 323
Olfactory Nerve (CN I) 329
Optic Nerve (CN II) 329
Oculomotor Nerve (CN III) 329
Trochlear Nerve (CN IV) 331
Clinical Considerations 331
Trigeminal Nerve (CN V) 334
Clinical Considerations 340
Abducens Nerve (CN VI) 341
Clinical Considerations 342
Facial Nerve (CN VII) 346

CONTENTS **•••** IX

Vestibulocochlear Nerve (CN VIII)	348
Glossopharyngeal Nerve (CN IX)	349
Clinical Considerations	349
Clinical Considerations	353
Vagus Nerve (CN X)	353
Clinical Considerations	354
Accessory Nerve (CN XI)	354
Clinical Considerations	357
Hypoglossal Nerve (CN XII)	357
Clinical Considerations	358
Synonyms and Eponyms of the Cranial Nerves	359
Follow-up to Clinical Case	360
Questions to Ponder	360

Chapter 18 Visual System 361
Clinical Case	361
Eyeball	361
Central Visual Pathways	367
Visual Reflexes	377
Clinical Considerations	384
Synonyms and Eponyms of the Visual System	387
Follow-up to Clinical Case	387
Questions to Ponder	387

Chapter 19 Auditory System 388
Clinical Case	388
Ear	389
Auditory Transmission	393
Central Auditory Pathways	394
Clinical Considerations	403
Synonyms and Eponyms of the Auditory System	404
Follow-up to Clinical Case	404
Questions to Ponder	404

Chapter 20 Vestibular System 405
Clinical Case	405
Vestibular Apparatus	406
Vestibular Nerve (CN VIII)	411
Central Pathways of the Vestibular System	413
Control of Ocular Movements	417
Vestibular Nystagmus	421
Caloric Nystagmus	421
Synonyms and Eponyms of the Vestibular System	423
Follow-up to Clinical Case	423
Questions to Ponder	423

Chapter 21 Olfactory System 424
Clinical Case	424
Olfactory Receptor Neurons	425
Olfactory Transduction	427
Olfactory Nerve (CN I)	428
Central Connections of the Olfactory System	429
Blood Supply and Drainage	431
Clinical Considerations	432
Synonyms and Eponyms of the Olfactory System	433
Follow-up to Clinical Case	433
Questions to Ponder	433

Chapter 22 Limbic System 434
Clinical Case	434
Limbic Lobe	435
Brainstem Centers Associated with Limbic System Function	446
Pathways of the Limbic System	446
Limbic Association Cortex	449
Limbic System Input to the Endocrine, Autonomic, and Somatic Motor Systems	449
Clinical Considerations	450
Synonyms and Eponyms of the Limbic System	451
Follow-up to Clinical Case	451
Questions to Ponder	452

Chapter 23 Hypothalamus 453
Clinical Case	453
Borders	454
Hypothalamic Zones and Component Nuclei	455
Hypothalamic Regions (Areas) and Component Nuclei	457
Connections of the Hypothalamus	462
Pathways of the Hypothalamus	463
Functions of the Hypothalamus	466
Hypothalamohypophyseal Connections	469
Clinical Considerations	475
Synonyms and Eponyms of the Hypothalamus	477
Follow-up to Clinical Case	478
Questions to Ponder	478

Chapter 24 Thalamus 480
Clinical Case	480
Borders	480
Anatomy	482
Internal and External Medullary Laminae	483
Thalamic Nuclei	484
Clinical Considerations	491
Synonyms and Eponyms of the Thalamus	492
Follow-up to Clinical Case	492
Questions to Ponder	493

Chapter 25 Cerebral Cortex 494
Clinical Case	494
Cells of the Cerebral Cortex	496
Types of Cortex	498
Cell Layers of the Neocortex	499
Vertical Columnar Organization of the Cerebral Cortex	500
Afferents (Input) to the Cerebral Cortex	500
Efferents (Output) from the Cerebral Cortex	501
Internal Capsule and Corona Radiata	504
Lobes of the Cerebral Cortex	504
Functional Organization of the Cerebral Cortex	505
Cerebral Dominance	512
Clinical Considerations	515
Synonyms and Eponyms of the Cerebral Cortex	519
Follow-up to Clinical Case	520
Questions to Ponder	521

X ••• CONTENTS

Chapter 26 Evolution of the Human Brain	**522**	Questions to ponder: answers to odd questions	524
Evolutionary Biology Fundamentals	522	Questions to ponder: answers to even questions	533
Early Hominins	522		
Non-Human Primates	522	Answers to clinical case margin questions	541
Modern Human Brains	523	Index	545
Prenatal and Postnatal Development	523		

Preface to the Third Edition

For *A Textbook of Neuroanatomy* by Maria A. Patestas, Amanda J. Meyer, and Leslie P. Gartner.

Unlike most prefaces, this preface to the third edition of *A Textbook of Neuroanatomy* is presented in two parts because we (MAP and LPG) wish to announce how pleased we are to be able to present our new coauthor, Dr. Amanda J. Meyer, who was kind enough to accept our invitation to join us in this revision of the second edition. Dr. Meyer is a distinguished faculty member of the College of Medicine and Dentistry, James Cook University in Townsville, Queensland, Australia. Her many additions to the current edition are invaluable, and we thank her for all the improvements that she made.

Maria A. Patestas
Leslie P. Gartner

It is always gratifying to publish the third edition of a textbook because it is an indication that our efforts were well received, that our colleagues endorsed our work, and that students have been well served by it. This new edition of *A Textbook of Neuroanatomy*, as indeed the first and second editions, was written with students in mind in full knowledge of the trepidation with which they face the prospect of learning a subject matter that appears to be not just daunting, but actually intimidating. The material, in part, appears overwhelming because of the many eponyms that have been retained, which add to the inherent complexity of studying the nervous system. Happily, for the students, the reports of being an overwhelming subject is an unjust reputation for neuroanatomy; although challenging, it is truly a fascinating topic. Its comprehension does not require amazing brilliance, but merely a willingness to learn new words, an aptitude for reading maps, and the ability to follow pathways from one place to another. In order to help the student achieve the goal of learning neuroanatomy, we made this textbook concise, yet complete and easy to read. We included many illustrations and schematic diagrams that clarify and elucidate the concepts being discussed. Presenting the exquisite structure and function of the nervous system in a clear and relevant context makes the learning experience simpler, more enjoyable, and quite memorable.

Every chapter of this edition was revised to include current material and to make it even easier to read than the previous two editions. The terminology has been updated based on the new international neuroanatomical terminology derived from the Terminologia Neuroanatomica (TNA) for the human central and peripheral nervous systems, eliminating *repetitions* of many of the older eponyms, but identifying them at least once in the text.

One new chapter (Chapter 26) was added at the end of the book on the Evolution of the Human Brain, which provides a general and broad overview of how the brain evolved over millions of years.

We continued to highlight the interrelationships between neural systems/structures and the rest of the body as we described the various regions of the brain. We firmly believe that a broad understanding of neuroanatomy, supported with a basic understanding of its physiology, is critical so that students, instead of merely memorizing structures, learn the principles thus establishing the foundation for future studies in the health sciences and for the logical forces that help drive it.

As with the first two editions, this textbook is divided into two sections. The first 11 chapters provide an overview of neuroanatomy that introduces terms and should be viewed as the vocabulary lessons that are necessary evils in the mastering of a new language. For no matter how well one understands the grammar of a foreign language, it is the possession of a rich vocabulary that permits one to communicate with speakers of that tongue. The second part of this textbook, Chapters 12–26, utilizes the vocabulary of the first part to detail information concerning the various pathways and discrete systems that act in concert to perform the myriad functions of the human nervous system. Pieces of the puzzle are assembled and fit together to create a composite design of the whole brain.

We retained in this edition the well-liked special features, which include:

- Chapter opening outlines that provide a quick overview of the chapter content and organizational logic;
- Clinical cases in the beginning of each chapter, setting the stage for the clinical relevance of that chapter's context;
- Key points in the chapter that are highlighted in the text;
- Clinical case questions that emphasize the relevance of the chapter opening case at key points in the chapter – the answers appear in the related website;
- Summary tables within each chapter that function as study guides to assist students in memorization that leads to learning;
- Clinical considerations that underscore the medical conditions relevant to the topics;
- Synonym/eponym tables that help organize the many possible terms for each vocabulary word;
- A follow-up to each clinical case at the end of each chapter that discusses the opening case, helping to correlate the text and its medical application;

XII ● ● ●   **PREFACE TO THE THIRD EDITION**

- Questions to ponder at the end of each chapter to reinforce the relevance of the material, with the answers to even-numbered and odd-numbered questions appearing at the end of the hard copy of the book;
- An accompanying website that includes all the illustrations, the odd and even numbered answers for questions to ponder, and all the answers for clinical case questions. It also features animations of key processes and links to useful sources. The site can be found at www.wiley.com. go/patestas/neuroanatomy3e.

Acknowledgments

We would like to thank the many individuals who helped us bring this project to fruition, including our editors at Wiley Blackwell Publishing: Julia Squarr, our original Editor, who signed and oversaw the beginning of the publishing process; Frank Weinreich, our new Publisher, who took over where Julia left off; our Managing Editor, Neena Ganjoo, who scrutinized the fine detail and helped us proceed through the intricacies of manuscript submission; Our artist, Todd Smith, worked closely with us to ensure a first-rate art program, and Dr Scott Thompson, who wrote the clinical cases that begin and close each chapter.

Additionally, we would like to thank the anonymous (to the three authors) reviewers who commented on the manuscript prior to its development. Their help was instrumental in the crafting of this book. And finally, we would like to thank our respective families for all the time we took away from them in the writing of this textbook.

While we appreciate all of the assistance that we received from our editors and colleagues, the responsibility for errors, omissions, and shortcomings is ours. In view of that fact, we welcome criticisms and suggestions for improvement of this text.

Maria A. Patestas
Amanda J. Meyer
Leslie P. Gartner

About the Companion Website

This book is accompanied by a companion website:

www.wiley.com.go/patestas/neuroanatomy3e

The website includes:

- Answers to the odd-numbered questions at the end of each chapter
- Answers to the even-numbered questions at the end of each chapter
- Answers for the clinical case questions
- Animations of key processes
- Powerpoints of all figures from the book for downloading

PART I

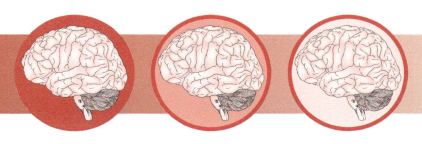

General Principles of the Nervous System

CHAPTER 1

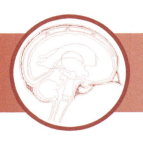

Introduction to the Nervous System

CELLS OF THE CENTRAL NERVOUS SYSTEM

CENTRAL NERVOUS SYSTEM

PERIPHERAL NERVOUS SYSTEM

QUESTIONS TO PONDER

The human nervous system is an extremely efficient, compact, fast, and reliable computing system that performs at an incredible capacity. In fact, at this point it has the capability of performing tasks that are far beyond the abilities of any artificial intelligence as yet devised. The present textbook discusses not only the anatomy of the central nervous system but also its function, making it a textbook dealing with the subject of **neurobiology**. In case the reader wonders why one should study this subject matter, it should be recalled that it is our central nervous system more than anything else about us that makes us what we are, human beings.

The nervous system is subdivided, *morphologically,* into two compartments, the **central nervous system** (**CNS**) and the **peripheral nervous system** (**PNS**). The CNS is composed of the brain and spinal cord, whereas the PNS, as a physical extension of the CNS, is composed of ganglia as well as cranial and spinal nerve fibers. *Functionally,* the nervous system is also subdivided into two components, the **somatic nervous system**, which is under the individual's conscious control, and the **autonomic nervous system**, which controls the myriad of unconscious activities in conjunction with the voluntary nervous system. The autonomic nervous system is a tripartite organization, in that it has a **sympathetic**, a **parasympathetic**, and an **enteric** component. Simply stated, the first initiates the "flight or fight or freeze or fornicate" response, the second is concerned with the body's vegetative activities, whereas the enteric nervous system is involved in regulating the processes of digestion. It must be understood, however, that the interplay of these three systems maintains **homeostasis**. The autonomic nervous system acts upon three cell types to perform its functions, namely cells of glands, smooth muscle cells, and cardiac muscle cells. The nervous system has two other functional components, **sensory** and **motor**. The sensory component collects information and transmits it to the CNS (and is therefore called **afferent**), where the information is sorted, analyzed, and processed. Generally speaking, the motor component delivers the results of the analyses away from the CNS (and is therefore called **efferent**) to the effector organs, that is, muscles (all three types: skeletal, smooth, and cardiac) and glands, resulting in a response to the stimulus.

Discussion of the topics of neuroanatomy requires that the student be familiar with some of the specialized terminology

---

A Textbook of Neuroanatomy, Third Edition. Maria A. Patestas, Amanda J. Meyer, and Leslie P. Gartner.
© 2025 John Wiley & Sons, Inc. Published 2025 by John Wiley & Sons, Inc.
Companion website: www.wiley.com.go/patestas/neuroanatomy3e

4 ● ● ● **CHAPTER 1**

Terms	Definition
Arcuate	Arc-like, resembles a segment of a circle
Bilateral	On both sides
Bipolar neuron	A neuron with a single axon and a single dendrite; usually arising on opposite sides of the soma (e.g., neurons in the olfactory mucosa)
Brainstem	Originally this term referred to the entire brain with the exception of the telencephalon; currently, most neuroanatomists refer to the medulla oblongata, pons, and midbrain as the brainstem
Inferior	Toward the tail (proceeding to a lower position; the opposite of cranial)
Column	A large bundle (funiculus) of ascending or descending nerve fibers in the CNS, composed of several different fasciculi (e.g., posterior column of the white matter of the spinal cord)
Commissure	A bundle of nerve fibers that run horizontally, connecting the right and left sides of the central nervous system (e.g., anterior commissure)
Contralateral	The opposite side (e.g., in many instances the right side of the brain receives information from and controls the left side, *contralateral*, of the body)
Cortex	The periphery of a structure; the superficial layer, the opposite of the deep medulla
Decussation	The level in the central nervous system where paired fiber tracts cross from one side of the body to the other (e.g., pyramidal decussation)
Exteroceptor	A sensory receptor that provides information to the central nervous system concerning the external environment
Fasciculus	A bundle (tract) of ascending or descending nerve fibers within the central nervous system (e.g., cuneate fasciculus)
Fiber	A long, thin structure; refers to an axon or a collection of axons
Fovea	A depression or a pit (e.g., fovea centralis of the retina)
Funiculus	A large bundle (column) of ascending or descending nerve fibers in the CNS, composed of several different fasciculi (e.g., posterior funiculus of the white matter of the spinal cord)
Ganglion	A collection of nerve cell bodies in the peripheral nervous system (e.g., spinal ganglion [previously named "dorsal root ganglion"]; although it is used occasionally in reference to a collection of nerve cell bodies in the central nervous system, e.g., basal ganglia – not used in this textbook)
Glomerulus	Structures with a spherical configuration (e.g., synaptic glomeruli in the olfactory bulb)
Infundibulum	A funnel-like structure (e.g., infundibulum of the hypophysis)
Interoceptor	A sensory receptor that provides information to the central nervous system concerning the body's internal environment
Ipsilateral	The same side (e.g., in some instances the right side of the cerebellum receives information from and controls the right side, *ipsilateral*, of the body)
Lamina	A layer of a specific material such as the layering of nerve cell bodies in the spinal cord
Multipolar neuron	A neuron with a single axon and multiple dendrites (e.g., motor neuron)
Myelin	A fatty substance that surrounds certain axons; composed of spiral layers of the cell membranes of neurolemmocytes (eponym: Schwann cells) in the peripheral nervous system and of oligodendroglia in the central nervous system
Neurite	A collective term for axons and dendrites
Neuropil	A complex of axons, dendrites, and processes of neuroglia that form a web-like network in which nerve cell bodies of the gray matter are embedded
Nucleus	Core; in a cell it is the region of the cell that houses the chromosomes; in the central nervous system it is a collection of nerve cell bodies
Operculum	A cover or lid (e.g., parietal operculum of the cerebrum that overhangs and partially masks the insula)
Peduncle	A massive collection of nerve fiber bundles that connect the cerebrum and the cerebellum to the brainstem
Perikaryon	The cell body of a neuron (i.e., a neuron without its dendrites and axon); also referred to as soma
Plexus	The interwoven arrangements of nerve fibers that serve a specific region (e.g., brachial plexus, L. "braid")
Project	When one group of nerve cells relay their information to a second group of nerve cells, it is said that the first group "projects" to the second group (e.g., the hippocampal formation projects to the hypothalamus)
Proprioceptor	Sensory nerve endings in muscles, joints, and tendons that inform the central nervous system concerning the position and movements of the regions of the body in space (e.g., muscle spindle)
Pseudounipolar neurons	See unipolar neuron
Raphe	A seam or midline structure (e.g., raphe nuclei of the reticular formation)
Rostral	Toward the nose (proceeding toward a anterosuperior position; the opposite of inferior or caudal)
Tract	A bundle (fasciculus) of ascending or descending nerve fibers within the central nervous system (e.g., solitary tract)
Unipolar neuron	A neuron with an axon only, no dendrites. These are very rare in vertebrates. In fact, the term is being used to identify a special case of bipolar neurons (known as pseudounipolar neurons) whose axon and dendrite fuse with each other and form a single short process that bifurcates forming a central process (that enters the CNS) and a peripheral process that goes to a sensory receptor in the body

Table 1.1 ● Common terms in neuroanatomy.

of the subject matter. One of the problems that students have in studying neuroanatomy is that there may be numerous terms applied to the same or similar structures. It is important, therefore, to begin the discussion of this subject matter by listing and defining in Table 1.1 some, but not all, of the terminology the student will encounter.

CELLS OF THE CENTRAL NERVOUS SYSTEM

Neurons

Neurons are the functional units of the central nervous system

Neurons, the functional units of the nervous system, although of several types (unipolar [pseudounipolar], bipolar, and multipolar, detailed in Chapter 3), all have similar structures and functions. Their most important property is that they are capable of receiving, conducting, and transmitting electrochemical information to each other as well as to muscle cells and acinar cells of glands. Neurons (Fig. 1.1) have one or more processes, collectively known as **neurites**, which are of two types, one or more **dendrites** through which they receive information and a single process, known as an **axon**, through which they transmit information to other neurons and to effector organs. Hence, dendrites conduct information toward the **cell body**, whereas axons conduct information away from the cell body. Neurons usually communicate with each other as well as with other cells at **synapses** of which there are three types. If the axon of the first (presynaptic) neuron synapses with the dendrite of the second (postsynaptic) neuron, it is called an axodendritic synapse which is the most common. Synapses can also be axoaxonic (axon to axon) or axosomatic (presynaptic axon to postsynaptic cell body). **Neurotransmitter substances** are released from the axon terminal of the first neuron into the synaptic cleft and bind to receptor molecules on the surface of the second neuron (or muscle/myoepithelial cell/acinar cell). Neurotransmitter substances are recycled so that they may be used numerous times without having to resynthesize them. Neurons may also communicate with each other via **gap junctions** which are exceptionally small aqueous pores in the membranes of the soma and/or neurites of each of two adjacent cells. These pores are aligned with each other in such a way that they permit the movement of small secondary messenger molecules from the cytoplasm of one cell (the **signaling cell**) into the cytoplasm of the neighboring cell (the **target cell**), initiating a requisite response in the target cell.

Neuroglia

Neuroglia constitute several categories of non-neuronal cells, namely microglia, macroglia, and ependymal cells

Additional cells, known as **neuroglia**, constitute several categories of non-neuronal supporting cells. Those in the central nervous system are known as **ependymal cells**, **macroglia**, and **microglia**. The first two are derived from cells of the neural tube, whereas microglia are **macrophages** whose origins are monocyte precursors of the bone marrow.

Ependymal cells form a simple ciliated cuboidal epithelium that lines the central canal of the spinal cord and the ventricles of the brain. Additionally, these cells also participate in the formation of the choroid plexus, vascular tufts of tissue that manufacture cerebrospinal fluid (Fig. 1.2). **Macroglia** is a collective term for the protoplasmic astrocytes, fibrous astrocytes, and oligodendroglia. **Protoplasmic astrocytes** support neurons in the gray matter, form a subpial barrier, and envelop capillaries of the CNS (Fig. 1.3). Due to their

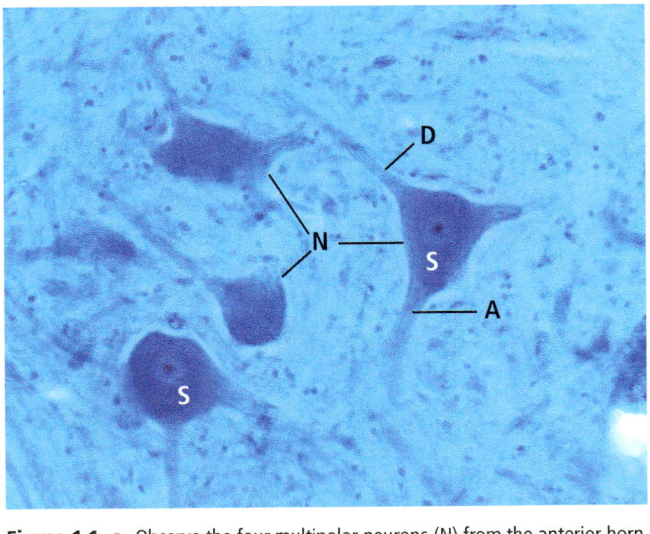

Figure 1.1 ● Observe the four multipolar neurons (N) from the anterior horn of the spinal cord. Note that each neuron has a cell body, also known as soma (S), and a number of processes, a single axon (A), and several dendrites (D). Collectively, axons and dendrites are known as neurites. Observe that the two neurons labeled with S display their large nucleus with prominent nucleoli. (Methylene blue, ×270).

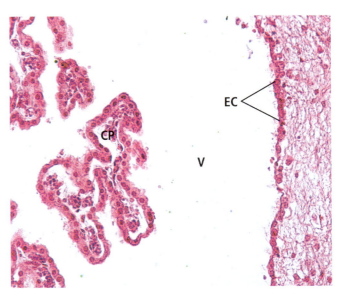

Figure 1.2 ● Note that ciliated cuboidal ependymal cells (EC) line the ventricle (V) of the brain. The structures on the left-hand side of the image are portions of choroid plexus (CP) which is responsible for manufacturing the cerebrospinal fluid. (Hematoxylin and Eosin; ×270).

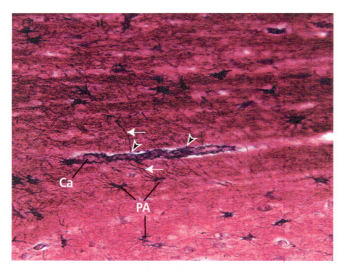

Figure 1.3 ● Note that the capillary (Ca), running horizontally through the middle of the image, is surrounded by a large number of protoplasmic astrocytes (PA) whose processes (arrows) extend to the blood vessel where they expand to form vascular feet (arrowheads), also known as pedicels. (Cerebral cortex, Cajal gold sublimate; ×270).

close relationship with blood vessels of the brain, they are responsible for the ability of endothelial cells of blood vessels to maintain the blood—brain barrier. **Fibrous astrocytes** are located in the white matter and appear to function in a similar fashion to protoplasmic astrocytes. Astrocytes also function in scavenging ions and neurotransmitter substances from the extracellular spaces. They also assist neurons in their role of transmitting impulses from one neuron to the other by regulating the presence of neurotransmitter substances as well as releasing **gliotransmitter substances** (e.g., adenosine triphosphate [ATP] and glutamate that appear to restrain synaptic transmission) in the vicinity of synapses.

Oligodendroglia form myelin sheaths around axons and also surround dendrites and cell bodies of neurons in the CNS. A single oligodendroglion can myelinate a single segment of several axons. **Neurolemmocytes (eponym: Schwann cells)** are located in the PNS, and they function in forming myelin around axons of the PNS. Unlike oligodendroglia, a single neurolemmocyte can myelinate a segment of only a single axon. It should be noted that the degree of myelination depends on the type of nerve fiber that is being myelinated, and the thicker the myelin sheath, the faster the axon can propagate the impulse.

CENTRAL NERVOUS SYSTEM

The central nervous system is composed of the large, superiorly situated brain and smaller, cylindrically shaped, inferiorly positioned spinal cord.

The **central nervous system** begins as a complex, hollow tube, whose superior end, the **brain**, is enlarged and folded in an elaborate manner, whereas its inferior end, the **spinal cord**, is a long, tubular structure (Fig. 1.4). The brain is housed in the cranial cavity and, at the foramen magnum, is continuous with the spinal cord, which is housed in the vertebral canal. The **posterior** surface of the spinal cord is closer to the lamina of the vertebral arch, whereas its **anterior** surface is closer to the bodies of the vertebrae. Since the CNS, as well as most of the body, is bilaterally symmetrical, the **sagittal** (midsagittal, according to some) plane bisects it into right and left halves. Positioning toward the sagittal plane is considered to be the **medial** direction, and away from the sagittal plane is the **lateral** direction.

Brain

The brain is subdivided into five regions: the telencephalon, diencephalon, mesencephalon, metencephalon, and myelencephalon

The **brain** is subdivided into five major regions, the largest being the **telencephalon**, which is composed of the cerebral hemispheres. The other four regions are: the **diencephalon**, whose component parts are the epithalamus, thalamus, hypothalamus, and subthalamus; the **mesencephalon**, consisting of the cerebral peduncles (tegmentum and crus cerebri) and the tectum (superior and inferior colliculi); the **metencephalon**, including the pons and cerebellum; and the **myelencephalon** (medulla oblongata). Frequently the medulla oblongata, mesencephalon, and the pons are collectively termed the **brainstem**. The lumen of the CNS is a narrow slit, which is known as the **central canal** in the spinal cord. However, the lumen is expanded into a system of dilated spaces, called **ventricles** in the brain, and these ventricles are filled with cerebrospinal fluid. Nine of the 12 pairs of **Cranial Nerves (CN)** emerge from the brain to supply motor, sensory, parasympathetic innervation, and proprioceptive information for the head and neck and much of the viscera of the body. Of the other three cranial nerves, one, the **optic nerve (CN II)**, arises from the retina; one, the **olfactory nerve (CN I)**, originates from the superior aspect of the nasal cavity; and one, the **accessory nerve (CN XI)**, ascends from the superior cervical spinal cord.

Spinal cord

The spinal cord is a cylindrical structure whose neurons are arranged in such a fashion that the motor functions are anteriorly positioned and the sensory functions are posteriorly positioned

The **spinal cord** (Fig. 1.5) is concentric more or less cylindrical aggregate of nervous tissue, where a cylinder of white matter (neuronal cell bodies, myelinated axons and glia) surrounds a central cylinder of gray matter (neuronal cell bodies, myelinated axons and glia). Nerves emerging from the spinal cord supply motor, sensory, and autonomic (sympathetic, parasympathetic) fibers to the entire body below the head. The cell bodies of neurons of the spinal cord

INTRODUCTION TO THE NERVOUS SYSTEM ● ● ● 7

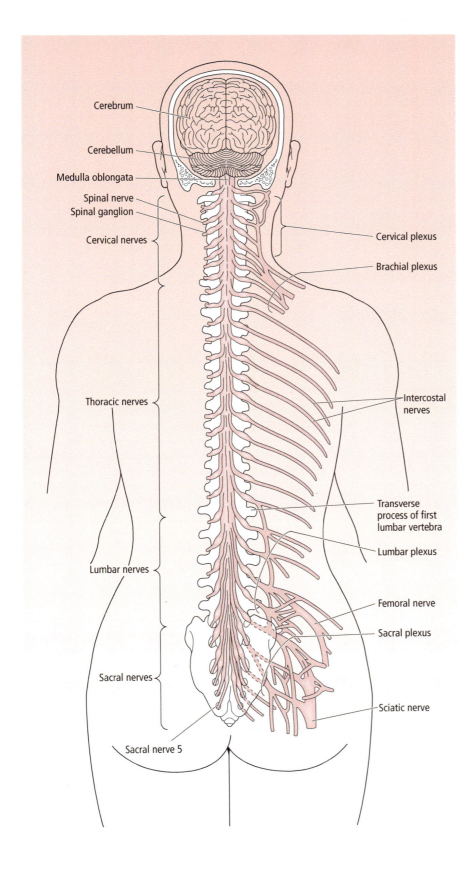

Figure 1.4 ● The brain, spinal cord, spinal nerves, and major somatic plexuses. Note that the back of the skull as well as the spinal processes of the vertebrae have been removed and that the dura mater and the arachnoid mater have been opened up so that the spinal cord may be viewed in its entire length.

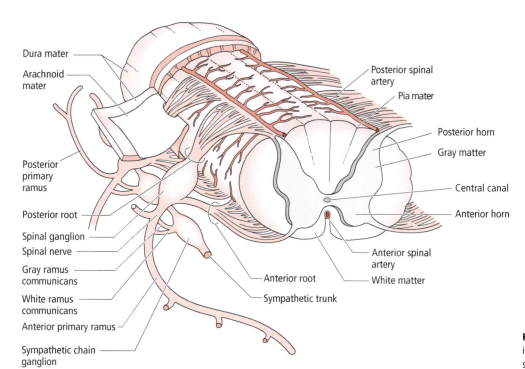

Figure 1.5 ● The spinal cord, its meninges, spinal nerves, and sympathetic chain ganglia.

are arranged in such a fashion that those concerned with somatic motor function are located in the **anterior horn**, and their axons leave via the anterior rootlets. These are accompanied by axons of the preganglionic sympathetic neurons, located in the **lateral horn** of the spinal cord gray matter in all the thoracic and upper two or three lumbar segments, and axons of preganglionic parasympathetic neurons located in the lateral horn of gray matter between the second and fourth sacral spinal cord levels. The **posterior horn** of the spinal cord is the location where central processes of unipolar neurons of posterior root ganglia enter the spinal cord, via posterior rootlets, bringing sensory information to the CNS. **Interneurons** connect two types of neurons to each other (e.g., unipolar sensory neurons of the posterior root ganglia to motor neurons of the anterior horn). Thus, interneurons have the capability of modulating (facilitating or inhibiting) a motor response to a sensory stimulus. For example, if one pricks one's finger the reflex response is to pull the finger away from the offending stimulus; however, if a health professional sticks one's finger for a blood test, the interneuron, receiving that information from higher levels in the central nervous system, inhibits the withdrawal of the finger.

The white matter of the spinal cord is composed of ascending and descending tracts of nerve fibers that connect regions of the CNS to each other. Anterior and posterior roots at each level of the spinal cord join one another to form the spinal nerves that leave the spinal cord at regular intervals, a condition that is indicative of its segmentation. Each posterior root displays a swelling, a spinal ganglion, housing the soma (cell bodies) of the unipolar (pseudounipolar) neurons.

Gray matter and white matter

Gray matter is composed of neuron cell bodies, clusters of which within the CNS are known as nuclei; white matter is recognized by the presence of myelinated axons

Gray matter is composed of neuronal cell bodies and their associated ganglia and can be arranged in layers (e.g., cortical gray matter) or in aggregations (e.g., subcortical nuclei). Neuronal cell bodies can be located in the CNS (brain and spinal cord gray matter) or in ganglia outside the CNS (e.g., spinal ganglia). There are two major categories of neurons, (1) those whose axons leave the CNS and (2) those whose axons remain within the CNS. The first group, called **principal cells** by some neuroanatomists, are generally motor neurons (somatic or autonomic); the second group, known as **interneurons** relay information from one neuron (or one group of neurons) to a second neuron (or second group of neurons) within the CNS (e.g., the interneuron of a reflex arc).

White matter is composed of neuroglia and processes of neurons, many of whose axons are wrapped in a myelin sheath, which in a living individual has a white color. These axons are collected into small bundles, known as **fasciculi**, or large bundles, called **funiculi** (L. *funiculus*, "cord"). Certain larger fiber bundles are named **tracts**, **brachia** (L. arm-like processes), **peduncles** (foot-like processes), or **capsules**, whereas axons that cross the midline to connect identical structures on opposing sides are known as **commissures**. The largest commissure of the CNS is the **corpus callosum**, which connects the two cerebral hemispheres of the brain. Axons that travel up or down the CNS and cross the midline

INTRODUCTION TO THE NERVOUS SYSTEM ● ● ● 9

from one side to the other are said to **decussate** at the point of crossing over.

PERIPHERAL NERVOUS SYSTEM

The peripheral nervous system is a continuation of the CNS; it is composed of clusters of nerve cell bodies, known as ganglia, as well as of bundles of axons and central processes, known as nerves

The **peripheral nervous system** is composed of cranial nerves, spinal nerves, their associated ganglia, and nerve fibers and ganglia of the autonomic nervous system. It must be understood that the PNS is in physical continuity with the CNS; in fact, cell bodies of many of the nerve fibers (axons) of the PNS are located in the CNS.

Somatic nervous system

The somatic nervous system is composed of the 12 pairs of cranial nerves and their ganglia as well as of the 31 pairs of spinal nerves and their spinal ganglia

There are 12 pairs of cranial nerves, identified both by name as well as by Roman numerals I through XII. All cranial nerves, with the exception of the vagus (CN X), serve structures in the head and neck only. The vagus nerve innervates structures in the head and neck, but also serves many of the thoracic and abdominal viscera, for example, the heart and the gastrointestinal tract. Those cranial nerves that have sensory components possess sensory ganglia housing the cell bodies of unipolar (pseudounipolar) neurons whose single process bifurcates into a central and a peripheral process. The central process of a unipolar neuron enters the brain, whereas its peripheral process goes to a sensory receptor. There are no synapses occurring in these sensory ganglia.

There are 31 pairs of spinal nerves (8 cervical, 12 thoracic, 5 lumbar, 5 sacral, and 1 coccygeal), attesting to the segmentation of the spinal cord (see Fig. 1.4). The cell bodies of sensory neurons (unipolar neurons) are located in the spinal ganglia (sensory ganglia that were named "posterior root ganglia"). Again, it must be remembered that just as in the sensory ganglia of the cranial nerves there are *no synapses* occurring in the spinal ganglia. The single process of each neuron bifurcates, and the short central process joins other central processes to form posterior rootlets that enter the spinal cord. The peripheral process goes to a sensory receptor, which, when stimulated (see Chapter 3 for a thorough discussion of this progression), causes a change in the peripheral process that, as a consequence, conveys the information to the central process. This information is conveyed from the central process either to an interneuron (in a three-neuron reflex arc) or to a motor neuron in a two-neuron reflex arc, such as the patellar reflex. This reflex is activated when a gentle strike of the patellar tendon with a rubber hammer causes contraction of the muscle that kicks the foot forward. Although this description is true for reflex arcs, it must be realized that in most instances the incoming information is transmitted to higher levels in the brain and is processed either consciously or subconsciously, or both, rather than just relying on simple spinal reflex phenomena, which do not involve the brain. These motor neurons are multipolar neurons whose cell bodies are located in the anterior horn of the spinal cord and serve *skeletal muscle cells only*. Their axons leave via the anterior rootlets that join the posterior rootlets to form the **spinal nerve**.

Each spinal nerve bifurcates to form a smaller **posterior primary ramus (dorsal primary ramus)** and a larger **anterior primary ramus (ventral primary ramus)**. Posterior primary rami supply sensory and motor innervation to a small section of the back, whereas anterior primary rami supply the lateral and anterior portion of the neck, trunk, and limbs. Anterior rami that supply the thorax and abdomen usually remain as separate nerves, whereas those of the cervical and lumbosacral regions join each other to form **plexuses** from which individual nerve bundles arise to serve the head, neck (cervical plexus), and upper and lower extremities (brachial and lumbosacral plexuses, respectively). Each spinal nerve receives sensory information from the skin of the segment, or **dermatome**, of the body that it serves (see Chapter 5). The entire body is mapped into a number of dermatomes; however, there are overlaps in the innervation, so that a single dermatome is supplied by more than one spinal nerve. Such overlaps prevent the total anesthesia of a particular dermatome if the posterior rootlets of the spinal nerve supplying it are damaged.

Autonomic nervous system

The autonomic nervous system, controlled by the hypothalamus, regulates the activities of smooth muscle, cardiac muscle, and acinar cells of glands, and is divided into three components: the sympathetic, parasympathetic, and enteric nervous systems

The autonomic nervous system, controlled by the hypothalamus, is a motor system, but unlike somatic motor neurons, it does *not* serve skeletal muscle cells; instead, it innervates visceral muscle (cardiac muscle cells, smooth muscle cells) and secretory cells (acinar cells) of glands (see Chapter 10, Autonomic Nervous System). Additionally, whereas a somatic motor neuron **directly** innervates its muscle cell (Fig. 1.6), in the autonomic nervous system the neuron whose cell body is located in the CNS (**preganglionic** also known as **presynaptic neuron**) synapses with a second neuron (**postganglionic** also known as **postsynaptic neuron**) located in a ganglion in the PNS. It is the axon of the *postganglionic neuron* that synapses with the cardiac muscle cell, smooth muscle cell, or secretory cell of a gland. Thus, the autonomic nervous system is said to be a two-cell system, and synapses *always occur* within an autonomic ganglion (Fig. 1.6). There is an exception to that rule where the presynaptic fibers synapse with secretory cells in the medulla of the suprarenal gland, instructing them to release catecholamines. Those secretory cells act as postsynaptic neurons. The axon of the presynaptic neuron is myelinated and is known as the **presynaptic fiber**.

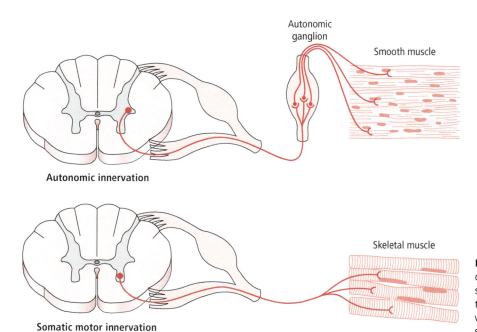

Figure 1.6 • Diagram demonstrating the difference between autonomic innervation (top) and somatic motor innervation (bottom). Observe that two neurons are present in the autonomic supply, whereas a single motor neuron is present in the somatic motor system.

The axon of the postsynaptic neuron is not myelinated and is known as the **postsynaptic fiber**.

The autonomic nervous system is responsible for the maintenance of **homeostasis**, and is composed of three functional components: sympathetic, parasympathetic, and enteric. The **sympathetic** component prepares the body for the "four F's" (fight, flight, freeze, or fornicate), whereas the **parasympathetic** component prepares the body for a vegetative state (e.g., rest). The **enteric nervous system** is situated completely within the wall of the digestive tract and controls the entire process of digestion. Although the sympathetic and parasympathetic components of the autonomic nervous system do modulate its activities, the enteric nervous system can function quite well on its own if the connections of the sympathetic and parasympathetic components are cut off from it.

Cell bodies of **spinal sympathetic motor neurons (preganglionic sympathetic neurons)** are located in the lateral horn of the spinal cord gray matter in thoracic and upper lumbar spinal cord (T1 to L2, L3), whereas those of the **cranial parasympathetic motor neurons (presynaptic parasympathetic neurons)** are located in the brain (and their axons travel with CN III, VII, IX, and X) and the **spinal parasympathetic motor neurons** are in the lateral horn of gray matter in the sacral spinal cord (S2–S4). Postganglionic cell bodies of sympathetic neurons are usually located near the spinal cord, just lateral to the vertebral column, within the **paravertebral sympathetic ganglia**; or a little farther away, in **prevertebral ganglia**. The cell bodies of *postganglionic parasympathetic neurons*, however, are located in ganglia that are in the walls of the viscera being innervated, and are, therefore, said to be **intramural**.

The cell bodies of the sensory neurons that supply the viscera are located in the spinal ganglia of spinal nerves or in the sensory ganglia of cranial nerves, along with the somatic sensory neurons. However, their peripheral processes accompany the preganglionic autonomic fibers into their respective ganglia, but do *not* synapse in those ganglia. Moreover, these peripheral fibers continue to accompany the postganglionic autonomic fibers to the same destinations. In spite of their route, these sensory neurons are *not* considered to be a part of the autonomic nervous system. Sensory information relayed by these autonomic sensory nerves are not registered as part of the conscious experience, and even pain sensations are experienced as "referred pain" in somatic regions of the body (e.g., **angina pectoris**, where pain sensations arising in the heart muscle are experienced as pressure in the left side of the chest, neck, back, and arm, regions served by the same segmental spinal nerve).

QUESTIONS TO PONDER

1. What is the relationship between the central and peripheral nervous systems?

2. What is meant by the "fight, flight, freeze, or fornicate" response?

3. What single characteristic is the major difference between microglia and the other neuroglia of the central nervous system?

4. What is the major difference between a two-neuron reflex arc and a three-neuron reflex arc, aside from the simplistic fact that one has an extra neuron associated with it?

CHAPTER 2

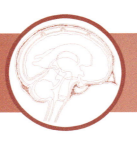

Development of the Nervous System

CLINICAL CASE

EARLY DEVELOPMENT

NEURULATION

EARLY DEVELOPMENT OF THE SPINAL CORD AND BRAIN

DEVELOPMENT OF THE SPINAL CORD

DEVELOPMENT OF THE BRAIN

CLINICAL CONSIDERATIONS

SYNONYMS AND EPONYMS OF THE NERVOUS SYSTEM

FOLLOW-UP TO CLINICAL CASE

QUESTIONS TO PONDER

CLINICAL CASE

A 22-year-old woman presents to the emergency room because she has not felt her baby kicking even though she is near term. There has been no prenatal care. A uterine ultrasound is performed that detects an abnormality. She is estimated to be 37 weeks pregnant. The obstetrician on call wishes to perform an emergency Caesarian section delivery because of the abnormality that was detected. This was performed without incident.

Examination of the newborn shows a fist-sized mass in the midline lumbosacral region. This protrudes above the skin surface and is dusky gray in appearance. The newborn has a normal cry and arms seem to move normally. Skin color is normal and temperature is normal. The legs do not move and are flaccid. There is deformity noted of the hips, ankles, and feet. Subsequent examination reveals a lack of cry or any physical response to sensory stimulation of the legs.

Embryogenesis, the development of the future individual, begins with the fusion of the spermatozoon with the secondary oocyte within the lumen of the uterine tube (oviduct). The process of fusion is known as **fertilization**, subsequent to which the fertilized ovum is referred to as a **zygote** (G., "yoked"), the very beginning of a new individual (Fig. 2.1). The mitotic activity of the zygote increases in the number of cells to allow for the construction not only of the future individual but also for the embryonic component of the supporting structures that will nourish the future embryo. The location of the cells as they proliferate and their new conditions will trigger the expression of different genes in different cells, resulting in the regulation of certain activities required for normal development. In the postembryonic period, the same or different genes act to insure normal functioning of the cell, tissue, organ, and/or body. Some of the proteins

---

A Textbook of Neuroanatomy, Third Edition. Maria A. Patestas, Amanda J. Meyer, and Leslie P. Gartner.
© 2025 John Wiley & Sons, Inc. Published 2025 by John Wiley & Sons, Inc.
Companion website: www.wiley.com.go/patestas/neuroanatomy3e

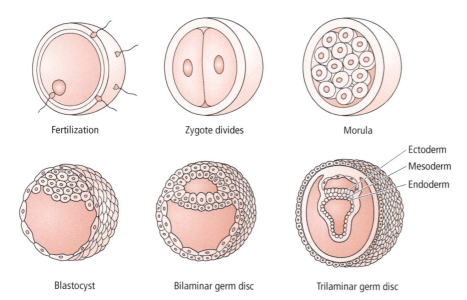

Figure 2.1 ● The early development of a human embryo. Fertilization, the fusion of the haploid sperm nucleus with the ovum's haploid nucleus, results in the formation of a diploid cell, known as a zygote. As the zygote undergoes mitosis, a solid cluster of cells is formed, the morula. Continued cell division and rearrangement of the newly formed cells results in the formation of a hollow sphere of cells, the blastocyst, whose cells form the bilaminar germ disc and later the trilaminar germ disc.

coded for by these genes control cellular metabolic activities, others migrate to the cells' membranes to act as receptors or cell signals, still others are released into the extracellular space to act as **signaling molecules**, **growth factors**, or **ligands** (L., "to bind"), and they reach their target cells by traveling in body fluids.

Because a particular signaling molecule may meet a number of other cells, it is important that only the target cells become influenced by that particular molecule, and this process is ensured by the presence of **receptor molecules** on the surfaces of cells. Usually, a specific receptor molecule recognizes only a particular signaling molecule. This is similar to a lock-and-key concept, where a specific lock can be opened only by a particular key.

If physical contact between cells is required, it is important that the two cells recognize each other. Cell recognition is also a function of a series of cell surface receptor molecules, some of which act as the keys and the others as locks. Usually, several molecules are involved on each cell so as to ensure that only the intended target cell is influenced by the signaling cell.

Once the cell surface receptors become activated by contact with the proper signaling molecule, or with the proper molecules on the signaling cell's surface, a sequence of intracellular events is initiated resulting in the activation (or suppression) of a single gene or a series of genes of the target cell. The expression of these genes may cause the release of further signaling molecules, may alter the activities of the target cell, may prompt the target cell to differentiate into another cell type, may compel the cell to proliferate, or may direct the cell to undergo a preprogrammed series of intracellular events resulting in the death of the cell (**apoptosis** [G., "dropping off"]). Each of these possible events is necessary for the progression of normal development. The process whereby one cell causes its target cell to differentiate – that is, become transformed into a different cell type – is known as **induction**.

It is interesting to note that there is a relatively small number of growth factors that regulates development. The reason for the paucity of their number is that they act in combination with each other, and by simple permutations and combinations of just a few of these factors, a very large number of signals may be generated.

Other molecules, referred to as **transcription factors**, enter the cell's nucleus, bind to the DNA of the chromosomes, thereby activating or repressing the expression of specific genes.

The normal development of the embryo is divided into a number of stages, and the interested reader is referred to a textbook of human embryology. In the present textbook, human embryology is portrayed from the perspective of neuroanatomy, and only those aspects of human development are described that will assist the student in appreciating the complexity and beauty of how the nervous system progresses from an ill-defined cluster of cells to an amazingly intricate functional unit. In order for the reader to be able to grasp the manner in which this process occurs, it is advisable to show not only the development of the nervous system *per se*, but also that of the structures that envelop and surround the nervous system. Therefore, this chapter includes a description of early embryology, the development of the pharyngeal (branchial) arches, formation of the face, and some of the molecular events that appear to govern this entire process.

EARLY DEVELOPMENT

During early development the zygote undergoes mitotic division to form a cluster of cells, known as the morula, whose cells rearrange themselves to form the hollow blastocyst

The zygote undergoes a series of mitotic divisions, known as **cleavage**, forming a solid cluster of cells, where each cell is smaller than the original zygote. This cluster of cells resembles a mulberry and hence is known as the **morula** (L., "mulberry"). The cells of the morula secrete a viscous fluid that creates a central cavity by pushing the cells to the periphery until a hollow ball of cells is formed, known as the **blastocyst** (G., "germ" + "bladder," Fig. 2.1).

Although most of the cells of the blastocyst are at the periphery, a few of the cells are trapped inside, adhering to one of the poles of this hollow ball of cells. These trapped cells are the **embryoblasts** (**inner cell mass**), whereas the peripherally positioned cells are the **trophoblasts** (G., "nourishment" + "germ").

The embryoblasts will give rise to the **embryo**, whereas the trophoblasts are responsible for the formation of the embryonic portion of the **placenta** (L., "flat cake"). The trophoblasts appear to be able to do this because they express a gene that codes for the signaling molecule known as **L-selectin** that binds to **receptors** located on the surface of the uterine endometrium. This interaction of the trophoblasts and the uterine endometrium triggers additional responses from both the trophoblasts and the endometrium that are responsible not only for the formation of the placenta but also for the **implantation** of the developing embryo into the endometrium of the anterior or posterior uterine wall.

Bilaminar germ disc

The bilaminar germ disc is composed of two cell layers: the epiblast and the hypoblast

The cells of embryoblasts rearrange themselves to form two layers, the **epiblast** (G., "upon" + "germ") and the **hypoblast** (G., "under" + "germ"), and the embryo becomes known as the **bilaminar germ disc**. The epiblast is closer to the trophoblast cells than is the hypoblast, and the forming embryo is about 7–8 days old (Fig. 2.1). The blastocyst entered the uterine cavity, where it burrowed into the wall of the uterus. This process of **implantation** begins on the late sixth or early seventh day after fertilization and is completed four or five days later.

During the period of implantation some of the cells of the epiblast delaminate, forming a membrane over the remaining epiblast. This new layer of cells is composed of the **amnioblasts** (G., "lamb" + "germ"), which will give rise to the **amniotic sac**, and the space between the amnioblasts and the epiblasts, known as the **amniotic cavity**, will be filled with amniotic fluid later during pregnancy.

Trilaminar germ disc

The trilaminar germ disc is responsible for the establishment of the three primary germ layers: the ectoderm, mesoderm, and endoderm

At about day 14 postfertilization, at the anterior (cranial) end of the bilaminar germ disc a few cells of the epiblasts form desmosomal contacts with a few cells of the underlying hypoblast, thus forming the **prochordal plate (prechordal plate)**, which will be the future **buccopharyngeal membrane**, a thin tissue between the future oral cavity and the pharynx. Approximately one day later a longitudinal furrow, the **primitive groove**, develops in the epiblast on the posterior aspect of the bilaminar germ disc. Because the anteriormost

extent (cranial end) of the primitive groove is deeper, it is referred to as the **primitive pit**. Bordering both sides of the primitive groove is an elevation of cells, the **primitive streak**, whereas the anterior border of the primitive pit is a small elevation of cells, the **primitive node** (eponym: **Hensen's node**).

Formation of ectoderm, mesoderm, and endoderm

Cells of the epiblast undergo active mitosis along most of its surface, and the newly formed cells migrate to the primitive streak. The ability of these cells to migrate is due to the cells of the primitive streak, which manufacture and release the signaling protein, **fibroblast growth factor 8**, which facilitates the movement of epiblast cells into the primitive groove thus presaging the formation of the other two cell layers. Hence, the newly formed epiblast cells enter the primitive groove and pass into the space between the epiblast and the hypoblast, thus forming a new intermediate layer of cells, the **mesoderm.** Some of these cells do not stop in the mesoderm space, but displace most of the original cells of the hypoblast, pushing the original cells of the hypoblast laterally. The cells of the hypoblast become incorporated into an extraembryonic region, known as the umbilical vesicle (formerly known as the yolk sac). As this process is occurring, the epiblast is renamed the **ectoderm** and the new cell layer that replaces the hypoblasts is referred to as the **endoderm**. Thus, as a result of the conversion of the embryoblast of the blastocyst into the trilaminar germ disc, a process known as **gastrulation** (G., gastrula, "belly"), the embryo is composed of three layers (ectoderm, mesoderm, and endoderm), and is known as the **trilaminar germ disc** (Fig. 2.1). The entire embryo, at this time, is approximately 1 mm in length. During the period of gastrulation, cells at the future anterior end of the embryo express homeobox (see below) and other genes that code for transcription factors as well as for signaling molecules (such as **cerberus** and **lefty**) that ascertain cephalization (that is head formation) as well as right- and left-side determination. Additionally, after gastrulation is completed another transforming growth factor, known as **Nodal**, assists in the maintenance of the establishment of cranial–caudal and posterior–anterior axes of the developing embryo, but it requires the presence of **BMP4** (**bone morphogenic protein 4**) as well as **fibroblast growth factor** (**FGF**) in order to achieve that. Additionally, high concentration of BMP4 promotes the formation of **epidermis** from cells of the **epiblast** as well as lateral plate and intermediate mesoderm from mesoderm (see below).

Genes known as **homeobox genes** (**HOX genes**) play major roles in the establishment of segmentation and the formation of the cranial–caudal axis. They code for proteins, known as **transcription factors**, which activate a cascade of other genes. In humans the HOX genes are actually four clusters of genes, and each cluster is assigned a capital letter, so there are HOXA (on chromosome 7), HOXB (on chromosome 17), HOXC (on chromosome 12), and HOXD (on chromosome 2). The clusters are said to be paralog clusters because many of the genes share the same function; thus there is a built-in redundancy, whereas other genes of the cluster possess different functions.

Interestingly, the HOX genes controlling the development of the spinal cord and rhombencephalon (hindbrain) belong to the oldest of the HOX genes known as the **antennapedia class** of genes, named after the studies in the fruit fly (*Drosophila melanogaster*) where they were first discovered. These HOX genes control the development of the head segment and thorax in *Drosophila*, but in the human, they control, among other things, the development of rhombomeres and the spinal cord. The development of the mesencephalon (midbrain) and prosencephalon (forebrain), however, is controlled by a newer set of homeobox genes **LIM1** (named after the initials of three proteins in which the protein product of the double zinc finger LIM protein was discovered) and **OTX2** (**orthodenticle homeobox 2 gene**) with the assistance of additional homeobox genes and growth factors (described below in the section entitled "Molecular basis of midbrain and forebrain development").

The time sequence of the events controlled by products of these homeobox genes and by **growth factors** is essential for normal development because certain genes can be switched on only if some other genes have already been activated and if still other genes have not as yet been switched on. In other words, there are "windows of opportunity" for the occurrence of certain events, and if that timeframe is missed then development will not proceed normally. Therefore, these homeobox genes become activated in a specific order, and their sequential expression establishes a pattern of developmental events. It appears that the correct temporo-sequential expression of these homeobox genes is probably regulated by **retinoic acid**. Many other macromolecules are involved in the control of embryonic development, but their description and the discussion of their various functions are beyond the scope of this textbook.

It should be noted that these homeobox genes act during different stages of development and they may be silent in some areas, active at a later stage in other regions of the developing embryo, and they may become silent again. In this chapter, only their roles in the development of the central nervous system are discussed. Therefore, the reader should not assume that the functions of a given homeobox gene, or of a transcription factor, or of a growth factor are limited to the embryology of the central nervous system.

Cells of the mesoderm cannot penetrate the contact of the ectoderm and endoderm at the prochordal plate, but they can migrate around this region of adhesion and some mesoderm cells will migrate anterior to this structure. These cells are **cardiogenic cells** and begin to form the future heart. An additional group of cells migrate from the primitive node into the primitive pit and proceed anteriorly (cranially), *en masse*, between the ectoderm and the endoderm until they reach the prochordal plate, where their progress is blocked. This pencil-shaped column of cells is known as the **notochordal process**, and it possesses special *inductive* capabilities. The notochord (G., back + string) induces the cells of the ectoderm lying above it to proliferate and form the flat **neural plate**, the beginning of the primitive nervous system of the embryo. This inductive capacity is due to the increased **fibroblast growth factor** (**FGF**) formation by cells of the notochord that suppresses the availability of **bone morphogenic protein 4** (**BMP4**). Additional factors, **follistatin**, **chordin**, and **noggin**, produced by the notochord cells as well as by the cells of the primitive node also suppress the availability of BMP4. The very low concentration of BMP4 in the area of the anterior (head region) epiblast permits the epiblast to differentiate into those cells of the **neural plate** that will form the forebrain and the midbrain. Other factors, namely **FGF** and **WNT3A** (**wingless type 3A** is a gene conserved from the fruitfly), suppress the availability of BMP4 in the more caudal region of the neural plate so that the hindbrain and the spinal cord can develop in a normal fashion.

NEURULATION

Neurulation is the process whereby the embryo internalizes its developing nervous system

Because in the trilaminar germ disc stage the primitive nervous system is on the external aspect of the developing embryo, it has to be internalized. The process of internalization, known as **neurulation**, is accomplished by an alteration in cell shape and by increased mitotic activity, especially of the lateral edges of the neural plate. These activities begin to fold the neural plate into a longitudinal furrow, the **neural groove**, whose two walls are the **neural folds**. The folding of the neural plate is believed to be initiated by the signaling molecule **sonic hedgehog** (**SHH**) manufactured and released by those cells of the notochord that are in close approximation of the cells of the neural plate. As the neural plate begins to fold its cells also secrete SHH which continues to facilitate the process of neurulation. Further cell division, and continued change in cell morphology, causes the neural folds of the two sides to approximate and fuse with each other in the midline, forming the closed **neural tube** (Fig. 2.2). The process of fusion starts at the midcervical level and proceeds anteriorly (cranially) and posteriorly (caudally) and is completed by the end of the fourth week of gestation. The cells of the two lips of the neural tube that contact each other stop expressing **E-cadherins** and, instead, express **N-cadherins** and **neural cell adhesion molecules** (**N-CAMs**) that permit the opposing lips to contact and adhere to each other thus forming the neural tube. The final regions of the neural tube to be closed are at the **anterior** and **posterior neuropores**. The cells of the neural tube just above the notochord differentiate to form the **floor plate**. These cells induce the differentiation of neuroblasts into motor neurons and establish the polarity of the neural tube.

Note that the clinical case at the beginning of the chapter refers to an embryonic abnormality detected during an ultrasound examination of the fetus.

1 Where is the lesion in relation to the innervation of the upper extremities?

2 Where is the lesion in relation to the motor innervation of the legs?

3 Where is the lesion in relation to the sensory innervation of the legs?

DEVELOPMENT OF THE NERVOUS SYSTEM • • • 15

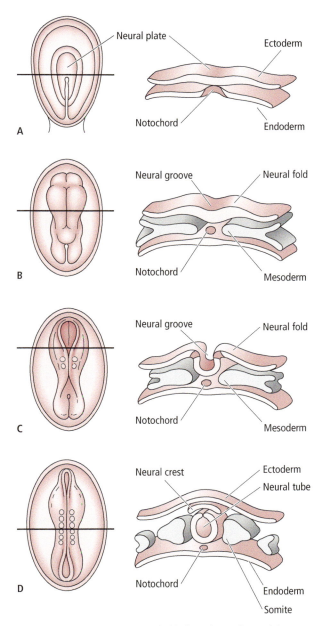

Figure 2.2 • The notochord is responsible for inducing the overlying ectodermal cells to form the neural plate (A). As the embryo continues its development, it enters the stage of neurulation, the process whereby the forming nervous system is brought into the body by the formation of an intermediary neural groove, and finally a neural tube, the precursor of the brain and spinal cord (B, C). Note that the neural crest, initially the lateral aspect of the neural plate, becomes separated as the neural tube is formed. Cells of the neural crest give rise to all of the ganglia of the peripheral nervous system as well as to numerous additional structures of the developing embryo (D).

Fusion of the lateral edges of the right and left sides of the central nervous system (CNS) is accompanied by fusion of the ectoderm, whose edges became approximated during neural tube formation. The simultaneous fusions permit a separation of the neural tube from the overlying ectoderm and complete internalization of the nervous system. Incomplete fusion results in a relatively common developmental anomaly known as **spina bifida**, a condition where not only is the spinal cord incomplete but also the dorsum of the vertebral canal remains open to the external environment. The nervous system, including the organs of special senses, originate from three sources: the **neural tube** (CNS, somatic motor neurons, and preganglionic autonomic neurons); the **neural crest** (see next section); and the **ectodermal placodes** (lens of the eye, internal ear, hypophysis (pituitary gland), and contribution to the formation of somatic sensory ganglia of the cranial nerves).

Neural crest

The neural crest is a narrow strip of cells at either edge of the developing neural plate

As the edges of the neural folds meet each other, some of their cells do not participate in the formation of the neural tube; instead, they form a narrow strip of cells, the **neural crest** (Fig. 2.2), initially located just above and later splitting into two longitudinal columns of cells, to be located on the right and left dorsolateral aspects of the neural tube. The cells of the neural crest will *migrate* throughout the body to form almost the entire peripheral nervous system, including the enteric nervous system, spinal ganglia, sensory ganglia of the cranial nerves, all postganglionic autonomic neurons, as well as melanocytes, parenchymal cells of the suprarenal gland medulla (pheochromocytes (chromaffin) and ganglion cells), neurolemmocytes (Schwann cells) and satellite cells of the peripheral nervous system, and most of the mesenchymal cells of the head and anterior neck. Therefore, most of the mesenchymal cells of this region are not derived from mesoderm; instead, they are neuroectodermal in origin and are referred to as **ectomesenchyme**. Derivatives of the neural crest, neural plate, and ectodermal placodes are summarized in Table 2.1. The transformation of neural tube cells into neural crest cells depends on a concentration of BMP4 that is *intermediate* between the high concentrations required for epidermis formation and the low concentration that triggers neural tube formation. In addition to BMP4, the transcription factors **FOXD3 (forkhead box 3D protein)** and the zinc finger transcriptional repressor protein **Snail** (Snail because it was first isolated in snails) are required to commit these cells to becoming neural crest cells. The cells of the neural crest require the presence of another transforming factor, **SLUG** (a member of the SNAIL family of genes but first isolated in a slug), in order to begin their migrations and then follow one of four general migratory pathways marked by signaling molecules manufactured and released by neighboring cells. Examples of these molecules are retinoic acid, transforming growth factor, and fibroblast growth factor. These signaling molecules contact cell surface receptors of target cells, activating intracellular molecular systems that cause specific responses within those cells. Some of these responses activate cytoplasmic enzyme systems, whereas other responses include the activation of **transcription factors** that regulate specific gene expressions, each resulting in the activation of specific inductive processes necessary for the formation of the nervous system.

Table 2.1 ● Derivatives of the neural crest, neural plate, and ectodermal placodes.

Neural crest	Neural plate	Ectodermal placodes
Neurons of: Posterior (dorsal) root ganglia; ganglia of CN V, VII, VIII, IX, and X; enteric plexuses; sympathetic ganglia; parasympathetic ganglia Satellite cells of: sensory ganglia; sympathetic ganglia; parasympathetic ganglia Glia cells of: Enteric ganglia Suprarenal medulla: pheochromocytes (chromaffin cells); neurons Carotid body parenchyma Melanocytes; (neurolemmocytes) Schwann cells Head: Most connective tissue elements; dentin, pulp, cementum; choroid, sclera Meninges: dura mater, arachnoid mater, pia mater Tunica media of: great vessels from heart	Prosencephalon: cerebral hemispheres; basal nuclei; Neurohypophysis Mesencephalon: cerebral peduncles; tegmentum; tectum Metencephalon: cerebellum; pons Myelencephalon: medulla oblongata Spinal cord All neurons of the CNS All oligodendroglia and astrocytes Ependymal cells (including those covering the choroid plexus) Optic nerve Eye: Retina Epithelium of: Ciliary body; ciliary processes; iris	Sensory ganglia: CN V, VII, VIII, IX, and X Lens of the eye Olfactory system: Receptor cells; epithelium Ear: Spiral organ (of Corti); epithelium of membranous labyrinth Adenohypophysis Teeth: enamel

Paraxial mesoderm

The paraxial mesoderm lies lateral to the developing neural tube and becomes segmented into clustered blocks of tissue known as somites

The mesoderm lying lateral to the neural tube and notochord in the region of the future trunk is known as the **paraxial mesoderm**, and it becomes segmented into paired blocks of mesodermal tissue, known as **somitomeres**. In the head region they remain as small, incomplete blocks of tissue on either side of the neural tube, and their mesoderm will form the skeletal muscles in the region and will make a minor contribution to the connective tissue components of the head and anterior portion of the neck; instead, most of their connective tissue elements are derived from neural crest cells. Somitomeres caudal to the occipital region form larger complete blocks of tissue known as **somites** (Fig. 2.3). The number of somites and the timing of their appearance vary by species, and in the human approximately 40 somites are formed, beginning at approximately the twentieth day of gestation with the formation of the first somite and ending about ten days later with the formation of the 40th somite; therefore, in the human, a new somite is formed approximately every six hours. A similar segmentation of the caudal aspect of the neural tube begins around the 29th day of

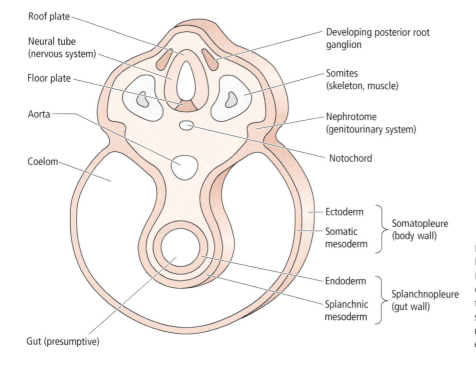

Figure 2.3 ● Cross-section through a developing human embryo during the process of neurulation. Note the presence of somites, gut, and intraembryonic coelom (body cavity). It is interesting to realize that the precursors of many of the future organ systems, such as the digestive, respiratory, genitourinary, musculoskeletal, and nervous systems, are being established at this early stage of development.

DEVELOPMENT OF THE NERVOUS SYSTEM ● ● ● **17**

development, where eight constrictions, known as **rhombomeres**, develop (see below "Development of the Brain") to form the hindbrain.

Each somite is composed of three elements. These are the **sclerotome**, responsible for the formation of two adjacent vertebrae and their intervening intervertebral disc; **myotome**, responsible for the formation of the muscle masses associated with that segment of the trunk; and **dermatome**, responsible for the formation of the dermis of the skin in its particular area of the trunk.

The region of the paraxial mesoderm lateral to the somites is known as the **intermediate mesoderm**, and it forms the genitourinary system. The most lateral aspect of the paraxial mesoderm, known as the **lateral plate mesoderm,** splits into two sheaths and the intervening space between these two sheaths is known as the **coelom** (G., "cavity") or body cavity (Fig. 2.3). The posterior of the two sheaths of the lateral plate mesoderm becomes known as the **somatic mesoderm**, and together with its associated ectoderm this is referred to as the **somatopleure** (G., "body" + "side"). The anterior sheath is known as the splanchnic mesoderm and, together with its associated endoderm, is referred to as the **splanchnopleure** (G., "viscera"). The somatopleure gives rise to the body wall, whereas the splanchnopleure forms the wall of the gut. The coelom becomes subdivided into the peritoneal, pleural, and pericardial cavities.

Pharyngeal arches

Pharyngeal arches are pairs of ectodermally covered mesenchymal thickenings responsible for the formation of much of the head and neck

As the neural tube becomes internalized, its anterior region, the future brain – whose development is detailed below – is developing so rapidly that it grows above and anterior to the prochordal plate. It overlies the cardiogenic region and approximates the amniotic sac. The anterior end of the embryo, possibly to provide more room for growth, begins to bend in an inferior direction (**cephalic flexure**). As that occurs, the future heart becomes neatly tucked beneath the embryo, and the prochordal plate, now referred to as the **buccopharyngeal membrane**, becomes positioned superiorly to the developing heart, separating the **stomodeum**, the (G., "mouth" + "on the way") primitive oral cavity, from the developing foregut (**pharynx** [G., "throat"]).

Note that the stomodeum is in fact a space captured by the embryo from the amniotic cavity. The formation of the stomodeum is indicative of the initiation of the development of the face.

As the buccopharyngeal membrane degenerates during the fifth week of development and communication is established between the future oral cavity and the pharynx, the stomodeum deepens and begins to be surrounded by ectodermal-covered mesenchymal thickenings, the **pharyngeal arches** (previously named *branchial arches*). These bilaterally positioned arches develop in an anteroposterior direction. The first arch is the **mandibular arch**, the second is the **hyoid arch**, whereas successive arches are numbered 3, 4, and 6, with the notable absence of the fifth pharyngeal arch, which in humans is rudimentary. Between neighboring arches an external depression – the ectodermally lined **pharyngeal groove** – is encountered, with its anatomical counterpart on the inside, the **pharyngeal pouch**, an **evagination** of the endodermally lined pharynx

Genetic influences of pharyngeal arch development

Ectomesenchymal cells express the homeobox gene products

As indicated above, the connective tissue elements of the pharyngeal arches, especially those associated with the developing musculature, arise from somitomere mesoderm (**mesenchyme**), whereas the majority of the mesenchyme, particularly in the anterior regions of the arches, arise from neural crest material and is referred to as **ectomesenchyme**. These neural-crest-derived cells, known as **ectomesenchymal cells**, express the homeobox gene products (**HOXA2, HOXA3, HOXB3,** and **HOXD3 genes** as well as the other homeobox genes, **LIM1** and **OTX2**) that reflect their neural origin. The recruitment of the neural crest cells has been shown to be directed by cells of the endodermal lining of the pharyngeal pouches. These endodermal cells express factors that are site-specific, and thereby they can modulate the differentiation of the ectomesenchymal cells to form structures that develop in and from each pharyngeal arch (see next section). The genes expressed by these endodermal cells of the pharyngeal pouches that play major roles in the development of the face are **bone morphogenic protein 7, fibroblast growth factor 8, sonic hedgehog,** and **paired box protein 1 (PAX1).**

Pharyngeal arch derivatives

Each pharyngeal arch possesses its own vascular and neural supply as well as supporting cartilaginous skeleton and associated skeletal muscles

Each pharyngeal arch possesses its own cartilage, nerve, vascular, and muscular components (see Table 2.2). The first two arches become developed to the greatest extent, whereas the last one is the least developed. Only those derivatives that are pertinent to the topic of this textbook will be discussed.

First pharyngeal arch (mandibular arch)

The first pharyngeal arch is responsible for the formation of the maxillary and mandibular arches and the muscles of mastication

The cartilage of the first (mandibular) arch is a horseshoe-shaped structure, known as **Meckel's cartilage**. Most of this cartilage will disappear, except for its anteriormost extent which participates in the formation of the **mental symphysis** (L., mentum = "chin") and its posteriormost extent which gives rise to the **malleus** (L., "hammer") and **incus** (L., "anvil"), two of the three **ossicles** (L., "little bones") of

Arch	Skeletal	Ligaments	Muscle	Nerve

Mandibular (I)	Meckel's cartilage
	Maxillae
	Mandible
	Malleus
	Incus
		Sphenomandibular
		Anterior ligament
		of malleus
			Muscles of mastication:
			  Temporalis
			  Masseter
			  Medial and lateral pterygoids
			Tensor veli palatini
			Tensor tympani
			Digastric (anterior belly)
			Mylohyoid
				Trigeminal (CN V)

Hyoid (II)	Reichert's cartilage
	Hyoid bone:
	  Lesser cornu
	  Superior part of
	  body
	Styloid process
	Stapes
		Stylohyoid	Muscles of facial expression;
			Stapedius
			Digastric (posterior belly)
			Stylohyoid
				Facial (CN VII)

III	Hyoid bone:
	  Greater cornu
	  Inferior part of body
	  Superior cornu of
	  thyroid cartilage
			Stylopharyngeus	Glossopharyngeal
				(CN IX)

IV and VI	Thyroid cartilage
	Laryngeal cartilages:
	  Cricoid
	  Arytenoid
	  Corniculate
	  Cuneiform
	  Body of hyoid bone
			Pharyngeal muscles
			Pharyngeal constrictors
			Laryngeal muscles
				CN IX via the
				pharyngeal branch
				of CN X and the
				pharyngeal plexus
				Recurrent laryngeal
				branch of CN X

Source: Adapted from Hiatt, JL, Gartner, LP (2001) *Textbook of Head and Neck Anatomy*, 3rd edn. Lippincott, Williams & Wilkins, Baltimore.

Table 2.2 ● Pharyngeal arch derivatives and their innervation.

the middle ear. The perichondrium of Meckel's cartilage becomes the **sphenomandibular** (G., "wedge" and L., "jaw") **ligament** and the **anterior ligament of the malleus.**

Mesenchyme of the first pharyngeal arch forms the muscles of mastication (temporalis, masseter, and lateral and medial pterygoid muscles), as well as the following muscles: the tensor veli palatini, tensor tympani, anterior belly of the digastric, and mylohyoid. Since these muscles are derived from the mandibular arch, they are innervated by the nerve of this arch, the **trigeminal** (L., "threefold") **nerve** (**CN V**). The mandibular arch, as will be evident later in this chapter, is responsible for the formation of much of the face.

Second pharyngeal arch (hyoid arch)

The second pharyngeal arch overgrows the remaining arches and is responsible for the formation of the muscles of facial expression and part of the hyoid bone

The cartilage of the second or hyoid arch is known as **Reichert's cartilage**. Most of this cartilage will also disappear, but parts of it will be responsible for the formation of the **stapes** (L., "stirrup"), the third ossicle of the ear, as well as the **styloid** (G., "post") **process**, and the lesser cornu and superior aspect of the body of the hyoid bone. Additionally, the **stylohyoid ligament** is derived from the perichondrium of Reichert's cartilage.

The muscles of facial expression take their origin in the hyoid arch mesoderm, along with the stapedius, stylohyoid, and the posterior belly of the digastric muscles. All of these muscles are, therefore, innervated by the nerve of the second arch, the **facial nerve** (**CN VII**).

The homeobox gene *HOXA2* is responsible for the normal development of the hyoid arch and its derivatives. In the absence of *HOXA2* the formation of second arch derivatives does not occur; instead, first arch derivatives develop in the hyoid arch. The implication of this switch is that mandibular arch substances form by default, and *HOXA2* gene products prevent the default condition from taking place.

Third pharyngeal arch

The third pharyngeal arch is responsible for the formation of the stylopharyngeus muscle and part of the hyoid bone

The cartilage of the third arch is unnamed, but it is responsible for the formation of the greater horn and the inferior half of the body of the hyoid bone. The only muscle to be derived from the mesoderm of the third arch is the stylopharyngeus muscle, innervated by the **glossopharyngeal** (G., "tongue" + "pharynx") **nerve** (**CN IX**), the nerve of the third arch. **HOXA3** is mostly responsible for the normal development of the third pharyngeal arch.

Fourth and sixth pharyngeal arches

The fourth and sixth pharyngeal arches participate in the formation of the larynx and its muscular apparatus

The cartilages of the fourth and sixth pharyngeal arches are also unnamed. They participate in the formation of the skeleton of the larynx. It is believed that the muscles associated with the larynx and pharynx are derived from the mesenchyme of these arches. The innervation of these muscles is somewhat confusing, for they receive their nerve supply from the **pharyngeal plexus** (composed of fibers from CN IX and X). The nerves of the fourth and sixth arches, however, are the superior laryngeal and recurrent laryngeal branches of the vagus nerve (CN X), respectively. The fifth pharyngeal arch begins to form, but then it disappears by becoming overgrown by and incorporated into pharyngeal arches 4 and 6. *HOXB3* and *HOXD4* genes appear to be responsible for normal development of the fourth and possibly the sixth pharyngeal arches.

Pharyngeal groove derivatives

Pharyngeal grooves are external depressions located between succeeding pharyngeal arches; the first pharyngeal groove forms the external ear canal, while the others disappear during development

Only the first pharyngeal groove, that lies between the mandibular and hyoid arches (first and second pharyngeal arches), gives rise to a definitive structure. This groove becomes deeper and approximates the first pharyngeal pouch, so much so, that only a thin membrane, the **pharyngeal membrane** (**closing plate**), separates the groove from the pouch. This closing plate becomes the tympanic (G., "drum") membrane whose external surface is covered by the ectodermal derivative of the first pharyngeal groove. The remainder of the first pharyngeal groove becomes the **external auditory canal**.

The subsequent pharyngeal grooves are submerged by the sudden spurt of growth occurring in the second pharyngeal arch, which grows in an inferior direction, to form the future neck. Occasionally, the submerged pharyngeal grooves do not become obliterated, and thus remain as **cervical sinuses**, which may result in the formation of cervical cysts.

The message for this sudden growth spurt arises from the epithelium covering the tip of the second arch. These ectodermal cells express **bone morphogenic protein 7**, **paired box protein 1 (PAX1)**, **sonic hedgehog**, and **fibroblast growth factor 8**, which are responsible for the proliferation of the mesenchymal cells of the hyoid arch.

Pharyngeal pouch derivatives

Pharyngeal pouches are internal depressions located between succeeding pharyngeal arches

The **first pharyngeal pouch**, lying between the mandibular and hyoid arches, forms the auditory tube (Eustachian tube), the tympanic cavity, and the endodermal lining of the eardrum. The **second pharyngeal pouch** forms the palatine tonsils. The **third pharyngeal pouch** gives rise to the thymus and the inferior parathyroid glands. The **fourth pharyngeal pouch** forms the superior parathyroid gland, and perhaps part of the thymus and the **fifth pharyngeal pouch** gives rise to the parafollicular cells of the thyroid gland (see Table 2.3). It is interesting to note that endoderm lining these pouches respond to four transcription factors:

- **Fibroblast growth factor 8**: anterior region of pouches 1 through 4
- **Bone morphogenic protein 7**: posterior region of pouches 1 through 4
- **Paired box protein 1**: posterior-most areas of pouches 1 through 4
- **Sonic hedgehog**: Posterior region of pouches 2 and 3

EARLY DEVELOPMENT OF THE SPINAL CORD AND BRAIN

During its early development the neural tube is subdivided into three regions: the prosencephalon, mesencephalon, and rhombencephalon

The neural tube, at its inception, is composed of a single layer of columnar **neuroepithelial cells**, whose proliferation forms a thickened tubular

Table 2.3 ● Derivatives of the pharynx and pharyngeal pouches.

Region	Pouch I level	Pouch II level	Pouch III level	Pouch IV level	Pouch V level
Roof		Pharyngeal tonsils			
Lateral walls	Tympanic cavity and lining of tympanic drum Mastoid air cells Auditory tube	Palatine tonsils and fossa	*Posterior:* Inferior parathyroid *Anterior:* Thymus	*Posterior:* Superior parathyroid *Anterior:* Thymus	Ultimopharyngeal body Incorporated into the thyroid gland as parafollicular cells (secrete calcitonin)
Floor (pharyngeal endoderm related to pharyngeal arch)	Tongue (anterior two-thirds) Foramen cecum (rostral end of the thyroglossal duct)	Tongue (posterior one-third) Lingual tonsil	Tongue (part of the base)	Tongue (part of the base) Epiglottis	

Adapted from Hiatt, JL & Gartner, LP (2010) *Textbook of Head and Neck Anatomy,* 3rd edn. Lippincott, Williams & Wilkins, Baltimore.

structure, the cephalic region of which forms three enlargements: the **prosencephalon** (forebrain), **mesencephalon** (midbrain), and **rhombencephalon** (hindbrain). The inferior extent of the neural tube forms the **spinal cord**. As the three vesicles are developing, the **cephalic flexure** in the region of the mesencephalon and the **cervical flexure**, between the rhombencephalon and the future spinal cord, are also developing.

By the fifth week of gestation the prosencephalon presents two regions: the **telencephalon** with its two lateral swellings, the future cerebral hemispheres, and the **diencephalon**, whose optic vesicles are in the process of development. The rhombencephalon is also subdivided, by the **pontine flexure,** into two regions: the **metencephalon**, which will form the pons and the cerebellum, and the **myelencephalon**, which will develop into the medulla oblongata. These flexures provide increased space for the folding and three-dimensional organization of the brain.

The lumen of the developing CNS will form the ventricles of the brain and the central canal of the spinal cord. The **lateral ventricles** of the cerebral hemispheres communicate with the **third ventricle** of the diencephalon via the **interventricular foramina** (of Monro). The **fourth ventricle**, located in the rhombencephalon, communicates with the third ventricle via the **mesencephalic aqueduct (cerebral aqueduct** [of Sylvius])**,** a short canal, that pierces the mesencephalon.

Neuroepithelial cells

The neuroepithelial cells form a three-layered neural tube, composed of ventricular, intermediate, and marginal zones

As the **neuroepithelial cells** continue to proliferate, they establish a thicker wall composed of a pseudostratified epithelium, each of whose cells initially extends the entire thickness of the developing neural tube. As development progresses, some of these neuroepithelial cells remain adjacent to the lumen, whereas most cells migrate away from it, resulting in a three-layered neural tube. The layer adjacent to the lumen is known as the **ependymal layer (ventricular zone)**, some of whose cells develop into **ependymal cells** that line the ventricles of the brain as well as the central canal of the spinal cord. The middle region of the developing neural tube is the **mantle layer (intermediate zone)**. The third layer of the developing neural tube, located the farthest from the lumen, is the **marginal layer (marginal zone)**.

Neuroepithelial cells are quite long; initially, they extend from the lumen of the neural tube to its periphery, spanning all three layers. Hence one may name regions of these cells according to the zone in which they are located. When a neuroepithelial cell is ready to undergo mitosis, its nucleus migrates from the ventricular zone to the region of the cell occupying the marginal layer, and the cell enters the S phase (DNA synthesis) of the cell cycle. Upon completion of the S phase, the nucleus returns to the ventricular zone, and the cell shortens and becomes round so that the cell is located completely in the ventricular zone. It is here that the cell divides to give rise to two daughter cells, which may remain in the ventricular zone or migrate to the mantle layer or into the marginal layer (Fig. 2.4).

There appears to be at least two types of neuroepithelial cells: those that can be stained for **glial fibrillary acidic (GFA)** proteins (GFA positive) and those that lack GFA proteins (GFA negative). GFA-negative cells give rise to an enormous number of daughter cells that eventually differentiate into **migratory neuroblasts** that are no longer able to divide. GFA-positive cells also give rise to an immense number of daughter cells that will differentiate into **glioblasts** and **ependymal cells.** Neuroblasts arise first and glioblasts and ependymal cells originate later. Neuroblasts migrate from the ventricular zone into the mantle layer where they differentiate into **neurons**. Glioblasts give rise to all macroglia (**astrocytes** and **oligodendroglia**), and **ependymal cells** lining the ventricles of the brain and the central canal of the spinal cord.

As the neuroblasts are differentiating into neurons within the mantle layer they develop **axons** that grow into the marginal layer. Concomitantly, many glioblasts will migrate into the marginal layer, and those that become oligodendroglia form a protective cellular sheath around the developing axons, dendrites, and cell bodies within the mantle layer. Many of the axons will become myelinated by the

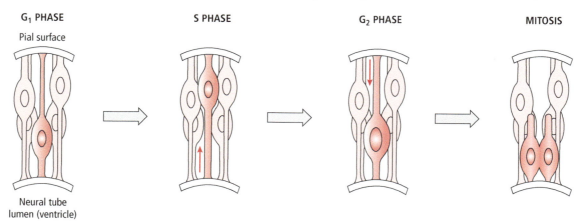

Figure 2.4 • Neuroepithelial cells are long cells that initially extend from the pial (marginal) to ependymal surface of the neural tube. As they enter the cell cycle to form new cells, their nuclei migrate along the length of the cell. G1 phase: the nucleus is in the vicinity of the ventricular surface and begins to migrate to the pial surface. S phase: the nucleus is in the pial surface and at the end of the S phase the nucleus begins its return to the ventricular surface. G2 phase: the nucleus reaches the ventricular surface and the cell begins to shorten. Mitosis (M phase): the cell divides to give rise to two daughter cells in the ventricular zone.

oligodendroglia. Thus, in the spinal cord, the mantle layer is the future **gray matter**, whereas the marginal layer, composed of numerous myelinated axons and attendant neuroglia, becomes the **white matter**.

Myelination in the CNS begins around the sixteenth week of gestation and continues until the individual is about 3 years old, although in the frontal lobes of the brain this process may continue into the early twenties and even later. It is interesting to note that there is a phylogenetic component to the sequence of myelination, in that the phylogenically older pathways are myelinated before the phylogenically newer pathways; also motor roots of spinal nerves are myelinated before sensory roots.

DEVELOPMENT OF THE SPINAL CORD

Basal plates, alar plates, and spinal ganglia

The basal plates and alar plates of the mantle layer are separated from each other by the sulcus limitans, a longitudinal furrow extending the entire length of the future spinal cord

The mantle layer increases in thickness in a disproportionate fashion, causing an anterior and posterior thickening (**basal plate** and **alar plate**, respectively) on each side of the entire length of the wall of the future spinal cord. These two thickenings are separated from each other by a longitudinal furrow, known as the **sulcus limitans**. The neuroblasts of the basal plate differentiate into motor neurons, responsible for the motor function of the spinal cord (**anterior gray column**), whereas neuroblasts of the alar plate differentiate into **interneurons**. Some of these interneurons will receive sensory information from primary sensory neurons (including those of the spinal ganglia). In the thoracic and upper lumbar regions (T1 through L2–L3) of the future spinal cord an intermediate thickening is discerned. Neuroblasts of this region, derived from the basal plate, will give rise to preganglionic sympathetic neurons (forming the **lateral gray column**) of the autonomic nervous system. It should be noted that various synonyms are in common use among neuroanatomists, and some of these are given at the end of the chapter.

The central midline of the floor (**floor plate**) and roof (**roof plate**) of the neural tube has few, if any, neuroblasts, and is devoid of nerve cell bodies. However, neuronal processes will be present in these regions, conveying information between the two halves of the spinal cord.

Neurons of the **spinal ganglia** arise from cells of the **neural crest**. These neurons are responsible for delivering information from the sensory receptor organs to the spinal cord for processing.

Histodifferentiation of neuroblasts

Neuroblasts are more or less spherical cells that differentiate to form the various classes of neurons: unipolar (pseudounipolar), bipolar, and multipolar

Neuroblasts, which begin their differentiation as relatively spherical cells, form two processes at opposite poles. Usually, one of the processes,

the **dendrite**, begins to arborize, whereas the other process, the **axon**, remains unbranched. The manner in which the axon and dendrite are established and modified, permits neurons to be classified as unipolar (pseudounipolar), bipolar, and multipolar (Fig. 2.5).

Unipolar (pseudounipolar) neurons are located in sensory ganglia. The two processes of each of these cells begin to grow toward and fuse with one another, forming a single process. This unified process then divides into two processes that grow in opposing directions. One of the processes, the **central process**, enters the posterior horn of the spinal cord where it may terminate or ascend to higher levels. Collections of central processes form the posterior roots of the spinal nerves. The other process, the **peripheral process**, is much longer. Collections of **peripheral processes** join the anterior root fibers to form the **spinal nerve**.

Bipolar neurons retain their two processes at opposing poles of the cell body. Dendrites of these cells collect information from the periphery of the body, whereas their axons deliver information to the CNS for processing. Bipolar neurons are associated only with the olfactory epithelium, cerebral cortex, retina, cochlear nucleus, and the vestibular nucleus.

Multipolar neurons, instead of having only two processes, develop several. One of these is the axon, whereas the remaining processes are dendrites. Collections of axons of multipolar neurons of the basal plate grow through the marginal zone to form the anterior root of the spinal nerve. As mentioned previously, they join with collections of peripheral processes of unipolar neurons of the spinal ganglia to form the spinal nerve.

Further differentiation of the basal and alar plates

Neuroblasts of the basal plate differentiate into multipolar neurons, whereas those of the neural-crest-derived spinal ganglia differentiate into unipolar (pseudounipolar) neurons

As neuroblasts of the **basal plate** differentiate they become multipolar neurons whose axons grow not only into but through the marginal layer and pierce the external boundary of the neural tube. Collections of these axons travel together forming the **anterior rootlets** of the spinal cord. These anterior rootlets join others in their vicinity to form the **anterior root** of a spinal nerve.

The central processes of the unipolar neurons of the spinal ganglia pierce the external aspect of the neural tube at the **posterolateral sulcus** and enter the marginal zone or the alar plate. Central processes that enter the alar plate form synapses with the **interneurons** whose axons form synaptic contacts with motor neurons of the basal plate and in this fashion form simple reflex arcs. Those that enter only the marginal layer travel to higher levels in the CNS. Many of these central processes travel *en masse* to reach their specific destinations. Collections of fibers going to the same destination are named accordingly.

As described previously, collections of the peripheral processes of unipolar neurons join the anterior roots to form the mixed (motor and sensory) **spinal nerves**. Motor nerves effect muscle contraction as well as contraction of

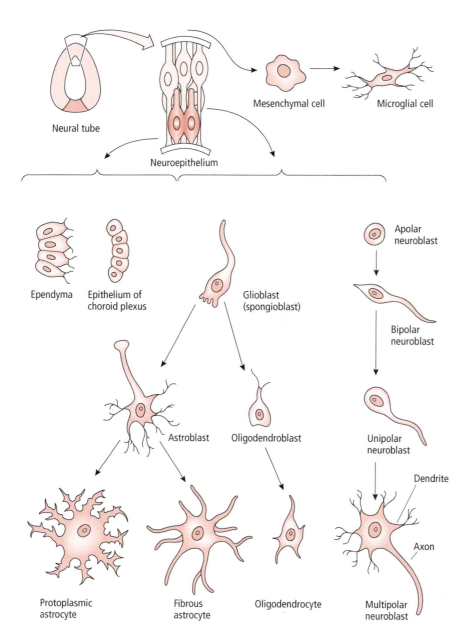

Figure 2.5 ● The origin of the developing cells of the CNS. Note that all of the cells are derived from the original neural tube, with the notable exception of the microglial cells, which are phagocytes of the CNS and originate from mesenchymal cells. Ependymal cells will line the ventricles of the brain as well as the central canal of the spinal cord; they also participate in the formation of the choroid plexus. Glioblasts will give rise to macroglia, namely the protoplasmic and fibrous astrocytes, as well as to oligodendrocytes; these cells are supporting cells of the CNS and, in the case of oligodendroglia, form myelin sheaths of CNS neurons. Neuroblasts are responsible for the formation of the neurons of the CNS.

myoepithelial cells that surround secretory acini, thereby inducing glandular secretion. Sensory nerves transmit information from outside or inside the body conveying it to the central nervous system. Motor nerves propagate information that goes *away* from the CNS to a muscle (or gland) and are called **efferent nerve fibers**. Sensory nerves propagate information *toward* the CNS and are called **afferent nerve fibers**.

Molecular bases of spinal cord differentiation

Factors secreted by the secondary signaling centers in the roof and floor plates of the neural tube are responsible for establishing the sensory and motor regions of the spinal cord

The division of the spinal cord into a posterior sensory and an anterior motor region is due to the differences in the concentration gradients of factors secreted by the secondary signaling centers located in the roof plate and the floor plate of the neural tube. These secondary signaling centers become established by the presence of bone morphogenic proteins 4 and 7 (BMP4 and BMP7) secreted by cells of the ectoderm anterior to the neural tube and by the presence of sonic hedgehog (SHH) secreted by cells of the notochord lying anterior to the neural tube. BMP4 and BMP7 trigger cells of the roof plate to secrete the following factors: BMP5, BMP7, dorsalin, and activin. Since the roof plate is exposed to a high concentration of these factors, they will induce the secretion of factors responsible for the formation of the sensory portion of the spinal cord. The high SHH concentration in the region of the floor plate triggers the release of factors that are responsible for the differentiation of the neuroblasts into motor neurons. Additionally, the anterior and posterior interneuron development and their migration to their proper locale are dependent on the presence of DOT1L (Disruptor of Telomeric 1 Like).

Readers interested in more specific information concerning these molecular events at work are referred to textbooks of human embryology and molecular neuroembryology.

Development of modalities

A typical spinal nerve possesses four functional components, known as modalities, whereas the typical cranial nerve possesses the same four plus three additional modalities

Motor fibers of spinal nerves that innervate skeletal muscle are said to be **general somatic efferent** (**GSE**) fibers, whereas motor fibers that innervate smooth muscle, cardiac muscle, or myoepithelial cells of glands are said to be **general visceral efferent** (**GVE**) fibers. Sensory fibers that bring information from somite-derived structures including the skin, skeletal muscle, tendons, and joints are called **general somatic afferent** (**GSA**) fibers, whereas those that bring information from the viscera (membranes, glands, and organs) are named **general visceral afferent** (**GVA**) fibers. The visceral components (GVE) belong to the **autonomic nervous system** (sympathetic and parasympathetic); however, GVA fibers are not autonomic fibers, even though they bring information destined for autonomic nervous system response. A typical spinal nerve has all of these four **functional components** (also known as **modalities**). It should be noted that cranial nerves carry the same four plus an additional three modalities, namely **special visceral efferent** (**SVE**) fibers (CN V, VII, IX, and X) which are motor fibers; **special somatic afferents** (**SSA**) fibers (CN VIII); and **special visceral afferents** (**SVA**) fibers (CN VII, IX, and X), which are sensory fibers. These modalities develop in specific regions along the alar and basal plates, forming columns of recognizable gray matter along the length of the developing nervous system. Those developing in the alar plate, in vertical rows, in posterior to anterior position are the SSA, GSA, SVA, and GVA, whereas those developing in the basal plate, again in an anterior to posterior direction, are GVE, SVE, and GSE (Fig. 2.7C).

As discussed above, during early embryogenesis, clusters of cells, known as **somites**, form on either side of the developing spinal cord. These somites form the skeletal muscles, skin (dermatomes), and connective tissue of the body. Skin and skeletal muscle derived from a particular somite are innervated by neurons arising from the segment of the spinal cord (and spinal ganglion) in its vicinity, giving rise to segmental innervation of the embryo that is conserved in the adult. Since, in the early embryo, the length of the vertebral canal equals the length of the spinal cord, the segment of the spinal cord is at the same level as the body segment. However, because growth in body length surpasses the lengthening of the spinal cord, in the adult the spinal cord is much shorter than the vertebral canal and thus (especially caudally), the body segment innervated lies farther away than it does in the embryo. The region of the vertebral canal that is devoid of spinal cord in the adult is occupied by the accumulation of posterior and anterior rootlets, collectively known as the **cauda equina** (as well as the non-nervous pial filament, the **filum terminale**). The space occupied by these structures is an enlarged subarachnoid space, known as the **lumbar cistern**.

DEVELOPMENT OF THE BRAIN

During early development the brain is subdivided into three morphologically recognizable components: the prosencephalon, mesencephalon, and rhombencephalon

During early development the primary brain divisions – the rhombencephalon, mesencephalon, and prosencephalon (from caudal to rostral) – were established, as were the three flexures, the cephalic, cervical, and pontine flexures. Since the development of the rhombencephalon is similar to that of the spinal cord discussed previously, the embryology of the brain will be detailed in the reverse order to permit a more logical approach to its study. It should be noted that the organization of the basal and alar plates noted in the spinal cord is somewhat altered in the brain, in that the role of the basal plate becomes diminished and the role of the alar plate becomes heightened. Table 2.4 summarizes the adult structures derived from the three primary divisions (Fig. 2.6).

Table 2.4 ● Derivatives of the primary brain divisions.

Primary brain divisions	Derivatives of the divisions
Prosencephalon (forebrain)	
Telencephalon	Cerebral hemispheres
	Cerebral cortex
	Corpus striatum
	Hypothalamus (superior aspect)
	Third ventricle (superior aspect)
	Lateral ventricles
Diencephalon	Epithalamus
	Thalamus
	Hypothalamus (caudal aspect)
	Third ventricle (caudal aspect)
Mesencephalon (midbrain)	Cerebral peduncles
	Tectum
	Tegmentum
	Cerebral aqueduct (of Sylvius)
Rhombencephalon (hindbrain)	
Metencephalon	Pons
	Cerebellum
	Middle and superior cerebellar peduncles
	Fourth ventricle (rostral aspect)
Myelencephalon	Medulla oblongata
	Inferior cerebellar peduncles
	Fourth ventricle (Inferior aspect)

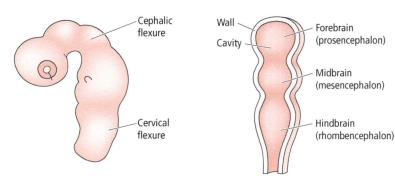

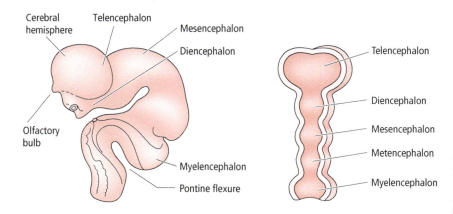

Figure 2.6 ● Three-dimensional representation of the human brain (left) and its longitudinal section (right; as if the brain were stretched out) at 6 and 27 mm (5 and 7½ weeks) of development. Note that the three primary brain divisions of the 6 mm embryo give rise to the five divisions of the 27 mm embryo.

Myelencephalon

The posteriormost region of the developing brain is the myelencephalon

The **myelencephalon** gives rise to the medulla oblongata, the rostral continuation of the spinal cord. During the development of the myelencephalon, the pontine flexure arises, and appears to stretch the roof plate, which becomes a thin epithelial layer. The alar plates of the two sides move farther away from one another, and the intervening cavity of the hindbrain becomes the caudal aspect of the **fourth ventricle** (Fig. 2.7). The junction between the anterior edge of the alar plates and the roof plate is known as the **rhombic lip**, composed of an inferior and a superior rhombic lip (located in the myelencephalon and metencephalon, respectively). Nuclei of cranial nerves VIII through XII as well as part of the nucleus of the trigeminal nerve (CN V) are formed from the alar plates (sensory neurons) and basal plates (motor neurons). Also, neuroblasts from the alar plate migrate anteriorly to become the neurons constituting the **inferior olivary nuclei** (as well as the **gracile nucleus** and **cuneate nucleus**. Subsequent to the establishment of the nuclei, fiber tracts will appear, the most pronounced of which are the **pyramids**, collections of corticonuclear and corticospinal axons on the anterior surface of the medulla oblongata.

Metencephalon

The rostral-most portion of the rhombencephalon is the metencephalon

The nuclei of cranial nerves VI and VII as well as the remaining portion of cranial nerve V arise from the alar and basal plates of the **metencephalon**. Moreover, two important, large structures develop from cells migrating from the alar plates of the metencephalon. One group of cells migrates anteriorly to form the **pontine nuclei**, whereas the other group migrates to the superior rhombic lip to form the **cerebellum**.

Three groups of motor neurons develop from the **basal plate** of the metencephalon: the **general somatic efferent** (gives rise to the nucleus of the abducent nerve), **general visceral efferent** (parasympathetic fibers for the facial nerve), and **special visceral efferent** (gives rise to the motor nuclei of the trigeminal and facial nerves, which will innervate muscles of branchiomeric origin: the first and second pharyngeal arches).

Three groups of sensory nuclei develop from the **alar plates** of the metencephalon: the **somatic afferent** (neurons of the trigeminal and some neurons of the vestibulocochlear nerves), **general visceral afferent**, and **special visceral afferent**.

DEVELOPMENT OF THE NERVOUS SYSTEM

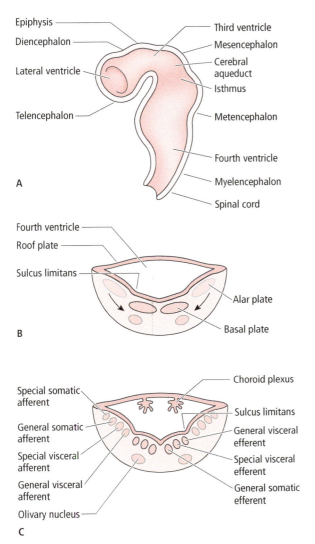

Figure 2.7 ● Diagrams of the developing nervous system. (A) Longitudinal section of the brain and spinal cord of an 11 mm (6 weeks) embryo. (B) Transverse section of the myelencephalon of an early embryo depicting the alar and basal plates. The arrows indicate the migration of cells from the alar plates to form the future olivary nuclei. (C) Transverse section of a later stage of the myelencephalon depicting the development of nuclei from the alar and basal plates.

Rhombomeres

The neural tube, similarly to the paraxial mesoderm, begins to form segmentations, but these segments, known as **neuromeres**, are not only incomplete but they are also transient. The term neuromere is a generalized name and, depending on the region of the developing brain, they are named prosomeres (in the forebrain), mesomeres (in the midbrain), and **rhombomeres** (in the hindbrain). There is a great deal of controversy concerning the existence of prosomeres and mesomeres, because many neuroembryologists do not see signs of segmentation in the forebrain or in the midbrain; however, there is almost total agreement concerning the presence of rhombomeres. Therefore, this textbook will not discuss prosomeres or mesomeres. As in the region of the neural tube responsible for the formation of the spinal cord, the process of delineation of the forming brain into its three regions, **prosencephalon (forebrain)**, **mesencephalon (midbrain)**, and **rhombencephalon (hindbrain)** is under the control of various **homeobox genes (HOX genes)** whose expression appears to be regulated by **retinoic acid**.

Four sets of **HOX genes** are known, HOXA, HOXB, HOXC, and HOXD, each located on separate chromosomes. Each set is composed of **13 genes** and they are named accordingly, for the sake of example HOXD6. The same numbered gene on each of the four sets, such as HOXA6, HOXB6, HOXC6, and HOXD6, are considered to be **paralogous groups**. These HOX genes are essential for normal development because they govern anterior to posterior arrangement.

Caudal to the **cephalic flexure** eight **rhombomeres** (Fig. 2.8) appear, temporarily, in the neural tube and these

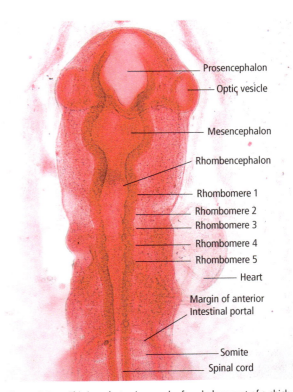

Figure 2.8 ● This is a photomicrograph of a whole mount of a chick embryo at about 30 hours of incubation stained with Eosin. The prosencephalon with the optic vesicles and the mesencephalon are clearly evident as is the rhombencephalon, which is beginning to display segmentations, known as rhombomeres. At 30 hours of development the chick embryo displays only five pairs of the eight pairs of rhombomeres, and the last three pairs will develop a few hours later. It should be noted that in humans, the rhombomeres appear by the 29th day of gestation. At this time there are about 10 pairs of somites that have developed from the mesoderm. The margin of the anterior intestinal portal is represented by the faint double line going across the cranial region of the first somite ×56.

26 ● ● ● **CHAPTER 2**

will be responsible for the formation of the **rhombencephalon** (hindbrain). Each rhombomere will give rise to ganglia and nerves that are specific to its location as well as to cranial neural crest cells. The development of the rhombomeres two through eight is controlled by HOX genes and is also regulated by retinoic acid. The development of rhombomere 1 is not controlled by HOX genes; instead, it is governed by genes that control the development of the mesencephalon and the prosencephalon.

Rhombomeres appear, in an antero-posterior linear sequence, relatively early during development. Initially, there may be some intermingling of cells of neighboring rhombomeres; however, very soon after their formation, the boundaries of adjacent rhombomeres become impenetrable, so that each rhombomere becomes a specialized compartment whose cells are restricted from entering neighboring rhombomeres. It is important to realize that these eight rhombomeres are responsible for providing both sensory and motor nerve supplies to the entire head and neck and also are responsible for managing the necessary functioning of heart contraction, respiration, swallowing, digestion, balance, hearing, touch, and taste. Additionally, it is the progeny of the cells of these rhombomeres that manage the movements of the eye, tongue, and mandible and even the muscles of the face that produce facial expression. Moreover, the processes of sleep, alertness, and the maintenance of normal posture also depend on the cells that have their origins in these rhombomeres. Many of the cranial motor nerves possess a dual rhombomeric origin, first appearing in the even numbered rhombomere and later receiving axons from their posteriorly positioned odd-numbered rhombomere. For example, the motor neurons of the trigeminal nerve (CN V) originate in the second rhombomere and their axons are shortly joined by axons whose cell bodies originate, and remain in, the third rhombomere. In a similar fashion, the facial nerve motor neurons (CN VII) originate in rhombomeres 4 and 5; the glossopharyngeal nerve motor neurons (CN IX) arise from rhombomeres 6 and 7. Motor nerves of the trochlear nerve (CN IV) originate in the first rhombomere. Although it was supposed that the vestibulocochlear nerve (CN VIII) originated in the fourth rhombomere, it is now believed that it arises from a placode, rather than from a rhombomere. The vagus nerve (CN X) originates from rhombomere 7, and the accessory nerve (CN XI) originates from rhombomere 8.

Migrating neural crest cells arising from the eight rhombomeres also form non-neural progeny. Three streams of neural crest cells migrate into the forming pharyngeal arches to differentiate into a special type of mesenchymal cells, known as **ectomesenchymal cells.** Although of neural crest origin, they act as if they were mesodermally derived mesenchymal cells, and form much of the connective tissue elements, including cartilage and bone of the developing pharyngeal arches. Other ectomesenchymal cells that migrate into the first pharyngeal arch are responsible for the formation of all tissues of teeth and their supporting apparatus, with the exception of enamel. The majority of the three streams of migrating neural crest cells arise from rhombomeres 1 and 2,

destined for the first pharyngeal arch; rhombomere 4, migrating into the second pharyngeal arch; and rhombomeres 6 and 7, destined for pharyngeal arches 4 and 6.

Still other neural crest cells remain near their site of formation and associate with the epithelial tissue in their vicinity, becoming cranial **ectodermal placodes**. They can form various structures, such as the lens placode that develops into the lens of the eye, otic placode that will form the structures responsible for equilibrium and hearing, nasal placode that is responsible for the formation of the olfactory epithelium, sensory ganglia of CN V, VII, IX, and X, as well as the adenohypophyseal placode, that will form the adenohypophysis (anterior pituitary gland).

Cerebellum

The inferior aspect of the two rhombic lips thickens to form the right and left cerebellar plates that will fuse with one another to develop into the two cerebellar hemispheres

As neurons of the alar plate undergo intense mitotic activity, their proliferation causes an expansion of each rhombic lip (the dorsolateral aspect of the neural tube). As the inferior aspect of the rhombic lip forms a bulge in the wall of the fourth ventricle, its superior aspect increases greatly in size and these two thickenings are known as the **cerebellar plates** (Fig. 2.9). The cerebellar plates of the right and left sides grow toward each other and fuse in the midline, each half forming the future cerebellar hemisphere of its own side. The inferior aspect of the cerebellar plates will become greatly reduced in size and remain as the roof over the ventricle; in the adult it forms the **vermis**. During the development of the cerebellum, neuroblasts enter the marginal layer and form the cerebellar cortex. The huge number of cells being formed triggers the folding of the cortex into leaf-like structures called **folia**. Neuroblasts that remain in the mantle layer of the forming cerebellum also proliferate to form the **dentate** and **deep cerebellar nuclei**. Axons to and from the cerebellar cortex form the very large tracts, the superior, middle, and inferior cerebellar peduncles as well as the corticopontocerebellar pathway.

Mesencephalon

The mesencephalon is the midbrain and surrounds the mesencephalic aqueduct (cerebral aqueduct of Sylvius)

The **mesencephalon** is situated rostral to the pons and is the region of the midbrain flexure. Its wall becomes quite thick, obliterating much of the central lumen, which becomes a narrow cleft – the **mesencephalic aqueduct** (**cerebral aqueduct [of Sylvius]**) connecting the third and fourth ventricles. The alar plates give rise to the **tectum**, composed of the **superior** and **inferior colliculi**, which are reflex centers for visual and auditory signals, respectively. Additionally, the alar plate is believed to give rise to the **periaqueductal gray**. The basal plate is responsible for the formation of the **tegmentum**, whose **red nucleus** and **substantia nigra**, along with the nuclei of the **oculomotor** and

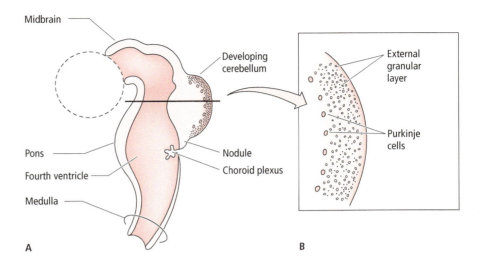
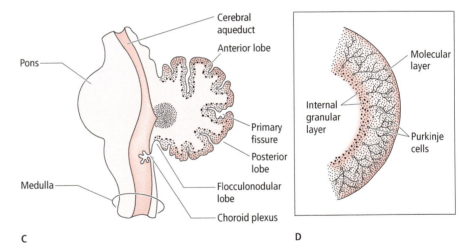

Figure 2.9 ● The developing cerebellum: (A, C) longitudinal sections at 6 and 17 weeks, respectively; (B, D) cellular morphology of the developing cerebellum at 6 weeks and shortly after birth. At 11 weeks the cerebellar plate forms and even Purkinje cell precursors are evident. By the 12th week the right and left cerebellar hemispheres may be observed, separated from each other by the future vermis. By the 17th week of development, folia are apparent and the primary fissure is clearly recognizable.

trochlear nerves, arise from its neuroblasts. With continued development of the cerebral cortex the anterior aspect of the mesencephalon thickens considerably, forming the **cerebral peduncles**, fiber tracts leading to and from the cerebral cortex.

Prosencephalon

The **prosencephalon** is the forebrain and is divided into two regions, the diencephalon and the telencephalon.

Diencephalon

The diencephalon, the caudal part of the forebrain, gives rise to the epithalamus, thalamus, and hypothalamus

The three regions of the **diencephalon** – the epithalamus, thalamus, and hypothalamus – form as three separate bulges in the walls of the **third ventricle**. The habenular nuclei and the pineal body form the **epithalamus**. The largest portion of the diencephalon, the **thalamus**, is separated from the underlying hypothalamus by the hypothalamic sulcus, a transverse groove in the wall of the third ventricle. The close functional association of the thalamus with the cerebral cortex necessitates a synchronous developmental association between these two structures. As nuclei within the thalamus (medial and lateral geniculate bodies, as well as the medial, lateral, and anterior thalamic nuclei) increase in size, they bulge into the third ventricle, decreasing its size. Frequently, the nuclei of the right and left thalami fuse at a small region across the third ventricle, forming the **interthalamic adhesion (massa intermedia)**. Nuclei within the **hypothalamus** form anterior to the hypothalamic sulcus. These nuclei will be responsible for monitoring and controlling various endocrine and autonomic functions. Additionally, an anterior evagination of the hypothalamus is responsible for the formation of the **median eminence** and the **infundibulum** (see below).

As the diencephalon develops, the roof of the third ventricle becomes extremely thin, so that it is formed only of a layer of ependymal cells and the vascular pia mater. This vascular roof invaginates into the third ventricle, forming the choroid plexus. Although choroid plexuses develop in all of the ventricles of the brain, their developmental process is identical to that occurring in the third ventricle and is not treated again during the discussions of the other ventricles. Concomitant with the development of this choroid plexus, three bulges become evident on the floor of the third ventricle, and these form the mammillary bodies, the infundibulum, and the optic chiasm.

Hypophysis (Pituitary gland)

The hypophysis (pituitary gland) is an important endocrine gland that develops from the diencephalon as well as from an evagination, the adenohypophyseal pouch (Rathke's pouch), derived from the roof of the oral cavity

The **infundibulum** of the diencephalon is a downward evagination of the floor of the third ventricle. As the infundibulum grows down, it meets an ectodermal diverticulum of the oral cavity, known as the **adenohypophyseal pouch (Rathke's pouch)**. These two structures fuse with each other to form the **hypophysis (pituitary gland)**. The **neurohypophysis** (median eminence, stalk, and pars nervosa) arises from the infundibulum, whereas the **adenohypophysis** (pars tuberalis, pars distalis, and pars intermedia) derives from the **adenohypophyseal pouch** (Fig. 2.10). A small colloid-filled cleft frequently remains as a vestige of the lumen of Rathke's pouch, in the pars intermedia.

Eye development

The nervous and pigmented layers of the retina of the eye develop from the evagination of the ventrolateral wall of the diencephalon

During the third week of gestation, the anterolateral wall on each side of the future diencephalon gives rise to an **optic vesicle**. These hollow vesicles grow in a lateral direction, reach the basal lamina of the head ectoderm, and induce the ectoderm to form the **lens placode**, the precursor of the lens of the eye. The optic vesicle invaginates to form a two-layered structure, known as the **optic cup**, which gives rise to the nervous and pigmented layers of the retina; the connection between the optic cup and the forming brain becomes known as the **optic stalk** (Fig. 2.11). As axons growing out of the developing retina increase the thickness of the wall of the optic stalk, its lumen becomes obliterated, and the optic stalk becomes known as the **optic nerve**. The optic nerves of the right and left sides join in the midline to form the **optic chiasm**, where axons arising from the medial (nasal) half of each retina cross over to the opposite side.

As the rim of the optic cup, in contact with the lens placode, encircles the prospective lens to form the **iris**, a slight cleft (**choroid fissure**) remains, permitting the future central artery and vein of the retina to access the retina.

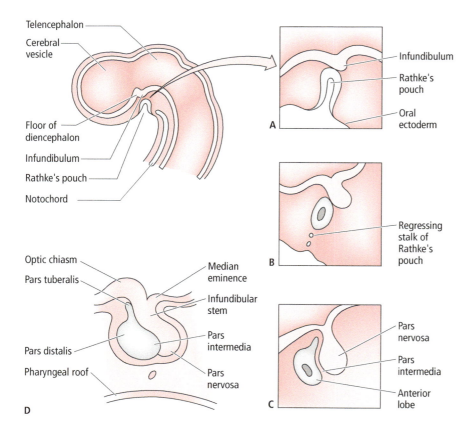

Figure 2.10 ● Development of the hypophysis (pituitary gland). Sagittal section of a 5-week-old embryo displaying the floor of the diencephalon with the beginning of the infundibulum and the roof of the stomodeum, showing its outpocketing, known as the adenohypophyseal pouch. (A) By the seventh week of development, the adenohypophyseal pouch contacts the infundibulum. (B) Around the ninth week of development, the adenohypophyseal pouch loses contact with the stomodeum. (C) Around the tenth week of development, cells of V proliferate to form the adenohypophysis. (D) The formed hypophysis.

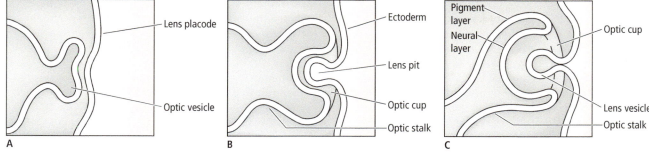

Figure 2.11 ● Development of the eye begins with the outgrowth of the diencephalon, known as the optic vesicle, around the time that the embryo is 4.5 mm long from crown to rump. (A) The optic vesicle induces the ectoderm to form the lens placode. (B) By the time the embryo is 5.5 mm long, the optic vesicle invaginates to form the optic cup and the lens placode differentiates to form the lens pit, which at this point is wide open to the outside. (C) The lens pit begins to form the lens vesicle around the time the embryo is 7.5 mm long and the optic cup forms a two-layered structure, the neural and pigment layers. The optic stalk, which will form the optic nerve, is also being established.

Telencephalon

The anteriormost portion of the prosencephalon, the telencephalon, is mainly responsible for the formation of the cerebral hemispheres

The rostral extent of the neural tube is the **telencephalon**, which extends from the diencephalon to the **lamina terminalis**, the anteriormost boundary of the neural tube. The telencephalon undergoes a tremendous growth spurt, forming the two telencephalic vesicles that will overgrow much of the brain by developing into the two **cerebral hemispheres**. The lumina of the telencephalic vesicles remain as the large, cerebrospinal fluid-filled lateral ventricles. These two ventricles will communicate with each other as well as with the third ventricle via the **interventricular foramen** (of Monro).

Growth of the telencephalic vesicles is asymmetric, in that the lateral expansion, forming the prospective **insula** (L., "island"), enlarges less than do the anterior, posterior, and medial regions. In fact, some of these regions overgrow and cover the insula and, in the adult, they are named the temporal, frontal, and frontoparietal opercula (L., "lid"; singular: operculum). The anterior expansion is also slight, forming the future **olfactory bulb**. Continued growth of the telencephalic vesicles within the limited space of the cranial cavity is responsible for the folding of the developing cerebral cortex into elevations called **gyri** (L., circle, singular: gyrus) and the intervening grooves, known as **sulci (singular: sulcus)**. Some of the deeper and constant grooves are named **fissures**; these separate lobes of the cerebral hemispheres from each other.

Major fiber tracts – the **optic chiasm**, **anterior commissure**, and the **corpus callosum** – develop in the region of the lamina terminalis. These fiber tracts connect similar areas of the right and left cerebral hemispheres to each other.

The superior aspect of the telencephalic vesicle, known as the **pallium**, is responsible for the formation of the **cerebral cortex**, a thin layer of gray matter covering the great mass of white matter of the cerebral hemispheres. Neuroblasts of the inferior aspect of the telencephalic vesicles undergo intense mitotic activity, forming a highly cellular region. Due to the large number of axons that pass through this cell-rich area, connecting the cerebral cortex to other regions of the brain and to the spinal cord, this area appears striated and is thus referred to as the **corpus striatum**. Many of these fibers join to form a major fiber tract, the **internal capsule**, which will isolate clusters of cells of the corpus striatum from each other; these cell clusters develop into the **lentiform nucleus** (putamen and globus pallidus) and the **caudate nucleus**.

Molecular bases of mesencephalon and prosencephalon development

Homeobox genes control development of the midbrain and forebrain

Unlike the developing spinal cord and hindbrain, the development of the first rhombomere (rhombomere 1), midbrain (mesencephalon), and forebrain (prosencephalon) are controlled by the products of **homeobox genes LIM1**, released from the prochordal plate, and the products of the **homeobox gene OTX2 (orthodenticle homeobox 2)**, released from the neural plate. These gene products activate the expression of additional **homeobox genes**, such as **EMX1 (empty spiracles-like 1 gene)**, **EMX2**, and **OTX1** that are responsible for the determinations of the specific regions of the mesencephalon and prosencephalon.

Therefore, the boundaries among the various regions of the brain become established due to the influences of homeobox genes and their transcription factors. Accordingly, the anteriormost region of the rhombencephalon belongs not to the hindbrain but to the midbrain. Subsequent to the determination of the boundaries, the embryonic brain establishes a slight swelling at the border of the mesencephalon and rhombencephalon, known as the **anterior neural ridge**, in the floor of the anterior region of the prosencephalon. Both the narrowed margin (isthmus) on the boundary between the mescencephalon and rhombencephalon and the anterior neural ridge are **organizing centers** that, under the influence of **fibroblast growth factor 8 (FGF8)**, release transcription factors that facilitate the expression of other homeobox genes.

The **isthmus** is responsible for the development of the cerebellum and the tectum and in order to accomplish that FGF8 stimulates the expression of two homeobox genes, **engrailed 1 (En1)** and **engrailed 2 (En2)** as well as the expression of the proto-oncogene known as **wingless type 1 (WNT1)**. The expression of En1 lasts only until En2 is expressed; therefore, it has a very short window of opportunity and it supports the development of the tectum (the four colliculi), cranial nerves III and IV, as well as most of the cerebellum. En2 is responsible for the development of the remainder of the cerebellum. Both En1 and En2 require the support of WNT1 to fulfill their functions. It should be noted that En1 expression does not shut off permanently because it will be expressed in somite, sternum, and limb development subsequent to its function in the midbrain.

Very early in development, FGF8 in the **anterior neural ridge** stimulates the expression of the transcription factor **forkhead box G1 (FOXG1)**, which is responsible for the formation of the cerebral hemispheres of the telencephalon and the formation of the optic vesicles that will form the retina and the optic nerves of the eyes.

As noted previously, the prochordal plate expresses the **sonic hedgehog protein (SHH protein)**, the protein product of the **sonic hedgehog gene**, which stimulates the expression of the homeobox gene **NK2 homeobox 1 (NKX2-1)**. It is this gene that plays a major role in the development of the hypothalamus, thus patterning the anterior aspect of the developing neural tube.

The **posterior aspect** of the neural tube is under the influence **of bone morphogenic proteins (BMP4** and **BMP7)**. These proteins are produced by the epithelial cells of the roof plate and act as a function of the concentration gradient differences of the SHH protein (see above, in the section on "Molecular basis of spinal cord differentiation").

CLINICAL CONSIDERATIONS

Spinal cord

The most common defect of the neural tube is incomplete fusion of the neural groove, resulting in **spina bifida cystica**. This involves a protrusion of the spinal cord tissue through a bony defect, so that a portion of the nervous tissue is visible on the individual's back as a cyst-like structure. There are various degrees of spina bifida, the worst of which involves mental retardation as well as paralysis below the level of the lesion. Although the cause of this anomaly is not known, there are several agents, such as an excess maternal intake of vitamin A and hyperthermia, which increase its incidence. Newer diagnostic methods, such as ultrasound examination during the 11th and subsequent weeks of gestation, as well as measuring fetoprotein levels in the amniotic fluid, can indicate and detect the presence of spina bifida cystica.

Spinal dermal sinus is a condition that frequently displays the presence of a slight depression, similar to a dimple, in the midline of the sacral region of the back. This slight depression is indicative of the final fusion of the sacral spinal cord level, where the separation of the neural tube from the surface ectoderm was almost incomplete. In fact, occasionally a fibrous cord persists between the dura mater and the dermis just beneath the dimple.

Brain

Hydrocephalus is a condition where the ventricles of the brain or the subarachnoid spaces (in external hydrocephalus) accumulate an excessive amount of cerebrospinal fluid (CSF). In the neonate this is most frequently a result of occlusion of the mesencephalic aqueduct that hinders the flow of CSF from the third into the fourth ventricle. In very extensive cases, the continued manufacture of CSF by the choroid plexuses of the two lateral and third ventricles increases the fluid volume of the ventricles, resulting in enlargement of the head, reduction in the thickness of the bones of the cranium, and compression of the walls of the ventricles. Such compression may cause lamination of the cerebral cortex resulting in severe mental retardation. A type of hydrocephalus, known as **communicating hydrocephalus**, is caused by the **Arnold–Chiari malformation**. This anomalous condition is characterized by herniation of the cerebellum and medulla oblongata into the foramen magnum, which impedes absorption of CSF, leading to the expansion of the ventricles with subsequent enlargement of the head. Arnold–Chiari malformation usually accompanies severe cases of spina bifida.

Anencephalus is due to the lack of fusion of the anterior neural groove. It is thus similar to spina bifida, in that the nervous tissue (in this case the brain) is open to the environment rather than being enclosed in the bony cranium. Usually, the nervous tissue is directly continuous with the skin of the scalp. Frequently vascular agenesis accompanies this condition, and much of the forming brain becomes necrotic a few weeks prior to birth.

Microcephaly is a condition with varied causes, such as maternal pelvic X-irradiation, a recessive genetic trait, or infection of the fetus with toxoplasmosis. As the name implies, microcephaly is characterized by a small head and small brain. The cerebral cortex is usually hypocellular with a resultant mental retardation and predisposition to motor disorders and seizures.

Agyri (lissencephaly) is characterized by the absence of fissures and sulci and an indiscriminate arrangement of neurons. This occurs as a result of decreased cellular activity and a consequent diminishing of brain growth.

Corpus callosum agenesis, an infrequent but interesting condition, is characterized by the absence of the corpus callosum, the major communicating pathway between the two cerebral hemispheres. Occasionally the condition is asymptomatic and individuals may be of average IQ. Frequently, however, mental deficiency and the occurrence of tremors and seizures are present.

Holoprosencephaly, defective development of the prosencephalon, may be of varying severity, extending from major malformation of the calvaria, forebrain, and midfacial skeleton to minor malformations involving these regions. The primary etiology of this condition is the consumption of alcohol during the first month or so of pregnancy. It should be stressed that even small quantities of alcohol consumed during this time and even a single instance of "binge" drinking may cause severe retardation and deformations of the midface. The association of alcohol consumption and holoprosencephaly is also referred to as the **fetal alcohol syndrome**.

SYNONYMS AND EPONYMS OF THE NERVOUS SYSTEM

Name of structure or term	Synonym(s)/eponym(s)
Alar plate	Dorsolateral lamina
Basal plate	Ventrolateral lamina
Mesencephalic aqueduct	Aqueduct of Sylvius
Embryoblast	Inner cell mass
Floor plate	Ventral lamina
Hyoid arch	Second pharyngeal arch
Interventricular foramen	Foramen of Monro
Mandibular arch	First pharyngeal arch
Mantle layer	Intermediate zone
Marginal layer	Marginal zone
Mesencephalon	Midbrain
Pharyngeal arches	Branchial arches
Pharyngeal grooves	Branchial grooves
Pharyngeal membrane	Closing plate
Pharyngeal pouches	Branchial pouches
Primitive node	Hensen's node
Prosencephalon	Forebrain
Rhombencephalon	Hindbrain
Roof plate	Dorsal lamina
Unipolar neurons	Pseudounipolar neurons
Ventricular zone	Ependymal layer (matrix cell layer)

FOLLOW-UP TO CLINICAL CASE

This newborn has a myelomeningocele, generically known as **spina bifida**. A myelomeningocele is one of the disorders in a group of conditions known as neural tube defects, otherwise known as dysraphism. A neural tube defect can occur along any part of the developing CNS. When it affects the region that should eventually become the brain it is known as anencephaly, meaning the brain is literally absent. However, it most commonly affects the most caudal portion and becomes known as spina bifida. There is failure of closure and neurulation of the neural tube. This occurs during the first few weeks of gestation.

Spina bifida cystica refers to an open spinal defect where nervous tissue and/or meninges protrude from a defect in the vertebral column and are exposed to the outside environment on the surface of the back. A myelomeningocele is an open defect where the spinal cord and meninges are exposed and often form a tumor-like mass in the midline just above the buttock. **Spina bifida occulta** refers to a closed spinal defect where skin covers the defect, and the spinal cord and meninges are within the defective vertebral column. A closed spinal defect is obviously milder than an open one. An overlying cutaneous abnormality, such as a tuft of hair or hemangioma, or a subcutaneous lipoma (a benign fatty tumor), may betray an underlying closed spinal defect. Both give rise to neurologic deficits, though those with open spinal defects are already present at birth and tend to be severe. Neurologic deficits, such as leg weakness, sensory deficits in the legs, and urinary incontinence, can occur much later in a closed spinal defect. This can present any time from early childhood to adulthood. It often occurs secondary to a tethered cord syndrome (a low-lying conus medullaris related to anchoring of the spinal cord or filum terminale to surrounding tissues).

A myelomeningocele needs to be surgically repaired within 48 hours of birth. The mortality rate from infection or other complications would otherwise be near 100%. A downward herniation of the cerebellum through the foramen magnum is always present (Chiari II malformation). Hydrocephalus may also be present, and both of these conditions may require surgery. Orthopedic malformations of the hip and/or feet often occur and may need to be corrected surgically.

Folic acid supplement very early in pregnancy decreases the incidence of neural tube defects, although the reason is unclear. The incidence is decreasing in the Western world, though it is still a common fetal malformation.

QUESTIONS TO PONDER

1. What is the importance of the transformation of the embryoblast of the blastocyst into the trilaminar germ disc?

2. What is the significance of Hensen's node?

3. What is the significance of homeobox genes?

4. What is the significance of the cellular arrangement of the developing neural tube?

5. What is the similarity between the development of the pituitary gland and the eye?

CHAPTER 3

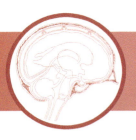  

Histophysiology of the Nervous System

CLINICAL CASE

NEURONS

NEUROGLIA

GENERATION AND CONDUCTION OF NERVE IMPULSES

CLINICAL CONSIDERATIONS

SYNONYMS AND EPONYMS OF NERVOUS SYSTEM HISTOPATHOLOGY

FOLLOW-UP TO CLINICAL CASE

QUESTIONS TO PONDER

CLINICAL CASE

A 32-year-old patient describes numbness and tingling in the left arm that began 5 days ago. The patient has also noted some recent "staggering" and being unbalanced when walking. There is no dizziness or weakness. The patient knows of no precipitating event and has no other complaints. Upon asking about past history, the patient recalls an episode of bilateral leg paralysis and poor sensation in the legs about 3 years ago. This improved over the course of 2 months, and the legs got back to normal. No specific etiology was found at that time, and the patient did not follow up with a doctor.

Examination shows sensory loss and altered sensation to stimulation in the left distal upper extremity and hand. Finger movements in both hands show poor coordination, somewhat more so on the left, but there is no weakness found in the face, arms, or legs. Gait is mildly ataxic. The patient cannot walk heel to toe in a straight line. Reflexes in the upper limbs are normal and are brisk at the knees and ankles but not necessarily pathologically so. Neurologic exam is otherwise unremarkable.

Aside from the meninges, blood vessels, and minor connective tissue elements, the central nervous system (CNS) is composed only of cells. These are categorized into two types: **neurons**, which receive and transmit impulses, and **neuroglia**, which support and facilitate the proper functioning of the neurons. A collection of neuronal cell bodies within the CNS is referred to as a **nucleus** (not to be confused with the nucleus of a cell), whereas a collection of neuronal cell bodies in the peripheral nervous system (PNS) is known as a **ganglion**. It was believed that **neurogenesis**, the formation of new neurons, does not occur in adults; however, it has been shown that in the brain, especially in the hippocampus, and possibly in the spinal cord, **neural stem cells**, that possess **centrioles**, have the ability to differentiate and form new neurons throughout an individual's life.

Neurons are excitable cells that are capable of conducting electrical impulses, by varying the voltage gradient across their cell membranes. They form **reflex arcs**, where

---

A Textbook of Neuroanatomy, Third Edition. Maria A. Patestas, Amanda J. Meyer, and Leslie P. Gartner.
© 2025 John Wiley & Sons, Inc. Published 2025 by John Wiley & Sons, Inc.
Companion website: www.wiley.com.go/patestas/neuroanatomy3e

information from the periphery is carried by a sensory neuron to the CNS and is directly transmitted to a motor neuron for a motor response. This is the simplest reflex arc because it involves only two neurons. An example of a **two-cell reflex arc** is the patellar reflex (knee jerk). Most reflex arcs are **three or more cell reflex arcs**, since they include at least a third neuron, the **interneuron**, interposed between the sensory neuron and the motor neuron. The reflex response associated with a three-cell reflex may be overridden when interneurons receive an input from higher centers of the brain that prevent them from transmitting the information from the sensory neurons to the motor neurons. An example of such a reflex override is when one picks up a hot cup of coffee and instead of dropping it on the dining room floor the person quickly places it back on the table. Information from the brain inhibited the interneurons from transmitting the information to the motor neurons that would have opened the fingers of the hand so that the hot cup could be dropped.

Neurons communicate with each other as well as with other target cells through specialized cell junctions, known as **synapses**. There are two types of synapses:

1. **electrical synapses** (**gap junctions**), where ions or small molecules may go from one cell into another cell by traversing small, contiguous channels present in the cell membranes of the two cells;
2. **chemical synapses**, where one cell releases ligands (e.g., neurotransmitters) into a narrow intercellular space, known as the **synaptic cleft**, and the ligand binds to receptors of the other cell's membrane, initiating the desired result.

Neuroglia are located both in the CNS and in the PNS. They act to support the neuron, both physically and nutritionally, to protect the neuron by forming an insulating sheath around it and its processes, aid neurons in establishing synapses with other neurons, phagocytose invading pathogens and defunct neurons, modulate the neuronal environment by removing excess neurotransmitters from the synaptic cleft, facilitate repair of damaged neural elements, as well as to isolate the neuron and its processes from the external milieu. It is interesting to note that the quantity of neuroglia in rats greatly exceeds the number of neurons, but, researchers showed that, in humans, there is a rough equivalence in neuronal and glial numbers.

NEURONS

A neuron is composed of a cell body, also known as a soma or perikaryon, and one or more processes, known as dendrites as well as a single axon

Usually, neurons have two regions, the **cell body** (**perikaryon**, **soma**), and neuronal processes called **dendrites** and **axons**, collectively known as **neurites** (Fig. 3.1). A neuron may possess several dendrites but it has only a single axon, although some exceptionally specialized neurons may present with no dendrites

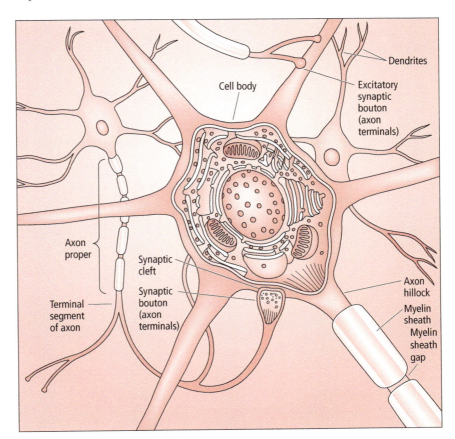

Figure 3.1 • Diagram of a multipolar neuron. Note that the processes of other neurons make synaptic contacts with it. Synapses may be formed, as illustrated, with the soma or with the dendrites, although other types of synapses also occur.

and/or no axons. Dendrites and axons frequently possess a higher percentage of the neuron's cytoplasm than does the soma, due to the large number of dendrites per neuron as well as due to the, possibly, great length of the axon.

Cell body

The cell body houses the nucleus and many of the organelles of the neuron

The size and morphology of the cell body vary and are specific to the particular region of the nervous system; they may vary from 5 to over 100 μm in diameter. The most prominent feature of the **cell body** is its **nucleus**, which is usually round to oval in shape, centrally placed, no more than 3–20 μm in diameter, with a fine chromatin network, and a well-defined **nucleolus**. It is the nucleolus that moderates protein production, organizes the chromatin material, and facilitates neuronal survival or neuronal death via apoptosis. The **cytoplasm** of the cell body is rich in **free ribosomes** and **rough endoplasmic reticulum** (**RER**), whose cisternae form numerous close, parallel configurations which, when viewed with light microscopy, stain dark blue with acidophilic stains and are frequently referred to as tigroid substance (eponym: Nissl bodies). Protein synthesis occurs on both free ribosomes for use in the cytosol and on the RER for eventual packaging. The **Golgi complex** is usually near the nucleus and is responsible for the modification and packaging of the various proteins, enzymes, and chemical messenger molecules manufactured on the RER and delivered to it. Numerous profiles of **smooth endoplasmic reticulum** (**SER**) are also present in the neuronal cytoplasm. Many of these are located just deep to the plasmalemma as **hypolemmal cisternae**, believed to be responsible for the sequestering of calcium, a key ion in triggering the exocytosis of neurotransmitter molecules, contained in synaptic vesicles, into the synaptic cleft. The energy requirement of the neuron is met by the presence of numerous **mitochondria**, which are distributed throughout the cell body. Therefore, protein synthesis, respiration, and many of the essential cellular functions occur in this region of the cell (Fig. 3.2). The soma is also rich in endosomes and lysosomes, as well as various types of vesicles that are being ferried to and from the neurites.

The cell body also contains numerous **inclusions**, nonliving substances such as melanin pigments, lipofuscin, and lipid droplets. Melanin pigments are present in granules in the cytoplasm of cells only in certain locations, such as the locus caeruleus and the substantia nigra, but their function in these cells is not understood, although it has been suggested that they may prevent oxidation and modulate metal ion activity. Neuromelanin is absent in infants and increases in quantity as the individual ages. Lipofuscin is believed to be the indigestible remains of lysosomal activity and, since it has been demonstrated to increase as a function of age, it is frequently referred to as "age pigment." Lipid droplets are present in the cytoplasm of neurons, as in other cells, as a form of energy reserve.

Cytoskeletal components of neurons consist of microfilaments, neurofilaments, and microtubules. **Microfilaments**, composed mostly of filamentous actin, are associated mainly with the cell membrane and aid in axon and dendrite

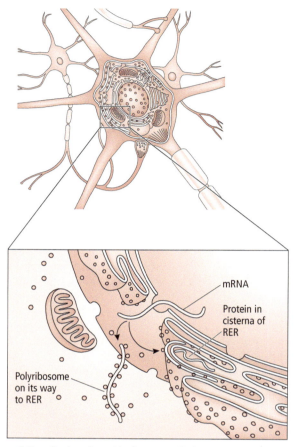

Figure 3.2 ● Section of a multipolar neuron displaying a messenger RNA, transcribed from the DNA, leaving the nucleus, picking up ribosomes to form a polysome, and proceeding to the rough endoplasmic reticulum (RER). On the surface of the RER the mRNA is translated and proteins, destined to be packaged into vesicles, are synthesized and funneled into the lumen of the RER.

elongation, in the movement of the plasmalemma, and as an essential component of the cytoskeleton. **Neurofilaments** are intermediate filaments that frequently become clumped during preparation for light microscopy, forming long filamentous threads known as **neurofibrils** (less than 2 μm in diameter). Neurofilaments assist in the regeneration and development of axons and dendrites. **Microtubules** are long, turgid structures that possess **microtubule-associated proteins** (such as **MAP-2** in the cytoplasm and dendrites and **MAP-3** in the axon). Microtubules provide a scaffold-like structure to maintain cell shape and aid in the transport of materials to and from the neurites.

Dendrites

Dendrites are short, branching processes of neurons that may house some organelles and transmit impulses toward the cell body

Dendrites, whose proximal region may house organelles, such as RER, SER, mitochondria, and vesicles, are relatively short, branching extensions of the cell body. Each neuron may have a number of dendrites, each of which receives signals from specialized receptors, axons, or other neurons and transmits these signals *toward* the cell body. Often the surface area of some dendrites is increased due to the presence of **dendritic spines**, permitting the presence of a greater number of synapses. Moreover, since dendrites are usually highly branched structures, they may receive information concurrently from many different sources. Dendrite orientation, branching, and distribution depend on extracellular proteins that can be adhesive, chemo-attractant, chemo-repellant, or trophic.

Axons

Each neuron has only a single axon, which originates from the organelle-free region of the cell body, known as the axon hillock that appears unstained in light microscopic preparations; axons transmit impulses away from the cell body

The axon transmits impulses away from the cell body and is divided into three regions: the initial segment, the axon proper, and the axon terminals. The **initial segment** is the short transition (30–50 µm in length) between the **axon hillock** and the axon proper. A characteristic feature of the initial segment is the thickening on the axoplasmic aspect of the axolemma (axon membrane) that is believed to be indicative of the presence of voltage-gated sodium channels. The initial segment is the primary region of the axon responsible for the *initiation and propagation of action potentials*.

The **axon proper** is a long process (in some instances exceeding 100 cm in length), whose diameter, which may be as great as 15–20 µm, remains constant for much of its length. Axons may possess a **myelin** covering (see below) that begins at the axon proper and stops at the axon terminal. Myelin covering (myelin sheath) is formed by **oligodendrocytes** within the central nervous system and by **neurolemmocytes** (eponym: Schwann cells) in the peripheral nervous system. The myelin sheath is elaborated in discontinuous segments, and each myelinated segment is called an **internode**. The discontinuities of myelin between adjacent internodes are referred to as **myelin sheath gaps** (eponym: **nodes of Ranvier**). Occasionally the axon proper has branches, known as **collaterals**, that diverge from the main axon at right angles. At its terminus the axon may arborize, forming numerous **axon terminals** (also known as **bouton terminaux** and **end feet**), which permits a single axon to make synaptic contact (see below) with numerous other neurons (or muscle or gland cells). The axoplasm does not possess ribosomes or RER, although mitochondria, SER, vesicles, neurofilaments, and microtubules are present.

Material produced in the cell body is transported, via the assistance of microtubules and their microtubule-associated proteins, to the axon for its use. Such transport is referred to as **anterograde transport**. Material, such as horseradish peroxidase that is injected into the synaptic region, may be conveyed in the opposite direction, toward the cell body, and this process is referred to as **retrograde transport**. Collectively, anterograde and retrograde transports are referred to as **axonal transport**, a process that is dependent on the presence of the ATPase molecular motors dynein and kinesin. The former is responsible for retrograde transport, whereas the latter functions in anterograde transport. Table 3.1 details the types and velocity of transport and the substances being transported. Axonal transport is essential for the health of the muscle cell or the parenchymal cells of the gland that is being innervated. If this **trophic relationship** is interrupted, the muscle cell or parenchymal cell atrophies.

Synapses

The synapse is a specialized intercellular junction that permits communication between neurons in the CNS or a neuron and an effector cell (another neuron, muscle cell, or gland cell) in the PNS

A **synapse** may be **electrical** with a synaptic cleft width of 2–4 nm or **chemical** with a synaptic cleft width of 20–30 nm. In an electrical synapse, ions are exchanged via **gap junctions** that permit cytoplasmic continuity between the two cells, whereas in a chemical synapse **messenger molecules** (**ligands**) that are housed in **synaptic vesicles** are released into the synaptic cleft, because there is no cytoplasmic continuity between the presynaptic and postsynaptic cell. These messenger molecules are of two types, namely **neurotransmitters** and **neuromodulators** (see Chapter 4, Neurotransmitter Substances), and they are released from the axon terminal, diffuse through the synaptic cleft to reach the target cell membrane where they contact receptors that cause ion channels to open (or close) in order to depolarize (or hyperpolarize) the membrane of the effector cell. Since electrical

Transport type	Transport rate (mm/day)	Substance being transported
Anterograde (slow component a)	0.1–1.0	Tubulins, microtubule-associated proteins, neurofilament proteins
Anterograde (slow component b)	1.0–10	Enzymes, proteins, clathrin, myosin, actin
Anterograde (fast, using kinesin)	50–500	Organelles (vesicles, mitochondria, secretory granules)
Retrograde (fast, using dynein)	200–300	Enzymes, proteins, lysosomes, horseradish peroxidase

Table 3.1 • Axonal transport.

synapses are seldom present in mammals (except in the retina, cerebral cortex, and brainstem), only chemical synapses will be discussed in this textbook, and the reader is referred to a textbook of histology or cell biology to become familiar with the structure of gap junctions.

There are several types of possible synaptic configurations within the CNS (Fig. 3.3), depending upon the regions of the neurons involved (e.g., axoaxonic, axosomatic, axodendritic, dendrodendritic), the most common being the **axodendritic**, where the axon of one neuron synapses with the dendrite of another neuron. In all chemical synapses, the neurotransmitter/neuromodulator travels across the synaptic cleft to the effector cell, giving rise to the terminology: presynaptic cell (the neuron releasing the messenger molecules) and postsynaptic cell (the effector cell). Thus, there are three major functional components of a synapse, namely the **presynaptic membrane**, the **synaptic cleft**, and the **postsynaptic membrane** (although usually there is an additional functional component, **astrocytes** that surround synapses and, to a certain extent, adjust the activity occurring there). These membranes are specialized so that the messenger molecules contained in the synaptic vesicles are released at the proper sites (**active sites**) on the presynaptic membrane, and so that they bind to receptor molecules located at specific sites of the postsynaptic membrane. When viewed with an electron microscope, the cytoplasmic aspect of either one of these membranes frequently displays a filamentous electron density known as the **presynaptic density** or **postsynaptic density**. Depending on the arrangement of these submembrane densities, the synapses are said to be either **type I** or **type II** (Table 3.2). Type I synapses are also known as **asymmetric synapses** since their postsynaptic density is thicker than their presynaptic density. Type II synapses are also referred to as **symmetric synapses** since the two densities are of equal thickness. Usually, type I synapses possess round synaptic vesicles, whereas type II synapses possess flattened synaptic vesicles. Generally, type I synapses are excitatory, so that the membrane of the effector cell becomes depolarized, initiating an **excitatory postsynaptic potential** (**EPSP**). Conversely, type II synapses are usually inhibitory, initiating an **inhibitory postsynaptic potential** (**IPSP**) so that the membrane of the effector cell becomes hyperpolarized, causing it to be less responsive to attempts at depolarization. It is important to note at this point that the process whereby a presynaptic neuron (signaling neuron) signals a postsynaptic neuron (target neuron) is extremely quick and accurate.

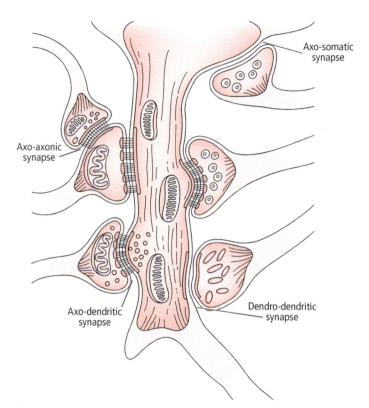

Figure 3.3 ● Synapses along a dendrite. Note that one of the axon terminals (terminal boutons, end-feet) on the top left, has an axon terminal impinging on it, probably acting in an inhibitory capacity. Source: Modified from Williams, PL, ed. (1998) *Gray's Anatomy*, 38th edn. Churchill Livingstone, London, fig. 7.19b.

Axon terminal and its presynaptic membrane

The axon terminal is the region that participates in synapse formation; the area of the plasmalemma of the axon terminal that contacts the synaptic cleft is known as the presynaptic membrane and the area of the effector cell bordering on the opposite side of the synaptic cleft is known as the postsynaptic membrane

The cytoplasm of the **axon terminal** is rich in cytoskeletal elements, such as microtubules, neurofilaments, spectrin, and microfilaments. Additionally, it also houses cisternae of SER, coated vesicles, occasional endosomes, a number of mitochondria, and numerous synaptic vesicles. **Synaptic vesicles** are membrane-bound vesicles whose membranes

Characteristic	Type I synapse	Type II synapse
Submembrane density	Asymmetric (thin presynaptic and thick postsynaptic density)	Symmetric (thin presynaptic and thin postsynaptic density)
Synaptic vesicle morphology	Round (50 nm in diameter)	Flattened (15–40 nm)
Synaptic cleft width	30 nm	20 nm
Functional type	Excitatory	Inhibitory

Table 3.2 ● Types of synapses.

contain numerous unique integral proteins that assist in binding the vesicles to specific sites within the axon terminal.

These synaptic vesicles, about 40 nm in diameter, contain chemical messengers (ligands, neurotransmitter molecules), ATP, and proteoglycans, all of which are destined to be released into the synaptic cleft. Some of these synaptic vesicles are held in reserve away from the active site, whereas others are located at the active site.

The submembrane density on the cytoplasmic aspect of the **presynaptic membrane** is composed of globular-to-oblong electron-dense subunits that are arranged in a hexagonal array at the active site (Fig. 3.4). These oblong structures are protein complexes, known as **docking complexes**, which form a **presynaptic grid** whose interstices are just large enough to accept synaptic vesicles but prevent these vesicles from contacting the presynaptic membrane.

Most synaptic vesicles are held in reserve, away from the active site. However, some are bound (*docked*) to the docking complex so that when an action potential arrives at the axon terminal, they are ready to fuse with the presynaptic membrane, open into the synaptic cleft, and release their messenger molecules.

The arrival of an action potential at the presynaptic membrane opens **voltage-gated calcium channels** located at the active site (possibly as a part of the docking complex), permitting the influx of Ca^{2+} ions into the axon terminal (Fig. 3.5). The presence of Ca^{2+} ions permits the fusion of the membrane of the *docked* synaptic vesicle with the presynaptic membrane and the subsequent release of its messenger molecules into the synaptic cleft. Precise control of calcium availability is necessary to ensure that only the required amount of messenger molecules are released as the consequence of a single action potential.

The sequestering of Ca^{2+} ions is performed by at least three mechanisms:

1. A cytosolic protein binds Ca^{2+} in the vicinity of the presynaptic membrane.
2. Calcium storage vesicles located in the axon terminal possess Ca^{2+} pumps that remove Ca^{2+} from the end-foot cytosol.
3. Na^+/Ca^{2+}-coupled transporters in the axon terminal membrane exchange extracellular Na^+ for cytosolic Ca^{2+}.

These three mechanisms act in conjunction with each other to control Ca^{2+} availability at the active site (Fig. 3.6).

There are two types of fusions of the synaptic vesicle with the presynaptic membrane: the "**kiss-and-run**" **fusion** and the "**collapse**" **fusion**. As the name implies, the "kiss-and-run" fusion is a temporary fusion where the synaptic vesicle contacts the presynaptic membrane, releases some of its contents into the synaptic cleft, detaches itself from the presynaptic membrane, returns to the active site of the axon terminal, and remains docked there until the next impulse arrives. The "collapse" fusion is a "permanent" fusion. Permanent in the sense that the fused vesicle empties its total contents into the synaptic cleft and the vesicle membrane becomes incorporated into the presynaptic membrane.

Continued synaptic vesicle fusion with and incorporation into the presynaptic membrane would enlarge the axon terminal; therefore, the excess membrane is recaptured by the formation of **clathrin-coated pits**. The pits form **clathrin-coated vesicles** and the captured membranes are returned to the cytosol. They travel as **endocytic vesicles**, possibly along microtubule-marked pathways. These vesicles lose their clathrin coat before fusing with the elements of SER to be recycled in the formation of new synaptic vesicles.

Since the flow of calcium ions into the axon terminal is essential for the release of the messenger molecules into the synaptic cleft, a method of diminishing its entry into the end-foot has evolved in mammals, a process known as **presynaptic inhibition**. This process occurs in areas where an axon forms a synapse with the axon terminal of another axon (Fig. 3.7). Hence, there are three neurons involved at this location:

1. The axon terminal of neuron A contacting the axon terminal of neuron B.
2. The axon terminal of neuron B that contacts neuron C.
3. The dendrite (or cell body) of neuron C.

The axon terminal of neuron A releases a chemical messenger destined for chloride ion channels of the axon terminal of neuron B. The binding of the ligands opens Cl^- channels causing an influx of Cl^- into the axon terminal of neuron B, hyperpolarizing its presynaptic membrane and making it much more difficult to open its voltage-gated Ca^{2+} channels. This event makes it less likely that neuron B will be able to release its chemical messengers intended for neuron C,

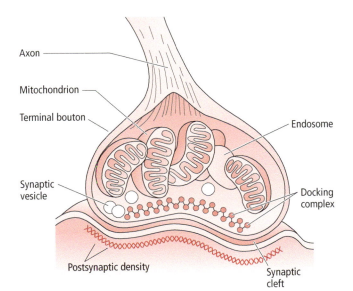

Figure 3.4 ● An axon terminal (terminal bouton) forming a synapse. Note that the synaptic vesicles are in close relationship with the docking complexes of the presynaptic membrane. The axon terminal also houses mitochondria and endosomes.

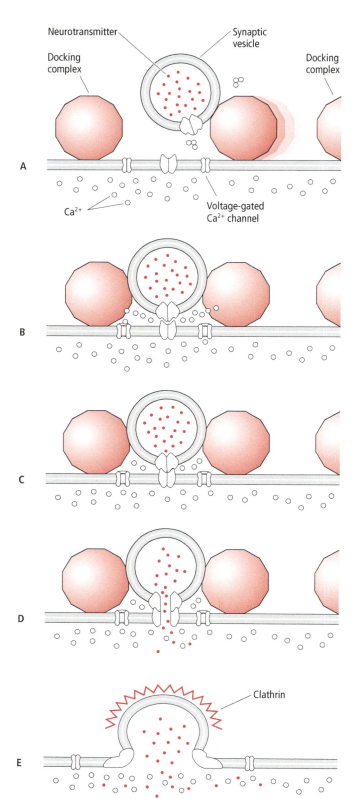

Figure 3.5 • Synaptic vesicles are held in reserve by docking complexes near the active sites of the presynaptic membrane. (A) Voltage-gated calcium channels of the presynaptic membrane are closed, and the synaptic vesicle does not contact the presynaptic membrane. (B, C) The arrival of an action potential opens the voltage-gated calcium channels; the influx of calcium ions permits the proper alignment of the synaptic vesicle with the presynaptic

and it is this inhibition of the activity of neuron B that is referred to as presynaptic inhibition.

Synaptic cleft and the postsynaptic membrane

The space located between the presynaptic membrane and the postsynaptic membrane is known as the synaptic cleft; the postsynaptic membrane is that region of the effector cell that borders the synaptic cleft

The **synaptic cleft** is a 20–30 nm wide extracellular space between the presynaptic and postsynaptic membranes. This space contains diverse families of macromolecules, some of which appear to be filamentous and span the entire width of the synaptic cleft. Many of these macromolecules are believed to be the extracytoplasmic aspects of the various transmembrane proteins of the presynaptic and postsynaptic membranes, whereas others are molecules of the extracellular matrix. It has been demonstrated that many of these synaptic cleft proteins are necessary to maintain a rigorous alignment of the active sites of the presynaptic membrane with those regions of the postsynaptic membrane that house receptor molecules designed to bind the released chemical messengers. These chemical substances, released in predetermined quantities, known as **quanta** (as many as 20,000 molecules), cross the synaptic cleft and bind to specific receptor molecules of the postsynaptic membrane. Each synaptic vesicle contains one quantum of chemical messenger molecules, which is the minimum amount necessary to precipitate a single postsynaptic event (**depolarization** or **hyperpolarization**).

The **postsynaptic membrane** is a region of the effector cell plasmalemma that is modified to receive and react to the chemical messenger (ligand) released into the synaptic cleft. Binding of the ligand results in the opening of ion channels of the postsynaptic membrane, resulting in subsequent depolarization or hyperpolarization of the effector cell plasmalemma. The submembrane density of the postsynaptic membrane is connected, via thin filamentous material, to an abundance of intertwined cytoskeletal elements, the subsynaptic **web**. The cytosol in the immediate vicinity of the subsynaptic web houses, among others, microtubules, microfilaments, SER, and, frequently, RER along with numerous mitochondria. The cytosol also contains enzymes

membrane by inducing slight rearrangements of both the docking complexes and the synaptic vesicle. (D) Once aligned properly, the synaptic vesicle fuses with the presynaptic membrane, opens, and releases its contents into the synaptic space. (E) To prevent an increase in the area of the presynaptic membrane, the excess membrane is recaptured via endocytosis either as clathrin-coated pits or as clathrin-independent pits; in either case the membranes are recycled in the cytosol. As these pits form, they are pinched off so that there is no longer a communication between the lumina of the pits and the synaptic cleft. This process of pinching off is performed by the protein dynamin that acts as the cord that closes the mouth of a duffel bag. Other dynamin molecules seal the opening in the presynaptic membrane. It should be mentioned that clathrin-dependent endocytosis is about 10–15 times slower than the clathrin-independent endocytosis which occurs in just a few seconds. Source: Adapted from Kingsley, RE (1996) *Concise Text of Neuroscience*. Williams & Wilkins, Baltimore, p. 96.

HISTOPHYSIOLOGY OF THE NERVOUS SYSTEM   39

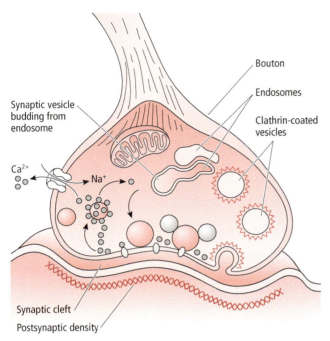

Figure 3.6 ● Some synaptic vesicles are formed in the soma and are transported into the axon terminal, whereas others are formed locally in the end-foot by budding from endosomes, and still others form via endocytosis of the presynaptic membrane. In order to prevent a constant release of neurotransmitters, the calcium level is reduced in the axon terminal by: 1. being actively exchanged for Na+, 2. being sequestered in the smooth endoplasmic reticulum, as well as by 3. being bound to cytoplasmic proteins. Furthermore, to maintain a constant size of the presynaptic membrane as it is enlarged by fusion of the synaptic vesicles with it, it is reduced by endocytosis via the formation of clathrin-coated vesicles or clathrin-independent vesicles that will then either join the endosomes or will form new synaptic vesicles.

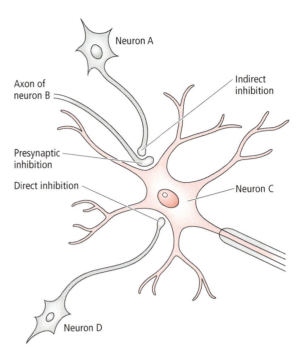

Figure 3.7 ● Diagram depicting the difference between direct and indirect inhibition. Note that in direct inhibition, neuron D forms a synapse with the large multipolar neuron C, permitting transmission of information directly. In indirect inhibition, neuron A forms a synapse with the axon of neuron B, and it is that other neuron that synapses with the large multipolar neuron C; thus, the information is transmitted indirectly.

such as protein kinase C and calmodulin-dependent kinase II. The postsynaptic membrane houses two major types of receptors: ionotropic and metabotropic receptors.

1 **Ionotropic receptors** are ligand-gated ion channels; they are ATP-independent, and the chemical messengers that bind to them are known as **neurotransmitters**. The binding of a neurotransmitter alters the conformation of the protein constituting the *ion channel*, opening it and permitting the flow of ions through it. Thus, a neurotransmitter is a chemical messenger that has a *direct* effect on the ion channel and the effect is *rapid*.

2 **Metabotropic receptors** are ATP-dependent and bind chemical messengers known as **neuromodulators** or **neurohormones**. These molecules also interact with a receptor molecule, causing a conformational change in its protein(s); however, this receptor molecule is *not an ion channel*. Instead, this receptor molecule interacts with other protein complexes, such as **G-proteins**. G-proteins in turn interact with one of the cell's secondary messenger systems, thus modulating the metabolic activities of the effector cell. In this fashion the required response occurs as an *indirect* function of the coupling of the ligand to the receptor and the effect is *slow*.

These two points are true in most instances, but there are exceptions. The following generalization is applicable in most instances, namely that *neurotransmitters have a direct influence*, whereas *neuromodulators and neurohormones have an indirect influence* in effecting the desired result. *It is also important to realize that the same chemical messenger may elicit different responses depending on the type of ionotropic or metabotropic receptor with which it is associated*. The various types of neurotransmitters and neuromodulators are discussed in Chapter 4, "Neurotransmitter Substances."

Classification of neurons

The number of processes a neuron possesses permits a classification of these cells into the following categories: unipolar, bipolar, multipolar, and pseudounipolar neurons (Fig. 3.8 I Top. Diagrams, II Bottom. Photomicrographs). True **unipolar neurons** are rare in vertebrates; they possess no dendrites, only an axon. **Bipolar neurons**, such as the ones in the olfactory mucosa, possess an axon and a dendrite, whereas **multipolar neurons**, such as the motor neurons of the spinal cord (Fig. 3.8 I. Top), have a single axon and two or more dendrites. An additional type of neuron, the **pseudounipolar neuron** (now referred to as **unipolar neurons** in vertebrates), starts out, during embryogenesis, as a bipolar neuron. The axon and dendrite of this neuron fuse into a single process that later bifurcates.

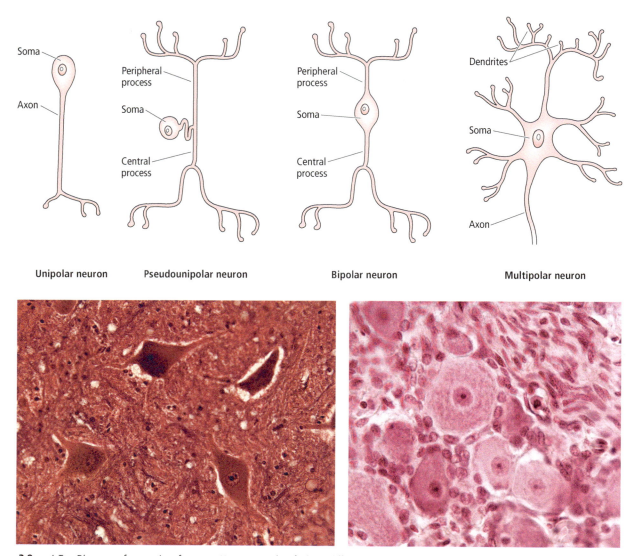

Figure 3.8 • *I. Top. Diagrams of categories of neurons:* Neurons are classified into different categories depending on the number of processes they possess. Note that true unipolar neurons are rare in humans and that pseudounipolar neurons are also referred to as unipolar neurons. *II. Bottom. Photomicrographs of categories of neurons:* (left) Low-power light micrograph of multipolar neurons from the anterior horn of a human spinal cord. (Right) Unipolar (pseudounipolar) neurons from a human spinal ganglion.

Neurons of the spinal ganglia and neurons of the ganglia of cranial nerves are representatives of this category. The two photomicrographs (Fig. 3.8 II. Bottom) present multipolar neurons (left) and unipolar (pseudounipolar) neurons (right).

NEUROGLIA

The six types of neuroglia are the supporting cells of the central and peripheral nervous systems

There are six types of supporting cells associated with the nervous system, five in the CNS (astrocytes, oligodendrocytes, microglia, polydendrocytes [NG2-glia], and ependymal cells) and one in the PNS (neurolemmocytes [eponym: Schwann cells]) (Table 3.3). Neuroglia of the CNS are usually smaller, and their population most probably equals those of neurons.

Astrocytes are stellate-shaped cells with numerous, long processes that extend between and among the neurons and vascular elements occupying much of the extracellular spaces of the CNS. Their processes surround all neural elements and also form gap junctions with one another. Additionally, astrocytes form a **perivascular covering** around blood vessels as well as a **glial-limiting membrane** at the interface of the brain and pia mater. There are two types of astrocytes: **fibrous astrocytes**, located in white matter, and **protoplasmic astrocytes**, located in gray matter. Both types of astrocytes are characterized by the unique **glial fibrillary acidic protein** (**GFAP**) intermediate filaments contained in their cytoplasm. Protoplasmic astrocytes surround synapses and, usually, are less than 1 μm away from the synaptic cleft; their cell membranes possess neurotransmitter/neuromodulator **receptors** as well as **carrier proteins** that are specific

Neuroglia	Origin	Location	Function
Protoplasmic astrocyte	Neural tube	CNS, gray matter	Support and isolate neurons; sequester Na^+; metabolize extracellular neurotransmitters; repair damage to CNS by forming glial scar tissue; surround unmyelinated axons; in the embryo they establish pathways along which neurons migrate
Fibrous astrocyte	Neural tube	CNS, white matter	Same as protoplasmic astrocyte
Oligodendrocyte	Neural tube	CNS, mostly white matter	In white matter they form myelin sheaths around axons of the CNS; in gray matter they surround neurons
Microglia	Bone marrow	CNS	Phagocytose damaged nervous tissue in the CNS
Polydendrocytes (NG2 cells)	Neural tube	CNS, white matter and gray matter	Give rise to oligodendroglia and, perhaps in the embryo and maybe even in the adult, also to protoplasmic astrocytes
Ependymal cells	Neural tube	Lining ventricles of the brain and central canal of spinal cord	Promote movement of cerebrospinal fluid; forms simple cuboidal ciliated epithelium of the choroid plexus; in third ventricle differentiates to form tanycytes
Neurolemmocytes (Schwann cells)	Neural crest	Peripheral nervous system only	Myelinate axons and surround unmyelinated axons of the PNS; function during degeneration and regeneration of axons

Table 3.3 ● Properties of neuroglia.

for the neurotransmitters/neuromodulators released by the presynaptic neuron. Astrocytes are believed to:

- provide physical and nutritional support to the neurons of the CNS;
- communicate with neurons;
- sequester Na^+ and K^+ from the extracellular space thus controlling the concentration of ions in the extracellular space;
- insulate neurons from each other;
- metabolize extracellular neurotransmitters, such as serotonin, gamma-aminobutyric acid (GABA), and glutamate;
- release **gliotransmitter substances** (principally, D-serine, glutamate, and ATP) that function in assisting neurons and neuroglia in communicating with each other;
- regulate the development, stabilization, and functions of synapses;
- surround blood vessels in the central nervous system and form an effective barricade that prevents molecules that escaped the blood–brain barrier from gaining access to the parenchyma of the central nervous system;
- are necessary for the preservation of the blood–brain barrier by assisting in the maintenance of zonula occludens-1 protein expression in the endothelial cell tight junctions;
- form glial scar tissue in an attempt to repair damage to the CNS; and
- form a framework along which developing neurons migrate to their destination during embryogenesis.

Oligodendrocytes are small cells that possess only a few processes, and their cytoplasm is devoid of GFAP. They are present in gray matter, where they abut neurons and are sometimes referred to as **satellite oligodendrocytes**. They are most numerous in white matter, where they form myelin sheaths around axons. Each oligodendrocyte may myelinate a single internode of as many as 40 axons.

Microglia are the smallest of the neuroglia and, unlike the others, they are derived from monocytes that, as they leave the bloodstream and enter the substance of the CNS, differentiate into small macrophages. Microglia function in the phagocytosis of damaged tissues in the central nervous system and participate in inflammatory and degenerative reactions.

Polydendrocytes (**NG2 cells**) are precursors of oligodendrocytes as well as – at least during the embryonic stage and maybe even in adults – of protoplasmic astrocytes. NG2 cells are present both in white matter and in gray matter where they undergo clonal expansion. It has been suggested that they may be able to differentiate and give rise to neurons not only during embryonic stages but, perhaps, also in the adult.

Ependymal cells form the simple cuboidal to simple columnar epithelial lining of the ventricles of the brain and the central canal of the spinal cord. These cells possess cilia in some regions of the spinal canal and ventricles facilitating the circulation of cerebrospinal fluid (CSF). Ependymal cells in certain regions of the ventricles become modified and participate in the formation of the choroid plexus, the structures that are responsible for the production of the CSF. Some of the ependymal cells of the third ventricle are modified and are known as **tanycytes**. These modified ependymal cells possess long processes that approximate capillaries and certain neurosecretory cells of the hypothalamus. It is speculated that they may transport CSF to these cells.

Neurolemmocytes (eponym: **Schwann cells**) are present only in the peripheral nervous system (PNS). They are derived from neural crest cells that migrate during development to the sites of axon formation and elongation. These cells envelop the forming axons, isolating them from the surrounding environment. Neurolemmocytes invest all unmyelinated axons as well as forming the myelin sheath of all myelinated axons of the PNS. Unlike oligodendrocytes, a neurolemmocyte is capable of myelinating only a single internode of a single axon. These cells form their own basal lamina and extracellular matrix; hence, they are capable of synthesizing type IV collagen, the glycoproteins fibronectin and laminin, as well as various proteoglycans. Neurolemmocytes also function during degeneration and regeneration of axons.

Myelin

Myelin is formed from the layers of oligodendrocytes and neurolemmocyte plasmalemma wrapped around axons of certain neurons

Each axon is served by a large number of oligodendrocytes or neurolemmocytes, since a single one of these cells is capable of myelinating only a short segment of an axon. These short, myelin-wrapped segments are known as **internodes** (200–1,000 μm in length), and the unmyelinated regions of the axon located between adjacent internodes are known as **myelin sheath gaps** (Fig. 3.9). As described above, oligodendrocytes myelinate in the CNS only and neurolemmocytes myelinate in the PNS only. Moreover, a single oligodendrocyte is able to myelinate a single internode of as many as 40 different axons (Fig. 3.10), whereas a single neurolemmocyte is capable of myelinating only one internode of one axon.

Note that the clinical case at the beginning of the chapter refers to a sudden appearance of sensory loss and motor disturbances.

1 Is a tumor usually responsible for bilateral paralysis of the leg?

2 Is a single CNS lesion usually responsible for a relapsing and remitting course of events?

3 Does a single lesion in the PNS usually affect both sensory and motor functions?

Although the mechanism of myelination is not well understood, it is believed that the process is similar whether it occurs in the peripheral or the central nervous system; therefore, only neurolemmocyte myelination will be discussed here. As the axon reaches 1 μm in diameter, neurolemmocytes proliferate, forming a continuous chain of cells along the length of the axon. As they envelop the axon, they begin to layer their membranes (as many as 50 or more layers, similar to wrapping a nearly empty toothpaste tube

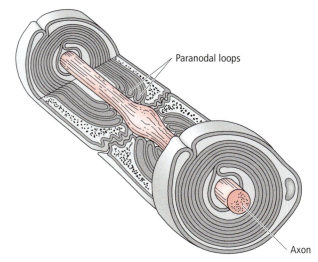

Figure 3.9 ● Longitudinal section of a nerve fiber at the myelin sheath gap. Note that as the two neurolemmocyte membranes approach each other they form cytoplasm-filled regions known as paranodal loops.

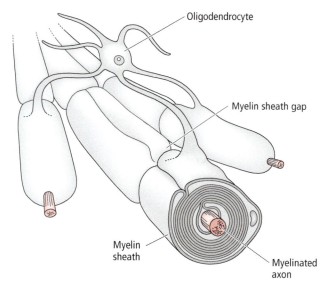

Figure 3.10 ● Myelination is very similar in the central nervous system where a single oligodendrocyte is capable of myelinating a single internode of numerous axons.

around a segment of a pencil) in a spiral fashion around the axon (Fig. 3.11). As the developing myelin sheath becomes more tightly wound, the cytoplasm of the wrapped region retracts into the neurolemmocyte cell body, and the cytoplasmic aspects of the apposing cell membranes approximate each other. Moreover, the extracytoplasmic aspects of the cell membranes also approach one another.

Viewed with the electron microscope the myelin sheath presents the appearance of concentric, wider, electron-dense lines alternating with concentric, narrower, less electrondense lines. The wider lines (3 nm in width) represent the apposition of the cytoplasmic surfaces of the neurolemmocyte plasmalemma and are called **major dense lines**. The narrower lines

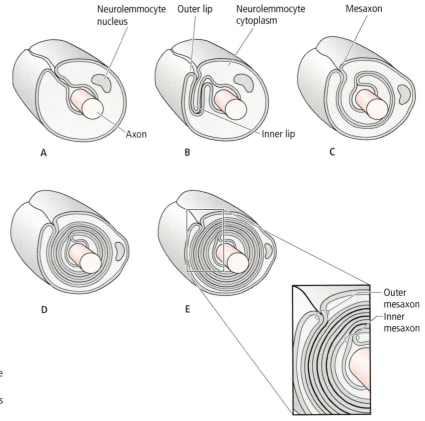

Figure 3.11 • Myelin formation and structure in the peripheral nervous system. (A) The axon is surrounded by a neurolemmocyte. (B) The neurolemmocyte begins to rotate around the axon. (C) As rotation progresses, the mesaxon is being formed. (D) Continued rotation of the neurolemmocyte squeezes the cytoplasm out of the myelin sheath. (E) Higher magnification, displaying the internal and external mesaxons as well as the major dense lines and minor dense lines.

represent the apposition of the external (extracellular) surfaces of the neurolemmocyte plasmalemma and are known as **intraperiod lines**. Although these lines *appear* to be concentric, they are actually spiral lines. The external surfaces of the neurolemmocyte plasmalemma do not contact each other, necessitating a change of terminology to the **fissure of intraperiod line** (synonym: **intraperiod gap**). This intraperiod gap is believed to be a conduit so that small molecules can reach the axon. The beginning of the internal aspect (closest to the axolemma) of the intraperiod gap is known as the **internal mesaxon**, whereas the distal aspect of the intraperiod gap (at the external surface of the myelin sheath) is referred to as the **external mesaxon**.

At the edges of the neurolemmocyte, in the region of the myelin sheath gap, bubbles of cytoplasm remain that separate the major dense line. Additionally, some of the cytoplasm may remain in the region of the internode, forming oblique arrays of bubbles of cytoplasm. When whole mounts of teased myelinated nerve fibers are viewed with the light microscope these areas appear as clear slanted lines in the myelin sheaths and are known as **myelin clefts** (synonym: **myelin incisures**; eponym: **Clefts of Schmidt–Lantermann**).

The process of myelination occurs over a period of several years and, as such, some axons are myelinated much sooner than others. Motor nerves are the first to be myelinated, whereas sensory nerves are not myelinated until several months after birth. Certain regions of the CNS, such as the prefrontal cortex, do not become completely myelinated until adolescence.

Nonmyelinated nerve fibers do not have several layers of cell membrane surrounding them, however, still possess a membranous sheath formed by a single layer of the cell membrane of an oligodendrocyte, in the CNS, or of a neurolemmocyte in the PNS.

Myelin is composed of approximately 70–80% lipids, and the remainder consists of proteins and water. The preponderant lipid constituent is cholesterol, followed by phospholipids and glycosphingolipids. There are several proteins whose roles in the structural integrity of the myelin sheath are beginning to be understood (Table 3.4). P_0 and P_2 are present mostly in PNS myelin, P_1 is located in both PNS and CNS myelin, and **proteolipid protein** (**PLP**) is the major structural protein that is present only in CNS myelin. All of these proteins are transmembrane proteins and are believed to play a role in the compaction of the major dense line, since their cytoplasmic aspects are composed of highly charged amino acids that attract their counterparts on the other side of the membrane. Another protein, **myelin-associated glycoprotein** (**MAG**), has globular subunits that extend into the extracellular space (into the intraperiod gap) of the myelin sheaths of both the CNS and PNS. Here they contact MAGs from the other side, thus establishing adhesive forces that hold the external aspects of the neurolemmocyte plasmalemma in close proximity. Additionally, the adhesion of the external surfaces of the neurolemmocyte plasma membrane is reinforced by the presence of **tight junctions** whose major components are not only the expected transmembrane

44 ••• CHAPTER 3

Protein	Location	Function
P_0	Mostly in PNS	Compaction of major dense line
P_1	Both in PNS and CNS	Compaction of major dense line
P_2	Mostly in PNS	Compaction of major dense line
PLP (proteolipid protein)	CNS only	Compaction of major dense line
MAG (myelin-associated glycoprotein)	Both PNS and CNS	Compaction of fissure of intraperiod line
CNS, central nervous system; PNS, peripheral nervous system.		

Table 3.4 • Proteins of myelin.

proteins **claudins** and **zonula occludens proteins**, but also **connexin 32**. The cytoplasmic aspects of these transmembrane proteins are associated with the cytoskeletal components spectrin and actin, thus reinforcing the rigorous architecture of the myelin sheath.

GENERATION AND CONDUCTION OF NERVE IMPULSES

Nerve fibers transmit information along their plasma membranes, which become depolarized subsequent to appropriate stimulation. Modulation of the transmembrane **resting potential**, due to the actions of voltage-gated ion channels, results in the transmission of an electrical signal along the dendrites, cell body, and/or axon either as an **action potential** or as a **postsynaptic potential**.

Resting potential, Nernst equation, and Goldman equation

The resting potential is the voltage difference between the cytoplasmic and extracellular sides of the plasmalemma; it can be calculated using the Nernst or Goldman equations

The **resting potential** is the voltage difference existing between the two sides (cytoplasmic and extracellular) of the semipermeable plasmalemma of any cell. This voltage difference is present only in the *immediate vicinity* of the membrane, since deeper in the cytoplasm and farther out in the extracellular space the electric charges are neutral. Neurons possess constant permeability (potassium leak) **K+ channels** that establish a specific charge difference (**potential difference**) between the two sides of the membrane, so that its cytoplasmic aspect *appears* negative with respect to its extracellular aspect. In order to understand the establishment of a resting potential one must realize that there are two factors operating on this semipermeable membrane, the concentration gradient and the electrical gradient.

Since the intracellular concentration of K+ is greater than the extracellular concentration, potassium is being driven by a concentration gradient from the cell into the extracellular space. Potassium ions travel through ion channels that are not gated and are known as **K+ leak channels**.

Because only K+ are permitted through these channels an electrical imbalance is established (more positive ions are present on the extracellular aspect of the membrane than on the intracellular aspect). It should be understood that a K+ may travel in either direction through the same K+ leak channel. At the point where the concentration gradient's "desire" to push potassium out of the cell and the electrical imbalance's "desire" to return potassium into the cell *equal each other*, the net flow of K+ becomes zero. This does not mean that there is no flow of K+ across the plasma membrane; it simply means that the same number of K+ flow into the cell as out of the cell. The establishment of this equilibrium potential (modified for K+) is described by the **Nernst equation**:

$$V_m = RT/ZF(\ln[K^+]out/[K^+]_{in})$$

where V_m is the membrane potential in volts; R is the gas constant; T is the temperature in Kelvin; F is Faraday's constant; Z is the ion valence; and $(\ln[K^+]_{out}/[K^+]_{in})$ is the natural logarithm of the ratio of the cytoplasmic to extracellular concentration of K+.

The Nernst equation can be applied for all ions and is solved for the most common ions in Table 3.5. However, the Nernst equation assumes that only the particular ion in question is permitted to traverse the cell membrane. But this is not a true situation since several types of ion channels of the plasma membrane, particularly of the neuron's membrane, are open simultaneously, thus permitting several types of ions to cross the plasmalemma. This state of affairs is better described by the Goldman equation.

The **Goldman equation** describes the membrane potential as a function of the relative permeability of the cell membrane to more than a single species of ions. Since the neuron membrane is influenced by at least three ions, Na+, K+, and Cl−, all three are considered in the Goldman equation:

$$V = RT/F \log\{(PK^+[K^+]_{out} + P_{Na^+}[Na^+]_{out} + P_{Cl}{}^-[Cl^-]_{in})/$$
$$(P_{K^+}[K^+]_{in} + P_{Na^+}[Na^+]_{in} + P_{Cl}{}^-[Cl^-]_{out})\}$$

As is evident, the Goldman equation is very similar to the Nernst equation, but it predicts the membrane potential from the concentration of several types of ions and the permeability of the membrane to those ions. It is interesting to note

Ion	Intracytoplasmic concentration (mmol/L)	Extracytoplasmic concentration (mmol/L)	Nernst equilibrium potential (mV)
Na^+	12	120	+58
K^+	5	125	−80
Ca^{2+}	10^{-4}	1–2	+178
Cl^-	5–15	110	−80
Organic ions that cannot leave cell	130	1–2	NA
NA, not applicable.			

Table 3.5 ● Ion concentration and Nernst equilibrium potential in mammalian neurons.

that the two equations would be identical if the permeability of two of the three ions were zero. Therefore, the Goldman equation is the Nernst equation expanded for multiple ions, and consequently it calculates a more realistic resting membrane potential for neurons.

The resting potential of a general neuron plasma membrane is approximately **−71 mV** when the ion concentrations inside and outside the cell equal the values listed in Table 3.5. The chloride ion concentration does not have much of an effect on the resting potential, indicating that it is the K^+ gradient that has the greatest role in establishing the resting potential. The following mental exercise should help in clarifying this statement. If we assume that originally the potential difference across the membrane is zero but the *intracellular* concentration of K^+ is greater than on the *outside*, K^+ will follow its concentration gradient and leave the cell via the K^+ leak channels. As these positive ions leave the cell, the inside of the membrane will become more negative (actually less positive), the outside of the membrane will become more positive, and this electrical imbalance will begin to drive K^+ back into the cell. At the same time the concentration gradient will continue to drive K^+ out of the cell, and there will come a point where the electric force and the concentration gradient force will equal each other. It is at that point that the same number of K^+ will leave the cell as enter the cell and, therefore, the net flow of K^+ equals zero, thus establishing the resting membrane potential. Simultaneously, Cl^- will be unable to enter the cell because the negative charges oppose their inward movement, thus preventing Cl^- from having much of an influence on the resting membrane potential. A further point to consider in this mental exercise is the number of ions that have to be involved in the establishment of the membrane potential. Since there are a far greater number of ions inside the cell than just those in the immediate vicinity (that is within 1–2 nm) of the cell membrane, the actual number of K^+ that have to leave the cell per unit area of membrane is miniscule when compared to the concentration of K^+ inside the cell.

A final point of interest is the effect of the Na^+/K^+-ATPase pump on the resting membrane potential. It should be recalled that this pump uses energy to move three Na^+ out of the cell into the extracellular space in exchange for two K^+ that it brings into the cell from the extracellular space. If the Na^+/K^+-ATPase pump is inactivated, the membrane resting potential will become somewhat less negative, and if the pump remains inactive for a longer period of time (a number of minutes), the resting membrane potential becomes disturbed. However, some time before that happens, cations, especially Na^+, slowly enter the cell and the intracellular Na^+ concentration becomes greater. As the concentration of Na^+ increases, the ion concentration inside the cell will also increase, driving water into the cell. Thus, the major function of the Na^+/K^+ ATPase pump in the cell is the maintenance of the cell's osmotic balance.

When the resting potential is disturbed at a particular point, **voltage-gated ion channels** open, and the passage of specific ions through the membrane generates a small, localized electric current. If the current is the result of the inward movement of cations (i.e., Na^+), the resting potential is decreased to such an extent that the **polarity** of the membrane is reversed (**depolarization**). This local current is very small and its ability to evoke an **action potential** diminishes in two ways:

1 as a function of its distance from the **initial segment** of the axon; and
2 as a function of the time elapsed between the initiation of the current and its arrival at the initial segment.

If the current is the result of the inward movement of anions through a voltage-gated channel (i.e., Cl^-) the resting potential is increased (**hyperpolarization**), leading to an increased resistance to depolarization (that is resistance to the formation of an action potential).

Action potentials

An action potential is the change in resting membrane potential when enough Na^+ enter the cytoplasm to reach a threshold level great enough to cause excitation

Action potentials last only a few milliseconds and occur due to the influx of **Na^+** into the neuron through **voltage-gated Na^+ channels**. These channels exist in one of three configurations: **closed, open,** or **refractory.** When the resting potential reaches a threshold level (about **−55 mV**) the voltage-gated Na^+ channel opens instantaneously (within 10 ms) and remains open, permitting a rapid influx of Na^+. As the number of Na^+ increases on the *cytoplasmic aspect* of the membrane, the accumulation of positive charges internally

46 ● ● ● **CHAPTER 3**

(at the interface of the cytoplasm and the plasmalemma) becomes greater than externally (at the interface of the external aspect of the plasmalemma and the extracellular space). Thus, the internal aspect appears "more" positive than the external aspect, and the membrane is said to have become depolarized (about +40 mV).

At this point we must revise our initial image of the voltage-gated sodium channel. Instead of assuming a single gate on the channel we must envision two gates. The first is the **activation gate** (at the extracellular aspect of the membrane), whereas the second is the **inactivation gate** (at the cytoplasmic aspect of the cell membrane), and both gates must be open for Na^+ to pass through the ion channel. At resting potential, the activation gate is closed, the inactivation gate is open, and Na^+ cannot pass through. When the threshold level (−55 mV) is reached the activation gate opens, the inactivation gate remains open, and Na^+ enter the cell through the ion channel. As the positive charge on the internal aspect builds up and the membrane is depolarized (+40 mV), the inactivation gate slowly closes, and Na^+ can no longer pass through the Na^+ channel. For 1–2 ms thereafter the inactivation gate remains closed, and the Na^+ channel *cannot* respond to alteration in the resting potential. Thus, the Na^+ channel is *inactivated*; it cannot be opened, and the axon is said to be in the **absolute refractory period**. Once the cell membrane is repolarized the activation gate shuts, the inactivation gate slowly opens, and, although Na^+ still cannot pass through the Na^+ channel, the Na^+ channel once again can respond to alterations in the resting potential.

In response to the sodium influx-induced depolarization of the membrane, **voltage-gated K^+ channels** slowly open and remain open for a longer period of time than do Na^+ channels. *It should be noted that voltage-gated K^+ channels are different from K^+-leak channels.* Voltage-gated K^+ channels open only during depolarization, whereas K^+-leak channels are always open. The increased positive charge (due to the influx of Na^+) on the internal aspect of the membrane drives K^+ out (*through the voltage-gated K^+ channels*) to the external aspect of the cell membrane. This efflux of K^+ ions is so intense that it hyperpolarizes the membrane, making the initiation of a second depolarization more arduous, until the resting membrane potential is re-established. This period of time is known as the **relative refractory period** and contributes to the **total refractory period**.

Propagation of an action potential

The advancement of an action potential (i.e., its constant regeneration) in a single direction, along the length of a membrane such as the axolemma, is called propagation of the action potential

An action potential is a local event involving a short segment of an axon, and it lasts merely a few milliseconds. In order to utilize action potentials for the transmission of information they must be able to advance along the axon in a single direction, until they reach the axon terminal.

The following points review the previous discussion of events occurring at a *voltage-gated Na^+ ion channel*:

1 The influx of Na^+ ions reversed the polarity of the membrane *at that particular point*.

2 The positive ions flowed away from the original site of entry, dispersing both toward the soma and toward the axon terminal.

3 As the positive charges reached the neighboring Na^+ ion channel (located closer to the axon terminal), the resting membrane potential was disrupted, that is the internal aspect of the axon membrane became more positive due to the presence of excess Na^+ ions.

4 Once the threshold level was reached the Na^+ ion channel opened permitting the entry of more Na^+ ions, resulting in the depolarization of the membrane at the new point and a new action potential was generated.

5 In this fashion the action potential is constantly regenerated (**and propagated**) at sequential Na^+ ion channels until the wave of depolarization reaches the axon terminal.

6 Remembering the discussion concerning the refractory period, it should be clear that those Na^+ ions that entered at the Na^+ ion channel and dispersed toward the cell body encountered Na^+ ion channels that were still inactivated, and their inactivation gates could not be opened.

7 Since the inactivation gate remained closed, Na^+ ions could not pass through the Na^+ ion channel, and the membrane could not be depolarized at the old point.

8 **Therefore, the action potential in a normal *in vivo* situation can travel only in a single direction, away from the cell body toward the axon terminal.**

The conduction velocity of an action potential is directly dependent on the diameter of the axon. The conduction velocity is further increased due to axon myelination (Fig. 3.12). As indicated previously, in an unmyelinated axon the propagation of the action potential is dependent on the sequential activation of successive Na^+ ion channels. In a myelinated axon voltage-gated Na^+ ion channels are much farther apart, located mostly at the myelin sheath gaps, requiring that the generation of action potentials occurs only at the myelin sheath gaps. Hence, the current must "jump" from node to node (**saltatory conduction**), a much faster process than that occurring in unmyelinated axons.

Postsynaptic potentials

The release of neurotransmitters at the synaptic cleft results in the formation of small, local currents at the postsynaptic membrane, known as postsynaptic potentials

Small local currents, initiated by the release of chemical messengers from synaptic vesicles, are spread throughout the cell body of the neuron and are known as **postsynaptic potentials** (**PSPs**). As described above, the ability of these postsynaptic potentials to evoke an action potential diminishes as a function of:

• their distance from the initial segment of the axon; and
• the amount of time that has elapsed from the time that the depolarization occurred at the **initial segment** of the axon.

The reason for this is that the dendrites and cell body have relatively few voltage-gated channels – the abundance of such channels is a specific requirement for the formation and propagation of action potentials. An increase in the number of postsynaptic potentials (e.g., by the rapid release of multiple quanta of chemical messengers from a single axon terminal or from many axon terminals on the same neuron), however, results in their summation. Hence, hundreds of nearly simultaneous, small postsynaptic potentials may be necessary to achieve the threshold required for the formation of an action potential when those currents arrive at the initial segment of the axon.

Generally, **inhibitory synapses** are closer to the initial segment of the axon than are excitatory synapses. Since inhibitory synapses function to hyperpolarize the postsynaptic membrane, they tend to diminish the flow of currents formed at the excitatory potential making it more difficult to reach threshold levels at the initial segment of the axon. This sequence of events is referred to as the **summation of IPSPs and EPSPs** (inhibitory postsynaptic potentials and excitatory postsynaptic potentials).

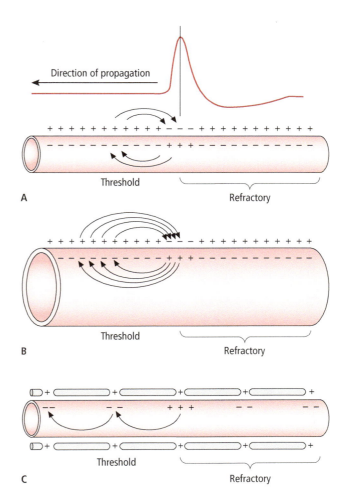

Figure 3.12 ● Propagation of the action potential depicted in three axons: (A, B) unmyelinated axons of different diameters, and (C) a myelinated axon. (A) Because the axon of smaller diameter offers greater axoplasmic resistance, a given localized current can hypopolarize the membrane to a shorter distance than an axon of larger diameter (indicated by the curved arrows) (B). Therefore, the propagation of action potentials occurs much faster in an axon of larger diameter than one of smaller diameter. (C) Myelinated axons offer less resistance than either of the two examples given, since the local current can only leave at the myelin sheath gap. Therefore, the propagation of the action potential is saltatory, jumping from node to node, and the impulse travels much faster along a myelinated axon than along a nonmyelinated one.

CLINICAL CONSIDERATIONS

Leigh's syndrome (subacute encephalomyelopathy), a childhood disease, is due to mutations in the mitochondrial DNA as well as mutations in chromosome 9 (surfeit-1 gene) of neurons located in the basal ganglia (basal nuclei). Children afflicted with this disease initially lose their verbal abilities and muscular coordination, and later may exhibit dementia, seizures, and optic atrophy; many die due to respiratory failure.

Tay–Sachs disease is an inherited lysosomal storage disease where specific neurons are unable to synthesize the enzyme hexosaminidase A. These defective lysosomes are unable to digest sphingolipids, which continue to accumulate and impede the normal functioning of the cell, leading to cell death. This autosomal recessive trait is especially prominent in families of Eastern European Jewish descent. Its symptoms – hyperacusis, irritability, hypotonia, reduced motor skills, blindness with red spots in the macula, and seizures – appear before the first year of life and result in the death of the child between the second and fourth year of age.

Multiple sclerosis is characterized by progressive demyelination of the white matter in the brain and spinal cord. This condition, believed to be an autoimmune disease, usually appears in the young adult and is distinguished by periods of deterioration followed by periods of remission that may continue for decades. Recently it has been proposed that oligodendrogliopathy is the actual cause of this condition, and the autoimmune response is simply a reaction to the pathologic state of the oligodendrocytes. Corticosteroids or interferon beta or monoclonal antibodies are administered to suppress the immune response and thereby ameliorate the disease. As the disease progresses the periods of remission become shorter and the disability becomes greater due, among other things, to the decreased velocity of impulse propagation in the affected axons. Any one of the demyelination episodes may lead to death within a short period of time. Current research is progressing in finding new treatment modalities to arrest and, perhaps even cure, this debilitating condition.

Neuroglial tumors account for approximately half of intracranial tumors. They range in severity from the relatively mild oligodendroglioma to the highly malignant, and fatal, **glioblastoma**. The latter disease originates from very rapidly proliferating astrocyte-derived neoplastic cells that invade the cerebrum.

SYNONYMS AND EPONYMS OF NERVOUS SYSTEM HISTOPHYSIOLOGY

Name of structure or term	Synonym(s)/eponym(s)
Axon terminal	Bouton terminaux
Cell body	Perikaryon; soma
Gap junction	Nexus
Myelin clefts	Myelin incisure/Clefts of Schmidt–Lantermann
Myelin sheath gaps	Nodes of Ranvier
Neurofilaments	Intermediate filaments
Neurolemmocytes	Schwann cells
Tigroid substance	Nissl bodies
Unipolar neuron	Pseudounipolar neuron

FOLLOW-UP TO CLINICAL CASE

The patient underwent an MRI examination of the brain and CSF examination that confirmed the diagnosis of multiple sclerosis (MS). MS is a neurologic disease that can vary widely in symptomatology, since it can cause lesions in any part of the CNS. However, it only affects the white matter. More specifically, prototypical MS causes demyelination of axons in the CNS by death or injury to oligodendrocytes. In most cases there is relative preservation of the axons. The understanding of the pathophysiology is incomplete, but it is clear that there is an immunologic attack causing loss of or dysfunction of oligodendrocytes. However, it is not a typical autoimmune condition since most of the medications used for autoimmune diseases such as rheumatoid arthritis do not work (or do not work well).

The diagnosis can be confirmed by MRI of the appropriate area of the CNS. The patient in the case study had multiple demyelinating plaques in the cerebellum and brain, including the corpus callosum (which is characteristic). The lesions of MS are often periventricular in location. There are often lesions seen that are clinically "silent." Using CSF and serum electrophoresis, synthesis of antibodies within the CNS and not the serum is often demonstrated. This is pathologic.

One definition for MS is that it causes multiple lesions in the CNS in space and time. Multiple focal white matter plaques of demyelination, easily seen by MRI, are present in different parts of the brain and/or spinal cord at different times. Peripheral nerve myelin is not affected at all since it is produced differently. The most common syndrome is a relapsing/remitting course. Over the course of many years, symptoms occur in episodes lasting typically weeks or months, and then symptoms spontaneously improve. Over the course of years, after a number of relapses, this can lead to disability.

Common symptoms of relapses include, but are certainly not limited to, focal numbness/tingling or weakness, gait ataxia, visual loss (optic neuritis), slurred speech, paraplegia and sensory disturbance up to a thoracic dermatome level, and ocular motility problems (producing double vision). One characteristic of MS is that heat or a rise in body temperature can temporarily bring about symptoms typical of that particular patient's MS. This is not due to any new lesions happening within the CNS. A rise in temperature actually reduces the efficiency of conduction along axons. The effects are not seen in people with normal CNS function because there is a large reserve capacity for conduction, but this is not so in the old plaques of demyelination in patients with MS. It is common for MS patients to complain of symptoms surfacing temporarily with fevers, hot flashes, hot baths or showers, prolonged exertion, or sunbathing.

QUESTIONS TO PONDER

1. Why is the neuron cell body so rich in organelles?

2. What is the functional difference between a node and an internode?

3. Why are results obtained by the use of the Goldman equation a more accurate prediction of resting membrane potential than the results obtained by the use of the Nernst equation?

4. Although action potentials can travel in both directions along an axon *in vitro*, why is it that they travel only in the anterograde direction *in vivo*?

5. Why are the locations of inhibitory synapses and excitatory synapses important?

CHAPTER 4

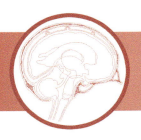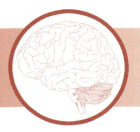

Neurotransmitter Substances

CLINICAL CASE

CLASSIFICATION OF NEUROTRANSMITTER SUBSTANCES

CLINICAL CONSIDERATIONS

FOLLOW-UP TO CLINICAL CASE

QUESTIONS TO PONDER

CLINICAL CASE

A 60-year-old patient presents to your office with imbalance and a few falls resulting in some severe bumps and bruises from the falls, which prompted the patient to come in to see the doctor. Symptoms seemed to begin about 6 months previously and are slowly getting worse. Also noted is a tremor that primarily involves the right hand and is most prominent when at rest and when walking. In fact, the patient barely notices it, but it is very bothersome to the patient's spouse, who notes that the patient's movements seem very slow and stiff.

The patient does not make good eye contact when being spoken to, and gaze is directed downward. Facial expression is limited, and there seems to be a paucity of spontaneous movements. All movements, including walking, speaking, and reaching, are very slow but performed accurately. There is a tremor, most prominent in the right hand, which is most noticeable at rest and goes away when reaching for objects. The tremor is also enhanced when walking, and handwriting is poor. Walking is slow with short, shuffling steps.

Cells of the nervous system communicate with each other at synapses either via electrical signals or by the release of relatively small messenger molecules, known as **neurotransmitters** (neurotransmitter substances). The morphology and physiology of electrical and chemical synapses were presented in Chapter 3, whereas the current chapter discusses the most common and significant of the neurotransmitter substances [of which there are more than 100] (Table 4.1). These messenger molecules may be classified into two major types, those that bind directly to **ion channels** (**ligand-gated receptors**) and those that bind to **G-protein-coupled receptors**. The neurotransmitter substances that bind specifically to ligand-gated ion channels cause those channels to open directly; thus, the target neuron responds immediately, resulting in an **ionotropic effect** (Fig. 4.1). Those neurotransmitter substances that bind to G-protein-coupled receptors exert an indirect effect on ion channels. Since the G-proteins act as intermediaries, the effect on the target neuron is slower than in the previous instance. Thus, neurotransmitter substances that act on G proteins are frequently known as **neuromodulators**, and they are said to exert a **metabotropic effect** (Fig. 4.2).

A Textbook of Neuroanatomy, Third Edition. Maria A. Patestas, Amanda J. Meyer, and Leslie P. Gartner.
© 2025 John Wiley & Sons, Inc. Published 2025 by John Wiley & Sons, Inc.
Companion website: www.wiley.com.go/patestas/neuroanatomy3e

50 ●●● **CHAPTER 4**

Table 4.1 ● Properties of the principal neurotransmitters (organized alphabetically).

Neurotransmitter	Major function	Precursor	Enzyme	Location in nervous system	Additional pertinent information
Acetylcholine	Excitatory/inhibitory	Acetyl CoA and choline	Choline acetyl-transferase	Neuromuscular synapse; autonomic nervous system; striatum	Removed by the enzyme acetylcholinesterase; cholinergic neurons degenerate in Alzheimer disease
Adenosine triphosphate (ATP)	Excitatory	ADP	Oxidative phosphorylation; glycolysis	Motor neurons of the spinal cord; autonomic ganglia	Also co-released with numerous neurotransmitters
Beta-endorphin	Inhibitory	Amino acids	Protein synthesis	Arcuate nucleus of hypothalamus; nucleus of solitary tract	Least numerous of the opioid neurotransmitter-containing cells; function in pain suppression
Dopamine	Excitatory/Inhibitory	Tyrosine (L-DOPA)	Tyrosine hydroxylase	Neurons of the substantia nigra, arcuate nucleus, and tegmentum	Associated with Parkinson disease; inhibition of prolactin release; schizophrenia
Dynorphin	Inhibitory	Amino acids	Protein synthesis	Hypothalamus; amygdala; limbic system	More numerous than beta-endorphin-containing cells; function in pain suppression
Enkephalins	Inhibitory	Amino acids	Protein synthesis	Raphe nuclei; striatum; limbic system; cerebral cortex	More numerous than beta-endorphin and enkephalin-containing cells; function in pain suppression
Epinephrine	Excitatory	Norepinephrine	Phenylethanolamine-N-methyltransferase	Rostral medulla oblongata	Not commonly present in the CNS
Gamma aminobutyric acid (GABA)	Inhibitory	Glutamate	Glutamic acid decarboxylase	Mostly local circuit interneurons	Decreased GABA synthesis in vitamin B_6 deficiency
Glutamate	Excitatory	Glutamine	Glutaminase	Most excitatory neurons of the CNS	Glutamate–glutamine cycle; excitotoxicity
Glycine	Inhibitory	Serine	Serine hydroxymethyl-transferase	Neurons of the spinal cord	Activity blocked by strychnine
Histamine	Excitatory	Histidine	Histidine decarboxylase	Tuberomammillary nucleus of hypothalamus	Controls neuronal excitability and synaptic transmission
Nitric oxide (NO)	Inhibitory	L-arginine	Nitric oxide synthase	Cerebellum; hippocampus; olfactory bulb	Smooth muscle relaxant, thus strong vasodilator
Norepinephrine (noradrenaline)	Excitatory	Tyrosine (dopamine)	Dopamine beta-hydroxylase	Postganglionic sympathetic neurons; locus caeruleus	Associated with mood and mood disorders (mania, depression, anxiety, and panic)
Serotonin (5-hydroxytryptamine)	Excitatory/inhibitory	Tryptophan	Tryptophan-5-hydroxylase	Pineal body; raphe nuclei of midbrain, pons, and medulla oblongata	Associated with sleep modulation; arousal, cognitive behaviors
Somatostatin	Inhibitory	Amino acids	Protein synthesis	Amygdala, small spinal ganglion cells, and hypothalamus	Also known as somatotropin release-inhibiting factor
Substance P	Excitatory	Amino acids	Protein synthesis	Spinal and trigeminal ganglia (C and Aδ fibers)	Composed of 11 amino acids; associated with transmission of pain

NEUROTRANSMITTER SUBSTANCES • • • 51

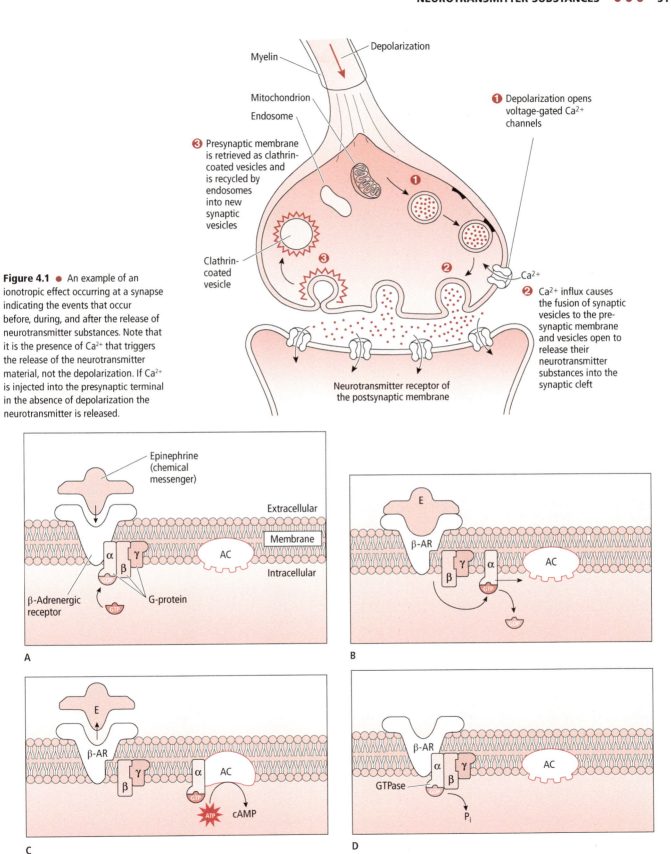

Figure 4.1 ● An example of an ionotropic effect occurring at a synapse indicating the events that occur before, during, and after the release of neurotransmitter substances. Note that it is the presence of Ca^{2+} that triggers the release of the neurotransmitter material, not the depolarization. If Ca^{2+} is injected into the presynaptic terminal in the absence of depolarization the neurotransmitter is released.

Figure 4.2 ● A diagram of G-protein action during a metabotropic event. (A) Binding of epinephrine (E) to its beta-adrenergic receptor (β-AR) activates the replacement of GDP by GTP on the alpha-subunit of the G-protein. (B, C) The alpha-subunit dissociates from the G-protein and activates adenylate cyclase (AC) to convert ATP to cAMP, and epinephrine dissociates from its beta-adrenergic receptor. (D) GTPase cleaves an inorganic phosphate molecule from GTP, converting it into GDP and the alpha-subunit rejoins its G-protein.

52 ● ● ● **CHAPTER 4**

In order for a substance to be considered a neurotransmitter (or neuromodulator) it must possess five basic characteristics; it:

1 must be synthesized by presynaptic neurons;
2 has to reside within the synaptic terminals (enclosed in synaptic vesicles);
3 must be able to elicit the same effect when tested under *in vitro* conditions;
4 has to bind to specific receptors on the postsynaptic membrane; and
5 must be inactivated in or removed from the synaptic cleft.

Each neuron has at least one, but usually a number of neurotransmitter substances – **co-transmitters** – in synaptic vesicles within the presynaptic terminal (frequently a classic neurotransmitter as well as neuropeptides). These co-transmitters may be released individually or together, and they are frequently housed in different synaptic vesicles. It is interesting to note that the "classic" neurotransmitters (e.g., acetylcholine and monoamines), known also as **small-molecule neurotransmitters**, are stored in small, electron-lucent synaptic vesicles, approximately 40 nm in diameter that are able to bind to the active sites of presynaptic terminals, whereas **neuropeptides** are stored in somewhat larger, dense-cored secretory granules, about 70–250 nm in diameter, which are unable to bind to the active sites of the presynaptic terminals.

Once neurotransmitter substances are released into the **synaptic cleft** they act on the postsynaptic cell. In order to ensure that a single release will not continue to exert an influence on the postsynaptic membrane one of three possibilities occurs to the neurotransmitter substance:

1 it is endocytosed by the endfoot of the releasing neuron,
2 it is digested by enzymes localized in the synaptic cleft (a process called catabolism), or
3 it diffuses from the synaptic cleft to be endocytosed by astrocytes protecting the synapse.

CLASSIFICATION OF NEUROTRANSMITTER SUBSTANCES

Neurotransmitter substances may be classified into four different groups: biogenic amines, neuropeptides, small-molecule neurotransmitters, and other ligands. Some authors classify neurotransmitters into only two groups: small-molecule neurotransmitters and neuropeptides; thus, all neurotransmitters other than neuropeptides are small-molecule neurotransmitters in their classification; and still other authors classify them according to other qualifications.

It should be noted that this chapter will indicate the locations, whether in the central nervous system or in the peripheral nervous system, where these neurotransmitters perform their functions. The reader is not expected to know where these specific structures are located or what their function(s) happen to be; however, as the reader progresses in this textbook, the locations and the functions of those structures will be elucidated and the reader will be reminded of the neurotransmitters that are pertinent to those areas.

Biogenic amines

Biogenic amines are derived from amino acids and exert a metabotropic effect

Biogenic amines, amino acid derivatives, bind to G-protein-coupled receptors and activate the second messenger system within the target neuron, which in turn will open ion channels. Biogenic amines include dopamine, norepinephrine, and epinephrine (all considered under the heading **catecholamines**), and serotonin and histamine (even though histamine is an imidazole). Serotonin and histamine are produced by enzymatic reactions on essential amino acids (tryptophan and histidine, respectively), which are amino acids humans *must get from diet because we are unable to manufacture them*. Biogenic amine manufacture is under strict control, and if an increased amount is required the rate of their synthesis can be greatly enhanced. Upon release from the presynaptic terminal, biogenic amines are endocytosed by the nerve terminals and glial cells, and are digested by the enzymes **cathecol O-methyltransferase** and **monoamine oxidase** (**MAO**). In this fashion, each release of these neurotransmitters is responsible for only a limited number of depolarizations/hyperpolarizations. **Monoamine oxidase inhibitors** (**MAOIs**) are drugs that act within the presynaptic neuron to inactivate catecholamines and indoleamines (such as serotonin and melatonin). They also act to increase the concentration of other neurotransmitters such as serotonin, dopamine, and norepinephrine, and, by some unknown mechanism, relieve depression.

Catecholamines

The three major catecholamines, dopamine, norepinephrine, and epinephrine, have similar characteristics. They are all derivatives of the amino acid **tyrosine**, they have a **catechol** component, and the enzyme **tyrosine hydroxylase** participates in the conversion of tyrosine to **dihydroxyphenylalanine** (L-**DOPA**) (Fig. 4.3).

Dopamine

Dopamine, a mostly excitatory neurotransmitter (derived from the decarboxylation of L-**DOPA**), is present in neurons of various regions of the central nervous system (CNS). They include the neurons of the arcuate nucleus, a group of nerve cell bodies that project to the infundibulum of the hypophysis, the neurons of the anterior tegmentum whose axons project to the limbic system, and the neurons of the substantia nigra whose axons project to the striatum.

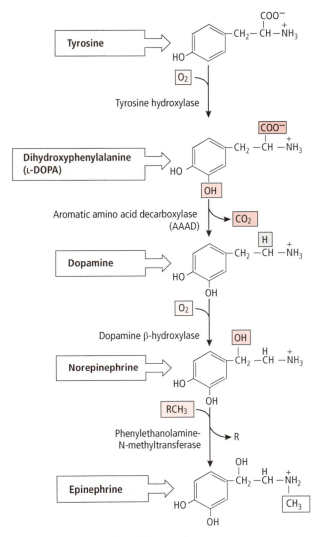

Figure 4.3 • Synthesis of catecholamines from tyrosine.

Note that the clinical case at the beginning of the chapter refers to a patient with imbalance and tremors.

1 What region of the brain should be suspected when a patient exhibits movement disorders?

2 What condition should a physician suspect when the symptoms include tremors of the hand at rest but not during the process of reaching for something?

3 What condition should a physician suspect when the patient's actions are slowed, and display limited facial expressions?

Norepinephrine

Norepinephrine (**noradrenaline**), an excitatory neurotransmitter located in postsynaptic sympathetic neurons, is generated from the hydroxylation of dopamine. The conversion occurs in the presynaptic terminals of adrenergic neurons and is facilitated by the enzyme **dopamine β-hydroxylase**. An additional site where norepinephrine is the neurotransmitter is the **locus caeruleus**. Neurons of the locus caeruleus project to the reticular formation as well as to various other regions of the CNS.

Higher-than-normal levels of norepinephrine in the CNS have been associated with mania, whereas lower-than-normal levels have been associated with depression. Additionally, panic disorders have also been associated with abnormal levels of norepinephrine in the CNS.

Epinephrine

Epinephrine (**adrenaline**) is an excitatory neurotransmitter derived from norepinephrine. The conversion is catalyzed by the enzyme **phenylethanolamine-N-methyltransferase**. Very few neurons of the CNS use epinephrine as a neurotransmitter substance, and those that do are restricted to the cranial medulla oblongata.

Serotonin

Serotonin is a derivative of tryptophan, an essential amino acid that is converted into 5-hydroxytryptophan by the rate-limiting enzyme **tryptophan-5-hydroxylase**. Decarboxylation of 5-hydroxytryptophan results in the formation of 5-hydroxytryptamine (serotonin), which is an excitatory neurotransmitter (Fig. 4.4). Neurons that use serotonin as their neurotransmitter in motor functions and inhibitory in sensory functions are known as **serotonergic neurons**, and they are present in the pons, medulla oblongata, raphe nuclei of the midbrain, and pineal body. These neurons function in arousal and sleep modulation as well as in regulating certain higher cognitive functions. Moreover, these neurons may exert modulatory effects on catecholamine levels. Interestingly, at least 90% of serotonin is manufactured by the enterochromaffin cells of the stomach, small intestine, and the colon. It has been demonstrated that low levels of the last three neurotransmitters – dopamine, noradrenaline, and serotonin – all produce depression and, if their uptake is blocked, the severity of the depression is greatly alleviated. Low levels of serotonin and dopamine can produce clinical depression, and the use of selective serotonin reuptake inhibitors (SSRIs) is often used to treat this condition.

Histamine

Histamine is an inflammatory pharmacologic agent peripherally released by mast cells, basophils, and enterochromaffin-like cells. It is responsible for causing capillaries to become

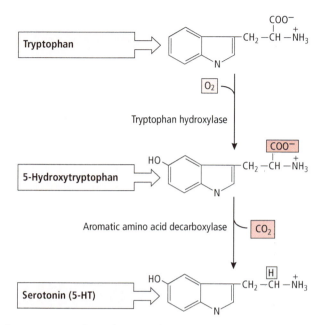

Figure 4.4 • Synthesis of serotonin from tryptophan.

more permeable to white blood cells and also for contraction of bronchiolar smooth muscles. In the nervous system it is located mostly in the hypophysis, and the tuberomammillary nucleus of the hypothalamus and is generated by the decarboxylation of histidine, an essential amino acid. Modulation of histamine's effect occurs through four subtypes of histamine receptors: H1, H2, H3, and H4. In the nervous system, H1 receptors have been localized in the supraoptic nucleus (to modulate antidiuretic hormone and oxytocin production), H2 receptors in the striatum (to modulate motor activity), H3 receptors on cell bodies (for autoregulation of histamine levels), and H4 receptors most strongly expressed in the spinal cord laminae I & II (to modulate nociception). The use of antihistamine (H_1 receptor antagonists), commonly used in treating allergies, often causes drowsiness due to CNS depression.

Neuropeptides

Neuropeptides are small polypeptides cleaved from their larger propeptides on the RER immediately after synthesis, then packaged in the Golgi apparatus, and transferred to the presynaptic terminal

Initially, many of the **neuropeptides** were known as hormones in the digestive and respiratory systems, but subsequently they were observed to function also as neurotransmitters in the nervous system. Neuropeptides are small polypeptides, cleaved, post-translationally before they reach the Golgi apparatus, from larger **propeptide molecules** synthesized on the rough endoplasmic reticulum (RER) in the cell body. The cleaved neuropeptides are packaged into secretory vesicles in the *trans*-Golgi network and transferred, via the rapid anterograde transport, to the presynaptic terminal. Often, a presynaptic terminal may possess and release several different neuropeptides simultaneously, but once released, instead of being recycled, they are destroyed by peptidases. Thus, unlike neurotransmitters such as acetylcholine, which is cleaved by acetyl-cholinesterase in the synaptic cleft and taken up at the presynaptic cell membrane, neuropeptides are degraded in the synaptic cleft and are not taken up by the presynaptic cell membrane. Instead, more neuropeptides are synthesized in the presynaptic cell body and transported into the presynaptic terminal. Because the released neuropeptide is degraded in the synaptic cleft by peptidases single release of a quantum of neuropeptides cannot elicit *multiple* responses from the postsynaptic cell. Depending on the specific neuropeptides and/or their receptors, these small peptides may be excitatory or inhibitory neurotransmitters or neuromodulators. Although there are numerous neuropeptides, only the most common ones will be detailed in this textbook, namely somatostatin, substance P, and the opioid neuropeptides (endorphins, enkephalins, and dynorphins). It should also be reaffirmed that neuropeptides are synthesized only in the cell body of the neuron, whereas some other neurotransmitter substances are synthesized both in the cell body as well as at the synaptic terminal of the neuron.

Somatostatin

Somatostatin (also known as **somatotropin release-inhibiting factor**) is a small polypeptide with two active forms somatostatin-14 and somatostatin-28, referring to the number of amino acids composing them. It was first observed in the digestive tract as one of the paracrine hormones manufactured and released by one of the **diffuse neuroendocrine system (DNES) cells** and functioned to inhibit the release of other paracrine and endocrine hormones manufactured by nearby DNES cells. In the CNS, somatostatin is an inhibitory neurotransmitter and is localized in the hypothalamus (preoptic area and arcuate nucleus), retina, hippocampal formation, amygdaloid body, the nucleus of the solitary tract, and the spinal ganglion cells. To date, five somatostatin receptor subtypes have been characterized, and activation of all of them hyperpolarizes the pre-synaptic membrane potential by inhibiting adenylyl cyclase, lowering intracellular cAMP and calcium ions, thereby inhibiting neurotransmitter release. Somatostatin is often co-released with GABA; however, its fate in the synaptic cleft is different because GABA is rapidly endocytosed, whereas somatostatin must be cleaved by local peptidases before being endocytosed; thus, somatostatin has a much longer duration of action.

Substance P

Substance P is a small polypeptide, composed of 11 amino acids, that was first discovered in the digestive system as a product of the DNES cells. Subsequently, it was observed in the spinal cord, hippocampus, and the neocortex, as well as in the unipolar neurons of cranial sensory (e.g., trigeminal ganglia) and spinal ganglia. Substance P is an excitatory neuropeptide that is one of the most important neurotransmitters in nociception as its release by unmyelinated C-type fibers is required to produce moderate-to-intense pain. Activation of opiate receptors inhibits the release of Substance P. It also elicits an inflammatory action by binding to neurokinin receptors 1–3 which causes microvascular permeability, arteriolar vasodilation, leukocyte chemoattraction, and mast cell degranulation which increases histamine and serotonin levels. Substance P is degraded by neutral endopeptidase or enkephalinase enzymes which are expressed on neurons, endothelial cells, leukocytes, and smooth muscle cells. Interestingly, the release of substance P in the vomiting center (the area postrema) initiates the vomiting reflex. Moreover, it has been recognized that substance P levels become elevated during magnesium deficiency which appears to promote the synthesis of inflammatory messenger molecules such as tumor necrosis factor-α (TNF-α).

Opioid neuropeptides

The **opioid neuropeptides** constitute a major subgroup of neuropeptides, composed of at least 20 neurotransmitter substances. Although they may be subdivided into three categories – enkephalins, endorphins, and dynorphins – they have three major characteristics in common in that they are mostly inhibitory, they bind to opium receptors of the postsynaptic membrane, and they serve as substance P antagonists, thereby inducing analgesia.

Enkephalins

Enkephalins are of two types, methionine enkephalins (met-enkephalins) and leucine enkephalins (leu-enkephalins). **Met-enkephalins** are derived from **proenkephalin** and **pro-opiomelanocortin (POMC)**, a large propeptide precursor that is cleaved to form the following: beta-endorphins, adrenocorticotropic hormone (ACTH), beta lipoprotein, and melanocyte-stimulating hormone (Fig. 4.5). **Leu-enkephalin** is a derivative of the propeptides **prodynorphin** and **proenkephalin**. Enkephalins are the neurotransmitter substances used by many interneurons as well as at synapses in the posterior horn of the spinal cord. Enkephalins are also used as neurotransmitters in the limbic system, the cerebral cortex, the striatum, and in the raphe nuclei of the brainstem. Moreover, they function to regulate nociception (pain sensation).

Endorphins (endogenous morphine)

The most common form of the endorphins, **beta-endorphins**, are among the pro-opiomelanocortin (POMC) derivatives. They are present in the hypothalamus where they exert an analgesic effect. As with most other opioid neuropeptides, beta-endorphins are inhibitory neurotransmitter substances. Specifically, they are released from presynaptic terminals and inhibit the release of gamma-aminobutyric acid (GABA)

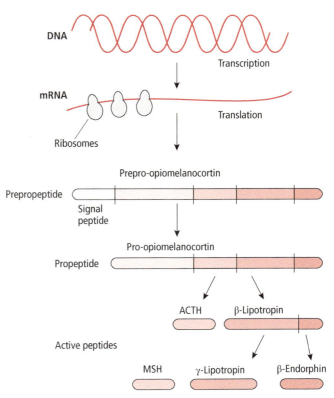

Figure 4.5 • Synthesis of enkephalins from prepro-opiomelanocortin. ACTH, adrenocorticotropic hormone; MSH, melanocyte-stimulating hormone.

whose absence permits the increased release of dopamine, the likely cause of drug dependency. Interestingly it has been demonstrated that acupuncture appears to cause an elevated endorphin level, possibly accounting for the ability of acupuncture to relieve pain.

Dynorphin

Dynorphin is an inhibitory neurotransmitter substance derived from the propeptide **prodynorphin**. This neurotransmitter substance is inhibitory and functions in pain suppression. Dynorphin is localized in the amygdaloid body, limbic system, striatum, substantia nigra, and in the hypothalamus. It appears that cocaine addiction may be decreased in individuals who produce an increased level of dynorphin because of dynorphin's ability to inhibit dopamine release.

Small-molecule neurotransmitters

Small-molecule neurotransmitters are low-molecular-weight substances that are synthesized in the presynaptic terminals

Small-molecule neurotransmitters, such as acetylcholine, glycine, glutamate, and gamma aminobutyric acid (GABA), are low-molecular-weight substances that are usually synthesized in the presynaptic terminals. The enzymes required for their synthesis are manufactured in the cell body and reach the presynaptic terminals via anterograde axonal transport.

Acetylcholine

Acetylcholine is the neurotransmitter at neuromuscular junctions, reticular activating system, presynaptic and postsynaptic terminals of the parasympathetic nervous system, and presynaptic terminals of the sympathetic nervous system, as well as in various regions of the CNS

Acetylcholine, a neurotransmitter substance localized both in the central and peripheral nervous systems, is synthesized in the presynaptic terminal from **acetyl coenzyme A (Acetyl CoA)** and **choline** by **choline acetyltransferase**, the rate-limiting enzyme of its synthesis (Fig. 4.6). Therefore, unlike the other small-molecule neurotransmitters, acetylcholine does not originate from an amino acid. Moreover, choline is not synthesized by the body; instead, it must be absorbed from the diet. Acetylcholine, as with many other neurotransmitters, has the ability to bind to both **ionotropic receptors** and **metabotropic receptors**. In skeletal muscle it is an **excitatory neurotransmitter substance**, binding directly to ligand-gated sodium (nicotinic receptors) as well as to potassium ion channels causing them to open. In cardiac muscle it is an **inhibitory neurotransmitter substance**, binding to G-protein-linked receptors (muscarinic receptors) that facilitate the opening of potassium ion channels.

Acetylcholine is present in the reticular activating system as well as in the autonomic nervous system, specifically at the presynaptic sympathetic, presynaptic parasympathetic, and postsynaptic parasympathetic synapses. Additionally, acetylcholine is located in other areas of the CNS such as the projection neurons of the tegmentum, interneurons of the striatum, projection neurons of the forebrain, and motor neurons of the brainstem and spinal cord.

The enzyme **acetylcholinesterase** is present in the synaptic cleft at the postsynaptic membrane. This enzyme cleaves the released acetylcholine into its two component molecules, choline and acetyl CoA, preventing multiple depolarizations from the release of a single quantum of acetylcholine. The cleaved moieties, acetyl CoA and choline, are transported, individually, into the presynaptic terminal where they are reassembled into acetylcholine and stored in synaptic vesicles.

Glycine

Glycine, one of the most common inhibitory neurotransmitters of the spinal cord, is synthesized by the enzyme **serine hydroxymethyl transferase** from the amino acid serine (Fig. 4.7A). Usually, glycine binds to ligand-gated chloride channels. Subsequent to its release into the synaptic cleft, glycine is recaptured by glycine-specific, membrane-bound carrier proteins (transport proteins), ferried into the presynaptic terminal, and is transported into synaptic vesicles for future use. The binding of glycine to the postsynaptic terminal opens chloride channels that result in an inhibitory postsynaptic potential (see the Chapter 3 section on Synapses).

Glutamate

Glutamate is present in almost every region of the brain as well as in the spinal cord at presynaptic terminals of the central processes of Aδ and C fibers of the posterior root ganglion

Glutamate, probably the most common excitatory neurotransmitter, is synthesized in the presynaptic terminal from glutamine, catalyzed by the enzyme **glutaminase**. Once glutamate is released into the synaptic cleft within the CNS, it must be quickly removed, or it will cause the postsynaptic neuron to undergo repeated excitations, resulting in neuronal degeneration and subsequent death. This process, known as **excitotoxicity**, results from repeated stimulation of the postsynaptic membrane receptors that open Ca^{2+} channels. The increased influx of Ca^{2+} into the cell body results in free radical formation and damage to the neuron and its system of membranes.

To prevent excitotoxicity, the free glutamate in the vicinity of the synapse is endocytosed by the presynaptic terminal as well as by local **neuroglia cells**. The glutamate endocytosed by the presynaptic terminal is converted, by the enzyme **glutamine synthetase**, to glutamine, which then is again converted to form glutamate. The same reaction occurs within the glia cells; however, these cells release the newly formed glutamine into the vicinity of the presynaptic terminal that endocytoses the glutamine and, as before, converts it into

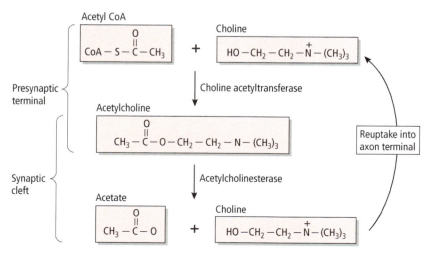

Figure 4.6 • Acetylcholine synthesis and degradation.

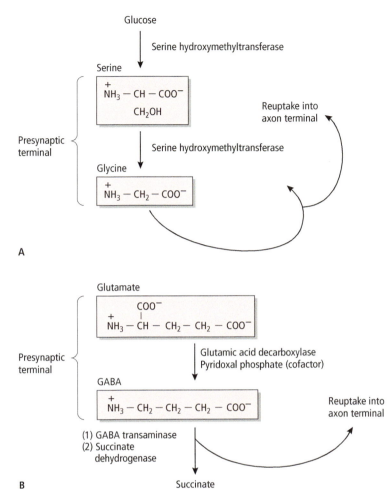

Figure 4.7 ● (A) Synthesis of glycine from glucose. (B) Synthesis of GABA from glutamate.

glutamate (Fig. 4.8). This entire sequence is known as the **glutamine–glutamate cycle**.

Experimental evidence indicates that when glutamate is injected into neurons *in vitro* the neurons undergo spontaneous, consecutive depolarizations every second or so, a situation reminiscent of epileptic seizures. These results suggest a possible relationship of glutamate to epileptic seizures.

Gamma-aminobutyric acid (GABA)

Approximately 30% of the neurons of the CNS use **GABA (gamma-aminobutyric acid)** as an inhibitory neurotransmitter substance. GABA is derived from glucose by way of glutamate, whose conversion to GABA is catalyzed by the enzyme **glutamic acid decarboxylase**, utilizing **pyridoxal phosphate** (a vitamin B_6 derivative) as a cofactor (Fig. 4.7B). Once GABA is released from the presynaptic terminal, the free neurotransmitter molecules (those not bound to postsynaptic receptors) are rapidly reabsorbed by neuroglia and by the presynaptic terminals by using GABA-specific membrane-bound protein transporters.

GABA is present in local circuit interneurons and Purkinje cells of the cerebellum. Depending on the receptor, the binding of GABA either opens chloride or potassium ion channels or closes calcium ion channels resulting in hyperpolarization. Interestingly, the role of GABA is different during brain development where its role is reversed, and it is an excitatory neurotransmitter even though it functions in the same way as in the adult brain; that is, it causes the opening of chloride channels. However, developing neurons have a higher intracellular Cl^- concentration than that of the extracellular milieu, and opening of the chloride channels results in a flow of Cl^- ions out of the cell thereby depolarizing the neuron rather than hyperpolarizing it. The higher intracellular Cl^- concentration in immature neurons is due to their higher concentration of Na^+-dependent NKCC1 co-transporter (which transports Na^+, K^+, and Cl^- into the cell [in a ratio of 1 : 1 : 2]) than their Na^+-independent KCC2 co-transporters (which transports Na^+ and Cl^- out of the cell [in a ratio of 1 : 1]).

Other neurotransmitters

Although there is a plethora of other neurotransmitter and neuromodulator substances, many of them are sufficiently rare in the nervous system that they will not be addressed in this textbook. The only additional agents that will be discussed are adenosine triphosphate and nitric oxide.

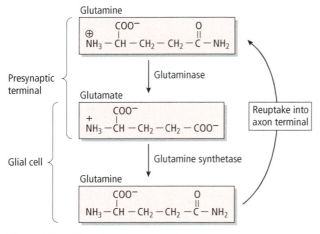

Figure 4.8 • Glutamine–glutamate cycle.

Adenosine triphosphate

Adenosine triphosphate (ATP) is a high-energy phosphate molecule that is customarily associated with energy storage in cells. Not much is understood concerning its functions as a neurotransmitter substance, but it is known to be co-released with other neurotransmitters, although it may also be released by itself. ATP has an excitatory effect by causing depolarization of the postsynaptic membrane in neurons whose postsynaptic membranes possess purinergic receptors (especially P2X receptors). Subsequent to being released, ATP is degraded to adenosine. It is present in autonomic postsynaptic neurons as well as in motor neurons of the spinal cord.

Nitric oxide

Nitric oxide (NO) is a gas, derived from the amino acid **L-arginine** by the action of the enzyme **nitric oxide synthase**. It is a very reactive molecule and, therefore, has a very short half-life. This inhibitory neurotransmitter is unusual not only because it is a gas but also due to the fact that it does not bind to membrane receptors. It may function in memory formation since it is located in the hippocampal formation. Additionally, NO is also present in the olfactory system, the cerebellum, striatum, cerebral cortex, and the hypothalamus. NO also causes relaxation of smooth muscle cells.

CLINICAL CONSIDERATIONS

Parkinson disease
Parkinson disease is a neurologic disorder characterized by progressive deterioration with time. The chief symptoms are rigidity of movement, increased tremors and trembling, being slow in initiating movement, muscle stiffness, stooping gait, and difficulties in maintaining balance. This disease is the result of the degeneration and death of dopaminergic neurons in the compact part of the **substantia nigra**, which causes a decrease in the availability of the neurotransmitter substance **dopamine**. The reason for the degeneration of the dopaminergic neurons is not understood. It is interesting to note that excess levels of dopamine in the anterior tegmentum have been associated with schizophrenia.

Eaton–Lambert syndrome and myasthenia gravis
Eaton–Lambert syndrome is an autoimmune disorder affecting mostly men older than 50 years of age, many of whom are long-time smokers who are also suffering from small-cell lung carcinoma. The symptoms of this syndrome include weakness of the axial and limb muscles. Additional symptoms include xerostomia (dry mouth), impaired accommodation of the lenses of the eyes, impotence, and the inability to elicit deep tendon reflexes. Since some of these symptoms also appear in **myasthenia gravis**, the two conditions resemble one another. However, in the Eaton–Lambert syndrome the ocular and facial and pharyngeal muscles are rarely affected, whereas ocular paresis is normally present in myasthenia.

The two diseases affect different regions of the neuromuscular synapse. In the Eaton–Lambert syndrome, the autoimmune response is against the calcium ion channels of the presynaptic membrane, interfering with the flow of calcium ions into the presynaptic terminal. The diminished number of calcium ions prevents the adherence of the normal number of synaptic vesicles to the presynaptic membrane, decreasing the release of acetylcholine with each stimulus, and resulting in a weak muscle contraction.

In myasthenia gravis, the autoimmune reaction is against the acetylcholine receptors of the postsynaptic membrane. Therefore, even though the release of acetylcholine from the presynaptic membrane is normal, the reduced number of normal acetylcholine receptors of the postsynaptic membrane results in weak muscle contraction. The muscles of the head, and especially of the eye, are the first ones to be affected, but as the disease progresses the respiratory muscles also become involved, and the patient dies of respiratory insufficiency.

Alpha latrotoxin (black widow spider venom)
Alpha latrotoxin, the venom of the black widow spider, is inserted into the presynaptic membrane of the neuromuscular junction, where it acts as an ungated, nonselective ion channel, permitting the entry of Na+, K+, and Ca^{2+} ions into the presynaptic terminal. The altered intracellular univalent cation concentration results in constant depolarization of the presynaptic terminal. Moreover, the elevated Ca^{2+} levels facilitate synaptic vesicle docking and the release of copious amount of acetylcholine into the synaptic cleft. The continuous release of acetylcholine initially results in muscle rigidity and then in paralysis of the affected muscles.

Huntington disease
Huntington disease is a genetic disorder resulting in degeneration of GABAergic neurons in the caudate nucleus and putamen. The reduction in GABA results in increased release of dopamine and a corresponding decrease in acetylcholine release in the striatum, causing an imbalance in the ratio of these two neurotransmitter substances. It is this imbalance that is responsible for the symptoms of Huntington disease, whose early signs include the incapacity to form new memory, the inability to make decisions, memory loss, depression, irritability, and mood swings. As the disease progresses the patient becomes emotionally disturbed, loses cognitive faculties, and will display uncontrolled, erratic movements (Huntington chorea).

FOLLOW-UP TO CLINICAL CASE

The diagnosis is Parkinson disease, which is very common and usually easy to diagnose. It is a purely clinical diagnosis. All testing is either normal or unrevealing.

This patient was given levodopa three times a day, and the symptoms improved dramatically within a half hour of taking the first dose. The patient noted that symptoms were bad in the mornings before he took his morning dose of medicine. Both the patient and the patient's spouse stated that the patient seemed "normal" with the medicine.

Parkinson disease is the most common of the class of diseases known as movement disorders. Movement disorders are caused by pathology somewhere within the basal nuclei (basal ganglia). In the case of Parkinson disease there is a degeneration of neurons in the midbrain's substantia nigra, which is composed of pigmented cells. Dopamine is the neurotransmitter released by these neurons, to communicate with other components of the basal nuclei. In Parkinson disease there is a relative lack of dopamine in the basal nuclei due to the degeneration of neurons that synthesize it. Levodopa is a precursor to dopamine. It is taken up by the remaining substantia nigra neurons, is converted to dopamine, and the relative deficiency of dopamine in the basal nuclei is temporarily corrected. Dopamine cannot be used as a medicine since it will not cross the blood–brain barrier to enter the neural tissue of the brain.

Medications that contain levodopa are very effective treatments for the symptoms of Parkinson disease. They readily relieve symptoms, but the effects typically wear off in a few hours. Dopamine agonists are used in Parkinson disease, and anticholinergic medicines are also sometimes used. However, none of these medicines alters the underlying pathology of the disease. They only reduce symptoms.

QUESTIONS TO PONDER

1. What is the difference between ionotropic and metabotropic effects?

2. What are the five basic characteristics of neurotransmitter substances?

3. Biogenic amines are one type of neurotransmitters. How are they defined?

4. Neuropeptides are neurotransmitters. What are their major characteristics?

5. Why is acetylcholine sometimes an excitatory and sometimes an inhibitory neurotransmitter substance?

CHAPTER 5

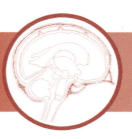  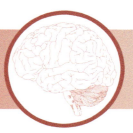

Spinal Cord

- CLINICAL CASE
- MORPHOLOGY OF THE SPINAL CORD
- INTERNAL MORPHOLOGY OF THE SPINAL CORD
- VASCULAR SUPPLY OF THE SPINAL CORD
- CLINICAL CONSIDERATIONS
- SYNONYMS AND EPONYMS OF THE SPINAL CORD
- FOLLOW-UP TO CLINICAL CASE
- QUESTIONS TO PONDER

CLINICAL CASE

A 45-year-old white patient describes back pain, weakness, and numbness of both legs. The back pain started about a month ago, and the other symptoms began 1 week ago as slight imbalance, tingling feet, and "funny" sensations around her abdomen. This has rapidly progressed to the point where the patient can no longer walk without help. The patient could not feel that the water was too hot and suffered burned legs. Today the patient had an episode of urinary incontinence. Sensation seems reduced or altered from the toes all the way up to the breast. The weakness and numbness are in both legs but worse on the left. The arms are normal. The back pain is located in the middle of back, over the spine, as opposed to the much more common lower back pain.

Exam shows moderate-to-severe weakness in both legs, which is worse on the right. Arm strength is normal. Pinprick sensation and touch are reduced bilaterally in the legs and abdomen, up to a fairly distinct level at the nipple line. Percussion of the back indicates tenderness over the spine in the upper thoracic region. Reflexes at the knees and ankles are pathologically brisk, and extensor (Babinski) responses are present bilaterally. Reflexes in the arms are normal. The patient cannot walk without assistance.

The **spinal cord**, the grayish-white oblong cylindrical continuation of the medulla oblongata of the brain, begins at the **foramen magnum** (L., "large hole") of the skull and extends within the vertebral canal to terminate as the cone-shaped **conus medullaris** (L., "medullary cone"). The caudal tip of the conus medullaris, in the adult, is located between vertebral levels **L1** and **L2**. Thus, the adult spinal cord is approximately 45 cm (17.7 in.) in length with an average diameter of 1–1.5 cm (0.4–0.6 in.) and an average weight of about 30 g (1.1 oz).

The spinal cord, a soft, gelatinous substance, is protected from injury by being encased in the bony vertebral column. Additional safeguard is provided by the **meninges** (G., "membrane" singular: "meninx"), composed of the three concentric sheaths (listed from superficial to deep): the dura mater (L., "hard mother"), arachnoid mater (G., "spider web-like"), and pia mater (L., "tender mother"). There is only a potential space between the outermost dura and the arachnoid mater, known as the subdural space, but the subarachnoid space, located between the arachnoid mater and pia mater, houses **cerebrospinal fluid** (**CSF**). The meninges surround the entire spinal cord and, in turn, are enveloped by epidural fat and a venous plexus (L., "braid") (for more detailed information, see Chapter 8, Meninges and Cerebrospinal Fluid).

A Textbook of Neuroanatomy, Third Edition. Maria A. Patestas, Amanda J. Meyer, and Leslie P. Gartner.
© 2025 John Wiley & Sons, Inc. Published 2025 by John Wiley & Sons, Inc.
Companion website: www.wiley.com.go/patestas/neuroanatomy3e

Along its length, the lateral aspect of the spinal cord is affixed by approximately 21 pairs of **denticulate ligaments** (L., "serrated") that are triangular extensions of the pia to the dura mater. The spinal cord is a two-way conduit to and from the brain. It functions as a *central relay station*, receiving incoming information (that is incoming into the spinal cord) from the body and the brain, and as a *central processing station*, conveying outgoing information (that is outgoing from the spinal cord) to the body and the brain.

MORPHOLOGY OF THE SPINAL CORD

General structure of the spinal cord

The spinal cord is a cylindrical structure whose circumference varies along its length; in the adult it is shorter than the length of the bony vertebral canal

The circumference of the spinal cord varies along its length. It presents two regions that are broader in diameter – the **cervical enlargement** (C5–T1) and the **lumbosacral enlargement** (T11–S1). These enlargements are due to the origins of the spinal nerves that are destined for the upper and lower limbs, respectively. Each spinal nerve emanates from the spinal cord as posterior and anterior rootlets, which eventually join each other within the intervertebral foramen to form that particular spinal nerve.

The spinal cord equals the length of the vertebral canal only until the end of the first trimester of prenatal life. Thereafter the trunk elongates much faster than does the spinal cord, so that by the time of birth the conus medullaris is only at the level of the third lumbar vertebra and, in the adult, it extends only as far as the caudal aspect of the first lumbar (L., "loin") vertebra. Therefore, since lower levels of the spinal cord segments are displaced superiorly in relation to their corresponding vertebral levels, spinal cord levels do not necessarily correspond to vertebral column levels (Table 5.1). Because of this differential growth, the subarachnoid space caudal to the conus medullaris, known as the **lumbar cistern** (L., "box"), which extends to the level of the second sacral vertebra, and is devoid of spinal cord (Fig. 5.1).

As the posterior and anterior rootlets of the lumbar and sacral segments elongate, they descend various distances and form a loose conglomeration of nerve fibers within the lumbar cistern. Since these fibers resemble a horse's tail, they are known as the **cauda equina** (L., "horse's tail"). However, the

Table 5.1 • Relationship of the spinal segment to the adult vertebral column.

Spinal cord level	Spinous process of vertebrae
C6	C5
T5	T4
T10	T8
L3	T11
L5	T12
S1, 2	L1
S4	L2

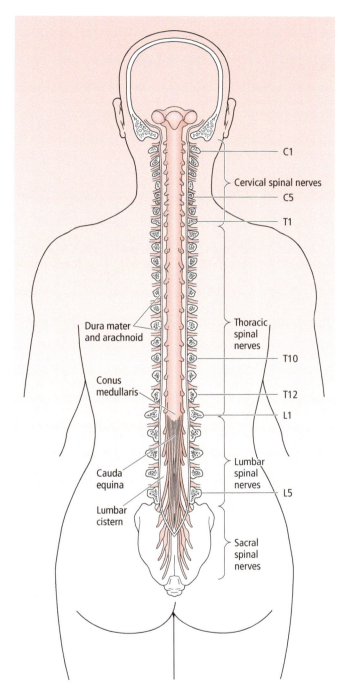

Figure 5.1 • The spinal cord in a human. Note that the spinal processes of the vertebrae have been removed and that the dura mater and the arachnoid mater have been opened up so that the spinal cord may be viewed in its entire length. It should be evident that the spinal cord ends at L1, L2, and the spinal nerves continue as the cauda equina within the lumbar cistern. It is the lumbar cistern that is accessed during a lumbar puncture to withdraw cerebrospinal fluid for laboratory examination. Note that some of the vertebral levels are shown by their abbreviations. Observe that the width of the spinal cord is enlarged in comparison to that of the outline of the body.

Note that the clinical case at the beginning of the chapter refers to a rapidly progressing deterioration of sensory and motor functions.

1 Why does the acute onset and rapid progression mandate immediate action?

2 What region of the central nervous system is responsible for the innervation of the upper extremity?

3 Why is the lesion suspected to be in the spinal cord and not in the brain?

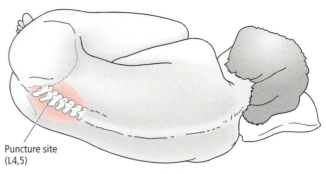

Figure 5.2 ● The preferred method of performing a lumbar puncture is to have the patient assume the lateral decubitus position with the needle piercing the intervertebral space between L4 and L5. Observe that the width of the spinal column is enlarged in comparison to that of the outline of the body.

pial covering of the spinal cord continues beyond the conus medullaris as a thin, non-nervous filament, the **filum terminale** (L., "terminal filament"), to the second sacral vertebra where it is covered by the continuation of the dura mater and arachnoid mater and is anchored into the first coccygeal vertebral segment. The CSF, by filling the **subarachnoid space**, surrounds, suspends, cushions, and thus further protects, the spinal cord from possible mechanical injury.

Since the spinal cord of the adult occupies only the upper two-thirds of the spinal column, the **lumbar cistern** is a CSF-filled chamber that is accessible for epidural anesthesia and lumbar puncture (spinal taps) procedures, avoiding damage to the spinal cord (Fig. 5.2). The CSF-filled **central canal of the spinal cord** is continuous superiorly with the fourth ventricle of the brain and, in the young adult, extends along the entire length of the cord, terminating within the filum terminale. However, by the fourth decade of adult life the central canal ends just superiorly to the conus medullaris. Frequently, in the over 40-year-old person, the central canal is partially occluded by ependymal cells. A study of the degree of stenosis in autopsies of more than 200 individuals of various ages was noted to be greatest in the thoracic region. With increasing age, the degree of occlusion and the length of the occluded regions increased.

External morphology of the spinal cord

The spinal cord is a bilaterally symmetric structure; it has an anterior median fissure and a posterior median sulcus along its entire length

The spinal cord possesses a bilateral symmetry, and the ordered presence of the anterior median fissure and the posterior median sulcus aid in the recognition of this symmetry. The prominent **anterior median fissure** is about 3 mm deep and extends along the entire anterior (ventral) length of the spinal cord. The **posterior median sulcus** is not as deep as its anterior counterpart, but it also occupies the entire posterior (dorsal) length of the spinal cord. It is interesting to note that a glial septum, the **posterior (dorsal) median septum**, is continuous with the deep aspect of the posterior median sulcus, and assists in separating the spinal cord into the right and left halves (Fig. 5.3A).

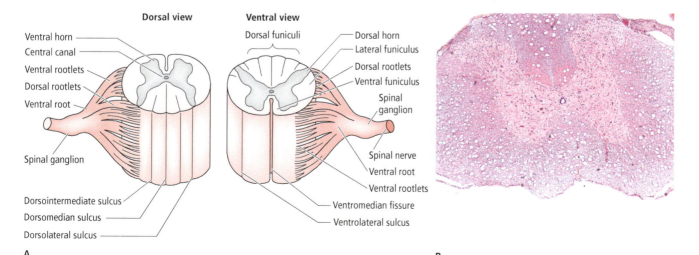

Figure 5.3 ● (A) Posterior and anterior views of the spinal cord, depicting the gray matter as well as the fissures and funiculi of the white matter. Observe also that the posterior and anterior roots join to form the spinal nerve. Note that the posterior view is the clinical, whereas the anterior view is the anatomical orientation. (B) A very low-power light micrograph of the human thoracic spinal cord.

Approximately 2 mm (0.08 in.) on either side of the posterior median sulcus is the **posterolateral (dorsolateral) sulcus** that extends the entire length of the spinal cord. It is distinguished by the presence of the **posterior rootlets of the spinal nerves** as they penetrate the substance of the spinal cord. The **dorsointermediate sulcus**, located between the posterior median and posterolateral sulci, is present only at the cervical and upper thoracic levels. The **anterior rootlets of the spinal nerves** emerge at the **anterolateral sulcus**, between the posterolateral sulcus and the anterior median fissure.

Segmentation of the spinal cord and the roots of the spinal nerves

The segmental design of the spinal cord is evident due to the regular arrangement of the spinal nerves The segmental design of the spinal cord is displayed by the presence of 31 pairs of spinal nerves and their associated spinal ganglia. Each spinal nerve emanates from the spinal cord as a number of **sensory** and **motor rootlets**, leaving the spinal cord's posterior and anterior surfaces, respectively. Each rootlet is composed of neuronal processes (synonym: neurites) of a large number of neurons.

The anterior rootlets, usually no more than eight in number, join to form a single **anterior root**, housing the axons of motor (and in certain regions, autonomic) neurons; therefore, they carry GSE/GVE axons. The posterior rootlets, also no more than eight in number, join and enter the slight swelling, known as the **spinal ganglion** housing the unipolar (pseudounipolar) nerve cell bodies (Fig. 5.4). These posterior rootlets thus contain the **central processes** of these unipolar cells; therefore, posterior rootlets carry GSA/GVA axons. The **peripheral processes** of the unipolar cells, which bring sensory information from sensory receptors distributed throughout the body, leave the spinal ganglion to join the anterior root of the same segment to form the mixed **spinal nerve**, which thus contains both sensory and motor fibers.

Each spinal nerve leaves the vertebral column via its **intervertebral foramen** and is named accordingly. The first eight spinal nerves, C1–C8, exit the vertebral canal *above* the correspondingly numbered cervical vertebrae (note that spinal nerve C8 leaves the vertebral canal above vertebra T1, because there are only seven cervical vertebrae and eight cervical spinal nerves), whereas all subsequent spinal nerves (T1–T12, L1–L5, S1–S5, and Co1) exit *below* the correspondingly named vertebrae (Fig. 5.5). Hence, there are 31 pairs

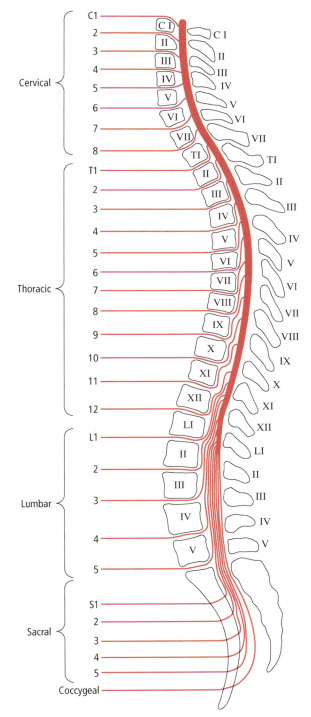

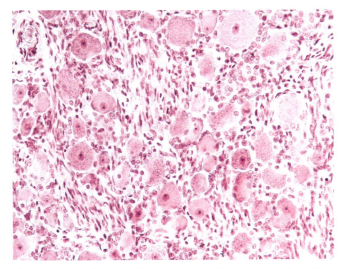

Figure 5.4 ● Medium magnification histological image of a spinal ganglion. Note the presence of small-to-large soma of the unipolar (pseudounipolar) neurons) each surrounded by much smaller, cuboidal-shaped satellite cells.

Figure 5.5 ● The spinal cord and the relationship between the numbered spinal nerves and the bodies of the associated vertebrae.

of spinal nerves emerging from the spinal cord. The region of the spinal cord associated with a particular pair of spinal nerves is called a **spinal segment**.

Spinal nerves are only a few millimeters in length. They give rise to a slender recurrent meningeal branch and then divide into two components, a smaller **posterior ramus** (L., "branch") and a larger **anterior ramus**. Every anterior primary ramus receives a **gray rami communicans** (L., "to share with someone"; plural: communicantes) from its corresponding sympathetic ganglion. Additionally, *but only* from T1 through L2, L3, each anterior ramus also provides a **white rami communicans** to its corresponding sympathetic ganglion (Fig. 5.6). These rami communicantes carry postganglionic and preganglionic **sympathetic fibers** from and to the sympathetic ganglia, respectively. White rami communicantes (carrying preganglionic sympathetic fibers) have myelinated fibers, whereas gray rami communicantes (carrying postganglionic sympathetic fibers) house nonmyelinated fibers. Additionally, anterior rami S2–S4 also provide spinal preganglionic **parasympathetic fibers** that pass directly to their sites of destination.

Table 5.2 ● List of clinically important dermatomes.

Spinal nerve	Distribution
C2	Occipitum of the head
C3–C4	Neck
C5	Shoulder
C6	Thumb
C7, 8	Elbow
C8	Little finger
T4, 5	Nipple
T10	Umbilicus
L1, 2	Groin
L3	Skin over patella
L5	Big toe (hallux)
S1	Heel
S1	Little toe
S5	Anus

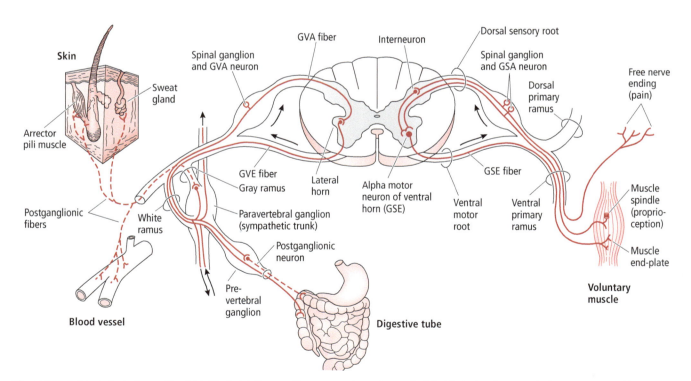

Figure 5.6 ● A typical spinal nerve. The right side depicts the somatic nervous system, whereas the left side depicts the sympathetic nervous system. The preganglionic fibers are shown as solid lines, and the postganglionic fibers are displayed as dashed lines. Note that the left side of the figure shows that the preganglionic sympathetic cell body is in the lateral horn of the thoracic and upper lumbar spinal cord levels (T1–L1, L2). The preganglionic fiber travels along the anterior rootlets, enters the spinal nerve, travels for a very short distance, and then leaves the spinal nerve via the white ramus communicans, which is the connection to the sympathetic chain ganglion. Once in the ganglion, there are three possibilities: (1) it may synapse with the postganglionic cell body located there; (2) it may travel up or down the sympathetic trunk until it reaches another ganglion and synapse there with a postganglionic sympathetic cell body; or (3) it may leave the sympathetic trunk altogether and travel to a prevertebral sympathetic ganglion (as part of a splanchnic nerve) and synapse there with a postganglionic sympathetic nerve cell body. The text describes the fate of the postganglionic fiber.

Sensory innervation of the skin is determined by its developmental origin, and the strip of skin for which a particular spinal nerve is responsible is referred to as a **dermatome**. Along the length of the trunk, each dermatome forms an ordered series of bands, whereas along the limbs the ordering is not as evident. These bands overlap each other, since the sensory nerves of one dermatome are also responsible for innervating regions of adjacent dermatomes (Fig. 5.7). It is interesting to note that the overlap is greater for light touch than for pain. Clinicians should remember the distribution of certain dermatomes (Table 5.2), especially, that C3, C4 serve the neck; C6 serves the thumb; C7, C8 serve the elbow; T4, T5 serve the nipple; T10 serves the umbilicus; L3 serves the region of the patella; L5 serves the big toe; S1 serves the little toe and the heel; and that S5 serves the anus. Additionally, all of the skeletal muscle cells that are innervated by the fibers of a particular spinal nerve are referred to as a **myotome**. Furthermore, a **sclerotome** is defined as the ligaments and bones that receive their innervation from the fibers of the same spinal nerve. Each dermatome, myotome, and sclerotome are derived from the same embryological somite.

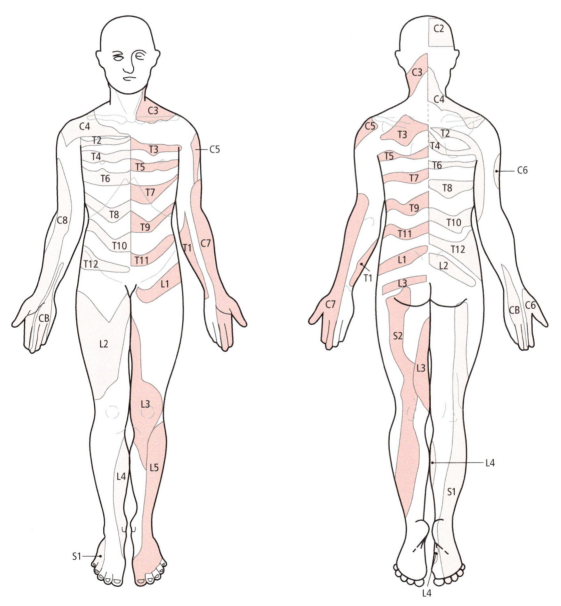

Figure 5.7 ● The evidence-based dermatome map representing the most consistent tactile dermatomal areas for each spinal posterior nerve root found in most individuals, based on the best available evidence. The dermatomal areas shown are NOT autonomous zones of cutaneous sensory innervation since, except in the midline where overlap is minimal, adjacent dermatomes overlap to a large and variable extent. Blank regions indicate areas of major variability and overlap. S3, S4, and S5 supply the perineum but are not shown for reasons of clarity.
Source: Lee, MWL, McPhee, RW, and Stringer, MD (2008) An Evidence-Based Approach to Human Dermatomes, Clinical Anatomy 21:363-373/University of Otago.

CHAPTER 5

Modalities of the spinal nerves

Spinal nerves possess functional components known as modalities: general somatic afferent, general visceral afferent, general somatic efferent, and general visceral efferent components

Neurons are classified by function into three categories: sensory (afferent), intercalated (connecting), and motor (efferent). The afferent (sensory) and efferent (motor) components are further categorized into more specific divisions, namely somatic and visceral. Thus, a typical spinal nerve has four functional components, known as **modalities**: general somatic afferent (GSA), general visceral afferent (GVA), general somatic efferent (GSE), and general visceral efferent (GVE) components. Cell bodies of the afferent fibers of a spinal nerve reside in the spinal ganglia, whereas the cell bodies of the efferent fibers are located within the gray matter of the spinal cord.

Visceral efferent neurons belong to the autonomic nervous system, and they may be part of its sympathetic or parasympathetic component. The cell bodies of the preganglionic sympathetic visceral efferent fibers are located in the lateral gray column of levels T1 through L2, L3 of the spinal cord. The cell bodies of the preganglionic parasympathetic visceral efferent fibers are located in the lateral gray column of sacral spinal levels 2–4. The cell bodies of visceral afferent neurons are located in the spinal ganglia. Their peripheral processes bring information from the viscera, using white rami communicantes to enter the spinal nerve. The central process of the visceral afferent neuron uses the posterior rootlets to enter the substance of the spinal cord. Once there it will either complete a reflex path by synapsing with sympathetic or somatic efferent neurons, usually by way of an interneuron, or will synapse with other neurons in the spinal cord or brainstem to deliver the information to higher centers.

General somatic afferent (GSA) is sensory information, such as touch, pressure, pain, temperature, and proprioception that is perceived in the body and is transmitted to the spinal cord. **General visceral afferent (GVA)** is sensory information, such as pressure or pain information that is perceived in the viscera – the organs, glands, and membranes – that is transmitted to the spinal cord. The **GSE (general somatic efferent)** fibers provide motor innervation to skeletal muscles of somatic origin. The **GVE (general visceral efferent)** fibers provide visceral motor innervation to the myoepithelial cells of glands, cardiac muscle, and smooth muscle.

Classification of the fiber components of the spinal nerves

Nerve fibers are classified according to their modalities, whether they are afferent or efferent, and also classified based on their conduction velocities

The nerve fibers constituting the spinal nerve are subdivided into two categories, afferent (sensory) and efferent (motor). All afferent fibers *enter* the spinal cord via the posterior rootlets, and all efferent fibers *leave* the spinal cord via the anterior rootlets. **Afferent fibers** are classified according to their peak conduction velocities. Unfortunately, there are two classifications commonly in use, one using Roman numerals I–IV and a second using the letters A, B, and C, but they are easily reconciled, as evident in Table 5.3. Furthermore, they are also classified

Fiber type	Myelin sheath	Conduction velocity (m/s)	Fiber diameter (μM)	Function
Afferent				
Ia (Aα)	Yes	70–120	12–22	Primary sensory from muscle spindles
Ib (Aα)	Yes	70–120	10–20	Sensory fibers of tendon organs
II (Aβ)	Yes	30–70	6–12	Secondary sensory terminals of muscle spindles
III (Aδ)	Thin	10–30	1–6	Nociceptive; touch; pressure
IV (C)	No	0.5–2	>1.5	Nociceptive in muscle; temperature
Efferent				
Aα	Yes	70–120	12–22	Motor to skeletal muscle fibers (extrafusal)
Aγ	Yes	10–45	2–10	Fusimotor to intrafusal fibers of muscle spindles
Sympathetic	Thin	3–15	3	Preganglionic to cell bodies in ganglion
Sympathetic	No	0.5–2	>1.5	Postganglionic to glands and smooth and cardiac muscles

Table 5.3 ● Classification of afferent and efferent nerve fibers.

according to their modality, into GSA and GVA fibers. **Efferent fibers** are also classified as A, B, and C fibers as well as by a different system utilizing the Greek letters alpha (α) and gamma (γ). Furthermore, they are also classified according to their modality, into GSE and GVE fibers.

INTERNAL MORPHOLOGY OF THE SPINAL CORD

The spinal cord is composed of gray matter surrounded by white matter

The spinal cord is composed of a central column of gray matter surrounded by a sheath of white matter. **Gray matter** is composed of neurons, their processes, and neuroglia. It is the large number of nerve cell bodies that is responsible for the grayish appearance of the gray matter. **White matter** is composed mostly of myelinated but also unmyelinated processes of neurons, neuroglia, and blood vessels, and it is the white coloration of the myelin that gives white matter its name.

Inspection of cross-sections of the spinal cord displays a bilateral symmetry. The gray matter is arranged in the shape of the letter "H" and the appearance of this image, as well as the ratio of white matter to gray matter, varies with level of the segment being examined (Fig. 5.3A, B).

The actual cross-sectional area of gray matter is largest in the mid-to-lowest cervical and mid-to-lowest lumbar levels, which are the regions of the spinal cord responsible for the neural supply of the upper and lower limbs, respectively. The region of the spinal cord that displays the least actual area of gray matter is at the thoracic level. The cross-sectional area of the white matter is greatest at the cervical level and smallest at the sacral level of the spinal cord, since the total number of ascending and descending nerve fibers increases in a superior direction. (However, it should be noted that, looking at the ratio of gray matter to white matter, it is the largest in the sacral region). Because of these characteristics the segmental level of a cross-section of the spinal cord can be distinguished with relative ease.

It is interesting to note that generally speaking the processes of the neurons run *parallel* to the longitudinal axis of the spinal cord in white matter, but *perpendicular* to the longitudinal axis of the spinal cord in gray matter.

Gray matter

Gray matter, composed of neurons, their processes, and neuroglia, is subdivided into the anterior, posterior, and lateral columns

During examination of a cross-section of the gray matter it becomes evident that the two "vertical bars" of the letter "H" of the gray matter are connected to each other by a narrow strip of gray matter, known as the **gray commissure** (the crossbar of the letter "H"), the center of which is occupied by the **central canal**. Viewed in three dimensions, the lateral aspect of the gray matter is concave and displays an **anterior gray funiculus** and a **posterior gray funiculus**, as well as a small projection, the **lateral (intermediolateral) gray funiculus**, which appears only between spinal cord levels T1 and L2/-3 (Figs. 5.3A and 5.8). The lateral gray funiculus represents the location of the cell bodies of the preganglionic autonomic neurons. It should be noted that these three funiculi are also referred to as the **anterior horn**, **posterior horn**, and **lateral horn** when the spinal cord is examined in cross-sections, and therefore these terms are frequently used interchangeably. Although the gray matter is completely surrounded by white matter, the posterior horn approaches the limit of the spinal cord and is separated from the posterolateral (dorsolateral) sulcus by a small bundle of nerve fibers, known as the **posterolateral tract** (eponym: Zone of Lissauer). The narrow region of the posterior horn in contact with the posterolateral tract is subdivided into two regions, the outer **posterolateral nucleus** and the somewhat deeper situated spinal lamina II (synonym: gelatinous substance; eponym: substance of Rolando).

Reticular formation and central canal

The central canal of the spinal cord, lined by ependymal cells, is surrounded by gray matter in spinal area X, the central gelatinous substance, and anterior and posterior gray commissures

The interface between the gray matter and white matter is usually well defined, but at cervical levels the base of the anterior gray funiculus forms network-like interdigitations with the adjoining white matter. This interwoven structure, known as the **reticular formation**, is similar to other regions present in the brain and inferior spinal cord levels and thus is part of the **reticular system**, discussed later in this textbook (see Chapter 16). The central canal of the spinal cord contains cerebrospinal fluid (CSF) and is lined by ciliated cuboidal-to-columnar epithelium composed of **ependymal cells**. In some regions the canal may be partially or completely occluded

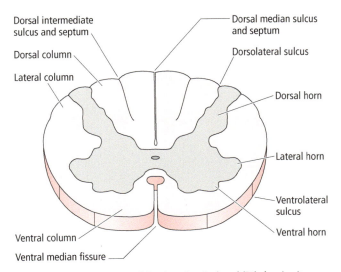

Figure 5.8 ● Cross-section of the thoracic spinal cord (T1) showing its major landmarks. Observe that the various regions of the white matter are described on the left side and the regions of the gray matter on the right side of the diagram.

by clusters of ependymal cells. Immediately surrounding the ependymal cells is the **central gelatinous substance**, a layer of neuroglia interspersed with occasional nerve cell bodies and fine nerve fibers. The remainder of gray matter surrounding the central gelatinous substance is referred to as the **anterior and posterior gray commissures of the spinal cord**. The white matter surrounding the gray commissure is called the **anterior** and **posterior white commissure of the spinal cord**.

Neuronal architecture

Processes of neurons are intertwined within the gray matter of the spinal cord; these form a complex web known as the neuropil

Much of the gray matter is composed of an intricate network of neuronal processes. These may be the central processes of unipolar neurons, the initial segments of motor neurons (somatic and autonomic), axons that are crossing over (decussating) from one side of the spinal cord to the other. These nonmyelinated processes form the highly complex and not well-understood **neuropil** whose extent is limited to the gray matter. Neurons of the spinal cord gray matter are **multipolar** and are collected in various sized clusters of cell bodies, which, for the sake of convenience, are described as belonging to the posterior gray funiculus, lateral gray funiculus, or anterior gray funiculus.

The gray matter of the spinal cord can be organized into nine layers plus the region surrounding the central canal, were named **Rexed laminae I–X**, after the Swedish neuroanatomist who mapped out their distribution during the 1950s; however, they are now known as **Spinal laminae I-X**. These laminae are numbered in a posteroanterior direction and generally correspond to the nerve cell clusters in the posterior gray funiculus (spinal laminae I–VI), lateral gray funiculus (spinal lamina VII), and anterior gray funiculus (spinal laminae VIII and IX) (Fig. 5.9; Table 5.4).

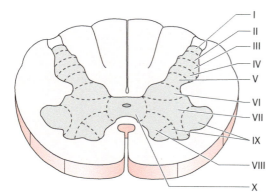

Figure 5.9 • Cross-section of the spinal cord displaying the divisions of the gray matter into nine regions, referred to as spinal laminae, as well as a tenth region, around the central canal of the spinal cord, known as the gray commissure (central gray, periependymal gray). These divisions are based on the cellular composition of the gray matter, and they bear some relationship to the nuclei of the spinal cord (see Table 5.4).

Neuronal groups of the posterior gray funiculus

The nuclei of the posterior gray funiculus are sensory in nature and are arranged in four groups: spinal lamina I, spinal lamina II, spinal laminae III and IV, and the posterior thoracic nucleus in spinal lamina VII

The nuclei of the posterior gray funiculus function in the reception and processing of sensory input from the central processes of the unipolar (pseudounipolar) neurons whose cell bodies are located in the spinal ganglia. These nerve cells are subdivided into four groups: spinal lamina I, spinal lamina II, spinal laminae III and IV, and the posterior thoracic nucleus in spinal lamina VII.

Spinal lamina I extends the entire length of the spinal cord, capping the posterior horn, and receives afferent fibers carrying pain, temperature, and light touch sensations. It also contributes fibers for the lateral and anterior spinothalamic tracts. It contacts the *posterolateral tract* of the white matter.

Spinal lamina II also extends the entire length of the spinal cord. It is densely packed with Golgi type II neurons that possess highly branched, unmyelinated axons. These cells receive sensory input from the central fibers of unipolar neurons of the spinal ganglia, delivering pain, temperature, and light touch information. Descending fibers from higher centers (such as the cerebral cortex) form excitatory and inhibitory synapses with the type II neurons, thus modifying the arriving pain and temperature sensations.

Spinal laminae III and IV also extend the entire length of the spinal cord. They are composed of densely clustered large nerve cell bodies, located just anterior to the substantia gelatinosa, and receive the central processes of the majority of the unipolar neurons of the spinal ganglia. The nucleus proprius receives pain, light touch, and temperature sensations and provides input to the lateral and anterior spinothalamic tracts.

Spinal lamina V (neck of the posterior horn) is located between spinal laminae IV and VI, bordering the base of the posterior horn. It is less cellular and houses thicker bundles of nerve fibers than most other of the laminae and its lateral aspect is referred to as the **reticular formation** (the **short propriospinal neurons** of spinal lamina V).

Spinal lamina VI (base of the posterior horn) is responsible for the removal of those regions of the extremities from harm's way that are exposed to painful stimuli. This lamina is not present throughout the entire length of the spinal cord; it is located only in the cervical (C5–T1) and lumbar (L1–S2) enlargements.

Spinal lamina VII (intermediate zone of the spinal cord) houses in its lateral aspect the **posterior thoracic nucleus** (eponymously known as **nucleus dorsalis of Clarke** and **Clarke's column**). Spinal lamina VII does not extend the entire length of the spinal cord; instead, it extends only from C8 to L3. It is located at the base of the posterior gray column and houses relatively large cell bodies that receive synapses from proprioceptive fibers, which bring information from tendon organs and muscle spindles. Some of the axons of these large nerve cell bodies travel in the posterior spinocerebellar tracts, conveying information from the lower limbs.

Table 5.4 ● Neuronal groups of the spinal cord gray matter and their relationship to spinal laminae.

Funiculus	Neuronal group	Extent	Spinal lamina
Posterior gray	Dorsomarginal nucleus	C1–S5	I
Posterior gray	Substantia gelatinosa (of Rolando)	C1–S5	II
Posterior gray	Nucleus proprius	C1–S5	III and IV
Posterior gray	Neck of the posterior horn	C1–S5	V
Posterior gray	Posterior thoracic nucleus (eponym: Clarke column)	C8–L3	VII
Posterior gray		C3–T2 and L1-S3	VI
Lateral gray	Lateral nucleus	T1–L2 (or L3)	VII
Lateral gray	Sacral parasympathetic nucleus (of Onufrowicz)	S2–S4	VII
Anterior gray		C1–S5	VIII
Anterior gray	Medial group	C1–S5 (except L5 and S1)	IX
Anterior gray	Central group	C1–C6 and L2–S2	IX
Anterior gray	Central group: phrenic nucleus	C3–C6	IX
Anterior gray	Central group: accessory nucleus	C1–C6	IX
Anterior gray	Central group: lumbosacral group	L2–S2	IX
Anterior gray	Lateral group	C4–T1 and L2–S3	IX
Pericentral canal	Gray commissure	C1–S5	X
Pericentral canal	Central gelatinous substance	C1–S5	X

Neuronal groups of the lateral gray funiculus

The **intermediolateral nucleus** (another region of spinal lamina VII) is composed of the relatively small multipolar cell bodies of **preganglionic sympathetic neurons**. These preganglionic sympathetic neurons are present only between T1 and L2, L3, and their axons pass through the anterior root of the spinal cord to enter the sympathetic trunk via the white rami communicantes.

A similar nucleus, the **sacral parasympathetic nucleus** is located at sacral levels 2–4. These **preganglionic neurons of the sacral outflow** of the parasympathetic nervous system are also considered to belong to spinal lamina VII.

Neuronal groups of the anterior gray funiculus

Neurons of the anterior gray funiculus include motor neurons and interneurons; they are subdivided into three groups: medial, central, and lateral

The nuclei of the anterior gray funiculus are composed of both large and small multipolar motor neurons whose axons leave the spinal cord via the anterior rootlets. The large motor neurons give rise to **alpha efferents** that supply skeletal muscles with motor innervation, whereas the smaller motor neurons supply **gamma efferents** to intrafusal muscle fibers of muscle spindles; an additional group of neurons, **interneurons**, are also located in the anterior gray funiculus. HOX patterning induces the formation of motor columns: HOX6 correlates with brachial, HOX9 with thoracic, and HOX10 with lumbar segments. Due to the location of their cell bodies, these nerve cells are subdivided into three major groups, medial, central, and lateral.

The **medial motor column (spinal lamina IX)** extends almost the entire length of the spinal cord (with the possible exceptions of C0). Between T1 and L4 it is subdivided into two components, for limb musculature, the **posterolateral (epaxial)** and **anteromedial (hypaxial) groups**. The motor neurons of the medial group also provide innervation for the skeletal muscles of the abdomen, the intercostal muscles, and the muscles of the neck. Proteins expressed by the medial motor column for epaxial muscles include Mnx1, IsL1/2, and Lhx and for hypaxial muscles include Mnx1, Isl2, Isl2^{low}, and Etv1.

The **central motor column (spinal lamina IX)** is the smallest of the three groups, and its distribution is not very extensive. The central group is present only in the cervical and lumbosacral segments of the spinal cord. Two regions of the cervical aspect of the central group have special names, the phrenic nucleus and the accessory nucleus. The **phrenic nucleus** (extending from C4 to C6) is responsible for the innervation of the diaphragm, and the **accessory nucleus** (extending from C1 to C5) is responsible for innervation of the sternocleidomastoid and trapezius muscles. Cells of the accessory nucleus provide fibers for the spinal root of cranial nerve XI (the accessory nerve). Proteins expressed by the phrenic nucleus include Mnx1, Isl1/2, Alcam, and Scip. Proteins expressed by the accessory nucleus are Isl1, Alcam, and Phox2b.

The **lateral group (Spinal lamina IX)** is present only in the regions of the spinal cord responsible for the motor innervation of the distal muscles of the upper and lower extremities (C5–T1 and L1–S2). The medial portion of this group expresses Foxp1 and Isl1/2, while the lateral group expresses Foxp1, Isl2, Lhx1, and Mnx1.

CHAPTER 5

Funiculi	Sulci/fissures	Sulci/fissures	Spinal cord levels
Posterior funiculus	Posterior median sulcus	Posterolateral sulcus	All cord levels
Gracile fasciculus	Posterior median sulcus	Posterointermediate sulcus	All cord levels
Cuneate fasciculus	Posterointermediate sulcus	Posterolateral sulcus	C1–T6
Lateral funiculus	Posterolateral sulcus	Anterolateral sulcus	All cord levels
Anterior funiculus	Anterolateral sulcus	Anterior median fissure	All cord levels
Anterior white commissure	Central canal	Anterior median fissure	All cord levels

Table 5.5 ● Subdivisions of the spinal cord white matter.

Neuronal groups of the gray commissure

The gray matter that surrounds the central canal of the spinal cord is known as **spinal lamina X** which is subdivided into a posterior gray commissure and an anterior gray commissure by the central canal. The gray commissures and the central gelatinous substance extend the entire length of the spinal cord and are believed to be associated with the autonomic nervous system. Frequently, spinal laminae VI, VII, and X are collectively known as the intermediate zone of the spinal cord gray matter.

White matter

White matter is also bilaterally symmetrical and is composed of myelinated and unmyelinated fibers; it is subdivided into three major funiculi: the posterior (dorsal), lateral, and anterior (ventral) funiculi

The **white matter** of the spinal cord is also bilaterally symmetrical. It surrounds the gray matter and is composed mostly of myelinated and unmyelinated nerve fibers. It has two main functions:

1 To bring information into the central nervous system (CNS) and transmit it to higher levels.
2 To transmit information from the higher levels to the spinal cord and to muscles and glands.

Based on the presence of sulci and fissures, the white matter is subdivided into three major funiculi: the *posterior*, *lateral*, and *anterior funiculi* (Table 5.5). The **posterior funiculus** is bounded by the posteromedian and posterolateral sulci; hence, it is between the posterior midline and the posterior nerve rootlets. The **lateral funiculus** is between the posterolateral and anterolateral sulci; hence, it is between the posterior and anterior rootlets. The **anterior funiculus** is between the anterolateral sulcus and the anterior median fissure, namely it is between the anterior rootlets and the anterior midline. The cervical and upper thoracic aspects of the posterior funiculus (C1–T6) are further subdivided by the posterior intermediate sulcus and septum into a medial **gracile fasciculus** and a lateral **cuneate fasciculus**. Moreover, the anterior funiculus houses the **anterior white commissure of the spinal cord**, the region of decussation for the spinothalamic tracts.

Within the funiculi, the nerve fibers that have similar destinations are arranged in bundles known as **tracts** (**fasciculi**). Some of these fiber bundles, especially in the posterior funiculus, are clearly defined into well-recognizable tracts that are separated from one another other by neuroglial sheaths, whereas most tracts appear to have overlapping boundaries. Tracts are classified into three categories: *ascending*, *descending*, and *intersegmental*. **Ascending tracts** transmit sensory information to higher centers, whereas **descending tracts** relay motor information originating at higher centers. The **intersegmental tracts** convey information between spinal cord segments, thus orchestrating intersegmental spinal reflexes. Although these tracts will be discussed in detail in later chapters of this book, the positions of these tracts are indicated in the transverse section of the spinal cord in Fig. 5.10 and Table 5.6.

VASCULAR SUPPLY OF THE SPINAL CORD

The spinal cord receives its blood supply from three longitudinally arranged vessels, the single anterior and paired posterior spinal arteries as well as from small, segmental radicular arteries

The **anterior spinal arteries**, direct branches of the vertebral arteries, join with each other to form a single median vessel, the **anterior spinal artery**, which occupies and follows the anterior median fissure of the spinal cord. This vessel extends from within the cranial cavity throughout the entire length of the spinal cord and provides small branches that penetrate and supply both the white matter and gray matter of the spinal cord. The anterior spinal artery may be quite small in the thoracic region (Fig. 5.11).

The **posterior spinal arteries** also arise from the vertebral arteries directly, or frequently indirectly, by way of the posterior inferior cerebellar branch of the vertebral artery. Each posterior spinal artery extends longitudinally from within the cranial cavity throughout the entire length of the spinal cord,

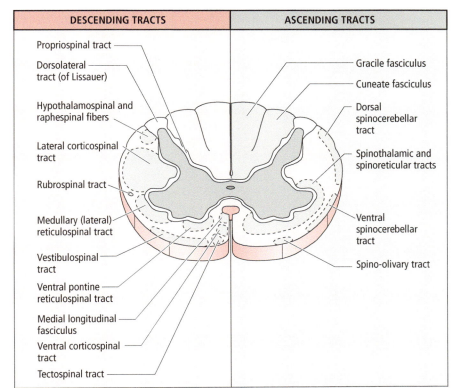

Figure 5.10 ● The white matter of the spinal cord is organized into fiber bundles known as tracts or fasciculi. These tracts have relatively well-defined boundaries and are subdivided into three major groups: descending, ascending, and intersegmental. The descending fiber tracts convey motor information from higher centers, ascending tracts carry sensory information to higher centers, and intersegmental tracts (not illustrated here) transmit information between spinal cord segments and are therefore restricted to the spinal cord.

Funiculi	Ascending tracts	Descending tracts	Intersegmental
Posterior	Gracile fasciculus		Posterior intersegmental
	Cuneate fasciculus		
Lateral	Posterior spinocerebellar	Lateral corticospinal	Lateral intersegmental
	Anterior spinocerebellar	Rubrospinal	
	Lateral spinothalamic	Lateral (medullary) reticulospinal	
	Spinotectal	Descending autonomic fibers	
	Posterolateral tract	Olivospinal	
	Spinoreticular		
	Spino-olivary		
Anterior	Anterior spinothalamic	Anterior corticospinal	Anterior intersegmental
		Vestibulospinal	
		Tectospinal	
		Reticulospinal	

Table 5.6 ● Tracts of the spinal cord white matter.

sandwiching the posterior rootlets between them. These two vessels provide small branches that penetrate and serve both the white matter and gray matter of the posterior spinal cord.

The 32 pairs of **radicular arteries** are small vessels that arise from arteries in the immediate vicinity of the spinal column. Each radicular artery enters the intervertebral foramen where it bifurcates, forming an **anterior** and a **posterior radicular artery**, which follow the anterior and posterior roots, respectively, to gain entrance into the vertebral canal. The anterior and posterior radicular arteries anastomose with branches of the anterior and posterior spinal arteries on the surface of the spinal cord, and arborize

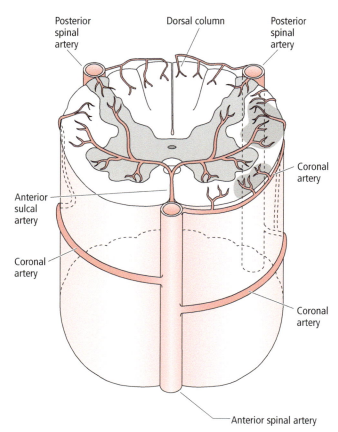

Figure 5.11 ● The vascular supply of the spinal cord. Note that there are two posterior spinal arteries but only a single anterior spinal artery. These three vessels, the coronal arteries, and branches of the anterior spinal artery vascularize the anterolateral aspect of the spinal cord, whereas the deep anterior aspect is vascularized by branches of the anterior sulcal artery. Much of the posterior aspect of the spinal cord is vascularized by branches of the posterior spinal arteries.

to supply the white and gray matter of the spinal cord. It should be stressed that the radicular arteries are extremely important for the vascularization of the spinal cord, because, with the exception of the cervical region, the anterior and posterior spinal arteries by themselves are unable to provide an adequate vascular supply to the spinal cord. Therefore, an injury to the spinal nerve damages not only the afferent and efferent fibers of a particular spinal cord level, but may also damage the segmental white and gray matter by producing ischemic conditions due to the reduction in blood supply from the radicular artery serving that region. It should be noted that the **great anterior radicular artery** (eponym: artery of Adamkiewicz) is the largest, albeit inconsistent, of the radicular arteries (Fig. 5.12). It usually arises on the left-hand side and serves much of the inferior half of the spinal cord, entering the vertebral canal between T8 and L1, and contributes to the formation of the inferior aspect of the anterior spinal artery.

Several longitudinally arranged tortuous veins of the pia mater are responsible for the venous drainage of the spinal

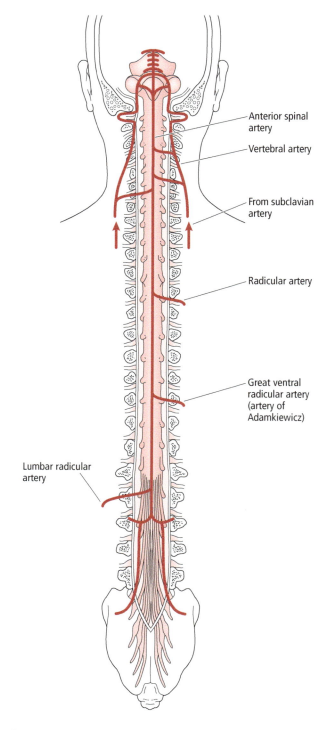

Figure 5.12 ● Anterior view of the spinal cord, drawn so that the bodies of all the vertebrae have been removed and the anterior spinal artery is displayed. Note that although there are 32 pairs of radicular arteries, only a few are displayed in this image and, to limit confusion, their anterior and posterior branches are not displayed. It should be noted that the anterior branches join the anterior spinal artery whereas their posterior branches join the posterior spinal arteries.

cord. These are the **anterior (ventral)** and **posterior (dorsal) spinal veins** that communicate with the segmentally arranged radicular veins. The **radicular veins** follow the paths of their companion radicular arteries to leave the vertebral canal via the intervertebral foramina. Along the way they also communicate with the **epidural venous network** embedded in the epidural fat of the vertebral canal. The epidural venous network delivers its blood into the cavernous sinus of the cranial cavity. Cranially, the anterior and posterior spinal veins drain into the veins and venous sinuses of the cranial cavity.

CLINICAL CONSIDERATIONS

Syringomyelia

Syringomyelia is a spinal cord disorder marked by the loss of the sense of pain and temperature at levels involving several consecutive spinal cord segments. Frequently, the other senses, such as touch and pressure, are unaffected, and then a **dissociated anesthesia** is said to be present. A number of conditions may cause this disorder, including the presence of a cleft within the spinal cord as well as an anomalous increase in the size of the central canal of the spinal cord.

Spinal shock

Spinal shock occurs infrequently and is characterized by complete anesthesia and paralysis, including the absence of all reflexes, somatic or autonomic, involving the body segments caudal to the site of damage. This condition may be permanent if the spinal cord is transected or of temporary duration if the spinal cord receives a sudden, but transient, injury, such as an unusually high dose of spinal anesthesia.

Tabes dorsalis

Tabes dorsalis is a late-stage effect of syphilis, arising about 10–20 years after the disease was contracted. The condition is characterized by infection of the posterior column and posterior roots causing sudden, sharp, shooting pain in the lower extremities. Occasionally, the condition is accompanied by ulceration of the feet, violent stomach cramps, and deterioration of the joints.

Brown–Séquard syndrome

Brown–Séquard syndrome is the result of unilateral spinal cord lesions characterized by contralateral pain and temperature loss and ipsilateral weakness and loss of proprioception.

Ependymomas

Ependymomas are tumors involving usually the caudal end of the spinal cord and the filum terminale. The unchecked cell division of the ependymal cells results in the presence of small-to-medium-sized growth in the affected area. These cells produce large quantities of proteins that elevate the protein levels in the CSF with a consequent reduction in the ability of the arachnoid granulations to deliver CSF into the lacunae lateralis/superior sagittal sinus. Therefore, the patient begins to suffer from hydrocephalus with headaches, and MRI examinations display the presence of enlarged cerebral ventricles.

Spina bifida

Spina bifida is a prenatally acquired developmental defect of incomplete fusion of one or more vertebral arches along with the incomplete fusion of the lips of the neural tube. It is a relatively frequent anomaly, whose incidence is about 1–2% of all births. The subtypes of spina bifida depend on the severity of the lack of fusion. The most common ones are spina bifida anterior, spina bifida cystica, and spina bifida occulta. **Spina bifida anterior** occurs along the anterior surface of the vertebral column. This incomplete closure of the vertebrae is often accompanied by anomalous development of the thoracic and abdominal viscera. **Spina bifida cystica** (also known as **spina bifida aperta**) (L., aperta, "bare," "exposed") is a more serious defect involving herniation of the spinal cord (myelocele) and/or of the meninges (meningocele) through the congenital defect in the vertebral column. The meningocele and myelocele may protrude far enough to reach the skin and may even rupture the skin with subsequent leakage of CSF and the possibility of bacterial infection resulting in meningitis. **Spina bifida occulta** is the most common of the spina bifida cases, being present in about 5% of the population. This is the mildest form of spina bifida and may not even be symptomatic. In cases like this the patient may even be unaware of the defect until for other reasons radiographs or MRI are taken of the region of the defect. Then it appears as an osseous defect without the accompanying herniation of the spinal cord or of the meninges. Frequently, the site of the defect is apparent on the overlying skin by the presence of a tuft of hair and/or of a dimple over the region of the osseous lesion.

SYNONYMS AND EPONYMS OF THE SPINAL CORD

Name of structure or term	Synonym(s)/eponym(s)
Afferent fibers	Sensory fibers
Anterior gray funiculus	Ventral gray column
Anterior median fissure	Ventral median fissure
Anterolateral sulcus	Ventrolateral sulcus
Posterolateral tract	Dorsolateral tract of Lissauer
Efferent fibers	Motor fibers
Gray commissure	Periependymal gray
Great anterior radicular artery	Artery of Adamkiewicz
Intermediolateral gray column	Lateral gray column
Lateral reticulospinal tract	Medullary reticulospinal tract
Posterior gray column	Dorsal gray column
Posterior median septum	Dorsal median septum
Posterior median sulcus	Dorsal median sulcus
Posterolateral sulcus	Dorsolateral sulcus
Substantia gelatinosa	Substantia gelatinosa of Rolando

FOLLOW-UP TO CLINICAL CASE

This patient has an acute myelopathy, meaning spinal cord pathology, which in this case seems to be in the upper thoracic region. This is located at T4 or above, but not below. The sensory level makes that very clear. The location of the back pain is also consistent. The brisk reflexes and Babinski response (external plantar response; a pathologic reflex of dorsiflexion of the big toe with a scratch on the lateral bottom of the foot) point to a CNS cause, as opposed to a peripheral nervous system cause. Bilateral weakness in the legs, and not the arms, also makes a thoracic cord pathology most likely. A lesion in the brain that causes leg weakness rarely leaves the arm or face unaffected, and is almost always unilateral (although there are exceptions).

This case represents a neurologic emergency. She should be admitted to the hospital without delay. The cause is not immediately clear, but the acute onset and rapid progression mandate immediate action. Work-up would include imaging of the thoracic spine, preferably with MRI. This would almost certainly reveal the pathology; probably in this case a mass of some sort. The mass would have to be rapidly growing, and the rapid deterioration and back tenderness indicates that the mass is most likely extra-axial (outside the cord). This could either be intradural or extradural. These types of masses can cause vascular compromise as well as physical compression of the cord.

Some causes of extra-axial masses include metastatic tumors (lung, breast, prostate [in a male patient], etc.), primary tumors (meningiomas, fibromas, etc.), epidural abscess, vascular malformations, or hemorrhage. Compression from vertebral fracture or intervertebral disc herniation can look similar. Treatment for any of these is surgical removal of the offending mass.

QUESTIONS TO PONDER

1. Why does the spinal cord have cervical and lumbar enlargements instead of having a uniform diameter throughout its length?

2. Why is the spinal nerve so important if it is only a couple of millimeters in length?

3. What is the significance of fasciculi (tracts) in the white matter of the spinal cord?

4. Why are the small radicular arteries important for the vascularization of the spinal cord?

5. What is the significance of the lumbar cistern?

CHAPTER 6

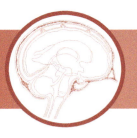

Gross Anatomy of the Brain

CLINICAL CASE

CEREBRUM

DIENCEPHALON

CEREBELLUM

BRAINSTEM

CLINICAL CONSIDERATIONS

SYNONYMS AND EPONYMS

FOLLOW-UP TO CLINICAL

CASE QUESTIONS TO PONDER

CLINICAL CASE

J.Z., a 28-year-old electrician, was struck in the back of the head by a steel beam while at work and lost consciousness for approximately 30 minutes. J.Z. was hospitalized due to confusion and disorientation while being examined in the emergency room. The following day J.Z. was felt to be back to normal except for a headache and was discharged. The patient had no memory of being in the emergency room and no memories of events just prior or subsequent to the injury until the next day. J.Z. also noted that, besides persistent headache, the injury caused a degree of forgetfulness. Neurologic examination was unremarkable.

The brain, a bilaterally symmetric, soft, gelatinous structure surrounded by its meninges and enclosed in its bony cranium, is continuous with the spinal cord at the foramen magnum located at the base of the skull. At birth the brain weighs less than 400 g, but by the beginning of the second year of life it has more than doubled in weight to 900 g. The adult brain weighs between 1,250 and 1,450 g, but there does not appear to be a relationship between brain weight and intelligence.

As detailed in Chapter 2, it is evident during embryogenesis that the brain is subdivided into five continuous regions, from cranial to inferior: the telencephalon (Gr., "end-brain"), diencephalon (Gr., "through-brain"), mesencephalon (Gr., mid-brain), metencephalon (Gr., "hind-brain), and myelencephalon (Gr., "after-brain"). As the brain grows in size and complexity, these regions fold upon and over one another, so that in the adult the evidence of these subdivisions is no longer clearly apparent.

The present chapter details the brain's basic morphology and architecture to provide an anatomical framework of reference for the chapters that follow. Much of this terminology should be memorized so that when, in later chapters, functional connections among regions of the brain are discussed the student has a visual image of the location of the various structures and the pathways the connections take.

A Textbook of Neuroanatomy, Third Edition. Maria A. Patestas, Amanda J. Meyer, and Leslie P. Gartner.
© 2025 John Wiley & Sons, Inc. Published 2025 by John Wiley & Sons, Inc.
Companion website: www.wiley.com.go/patestas/neuroanatomy3e

Note that the clinical case at the beginning of the chapter refers to a patient who loses consciousness subsequent to head trauma.

1 Why is it important to do a CT scan or MRI in the case of a head injury that involves loss of consciousness?

2 What happened to the brain as patient's head was struck by the steel beam?

3 Most tissues swell when they are exposed to blunt trauma. Does that happen to the brain?

Viewing the adult brain in three dimensions, only three regions are clearly visible: the **cerebrum, cerebellum,** and the lower part of the **brainstem.**

CEREBRUM

The cerebrum, observed from above, appears to be composed of two large, oval, cerebral hemispheres

Each **cerebral hemisphere** (cerebral, L., "brain") is narrower posteriorly, at the **occipital pole,** than anteriorly, at the **frontal pole.** The two hemispheres are large, oval structures that, viewed together, superficially resemble the surface of a shelled walnut (Fig. 6.1). The midline **longitudinal cerebral fissure,** occupied in life by the falx cerebri (L., "sickle, curved blade"), partially separates the two cerebral hemispheres from one another. The floor of the longitudinal cerebral fissure is formed by the **corpus callosum** (L., "tough body"), a large myelinated fiber tract that forms an anatomical and functional connection between the right and left cerebral hemispheres.

The surface few millimeters of the cerebral hemispheres is composed of a highly folded collection of gray matter, known as the **cerebral cortex** (L., "bark"). This folding increases the

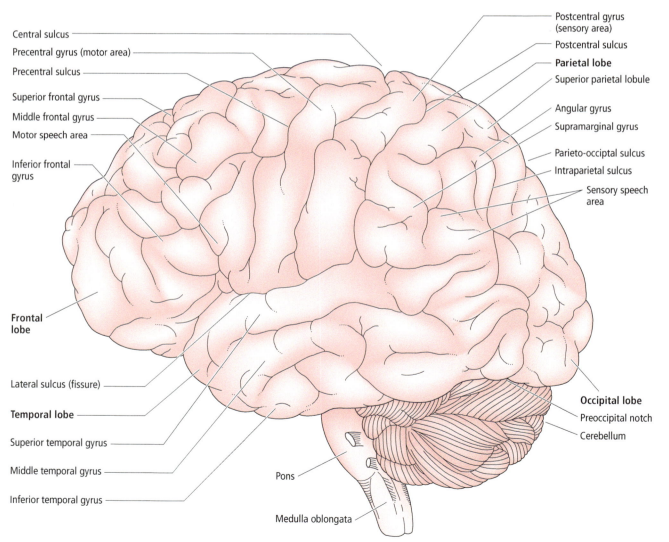

Figure 6.1 • Diagram of the brain from a lateral view.

surface area and volume of the cortex and presents elevations, **gyri** (gyrus, singular; L., "circle, ring"), and depressions, **sulci** (sulcus, singular; L.,"furrow"). Deep to the cortex is a central core of **white matter** that forms the bulk of the cerebrum and represents fiber tracts, supported by neuroglia, ferrying information destined for the cortex and cortical responses to other regions of the central nervous system (CNS). Buried within the mass of white matter are collections of neuron cell bodies, some of which are grouped together under the rubric of **basal nuclei** (frequently named **basal ganglia** even though, technically, they are *nuclei*). Large collections of gray matter are also present in the **diencephalons**, namely, the epithalamus, thalamus (Gr., "bedroom"), hypothalamus, and subthalamus.

Each **cerebral hemisphere** is a hollow structure, and the cavities are called the **right** and **left lateral ventricles** (L., "belly"), which communicate with the third ventricle, located in the diencephalon, via the **interventricular foramen** (L., "hole, opening") (Fig. 6.2A). The two lateral ventricles are separated from one another by two closely adjoined non-nervous membranes, each known as a **septum pellucidum** (L., "translucent wall"). Ependymal cells line each lateral ventricle, and protruding into each ventricle is a **choroid plexus** (Gr., "membrane-like network") that functions in the manufacturing of cerebrospinal fluid.

Lobes of the cerebral hemispheres

The five lobes of the cerebral hemisphere are the frontal, parietal, temporal, and occipital lobes, and the insula

Each cerebral hemisphere is subdivided into five lobes: frontal, parietal (L., "wall"), temporal (L., "temple"), occipital (L., ob caput "back of the head), and the insula (L., "island") (Table 6.1). Additionally, the cortical constituents of the limbic system are also considered to be a region of the cerebral hemisphere, and some consider it to be the sixth lobe, the **limbic lobe** (L., "hem, fringe").

Viewed from the side, each cerebral hemisphere resembles the shape of a boxing glove, where the thumb is the **temporal lobe** and is separated from the **parietal lobe** by the **lateral fissure** (eponym: lateral fissure) (Fig. 6.1). The floor of the lateral fissure is formed by the **insula** (eponym: island of Reil) that is hidden by the **frontal, parietal,** and **temporal opercula** (L., "lids"), extensions of the same named lobes. Although the geographic distributions of many of the sulci and gyri are relatively inconsistent, some are constant, occupy specific locations, are recognizable in most brains, and are named. The sulci are generally smaller and shallower than the fissures, and one of these, the **central sulcus** (eponym: of Rolando), separates the frontal lobe from the parietal lobe. The division between the parietal and occipital lobes is not readily evident when viewed from the lateral aspect because it is defined as the imaginary line between the **preoccipital** and the **parieto-occipital notches.** However, the division is clearly delimited on the medial aspect of the cerebral hemisphere, where the boundary between these two structures is the **parieto-occipital sulcus** and its continuation, the **calcarine** (L., "spur") **fissure** (Fig. 6.3).

Frontal lobe

The frontal lobe extends from the frontal pole to the central sulcus

The **frontal lobe** extends from the frontal pole of the cerebral hemisphere to the central sulcus (Fig. 6.1), thus forming the anterior one-third of the cerebral cortex. Its posteriormost gyrus, the **precentral gyrus,** consists of the primary motor area and is bordered anteriorly by the **precentral sulcus** and posteriorly by the **central sulcus.** The region of the frontal lobe located anterior to the precentral sulcus is subdivided into the **superior, middle,** and **inferior frontal gyri.** This subdivision is due to the inconsistent presence of two longitudinally disposed sulci, the superior and inferior frontal sulci. The inferior frontal gyrus is demarcated by extensions of the lateral fissure into three subregions: the pars triangularis, pars opercularis, and pars orbitalis. In the dominant hemisphere, a region of the inferior frontal gyrus, known as **Broca's area,** functions in the production of speech.

On its *inferior aspect,* the frontal lobe presents the longitudinally disposed **olfactory sulcus** (L., "to smell"). Medial to this sulcus is the **gyrus rectus** (also known as the **straight gyrus),** and lateral to it are the **orbital gyri.** The olfactory sulcus is partly occupied by the **olfactory bulb** and **olfactory tract** (Figs. 6.4 and 6.5). At its posterior extent, the olfactory tract bifurcates to form the lateral and medial olfactory striae (stria, sing. L., "channel, furrow"). The intervening area between the two striae is triangular in shape and is known as the **olfactory trigone,** and it abuts the **anterior perforated substance** (Fig. 6.5).

The frontal lobe is bordered, on its *medial aspect,* by the arched **cingulate sulcus** (L., "girdled"), which forms the boundary of the superior aspect of the cingulate gyrus. The quadrangular-shaped cortical tissue anterior to the central sulcus is a continuation of the precentral gyrus and is known as the **anterior paracentral lobule.**

Parietal lobe

The parietal lobe extends from the central sulcus to the parieto-occipital sulcus

The **parietal lobe,** located between the frontal and occipital lobes, is situated above the temporal lobe. The **postcentral gyrus,** its anteriormost gyrus, located on the lateral aspect of the parietal lobe, is the primary somesthetic area to which primary somatosensory information is channeled from the contralateral half of the body. The remainder of the parietal lobe, separated from the postcentral gyrus by the **postcentral sulcus,** is subdivided by the inconsistent intraparietal sulcus, into the **superior** and **inferior parietal lobules.** The former is an association area involved in somatosensory function, whereas the latter is separated into the **supramarginal gyrus,** which integrates auditory, visual, and somatosensory

78 ••• CHAPTER 6

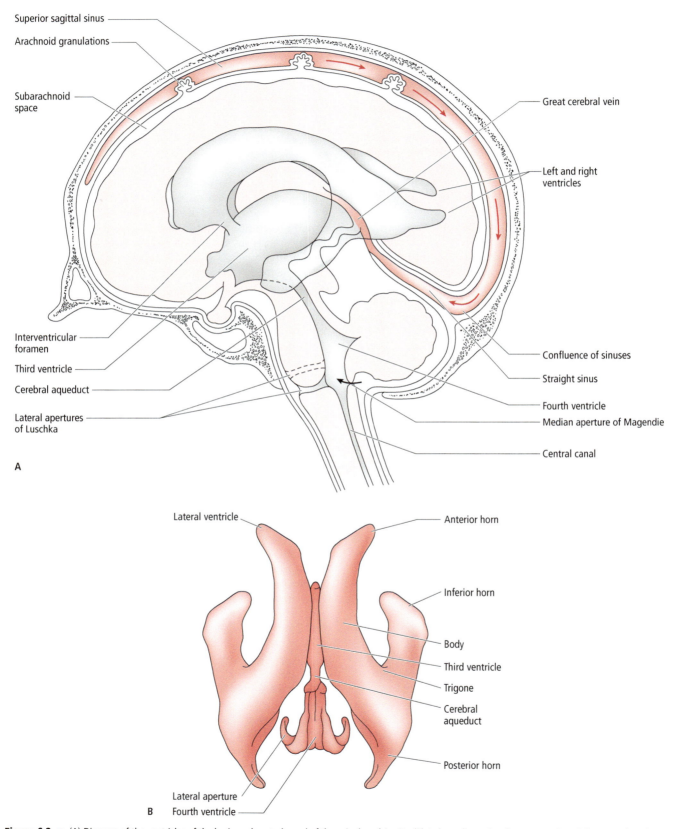

Figure 6.2 ● (A) Diagram of the ventricles of the brain and central canal of the spinal cord *in situ*. (B) A three-dimensional representation of the ventricles of the brain.

Table 6.1 ● Lobes of the cerebral hemispheres.

Lobe	Surface	Major gyri	Function/comment
Frontal	Lateral	Precentral gyrus	Primary motor area
		Inferior frontal gyrus	Broca's area in dominant hemisphere; functions in speech production
	Medial	Anterior paracentral lobule	Continuation of precentral/precentral gyrus
	Inferior	Gyrus rectus and orbital gyri	Olfactory bulb and tract in the olfactory sulcus
Parietal	Lateral	Postcentral gyrus	Primary somesthetic area
		Superior parietal lobule	Association area involved in somatosensory function
		Inferior parietal lobule	
		Supramarginal gyrus	Integrates auditory, visual, and somatosensory information
		Angular gyrus	Receives visual input and is involved in reading and other complex language skills
	Medial	Posterior paracentral lobule precuneus	Continuation of the precentral/postcentral gyrus
Temporal	Lateral	Superior temporal gyrus Middle and inferior temporal gyri	Wernicke's area in dominant hemisphere; ability to read, understand, and speak the written word
	Superior	Transverse temporal gyri (of Heschl)	Primary auditory cortex
	Inferior	Fusiform gyrus	Borders the parahippocampal gyrus of the limbic lobe
Occipital	Lateral	Superior and inferior occipital gyri	Separated from each other by the lateral occipital sulcus
	Medial	Cuneate gyrus (cuneus) Lingual gyrus	Separated from each other by the calcarine fissure; striate cortex (primary visual cortex) is on the banks of this fissure
Insula	Lateral	Short and long gyri	Forms the floor of the lateral sulcus; associated with taste
Limbic	Medial	Cingulate gyrus	Above the body of the corpus callosum and continues as the isthmus
		Parahippocampal gyrus	Anterior continuation of isthmus; ends in the uncus
		Hippocampal formation	Composed of hippocampus, subiculum, and dentate gyrus
		Subcallosal, paraolfactory, and preterminal gyri	Collectively known as the subcallosal area

information, and the **angular gyrus,** which receives visual input to integrate written and spoken forms of language.

The parietal lobe is separated from the occipital lobe on their *medial aspect* by the parieto-occipital sulcus and its inferior continuation, the calcarine fissure. This region of the parietal lobe is subdivided into two major structures, the anteriorly positioned **posterior paracentral lobule** (a continuation of the postcentral gyrus) and the posteriorly situated **precuneus** (Gr., "angle").

Temporal lobe

The temporal lobe, the "thumb of the boxing glove," is situated inferiorly to the lateral fissure and anterior to the parieto-occipital sulcus

The **temporal lobe** is separated from the frontal and parietal lobes by the lateral fissure and from the occipital lobe by an imaginary plane that passes through the parieto-occipital sulcus. The anteriormost aspect of the temporal lobe is known as the **temporal pole.** The temporal lobe exhibits three parallel gyri on its *lateral aspect*, the **superior, middle, and inferior temporal gyri,** separated from each other by the inconsistently present superior and middle temporal sulci. The superior temporal gyrus of the dominant hemisphere contains **Wernicke's area,** which is similar to Broca's area in

that it is also responsible for the individual's ability to speak and understand the spoken and written word.

Hidden within the lateral fissure is the *superior aspect* of the temporal lobe whose surface is marked by the obliquely running **transverse temporal gyri** (eponym: of Heschl), the primary auditory cortex.

The *inferior aspect* of the temporal lobe is grooved by the **inferior temporal sulcus** that is interposed between the inferior temporal gyrus and the **lateral occipitotemporal gyrus (fusiform gyrus).** The **collateral sulcus** separates the fusiform gyrus from the **parahippocampal gyrus** of the limbic lobe.

Occipital lobe

The occipital lobe extends from the parieto-occipital sulcus to the occipital pole

The **occipital lobe** extends from the occipital pole to the parieto-occipital sulcus. The occipital lobe presents, on its *lateral aspect*, the **superior** and **inferior occipital gyri,** separated from each other by the horizontally running **lateral occipital sulcus.**

The occipital lobe is subdivided, on its *medial aspect*, into the superiorly located **cuneate gyrus (cuneus)** and the inferiorly positioned **lingual gyrus,** separated from each other by the **calcarine fissure.** The cortical tissue on each bank of this

CHAPTER 6

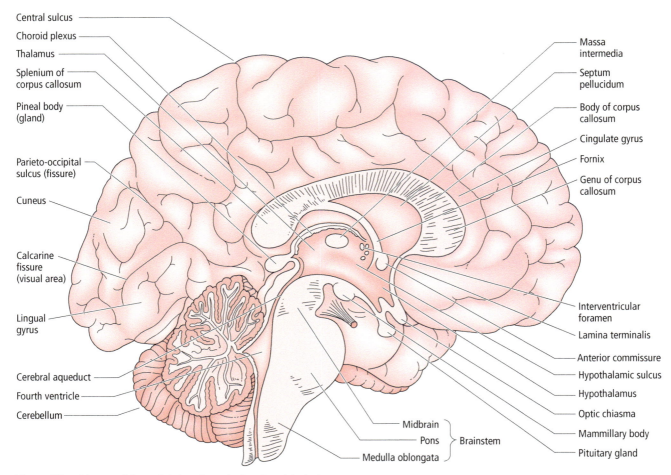

Figure 6.3 • Diagram of the medial view of a sagittal section of the brain.

fissure is known collectively as the **striate cortex (calcarine cortex)** and forms the **primary visual cortex.**

Insula

The insula forms the floor of the lateral sulcus

Pulling apart the frontal, temporal, and parietal opercula, displays the **insula**, since this lobe is submerged within and forms the floor of the lateral sulcus. Its lateral surface is subdivided into several short and long gyri, the most prominent of which is located posteriorly. The insula, which is believed to be associated with taste, and perhaps other visceral functions, is completely circumscribed by the **circular sulcus.**

Limbic lobe

The limbic lobe is a hemispherical region on the medial aspect of the cerebral cortex that surrounds the corpus callosum and the diencephalon

The **limbic lobe** is a complex region and includes the cingulate gyrus, parahippocampal gyrus, hippocampal formation, subcallosal gyrus, parolfactory gyrus, and the preterminal gyrus.

The following description is a view of the *medial aspect* of the hemisected brain and the various regions of the corpus callosum are obvious landmarks. Therefore, the corpus callosum will now be described, even though it is not a part of the limbic lobe. The anterior extent of the corpus callosum, known as the **genu** (G., "knee"), bends inferiorly and turns posteriorly, where it forms a slender connection, the **rostrum** (L., "beak, snout"), with the anterior commissure. The posterior extent of the corpus callosum is bulbous in shape, and is known as the **splenium** (G., "compress made of linen") (see Fig. 6.3).

The **cingulate gyrus** is located above the corpus callosum and is separated from it by the **callosal sulcus.** As the cingulate gyrus continues posteriorly, it follows the curvature of the corpus callosum and dips beneath the splenium to continue anteriorly as the **isthmus** (L., "narrow passage") of the cingulate gyrus. The anterior continuation of the isthmus is the **parahippocampal** (L., "seahorse"), whose anteriormost extent is known as the **uncus** (L., "hook"). Above the parahippocampal gyrus is the **hippocampal sulcus,** which separates the parahippocampal gyrus from the **hippocampal formation** (composed of the hippocampus, subiculum, and dentate gyrus).

Just beneath the rostrum of the corpus callosum is the **subcallosal gyrus.** The connection between the anterior

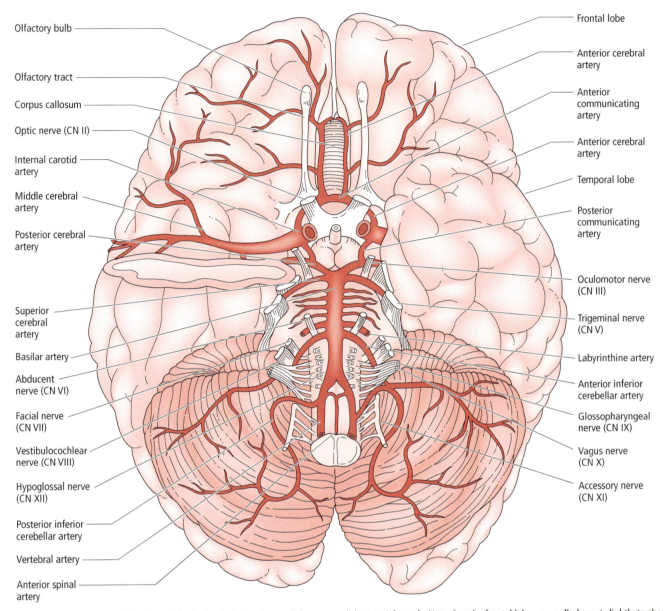

Figure 6.4 ● Diagram of the base of the brain displaying the cranial nerves and the arterial supply. Note that the frontal lobes are pulled apart slightly to show the corpus callosum and the anterior cerebral arteries; also, the right temporal lobe is sectioned to demonstrate the middle cerebral artery.

commissure and the optic chiasm is the lamina terminalis and the cortical tissue anterior to the lamina terminalis is the **parolfactory gyrus** and **preterminal gyrus**. The subcallosal, parolfactory, and preterminal gyri are referred to as the **subcallosal area**.

Brodmann's classification of the cerebral cortex

The best accepted system of functional regionalization of the cerebral cortex was developed by the neuroanatomist, K. Brodmann, who in the late nineteenth and early twentieth centuries mapped the cortex into 47 unique areas, each associated with specific morphological characteristics. Although later investigators refined and expanded his map into more than 200 areas and assigned functional characteristics to them, Brodmann's original classification is still widely used in the literature. The major areas, their location, and function are presented in Table 6.2.

Histology of the cerebral cortex

The histological organization of the cerebral cortex permits its subdivision into three regions: the archicortex, mesocortex, and neocortex

The cerebral cortex has a rich vascular supply that nourishes its exceptionally large number of neurons, neuroglia, and nerve fibers. The arrangement

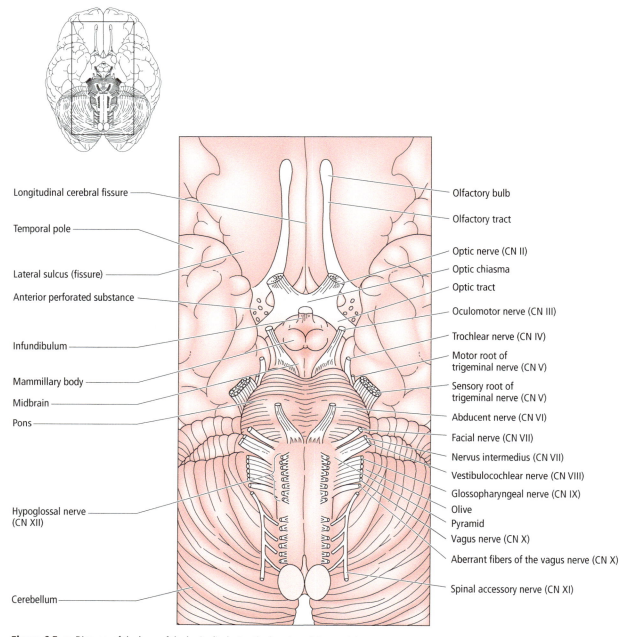

Figure 6.5 ● Diagram of the base of the brain displaying the location of the cranial nerves.

of the three types of neurons that populate the cortex – pyramidal cells, stellate neurons, and fusiform neurons – permits the classification of the cortex into: archicortex (allocortex), mesocortex (juxtallocortex), and neocortex (isocortex).

The **archicortex**, phylogenetically the oldest region, is composed of only three layers and is located in the limbic system. The **mesocortex**, phylogenetically younger, is composed of three to six layers, and is located predominantly in the insula and cingulate gyrus. The **neocortex**, phylogenetically the youngest region of the cerebral cortex, comprises the bulk of the cerebral cortex, and is composed of six layers.

The cerebral cortex is arranged in layers and superimposed upon this cytoarchitecture is a functional organization of **cell columns**, where each cell column is less than 0.1 mm in diameter, is arranged vertically, perpendicular to the superficial surface of the cortex. Each column passes through the six cortical layers and is composed of neurons with similar functions. All neurons of a single column respond to similar stimuli from the same region of the body.

The organization of the six layers of the neocortex is known as its **cytoarchitecture**, where each layer has a name and an associated Roman numeral (Table 6.3).

Area number	Area name	Location
1, 2, and 3	Primary somatosensory cortex	Parietal lobe
4	Primary motor cortex	Frontal lobe
5 and 7	Somatosensory association cortex	Parietal lobe
6	Supplementary motor area and premotor cortex	Frontal lobe
8	Frontal eye field	Frontal lobe
9–12 and 46–47	Prefrontal cortex	Frontal lobe
17	Primary visual cortex	Occipital lobe
18 and 19	Visual association cortex	Occipital lobe
22	Wernicke's speech area (dominant hemisphere)	Temporal lobe (and perhaps into the parietal lobe also)
	Auditory association cortex	
41 and 42	Primary auditory cortex	Temporal lobe
44 and 45	Broca's speech area (dominant hemisphere)	Frontal lobe

Table 6.2 ● Selected Brodmann's areas.

Name	Numeral	Components
Molecular layer	I	Afferent fibers from the thalamus or from the cerebral cortex
External granular layer	II	Stellate neurons
External pyramidal layer	III	Small pyramidal cells
Internal granular layer	IV	Fusiform neurons
Internal pyramidal layer	V	Larger pyramidal cells
Fusiform layer	VI	Fusiform neurons

Table 6.3 ● Layers of the neocortex.

White matter of the cerebral hemispheres

There are three categories of fiber components in the cerebral hemispheres: commissural, projection, and association fibers

The central core of white matter that forms the bulk of the cerebrum is composed of myelinated nerve fibers of varied sizes supported by neuroglia that may be classified into the following three categories: *commissural, projection,* and *association fibers.*

Commissural fibers

Commissural fibers are bundles of axons that connect the right and left cerebral hemispheres

The four bundles of **commissural fibers (transverse fibers),** the corpus callosum, anterior commissure, posterior commissure, and hippocampal commissure (see Fig. 6.3), interconnect the right and left cerebral hemispheres.

- The **corpus callosum,** is the largest group of the commissural fibers, and it comprises four regions: the anteriormost **rostrum,** the curved **genu,** the relatively flattened

body, and its posteriormost region, the **splenium.** The corpus callosum connects the **neocortex** of the right hemisphere with that of the left.
- The **anterior commissure** connects the right and left amygdaloid bodies (L., "almond"), the olfactory bulbs, and several cortical regions of the two temporal lobes.
- The **posterior commissure** connects the right and left pretectal regions and related cell groups of the mesencephalon.
- The **hippocampal commissure (commissure of the fornix)** joins the right and left hippocampi to one another.

Projection fibers

Projection fibers are restricted to a single hemisphere and connect the cerebral hemispheres with lower levels

Projection fibers are restricted to one side of the brain; they connect the cerebral cortex with lower levels, namely the corpus striatum, diencephalon, brainstem, and spinal cord. The majority of these fibers are axons of pyramidal cells and of fusiform neurons. These fibers are component parts of the **internal capsule,** which is subdivided into the anterior limb, genu, posterior limb, retrolentiform, and sublentiform

84 ●●● **CHAPTER 6**

regions. The projection fibers may be subdivided into corticopetal and corticofugal fibers.

- **Corticopetal fibers** are *afferent fibers* that bring information from the thalamus to the cerebral cortex. They consist of **thalamocortical fibers.**
- **Corticofugal fibers** are *efferent fibers* that transmit information from the cerebral cortex to lower centers of the brain and spinal cord. Their fibers belong to four groups, namely corticobulbar, corticopontine, corticospinal, and corticothalamic fibers.

Association fibers

Association fibers connect regions of a hemisphere to other regions of the same hemisphere

Association fibers, also known as **arcuate fibers,** are also restricted to one side of the brain; they are axons of pyramidal cells and fusiform neurons and are subdivided into two major categories, short arcuate fibers and long arcuate fibers.

- **Short arcuate fibers** connect adjacent gyri and do not usually reach the subcortical white matter of the cerebral cortex; most of them are confined to the cortical gray matter.
- **Long arcuate fibers** connect nonadjacent gyri and consist of the following fiber tracts (Table 6.4): the uncinate fasciculus, cingulum, superior longitudinal fasciculus, inferior longitudinal fasciculus, and fronto-occipital fasciculus.

Basal nuclei

The basal nuclei consist of some deep cerebral nuclei and brainstem nuclei that, when damaged, produce movement disorders

The **basal nuclei** (as indicated above, are frequently called basal ganglia even though they are nuclei) are large collections of cell bodies that are embedded deep in the white matter of the brain (Fig. 6.6). By definition, these cell bodies include those deep nuclei of the brain and brainstem which, when damaged, produce **movement disorders**.

The basal nuclei are composed of the caudate nucleus, lenticular nucleus (putamen and globus pallidus), subthalamic nucleus of the anterior thalamus, and the substantia nigra of the mesencephalon (the caudate nucleus and the putamen together are referred to as the striatum). These nuclei have numerous connections with various regions of the CNS:

- some receive input and are categorized as **input nuclei,**
- some project to other regions and are referred to as **output nuclei,** whereas
- some receive input, project to other regions of the CNS, and have local interconnections, and these are known as **intrinsic nuclei.**

DIENCEPHALON

The diencephalon is that portion of the prosencephalon that surrounds the third ventricle

The **diencephalon,** interposed between the cerebrum and the midbrain, has four regions: the epithalamus, thalamus, hypothalamus, and subthalamus. The right and left halves of the diencephalon are separated from one another by a narrow slit-like space, the ependymal-lined **third ventricle** (Fig. 6.6). Cranially, the interventricular foramina (eponym: of Monro) leads from the lateral ventricles into the third ventricle (Fig. 6.6), whereas inferiorly, the third ventricle is connected to the **fourth ventricle** by the **mesencephalic aqueduct** (eponym: **of Sylvius**).

The **epithalamus,** composed of the **pineal body, stria medullaris,** and **habenular trigone,** constitutes the posterior surface of the diencephalon. The right and left thalami compose the bulk of the diencephalon and form the superior aspect of the lateral walls of the third ventricle. The two thalami, structures composed of numerous nuclei, are connected to each other by a bridge of gray matter, the **interthalamic adhesion (massa intermedia).** Some of the nuclei of the thalamus form distinctive bulges on its surface, namely the **pulvinar** (L., "cushion") and the **medial** and **lateral geniculate bodies.** The boundary between the thalamus and the **hypothalamus** is marked by a groove, the **hypothalamic sulcus,** located along the lateral walls of the third ventricle. Structures associated with the hypothalamus are the **pituitary gland** and its **infundibulum,**

Fiber group	Extends from	Extends to
Uncinate fasciculus	Anterior temporal lobe and uncus	Motor speech area and orbital gyri of frontal lobe
Cingulum	Medial cortex below rostrum	Parahippocampal gyrus and parts of temporal lobe
Superior longitudinal fasciculus	Anterior frontal lobe	Occipital, parietal, and temporal lobes
Inferior longitudinal fasciculus	Posterior region of the parietal and temporal lobes	Anterior region of the occipital lobe
Fronto-occipital fasciculus	Ventrolateral regions of the frontal lobe	Temporal and occipital lobes

Table 6.4 ● Extent of the long arcuate fiber groups.

GROSS ANATOMY OF THE BRAIN ••• 85

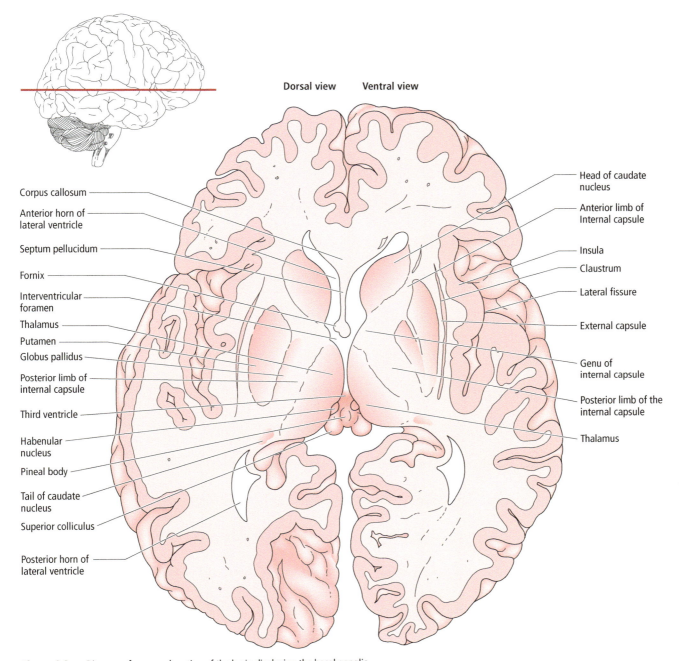

Figure 6.6 • Diagram of a coronal section of the brain displaying the basal ganglia.

the **tuber cinereum,** and the two **mammillary bodies.** The subthalamic nuclei and fiber tract form the **subthalamus.**

CEREBELLUM

The cerebellum is located below the occipital lobe of the cerebral hemispheres. It is connected to the brainstem via the superior, middle, and inferior cerebellar peduncles

The **cerebellum**, located in the posterior aspect of the brain just below the occipital lobes of the cerebrum (Figs. 6.7 and 6.8), is separated from the cerebrum via a horizontal dural reflection, the tentorium cerebelli. It is connected to the midbrain, pons, and medulla oblongata of the brainstem via three large pairs of fiber bundles, the **superior, middle,** and **inferior cerebellar peduncles,** respectively. The cerebellum is composed of the right and left **cerebellar hemispheres** and the narrow, intervening **vermis** (Fig. 6.7), which is subdivided into a superior and an inferior portion, where the superior portion is visible between the two hemispheres, whereas its inferior portion is buried between the two hemispheres of the cerebellum.

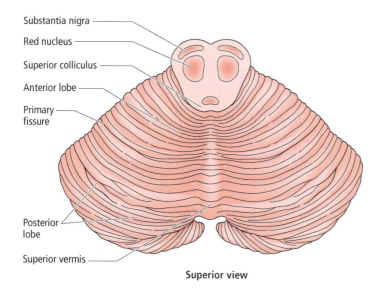

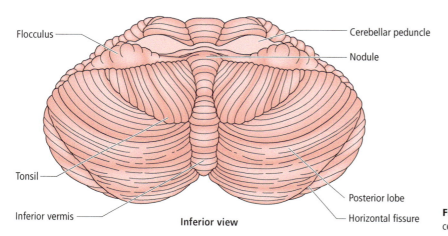

Figure 6.7 ● Superior and inferior views of the cerebellum.

The surface of the cerebellum has horizontal elevations, known as **folia,** and indentations between the folia, known as **sulci.** Some of these sulci are deeper than others, and they are said to subdivide each hemisphere into three lobes, the small **anterior lobe,** the much larger **posterior lobe,** and the inferiorly positioned **flocculonodular lobe** (formed from the **nodule** of the vermis and the **flocculus** of each cerebellar hemisphere). The anterior lobe is separated from the posterior lobe by the **primary fissure,** and the **posterolateral fissure** separates the flocculonodular lobe from the posterior lobe (Figs. 6.7–6.9).

Similar to the cerebrum, the cerebellum has an external rim of gray matter, the **cortex,** an internal core of nerve fibers, the **medullary white matter,** and the **deep cerebellar nuclei,** located within the white matter. The cortex and white matter are easily distinguished from each other in a midsagittal section of the cerebellum, where the white matter arborizes, forming the core of what appears to be a tree-like architecture, known as the **arbor vitae** (L., tree of life).

Histology of the cerebellar cortex demonstrates it to be a three-layered structure, the internal most granular layer, the middle Purkinje layer, and the external most molecular layer.

- The **granular layer** is composed of numerous, small cells.
- The **Purkinje layer**, composed of a single layer of large Purkinje cell perikaryons, is also easily recognizable.
- The **molecular layer** is rich in axons and dendrites as well as capillaries that penetrate deep into this layer.

Four pairs of nuclei are located within the substance of the cerebellar white matter. These are the **fastigial, dentate, emboliform,** and **globose nuclei.** The connections between the cortical regions and the deep nuclei permit the subdivision of the cerebellum into three zones – the **vermal, paravermal,** and **hemispheric**—where each zone is composed of deep cerebellar nuclei, white matter, and cortex.

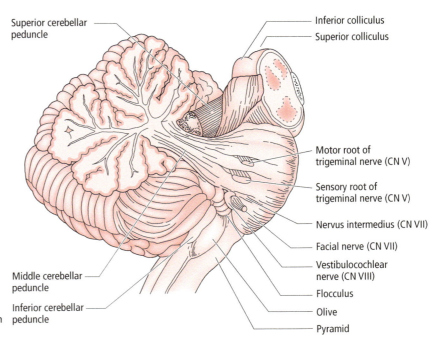

Figure 6.8 ● Diagram of a lateral view of the cerebellum and medulla oblongata.

BRAINSTEM

The brainstem is composed of the mesencephalon, metencephalon, and myelencephalon

The **brainstem,** the oldest part of the CNS, is composed of the **mesencephalon, metencephalon,** and **myelencephalon** (although some authors also include the diencephalon) (Figs. 6.9 and 6.10). Because these are embryologic terms, one may also state that the brainstem is composed of the mesencephalon, pons, cerebellum, and medulla oblongata. As parts of it have been overgrown by the cerebrum and the cerebellum, its posterior aspect is mostly hidden from view in the whole brain, whereas its anterior and lateral aspects are visible. Removal of the cerebral and cerebellar hemispheres exposes the brainstem in its entirety, and it is usually examined in that fashion as well as by hemisecting the entire brain (Fig. 6.3).

Mesencephalon

The mesencephalon is that region of the brainstem that surrounds the mesencephalic aqueduct

The **mesencephalon (midbrain)** is a relatively narrow band of the brainstem, extending from the diencephalon to the pons, surrounding the mesencephalic aqueduct. The posterior aspect of the midbrain is known as the **tectum** (L., "roof") and incorporates the paired **superior** and **inferior colliculi** (also known as the **corpora quadrigemina).** These structures are associated with the **lateral** and **medial geniculate bodies,** respectively, and they are all associated with visual and auditory functions. The **trochlear nerve (CN IV)** is the only cranial nerve that exits the posterior aspect of the mesencephalon just below the inferior colliculus; all other cranial nerves exit the anterior aspect of the brainstem. The region of the mesencephalon below the mesencephalic aqueduct is known as the **midbrain (mesencephalic) tegmentum** (L., "cover"). The cerebral hemispheres are connected to the brainstem by two large fiber tracts, the **cerebral peduncles,** and the depression between the peduncles is known as the interpeduncular fossa, the site of origin of the **oculomotor nerve (CN III).**

Metencephalon

The metencephalon is located below the cerebellum; its anterior bulge, the pons, is clearly visible

The cerebellum overlies and hides the posterior aspect of the **metencephalon,** but its anterior aspect, the **pons,** is clearly evident. Cranially, the **superior pontine sulcus** acts as the boundary between the metencephalon and the midbrain, and the **inferior pontine sulcus** as the boundary between the metencephalon and the myelencephalon. Part of the floor of the fourth ventricle is formed by the posterior aspect of the pons, and is known as the **pontine tegmentum,** the structure that houses the nuclei of the trigeminal, abducens, facial, and vestibulocochlear nerves. Cranial nerves VI, VII, and VIII leave the brainstem at the inferior pontine sulcus, whereas the trigeminal nerve exits the brainstem through the middle cerebellar peduncle.

Myelencephalon

The myelencephalon, the inferior-most portion of the brainstem, houses the fourth ventricle

The inferior-most portion of the brainstem, the **myelencephalon,** also known as the

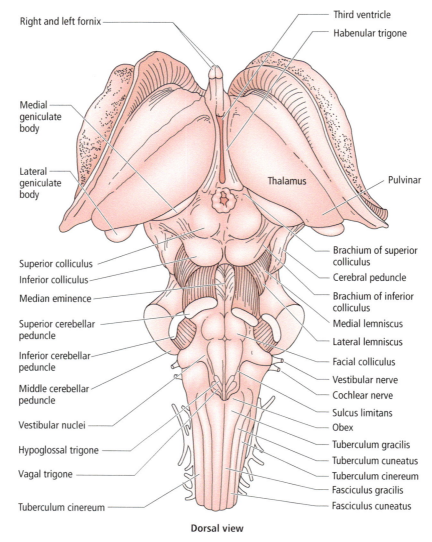

Figure 6.9 ● Diagram of the posterior view of the brainstem.

medulla oblongata, extends from the inferior pontine sulcus to the spinal cord. The boundary between them is the region where the lateral walls of the fourth ventricle converge in a V shape at the midline **obex** (at the level of the foramen magnum).

The anterior surface of the myelencephalon displays the **anterior midline fissure,** bordered on each side by the **pyramids** and crossed by the **pyramidal decussations,** connecting the two pyramids to oneanother. The **olives** are olivepit-shaped swellings lateral to each pyramid. The hypoglossal nerve is evident as a number of thin filaments on each side of the brainstem, arising from the **anterior lateral sulcus** between the pyramids and olives. The glossopharyngeal, vagus, and accessory nerves arise from the groove posterior to the olives.

The posterior surface of the myelencephalon displays the posterior median fissure, which is interposed between the right and left **tuberculum gracilis,** swellings formed by the nucleus gracilis. Just lateral to the tuberculum gracilis is another swelling, the **tuberculum cuneatus,** a bulge formed by the underlying nucleus cuneatus. The inferior continuation of the tuberculum gracilis is the **fasciculus gracilis,** and the continuation of the tuberculum cuneatus is the **fasciculus cuneatus.** Just lateral to the tuberculum cuneatus is another swelling, the **tuberculum cinereum,** formed by the **descending tract of the trigeminal nerve.**

CLINICAL CONSIDERATIONS

As indicated at the beginning of this chapter, most of the major topics discussed here are presented in detail in subsequent chapters. Therefore, the pertinent clinical considerations are presented in the chapters dealing with the specific topics.

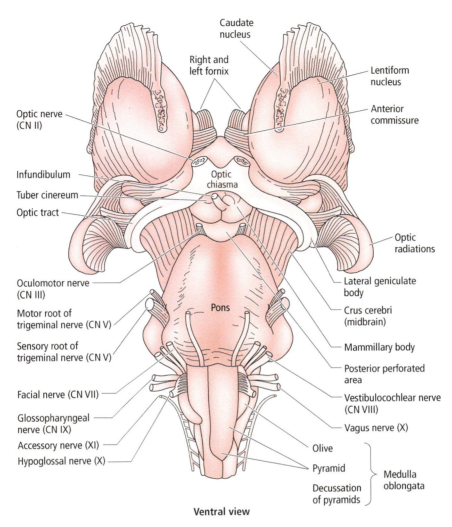

Figure 6.10 • Diagram of the anterior view of the brainstem.

SYNONYMS AND EPONYMS OF THE BRAIN

Name of structure or term	Synonym(s)/eponym(s)
Archicortex	Allocortex
Central sulcus	Central sulcus of Rolando
Commissural fibers	Transverse fibers
Cuneate gyrus	Cuneus
Hippocampal commissure	Commissure of the fornix
Interthalamic adhesion	Massa intermedia
Interventricular foramen	Foramen of Monro
Lateral fissure	Fissure of Sylvius
Lateral occipitotemporal gyrus	Fusiform gyrus
Mesocortex	Juxtallocortex
Myelencephalon	Medulla oblongata
Neocortex	Isocortex
	Neopallium
	Homogenetic cortex
Striate cortex	Calcarine cortex
Superior and inferior colliculi	Corpora quadrigemina
Transverse temporal gyri	Transverse temporal gyri of Heschl

FOLLOW-UP TO CLINICAL CASE

A review of the head CT that was performed in the emergency room revealed the extent of the patient's injuries. There were modest contusions of the bilateral anterior and orbital regions of the frontal lobes and the left anterior temporal lobe. Contusions are brain "bruises." The patient had a concussion, meaning any alteration of consciousness resulting from head trauma. Persistent headache and memory difficulties are often characteristics of "postconcussion syndrome," which can also include dizziness, poor balance, poor concentration, and other vague symptoms.

There are several ways in which head trauma can cause brain injury. As in the above case, contusion can result. The locations of these are often predictable. *Coup* and *contrecoup* are terms used to define the location of contusion related to the site of head trauma. A *coup* injury refers to a contusion that occurs to the part of the brain that directly underlies the site of head trauma. A *contrecoup* injury refers to a contusion that occurs in parts of the brain directly opposite the site of head trauma. The patient above suffered *contrecoup* contusions. The anterior and orbital regions of the frontal lobes, and the anterior temporal lobes are the most common locations of contusions. The skull is somewhat rigid and is the first to have contact with the object of injury. There is a very sudden acceleration or deceleration of the skull. The brain within is somewhat mobile. As the skull accelerates or decelerates, the brain crashes into the skull at certain points. The damage occurs mostly to those parts of the brain surface that are most angular and also in close proximity to the bones of the skull.

There are other locations and mechanisms of injury that are often more important, especially in cases where disability or death occur from head trauma. There is also a rotational component to brain movement inside the skull at the moment of impact. This causes shearing stresses, especially to the upper brainstem region, which can damage the reticular system, which maintains consciousness and awareness. This is probably the mechanism by which head injury causes loss or alteration of consciousness. Damage to the corpus callosum is also often demonstrated pathologically, secondary to its rotational momentum against the rigid falx. Diffuse axonal injury refers to diffuse damage to white matter, specifically to axons, which has been demonstrated pathologically. This is also believed to be due to shearing or stretching injury to the long and thin axons due to sudden and severe rotational and angular stresses.

QUESTIONS TO PONDER

1. How are the categories of the three fiber components of the cerebral white matter classified?

2. Why is it inaccurate to call the basal nuclei, "basal ganglia?"

3. What are peduncles?

4. Describe the reason why the trochlear nerve is an unusual cranial nerve.

CHAPTER 7

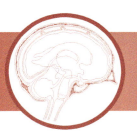

Brainstem

CLINICAL CASE

INTERNAL ORGANIZATION OF THE BRAINSTEM

MEDULLA OBLONGATA

PONS

MIDBRAIN

CLINICAL CONSIDERATIONS

SYNONYMS AND EPONYMS OF THE BRAINSTEM

FOLLOW-UP TO CLINICAL CASE

QUESTIONS TO PONDER

CLINICAL CASE

A 51-year-old patient with a history of treated hypertension presents with the sensation as though they are leaning to the left when walking. These symptoms become more intense the next day, at which time they also notice that they have some facial numbness in the left perioral region, right hemibody paresthesia, and in addition feel that their left leg is not working properly. Within 4 days they develop left retro-orbital pain, mild ptosis of the left eyelid and a left-sided headache, but no hiccups.

The neurological exam shows the patient to be alert and oriented in time, person, and place. Visual field testing and visual acuity are normal. There is mild ptosis of the left eyelid and mild left miosis but no proptosis. Extraocular movements are full with no objective evidence of diplopia. To light touch stimuli, subtle decreased sensation around the lips on the left side is documented. There is no hearing loss and no bulbar weakness. Strength in all four extremities is normal and reflexes are normal. The plantar responses are flexor bilaterally. Finger-to-nose testing is normal bilaterally. Heel-to-shin testing is slightly uncoordinated on the left side. Gait testing demonstrates a cautious pattern of ambulation which has decreased stride length but is not wide-based. The neuroanatomic issue is to determine the central nervous system site of origin of this constellation of symptoms and findings.

Non-contrast CT scan of the brain performed on the night of admission is normal. MRI scan of the brain done the following day shows a clearly defined 5 mm focus of diffusion abnormality in the left lateral medulla oblongata compatible with recent ischemic infarction. Echocardiogram identifies a mitral valve leaflet fibroelastoma thought to be the source of a cardiac embolus to the posterior inferior cerebellar artery.

A diagnosis of cardioembolic left lateral medullary ischemic stroke (Wallenberg syndrome) is made.

INTERNAL ORGANIZATION OF THE BRAINSTEM

The brainstem serves as a conduit for the ascending sensory and descending motor tracts to reach their destinations in the cerebral hemispheres or the spinal cord, respectively

The brainstem includes, in ascending order, the **medulla oblongata**, **pons**, and **midbrain** (Figs. 7.1 and 7.2). Rostrally, the brainstem is continuous with the diencephalon and inferiorly with the spinal cord. Similar to the spinal cord and cerebral hemispheres, the brainstem consists of gray matter and white matter. Gray matter is characterized by the presence of nerve cell bodies and their supporting neuroglia, whereas white matter is distinguished by the presence of myelinated axons and neuroglia and the notable absence of nerve cell bodies. In the brainstem, axons form fiber bundles that may be referred to as a tract, a fasciculus (L., "fascicle"), a lemniscus (L., "ribbon"), a peduncle (L., "little foot"), a commissure (consisting of crossing axons),

---

A Textbook of Neuroanatomy, Third Edition. Maria A. Patestas, Amanda J. Meyer, and Leslie P. Gartner.
© 2025 John Wiley & Sons, Inc. Published 2025 by John Wiley & Sons, Inc.
Companion website: www.wiley.com.go/patestas/neuroanatomy3e

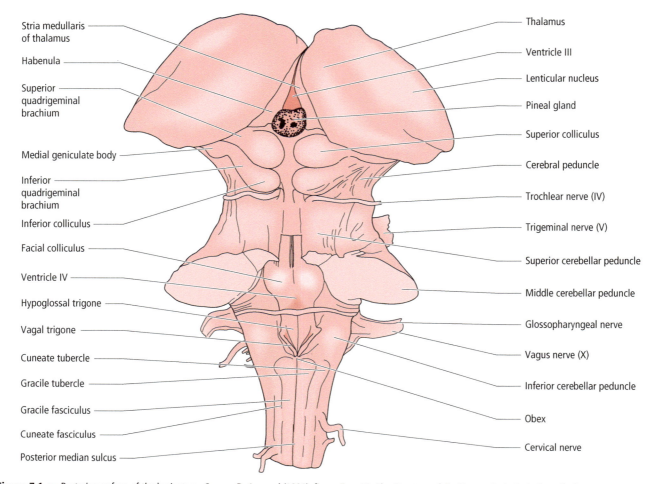

Figure 7.1 ● Posterior surface of the brainstem. Source: DeArmond (1989), figure 5, p. 10. *The Structure of the Human Brain*, 3rd edn/Oxford University Press.

or a brachium (L., "arm"-like structure). These axons are either local projections or form part of the long tracts of the descending motor or ascending sensory pathways traversing the brainstem on their way to their destinations. The core of the brainstem is known as the reticular formation (discussed in Chapter 16), a phylogenetically ancient, complex system of nuclei and network of fibers, forming a matrix with its myriads of connections and functions some of which are crucial for survival. Essential functions of the reticular formation include somatic and visceral sensation, control of consciousness, regulation of the respiratory and cardiovascular systems, control of muscle tone, of posture maintenance, as well as of movement. The brainstem is also the origin of most of the cranial nerves and the location of most of their nuclei.

The brainstem is subdivided internally into the **tectum** (L. "roof"), which is prominent only in the midbrain, the **tegmentum** (L. "covering"), and the **basis** (Fig. 7.3). The tegmentum is continuous and appears somewhat similar throughout its extent, unlike the basis, which exhibits different morphological characteristics at each of its three brainstem levels. The tegmentum serves not only as a conduit of the ascending and descending tracts, but it also contains cranial nerve nuclei and the nuclei of the reticular formation. The basis contains most of the descending tracts of the motor system.

Although a number of the ascending and descending tracts are present at all levels of the brainstem as they project to their nearby or distant terminations, others pass through only certain regions of the brainstem. The ascending sensory tracts that are present throughout the entire length of the brainstem, include the *spinothalamic*, *spinotectal*, and the *anterior trigeminothalamic tracts*, all of which are situated in the lateral aspect of the brainstem tegmentum. Other tracts, including the *medial lemniscus (ML)*, the *medial longitudinal fasciculus (MLF)*, and the *tectospinal tracts*, are located adjacent to the midline in the tegmentum as they pass through the brainstem. The descending tracts of the motor system that are present at all brainstem levels include the *corticospinal*, *corticonuclear*, and *corticoreticular*

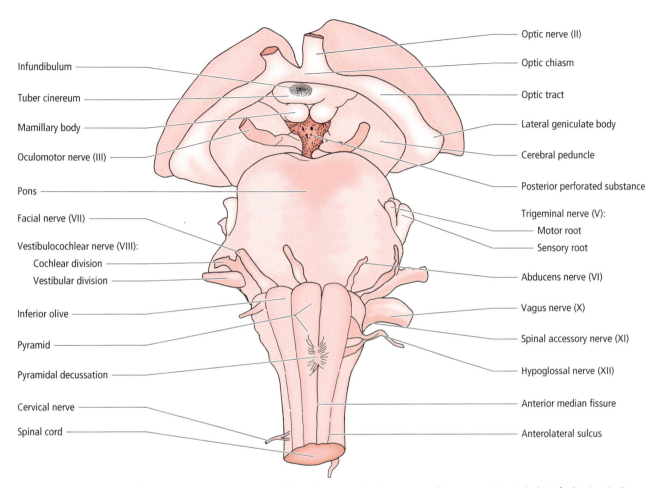

Figure 7.2 ● Anterior surface of the brainstem. Source: DeArmond (1989), figure 6, p. 12. *The Structure of the Human Brain*, 3rd edn/Oxford University Press.

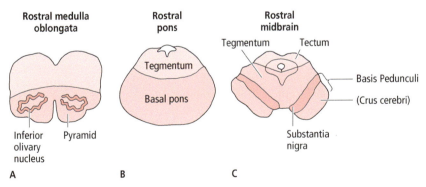

Figure 7.3 ● Regions of the brainstem at the (A) medulla oblongata, (B) pons, and (C) midbrain. Note that the tectum is only present in the midbrain, the tegmentum is present at all brainstem levels, and anteriorly, prominent structures at the level of the midbrain include the substantia nigra and the basis pedunculi, at the level of the pons the basal pons (basis pontis) with the corticonuclear and corticospinal tract fibers, and at the level of the rostral medulla oblongata, the pyramids, and the inferior olivary nuclei. Source: Smith (2002), figure 11.5, p. 268. Reproduced with permission of Elsevier. Nolte (2001), figure 11.5, p. 268. Reproduced with permission of Elsevier.

tracts, all of which occupy the anterior region of the brainstem. It should be noted that all tracts are somatotopically organized.

The brainstem will be discussed in ascending order, that is the medulla oblongata first, then the pons, and finally the midbrain.

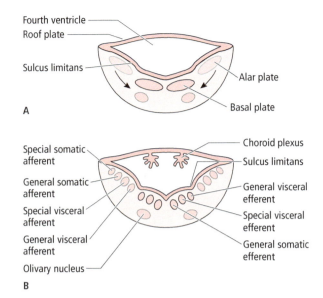

Figure 7.4 • (A) Transverse section of the myelencephalon of an early embryo depicting the alar and basal plates. The arrows indicate the migration of cells from the alar plates to form the future olivary nuclei. (B) Transverse section of a later stage of the myelencephalon depicting the development of nuclei from the alar and basal plates.

Embryonic derivation of cranial nerve nuclei

The embryological development of the cranial nerve nuclei provides an explanation of their organization, including their relative location and functional composition in the adult brainstem

The nuclei of cranial nerves III–X and of XII reside in the brainstem. Due to the complex nature of the cranial nerves, tracing the embryological development of their nuclei provides an explanation of their organization, including their relative location and functional composition in the adult brainstem. During the early development of the nervous system (described in more detail in Chapter 2), the cells of the neural tube become arranged into longitudinal cellular columns, the **basal plates** anteriorly, and the **alar plates** posteriorly, with the two plates being separated by the intervening longitudinal **sulcus limitans** (Fig. 7.4A, B).

During the early stages of nervous system development, the future spinal cord and brainstem follow a similar developmental pattern. The neuroblasts of the basal plates differentiate into motor and autonomic neurons (Table 7.1), whereas the neuroblasts of the alar plates differentiate into sensory neurons (Table 7.2). The embryonic arrangement of the cell clusters that develop into the cranial nerve nuclei exhibits a predictable pattern at the level of the future midbrain and medulla oblongata, but less so at the level of the pons (Fig. 7.4A, B). During early development, the lateral walls of the medulla oblongata move laterally, causing the alar plates to shift in a direction that is located lateral to the basal plates. Consequently, the pattern of organization of the cell clusters that will develop into the neurons of the cranial nerve nuclei is as follows: In the **medulla oblongata**, the three cell groups (derived from the basal plates) on each side of the midline and deep to the floor of the fourth ventricle will develop into motor nuclei (Table 7.1), whereas the four cell groups (derived from the alar plates) located posterolaterally will develop into the sensory nuclei. (Table 7.2). From medial to lateral, the arrangement of the **motor nuclei** and the modalities that they carry is as follows (Fig. 7.4A, B):

Table 7.1 • Basal plate derivatives are arranged by modality (fiber type).

Cranial nerve nucleus	Location	Modalities	Function
Oculomotor (CN III)	Rostral midbrain	GSE	Eye movement
Trochlear (CN IV)	Caudal midbrain	GSE	Eye movement
Abducens (CN VI)	Caudal pons	GSE	Eye movement
Hypoglossal (CN XII)	Medulla oblongata	GSE	Tongue movement
Motor nucleus of the trigeminal (CN V)	Midpons	SVE	Chewing, elevation of mandible, movement of mandible from side to side
Facial (CN VII)	Caudal pons	SVE	Facial expression
Nucleus ambiguus (CN IX, X)	Rostral ¾ of medulla oblongata	SVE	Swallowing, phonation
Accessory Oculomotor (Edinger–Westphal) (CN III)	Rostral midbrain	GVE	Pupillary constriction
Superior salivatory (CN VII)	Pons	GVE	Tear formation, salivation
Inferior salivatory (CN IX)	Medulla oblongata	GVE	Salivation
Dorsal motor nucleus of the vagus (CN X)	Rostral medulla oblongata	GVE	Stimulation of peristalsis and laryngeal, thoracic, and abdominal gland secretion; decrease of suprarenal gland secretion and heart rate

BRAINSTEM • • • 95

Table 7.2 • Alar plate derivatives are arranged by modality (fiber type).

Cranial nerve nucleus	Location	Modality	Function
Nucleus of solitary tract (Solitary nucleus) (caudal half) (CN VII, IX, X)	Medulla oblongata	GVA	Visceral sensation from carotid body and sinus, pharynx, tonsil, ear, posterior tongue
Nucleus of solitary tract (Solitary nucleus) (rostral half) (CN VII, IX, X)	Medulla oblongata	SVA	Taste
Mesencephalic nucleus of the trigeminal (CN V)	Midbrain – midpons	GP	Proprioception from muscles of mastication
Principal (main, chief) nucleus of the trigeminal (CN V)	Pons	GSA	Discriminative touch from orofacial region General proprioception from the extraocular muscles and TMJ
Spinal nucleus of the trigeminal (CN V)	Midpons – upper 2 cervical spinal cord levels	GSA	Pain, temperature, tactile sensation from orofacial region, nasal cavity, tonsil, pharynx, ear
Vestibular (CN VIII)	Rostral medulla oblongata Caudal pons	SSA	Balance
Cochlear (CN VIII)	Rostral medulla oblongata Caudal pons	SSA	Hearing

- general somatic efferent (GSE) – provide motor innervation to skeletal muscles derived from somites
- special visceral efferent (SVE) – provide motor innervation to skeletal muscles derived from the pharyngeal arches
- general visceral efferent (GVE) – provide visceral motor innervation to smooth muscle, cardiac muscle, or myoepithelial cells in the glands.

Lateral to these cell groups are the **sensory nuclei** derived from the alar plates and are arranged from medial to lateral as follows:

- general visceral afferent (GVA) – receive sensory input from the viscera
- special visceral afferent (SVA) – receive taste and olfactory fibers
- general somatic afferent (GSA) – receive sensory input from the superficial structures of the head
- special somatic afferent (SSA) – receive sensory input from the special senses of hearing and balance.

The cranial nerve sensory nuclei are collections of nerve cell bodies that receive sensory input from peripheral structures via the sensory branches of the cranial nerves that enter the brainstem. The cranial nerve motor nuclei are collections of motor neurons whose axons emerge from the nuclei, and course in the motor branches of the cranial nerves to reach their targets, providing them with motor innervation. The cranial nerve nuclei are positioned posteriorly in the floor of the ventricular system, but their roots emerge anterolaterally (except for the trochlear nerve, CN IV) which emerges from the posterior surface of the brainstem).

None of the cranial nerves carry all seven modalities (listed above).

Cranial nerves III, IV, VI, and XII have GSE nuclei. These cranial nerves are referred to as "somatic efferent nerves" and innervate the extraocular muscles (CN III, IV, VI) and the muscles of the tongue (CN XII), all of which are derived from somites. Their GSE motor nuclei are located most medially in a rostrocaudal direction of the brainstem.

Cranial nerves V, VII, IX, and X have SVE nuclei. They are referred to as the "branchiomeric motor nerves" and innervate muscles derived from the pharyngeal arches. Their SVE motor nuclei are located immediately lateral to the GSE nuclei.

Cranial nerves III, VII, IX, and X have GVE nuclei. They provide parasympathetic innervation to smooth muscle, cardiac muscle, as well as secretory and myoepithelial cells in the glands of visceral structures located in the head, neck, thorax, and abdomen. Their GVE nuclei are located immediately lateral to the SVE nuclei.

The cranial nerve sensory nuclei include the following functional components: GVA, SVA, GSA, and SSA.

Cranial nerves VII, IX, and X share the *caudal (inferior)* half of the nucleus of solitary tract (solitary nucleus) by sending the central processes of some of their sensory neurons to this nucleus. The nucleus of solitary tract (solitary nucleus), located in the medulla oblongata, contains GVA neurons in its caudal half that relay GVA input to the thalamus. Cranial nerves VII, IX, and X also share the *rostral* half of the nucleus of the solitary tract (solitary nucleus), which contains SVA neurons that relay taste to the ventral posterior nucleus of the thalamus.

The trigeminal nerve has three sensory nuclei that relay sensory input from the orofacial region: (1) the mesencephalic nucleus of the trigeminal, a general proprioception (GP) nucleus that relays proprioception (position sense of the mandible) from the muscles of mastication, and periodontal ligaments of the teeth (2) the principal (main, chief) sensory nucleus of the trigeminal, a GSA nucleus that relays discriminative touch, from the orofacial region and general

96 ••• CHAPTER 7

proprioception from the extraocular muscles and temporomandibular joint (TMJ), and (3) the spinal nucleus of the trigeminal, a GSA nucleus that relays pain, temperature, and tactile sensation that is transmitted by the branches of the trigeminal, facial, glossopharyngeal, and vagus nerves.

The vestibulocochlear nerve (CN VIII) has SSA nuclei. Four vestibular nuclei that process balance and two cochlear nuclei that process hearing.

MEDULLA OBLONGATA

The medulla oblongata consists of a caudal ("closed") portion and a rostral ("open") portion based on whether or not the lower part of the fourth ventricle extends into its posterior surface

The **medulla oblongata**, forming the caudal third of the brainstem, extends superiorly from the pyramidal decussation to the pontine sulcus. Viewing the fourth ventricle as a point of reference, the medulla oblongata consists of a caudal ("closed") portion and a rostral ("open") portion based on whether or not the lower part of the fourth ventricle extends into its posterior surface. The central canal of the spinal cord continuing superiorly from the spinal cord into the brainstem is embedded on the posterior aspect of the "closed" medulla oblongata, whereas approximately the lower third of the fourth ventricle is located on the posterior aspect of the "open" medulla oblongata. Three cranial nerves, namely the glossopharyngeal, vagus, and hypoglossal, emerge from the anterior or lateral surface of the medulla oblongata. The medulla oblongata also contains a segment of the spinal nucleus of the trigeminal nerve, and the nuclei of cranial nerves VIII through XII. As the **spinal nucleus of the trigeminal nerve** extends from the mid-pons caudally to the upper two cervical spinal cord levels, it occupies the lateral aspect of the entire length of the medulla oblongata with the same named tract descending on its lateral surface. The medulla oblongata contains a central core composed of numerous nuclei embedded in a complex fiber network, known as the **reticular formation** (Chapter 16). Furthermore, the medulla oblongata contains autonomic centers which control respiratory, cardiovascular, and gastrointestinal functions. It is divided posteriorly into the **tegmentum** and anteriorly into the **base** of the **medulla oblongata**. The following structures are located in the tegmentum *at all levels of the medulla oblongata*: the *medial longitudinal fasciculus (MLF), tectospinal tract, reticulospinal tracts, rubrospinal tract, spinal tract* and *nucleus of the trigeminal nerve, spinothalamic tract*, and the *anterior spinocerebellar tract*. The caudal half of the medulla oblongata contains two decussations which serve as distinctive landmarks of the caudal medulla oblongata, namely the *pyramidal (motor) decussation* situated most caudally at the medullo-spinal border and the *(sensory) decussation of the medial lemniscus (ML)* located rostral to it.

Structures present at all levels of the medulla oblongata

The **medial longitudinal fasciculus (MLF)** is a prominent, heavily myelinated fascicle, which consists of an ascending and a descending portion. The ascending component will be discussed in the appropriate sections on the pons and midbrain later, whereas the descending portion, which arises from the *medial vestibular nucleus* in the medulla oblongata and descends to the spinal cord, will be discussed here. The **descending MLF** is located near the midline and retains its position in its entire course within the medulla oblongata, as it descends to the spinal cord. In a cross-section of the upper three-fourths of the medulla oblongata, it appears to form a cap over the posterior aspect of the **medial lemniscus (ML)**. In both caudal and rostral medullary levels, the descending MLF contains the heavily myelinated axons of the **medial vestibulospinal tract**, which is involved in stabilizing head orientation.

The **tectospinal tract**, a narrow tract arising from the superior colliculus of the midbrain, accompanies the MLF near the midline of the medulla oblongata in its descent to the spinal cord. This tract is involved in the reflex turning of the head and upper trunk in the direction of startling cutaneous, auditory, or visual stimuli.

The medial (pontine) and lateral (medullary) **reticulospinal tracts** descend to the spinal cord where they modulate muscle tone and posture maintenance. Because these tracts originate from nuclei that belong to the reticular formation, they are obscured by the multitude of fibers surrounding them as they descend through the medullary reticular formation that forms the core of the medulla oblongata. The **medial (pontine) reticulospinal tract** terminates at all spinal cord levels where it has an excitatory influence on the motor neurons that innervate the extensor (antigravity) muscles of the neck, trunk, and lower limbs, in order to modulate muscle tone and control posture (whether the individual is standing or walking). The **lateral (medullary) reticulospinal tract** also terminates at all spinal cord levels, but it has an inhibitory influence on the motor neurons that innervate the extensor muscles (as when the individual is lying down).

The **rubrospinal tract** arises from the red nucleus of the midbrain and descends to the cervical spinal cord where it synapses with and modulates the motor neurons that innervate the upper limb flexor muscles of the forearm, hand, and fingers. In the medulla oblongata, it descends in the lateral tegmentum just medial to the spinocerebellar tracts and posterior to the spinothalamic tract.

Another set of structures present throughout the entire length of the medulla oblongata are the **spinal tract** and **nucleus** of the **trigeminal nerve** (discussed in more detail in the section on the pons of this chapter and in Chapter 17). Trigeminal fibers relaying tactile, nociceptive, and thermal input from the orofacial region enter the brainstem at the midpontine level and then descend in the spinal tract of the trigeminal to synapse in the spinal nucleus of the trigeminal. Note that since the facial, glossopharyngeal, and vagus nerves do not have a nucleus that processes nociception arising from the structures they innervate, they contribute their nociceptive fibers to the spinal tract of the trigeminal and terminate in the spinal nucleus of the trigeminal for further processing. Both tract and nucleus extend from the midpons inferiorly through the medulla oblongata to the upper two cervical spinal cord levels. The nucleus is continuous inferiorly with the substantia gelatinosa of the spinal cord, which is involved in the processing of nociception from the body. Axons that leave the spinal nucleus of the trigeminal form the **anterior trigeminothalamic tract**, which relays nociceptive, thermal, and tactile sensation from structures located in the anterior and lateral parts of the head to the ventral posteromedial (VPM) nucleus of the thalamus.

The **spinothalamic tract** (a component of the **anterolateral system**) transmits nociceptive, thermal, and some non-discriminative (light, crude) touch sensation from the contralateral side of the body to the ventral posterolateral (VPL) nucleus of the thalamus. This tract, located in the lateral medullary tegmentum, maintains its lateral position as it ascends from the spinal cord lateral funiculus. The **anterior spinocerebellar tract** relays unconscious position sense from the lower limb and lower trunk to the cerebellum.

Some of the tracts discussed here (the spinal tract and nucleus of the trigeminal, the spinothalamic, anterior spinocerebellar, and the rubrospinal tracts) occupy the lateral aspect of the medullary tegmentum, midway between the posterior and anterior aspects of the medulla oblongata. In addition to the list of structures discussed here, which are located in the *entire* length of the medulla oblongata, each of the four regions of the medulla oblongata discussed next, contain other tracts and nuclei, some of which are unique to a specific medullary level. They are now discussed in the corresponding region that they occupy.

Caudal (closed) medulla oblongata at the motor decussation

The pyramidal (motor) decussation (crossing of the corticospinal tracts) serves as the salient morphological characteristic of the caudal-most medulla oblongata, at the medullo-spinal junction

In addition to the structures listed previously, the *posterior spinocerebellar tract* is also located in the medullary tegmentum at the level of the motor decussation. At this level, the *gracile fasciculus* and *nucleus*, and the *cuneate fasciculus* and *nucleus*, are located posterior to the medullary tegmentum, whereas the *pyramidal (motor) decussation* is located anterior to the tegmentum (Fig. 7.5).

As the **gracile fasciculus** and **cuneate fasciculus** ascend in the posterior column of the spinal cord and enter the caudal medulla oblongata, they retain their posterior location by ascending posterior to the medullary tegmentum. Their fibers continue to ascend until they terminate and synapse in their respective nuclei, the **gracile nucleus** and **cuneate nucleus**. The gracile nucleus is present throughout the length of the

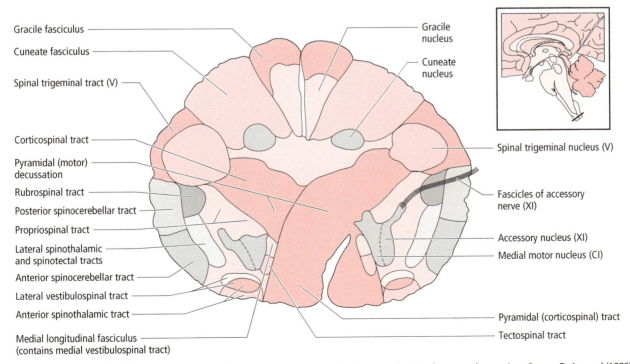

Figure 7.5 • Cross section of the brainstem at the level of the caudal (closed) medulla oblongata, showing the motor decussation. Source: DeArmond (1989) *The Structure of the Human Brain*, 3rd edn/Oxford University Press, figure 40, p. 80.

98 ••• **CHAPTER 7**

caudal (closed) medulla oblongata. In contrast, the cuneate nucleus begins to appear and is perceptible as a small pale structure deep to the cuneate fasciculus. The heavily myelinated, gracile, and cuneate fasciculi are located at the posterior aspect of the medulla oblongata. The pale gracile nucleus and cuneate nucleus are nestled just deep to their respective fasciculi. The gracile fasciculus and nucleus are located medially, close to the midline, whereas the cuneate fasciculus and nucleus are located immediately lateral to them. The gracile fasciculus carries somatosensory information (discriminative touch, vibratory sense, and conscious proprioception) from the ipsilateral lower limb and lower half of the trunk to its nucleus. The cuneate fasciculus relays the same type of somatosensory information but from the ipsilateral upper half of the trunk and upper limb to its nucleus.

The **posterior spinocerebellar tract** arises from the spinal cord, ascends to the medulla oblongata, and then passes into the cerebellum via the restiform body of the inferior cerebellar peduncle to relay unconscious proprioception (position sense sensory information) from the lower limb and lower trunk to the cerebellum, where it plays a role in coordination of movement.

Anterior to the tegmentum is the **pyramidal (motor) decussation** (crossing of the corticospinal tracts), which occupies almost the entire anterior half of the medulla oblongata and serves as the salient morphological characteristic of the caudal-most medulla oblongata, at the medullo-spinal junction. This is in contrast to the morphologic characteristic of the remaining, more rostral region of the medulla oblongata, where the pyramids (containing the corticospinal tracts) reside in the anterior medulla oblongata. Axons arising primarily from the motor cortex form the corticospinal tract, which descends in the corona radiata and the internal capsule of the cerebral hemispheres. This tract descends in the midbrain's crus cerebri, the basilar pons, and the anterior medulla oblongata. When these fibers reach the medulla oblongata they form the **pyramid**, a prominence on its anteromedial aspect. When the axons of the corticospinal tract reach the most caudal part of the medulla oblongata, the majority of these axons shift to a posterolateral position as they cross the midline in the **pyramidal decussation** to the opposite side and then descend in the lateral funiculus of the spinal cord as the *lateral corticospinal tract*. The remaining uncrossed axons descend ipsilaterally to the spinal cord in the anterior funiculus, as the *anterior corticospinal tract*. The lateral corticospinal tract is involved in the voluntary movement of the trunk and limbs. The anterior corticospinal tract controls movement of the neck, shoulder, and trunk. These tracts are discussed in more detail in Chapter 13.

Caudal (closed) medulla oblongata at the sensory decussation

The sensory decussation occurs in the caudal medulla oblongata to form the medial lemniscus (ML)

The structures located posterior to the medullary tegmentum at the level of the sensory decussation include the *gracile fasciculus* and *nucleus* and the *cuneate fasciculus* and *nucleus*. The structures located in the medullary tegmentum include the *solitary tract* and *nucleus*, *dorsal motor nucleus of the vagus*, *accessory cuneate nucleus* and its *cuneocerebellar tract*, *hypoglossal nucleus* and its *nerve rootlets*, *nucleus ambiguus*, *internal arcuate fibers*, *decussation of the medial lemniscus (ML)*, *posterior spinocerebellar tract*, and *central tegmental tract*. Anterior to the medullary tegmentum, both located in the base of the medulla oblongata, are the *inferior olivary nucleus* and the *pyramid* (Figs. 7.6 and 7.7).

The **gracile** and **cuneate fasciculi** continue their ascent in the posterior medulla oblongata where their ascending fibers continue to terminate and synapse in their respective nuclei. The **gracile nucleus** and **cuneate nucleus** form the **gracile** and **cuneate tubercles**, two small prominences on the posterior aspect of the caudal medulla oblongata just inferior to the inferior border of the fourth ventricle. The gracile nucleus is located medial and slightly inferior to the level of the cuneate nucleus. The gracile nucleus and cuneate nucleus contain the cell bodies of second-order neurons that receive synaptic contacts from the axons ascending in the gracile fasciculus and cuneate fasciculus, respectively. Since the gracile nucleus is located at a slightly lower level than the cuneate nucleus, synaptic contacts between ascending terminals and second-order neurons begin in the gracile nucleus first (at a lower level) followed by those occurring in the cuneate nucleus (at a higher level). The second-order neuron axons that exit from the nuclei emerge first from the gracile nucleus. As the axons emerging from the gracile and cuneate nuclei sweep anteromedially in the medullary tegmentum, they form an arc of ripple-like fibers from medial to lateral, referred to as the **internal arcuate fibers**. These fibers converge to decussate across the midline to the opposite side, and when they do so, they form a flat, ribbon-like ascending tract the **medial lemniscus** (**ML**: L., *lemniscus*, "ribbon") located next to the midline of the brainstem. This shift of the internal arcuate fibers from one side of the medulla oblongata to the opposite side is referred to as the **sensory decussation**, a rather conspicuous feature at this level of the medulla oblongata. Note that the sensory decussation occurs rostral to (above) the motor decussation. Viewed in a cross-section of the caudal medulla oblongata, the ML is located next to the midline, posterior to the pyramid from the site of its origin to the upper medulla oblongata. The ML ascends through the rostral medulla oblongata, and then through the pons and midbrain to the diencephalon where it terminates in the ventral posterolateral (VPL) nucleus of the thalamus relaying discriminative touch, vibratory sense, and conscious proprioception from the opposite side of the body. This pathway (Fig. 7.7) is discussed in more detail in Chapter 12. Anterior to the gracile nucleus are the solitary tract and nucleus as well as the dorsal motor nucleus of the vagus. The **solitary tract** consists of the central processes of pseudounipolar neurons whose cell bodies reside in the sensory ganglia of the facial, glossopharyngeal, and vagus nerves. These fibers terminate in the **nucleus of the solitary tract (solitary nucleus)** located alongside the tract. The rostral half of the nucleus of the solitary tract (solitary nucleus) receives taste

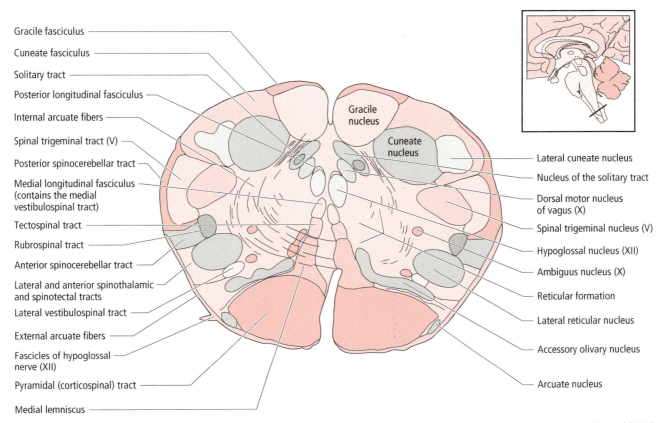

Figure 7.6 • Cross section of the brainstem at the level of the caudal (closed) medulla oblongata, showing the sensory decussation. Source: DeArmond (1989) *The Structure of the Human Brain*, 3rd edn/Oxford University Press, figure 41, p. 82.

sensations from the tongue, epiglottis, and adjacent pharyngeal wall, whereas its caudal half receives visceral sensation from the carotid body and sinus, pharynx, tonsils, ear, and posterior tongue. The **dorsal motor nucleus of the vagus** is the most prominent parasympathetic nucleus of the brain. It contains the cell bodies of *preganglionic parasympathetic* neurons. Their axons leave the nucleus and join the other fibers of the vagus nerve whose branches distribute them to parasympathetic ganglia located near or within the wall of thoracic, abdominal, and pelvic organs. The parasympathetic fibers regulate the contraction of smooth and cardiac muscles and the secretory activity of glands.

The **accessory cuneate nucleus**, which is located immediately lateral to the cuneate nucleus, gives rise to the **cuneocerebellar tract** that enters the ipsilateral cerebellum via the **restiform body** of the inferior cerebellar peduncle. The restiform body consists of proprioceptive fibers relaying unconscious position sense from the head and body to the cerebellum. The cuneocerebellar tract relays unconscious proprioception from the upper limb and upper trunk to the ipsilateral cerebellum.

The **hypoglossal nucleus** is close to the midline. It contains the cell bodies of motor neurons whose axons leave the nucleus to form multiple fascicles that course anteriorly to emerge from the anterior aspect of the medulla oblongata in about six distinct rootlets in the sulcus between the pyramid and the olive. The rootlets then gather to form the hypoglossal nerve that courses to the tongue to provide its intrinsic and extrinsic muscles with motor innervation.

The **nucleus ambiguus** is located lateral to the ML and posterior to the inferior olivary nucleus, approximately midway between the posterolateral margin of the inferior olivary nucleus and the hypoglossal nucleus. Since this nucleus provides motor innervation to muscles controlled by two cranial nerves, it is "shared" by these nerves. The nucleus ambiguus contains the cell bodies of motor neurons whose axons run in the branches of the glossopharyngeal and vagus nerves to distribute to skeletal muscles supplied by each of these nerves, including the stylopharyngeus muscle and the muscles of the pharynx and larynx, which are involved in swallowing and speaking, respectively.

The posterior spinocerebellar tract and the central tegmental tract are located on the anterolateral aspect of the tegmentum. The **posterior spinocerebellar tract** carries unconscious proprioceptive information from the lower limb and lower trunk to the cerebellum. At this level, the **central tegmental tract** carries two groups of fibers: one group of fibers that relays taste sensation from the nucleus of the solitary tract (solitary nucleus) ascends to the thalamus. The other group of fibers (rubroolivary fibers) descends from the red nucleus to

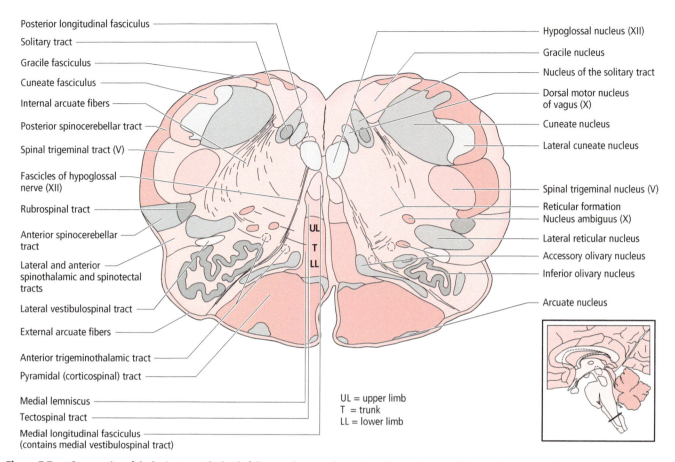

Figure 7.7 • Cross section of the brainstem at the level of the rostral sensory decussation. Source: DeArmond (1989) *The Structure of the Human Brain*, 3rd edn/Oxford University Press, figure 42, p. 84.

terminate in the inferior olivary nucleus. Both the red nucleus and the inferior olivary nucleus are involved in movement.

The two most conspicuous structures located anterior to the tegmentum in the anterior aspect (base) of the medulla oblongata are the inferior olivary nucleus and the pyramid. The **inferior olivary nucleus**, whose unusual appearance resembling a "crumpled bag," is located posterior to the pyramid and lateral to the ML. It has a hilus (G. "lip") surrounding what appears as the open part of the nucleus facing the ML through which fibers enter or leave the nucleus. The fibers terminating in the cerebellum are referred to as "climbing fibers." The inferior olivary nucleus receives afferent (input) *rubroolivary fibers* from the red nucleus and projects efferent (output) *olivocerebellar fibers* to the contralateral cerebellar cortex and cerebellar nuclei via the inferior cerebellar peduncle. The inferior olivary nucleus forms an oval swelling, the **olive** on the anterior surface of the medulla oblongata. At this medullary level, the **pyramid** contains the uncrossed corticospinal tract.

Note that the *motor* decussation is located in the caudal-most level of the medulla oblongata at the medullo-spinal junction. In contrast, the *sensory* decussation is located rostral to (above) the motor decussation.

Medulla oblongata at mid-olivary level

The following structures are located in the medullary tegmentum, at the mid-olivary level: *hypoglossal nucleus, dorsal motor nucleus of the vagus, solitary tract and nucleus, medial and inferior vestibular nuclei, inferior cerebellar peduncle, nucleus ambiguus,* the *ML*, and *central tegmental tract*. The structures anterior to the tegmentum in the anterior aspect (base) of the medulla oblongata include the *hypoglossal nerve rootlets* and the prominent *inferior olivary nucleus* and *pyramid*. To avoid redundancy, those structures that are present at the level of the sensory decussation and at the mid-olivary level that were discussed in the previous section are not mentioned in the discussion that follows, but the student should be aware of their presence. Only those new structures appearing at the mid-olivary level are discussed here (Fig. 7.8). The **medial vestibular nucleus** is the largest of the vestibular nuclei located in the posterolateral aspect of the lower part of the rostral medulla oblongata receives the central processes of the vestibular nerve relaying sensory information about the orientation and motion of the head. This nucleus is the origin of the **medial vestibulospinal tract** whose fibers become incorporated into the *descending* portion of the medial longitudinal fasciculus

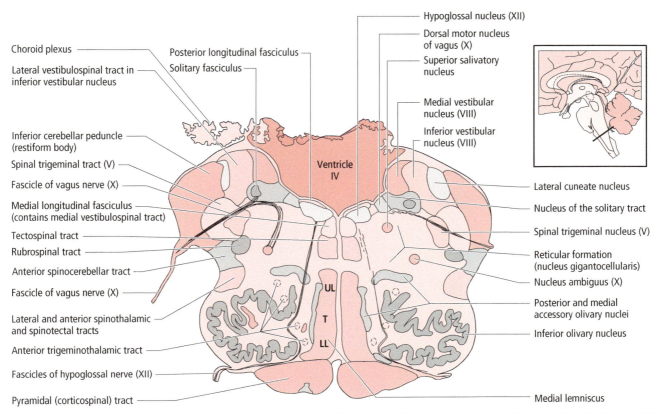

Figure 7.8 ● Cross section of the brainstem at the level of the rostral (open) medulla oblongata. Fibers transmitting sensory information from the UL (upper limb), T (trunk), LL (lower limb). Source: DeArmond (1989) *The Structure of the Human Brain*, 3rd edn/Oxford University Press, figure 44, p. 88.

(MLF) to terminate at cervical spinal cord levels where these fibers are involved in stabilizing head orientation. This nucleus also sends fibers via the *ascending* portion of the MLF to the cranial nerve motor nuclei and accessory nuclei that control eye movements to facilitate the coordination of eye movements in response to head movement.

The **inferior vestibular nucleus** located lateral to the medial vestibular nucleus in the lower part of the rostral medulla oblongata receives most of the first-order vestibular fibers bringing vestibular sensory input from the inner ear. It projects to the reticular formation and the cerebellum. Sensory input to the inferior vestibular nucleus plays a role in reflex activity related to muscle tone and upright posture.

The **inferior cerebellar peduncle** anchors the cerebellum to the medulla oblongata. It is a thick bundle of white matter consisting of heavily myelinated axons most of which arise in the spinal cord and brainstem to terminate in the cerebellum. It transmits proprioceptive fibers and vestibular fibers and consists of two divisions: the restiform body and the juxtarestiform body. The **restiform body** includes proprioceptive afferent fibers from the spinal cord and brainstem to the cerebellum. The **juxtarestiform body** next to it is composed of fibers that interconnect the vestibular system and the cerebellum and are involved in balance.

The **fascicles of the hypoglossal nerve** continue to emerge from the sulcus between the olive and the pyramid at the anterior surface of the medulla oblongata.

Rostral medulla oblongata

The rostral medulla oblongata houses part of the fourth ventricle in its posterior aspect

The structures located in the rostral medullary tegmentum include the *medial* and *inferior vestibular nuclei, posterior* and *anterior cochlear nuclei, inferior cerebellar peduncle, nucleus ambiguus, nucleus of the solitary tract (solitary nucleus), solitary tract, inferior salivatory nucleus,* and *ML*. Anteriorly, the structures in the rostral medulla oblongata include the *central tegmental tract,* the prominent *inferior olivary nucleus,* and the *pyramid* (Figs. 7.8 and 7.9).

The **posterior** and **anterior cochlear nuclei** are located in the posterolateral aspect of the rostral medulla oblongata. They receive the central processes of sensory neurons residing in the cochlear ganglion of the inner ear. These axon terminals relay auditory information from the cochlea, the sensory organ for hearing. The posterior cochlear nucleus relays monaural (from one ear) information to the inferior colliculus via the lateral lemniscus. The anterior cochlear nucleus projects bilaterally to the superior olivary nuclear complex which in turn projects binaural (from both ears) input to the inferior colliculus via the lateral lemniscus.

The **inferior salivatory nucleus** is a parasympathetic nucleus that encloses the cell bodies of *preganglionic parasympathetic* neurons. The axons of these neurons join the glossopharyngeal nerve which distributes them to the otic

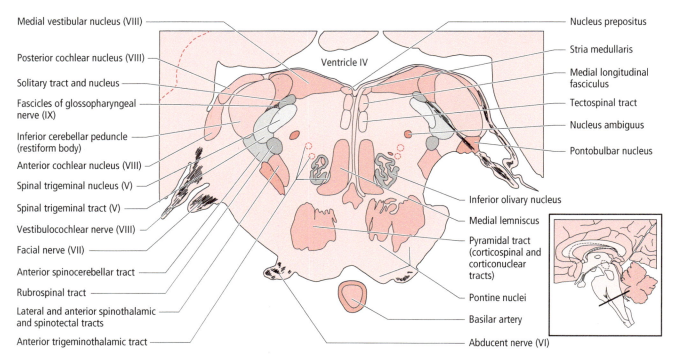

Figure 7.9 ● Cross section of the brainstem at the level of the pontomedullary border. Source: DeArmond (1989) *The Structure of the Human Brain*, 3rd edn/Oxford University Press, figure 45, p. 90.

ganglion. There, they synapse with postganglionic parasympathetic neurons whose axons terminate in the parotid gland to stimulate salivation.

The **medial lemniscus** (**ML**) ascends as a prominent tract in the medullary tegmentum, adjacent to the midline. This was discussed in more detail in the caudal ("closed") medulla oblongata at the sensory decussation section previously and is detailed in Chapter 12.

The **medial longitudinal fasciculus** (**MLF**), a prominent, heavily myelinated fascicle, located near the midline retains its position along its entire course in the brainstem, as it proceeds to the midbrain. Viewed in cross section of the upper three fourths of the medulla oblongata, it appears to form a cap over the posterior aspect of the ML. In both caudal and rostral medullary levels, the MLF contains the heavily myelinated axons of the **medial vestibulospinal tract** which is involved in stabilizing head orientation and coordination of head and eye movements.

PONS

The pons contains nuclei and tracts associated with the trigeminal, abducens, facial, and vestibulocochlear nerves

The pons (L., "bridge") appears to form a bridge between the two cerebellar hemispheres. Posteriorly, the pons consists of the more complex **pontine tegmentum** and anteriorly of the simpler **basis pontis**. The tegmentum forms part of the floor of the fourth ventricle and contains nuclei and tracts associated with the neighboring cranial nerves (*trigeminal, abducens, facial,* and *vestibulocochlear nerves*). Also in the pontine tegmentum are *reticular nuclei*, embedded in the complex fiber network of the reticular formation, and the major *ascending sensory pathways*. Due to the complex structure of the pons, it is divided, for descriptive purposes, into **rostral**, **mid-**, and **caudal pontine levels**. The structures residing in the posteriorly located pontine tegmentum and the anteriorly located basis pontis will be discussed for each level. Structures in the pons include cranial nerves, their nuclei and pathways, the cerebellar pathways traversing the pons, and the long ascending sensory and descending motor pathways.

Caudal pons

The caudal pons is the origin of the abducens, facial, and vestibulocochlear nerves

Structures located in the caudal pons include the *abducens nucleus* and *nerve*, the *corticonuclear tract*, the *facial nucleus* and *nerve*, the *internal genu of the facial nerve*, the *facial colliculus*, the *superior salivatory nucleus*, the *MLF*, the *superior and lateral vestibular nuclei*, the *juxtarestiform body (JRB)*, the *cochlear nuclei*, the *solitary tract and nucleus*, the *spinal tract and nucleus of the trigeminal*, the *anterolateral system (ALS)*, the *lateral lemniscus (LL)*, the *pontine nuclei*, and the *corticospinal tract* (Fig. 7.10).

The **abducens nucleus** is located in the caudal pons, deep to the floor of the fourth ventricle, adjacent to the midline. It receives cortical fibers from the contralateral frontal eye field (Brodmann's area 8) relaying motor information to the

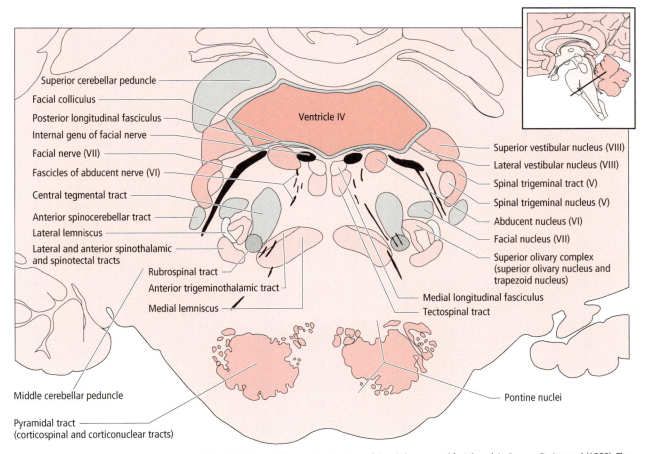

Figure 7.10 • Cross section of the brainstem at the level of the caudal pons, the location of the abducens, and facial nuclei. Source: DeArmond (1989) *The Structure of the Human Brain*, 3rd edn/Oxford University Press, figure 48, p. 96.

neurons of this nucleus. The abducens nucleus contains two distinct populations of neurons: *motor neurons* and *interneurons*. The axons of **motor neurons** exit the nucleus to form root fascicles of the **abducens nerve** in the tegmentum. These root fascicles course anteriorly in the pontine tegmentum and travel lateral to the corticospinal tract fibers in the basis pontis. They emerge at the pontomedullary junction on the anterior surface of the brainstem. The abducens nerve then courses to the orbit where it provides motor innervation to a single extraocular muscle, the lateral rectus muscle which mediates abduction (lateral rotation) of the eye. The interneurons connect the abducens nucleus to the contralateral oculomotor nucleus. The **interneurons** give rise to axons that leave the nucleus and immediately cross the midline to join the contralateral *ascending* MLF to reach the rostral midbrain where they terminate in the *contralateral oculomotor nucleus* and synapse specifically with the motor neurons that innervate the *medial rectus muscle*. The abducens nucleus is the **center for conjugate horizontal eye movement**; thus, simultaneous stimulation of both populations of neurons of this nucleus mediates abduction of the ipsilateral eye (via the abducens nerve) and adduction of the contralateral eye (via the interneurons and oculomotor nucleus) to execute lateral gaze ipsilateral to the side of the stimulated abducens nucleus.

In caudal pontine levels, the fibers of the descending **corticonuclear tract** terminate in the motor nucleus of the facial nerve. The corticonuclear fibers relay motor messages from the motor cortex to the facial nucleus, which controls the muscles of facial expression.

The **facial nucleus** contains motor neurons whose axons leave the nucleus, form the root of the facial nerve which courses posteromedially and inferiorly toward the floor of the fourth ventricle, to form a sharp loop (the **internal genu of the facial nerve**) over the abducens nucleus located deep to the floor of the fourth ventricle. The root fascicles then course ventrolaterally to emerge from the brainstem at the cerebellopontine angle as the **facial nerve proper**, the motor root of the facial nerve. The abducens nucleus and the overlying looping fascicles of the facial nerve form an elevation, the **facial colliculus** on the floor of the fourth ventricle. The facial nucleus provides motor innervation to the muscles of facial expression as well as to a few other muscles (stylohyoid, posterior belly of the digastric, stapedius, auricularis, and frontal belly of the occipitofrontalis muscles).

The **superior salivatory nucleus**, a parasympathetic nucleus of the facial nerve, is located anterolateral to the abducens nucleus. It contains the cell bodies of *preganglionic, parasympathetic* neurons whose axons exit the brainstem as part of the **nervus intermedius** next to the facial nerve proper. These fibers terminate in their target ganglia, the pterygopalatine, and submandibular ganglia where they synapse. The pterygopalatine ganglion provides parasympathetic secretomotor innervation to the lacrimal, nasal, and palatine glands. The submandibular ganglion provides parasympathetic secretomotor innervation to the submandibular and sublingual salivary glands.

The **medial longitudinal fasciculus (MLF)** is located medial to the abducens nucleus, adjacent to the midline, deep to the floor of the fourth ventricle. In its course through the brainstem, the MLF carries fibers that communicate messages from the vestibular nuclei to the cranial nerve nuclei that control eye movements. It is by way of these connections that the vestibular system, in conjunction with the nuclei that control eye movements (motor nuclei and a few accessory nuclei), orchestrates eye movements in response to head movement.

Two of the four **vestibular nuclei** are located in the posterolateral tegmentum of the caudal pons deep to the lateral floor of the fourth ventricle. They are the superior and lateral vestibular nuclei. The superior vestibular nucleus is located entirely in the caudal pons, whereas the lateral vestibular nucleus also extends from the caudal pons into the rostral medulla oblongata. The **superior vestibular nucleus** sends projections via the MLF to the motor and accessory nuclei that control the extraocular muscles. The **lateral vestibular nucleus** is the origin of the *lateral vestibulospinal tract* via which it plays a role in modulation of muscle tone, posture maintenance, and balance by facilitating the extensor (antigravity) muscles of the neck, trunk, and lower limb.

The **juxtarestiform body (JRB)** consists mostly of afferent fibers to the cerebellum but also possesses some efferent fibers that leave the cerebellum. The afferent fibers arise from the vestibular system and course in the JRB on their way to the vestibulocerebellum. The efferent fibers are projected back to the vestibular nuclei. These interconnections play a role in balance maintenance.

The **vestibulocochlear nerve** consists of two anatomically and functionally distinct nerves, the **vestibular nerve** and **cochlear nerve** that are enclosed in the same connective tissue sleeve to terminate in sensory structures for hearing and balance located in the internal ear. The vestibular nerve is involved in equilibrium and the cochlear nerve mediates hearing.

The **vestibular nerve** relays sensory information from the inner ear concerning the orientation and motion of the head. Sensory information from the vestibular apparatus of the internal ear is relayed to the **superior** and **lateral vestibular nuclei** in the pontine tegmentum and cerebellum. Via their projections to the oculomotor, trochlear, and abducens nuclei, the vestibular nuclei are involved in the coordination of eye movements in response to head movements. Furthermore, the vestibular nuclei via the *vestibulospinal tracts* control posture and balance. The vestibular nerve is the only cranial nerve that projects fibers to the flocculonodular lobe of the cerebellum, which is responsible for balance maintenance.

The **cochlear nerve** brings auditory information from the cochlea of the ear to the ipsilateral **posterior** and **anterior cochlear nuclei** located in the posterolateral aspect of the pontine tegmentum. The posterior cochlear nucleus projects its fibers via the lateral lemniscus directly to the inferior colliculus. The anterior cochlear nucleus projects its horizontally oriented fibers forming the **trapezoid body** (commissural fibers) in the pontine tegmentum to the **nuclei of the trapezoid body** and the **superior olivary nucleus**. Fibers from these nuclei ascend in the **lateral lemniscus** to the inferior colliculus. The lateral lemniscus, the principal ascending auditory pathway, begins in the pons and extends to the inferior colliculus of the midbrain. The cochlear nerve and cochlear nuclei together process auditory information from the ipsilateral ear. Nuclei and pathways distal to the cochlear nuclei process binaural auditory information. The superior olivary nucleus, a pontine relay nucleus of the auditory system, is involved in the localization of sound.

The **solitary tract**, which relays taste sensation from the tongue, epiglottis, and adjacent pharyngeal wall to the nucleus of the solitary tract (solitary nucleus) for further processing, and **nucleus of the solitary tract** (**solitary nucleus**) are located in the posterolateral tegmentum. The rostral portion of the **solitary tract** and **nucleus** reside anterior to the vestibular nuclei.

The **spinothalamic tracts** of the anterolateral system (ALS) and the **spinal tract** and **nucleus of the trigeminal nerve** are located in the anterior and lateral tegmentum, respectively. The spinothalamic tracts relay pain, temperature, and some touch sensation from the body, whereas the spinal tract of the trigeminal relays the same type of sensory information from the face.

The **superior cerebellar peduncle**, located lateral to the fourth ventricle, consists mostly of efferent (cerebellothalamic) fibers from the cerebellum most of which terminate in the thalamus. On their way to the midbrain, these peduncles pass through the posterior aspect of the pons. The **central tegmental tract** is located in the medial tegmentum. The **corticospinal**, **corticonuclear**, and **corticopontine fibers**, the **pontine nuclei**, and **pontocerebellar fibers**, are located in the **basis pontis**.

Midpons

The midpons is the origin of the trigeminal nerve and the location of the middle cerebellar peduncle (brachium pontis)

The salient features of the **midpontine tegmentum** include the *trigeminal nerve and its nuclei: the motor nucleus of the trigeminal, principal (main, chief) nucleus of the trigeminal* and *mesencephalic nucleus of the trigeminal*, and the rostral portion of the *spinal nucleus of the trigeminal nerve*. In addition, other conspicuous structures include the *pontine nuclei*, the *pontocerebellar fibers*, and the prominent *middle cerebellar peduncles* (Fig. 7.11). The trigeminal nerve consisting of motor and sensory roots is the only cranial nerve to emerge from the lateral surface of the pons.

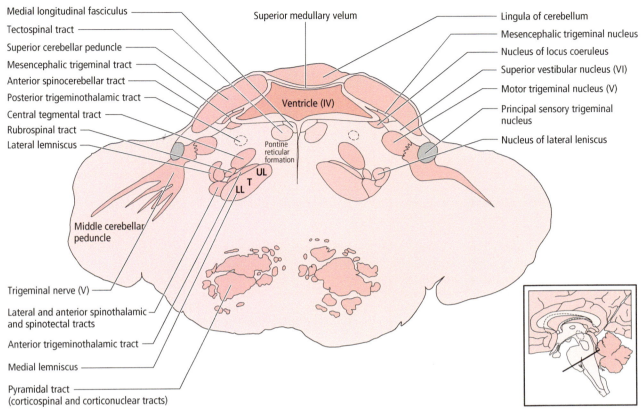

Figure 7.11 • Cross section of the brainstem at the level of the mid-pons, showing the motor nucleus, principal nucleus, and fascicles of the trigeminal nerve. Fibers transmitting sensory input from the UL (upper limb), T (trunk), LL (lower limb). Source: DeArmond (1989) *The Structure of the Human Brain*, 3rd edn/Oxford University Press, figure 49, p. 98.

In mid-pontine levels, some of the fibers of the **corticonuclear tract** terminate in the motor nucleus of the trigeminal nerve. As the corticonuclear tract fibers descend from the motor cortex, this is the first cranial nerve motor nucleus where they synapse. The corticonuclear fibers transmit motor information from the motor cortex to the trigeminal motor nucleus that controls the muscles derived from the first pharyngeal arch.

The **motor nucleus of the trigeminal nerve**, located in the lateral mid-pontine tegmentum, medial to the principal (main, chief) nucleus of the trigeminal nerve, contains the cell bodies of motor neurons whose axons leave the nucleus, course laterally and exit the lateral aspect of the pons as the **motor root** of the trigeminal nerve. This root later joins the mandibular division of the trigeminal nerve to distribute its fibers to the muscles derived from the first pharyngeal arch, namely the muscles of mastication (temporalis, masseter, and medial and lateral pterygoid muscles) and the mylohyoid, anterior belly of the digastric, tensor tympani, and tensor veli palatini, to provide them with motor innervation.

The **principal (main, chief) sensory nucleus of the trigeminal nerve** is located in the lateral pontine tegmentum and processes discriminative touch sensation from the orofacial region and proprioception from the extraocular muscles and TMJ. The central processes of pseudounipolar (unipolar) sensory neurons (whose cell bodies reside in the trigeminal ganglion) relaying sensory information from the orofacial region, enter the pons and terminate in the main sensory nucleus. Neurons of this nucleus give rise to one bundle of axons that form the ipsilateral **posterior trigeminothalamic tract** and another set of axons that join the contralateral **anterior trigeminothalamic tract**. Both trigeminothalamic tracts ascend to terminate in the ventral posteromedial (VPM) nucleus of the thalamus.

The **mesencephalic nucleus** and **tract of the trigeminal nerve** located lateral to the periaqueductal gray (PAG) matter relay proprioception from orofacial structures to the principal and motor nuclei of the trigeminal involved in sensory perception and reflexes, respectively.

The central processes of some of the pseudounipolar (unipolar) neurons whose cell bodies reside in the trigeminal ganglion relay tactile, thermal, and nociceptive input from the orofacial region. They enter the brainstem at midpons as part of the sensory root of the trigeminal nerve and begin their descent in the **spinal tract of the trigeminal nerve**. As the axons descend, their terminals synapse along the different levels of the spinal nucleus of the trigeminal nerve located immediately medial to the tract. The tract and nucleus are situated side by side along their entire length. The **spinal tract** and **nucleus of the trigeminal nerve** extend from the mid-pons inferiorly through the medulla oblongata to the upper two cervical spinal cord levels, overlying the **substantia gelatinosa** (**SG**) of the spinal cord. The caudal third of the spinal nucleus of the trigeminal nerve and the SG have a similar morphology and function – both process pain and temperature sensations. The SG processes pain and temperature sensation from the body, whereas the overlying spinal trigeminal nucleus, the head equivalent of the posterior horn of the spinal cord, processes pain and temperature sensation from the head. Similarly, the caudal extent of the spinal trigeminal tract overlies the structurally and functionally similar posterolateral tract (of Lissauer). The posterolateral tract contains the central processes of the sensory neurons relaying pain and temperature sensations from the body to the spinal cord. Axons leaving the spinal nucleus of the trigeminal nerve form the ipsilateral and contralateral **anterior trigeminothalamic tracts**, which relay pain, temperature, and some touch sensation from structures in the anterior and lateral parts of the head to the ventral posteromedial (VPM) nucleus of the thalamus.

Widespread areas of the cerebral cortex project *corticopontine fibers* to the **pontine nuclei**, scattered in the basilar pons. The pontine nuclei give rise to **pontocerebellar fibers** that form the most prominent afferent (input) projection to the cerebellum. As the pontocerebellar fibers leave the pontine nuclei, they cross the midline (they are commissural fibers) and pass into the cerebellum via the paired **middle cerebellar peduncles**. The massive middle cerebellar peduncles anchor the cerebellum to the posterior aspect of the mid to caudal pons and consist exclusively of pontocerebellar fibers which function in the coordination of voluntary movement.

The **medial longitudinal fasciculus** (**MLF**) is located in the midline, anterior to the fourth ventricle. The horizontally oriented **medial lemniscus** (**ML**) and the **spinothalamic tract** immediately lateral to it, are anteromedial to the trigeminal nuclei. The **rubrospinal fibers** have shifted to a position medial to the **spinothalamic tract** of the anterolateral system (ALS). The **anterior spinocerebellar fibers** enter the cerebellum via the superior cerebellar peduncle (brachium conjuctivum).

Rostral pons

The main structures located in the **rostral pontine tegmentum** include the *mesencephalic nucleus* and *tract* of the trigeminal nerve, situated just lateral to the periaqueductal gray (PAG) matter, the horizontally oriented *ML*, the *spinothalamic tract* adjacent to the ML, the *lateral lemniscus* located posterolaterally, and the *rubrospinal tract*, located medially. The *superior cerebellar peduncle* is ascending to the midbrain in the posterolateral tegmentum (Fig. 7.12).

The basis pontis contains the vertically oriented fiber bundles of the **corticospinal tracts**, **corticonuclear tracts**, and

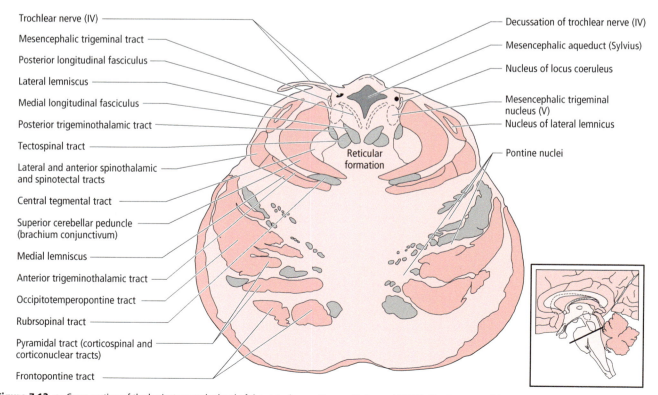

Figure 7.12 • Cross section of the brainstem at the level of the rostral pons. Source: DeArmond (1989) *The Structure of the Human Brain*, 3rd edn/Oxford University Press, figure 51, p. 102.

corticopontine fibers, all descending from the midbrain crus cerebri. As the vertically oriented **corticospinal tract** and **corticonuclear tract** fibers transition from the crus cerebri into the pons, they disperse into numerous distinct bundles of various sizes that occupy the basis pontis. The corticopontine fibers terminate in a group of nuclei, the **pontine nuclei**, scattered in the basis pontis, which give rise to the horizontally oriented **pontocerebellar fibers** that enter the cerebellum via the middle cerebellar peduncle. The corticospinal fiber bundles are surrounded by the pontine nuclei and the transversely oriented pontocerebellar fibers.

MIDBRAIN

Components of the midbrain function in visual and auditory reflexes

From posterior to anterior, the midbrain is subdivided into the **tectum** and the **cerebral peduncle**. The midbrain contains structures that function in visual and auditory reflexes (Fig. 7.13).

The **tectum** (L., "roof") is located posteriorly and contains the paired superior and inferior colliculi (L., *colliculus*, "little hill") collectively referred to as the *quadrigeminal plate (corpora quadrigemina)* (L., "quadruplet bodies").

The **cerebral peduncle** (L., peduncle "little foot") which forms most of the midbrain includes the *tegmentum* (L., "covering part") posteriorly, and the *basis pedunculi* anteriorly, which in turn consists of the *substantia nigra* (L., "black substance") and the *crus cerebri*. The **tegmentum** serves as a conduit for the ascending sensory tracts relaying sensory input from the body and head to higher brain centers, and numerous descending tracts some of which are motor tracts, that terminate in the brainstem or spinal cord. The **paramedian reticular formation**, which is involved in maintenance of consciousness, also resides in the midbrain tegmentum. Consequently, a lesion damaging this part of the midbrain results in coma. The anterior part of the midbrain is the prominent **basis pedunculi** whose location and morphology remains about the same at rostral and caudal midbrain levels. It consists of the **substantia nigra**, a massive nucleus involved in the subconscious control of movement, and the **crus cerebri** that conducts the descending axons of the corticospinal, corticonuclear, and corticopontine tracts. Although some structures may serve as landmarks that are unique to the rostral or caudal midbrain, others are constant and are present throughout its entire length. Structures present in both the rostral and caudal midbrain, include the *periaqueductal gray (PAG) matter*, the *mesencephalic tract* and *nucleus of the trigeminal*, the *mesencephalic aqueduct (cerebral aqueduct of Sylvius)*, the *MLF*, the *ML*, the *spinothalamic tract* (of the *anterolateral system (ALS)*, the *anterior* and *posterior trigeminothalamic tracts*, the *substantia nigra*, and the *crus cerebri*.

Caudal midbrain

The main structures that serve as landmarks of the caudal midbrain are the inferior colliculus, the trochlear nucleus, and the decussation of the superior cerebellar peduncle

The main structures that serve as landmarks of the caudal midbrain are the **inferior colliculus**, the **trochlear nucleus**, and the **decussation** of the **superior cerebellar peduncle** (Fig. 7.14).

Posteriorly, the tectum of the caudal midbrain contains the paired **inferior colliculi**, the midbrain relay nuclei of the ascending auditory pathway that receive binaural auditory input via the lateral lemniscus and function in sound localization. The inferior colliculus has an arm-like structure the **brachium** of the **inferior colliculus**, an **efferent** path extending laterally to the medial geniculate nucleus. The inferior colliculus uses this efferent path to project to the superior colliculus and to the medial geniculate nucleus of the thalamus. Projections to the superior colliculus elicit the reflex turning of the head in the direction of a startling sound, whereas the auditory information arriving at the medial geniculate nucleus is processed for sound intensity and frequency. The inferior colliculi are reciprocally connected via fibers that decussate in the **commissure** of the **inferior colliculus**.

The **lateral lemniscus** (not to be confused with the medial lemniscus) relays auditory information from both ears, from the auditory nuclei in the pons, as well as from the medulla oblongata to its termination in the inferior colliculus. Thus, the lateral lemniscus is not present in the rostral midbrain.

Just anterior to the inferior colliculi is the **PAG matter** and the **mesencephalic aqueduct (of Sylvius)**. These prominent landmarks of the midbrain form its core and are thus a transitional region between the tectum and tegmentum and extend from the caudal to the rostral midbrain. The PAG is associated with the endogenous modulation of nociception by ultimately influencing the neurons in the spinal nucleus of the trigeminal nerve and the posterior horn of the spinal cord. There nociception is attenuated from reaching higher brain centers and entering consciousness. The mesencephalic aqueduct (of Sylvius) allows cerebrospinal fluid (CSF) to flow from the third to the fourth ventricle. Embedded in the lateral border of the PAG is the mesencephalic nucleus and

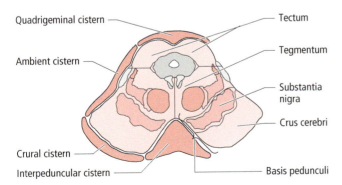

Figure 7.13 • General subdivisions of the midbrain. The tectum consists of the superior and inferior colliculi. The cerebral peduncle includes the tegmentum and the basis pedunculi. The basis pedunculi consists of the substantia nigra and the crus cerebri. Source: Haines (2013), figure 13.6, p. 171/with permission of Elsevier.

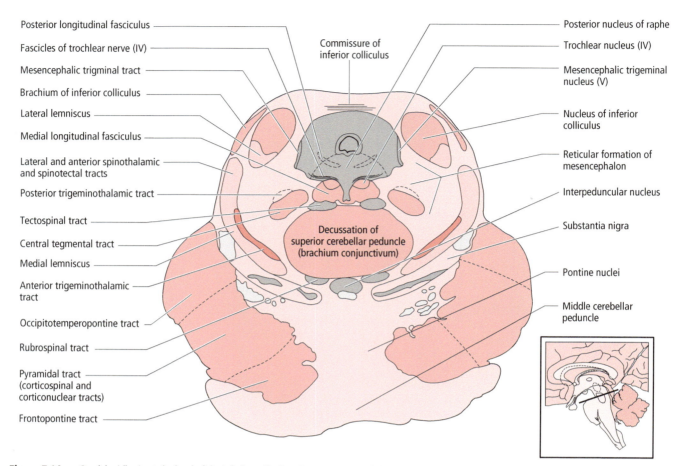

Figure 7.14 • Caudal midbrain at the level of the inferior colliculus. Source: DeArmond (1989) *The Structure of the Human Brain*, 3rd edn/Oxford University Press, figure 52, p. 104.

tract of the trigeminal nerve. The **mesencephalic nucleus of the trigeminal nerve** extends from the rostral pons to the midbrain. It processes proprioceptive impulses from the orofacial structures (muscles of mastication, and periodontal ligament of the teeth) relayed by the mesencephalic tract axon terminals. It is the only nucleus in the entire central nervous system that houses the cell bodies of pseudounipolar (unipolar) neurons that are characteristic of sensory ganglia and thus contains no synapses.

Continuing, anteriorly, is the **tegmentum** that in the caudal midbrain includes the *trochlear nucleus*, the *MLF*, the *ML*, the *spinothalamic tract* of the anterolateral system, the *trigeminothalamic tracts*, and the *central tegmental tract*. Following is the discussion of each structure.

The **trochlear nucleus**, located in the tegmentum of the caudal midbrain, just anterior to the PAG matter, contains the cell bodies of *motor neurons* whose axons leave the nucleus and then do something out of the ordinary for a cranial nerve: the axons curve around the PAG matter, course posteriorly, *cross within the midbrain*, and exit inferior to the inferior colliculus on the posterior aspect of the midbrain as the trochlear nerve. The trochlear nerve, the only cranial nerve that emerges from the posterior surface of the brainstem, courses anteriorly to the orbit to innervate a single extraocular muscle, the superior oblique which turns the eye inferolaterally ("down and out").

A prominent heavily myelinated fascicle, the *ascending* **medial longitudinal fasciculus (MLF)**, is located near the midline of the brainstem adjacent to the motor nuclei that innervate the extraocular muscles. In the caudal midbrain it is located immediately anterior to the trochlear nucleus. It contains heavily myelinated axons that interconnect the motor nuclei of the oculomotor, trochlear, and abducens nerves that innervate the extraocular muscles and also contains vestibular fibers involved in the coordination of eye movements in response to head movements. The MLF may be affected by demyelinating diseases such as multiple sclerosis, an autoimmune disease where T-lymphocytes attack myelin sheaths and in that way demyelinate axons of the central nervous system, although recent studies indicate the possibility that primary oligodendrogliopathy may be the actual cause of the disease and the immune reaction is merely a secondary response.

The **medial lemniscus (ML)** mediates discriminative (fine, detailed) touch, vibratory sense, proprioception (position sense), and some non-discriminative (light, crude) touch sensation from the contralateral side of the body to the ventral posterolateral (VPL) nucleus of the thalamus. As these fibers ascend from their origin (the gracile nucleus and cuneate nucleus) in the medulla oblongata through the pons to the midbrain, they shift from a medial, vertically oriented ribbon-like tract (in the medulla oblongata) to become horizontally orientated (in the rostral pons) to an obliquely oriented tract located posterior to the substantia nigra in the midbrain (Figs. 7.14 and 7.15). The **spinothalamic tract** (of the anterolateral system) mediates nociceptive, thermal, and some non-discriminative (light, crude) touch sensation from the contralateral side of the body to the ventral posterolateral nucleus (VPL) of the thalamus. The ML and the spinothalamic tract are located next to each other and form a wide "V-shaped configuration" on the lateral aspect of the midbrain tegmentum.

In addition to the **spinothalamic tract**, the anterolateral system (ALS) also includes the *spinoreticular, spinomesencephalic, spinotectal, spinoolivary,* and *spinohypothalamic* fibers, all of which terminate in the brainstem with the exception of the spinohypothalamic fibers that terminate in the hypothalamus of the diencephalon. The **spinoreticular fibers** terminate in the brainstem reticular formation of the midbrain, pons, and medulla oblongata. The reticular formation is involved in arousal and wakefulness that alerts the organism following injury. The **spinomesencephalic fibers** terminate in the PAG matter and in the midbrain raphe nuclei which are involved in the modulation of nociception. The **spinotectal fibers** terminate in the superior colliculus to mediate reflex turning of the head and upper trunk toward a noxious stimulus.

The **ML**, **spinothalamic tract** (of the **ALS**), and the **trigeminothalamic tracts** are located on the lateral aspect of the midbrain tegmentum. These three tracts are part of the major ascending sensory systems that carry somatosensory information from the body and head region to the thalamus.

The **central tegmental tract**, located lateral to the MLF, contains three types of fibers: (1) ascending fibers that arise from the nucleus of the solitary tract (solitary nucleus) to relay taste sensation to the ventral posteromedial (VPM)

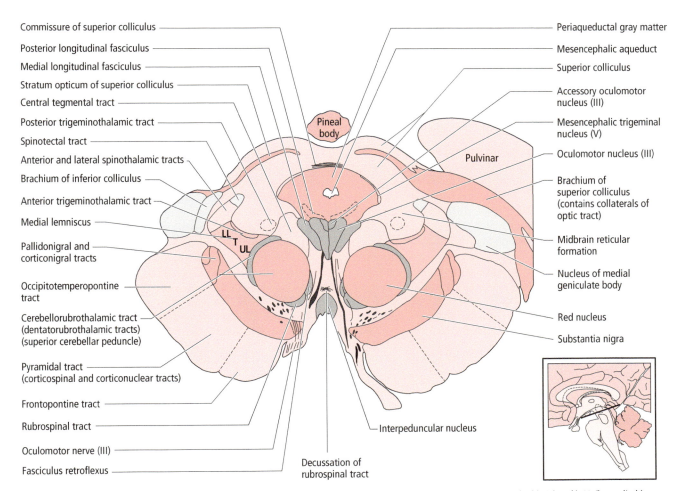

Figure 7.15 ● Rostral midbrain, at the level of the superior colliculus. Fibers transmitting sensory input from the UL (upper limb), T (trunk), LL (lower limb).
Source: FitzGerald et al. (2012) *Clinical Neuroanatomy and Neuroscience*, 6th edn/with permission of Elsevier Ltd.; DeArmond (1989) *The Structure of the Human Brain*, 3th edn/Oxford University Press, figure 54, p. 108.

110 ● ● ● **CHAPTER 7**

nucleus of the thalamus; (2) descending rubroolivary fibers that arise from the red nucleus to relay information about movement to the inferior olivary nucleus; and (3) reticulothalamic fibers to dorsal thalamic nuclei, which via their projections elicit an alerting response to nociception.

The paired **superior cerebellar peduncles** anchor the cerebellum to the posterior aspect of the midbrain. Their axons leave the cerebellum to ascend to the midbrain where they cross, forming the **decussation** of the **superior cerebellar peduncle**, the most prominent tegmental structure at the level of the inferior colliculus. Most of these cerebellar output fibers terminate in the thalamus. The **red nucleus** is not discernable at the level of the inferior colliculus in the caudal midbrain, since the decussation of the superior cerebellar peduncle occupies the tegmental region that is occupied by the red nucleus in the rostral midbrain.

Anterior to the decussation of the superior cerebellar peduncle near the midline and just posterior to the interpeduncular fossa is the **interpeduncular nucleus**, which is associated with the limbic system that processes emotions.

Similar to the situation in the rostral midbrain, the **substantia nigra** is located just posterior to the crus cerebri at the caudal midbrain level. Although present in the caudal midbrain, the substantia nigra is not as prominent or as well defined as it is in more rostral levels of the midbrain. It is discussed in more detail below, in the section on the rostral midbrain.

The **base** of the midbrain is formed by the **crus cerebri**, a massively myelinated fiber bundle on each side of the midbrain. This mass of white matter serves as a conduit for fibers arising from widespread areas of the cerebral cortex that descend in the corona radiata and the internal capsule of the cerebral hemispheres. In their descent, these fibers gather to occupy the crus cerebri, a structure that consists exclusively of *descending* axons, belonging to the *corticospinal, corticonuclear*, and *corticopontine* fibers. The corticospinal tract descends in approximately its middle third in tandem with the smaller corticonuclear tract which descends just medial to it. The remaining white matter on either side of the corticospinal and corticonuclear tracts consists of corticopontine (occipitotemperopontine and frontopontine) fibers that also arise from the cerebral cortex. The **corticospinal tract** terminates in the spinal cord to control voluntary movement in the contralateral side of the body, whereas the **corticonuclear tract** terminates in most cranial nerve motor nuclei located in the brainstem to control voluntary movement of the head and neck. The **corticopontine fibers** terminate in the pontine nuclei embedded in the basal pons, which in turn give rise to pontocerebellar fibers that enter the contralateral neocerebellum via the middle cerebellar peduncles, to participate in coordinating voluntary movement.

Rostral midbrain

The main structures that serve as landmarks of the rostral midbrain are the superior colliculus, the oculomotor nucleus, and the red nucleus

The main structures that serve as landmarks of the rostral midbrain are the **superior colliculus**, the **oculomotor nucleus**, and the **red nucleus** (Fig. 7.15). The two **superior colliculi** are part of the visual system and reside in the tectum of the rostral midbrain. Each superior colliculus receives: (1) visual information from the retina as it is propagated through the visual pathway; (2) corticotectal fibers from the frontal eye field (Brodmann's area 8) and the visual cortex (Brodmann's area 19) that function in the coordination of eye movements, particularly those mediating voluntary visual tracking of a moving object via conjugate deviation of the eyes, or changes in the direction of gaze; (3) corticotectal fibers from the visual cortex (Brodmann's area 19) involved in the accommodation reflex; and (4) auditory information from the inferior colliculus involved in audiovisual reflexes. The superior colliculi are reciprocally connected via commissural fibers that decussate in the **commissure** of the **superior colliculus**. The **brachium** of the **superior colliculus** relays **afferent** fibers arising from the retina and corticotectal fibers from the cerebral cortex to the superior colliculus. The superior colliculus via its tectospinal projections to the cervical and upper thoracic levels of the spinal cord functions in the control of reflex head and upper trunk movements in response to startling visual, auditory, or cutaneous stimuli.

Just cranial to the superior colliculus is the **pretectal area** that is associated with the pupillary light reflex in response to bright light. The **center for vertical conjugate gaze** is located in the posterior rostral midbrain. Intervening between the tectum and tegmentum is the **PAG matter** and **mesencephalic aqueduct** (of Sylvius).

Continuing, anteriorly, is the **tegmentum** which in the rostral midbrain includes the *oculomotor nuclear complex*, the *MLF*, the *ML*, the *spinothalamic tract of the anterolateral system*, the *trigeminothalamic tracts*, the *central tegmental tract* and the *red nucleus*. Each of these structures is discussed separately in the following sections.

The **oculomotor nuclear complex** includes the oculomotor and the accessory oculomotor (eponym: Edinger–Westphal) nuclei. The **oculomotor nucleus** houses the cell bodies of *motor neurons* whose axons exit the nucleus, sweep anteriorly in the tegmentum as numerous fascicles, pass through the red nucleus and the medial aspect of the substantia nigra, and then gather to form the **oculomotor nerve**. This nerve then emerges from the anterior aspect of the midbrain at the interpeduncular fossa (between the two crus cerebri). The oculomotor nerve courses anteriorly to the orbit where it subdivides to form muscular branches that innervate the levator palpebrae superioris muscle and four of the six extraocular muscles, namely the superior rectus, medial rectus, inferior rectus, and inferior oblique muscles. The oculomotor nucleus projects to the contralateral superior rectus muscle and via bilateral fibers to the levator palpebrae superioris muscle. The motor neurons of this nucleus mediate the elevation of the upper eyelid, and vertical and medial movement of the eyeball.

The **accessory oculomotor nucleus** located posterior to the oculomotor nucleus houses the cell bodies of two distinct populations of neurons: (1) those of the *accessory oculomotor preganglionic parasympathetic neurons* whose axons join the oculomotor nerve and course to the orbit where they synapse in the ciliary ganglion. This group of fibers is involved in two autonomic functions, pupillary constriction and accommodation of the lens by providing parasympathetic innervation to two of the three smooth muscles of the eye, the sphincter pupillae and ciliary muscles, respectively. (2) The other group of cell bodies within the accessory oculomotor nucleus is the *accessory oculomotor centrally projecting neurons* that send their axons to various brainstem nuclei including the gracile, cuneate, spinal trigeminal, parabrachial, and inferior olivary nuclei. Other fibers descend to the spinal cord. The projections of the *accessory oculomotor centrally projecting neurons* play a role in diverse functions including consumptive and stress-related behaviors. Several small nuclei that control eye movements reside near the oculomotor nucleus. They are the **elliptic nucleus** (eponym: nucleus of Darkschewitsch) located in the ventrolateral margin of the PAG matter, the **interstitial nucleus of the vestibular nerve** (eponym: interstitial nucleus of Cajal) located next to the MLF, and the **nucleus of the posterior commissure** (a pretectal nucleus). The latter two nuclei play a role in control of eye movements, especially those related to vertical gaze.

The **MLF** is present in the center, close to the midline of the rostral midbrain and occupies the same location as it does in the caudal midbrain. The *ML*, the *spinothalamic tract* of the anterolateral system, the *trigeminothalamic tracts*, and the *central tegmental tract* although present in the rostral midbrain, were discussed previously in the caudal midbrain section.

The **red nucleus**, a cylinder-shaped cell column that extends for most of the rostrocaudal length of the midbrain, earned its name from its pink appearance in fresh specimens due to its rich capillary network. On cross-section of the midbrain, it appears as a round structure posterior to the substantia nigra. It consists of the *rostral parvocellular* and *caudal magnocellular* portions. Afferent (input) projections include bilateral corticorubral fibers from the motor cortex and contralateral fibers from the cerebellar nuclei. Efferent (output) projections include the rubrospinal tract and the rubroolivary fibers. The **rubrospinal tract** fibers arise from the magnocellular portion, cross the midline at the level of the midbrain, and then descend to the cervical levels of the spinal cord. Although rudimentary in the human, the rubrospinal tract stimulates the motor neurons that innervate the distal flexors of the upper limb (the muscles that flex the hand and fingers). The **rubroolivary fibers** arise from the parvocellular portion and join the **central tegmental tract** to terminate in the ipsilateral inferior olivary nucleus located in the medulla oblongata in the lower brainstem. Via its connections to components of the motor system, the red nucleus is involved in the control of voluntary movement.

Adjacent to the red nucleus are the cerebellorubral and cerebellothalamic fibers that appear to form a parenthesis on its lateral surface. These fibers originate from the cerebellar nuclei and leave the cerebellum via the superior cerebellar peduncle to ascend to the caudal midbrain, where they cross in the decussation of the superior cerebellar peduncle. The **cerebellorubral fibers** terminate in the red nucleus, whereas the **cerebellothalamic fibers** continue their ascent to terminate in the ventrolateral (VL) nucleus of the thalamus. The red nucleus gives rise to the rubrospinal tract fibers which synapse with motor neurons that innervate the flexors of the hand and fingers. The VL nucleus of the thalamus projects to the motor cortex that controls the movement of the axial and proximal limb muscles of the upper and lower limbs. The cerebellum coordinates voluntary movement via the cerebellorubral and cerebellothalamic fibers.

Two important ascending sensory tracts, the medial lemniscus (**ML**) and the **spinothalamic tract**, ascend adjacent to one another on the lateral aspect of the midbrain tegmentum. Anterior to the tegmentum is the striking **substantia nigra**, the largest nucleus of the midbrain. It consists of two subdivisions, the *compact part* and the *reticular part*. Dopaminergic neurons in the compact part of the substantia nigra store neuromelanin, a byproduct of dopamine, giving the substantia nigra its characteristic dark appearance in a fresh or non-stained specimen. These dopaminergic *nigrostriatal fibers* project to the caudate nucleus and putamen of the basal nuclei, deep within the cerebral hemisphere. Degeneration of these neuromelanin-containing neurons is the putative cause of Parkinson's disease. The substantia nigra also projects *nigrothalamic fibers* to the ventral anterior, ventral lateral, and mediodorsal nuclei of the thalamus. The substantia nigra receives GABAergic *striatonigral fibers* from the caudate nucleus and the putamen. This large nucleus, via its connections, is involved in the subconscious control of eye movement (gaze) and general body movements. The location and appearance of the **crus cerebri** remains essentially similar to that at the level of the rostral midbrain.

CLINICAL CONSIDERATIONS

Brainstem vascular syndromes

Usually, damage to brainstem neural tissue results from interruption of normal blood flow as a consequence of a thrombus or an embolus. A thrombus is a blood clot which forms on the inner surface of a blood vessel (an artery), preventing normal blood flow to tissues. If a part of, or the entire, blood clot detaches from the vessel wall and flows in the blood stream to another part of the body where it occludes another vessel's lumen and thus obstructs blood flow to tissues, it is known as an embolus. Brainstem lesions result in disorders characterized by a set or pattern of symptoms (symptom complex) known as a syndrome. Only the classical brainstem syndromes resulting following brainstem focal vascular lesions are discussed in this section. Although when a particular arterial vessel that supplies the brainstem becomes occluded by an embolus, various other brain regions located in the vascular territory supplied by this same vessel are also affected, and more deficits may be apparent. However, localized vascular lesions involving only a

CLINICAL CONSIDERATIONS (continued)

specific region of the brainstem do occur and have been observed. The study of lesions here is to enhance understanding of the anatomy of the different segments of the brainstem. This can be accomplished by summarizing the location and functions of the different pathways traversing the brainstem and the nuclei residing at its various levels.

To localize a lesion, the student should be familiar with the organization of the brainstem including the location of important long ascending and descending tracts and other pathways within the brainstem that may be involved. These tracts include the corticospinal tract, corticonuclear tract, spinothalamic tract (which tends to maintain its lateral position as it ascends through the brainstem), the ML (which tends to shift position from medial to lateral as it ascends through the brainstem), the spinal tract of the trigeminal, and the superior and inferior cerebellar peduncles.

Structures that may be involved in *medial* brainstem lesions, since they are located close to the midline, include the corticospinal tract and the ML. For medial brainstem lesions, the level of the lesion may be determined by the following medially located cranial nerves: the oculomotor nerve located in the rostral midbrain, the trochlear nerve located in the caudal midbrain, the abducens nerve located in the caudal pons, and the hypoglossal nerve located in the medulla oblongata.

Structures that may be involved in *lateral* brainstem lesions, due to their lateral location, include the spinothalamic tract, and the spinal tract of the trigeminal. For lateral brainstem lesions, the level of the lesion may be determined by the following laterally located cranial nerves: the trigeminal nerve located in the mid-pons, the facial nerve located in the caudal pons, the vestibulocochlear nerve located in the pons/medulla oblongata, and the glossopharyngeal and vagus nerves located in the medulla oblongata. The level of the brainstem lesion is most often localized by the cranial nerve involved.

Medullary syndromes

Medial medullary syndrome

Medial medullary syndrome results following occlusion of the *paramedian branches of the vertebral or anterior spinal arteries* leading to infarction of neural tissue of the medial medulla oblongata (Fig. 7.16). The structures affected include the medially located corticospinal tract (descending in the pyramid, anteriorly), the ML (just posterior to the corticospinal tract), and the hypoglossal nerve root and nucleus (located near the midline). The **corticospinal tract** (a descending motor tract) consists of uncrossed axons at the level of the lesion (rostral to the motor decussation); thus, a lesion at this site produces weakness in the contralateral trunk and upper and lower limbs (contralateral hemiparesis). The **ML** (an ascending sensory tract) consists of crossed axons producing a contralateral decrease in conscious proprioception, vibratory sense, and discriminatory (fine, detailed) touch sensation from the opposite side of the body. Damage to the **hypoglossal nucleus**, root fascicles, and nerve results in weakness/paralysis and atrophy of the ipsilateral tongue muscles. Upon protrusion, the tongue deviates toward the weak side (side of the lesion).

Lateral medullary syndrome (Wallenberg syndrome)

Lateral medullary syndrome results following occlusion of the vertebral artery caused by a thrombosis, or less often by occlusion of the *posterior inferior cerebellar artery (PICA)* (Fig. 7.16).

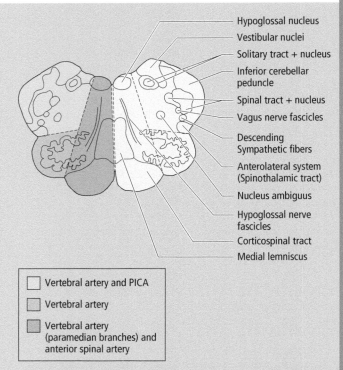

Figure 7.16 • Cross section at the level of the rostral medulla oblongata. Source: *Blumenthal/Neuroanatomy through Clinical Cases*, 2 edn, pp. 652–653. Reproduced with permission of Sinauer Associates, Inc. Blumenfeld (2002) *Neuroanatomy through Clinical Cases*, 1st edn, figure 14.20d, p. 611/Oxford University Press.

The structures affected include the inferior cerebellar peduncle, the nearby vestibular nuclei, the spinal tract and nucleus of the trigeminal nerve, the spinothalamic tract, the descending sympathetic fibers, the nucleus ambiguus, and the nucleus of the solitary tract (solitary nucleus). Since this syndrome involves the **lateral medullary tegmentum** the medially situated pyramid containing the corticospinal tract is usually not included in the lesion; thus, loss of motor function is minimal. Although numerous structures are involved in this medullary lesion, damage to the inferior cerebellar peduncle and vestibular nuclei produces the most debilitating symptoms. The inferior cerebellar peduncle relays proprioception from the spinal cord and medulla oblongata for the body and head, respectively. Damage to the **inferior cerebellar peduncle** produces *ipsilateral* cerebellar signs: ataxia (lack of coordination of movement) and damage to the **vestibular nuclei** produces unsteady gait, vertigo, nystagmus, nausea, and vomiting. The sensory systems relaying nociceptive and thermal sensations from the face and body may also be affected. Damage to the **spinal tract and nucleus of the trigeminal nerve** results in the diminution of nociceptive and thermal sense from the *ipsilateral* face; damage to the **spinothalamic tract** results in the diminution of nociceptive and thermal sensation from the *contralateral* side of the body. The **descending sympathetic fibers** arising from the hypothalamus as the *hypothalamospinal tract* and descending close to the spinothalamic tract are involved in the lesion, resulting in *ipsilateral* Horner syndrome characterized by ptosis (drooping) of the upper eyelid, miosis (constriction of

CLINICAL CONSIDERATIONS (continued)

the pupil), and anhidrosis (loss of sweating in the ipsilateral face). Damage of the **nucleus ambiguus** causes **dysphonia** (hoarseness due to *ipsilateral* vocal cord paralysis) and **dysphagia** (difficulty swallowing, and *ipsilateral* decrease of the gag reflex, due to involvement of the motor limb of the gag reflex arc formed by the vagus nerve). Lesion of the **nucleus of the solitary tract (solitary nucleus)** results in the decrease of taste sensation from the *ipsilateral* tongue and epiglottis. The level of a brainstem lesion is usually located by the cranial nerve involvement and resulting deficits. In lateral medullary syndrome, the level of the lesion is located by the dysphonia, dysphagia, and loss of taste as being in the medulla oblongata instead of the pons.

Pontine syndromes

Lesion in medial pontine basis

Dysarthria hemiparesis syndrome (pure motor hemiparesis syndrome)

The vessels involved in **dysarthria hemiparesis syndrome** are the *paramedian branches of the basilar artery* (Fig. 7.17). These branches stop at the midline (do not cross the midline or overlap the vascular territory of the vessels of the other side). Occlusion of the *paramedian branches of the basilar artery* supplying the medial pontine basis results in infarction of the **corticonuclear** and **corticospinal tracts**. Damage to the corticonuclear tract results in a *contralateral* lower face weakness and dysarthria (a motor speech disorder due to weakness/paralysis of the lips, tongue, and lower face). Damage to the corticospinal tract results in *contralateral* upper and lower limb weakness (motor hemiparesis).

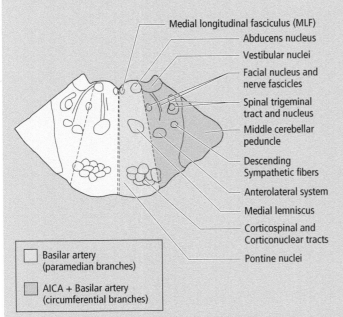

Figure 7.17 • Cross section at the level of the caudal pons. Source: Blumenthal/Neuroanatomy through Clinical Cases, 2 edn, pp. 652–653. Reproduced with permission of Sinauer Associates, Inc. Blumenfeld (2002) Neuroanatomy through Clinical Cases, 1st edn, figure 14.20c, p. 611/Oxford University Press.

Ataxic hemiparesis syndrome

The vessels involved in **ataxic hemiparesis syndrome** are the *paramedian branches of the basilar artery* (Fig. 7.17). The structures affected in this syndrome include the **corticonuclear tract**, the **corticospinal tract**, and the **pontine nuclei** and **pontocerebellar fibers**. Damage to the corticonuclear tract results in a contralateral lower face weakness and dysarthria. Damage to the corticospinal tract results in *contralateral* upper and lower limb weakness (motor hemiparesis). Damage to the **pontine nuclei** and **pontocerebellar fibers** results in *contralateral* ataxia. In order to understand why ataxia is *contralateral* to the side of the lesion may be understood by realizing that the (right) cerebral cortex gives rise to *corticopontine fibers* that descend ipsilaterally to the pons to synapse with the cell bodies of neurons in the (right) *pontine nuclei*. These neurons give rise to axons that form the *pontocerebellar fibers* that **cross** the midline in the pons to terminate in the contralateral (left) cerebellar hemisphere. Output fibers (*cerebellothalamic fibers*) arising from the (left) cerebellar hemisphere ascend to the (left) midbrain via the (left) *superior cerebellar peduncle* that **crosses** (to the right) in the caudal midbrain at the level of the inferior colliculus. From there the cerebellothalamic fibers continue their ascent to the (right) thalamus which in turn projects to the (right) motor cortex. Upper motor neuron axons (of the corticospinal tract) from the (right) motor cortex descend to the caudal medulla oblongata and **cross** to the opposite side (to the left) to descend in the left side of the spinal cord where they influence movement in the (left) side of the body. Thus, a lesion in the right side of the pontine basis involving the pontine nuclei and the pontocerebellar fibers arising from them, results in a *contralateral* ataxia due to the crossing of the fibers at three different levels.

Lesion in medial pontine basis and tegmentum

Foville syndrome

Occlusion of the *paramedian branches of the basilar artery* (perforating the pontine basis and the tegmentum) may result in an infarct that not only involves structures located in the medial pontine basis but extends posteriorly to include structures in the medial pontine tegmentum as well. The structures affected include the **corticonuclear** and **corticospinal tracts** (located in both rostral and caudal pontine levels) and the **facial colliculus** (located in the caudal pons) on the floor of the fourth ventricle. The deficits resulting from a lesion to the corticonuclear and corticospinal tracts include hemiparesis (weakness) in the *contralateral* lower face (with dysarthria) and the upper and lower limbs. Damage involving the facial colliculus includes the **abducens nucleus** (or **paramedian pontine reticular formation, PPRF**) and the **facial nerve root fascicles** as they curve over the underlying abducens nucleus. Deficits include *ipsilateral* horizontal (lateral) gaze palsy (due to damage of the abducens nucleus) and *ipsilateral* facial paralysis (due to damage of the facial nerve root fascicles) (Fig. 7.17).

Pontine wrong-way eyes syndrome

Occlusion of the *paramedian branches of the basilar artery* may cause infarction of the **corticonuclear** and **corticospinal tracts** located in the medial pontine basis and the **abducens nucleus** or **paramedian pontine reticular formation (PPRF)** located in the medial pontine tegmentum. The deficits resulting from a lesion to the corticospinal and corticonuclear tracts include hemiparesis (weakness) in the *contralateral* lower face (with dysarthria) and upper and lower

CLINICAL CONSIDERATIONS (continued)

limbs. Deficits resulting from a lesion to the abducens nucleus or PPRF include *ipsilateral* horizontal gaze palsy (inability to look to the side of the lesion, that is, abduct the ipsilateral eye and adduct the contralateral eye) (Fig. 7.17). A brief description of right-way eyes is provided here to clarify the meaning of wrong-way eyes. A lesion that damages the fibers arising from the eye fields in the cerebral hemisphere results in the inability to look away from the side of the lesion. The eyes preferentially gaze to the side of the lesion. If the corticospinal fibers are also damaged, producing a contralateral weakness (in the body), and since the *eyes gaze to the side of the lesion* (away from the side of the weakness in the body), this impairment is referred to as *right-way eyes*. In contrast, the **wrong-way eyes syndrome** is characterized by the inability to gaze to the side of the lesion, and instead, *look toward the side of the weakness* (Fig. 7.17).

Millard–Gubler syndrome

Occlusion of the *paramedian branches of the basilar artery* may damage the **corticonuclear** and **corticospinal tracts** anteriorly, and the more laterally located **fascicles of the facial nerve** in the tegmentum. The deficits resulting from a lesion to the corticonuclear and corticospinal tracts include hemiparesis (weakness) in the *contralateral* lower face (with dysarthria) and upper and lower limbs. Deficits from damage to the facial nerve fascicles include *ipsilateral* face paralysis.

Other structures located in the pontine tegmentum that may be variably involved in pontine tegmental infarcts include the **ML** and the **MLF**. A lesion to the ML will result in the diminution of discriminative touch, proprioception, and vibratory sense from the *contralateral* side of the body, whereas damage to the MLF results in internuclear ophthalmoplegia (discussed in Chapter 17) (Fig. 7.17).

Lesion in lateral caudal pons

AICA syndrome

Occlusion of the *anterior inferior cerebellar artery* (*AICA*) produces an infarct in the lateral caudal pons that will damage the laterally located structures and result in the deficits as follows: Damage of the **middle cerebellar peduncle** results in *ipsilateral* ataxia, the **vestibular nuclei**, vertigo, nystagmus, nausea, and vomiting, the **spinothalamic tract**, decrease in nociceptive and thermal sensation from the *contralateral* side of the body, the **spinal tract** and **nucleus** of the **trigeminal nerve**, diminution in nociceptive and thermal sensation from the *ipsilateral* face, and **descending sympathetic fibers**, *ipsilateral* Horner syndrome. Another region that may be variably involved if the *labyrinthine artery*, a branch of the AICA is affected (occluded), is the inner ear. This will result in *ipsilateral* hearing loss. Infarcts of the lateral rostral pons are not common (Fig. 7.17).

Lesion in posterolateral rostral pons

SCA syndrome

Occlusion of the *superior cerebellar artery* (*SCA*) will result in an infarct involving the superior cerebellar peduncle and the cerebellum. Deficits include *ipsilateral* ataxia of cerebellar origin. (Fig. 7.18).

Midbrain syndromes

The midbrain syndromes discussed next result from occlusion of the perforating branches emerging from the *top of the basilar artery* and the proximal *posterior cerebral arteries*.

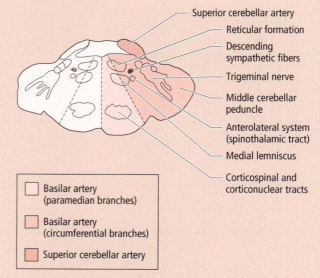

Figure 7.18 • Cross section at the level of the rostral pons. Source: Blumenthal/*Neuroanatomy through Clinical Cases*, 2 edn, pp. 652–653. Reproduced with permission of Sinauer Associates, Inc. Blumenfeld (2002) *Neuroanatomy through Clinical Cases*, 1st edn, figure 14.20b, p. 610/Oxford University Press.

Lesion in rostral midbrain basis

Medial midbrain syndrome (Weber syndrome)

Occlusion of the arterial vessels that supply the rostral midbrain basis will produce an infarct that will damage the oculomotor nerve fascicles as they exit the midbrain at the interpeduncular fossa to produce an *ipsilateral* **oculomotor nerve palsy**. Damage to the crus cerebri will result in *contralateral* hemiparesis (Fig. 7.19).

The **oculomotor nerve** innervates the ipsilateral eye. If the oculomotor nerve (containing **motor** and **parasympathetic fibers**) is damaged as it exits the midbrain, it will result in **oculomotor nerve palsy**, and the following deficits will occur in the ipsilateral eye: *ipsilateral* paralysis of the levator palpebrae superioris known as **ptosis** (drooping) of the upper eyelid, and paralysis of all of the extraocular muscles, except the lateral rectus and the superior oblique. The eye is deviated "down and out" (**lateral strabismus**), resulting in double vision (**diplopia**). The individual is unable to move the affected eye medially, or vertically. Since the parasympathetic innervation to the sphincter pupillae muscle and the ciliary muscle is interrupted, the *ipsilateral* pupil is dilated (**mydriasis**), and the lens cannot accommodate for near focus. For a more detailed description, see Chapter 17.

The **corticonuclear fibers** are damaged as they descend in the crus cerebri, prior to their distribution to the cranial nerve motor nuclei in the pons and medulla oblongata. The lesion includes the corticonuclear fibers terminating in the facial nucleus, the motor nucleus of the trigeminal, the nucleus ambiguus, and the hypoglossal nucleus. This will cause the following deficits: (1) weakness or paralysis of the muscles of the *contralateral* lower face (mouth) since the lower half of the facial nucleus receives mainly contralateral corticonuclear fibers; thus if these fibers are damaged, there are few, if any, back up corticonuclear fibers from the ipsilateral side to stimulate the neurons in the lower half of the facial nucleus which innervate the muscles in the lower half of the face; (2) weakness

CLINICAL CONSIDERATIONS (continued)

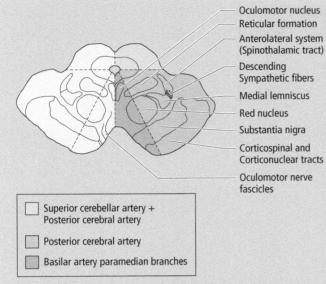

Figure 7.19 • Cross section of the rostral midbrain, at the level of the superior colliculus. Source: *Blumenthal/Neuroanatomy through Clinical Cases*, 2 edn, pp. 652–653. Reproduced with permission of Sinauer Associates, Inc. Blumenfeld (2002) *Neuroanatomy through Clinical Cases*, 1st edn, figure 14.20a, p. 610/Oxford University Press.

or paralysis of the *contralateral* lateral pterygoid muscle since the lower motor neurons that innervate the lateral pterygoid muscle receive only contralateral corticonuclear fibers. Normally, when the lateral pterygoid muscle contracts, due to the attachments and direction of its fibers, it swings the mandible toward the opposite side. Thus, if there is a lesion damaging the corticonuclear fibers at the level of the midbrain, on the left, the corticonuclear fibers that cross to the right side to stimulate the lower motor neurons that innervate the right lateral pterygoid are damaged. The right lateral pterygoid will be weak, and upon protrusion of the jaw, the jaw will deviate to the weak side, to the right; (3) *ipsilateral* deviation of the uvula. The lower motor neurons in the nucleus ambiguus that innervate the musculus uvulae receive contralateral corticonuclear projections. If the left corticonuclear projections that stimulate the lower motor neurons in the nucleus ambiguus that innervate the right musculus uvulae are damaged, the right musculus uvulae will be weak. Due to the attachment and direction of the musculus uvulae muscle fibers in the soft palate, when asked to say "Ahhh," the uvula will deviate to the side of the lesion (to the left), because the innervated musculus uvulae on the left will pull the right (weak) musculus uvulae to the left; and (4) *contralateral* deviation of the tongue upon protrusion. Normally, tongue protrusion is mediated by the joint action of the two genioglossus muscles. When the genioglossus muscle of one side contracts it causes the tongue to move medially and forward. The lower motor neurons that innervate the genioglossus receive only contralateral corticonuclear fibers. Thus, if there is a lesion damaging the corticonuclear fibers at the level of the midbrain, on the left, the corticonuclear fibers that cross to the right side to stimulate the lower motor neurons that innervate the right genioglossus muscle, are damaged. The right genioglossus will be weak, and upon protrusion of the tongue, the innervated (left) genioglossus will push the tongue to the right (toward the weak side). The corticonuclear fibers are discussed in more detail in Chapter 13.

The **corticospinal tract** will be damaged above the pyramidal decussation as it descends in the crus cerebri lateral to the corticonuclear fibers. This will result in *contralateral* hemiparesis (weakness) on the opposite side of the body. For a more detailed description, see Chapter 13.

Lesion in midbrain tegmentum (central midbrain)

Claude syndrome

Occlusion of the arterial vessels that supply the rostral midbrain tegmentum will produce an infarct that will damage the **oculomotor nerve fascicles** as they course anteriorly in the midbrain tegmentum to produce an *ipsilateral* **oculomotor nerve palsy** (see previous section for deficits). Other structures involved in the lesion include the **red nucleus** and the **cerebellothalamic fibers**. Damage to these structures will produce a *contralateral* **ataxia** and **tremor** of cerebellar origin (Fig. 7.19).

Cerebellothalamic fibers leave the cerebellum via the superior cerebellar peduncle. When these fibers ascend to the caudal midbrain, they decussate in the **decussation of the superior cerebellar peduncle**, which occurs in the caudal midbrain at the level of the inferior colliculus. Thus, a lesion in the (right) rostral midbrain tegmentum (rostral to the decussation of the superior cerebellar peduncle) will damage the cerebellothalamic fibers carrying information from the contralateral (left) cerebellum. The cerebellothalamic fibers arising from the (left) cerebellum normally decussate in the midbrain and then ascend to terminate in the (right) thalamus, which in turn projects to the (right) motor cortex. The (right) motor cortex gives rise to the (right) corticospinal tract, a descending motor tract most of which decussates in the caudal medulla oblongata to descend in the lateral funiculus of the (left) spinal cord. The cerebellum has an ipsilateral association with the body. That is, messages arising from the (left) cerebellum eventually control movement in the (left) side of the body. Therefore, deficits following a lesion in the (right) rostral midbrain tegmentum, will include a *contralateral* (left) ataxia, and tremor of cerebellar origin.

Lesion in rostral midbrain basis and tegmentum

Benedikt syndrome

Occlusion of the arterial vessels that supply the rostral midbrain basis and tegmentum may produce an infarct that will damage the **oculomotor nerve fascicles**, the **crus cerebri**, the **red nucleus**, the **substantia nigra**, and the **cerebellothalamic fibers**. Deficits include an *ipsilateral* **oculomotor palsy**, and *contralateral* hemiparesis, (as discussed previously), as well as a contralateral **ataxia, tremor** (as discussed previously), and **involuntary movements**. The involuntary movements are the result of damage to the substantia nigra. The substantia nigra is part of the basal nuclei which are involved in the subconscious control of movement. A lesion in the (right) rostral midbrain basis and tegmentum will damage the (right) substantia nigra, which normally sends messages to the ipsilateral (right) striatum of the basal nuclei that are then relayed to the ipsilateral (right) thalamus and then to the ipsilateral (right) motor cortex. The (right) motor cortex gives rise to the corticospinal tract, most of which decussates in the caudal medulla oblongata to descend in the (left) lateral funiculus of the spinal cord. Thus, if there is a lesion in the (right) rostral midbrain tegmentum, the involuntary movements due to the lesion of the substantia nigra, will appear on the left, *contralateral* to the side of the lesion, since unlike the cerebellum which has an *ipsilateral* association with the body, the basal nuclei have a *contralateral* association with the body (Fig. 7.19).

SYNONYMS AND EPONYMS OF THE BRAINSTEM

Name of structure or term	Synonym(s)/eponym(s)
Principal nucleus of the trigeminal	Main nucleus of the trigeminal
	Chief nucleus of the trigeminal
Medial reticulospinal tract	Pontine reticulospinal tract
Lateral reticulospinal tract	Medullary reticulospinal tract
Pyramidal decussation	Motor decussation
Decussation of the medial lemniscus	Sensory decussation
Corticonuclear tract	Corticobulbar tract (older term)
Basal nuclei	Basal ganglia (older term)
Trigeminothalamic tract	Trigeminal lemniscus
Red nucleus	Nucleus ruber
Lateral medullary syndrome	Wallenberg syndrome
Nucleus of the solitary tract	Nucleus solitarius (Solitary nucleus)
Rostral midbrain basis syndrome	Weber syndrome
Rostral midbrain tegmentum syndrome	Claude syndrome
Rostral midbrain basis and tegmentum syndrome	Benedict syndrome

FOLLOW-UP TO CLINICAL CASE

The patient is started on daily aspirin therapy, given increased antihypertensive medication and begun on statin therapy to decrease low-density lipoprotein cholesterol.

Within 6 weeks of the initial series of events the patient feels that they are 85% improved in terms of symptoms. No further episodes occur in the ensuing 6 months.

QUESTIONS TO PONDER

1. What are the deficits that would result following a lesion of the left rostral medial midbrain basis?

2. What are the deficits that would result following a lesion of the right rostral midbrain tegmentum?

3. What are the deficits that would result following a lesion of the right rostral midbrain basis and tegmentum?

4. What are the deficits that would result following a lesion in the rostral medial pontine basis?

5. What are the deficits that would result following a lesion at the level of the caudal pons, in the medial pontine basis on the left?

CHAPTER 8

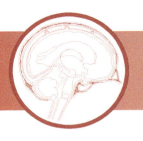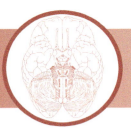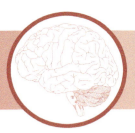

Meninges and Cerebrospinal Fluid

CLINICAL CASE

CRANIAL MENINGES

SPINAL MENINGES

VENOUS SINUSES OF THE CRANIAL DURA MATER

CEREBROSPINAL FLUID

VENTRICLES OF THE BRAIN

CLINICAL CONSIDERATIONS

SYNONYMS AND EPONYMS

FOLLOW-UP TO CLINICAL CASE

QUESTIONS TO PONDER

CLINICAL CASE

A 23-year-old patient, O.P., comes to the emergency room reporting of 2 days of fever, headache, stiff neck, and general ill feeling. Symptoms have been getting worse over the course of these 2 days. O.P. notes pain on eye movements and sensitivity to light. The patient's roommate who accompanies O.P. noted that the patient has been somewhat lethargic, but was quite well before the onset of the present symptoms and did not have a seizure.

Upon examination in the ER, O.P. appeared ill and lethargic but easily arousable and had a fever of 38.5°C. Blood pressure, heart rate, and respirations were normal. With the patient supine the neck was passively flexed. The neck was abnormally stiff (i.e., it did not flex to the normal degree). At the same time as the neck was flexed, O.P.'s knees flexed slightly. When eye movement was tested, the patient noted eye pain when they moved. Strength, sensation, reflexes, and all other parts of the neurologic and general exams were normal or unremarkable.

The central nervous system (CNS) is enclosed in the bony cranium and vertebral column. Protecting the brain and spinal cord from the calcified tissue are three more or less concentric membranes, the outermost, dense, irregular collagenous connective tissue, the **dura mater**, also known as the **pachymeninx**; the middle spiderweb-like **arachnoid**; and the very flimsy, innermost **pia mater**. The arachnoid and pia mater, together, are also known by the term **leptomeninges** and are separated from one another by the subarachnoid space. The dura mater forms reflections upon itself, some of which contain **dural venous sinuses**. The meninges surrounding the brain and spinal cord are compared in Table 8.1. Recently, the presence of a fourth meningeal membrane, known as the **subarachnoid lymphatic-like membrane** (**SLYM**), has been proposed which is discussed below in this chapter. Moreover, it was believed that the central nervous system does *not* have lymphatic vessels; however, in the mid-2010s lymphatic vessels have been detected in the meninges that convey material from the cerebrospinal fluid to lymph nodes of the scalp.

Cerebrospinal fluid (**CSF**) is a clear fluid whose composition and content is detailed below in this chapter. It is manufactured by the **choroid plexuses** (Gr., "braided membrane") of the ventricles of the brain; the CSF circulates not only in the ventricles but also in the subarachnoid space and in the unoccluded

---

A Textbook of Neuroanatomy, Third Edition. Maria A. Patestas, Amanda J. Meyer, and Leslie P. Gartner.
© 2025 John Wiley & Sons, Inc. Published 2025 by John Wiley & Sons, Inc.
Companion website: www.wiley.com.go/patestas/neuroanatomy3e

Skull (bone)	Vertebra (bone)
Periosteal layer of dura mater	Periosteum
Epidural space (potential space)	Epidural space (a real space that contains the venous plexus in a connective tissue surrounded by fat)
Meningeal layer of dura mater	Dura mater (meningeal layer only)
Subdural space (potential space)	Subdural space (potential space)
Arachnoid mater	Arachnoid mater
Subarachnoid space (real space; contains CSF)	Subarachnoid space (real space; contains CSF and cauda equina; expands into the lumbar cistern inferior to L2)
Pia mater	Pia mater (forms denticulate ligaments and filum terminale)
Subpial extracellular matrix	Subpial extracellular matrix
Astrocyte end-feet	Astrocyte end-feet
Brain tissue	Spinal cord tissue

CSF, cerebrospinal fluid.

Table 8.1 • Comparison of the layers of meninges around the brain and spinal cord.

portion of the central canal of the spinal cord. It is delivered into the lacunae lateralis, and from there into the vascular supply by structures, derived from both the pia mater and arachnoid, known as **arachnoid granulations**. As described earlier in this textbook (see Chapters 5 and 6) the four ventricles of the brain are fluid-filled, ependymal cell-lined cavities that are continuous with each other as well as with the central canal of the spinal cord. These ventricular spaces are the remnants of the lumen of the embryonic neural tube (see Chapter 2).

CRANIAL MENINGES

Note that the clinical case at the beginning of the chapter refers to a patient suffering with increasing severity of fever, headache, and stiff neck.

1 What should the physician suspect when the symptoms include fever accompanied by light sensitivity?

2 Why is a stiff neck an important clue in this case?

3 Should the fever be treated with antibiotics? Why or why not?

Cranial dura mater

The cranial dura mater is composed of an external periosteal dura mater and an internal meningeal dura mater

Although the **dura mater** surrounding the brain is continuous with the dura mater surrounding the spinal cord at the level of the foramen magnum, it is customary to discuss the two separately. The cranial dura mater has two layers (Fig. 8.1), an external **periosteal dura mater** that adheres to the internal table of the diploe and acts as a true periosteum, and an internal **meningeal dura mater** that is in intimate contact with the arachnoid mater. The periosteal and meningeal layers of the dura mater are tightly attached to each other throughout much of their extent; however, in certain regions the two layers are separated from each other to form **endothelially lined** venous channels, known as **dural sinuses**.

The periosteal dura mater, a coarse, dense irregular collagenous connective tissue interlaced with some elastic fibers, is a very tough, mostly inelastic tissue that is tightly attached to the surrounding bony vault of the skull. It is especially firmly attached at the sutures and to the floor of the cranial cavity. At the foramina of the skull the periosteal dura mater forms a connective tissue sheath around the cranial nerves as they leave the skull, and this dural layer is quickly replaced by the epineurium, derived from extracranial connective tissue. At the rim of the foramen magnum the periosteal dura becomes continuous with the periosteum of the vertebral canal.

The meningeal layer of the dura is also composed of dense irregular collagenous connective tissue. Its innermost aspect is lined by a single layer of flattened **fibroblasts** that form an **epithelioid sheath** that is in direct contact with, and separates the **arachnoid mater** from, the collagenous connective tissue component of the meningeal dura mater.

Vascular and nerve supply of the cranial dura mater

The periosteal layer of the dura mater has a very rich vascular supply, whereas the meningeal layer of the dura mater has no vascular supply

The periosteal layer of the cranial dura mater is richly supplied by blood vessels, whereas the meningeal layer has no vascular supply. The blood vessels of the periosteal layer include the **middle meningeal** and **accessory meningeal arteries** of the middle cranial fossa, as well as the **meningeal branches of the anterior** and **posterior ethmoidal arteries** and **meningeal branches of the internal carotid artery** of the anterior cranial fossa. **Meningeal branches of the vertebral, occipital,**

MENINGES AND CEREBROSPINAL FLUID

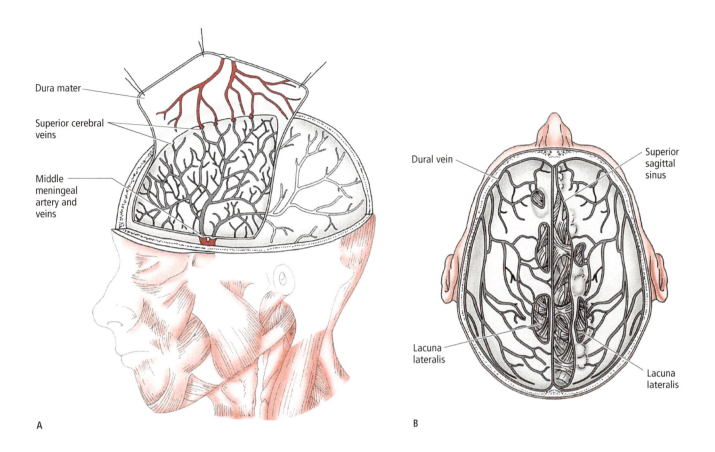

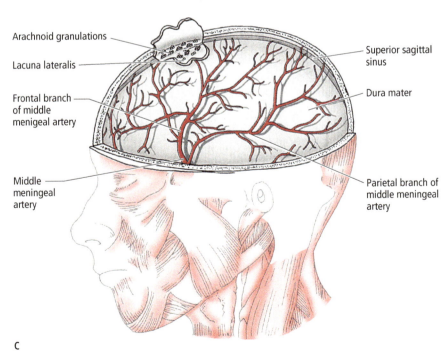

Figure 8.1 • Three views of the dura mater. (A) The periosteal layer of the dura is reflected to demonstrate the branches of the middle meningeal artery and the tributaries of the middle meningeal vein. (B) The dura is opened to display the superior sagittal sinus and several lacunae lateralis. (C) The dura is reflected to display the arachnoid granulations in a lacuna lateralis.

middle meningeal, and **ascending pharyngeal arteries** serve the periosteal dura of the posterior cranial fossa. Most of these vessels enter the cranial fossa via foramina and canals of the skull, such as the jugular and mastoid foramina, and the hypoglossal canal. Blood is drained from the dura mater by meningeal veins that empty their blood (indirectly through venous lacunae) into several of the venous sinuses as well as into nearby **emissary veins** and **diploic veins**.

The cranial dura mater possesses a very rich **sensory nerve** supply, composed mostly of pain fibers, derived mostly from **cranial nerve V** (**trigeminal nerve**) but also from the first three cervical spinal nerves that serve the dura mater of the posterior cranial fossa. Sympathetic fibers also reach the dura mater, arising from the vertebral and carotid plexuses, to serve the dural blood vessels.

Reflections of the meningeal layer of the dura mater

The meningeal layer of the dura mater is reflected upon itself to form dural folds

The meningeal layer of the dura mater forms several folds that are interposed between separate parts of the brain (Figs. 8.2 and 8.3). These folds are reflections of the meningeal layer of the dura upon itself and are known as the **cerebral falx** (**falx cerebri**), **cerebellar tentorium** (**tentorium cerebelli**), and **cerebellar falx** (**falx cerebelli**). Additionally, the meningeal layer of the dura also forms the **sellar diaphragm** (**diaphragma sella**) covering the hypophyseal fossa, as well as a roof over the superior surface of the trigeminal ganglion, thus forming a shallow housing known as the **trigeminal cave** (**cavum trigeminale**, eponym: **Meckel's cave**).

Cerebral Falx

The cerebral falx (falx cerebri) is a sickle-shaped fold of the meningeal layer of the dura mater

The **cerebral falx** (**falx cerebri**) (L., "sickle") occupies the midline **longitudinal cerebral fissure**, the space located between the two cerebral hemispheres. The narrow anterior aspect of the cerebral falx is attached to the **crista galli** (L., "rooster's comb") and its broader, posterior aspect adheres to the superior surface of the **cerebellar tentorium** (**tentorium cerebelli**; L., "tent"), extending posteriorly as far as the internal occipital protuberance. The superior, convex surface of

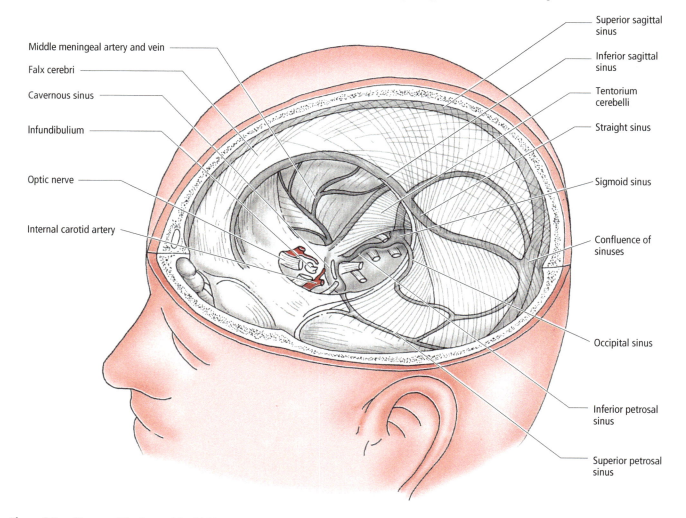

Figure 8.2 • Diagram of the dura and dural folds containing the venous sinuses.

MENINGES AND CEREBROSPINAL FLUID 121

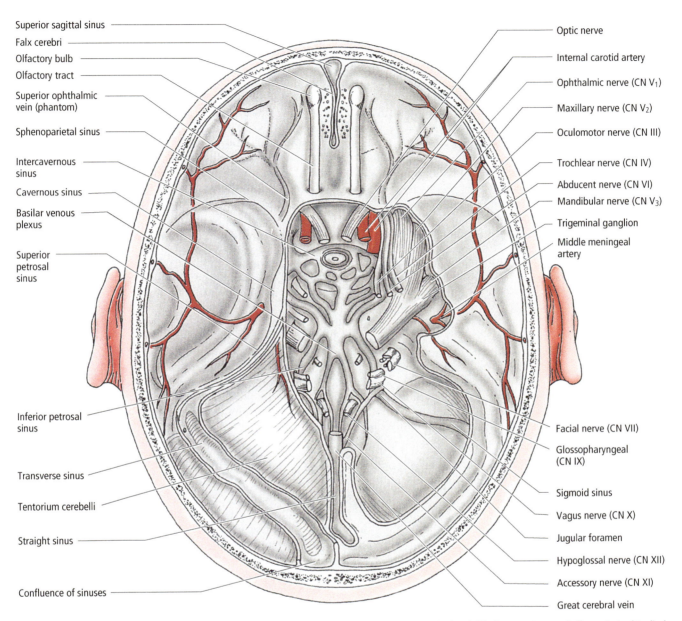

Figure 8.3 • Diagram of the dura and dural reflections housing the venous sinuses. Note that on the right-hand side the tentorium cerebelli was incised to display the trigeminal ganglion, the three divisions of the trigeminal nerve, and the contents of the cavernous sinus.

the falx cerebri is attached to the periosteal layer of the dura, leaving only a narrow, endothelially lined channel, the **superior sagittal sinus**. This sinus, the largest of the dural venous sinuses, begins at the **foramen cecum** (L., "blind opening") and terminates posteriorly at the **confluence of sinuses**. The inferior surface of the falx cerebri is a free, concave-shaped edge, occupied by the **inferior sagittal sinus**. At the junction where the cerebral falx joins and fuses with the cerebellar tentorium is another endothelially lined space, the **straight sinus**, that receives blood from the inferior sagittal sinus and from the **great cerebral vein**. Blood from the straight sinus also enters the confluence of sinuses.

Cerebellar tentorium

The cerebellar tentorium separates the occipital lobe of the cerebral hemispheres from the cerebellum

The **cerebellar tentorium**, a horizontal reflection of the meningeal layer of the dura mater, is housed in the space between the cerebellum and the occipital lobes of the cerebrum (Figs. 8.2 and 8.3); in fact, the occipital lobes are physically supported by it. The anterior sharp concave margin of the cerebellar tentorium is free and partially surrounds the midbrain. The lateral borders of the cerebellar tentorium extend much farther anteriorly than does its midline (where it is joined by the cerebellar falx). The superior surface of the central region is convex, being highest along the length of the **straight sinus**. The **transverse sinuses**, endothelially lined vascular spaces, sit in the lateral aspect of the tentorium cerebelli, where this dural reflection is attached to the lips of the **grooves** for the left and right **transverse sinuses of the occipital bone** (Fig. 8.3). Anteriorly, the lateral aspect of the tentorium is attached to the superior surface of the **petrous** portion of the **temporal bone** and forms endothelially lined spaces, the right and left **superior petrosal sinuses**, and continues anteriorly to attach to the **posterior clinoid processes of the sphenoid bone**. The free, medial edge of the cerebellar tentorium crosses over the lateral edge and attaches to the **anterior clinoid processes** of the sphenoid bone. Because of the separate attachments of the free and attached aspects of the cerebellar tentorium, an oval opening is created in the dura mater. This opening, known as the **tentorial notch** (**tentorial incisure**), surrounds the midbrain and permits the ascent of the **posterior cerebral arteries** to reach the **cerebral hemispheres**.

Cerebellar falx

The **cerebellar falx**, a relatively small reflection of the meningeal layer of the dura folded upon itself, is interposed between the right and left cerebellar hemispheres. The posterior border of the cerebellar falx meets the periosteal layer of the dura mater on the internal aspect of the occipital bone, where it is attached along the entire length of the **internal occipital crest**, and forms the endothelially lined **occipital sinuses**.

Sellar diaphragm (Diaphragma sella)

The **sellar diaphragm** (**diaphragma sella**; L., "diaphragm of the saddle"), a thin reflection of the meningeal layer of the dura mater, is attached, laterally, to the **clinoid processes** (Gr., "bed"), whereas its central aspect is open, thus forming an incomplete roof over the hypophyseal fossa that is penetrated by the **infundibulum of the hypophysis**. The anterior and posterior edges of the sellar diaphragm house the **anterior** and **posterior intercavernous sinuses** (Fig. 8.3).

Trigeminal cave (cavum trigeminale)

The **trigeminal cave** (**cavum trigeminale**, eponym: **Meckel's cave**) is a narrow, slit-like region located between the periosteal and meningeal layers of the dura mater positioned on the **trigeminal impression** of the petrous portion of the **temporal bone**. It is occupied by the **trigeminal ganglion** (Fig. 8.3) as well as by the sensory and motor roots of the trigeminal nerve.

Cranial arachnoid mater

The cranial arachnoid mater is a fine, spiderweb-like, nonvascular membrane that is interposed between the meningeal layer of the dura mater and the pia mater

There is a potential space between the meningeal layer of the cranial dura mater and the **cranial arachnoid** (Gr., "spider-like web") mater (Figs. 8.4 and 8.5), known as the **subdural space**, that may become filled with blood in case of hemorrhage due to a cerebrovascular accident. Similar to the dura mater, the arachnoid's surface is composed of a single layer of flattened **fibroblasts** that form an **epithelioid sheath**, from which trabeculae, known as **arachnoid trabeculae**, extend toward and attach to the external surface of the pia mater. These trabeculae are formed from highly attenuated fibroblasts that surround some collagen fibers. The space between the epithelioid sheath and the pia mater, known as the **subarachnoid space**, is occupied by **cerebrospinal fluid** (**CSF**) and is traversed by numerous arachnoid trabeculae. Blood vessels from the vascular meninges perforate the arachnoid to reach the pia mater; however, each vessel is completely surrounded by arachnoid fibroblasts and, therefore, the vessels do not actually enter the subarachnoid space. It should be recalled that the **blood–brain barrier**, established by the endothelial cells of the blood vessels, prevents large molecules from leaving the bloodstream (See Chapter 9, Vascular Supply of the Central Nervous System).

Arachnoid cisternae

The arachnoid resembles the dura mater in that it does not follow closely the contours of the brain; therefore, the subarachnoid space is much deeper over sulci than it is over the gyri. Moreover, in certain areas of the brain the arachnoid completely diverges from the pia mater, forming expanded subarachnoid spaces known as **subarachnoid cisterns** (L., "chest, box"). As the CSF percolates through the subarachnoid spaces, it also enters the subarachnoid cisterns, filling them. The major subarachnoid cisterns are the posterior cerebellomedullary cistern (cisterna magna), pontine cistern (L., "bridge, passage"), and interpeduncular (L., "little foot") and chiasmatic cisterns (cisterna basalis), as well as the superior cistern (cistern of the great cerebral vein). According to recent publications, a fourth meningeal membrane, the **subarachnoid lymphatic-like membrane** (**SLYM**), located between the arachnoid and the pia, envelops the brain and is usually fused with the subarachnoid barrier cell layer. However, in subarachnoid cisterns the two layers are separated, so that the SLYM forms a thin membrane, subdividing the CSF content of the cistern into two compartments, an **internal** and an **external subarachnoid spaces**, where the internal subarachnoid space is a perivascular space surrounding the major arterial vessels of the brain. It thus appears that the newly formed, "clean," CSF is shunted into the perivascular spaces so that it does not intermix with the "older" CSF that has already percolated through the substance of the brain tissue and contains waste products, such as tau, synuclein, and amyloid, among others that were washed out from the extracellular spaces. Therefore, only the clean CSF will percolate through the brain tissue, and the "older" CSF will be removed from the subarachnoid spaces. The major subarachnoid cisterns are the posterior cerebellomedullary cistern (cisterna magna), pontine cistern, interpeduncular and chiasmatic cisterns, and the superior cistern.

MENINGES AND CEREBROSPINAL FLUID • • • 123

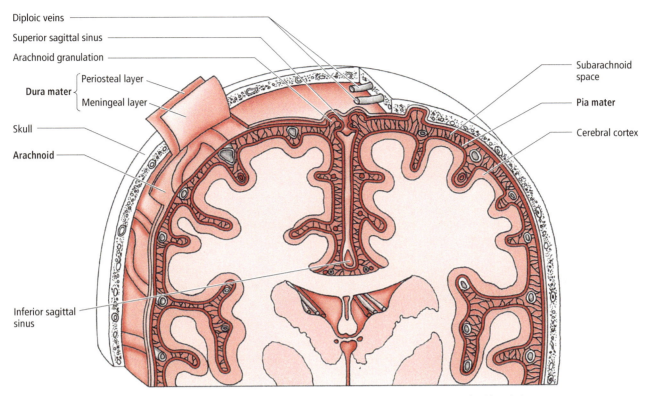

Figure 8.4 • Diagram of a frontal section of the skull and brain to display the three meninges: the dura mater, arachnoid, and pia mater.

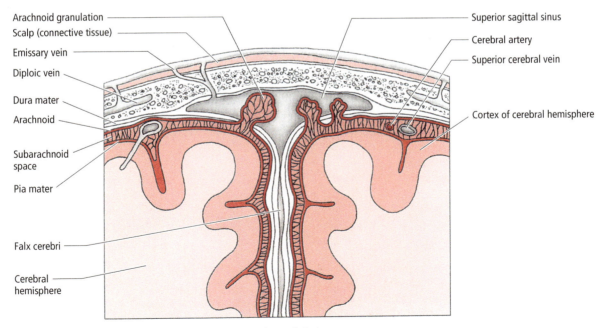

Figure 8.5 • Diagram of the superior sagittal sinus housing arachnoid granulations.

- The **posterior cerebellomedullary cistern (cisterna magna)** is the largest of the subarachnoid cisterns. It is situated between the inferior aspect of the cerebellum and the posterior surface of the medulla oblongata (L., "pith, marrow") as the arachnoid extends across these two structures rather than following the contour of the brain. The posterior cerebellomedullary cistern is the first subarachnoid space that receives CSF from the fourth ventricle

of the brain (via the **medial aperture** [eponym: **foramen of Magendie**]).
- The **pontine cistern**, a much smaller space than the posterior cerebellomedullary cistern, is located along the anterior surface of the pons and communicates with the subarachnoid space of the spinal cord inferiorly, and with the interpeduncular cistern cranially. It receives CSF from the right and left **lateral apertures** (eponym: **lateral foramina of Luschka**). The **basilar artery**, the **anterior inferior cerebellar** and **superior cerebellar arteries**, as well as the **cerebral arterial circle** traverse the pontine cistern and **CN VI (abducens nerve)** arises from the brainstem within this cistern.
- The **interpeduncular cistern (basal cistern)** is located between the right and left cerebral peduncles and receives CSF from the **chiasmatic cistern** with which it is continuous superiorly. Frequently, the two are considered to be a single cistern, the **cisterna basalis**, even though the **optic chiasma** is interjected between them. The **posterior cerebral artery**, part of the **basilar artery**, and the **basal vein** pass through the interpeduncular and chiasmatic cisterns.
- The **superior cistern (cistern of the great cerebral vein)** is located in the vicinity of the superior aspect of the cerebellum, the corpora quadrigemina, and the pineal body. As its alternate name implies, the great cerebral vein traverses the superior cistern as well as the posterior cerebral arteries.

Arachnoid granulations

Arachnoid granulations function in transporting CSF manufactured by the choroid plexuses of the ventricles of the brain into the vascular system

Arachnoid granulations (also known as **arachnoid villi**) are small, mushroom-shaped evaginations of the arachnoid protruding into the lumina of the dural sinuses (Fig. 8.6). They appear as very small granular substances that are visible with the unaided eye.

Interestingly, about 70% of the arachnoid granulations are easily separated from the dura, and only 30% of them form firm attachment to the dural structures. Most of the arachnoid granulations are associated with **lacunae lateralis**, diverticula of the **superior sagittal sinus**, although some jut directly into the lumen of the sinus. They have a varied morphology, where the simple cylindrical stalk will have a bulbous head, whose shape can range from being single-lobed to multi-lobed and where the cores of some of the lobes are attached to each other via bridge-like structures. Others, instead of forming lobes, have flattened, plaque-like heads that may extend to the dural roof or end free in the

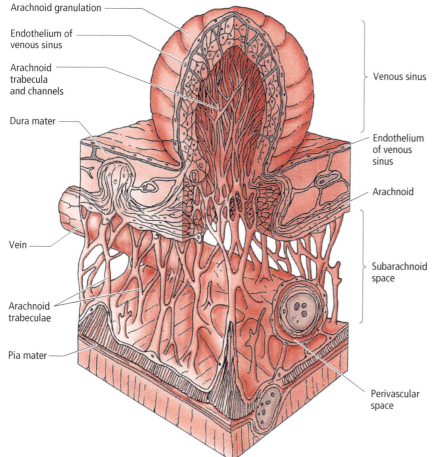

Figure 8.6 • Schematic diagram of an arachnoid granulation protruding into the superior sagittal sinus.

subdural space. The diameter of the stalk varies, becoming larger as a function of the individual's age. The core of an arachnoid granulation, composed of arachnoid trabeculae, is continuous with the subarachnoid space and is surrounded by the epithelioid layer of the arachnoid and of the dura, forming a membrane that is two cell layers thick. Interestingly, when the peripheral aspects of the bulbous portion and the core were stained for various substances, both the core and the periphery displayed positive staining for collagen, vascular endothelium, macrophagic, and arachnoidal markers but not for neuronal, aquaporin, astrocytic, lymphatic endothelial, or neutrophilic markers. Most arachnoid granulations were encapsulated by arachnoid cells only, and fewer than 5% had dural cells included in their capsule; however, as many as 7% of the arachnoid granulations were not encapsulated. Interestingly, the arachnoid granulations of young (90%) and middle aged (70%) individuals possessed capsules, whereas those of old individuals only 30% had capsules. The substance of the arachnoid granulations was composed of a sponge-like network of collagen fibers ensheathed by arachnoid epithelioid cells surrounding cavernous spaces that were smallest in the young and largest in old individuals. The contents of these cavernous spaces included macrophages, plasma cells, lymphocytes. Interestingly, younger individuals had more of these immune cells than the arachnoid granulations of older persons; but older individuals possessed more immunologically active B cells, T cells, and MHCII+ antigen-presenting cells, in their arachnoid granulations than did younger persons. Unfortunately, the significance of these recent findings, especially as related to the age-specific differences, is not clear.

Most of the arachnoid granulations protrude into the **lacunae lateralis** (L., "lateral lake"), diverticula of the **superior sagittal sinus**, although some jut into the lumen of the sinus itself. As the arachnoid granulation evaginates into the lacuna lateralis it is invested by some cellular and collagenous elements of the meningeal dura mater, which in turn is surrounded by the endothelial lining of the blood vessel. Cerebrospinal fluid from the subarachnoid space enters into the core of the arachnoid granulation and from there penetrates, probably by osmosis, the epithelioid layers of the arachnoid granulation and the endothelial lining to escape into the lacuna lateralis. Therefore, the function of the arachnoid granulations is transporting almost half of the CSF manufactured by the choroid plexuses of the ventricles of the brain into the vascular system. Most of the other half of the CSF leaves the subarachnoid space via the cribriform plate of the ethmoid bone, paralleling the olfactory nerve axons (the **perineural space**) into the nasal cavity to reach the lymphatic plexus present there. The remaining 1–2% of the CSF follows other cranial nerves out of the skull and into lymphatic vessels in the vicinity of the extracranial portions of the cranial nerves. This method whereby the CSF enters along the arterial portion of the cerebral vascular system, combines with the interstitial fluid, and exits as a pooled CSF and interstitial fluid at the venous aspect of the cerebral vascular system that removes deleterious proteins and other waste materials from the interstitial spaces of the brain

tissue to deliver them into the lymphatic system outside the cranial cavity has been termed the **glymphatic system**.

Combined, the average drainage of the entire CSF is approximately 20–25 mL per hour with a maximum capability of 60 mL per hour. Apparently, the arachnoid granulation drainage is reduced as a function of age which may have consequences for CSF drainage with a resultant decrease in the clearance of materials, such as beta amyloid and other proteins associated with neurodegenerative disorders, present in the extracellular spaces among neurons of the brain. Therefore, the function of the arachnoid granulations appears to be not only the transportation of CSF manufactured by the choroid plexuses of the ventricles of the brain into the vascular system, but also the clearing of unwanted materials from the interstitial spaces of the CNS.

An interesting series of experiments of the glymphatic system in rodents have demonstrated that during the time that the animals were awake the flow of CSF into the extracellular spaces, occupied by the cell bodies and processes of protoplasmic astrocytes bathed in a scant amount of interstitial fluid, is very low. However, when these rodents fall asleep, norepinephrine is released in the brain and causes the shrinkage of the astrocytes and their processes, thus increasing the volume of the extracellular spaces. Additionally, **aquaporin-4 water channels** (L., "water pore") in the end-feet of the astrocytes contacting the blood vessels open and interstitial fluid drainage increases, reducing further the amount of extracellular fluid. The increase in the volume of the extracellular spaces and the reduction in the amount of extracellular fluid decreases the pressure within the extracellular spaces thus encouraging the volume and velocity of the flow of cerebrospinal fluid from the subarachnoid space into the extracellular spaces. This augmented flow volume and flow rate increases the clearance of deleterious materials present in the extracellular spaces. Interestingly, when the animals woke up the astrocytes and their processes increased in size and the aquaporin-4 channels became closed, decreasing the volume of the extracellular spaces and reducing the flow rate and flow volume of the CSF from the subarachnoid space into the extracellular spaces. A similar mechanism of the glymphatic system has been shown to occur in most mammals, including humans.

Cranial pia mater

Cranial pia mater is a vascular tissue that closely invests the contours of the brain

The **cranial pia mater** is composed of a single layer (or occasionally two layers) of attenuated **fibroblasts** that form a transparent **epithelioid membrane**, which closely invests the contours of the brain. The pia mater, unlike the arachnoid or the dura mater, intimately follows the gyri and the sulci, maintaining an uninterrupted contact with the surface of the brain. Since the pia mater is vascular it has numerous blood vessels associated with it; however, because this layer is so thin, the vessels are in part surrounded by cells derived from the arachnoid trabeculae and, in part, by cells of the pia mater. Deep to the epithelioid sheath is a thin, discontinuous layer of collagen and elastic fibers. The end-feet of the astrocytes form a protective layer that underlies this subpial extracellular matrix, separating it from the neural tissue. In addition to separating the subarachnoid space from the brain tissue, the pia mater also serves to degrade neurotransmitter substances and to prevent material in the subarachnoid space from entering the nervous tissue, as evidenced by the inability of erythrocytes to cross the pia mater in cases of subarachnoid hemorrhages.

SPINAL MENINGES

Spinal dura mater

Spinal dura mater, unlike cranial dura mater, is composed only of a meningeal layer

The **dura mater** of the spinal cord differs from that of the cranial cavity because it is composed only of the meningeal layer (Fig. 8.7). The periosteal layer is the true **periosteum** of the vertebral canal and is separated from the meningeal layer by a loose connective tissue and fat-filled **epidural space**. A plexus of veins is embedded in the connective tissue and fat layer. At the level of the second lumbar vertebra the spinal cord ends in the conical-shaped **conus medullaris**, but the dura mater continues as a cylindrical sheath until it narrows into a cone-shaped structure at the second sacral vertebra. It then continues as a very narrow cylindrical sheath, which becomes anchored into the periosteum of the first and second coccygeal vertebrae.

Several investigators described the presence of slender tendon-like fibers passing through the posterior region of the atlanto-occipital as well as the atlanto-axial junctions of cadavers connecting the rectus capitis posterior minor muscle and the cervical dura mater. Manipulation of this muscle resulted in visually evident movement of the spinal dura mater extending through the entire cervical region (from C1 to T1). It has been suggested that some migraines and cervicogenic headaches may be due to pathological rhythmic contraction of the rectus capitis posterior minor muscle.

Spinal arachnoid mater

The **spinal arachnoid mater** is seamlessly continuous with the cranial arachnoid at the foramen magnum. It closely adheres to the spinal dura mater, and there is a potential subdural space between the two layers. The narrow subarachnoid space is filled with CSF that percolates through and around the arachnoid trabeculae. At the level of the conus medullaris the subarachnoid space becomes much larger, is referred to as the **lumbar cistern**, and is filled with the posterior and anterior rootlets of the spinal nerves, collectively referred to as the **cauda equina**. These nerve rootlets extend to the lumbar, sacral, and coccygeal intervertebral foramina through which they exit to form their respective spinal nerves. As the rootlets of the spinal nerves leave the vertebral canal, they are surrounded by a thin arachnoid sheath to be replaced by the proper connective tissue layers (endoneurium, perineurium, and epineurium) as they enter the realm of the peripheral nervous system.

Spinal pia mater

The spinal pia mater is continuous with the cranial pia mater at the foramen magnum and the two are essentially identical to each other

The **spinal pia mater** very closely invests the spinal cord and at the level of the conus medullaris it is gathered into a

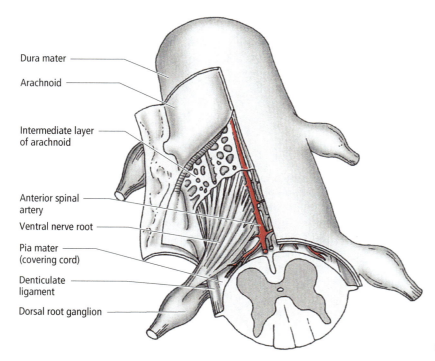

Figure 8.7 • Schematic diagram of the spinal meninges.

MENINGES AND CEREBROSPINAL FLUID ● ● ● 127

very thin, non-nervous filament, the **terminal filament** (**filum terminale**), that extends for about 20 cm from the tip of the conus medullaris to become attached to the periosteum of the first coccygeal vertebra. Interestingly, nerve fibers that appear to be stretch and pain fibers have been demonstrated entering the core of the filum terminale and occupying its entire length. It appears that the stretch fibers are associated with mechano-receptors within the terminal filament that can stimulate contraction of the paraspinal group of skeletal muscles.

Narrow, triangular, fibrous extensions of the pia mater extend laterally through the arachnoid to attach to the meningeal layer of the dura mater, helping to anchor the spinal cord along its entire length. These triangular extensions of the pia resemble shark's teeth and, therefore, are known as **denticulate ligaments**. There are 21 pairs of denticulate ligaments, positioned in such a fashion that they are situated about half way between successive spinal nerves. The first pair is located at the level of the foramen magnum, and the last pair is situated inferior to the 12th thoracic vertebra.

VENOUS SINUSES OF THE CRANIAL DURA MATER

Venous sinuses of the cranial dura mater are endothelially lined vascular spaces formed in reflections of the meningeal layer of the cranial dura mater

The **venous sinuses** of the dura mater, formed within reflections of the meningeal dura, are endothelially lined venous channels that are devoid of valves (Fig. 8.8). These sinuses collect blood from the brain and from emissary veins and also receive CSF from the subarachnoid spaces. They empty their contents into the superior jugular bulb of the **internal jugular vein** as that bulb sits in the jugular foramen. Based on their location and draining patterns the dural venous sinuses are grouped into two major categories, the **anterior inferior** and **posterior superior groups**.

Anterior inferior group

The **anterior inferior group** of dural sinuses include the cavernous, anterior and posterior intercavernous, sphenoparietal, and superior and inferior petrosal sinuses and the basilar plexus (Figs. 8.3–8.5, 8.8–8.10).

Cavernous sinus

The cavernous sinuses, located on either side of the body of the sphenoid bone, are associated with the internal carotid artery as well as with a number of cranial nerves

The two, rather large **cavernous sinuses** (Fig. 8.9), located on either side of the body of the sphenoid bone just inferior to the hypophyseal fossa, are connected to each other by the very small **anterior** and **posterior intercavernous sinuses**,

thus forming the **circular sinus**, encircling the infundibulum of the pituitary gland. The lumen of each cavernous sinus is criss-crossed by a spongy network of endothelially covered filamentous structures that reduce the luminal size and reduce the velocity of blood flow. The cavernous sinus receives its blood from a number of sources, including the **pterygoid plexus of veins** (via the emissary veins), the **angular vein** (via the inferior and superior ophthalmic veins), as well as the **middle** and **inferior cerebral veins**. There are numerous interrelationships of the cavernous sinus with the venous channels on the outside of the skull, which presents many possible pathways for the entry of infectious agents with a possibility of causing **meningitis**. The **superior** and **inferior petrosal sinuses** drain blood from the cavernous sinus into the transverse sinus and into the superior bulb of the internal jugular vein, respectively.

The cavernous sinus is associated with the internal carotid artery as well as a number of cranial nerves, four of which are embedded in its wall and one other that passes through the lumen of the sinus. Embedded in the lateral wall of the cavernous sinus are the **ophthalmic** and **maxillary divisions of the trigeminal nerve** and the **oculomotor** and **trochlear nerves**. The **internal carotid artery** and its associated **carotid sympathetic plexus** of fibers, as well as the **abducens nerve**, pass through the lumen of the cavernous sinus. It should be stressed, however, that the vessel and nerves travel in a sheath of endothelial cells that isolate them from the bloodstream.

Sphenoparietal sinus

The **sphenoparietal sinus**, a small sinus, is located on the inferior aspect of the edge of the lesser wing of the sphenoid bone (Fig. 8.3). It receives blood from several tiny meningeal veins and from the anterior temporal diploic vein; it drains into the cavernous sinus.

Superior petrosal sinus

The **superior petrosal sinus** is located in a shallow groove on the superior aspect of the petrous portion of the temporal bone (Fig. 8.8). It receives blood via small veins from the tympanic cavity, as well as from the cerebellar and inferior cerebral veins and drains into the transverse sinus just prior to its joining the sigmoid sinus.

Inferior petrosal sinus

The **inferior petrosal sinus**, located in the shallow groove formed between the petrous temporal and occipital bones (Fig. 8.8), receives blood from the cavernous and superior petrosal sinuses, the veins of the pons, medulla oblongata, and cerebellum, and from the labyrinthine veins. The inferior petrosal sinus drains into the superior bulb of the internal jugular vein.

128 ••• CHAPTER 8

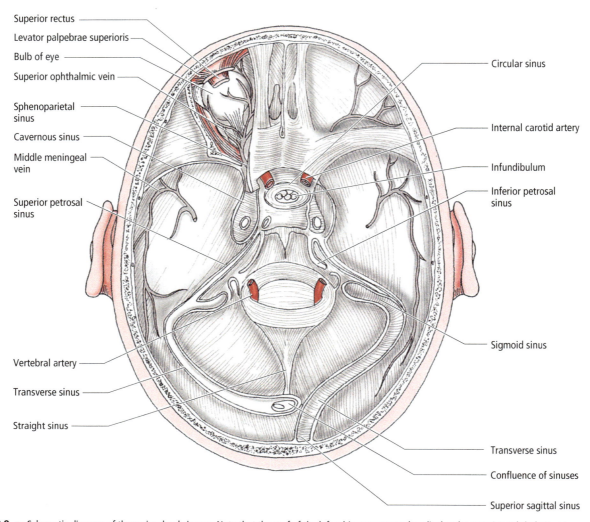

Figure 8.8 • Schematic diagram of the major dural sinuses. Note that the roof of the left orbit was removed to display the superior ophthalmic vein.

Basilar plexus

The **basilar plexus**, a group of slender venous channels located in the meningeal dura mater on the basilar portion of the occipital bone, acts as a venous connection between the right and left inferior petrosal sinuses. It drains into the anterior vertebral plexus of veins.

Posterior superior group

The **posterior superior group** of dural sinuses includes the superior sagittal, inferior sagittal, straight, transverse, sigmoid, petrosquamous, and occipital sinuses as well as the confluence of sinuses (see Figs. 8.1–8.3, 8.9, 8.10).

Superior sagittal sinus

The superior sagittal sinus begins at the foramen cecum and ends in the transverse sinus of the right-hand side

The **superior sagittal sinus** (Figs. 8.3–8.5, 8.10), located in the superior, convex aspect of the falx cerebri, begins at the **foramen cecum**, where blood from the **emissary vein of the foramen cecum** enters it and, usually, ends in the **transverse sinus** of the right side. Along its length the superior sagittal sinus grooves the internal aspects of the frontal bone, the suture line between the two parietal bones, and the squamous portion of the occipital bone, forming the **groove for the superior sagittal sinus**. This sinus receives blood from many of the cerebral veins, from the superior cerebellar veins, and the emissary veins traversing the parietal foramina. Additionally, **lacunae lateralis**, endothelially lined spaces housing arachnoid granulations, drain their blood mixed with CSF into the superior sagittal sinus.

Inferior sagittal sinus

The **inferior sagittal sinus** is much smaller, narrower, and shorter than the superior sagittal sinus. It is located in the

MENINGES AND CEREBROSPINAL FLUID ● ● ● 129

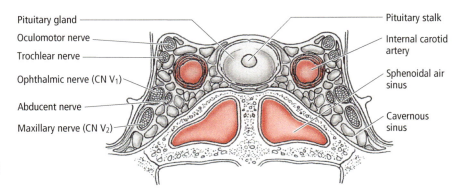

Figure 8.9 ● Diagram of the cavernous sinuses and their contents.

concave, inferior aspect of the **falx cerebri** occupying a little more than its posterior half (Fig. 8.4). Occasional veins from the medial aspect of the cerebral hemispheres and small veins of the falx cerebri deliver their blood into the inferior sagittal sinus, which, in turn, drains into the straight sinus.

Straight sinus

The **straight sinus** occupies the interface between the tentorium cerebelli and the falx cerebri (Fig. 8.3) and is formed by the junction of the great cerebral vein with the inferior

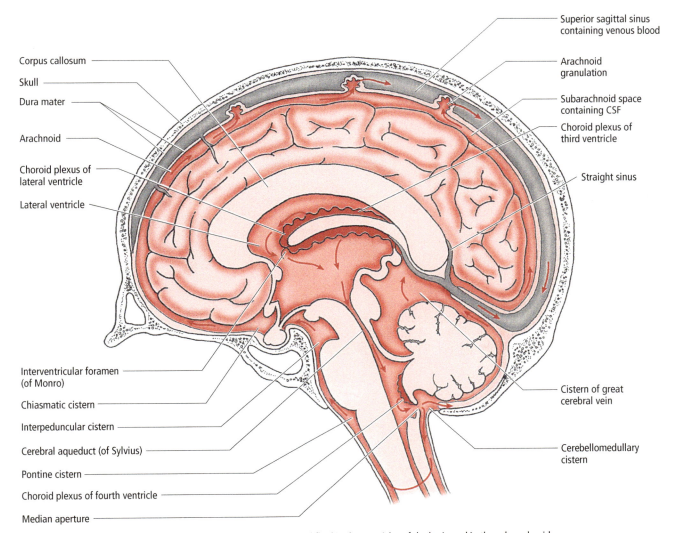

Figure 8.10 ● Hemisected skull demonstrating the flow of cerebrospinal fluid in the ventricles of the brain and in the subarachnoid spaces.

CHAPTER 8

sagittal sinus. It receives blood from those two structures as well as from some of the cerebellar veins and drains, usually, into the left transverse sinus.

Transverse sinuses

The paired **transverse sinuses** begin at the internal occipital protuberance of the occipital bone and are located in the lateral aspect of the tentorium cerebelli (Figs. 8.3 and 8.8). They are responsible for the formation of the grooves for the transverse sinuses. The right transverse sinus is usually larger and is the continuation of the superior sagittal sinus, whereas the left transverse sinus is the continuation of the straight sinus, although, occasionally the reverse occurs. Each transverse sinus ends in, and delivers its blood to, the sigmoid sinus; whereas each receives blood from the inferior cerebellar vein, the diploic veins from the temporal and occipital regions, the veins from the cerebral hemispheres, and from the petrosquamous and superior petrosal sinuses.

Sigmoid, petrosquamous, and occipital sinuses and the confluence of sinuses

Each **sigmoid sinus** – a continuation of each transverse sinus – follows an S-shaped curve as it grooves the temporal and occipital bones (Fig. 8.3) to terminate in the posterior aspect of the superior bulb of the internal jugular vein in the jugular foramen. The **petrosquamous sinus,** when present, is located in the meningeal reflection of the dura along the intersection of the petrous and squamous portions of the temporal bone. It receives blood from a diploic vein in its vicinity and drains into the transverse sinus. The smallest of the sinuses is the **occipital sinus** located in the posterior margin of the falx cerebelli. It delivers its blood into the transverse sinus or, with equal frequency, into the **confluence of sinuses** located just lateral to the internal occipital protuberance (Fig. 8.3). The confluence of sinuses receives blood from the superior sagittal, straight, and occipital sinuses. Blood flow in the confluence of sinuses is somewhat turbulent and pressures are established in such a fashion that, usually, blood arriving from the straight sinus is shunted into the left transverse sinus, whereas blood from the superior sagittal sinus reaches the right transverse sinus.

CEREBROSPINAL FLUID

Cerebrospinal fluid is manufactured by the choroid plexuses of the ventricles of the brain

Although the choroid plexuses located in every ventricle of the brain manufacture **cerebrospinal fluid**, the majority of this fluid is formed in the two lateral ventricles. The average rate of CSF production is approximately 14–35 mL per hour for a total daily production of less than 800 mL. However, only about 150 mL of CSF occupies the ventricles, central canal of the spinal cord, perivascular space, and the subarachnoid space and

Table 8.2 ● Composition of cerebrospinal fluid.

Constituent	Amount
White blood cells	0–5 cells/mL
Protein	Almost none
Glucose	2.1–4.0 mmol/L
Na+	135–150 mmol/L
K+	2.8–3.2 mmol/L
Cl-	115–130 mmol/L
Ca2+	1.0–1.4 mmol/L
Mg2+	0.8–1.3 mmol/L
pH	7.3

its dilated cisterna at any one time because there is a constant drainage of CSF into the superior sagittal sinus by the numerous arachnoid granulations and by the drainage of CSF alongside of the olfactory nerves through the cribriform plate of the ethmoid bone; there is also a very limited amount of drainage of CSF alongside of the other cranial nerves as they leave the interior of the skull.

CSF, a clear fluid with a low density, is rich in sodium, potassium, and chloride ions but has almost no protein and only occasional lymphocytes and a few desquamated epithelioid cells (Table 8.2). CSF forms a protective cushion for the brain and spinal cord and is also a recipient of the brain metabolites, which then reach the systemic circulation as the CSF is returned to the bloodstream.

The **blood–CSF barrier**, composed of zonulae occludentes—tight junctions formed by the fusion of cell membranes of contiguous cells of the simple cuboidal epithelium of the choroid plexus – maintains the chemical stability of the CSF. These tight junctions prevent paracellular movement of substances, thus requiring them to take the transcellular route via facilitated and active transports across the epithelium of the choroid plexus, resulting in differences in the composition of CSF and plasma.

VENTRICLES OF THE BRAIN

The four ventricles of the brain – the two lateral ventricles, the third ventricle, and the fourth ventricle – are lined by ependymal cells that isolate the CSF from the brain tissues

Although the ventricles of the brain have been described previously (see Chapter 6), they are addressed here again, for completeness of discussion of the cerebrospinal fluid.

The four ventricles and the central canal of the spinal cord are the remnants of the lumen of the embryonic neural tube.

The largest ventricles, the paired **lateral ventricles**, are horseshoe-shaped cavities separated from one another by the **septum pellucidum**, and are located in the right and left cerebral hemispheres. Due to their shape, each lateral ventricle is said to possess a body and four horns, the anterior, posterior, lateral, and inferior horns. The **anterior horn** hollows out the frontal lobe; the **inferior horn** is the cavity of the temporal lobe, whereas the occipital lobe houses the variable-sized **posterior horn** of the cerebral hemisphere. The **body** of each lateral ventricle is located mostly in the parietal lobe. The body and the inferior horn contain a relatively extensive **choroid plexus** that is responsible for the manufacture of most of the CSF. The two lateral ventricles communicate with the third ventricle via the right and left **interventricular foramina** (eponym: **of Monro**) (Fig. 8.10).

The **third ventricle**, the quadrilateral, slit-like, vertically positioned space whose walls are formed by the right and left thalami, is interrupted by a mass of gray matter, the **intermediate mass (massa intermedia)**, that forms a bridge between the two thalami. The roof of the third ventricle is formed by the tela choroidea (L., "woven choroid"), whereas its floor is formed by the hypothalamus, whose separation from the thalamus is indicated on the wall of the third ventricle by the **hypothalamic sulcus**. The third ventricle has several out-pocketings: the preoptic, infundibular, mammillary, and pineal recesses. It is drained by the **mesencephalic aqueduct (of Sylvius)** that conveys CSF into the fourth ventricle.

The **fourth ventricle** is an irregularly shaped space in the hindbrain, extending from the mesencephalic aqueduct (of Sylvius) posteriorly to the **obex** anteriorly. It is continuous with the central canal of the spinal cord. The lateral aspect of the fourth ventricle has two openings, the right and left **foramina of Luschka** and the single, median **foramen of Magendie**, all three foramina draining the CSF from the fourth ventricle into the **subarachnoid space**. Specifically, the foramina of Luschka lead to the **interpeducular cistern**, whereas the foramen of Magendie delivers the CSF into the **cerebellomedullary cistern**.

CLINICAL CONSIDERATIONS

Meningitis

Meningitis, a bacterial or viral inflammation of the meninges, is an exceptionally dangerous condition affecting about two or three people per 100,000 in the USA annually (the incidence in neonates is lower, about 5 in 1,000,000). The symptoms in neonates and babies include fever, lethargy, respiratory distress, poor feeding, vomiting, irritability, an unusual high-pitched cry when held and quiet when placed down in a stationary position, and occasionally bulging at the fontanelle. In children and adults, the symptoms are different from those of neonates and babies and include high fever and chills, severe headache, painful and stiff neck, nausea and/or vomiting, and in later stages, sleepiness, confusion, and difficulty in waking up. Meningitis has a high degree of morbidity and mortality, especially if not diagnosed properly.

Meningiomas

Meningiomas are benign, encapsulated tumors composed of fusiform cells originating in the leptomeninges, specifically the arachnoid. Usually, these tumors occur in adults and reach a fairly large size, on average 3 cm, before being diagnosed. Approximately 20% of all brain tumors and 10% of all spinal cord tumors are meningiomas, but, depending on their location, most can be treated with relative ease with good prognosis. The symptoms include headaches, seizures, weakness, and possibly paralysis or impairment of some brain functions due to pressure being applied to specific areas of the brain by the tumor. Diagnosis requires either radiographic or MRI techniques.

Blood–brain barrier

The integrity of the blood–brain barrier prevents many substances, including some neurotransmitters and drugs, from penetrating it. In order to permit the delivery of certain drugs through the blood–brain barrier, procedures were developed such as perfusion with a hypertonic solution, **mannitol.** This temporarily disables the fasciae occludentes of the capillary endothelial cells, allowing the delivery of certain therapeutic drugs.

Hydrocephalus

The constant production of CSF by the choroid plexus must be mirrored by its constant resorption by the arachnoid villi. If too little CSF is resorbed or if there is a blockage of CSF flow within the ventricular system of the brain, the result will be swelling of the brain tissue, a condition known as **hydrocephalus** (G., "water head"). This situation results in an increased head size in the neonate and fetus, and impaired muscular, cognitive, or other mental functions in the adult, possibly resulting in death if left untreated.

CHAPTER 8

SYNONYMS AND EPONYMS OF THE CRANIAL MENINGES

Name of structure or term	Synonym(s)/ eponym(s)
Arachnoid granulations	Arachnoid villi
Cavum trigeminale	Meckel's cave
Chiasmatic cistern	Cisterna basalis
Cisterna magna	Cisterna cerebellomedullaris
Superior cistern	Cistern of the great cerebral vein

FOLLOW-UP TO CLINICAL CASE

The immediate concern for any ER physician should be the possibility of meningitis (infection of the meninges) or possibly even encephalitis (infection within the brain itself). These are two examples of neurologic emergencies and require careful evaluation. Both of these conditions could result in death, depending on the particular infectious organism involved (among other things).

Work-up did reveal that this O.P. had **acute meningitis**. Because the initial presentation was so suggestive of meningitis, the patient should have been started immediately on appropriate antibiotics. Any acute infectious illness can be accompanied by dangerously low blood pressure (i.e., sepsis) or other abnormalities of vital signs. A head CT was performed to look particularly for bleeding or a mass lesion. This was normal (as it almost always is in cases of meningitis). A lumbar puncture was performed to look at the CSF. This

showed elevated protein (80 mg/L), elevated white blood cells (210 cells/mm^3), and no red blood cells. The white cells were mostly lymphocytes. Staining for bacteria, fungi, and parasites was negative. CSF cultures were negative (although these are not known until a few days after the CSF is drawn).

After several days, O.P. fully recovered. This case represents aseptic meningitis since no organism was identified. Most cases of aseptic meningitis are assumed to be viral. This is actually a fairly common infection, and very mild cases are certainly missed since patients may not even go to the doctor for it. Viral meningitis is usually a self-limiting infection and is often mild. However, bacterial meningitis is very serious and often results in death or permanent disability. Both viral and bacterial infections are acute, but other organisms cause subacute or chronic meningitis such as tuberculosis or fungi.

QUESTIONS TO PONDER

1. How is the cranial dura mater different from the spinal dura mater?

2. How does the vascular supply of the cranial periosteal layer of the dura mater differ from that of the meningeal layer?

3. It is well known that the brain itself does not "feel" pain, yet patients with epidural or subdural hematomas may experience severe pain.

4. How does cerebrospinal fluid located in the ventricles of the brain reach the subarachnoid cisterns?

5. How does the cerebrospinal fluid leave the subarachnoid space and where does it go?

CHAPTER 9

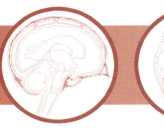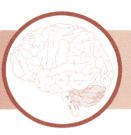

Vascular Supply of the Central Nervous System

CLINICAL CASE

VASCULAR SUPPLY OF THE SPINAL CORD

ARTERIAL SUPPLY OF THE BRAIN

VENOUS DRAINAGE OF THE BRAIN

CLINICAL CONSIDERATIONS

SYNONYMS AND EPONYMS

FOLLOW-UP TO CLINICAL CASE

QUESTIONS TO PONDER

CLINICAL CASE

J.G., a 68-year-old patient, presents to the emergency room with a sudden onset of right arm and leg weakness and altered speech. The right side of J.G.'s face is noted by the family to be "droopy," and they say that the patient is speaking "gibberish." This began very suddenly about 3 hours before presenting to the ER. The symptoms have been about the same since the onset. The history cannot be obtained from the patient since J.G.'s speech is not intelligible.

The exam shows the patient to be alert and attentive. J.G. can speak words but they are contextually inaccurate, and the patient cannot form understandable sentences or even phrases. J.G. cannot tell the names of simple objects that are presented and cannot follow direction appropriately. The right side of the patient's face and the right arm are very weak, while the right leg is only mildly weak. The rest of the neurologic exam is unremarkable.

The average brain weighs only about 1,250–1,450 g and occupies almost all of the limited space available in the cranial cavity; it should be noted that the volume occupied by the brain cannot expand because of the synarthrotic joints that hold the skull's component bones together. The brain is supplied by approximately 400 miles (~640 km) of blood vessels and receives about 750 mL of blood per minute, almost 20% of the total blood volume of an adult individual, indicating the brain's requirement for a large supply of oxygen and nutrients, especially glucose. In order to provide the brain with such a high volume of blood, two pairs of major arteries – the right and left **vertebral arteries** and the right and left **internal carotid arteries** – deliver all of their blood almost exclusively to the brain. These four vessels form a very effective anastomotic network, the **cerebral arterial circle** (eponym: of Willis), that is designed to ensure an uninterrupted blood supply to the brain. The brain's requirement for a constant blood flow is evidenced by the fact that interruption of the blood supply for a mere 10–15 seconds results in losing consciousness and after 5 minutes without blood flow to

---

A Textbook of Neuroanatomy, Third Edition. Maria A. Patestas, Amanda J. Meyer, and Leslie P. Gartner.
© 2025 John Wiley & Sons, Inc. Published 2025 by John Wiley & Sons, Inc.
Companion website: www.wiley.com.go/patestas/neuroanatomy3e

the brain irreversible brain damage occurs. With the notable exceptions of the basilar artery and the anterior communicating artery, all other arteries of the brain are paired.

Before discussing the blood supply of the brain, it should be noted that although the description that follows implies very specifically established vascular patterns, there are numerous variations even in the larger vessels. The vertebral arteries serve the occipital lobes of the cerebral hemispheres, brainstem, parts of the thalamus, as well as the cerebellum, whereas the internal carotid artery supplies blood to the remainder of the brain. One tends to think of larger vessels as having thick walls, but those of the brain are thin walled. Furthermore, since the arteries of the brain travel in the subarachnoid space, if there is a hemorrhage the blood accumulates in that space. Finally, it should be realized that there is only a limited amount of collateral circulation in the brain, even though the cerebral arterial circle forms anastomotic connections between the right and left sides of the brain.

Although the vascular supply of the spinal cord was discussed in Chapter 5, it will be repeated here for two reasons: so that the reader does not have to refer back to a previous chapter and so that the reader can appreciate the vascular supply of the entire central nervous system (CNS) in a single chapter.

VASCULAR SUPPLY OF THE SPINAL CORD

The spinal cord receives its blood supply from two pairs of longitudinally arranged vessels, the anterior and posterior spinal arteries as well as from small, segmental radicular arteries

The **anterior spinal arteries**, direct branches of the vertebral arteries, join with each other to form a single median vessel, the anterior spinal artery, which occupies and follows the anterior median fissure of the spinal cord (Fig. 9.1). This vessel extends from within the cranial cavity throughout the entire length of the spinal cord and provides small branches that penetrate and supply the white and gray matter of the spinal cord. Moreover, this vessel also supplies the medulla oblongata. The anterior spinal artery may be quite small in the thoracic region. It has been demonstrated in a small study of nine adult cadavers in 2008, that there is a great degree of variation in the origin of the anterior spinal artery. Most of the variability involved the surface of the vertebral artery which gave rise to the anterior spinal arteries; in slightly over 50% of the cases they originated from the medial surface, in approximately 25% from the anteromedial, and about 25% from the posteromedial surface.

The **posterior spinal arteries** also demonstrate variability in their site of origin. They arise from the vertebral arteries

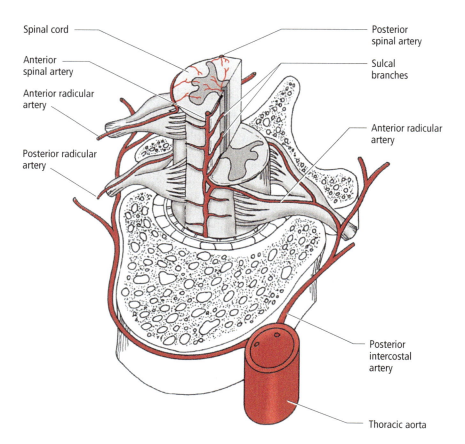

Figure 9.1 • Arterial supply of the spinal cord.

VASCULAR SUPPLY OF THE CENTRAL NERVOUS SYSTEM

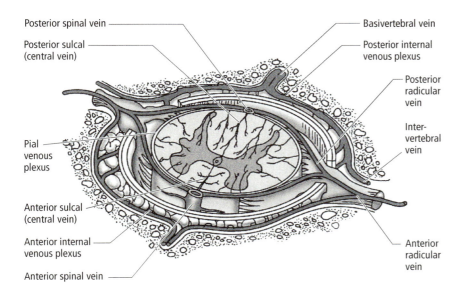

Figure 9.2 • Venous drainage of the spinal cord.

directly, or frequently indirectly, by way of the posterior inferior cerebellar artery. Moreover, it has been reported in 2018 that only approximately 30% of the cadavers studied, did the posterior spinal arteries arise symmetrically from their vessel of origin; in the other 70% of the cases, the posterior spinal arteries arose prior to or after the vertebral artery entered the base of the skull. Each posterior spinal artery extends from within the cranial cavity throughout the entire length of the spinal cord, sandwiching the posterior rootlets between them. These two vessels provide small branches that penetrate and serve the white and gray matter of the spinal cord as well as the medulla oblongata.

The 31 pairs of **radicular arteries** are small vessels that arise from arteries in the immediate vicinity of the spinal column. Each radicular artery enters the intervertebral foramen where it bifurcates, forming an **anterior** and a **posterior radicular artery**, which follow the anterior and posterior roots, respectively, to gain entrance into the vertebral canal. The anterior and posterior radicular arteries anastomose with branches of the anterior and posterior spinal arteries on the surface of the spinal cord, and arborize to supply the white and gray matter of the spinal cord. It should be stressed that the radicular arteries are extremely important for the vascularization of the spinal cord, because, with the exception of the cervical region, the anterior and posterior spinal arteries by themselves are unable to provide an adequate vascular supply to the spinal cord. Therefore, an injury to a spinal nerve damages not only the afferent and efferent fibers of a particular spinal cord level, but may also damage the segmental white and gray matter by producing ischemic conditions due to the reduction in blood supply from the radicular artery serving that region. It should be noted that the **great anterior radicular artery** (eponym: artery of Adamkiewicz) is the largest, albeit inconsistent, of the radicular arteries. It usually arises on the left-hand side and serves much of the inferior half of the spinal cord, entering the vertebral canal between T8 and L1, and contributes to the formation of the inferior aspect of the anterior spinal artery.

Several longitudinally arranged tortuous veins of the pia mater are responsible for the venous drainage of the spinal cord (Fig. 9.2). These are the **anterior** and **posterior spinal veins** that communicate with the segmentally arranged radicular veins. The **radicular veins** follow the paths of their companion radicular arteries to leave the vertebral canal via the intervertebral foramina. Along the way they also communicate with the **epidural venous network** (embedded in the epidural fat of the vertebral canal), an interconnected system of veins that delivers its blood into the cavernous sinus of the cranial cavity. Cranially, the anterior and posterior spinal veins drain into the inferior cerebellar veins and inferior petrosal sinuses.

ARTERIAL SUPPLY OF THE BRAIN

In order to protect the brain from certain substances carried in the bloodstream, the blood–brain barrier and neurovascular unit is established by endothelial cells in conjunction with vascular smooth muscle cells, pericytes, astrocytes, neurons, microglia, astrocytes, and the basal lamina matrix

Blood–brain barrier and neurovascular unit

The blood–brain barrier and the neurovascular unit prevent certain molecules and macromolecules from entering the substance of the brain by the formation of occluding junctions and adhering junctions by the endothelial cells

The basal lamina-lined endothelial cells of the capillaries of the brain and spinal cord form a **blood–brain barrier**. These endothelial cells not only form **fasciae occludentes** with each other, but they have only a limited ability to form **caveolae**, relying instead mostly on **receptor-mediated transport** to transfer material between the capillary lumen and the neural tissue of the CNS. Therefore, most macromolecules injected into the capillary lumen are unable to gain access into the intercellular spaces of the brain and spinal cord. Similarly, most macromolecules injected into the intercellular compartment of the CNS are unable to enter the capillary lumen, unless the endothelial cells possess specific receptors for them. Small, noncharged molecules and numerous small lipid-soluble materials can dissolve through the lipid membrane of the endothelial cells and thus are able to penetrate the blood–brain barrier. Molecules, such as some of the vitamins, amino acids, glucose, and nucleosides, can penetrate the blood–brain barrier via carrier proteins, specific to the various molecules that may be transported, with or without passive diffusion. Ion channels are also present in the endothelial cell membranes, and they are responsible for the transport of ions across the blood–brain barrier. The blood–brain barrier does not permit immune cells, especially lymphocytes, from leaving the blood vessels thus preventing them from entering the brain tissue whereas monocytes can readily pass through the tight and adhering junctions between endothelial cells to become microglia.

The end-feet of astrocytes contact and form a sheet around the basal lamina of the capillaries of the CNS. These end-feet are referred to as the **perivascular glia limitans**. These astrocytes function not only in maintaining the neurochemical balance of the intercellular compartment of the CNS by removing excess K^+ ions and neurotransmitters that were released into the intercellular spaces, but also in transferring metabolites from the capillary lumen to neurons in their vicinity.

The **neurovascular unit** (**NVU**), composed of the endothelial cells, vascular smooth muscle cells, pericytes, astrocytes, microglia, and the basal lamina matrix of the capillaries, functions in bolstering the ability of the blood–brain barrier to perform its functions. The barrier as well as the transport abilities of the endothelial cells are regulated by the non-endothelial cell components of the NVU. They do that by the release of various signaling molecules some of which relax the tight junctions between endothelial cells to permit the paracellular passage of certain substances and cells (e.g., permitting monocytes but not lymphocytes). Additionally, it has been demonstrated that certain gases, such as CO_2 and NH_3, do not diffuse through the plasma membranes by simple diffusion; instead, they pass through the NVU via aquaporin-1 and aquaporin-4 channels by way of facilitated diffusion. The NVU also regulates blood flow and oxygen levels by way of the astrocytic end-feet which can release vasoactive substances that can either dilate (e.g., NO) or constrict (e.g., tetraenoic acid derivatives) blood vessels depending on the prevailing situation. Neurovascular units also guard the "immune-privileged" status of the central nervous system, so that many types of immune cells, such as B and T lymphocytes, are prevented from entering the CNS. However, monocytes, cells that are much larger than lymphocytes, are permitted access to the central nervous system, where they differentiate into specialized macrophages,

known as microglia. Using a somewhat simplistic description, there are two phenotypes of macrophages/microglia, M1 and M2, where M1 are proinflammatory and neurotoxic, that is they release factors that foster inflammation that harm neurons and the "alternatively activated" M2 macrophages that are anti-inflammatory and have neuroprotective properties. It should be noted, however, that macrophages/microglia possess intermediary properties of both M1 and M2 cell types, depending on the health or pathology of the individual patient.

Microglia constantly monitor their environment by extending their many processes and, due to their numerous membrane-bound receptor molecules, communicate with other members of the neurovascular unit to respond to pathologic events in order to protect the central nervous system. They have specific roles in certain pathological conditions, such as strokes, neuroinflammation, meningitis, multiple sclerosis, Alzheimer disease, Parkinson disease, benign and malignant tumors. It should be noted that recent investigations of SARS-CoV-2 (Corona virus-19) have demonstrated that this virus is capable of disrupting the blood–brain barrier.

Circumventricular organs

Circumventricular organs (CVOs) are highly vascular specific secretory and sensory regions of the brain that display the absence of the blood–brain barrier. The secretory components of the CVOs are the pineal gland, neurohypophysis, and the median eminence, all of which are able to release neurohormones into their vascular supplies without being restricted by a blood–brain barrier. The sensory CVOs are the vascular organ of the lamina terminalis, the subfornical organ, and the area postrema that are able to detect the presence of specific blood-borne molecules. Neurons associated with these sensory components of the CVOs transmit that information to other regions of the brain. The area postrema houses the vomiting center and, if it detects the presence of certain toxins in the circulating blood, it sends neural impulses to induce vomiting. The other two sensory CVOs in the lamina terminalis and in the subfornical organ both act as sensors of fluid homeostasis.

Internal carotid artery

The internal carotid artery, one of the paired terminal branches of the common carotid artery, forms numerous branches supplying the brain and also participates in the formation of the cerebral arterial circle (eponym: of Willis)

The **internal carotid artery**, one of the paired terminal branches of the common carotid artery, has no branches in the neck. It ascends in the neck, enters the cranial cavity through the carotid canal of the petrous portion of the temporal bone, and reaches the cavernous sinus from below. It passes through, and subsequently pierces the roof of the cavernous sinus and enters the

VASCULAR SUPPLY OF THE CENTRAL NERVOUS SYSTEM ••• 137

cranial cavity where it is flanked by the oculomotor and optic nerves (Fig. 9.3). Occasionally, the anterior and posterior clinoid processes fuse to form a foramen, and the internal carotid artery uses that aperture to enter the cranial cavity. The internal carotid artery then approximates the anterior perforated substance where it terminates into its various branches. Along this path, from the neck to its termination, the internal carotid artery makes two almost 90° turns, thus reducing both the pressure and the velocity of blood it brings to the thin-walled vessels of the brain. It should be noted that were it not for the almost 90° turns the blood pressure in the intracranial portion of the carotid artery would damage and most probably destroy the vessel wall causing an intracranial hemorrhage. Although the internal carotid artery is said to be divided into four regions – the cervical, petrous, cavernous, and cerebral – only the **cavernous** and **cerebral portions** are discussed in this textbook. Readers interested in the distribution and paths of the cervical and petrous portions are urged to read any major textbook of Gross Anatomy, Head and Neck Radiology, or Head and Neck Anatomy.

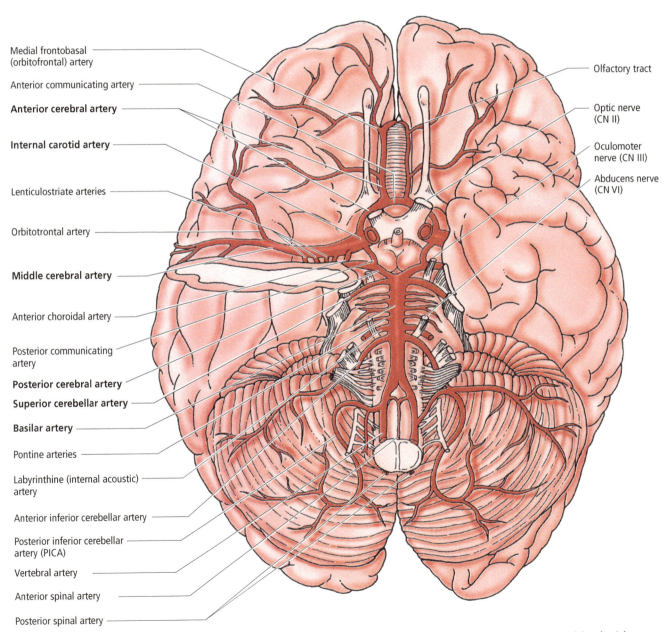

Figure 9.3 • Arterial supply of the brain. Note that the frontal lobes are spread apart somewhat to show the anterior cerebral arteries and that the right temporal lobe is severed to show the path and branches of the middle cerebral artery. The ophthalmic artery, a branch of the internal carotid artery, is not included in this drawing.

CHAPTER 9

Cavernous portion of the internal carotid artery

Branches of the cavernous portion of the internal carotid artery supply the hypophysis as well as the dura mater of the anterior cranial fossa

The **cavernous portion** of the internal carotid artery passes through the blood-filled cavity of the cavernous sinus, but is isolated from the blood of the cavernous sinus by being invested with a simple squamous epithelium, the lining of the cavernous sinus. The abducens nerve accompanies the internal carotid artery through the cavernous sinus. While in the sinus the artery describes an S-shaped curve and gives off small branches: the cavernous, superior, and inferior hypophyseal, ganglionic, and anterior meningeal arteries. As their names suggest, these vessels serve the areas of their namesake (Table 9.1).

Cerebral portion of the internal carotid artery

The cerebral portion of the internal carotid artery has the following branches: ophthalmic, anterior choroidal, posterior communicating, anterior cerebral, and middle cerebral arteries

The **cerebral portion** of the internal carotid artery is quite short (Fig. 9.3). Almost as soon as it pierces the roof of the cavernous sinus it gives rise to its branches, the ophthalmic, anterior choroidal, and posterior communicating arteries, as well as to its two terminal branches, the anterior cerebral and middle cerebral arteries. The ophthalmic artery serves the contents of the orbit, regions of the nasal cavity, and the superficial face and is not discussed in this textbook.

Anterior choroidal artery

The **anterior choroidal artery** is a rather small, narrow artery that arises from the posterior aspect of the internal carotid

Table 9.1 ● Branches of the cavernous portion of the internal carotid artery.

Branch	Major areas served	Additional pertinent information
Cavernous	Walls of the cavernous sinus Walls of the inferior petrosal sinus	Anastomoses with branches of the middle meningeal artery
Superior hypophyseal	Pars tuberalis and infundibulum	Forms primary capillary bed
Inferior hypophyseal	Posterior lobe of the hypophysis	Sends some branches to the anterior lobe
Ganglionic	Trigeminal ganglion	Composed of several small branches
Anterior meningeal	Dura mater of the anterior cranial fossa	Anastomoses with the meningeal branch of the posterior ethmoidal artery

Table 9.2 ● Regions supplied by the anterior choroidal and posterior communicating arteries.

Branch	Major area served	Additional pertinent information
Anterior choroidal	Choroid plexus of the inferior horn of the lateral ventricle thus providing the choroid plexus with blood whose ultrafiltrate serves to form cerebrospinal fluid	Also serve the optic tract, lateral geniculate body, optic radiation, hippocampus, posterior limb of the internal capsule, and the tail of the caudate nucleus
Posterior communicating	Connects the internal carotid system with the vertebral arterial system	Anastomoses with the posterior cerebral artery; right and left arteries are not identical and one may be absent; occasionally it may bypass the cerebral arterial circle
Central branches	Posterior limb of the internal capsule, medial aspect of the thalamus, and tissue forming the lateral border of the third ventricle	Pierce the region of the base of the cerebrum that is posterolateral to the infundibulum of the pituitary

artery and follows the path of the optic tract until it reaches the lateral geniculate body, where it enters the choroid fissure to reach and supply the choroid plexus of the inferior horn of the lateral ventricle (Table 9.2), providing blood whose ultra-filtrate serves to form the cerebrospinal fluid. The anterior choroidal artery gives off numerous branches along the way which serve structures in their vicinity, including the optic tract, lateral geniculate body, optic radiation, hippocampus, posterior limb of the internal capsule, and the tail of the caudate nucleus.

Posterior communicating artery

The **posterior communicating artery** connects the cerebral portion of the internal carotid artery to the posterior cerebral branch of the basilar artery (Table 9.2). The right and left arteries are not identical, in that one is frequently smaller than the other, and, in fact, one may be entirely absent or even doubled. Other variation of this vessel was the underdevelopment of either one or of both arteries and deviations in their terminations, such as instead of terminating in the posterior cerebral artery it terminates in the basilar artery thus still completing the cerebral arterial circle. Occasionally, the posterior communicating artery bypasses the cerebral arterial circle and supplies the posterior aspect of the cerebrum, directly. The main function of the posterior communicating artery is to ensure a viable blood supply to the brain in case the internal carotid artery or the vertebral artery becomes occluded. However, it does also have **central branches** – thin, unsubstantial vessels – that pierce the region of the base of the cerebrum that is posterolateral to the infundibulum of the hypophysis and serve part of the posterior limb of the internal capsule, the medial aspect of the thalamus, and the tissue forming the lateral border of the third ventricle.

Anterior cerebral artery

The anterior cerebral artery has the following branches: the anterior communicating artery and the central and cortical branches

The **anterior cerebral artery**, one of the two terminal branches of the internal carotid artery, passes between the frontal lobes in the anterior part of the longitudinal cerebral fissure, where it lies within a few millimeters of the anterior cerebral artery of the other side (Figs. 8.3 and 8.4). The two vessels follow the genu and then the superior border of the corpus callosum until they each anastomose with the posterior cerebral artery, a branch of the basilar artery. It also supplies the anterior perforated substance, the corpus callosum, the frontal and parietal lobes of the cerebrum, and fuses with the posterior cerebral artery. The branches of the anterior cerebral artery are the anterior communicating artery, the central branches, and the cortical branches (Table 9.3).

Anterior communicating artery

The **anterior communicating artery** is a very short vessel, no more than 5–6 mm in length that connects the right and left anterior cerebral arteries to each other, forming the anterior end of the cerebral arterial circle. It has two or more small branches, the **anteromedial branches**, which serve the region of the optic chiasm, globus pallidus, and the amygdaloid body. Occasionally, the artery exists only as a fusion of the right and left anterior cerebral arteries and, in some instances, there are two or even three anterior communicating arteries; the incidence of only a single anterior communicating artery occurs only in approximately 60% of individuals.

Central branches

A number of small, unnamed vessels arise from the initial part of the anterior cerebral artery, just prior to the origin of

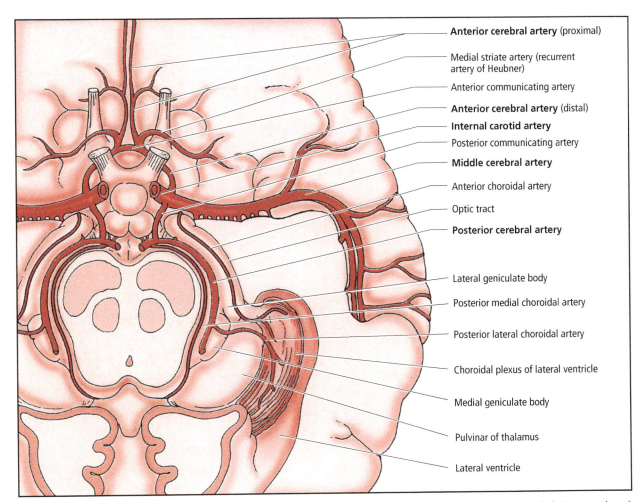

Figure 9.4 ● Region of the arterial cerebral circle displaying the blood supply to the choroidal plexus of the lateral ventricle. The ophthalmic artery, a branch of the internal carotid artery, is not included in this drawing.

Table 9.3 ● Regions supplied by the anterior cerebral artery and its branches.

Branch	Major area served	Additional pertinent information
Anterior communicating artery	Region of the optic chiasm via its anteromedial branches	Connects the right and left anterior cerebral arteries, forming the anterior boundary of the cerebral arterial circle (of Willis)
Central branches		
Unnamed branches	Septum pellucidum and anterior part of the corpus callosum	Arise just before the origin of the anterior communicating artery
Medial striate artery	Head of caudate nucleus, parts of putamen, globus pallidus, and anterior limb of internal capsule	Also known as the recurrent artery, it arises just after the origin of the anterior communicating artery
Cortical branches		
Orbital branches Frontal branches	Gyrus rectus, medial orbital gyrus, olfactory lobe	Serves area of the frontal lobe that lies on the intracranial roof of the bony orbit
Frontopolar artery	Medial surface of the superior frontal gyrus and lateral surface of the superior and middle frontal gyri	Originates at the level of the genu of the corpus callosum
Callosomarginal artery	Corpus callosum, cingulate gyrus, medial aspect of superior frontal gyrus, and precentral and postcentral gyri	Arises at the genu of the corpus callosum and parallels the remainder of the anterior cerebral artery as far posteriorly as the postcentral gyrus
Parietal branches	Medial aspect of precuneate gyrus and neighboring structures	Serve region of the brain as far posteriorly as the parieto-occipital sulcus

the anterior communicating artery; these supply the septum pellucidum as well as the anterior part of the corpus callosum. A larger branch, the **medial striate artery** (also referred to as the **recurrent artery**), arises just after the origin of the anterior communicating artery. It follows the anterior limb of the internal capsule to supply parts of the following regions of the basal nuclei: the head of the caudate nucleus, part of the putamen, the globus pallidus, and the anterior limb of the internal capsule.

Cortical branches

The cortical branches of the anterior cerebral artery are named by the regions that they serve; these are the orbital, frontal, and parietal arteries

The **cortical branches** of the anterior cerebral artery arise from that vessel as it courses along the genu and superior aspect of the corpus callosum. These branches, named by the areas that they serve, are the orbital, frontal, and parietal arteries.

The **orbital branches** distribute to the inferior aspect of the frontal lobes as they lie on the bony roof of the orbit, and serve the gyrus rectus, medial orbital gyrus, and olfactory lobe.

The frontopolar and callosomarginal branches constitute the **frontal branches** of the anterior cerebral artery. The **frontopolar artery** originates at the level of the genu of the corpus callosum and serves the medial surface of the superior frontal gyrus and the region of the lateral surface of the superior and middle frontal gyri. The **callosomarginal artery** branches from the anterior cerebral artery at the genu of the corpus callosum and parallels its parent artery as far posteriorly as the postcentral gyrus. It supplies blood to the corpus callosum, the cingulate gyrus, the medial aspect of the superior frontal gyrus, as well as to the precentral and postcentral gyri.

The remaining portion of the anterior cerebral artery, frequently renamed the **pericallosal artery**, extends as far posteriorly as the parieto-occipital sulcus, and gives rise to the **anterior** and **posterior parietal arteries**. These two vessels and their branches serve the medial aspect of the precuneate gyrus and neighboring structures.

Middle cerebral artery

The middle cerebral artery has central and cortical branches

The **middle cerebral artery**, the terminal branch and largest branch of the internal carotid artery, enters the lateral fissure (eponym: of Sylvius), passes along the free surface of the insula, and gives off several branches that ramify over the posterior and lateral surfaces of the cerebral cortex (Figs. 9.3–9.6). The middle cerebral artery has **central** and **cortical branches**, where the former supply blood to the deeper structures and the latter serve the cortical regions (Table 9.4).

VASCULAR SUPPLY OF THE CENTRAL NERVOUS SYSTEM

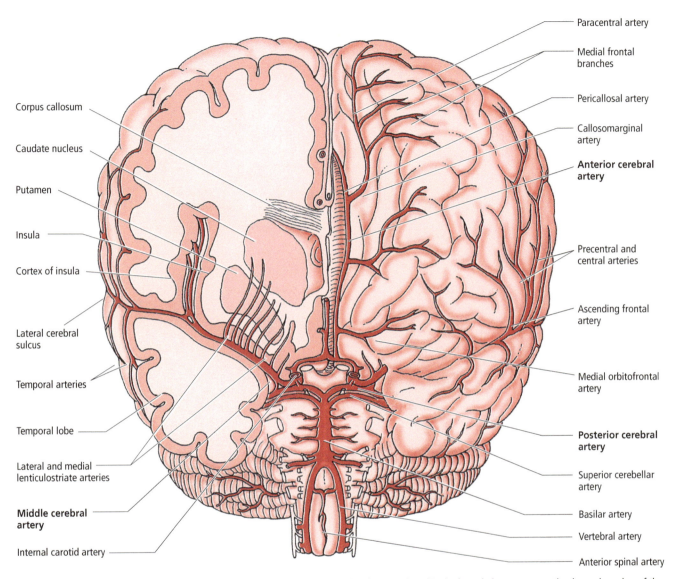

Figure 9.5 • Blood supply to the brain. Note that the right cerebral hemisphere has been sectioned in the frontal plane to expose the deeper branches of the middle cerebral artery.

Table 9.4 • Regions supplied by the middle cerebral artery and its branches.

Branch	Major area served	Additional pertinent information
Central striate arteries	Corpus striatum, head and body of caudate nucleus, lentiform nucleus, external and internal capsules	These vessels pierce the floor of the lateral sulcus to reach their points of destination located deep within the white matter of the cerebrum
Cortical branches		
Orbital branches	Inferior frontal gyrus, lateral aspect of orbital surface of frontal lobe	Orbitofrontal artery is the most prominent of the orbital branches
Frontal branches	Middle frontal and precentral gyri, part of inferior frontal gyrus	Anterior and posterior parietal branches
Parietal branches	Postcentral gyrus, part of superior and all of inferior parietal lobules, angular and supramarginal gyri	Parietotemporal and parieto-occipital arteries
Temporal branches	Lateral aspect of entire temporal lobe as far posteriorly as occipital gyri	Anterior, middle, and posterior temporal arteries

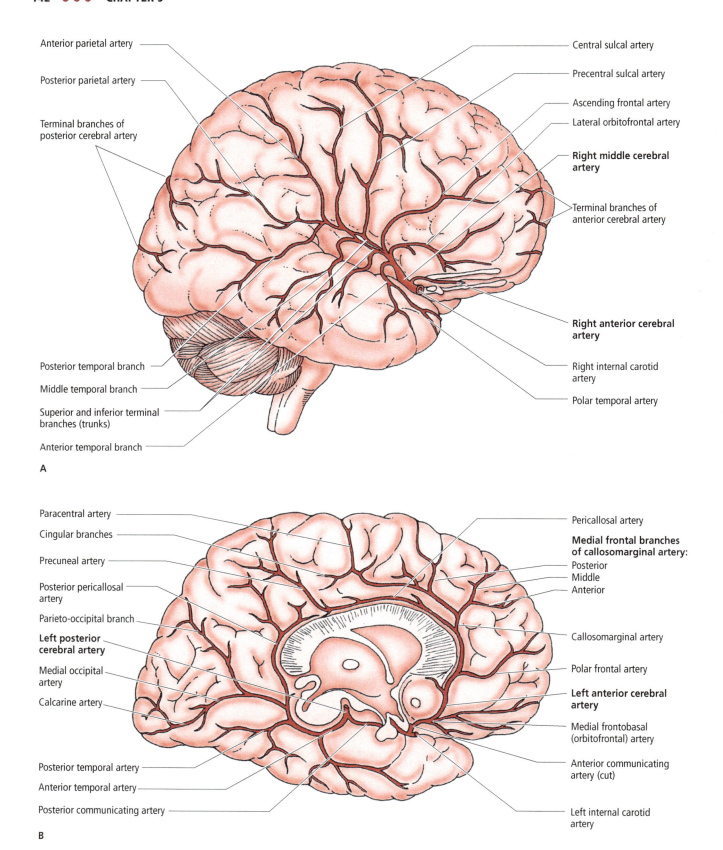

Figure 9.6 ● (A) Arterial supply to the lateral aspect of the brain. Note that the temporal lobe is partially reflected to permit a view of the vessels lodged within the lateral fissure (of Sylvius). (B) Arterial supply of the medial aspect of the brain.

VASCULAR SUPPLY OF THE CENTRAL NERVOUS SYSTEM • • • 143

Note that the clinical case at the beginning of the chapter refers to a patient with a sudden onset of unintelligible speech.

1 What should the physician suspect when there is a sudden onset of speech disorder?

2 What lesion of the CNS should be suspected when there is a sudden weakness on one side?

3 Why is the fact that "otherwise the patient is alert and attentive" is not a surprising finding in this case?

Central branches

There are 10–15 slender vessels, the **striate arteries (lenticulostriate arteries)**, that arise from the middle cerebral artery as it enters the lateral fissure (eponym: of Sylvius). These vessels pierce the floor of the lateral sulcus to supply structures deep within the white matter of the cerebral cortex, including the corpus striatum, much of the head and body of the caudate nucleus, and large portions of the lentiform nucleus and of the external and internal capsules, passing as deep as the external surface of the thalamus.

Cortical branches

Cortical branches of the middle cerebral artery are named by the regions that they serve: the orbital, frontal, parietal, and temporal arteries

The **cortical branches** originate from the middle cerebral artery as it courses along the lateral sulcus. These vessels, named according to their location and distribution, are the orbital, frontal, parietal, and temporal branches.

The **orbital branches**, the most prominent of which is the **orbitofrontal artery**, serve the inferior frontal gyrus and the lateral aspect of the orbital surface of the frontal lobe as it lies over the bony roof of the orbit. The **frontal branches** supply the middle frontal and precentral gyri as well as a region of the inferior frontal gyrus. The **anterior** and **posterior parietal branches** serve the postcentral gyrus as well as the inferior aspect of the superior parietal lobule and all of the inferior parietal lobule. Two other branches, the **parietotemporal** and **parieto-occipital arteries**, serve the remainder of the parietal lobe, terminating at the parieto-occipital sulcus, thus supplying the angular and supramarginal gyri. The temporal branches consist of three main vessels, the **anterior, middle**, and **posterior temporal arteries**. They serve the lateral aspect of the entire temporal lobe as far posteriorly as the occipital gyri.

Table 9.5 ● Regions supplied by the cranial portion of the vertebral artery and its branches.

Branch	Major area served	Additional pertinent information
Meningeal branch	Falx cerebelli	Ramify between periosteal and meningeal layers of the dura
Posterior inferior cerebellar artery		
Medial branch	Medial wall of cerebellum	Additional branches supply choroid plexus of the fourth ventricle
Lateral branch	Inferior aspect of cerebellum	Anastomoses with anterior inferior cerebellar and superior cerebellar arteries of the basilar artery
Medullary branches	Medulla oblongata	Several small branches of the vertebral artery

Vertebral artery

The right and left vertebral arteries enter the cranial cavity via the foramen magnum and join each other to form the basilar artery

The **vertebral artery**, the largest branch of the subclavian artery, arises from its first part and ascends in the neck, traveling through the **transverse foramina** of the sixth cervical through first cervical vertebrae. It describes an S-shaped curve along the posterior arch of the **atlas**, passes through the **foramen magnum** and enters the cranial cavity where it joins its counterpart of the other side to form the **basilar artery**. Only the cranial branches of the vertebral artery are described, and these are the meningeal branches, the posterior inferior cerebellar artery, and the medullary branches (Table 9.5).

Meningeal branches of the vertebral artery

The two or three **meningeal branches** of the vertebral artery serve the falx cerebelli. They ramify between the periosteal and meningeal layers of the dura mater of the cerebellar fossa.

Posterior inferior cerebellar artery

The **posterior inferior cerebellar artery** arises within the cranial cavity from the vertebral artery about 1–2 cm before the formation of the basilar artery (Fig. 9.7). The most common variations are that it may arise from the basilar artery, and it may be completely absent on one side. Usually, the posterior inferior cerebellar artery passes laterally from its origin from the vertebral artery, follows the contour of the medulla oblongata to pass between the rootlets of cranial nerves IX and X, and divides into a medial and a lateral branch.

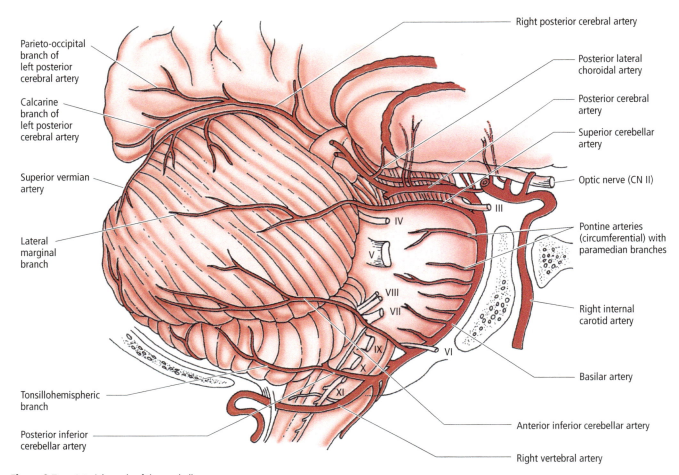

Figure 9.7 • Arterial supply of the cerebellum.

The **medial branch** supplies the medial aspect of the cerebellar hemispheres, whereas the **lateral branch** serves the inferior aspect of the cerebellum and anastomoses with the anterior inferior cerebellar artery and the superior cerebellar artery. Additional branches of the posterior inferior cerebellar artery supply the medulla oblongata as well as the choroid plexus of the fourth ventricle.

Medullary branches

Several small branches of the cranial portion of the vertebral artery, known as the **medullary branches**, serve the medulla oblongata. It should be noted that this important region is also served by the posterior inferior cerebellar artery (as indicated in the above paragraph), as well as by the anterior and posterior spinal arteries.

Basilar artery

> The basilar artery extends from the cranial end of the pyramids to the cranial end of the pons; it has several small branches and three larger ones

The **basilar artery**, formed by the fusion of the right and left vertebral arteries, lies in a shallow groove in the midline of the pons, extending from the cranial end of the pyramids to the cranial end of the pons (Fig. 9.7). It has several small branches, the pontine and labyrinthine arteries, and three larger branches, the anterior inferior cerebellar, superior cerebellar, and posterior cerebral arteries (Table 9.6).

Pontine arteries

The several small **pontine arteries** arise at right angles to the basilar artery and serve the pons and the midbrain, penetrating deep into the substance of these two regions.

Labyrinthine artery

The **labyrinthine artery (internal auditory artery)** is larger than the pontine arteries and originates somewhat cranial to the anterior inferior cerebellar artery (from which it occasionally arises). It reaches and enters the internal auditory meatus in the company of cranial nerves VII and VIII, and supplies the internal ear.

VASCULAR SUPPLY OF THE CENTRAL NERVOUS SYSTEM ● ● ● 145

Table 9.6 ● Regions supplied by branches of the basilar artery.

Branch	Major area served	Additional pertinent information
Pontine arteries	Pons and midbrain	Small vessels that arise at right angles from the basilar artery
Labyrinthine (internal auditory) artery	Internal ear	Enters the internal acoustic meatus in company with CN VII and VIII
Anterior inferior cerebellar artery		Its branches anastomose with branches of the posterior inferior cerebellar and superior cerebellar arteries
Pontine	Lateral aspect of pons	
Medial branches	Anteromedial aspect of cerebellum	
Lateral branches	Anterolateral aspect of cerebellum	
Superior cerebellar artery	Superior surface of cerebellum as well as pineal body, midbrain, and choroid plexus of third ventricle	Its branches anastomose with branches of the anterior inferior cerebellar artery
Posterior cerebral artery		It is connected to the internal carotid artery by the posterior communicating artery
Cortical branches		
Anterior temporal	Uncus and parahippocampal gyrus	
Posterior temporal	Lateral and medial occipitotemporal gyri	
Parieto-occipital	Cuneate and precuneate gyri	
Central branches		
Anterior thalamic	Optic tract, mammillary bodies, anterior and ventral medial nuclei of thalamus	
Posterior thalamic	Posterior thalamic nuclei, geniculate bodies	
Mesencephalic	Cerebral peduncles, interpeduncular region, corticospinal tracts, mesencephalic reticular formation, substantia nigra, tegmentum of midbrain	Has a small branch, the circumflex mesencephalic
Posterior choroidal branches		
Lateral choroidal	Choroid plexus of lateral ventricle, lateral geniculate body, pulvinar, dorsomedial nucleus of thalamus	Anastomoses with branches of the medial choroidal
Medial choroidal	Choroid plexus of third ventricle, pineal body, superior and inferior colliculi, dorsal medial nucleus of thalamus	Anastomoses with branches of the lateral choroidal

Anterior inferior cerebellar artery

The **anterior inferior cerebellar artery** arises from the basilar artery just cranial to the pyramids (although it may arise from the vertebral artery) and follows cranial nerve VI for a short distance. It loops around cranial nerves VII and VIII, just before they enter the internal acoustic meatus. Shortly before reaching the cerebellopontine angle, the artery gives rise to small **pontine branches** that pierce the lateral aspect of the pons. At the cerebellopontine angle the anterior inferior cerebellar artery bifurcates into **medial** and **lateral branches** that serve the anteromedial and anterolateral aspects of the cerebellum, respectively. The branches of the anterior inferior cerebellar artery anastomose with branches of the posterior inferior cerebellar artery and with those of the superior cerebellar artery.

Superior cerebellar artery

The **superior cerebellar artery** originates just inferior to the third cranial nerve. It curves around the cerebral peduncle to gain the superior surface of the cerebellum, which it vascularizes and as it arborizes, its branches provide vascular supply to the pineal body, the midbrain, and the choroid plexus of the third ventricle. Branches of the superior cerebellar artery anastomose with branches of the anterior inferior cerebellar artery. There are frequent duplications of the superior cerebellar artery, where one of the anomalous branches may place pressure on the trigeminal nerve, possibly causing trigeminal neuralgia in the patient. At times the superior cerebellar artery arises from the posterior cerebral rather than from the basilar artery.

Posterior cerebral artery

The posterior cerebral artery gives rise to the cortical, central, and posterior choroidal arteries

The right and left **posterior cerebral arteries** are the terminal branches of the basilar artery (Fig. 9.7). Each originates just cranial to the superior cerebellar artery and follows the course of that vessel as they both curve around the cerebral peduncle. It should be noted that the third cranial nerve is located between the posterior cerebral and superior cerebellar arteries. The posterior cerebral artery is connected to the internal carotid artery by the posterior communicating artery, forming the posterior limb of the cerebral arterial circle. There are only a few reported variations in the course of the posterior cerebral artery. Branches of the posterior cerebral artery are the cortical, central, and posterior choroidal arteries.

Cortical branches

The **cortical branches** of the posterior cerebral artery vascularize the medial aspect of the temporal and occipital lobes. They are usually four in number: the anterior and posterior temporal, parieto-occipital, and occipital (calcarine) branches. The **anterior temporal branch** serves the uncus and parahippocampal gyrus; the **posterior temporal branch** vascularizes the lateral and medial occipitotemporal gyri; the **parieto-occipital branch** serves the cuneate and precuneate gyri; and the **occipital (calcarine) branch** follows the calcarine sulcus to serve the lingual and cuneate gyri of the primary visual cortex.

Central branches

The **central branches** of the posterior cerebral artery are classified into two divisions: the thalamic and mesencephalic (as well as the circumflex mesencephalic) branches.

The **thalamic branches**, supplying regions of the thalamus and hypothalamus, are divided into the anterior and posterior thalamic branches, where the **anterior thalamic branches** serve the optic tract, the mammillary bodies, and the anterior and ventromedial nuclei of the thalamus. The **posterior thalamic branches** vascularize the posterior thalamic nuclei and the geniculate bodies.

The **mesencephalic** and **circumflex mesencephalic branches** serve the cerebral peduncles, the interpeduncular region, the corticospinal tracts, the mesencephalic reticular formation, the substantia nigra, and the tegmentum of the midbrain.

Posterior choroidal arteries

There are at least two **posterior choroidal arteries** – the lateral choroidal and medial choroidal arteries. The **lateral choroidal branches** serve the choroid plexus of the lateral ventricle, providing blood whose ultrafiltrate serves to form the cerebrospinal fluid. The lateral choroidal branches also serve the lateral geniculate body, the pulvinar, and the dorsomedial nucleus of the thalamus. The **medial choroidal branches** serve the choroid plexus of the third ventricle, providing blood whose ultrafiltrate serves to form the cerebrospinal fluid. Additionally, the medial choroidal branches vascularize the pineal body, the superior and inferior colliculi, the dorsal medial nucleus of the thalamus, and in many cases even vascularize the pulvinar, where they anastomose with the lateral choroidal branches.

Cerebral arterial circle (eponym: of Willis)

The cerebral arterial circle (eponym: of Willis) is an essential structure in maintaining the normal blood supply to the brain

As described above, classically, the **cerebral arterial circle** connects the branches of the two internal carotid arteries to the branches of the two vertebral arteries, forming a method of shunting blood from one side to the other in case of a blockage in any one of the vessels (Figs. 9.8–9.10). The anterior communicating artery forms the anterior connection, and the posterior communicating artery forms the posterior connection between the right and left sides. The arteries involved in the circle are: the right internal carotid, right anterior cerebral, anterior communicating, left anterior cerebral, left internal carotid, left posterior communicating, left posterior cerebral, right posterior cerebral, right posterior communicating, and back to the right internal carotid. The major vessels arising from the anterior, lateral, and posterior aspects of the cerebral arterial circle supply the cerebrum (via the anterior, middle, and posterior cerebral arteries). The arterial circle lies inferior to the hypothalamus and encompasses the mammillary bodies, infundibulum, optic chiasm, and tuber cinereum, among other structures.

VASCULAR SUPPLY OF THE CENTRAL NERVOUS SYSTEM ● ● ● 147

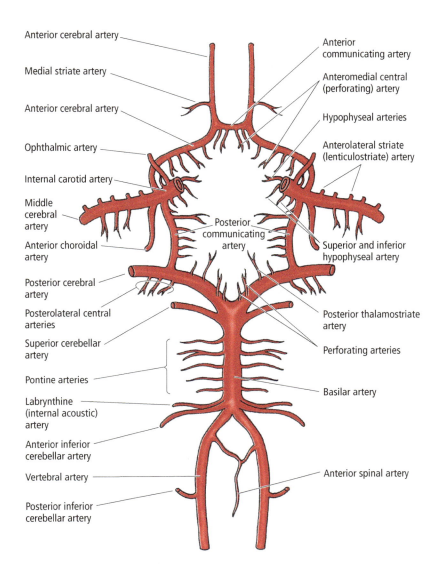

Figure 9.8 ● Diagram of the anterior view of the cerebral arterial circle (of Willis).

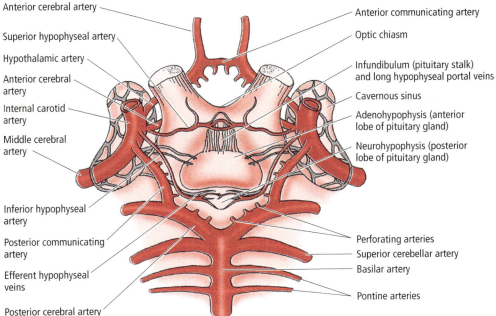

Figure 9.9 ● Close-up diagram of the anterior view of the cerebral arterial circle.

148 ••• CHAPTER 9

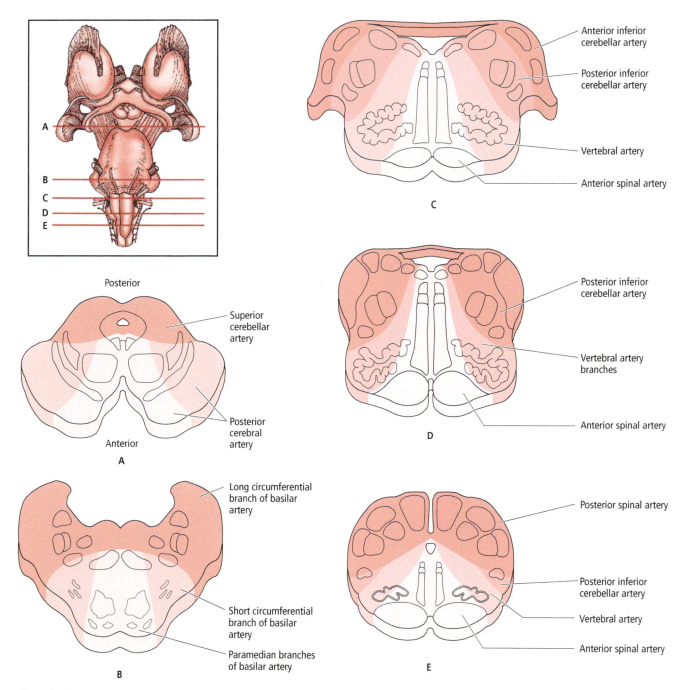

Figure 9.10 • Schematic diagrams of the distribution of the major branches of the arterial supply of the brainstem, showing five different cross-sections of the brainstem as indicated in the three-dimensional figure in the boxed area on the top left.

VENOUS DRAINAGE OF THE BRAIN

The venous drainage of the brain is accomplished by the cerebral veins, the cerebellar veins, and the veins of the brainstem

It should be noted that the veins of the brain do not usually possess a smooth muscle coat, nor do they have valves and they do not travel with the like named cerebral arteries. Blood from these thin-walled veins reaches the dural venous sinuses via vessels that penetrate both the arachnoid and the meningeal layer of the dura mater. The dural venous sinuses are discussed in Chapter 8.

Cerebral veins

The **cerebral veins** are subdivided into two categories: the external veins, which drain the superficial aspect of the cerebrum, and the internal veins, which drain the deep regions of the cerebrum (Figs. 9.11–9.13).

External veins

There are three veins in the external vein group, these are the superior, superficial middle, and inferior cerebral veins

There are three veins in the **external vein group**, the superior, superficial middle, and inferior cerebral veins. The 15 or so **superior cerebral veins** are lodged mostly in the sulci, though they occasionally pass along the surface of the gyri. These vessels drain blood from the medial and superolateral aspects of the cerebrum and empty into the superior sagittal sinus. The **superficial middle cerebral vein** (Fig. 9.12) is located within the lateral fissure (eponym: of Sylvius) and drains blood from the lateral aspect of the cerebral hemisphere. It delivers its blood into the superior sagittal sinus by way of the **superior anastomotic vein**, into the transverse sinus, via the **inferior anastomotic vein**, and continues posteriorly until it reaches, and empties into, the cavernous sinus. The small **inferior cerebral veins** (Fig. 9.12) drain blood from the inferior aspect of the cerebrum and deliver it into the superior sagittal sinus by way of the superior cerebral veins. The inferior cerebral veins that drain blood from the inferior aspect of the temporal lobe deliver their blood into the cavernous, transverse, sphenoparietal, and superior petrosal sinuses, via the basal and middle cerebral veins.

Internal veins

The internal vein group is composed of the great cerebral vein, its tributaries, and the basal vein

The **internal vein group** drains structures deeper within the brain and is composed of the great cerebral vein and its tributaries as well as the basal vein. The **great cerebral vein**

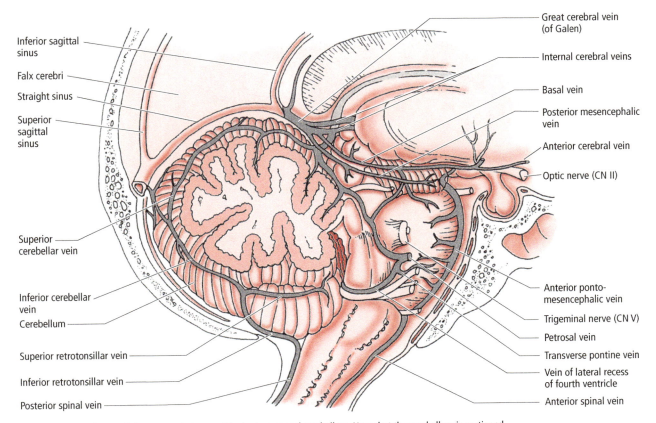

Figure 9.11 ● Lateral view of the venous drainage of the brainstem and cerebellum. Note that the cerebellum is sectioned.

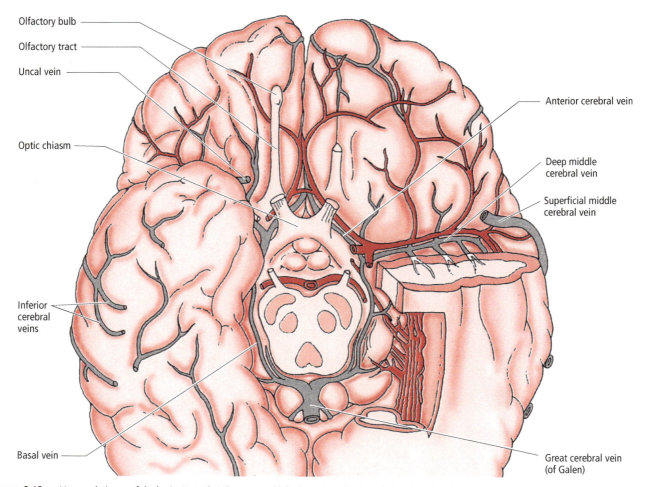

Figure 9.12 • Venous drainage of the brain. Note that the temporal lobe is sectioned and a window is cut into it to provide a view of the lateral ventricle.

(eponym: of Galen [Figs. 9.11–9.13]), a short vessel formed by the union of the right and left internal cerebral veins, drains into the straight sinus in common with the inferior sagittal sinus. The great cerebral vein receives blood from its tributaries that drain the thalamus, hypothalamus, basal nuclei, and midbrain.

The **thalamostriate vein** (Fig. 9.13) receives tributaries that drain the region of the thalamus and corpus striatum, whereas the **choroid vein** follows the choroid plexus of the lateral ventricle and receives tributaries from the fornix, hippocampus, and corpus callosum. At the level of the interventricular foramen (eponym: of Monro) the choroid vein joins the thalamostriate vein to form the **internal cerebral vein** (Fig. 9.13), which, shortly after receiving the basal vein, joins its counterpart from the other side to form the **great cerebral vein**.

The **basal vein** (Figs. 9.11 and 9.12), formed by its three tributaries at the anterior perforated substance, the anterior cerebral vein, the deep middle cerebral vein, and the inferior striate veins, receives blood from the parahippocampal gyrus, interpeduncular fossa, midbrain, and lateral ventricle.

Cerebellar veins

The **superior cerebellar veins** (Fig. 9.11) drain the superior region of the cerebellum and empty into the straight sinus, the great cerebral vein, and the transverse and superior petrosal sinuses. The **inferior cerebellar veins** (Fig. 9.11) drain the inferior aspect of the cerebellum and empty into the occipital, superior petrosal, and transverse sinuses.

VASCULAR SUPPLY OF THE CENTRAL NERVOUS SYSTEM ••• 151

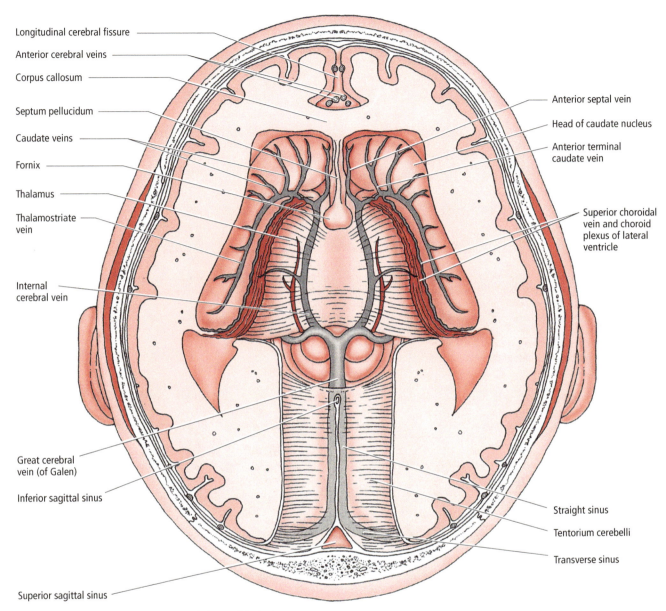

Figure 9.13 ● Diagram of the deep venous drainage of the brain.

Veins of the brainstem

The veins of the brainstem drain the medulla oblongata, pons, and midbrain by forming a superficial plexus of veins. These vessels deliver their blood to the vertebral veins, the veins of the spinal cord, and the basilar and inferior petrosal sinuses, as well as into the basal and great cerebral veins.

CLINICAL CONSIDERATIONS

Aneurysms
Aneurysms are local dilation or ballooning of a vessel due to the weakening of its wall. The most common location for an aneurysm in the brain is the posterior communicating artery as it originates from the internal carotid artery.

Miliary aneurysm (Eponym: Charcot–Bouchard aneurysm)
The largest of the slender **striate arteries (lenticulostriate arteries)** is frequently known as the **artery of cerebral hemorrhage** because its aneurysm is the most likely source of strokes in hypertensive patients. This vessel, a branch of the middle cerebral artery, passes between the external capsule and the lentiform nucleus and penetrates as deep as the caudate nucleus.

Ischemia and infarction
The brain receives approximately 750 mL of blood per minute from the right and left internal carotid arteries and from the basilar artery, where each of the three vessels supplies about the same volume of blood flow. This volume translates to about 55 mL of cerebral blood flow per minute per 100 g of brain tissue. **Ischemia** occurs if this blood flow is less than 35 mL/min/100 g of brain tissue, whereas a blood flow of below 15 mL/min/100 g of brain tissue for 5 minutes or longer is considered to be an **infarct** resulting in the death of brain tissue. The infarct may also be due to complete cessation of blood flow because of hemorrhage, thrombosis, embolism, or vasospasm.

Neurovascular unit
Due to the specificity of the **neurovascular unit,** certain drugs, antibiotics, and neurotransmitters, such as dopamine, are unable to penetrate the **blood–brain barrier**. In order to facilitate the entry of therapeutic substances, a hypertonic solution of **mannitol** is administered which temporarily disables the tight junctions of the capillary endothelial cells. Additionally, it is possible to couple antibodies that have been developed against capillary endothelial cell **transferrin receptors** to the drugs that need to be administered. The antibody permits the transport of the drug–antibody complex across the blood–brain barrier and into the CNS.

Stroke
A **stroke** (also known as a **cerebrovascular accident**) is a very serious event that results in irreversible damage to a region of the brain whose blood supply has been compromised to such an extent that the involved region can no longer be perfused adequately. The neurons of the region do not receive enough oxygen to sustain them, and they die. The severity of the stroke is a function of the region involved and the extent of the damage; therefore, the effect may range from minimal damage to death. In the USA, stroke is the third most common cause of death, preceded only by heart disease and cancer. Stroke has numerous risk factors, some of which, such as cigarette smoking and hypertension, can be controlled with relative ease, whereas other factors, such as family history and heart disease, cannot be controlled. There are two types of stroke, ischemic and hemorrhagic stroke. Approximately 85% of all stroke cases are due to **ischemic stroke,** which is the result of embolism of one of the cerebral arteries, vascular disease, or anomalies of the coagulation process. It has been shown that 15–20% of the patients with ischemic stroke die within 30 days of the event. **Hemorrhagic stroke,** which involves bleeding either into the meninges or the brain tissue, has a much higher mortality rate, with 40–80% of patients dying within 30 days of the stroke.

Lateral medullary syndrome
The **lateral medullary syndrome,** also known as the **posterior inferior cerebellar artery (PICA) syndrome**, involves occlusion of the branches of the vertebral artery or, less commonly, of the PICA. It is the most commonly occurring brainstem stroke, and it has a very good recovery rate. The symptoms of the lateral medullary syndrome include miosis, ptosis, vertigo, dysarthria, and ipsilateral hemiataxia.

Basilar artery thrombosis
Basilar artery thrombosis is a very serious stroke that involves loss of vascular supply to much of the brainstem, resulting in quadriplegia and frequently in death due to respiratory failure.

SYNONYMS AND EPONYMS OF THE VASCULAR SUPPLY OF THE CENTRAL NERVOUS SYSTEM

Name of structure or term	Synonym(s)/ eponym(s)
Cerebral arterial circle	Circle of Willis
Great anterior radicular artery	Artery of Adamkiewicz
Labyrinthine artery	Internal auditory artery
Medial striate artery	Recurrent artery
Occipital branch of the posterior cerebral artery	Calcarine branch of the posterior cerebral artery
Striate arteries	Lenticulostriate arteries

FOLLOW-UP TO CLINICAL CASE

This patient had a stroke in the distribution of the left middle cerebral artery (MCA). This artery is a common target for stroke which infrequently occurs in the full distribution of the MCA, but more commonly only part of the artery is affected. Some symptoms that are characteristic of MCA territory strokes include hemiparesis and/or hemisensory loss (contralateral), aphasia (usually only if the left MCA is affected), and sometimes hemineglect (contralateral, especially with right MCA stroke). This patient exhibits aphasia and hemiparesis. Some information cannot be ascertained in aphasic patients since they cannot talk about their symptoms.

The two common etiologies of strokes are embolism and cerebral thrombosis. Emboli commonly arise from a cardiac source. A thrombus forms within the heart and a piece or pieces break off, travel downstream, and become lodged within one or more cerebral arteries. A thrombus can also form within a cerebral artery, or one of the arteries leading to the brain (e.g., internal carotid). The blockage can lead to a reduction of blood flow that, if severe enough and prolonged, leads to infarction.

This patient had an urgent head CT in order to rule out hemorrhage. Other testing could include an echocardiogram as well as carotid ultrasound or cerebral magnetic resonance angiography (MRA) to look at the appropriate arterial circulation. Cardiac thrombus is often prompted by atrial fibrillation, so it would be important to monitor the cardiac rhythm. Treatment for stroke includes intravenous tissue plasminogen activator if it can be started within 3 hours of the stroke onset, intra-arterial thrombolysis if given within 6 hours of onset, carotid endarterectomy if there is severe carotid stenosis from the thrombus (this will only help prevent future strokes), aspirin or similar medications such as apixaban (Eliquis), heparin, or warfarin.

QUESTIONS TO PONDER

1. What is the significance of the radicular arteries that serve the spinal cord?

2. What is the most significant component of the blood–brain barrier?

3. What is the importance of the posterior communicating artery?

4. What is the significance of the cerebral arterial circle (of Willis)?

5. Where do the veins of the brain drain their blood?

CHAPTER 10

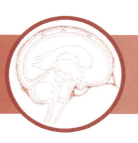

Autonomic Nervous System

- CLINICAL CASE
- SYMPATHETIC NERVOUS SYSTEM
- PARASYMPATHETIC NERVOUS SYSTEM
- ENTERIC NERVOUS SYSTEM
- NEUROTRANSMITTERS AND RECEPTORS OF THE AUTONOMIC NERVOUS SYSTEM
- PELVIC AUTONOMIC FUNCTIONS
- CLINICAL CONSIDERATIONS
- SYNONYMS AND EPONYMS
- FOLLOW-UP TO CLINICAL CASE
- QUESTIONS TO PONDER

CLINICAL CASE

A 55-year-old patient, J.H., presented with generalized weakness and fatigue. This had been ongoing for the past year. J.H. also noted a dry mouth and dry irritated eyes, revealed impotence and some mild constipation. The patient had been a heavy smoker for 40 years.

On examination there was mild-to-moderate weakness noted in the proximal muscles of the upper and lower limbs. J.H. had difficulty rising from a chair without the help of armrests on the chair. There was no sensory abnormality and J.H.'s reflexes were normal. The patient's blood pressure when supine was 150/90 with a heart rate of 80. When standing, the blood pressure was 118/76 and the heart rate was 78.

The role of the **autonomic nervous system** is to serve in the capacity of functional communication between the body and the external environment, thereby balancing the physiological and psychological functioning of the body to achieve not only a level of stability known as **homeostasis** but also in adjusting the body's internal milieu to the external **stresses** that the person encounters. Homeostasis refers to precise control of respiratory and cardiovascular status, regulation of body temperature, control of intestinal movement, defecation and micturition, directing reproductive urges, as well as regulating the metabolic and endocrine activities of the body. Adjusting the internal milieu refers to the "fight or flight or fornicate or freeze" response, which means that the body is prepared to fight an adversary or flee from an adversary or in very dire circumstances the body "freezes" and, in a manner of speaking, "plays dead" so that the adversary stops the attack, assumes that the individual is dead, departs the scene of the attack, and the individual survives. The autonomic nervous system also readies the individual for the reproductive function of fornication. These actions are accomplished by specialized neurons that are located in both the central and peripheral nervous systems. The neurons of the autonomic nervous system innervate smooth muscles, cardiac muscles, and glands and perform their functions below the

---

A Textbook of Neuroanatomy, Third Edition. Maria A. Patestas, Amanda J. Meyer, and Leslie P. Gartner.
© 2025 John Wiley & Sons, Inc. Published 2025 by John Wiley & Sons, Inc.
Companion website: www.wiley.com.go/patestas/neuroanatomy3e

conscious level. Functionally, the autonomic nervous system has three components, namely **sympathetic**, **parasympathetic**, and **enteric**. Although it is usually treated as a motor system – the **general visceral efferent** (GVE) system – the autonomic nervous system does not possess **afferent components**, but the **primary visceral sensory neurons** bring information to the autonomic nervous system. As a generalization it may be stated that the *sympathetic nervous system* functions in "fight or flight or freeze," the *parasympathetic nervous system* functions in "digest and rest," and, in concert with the sympathetic nervous system, readies the body for fornication, whereas the *enteric nervous system* functions in "overseeing the digestive process."

It is important to note from the outset that there are major differences in the neuronal arrangement of the somatic motor nervous system and the motor limb of the autonomic nervous system (Figs. 10.1 and 10.2). In the **somatic nervous system**, the cell body of the lower motor neuron is located in the **anterior horn** of the spinal cord. The axon of that neuron leaves the spinal cord via the anterior rootlets and passes directly to the muscle fibers it is destined to innervate. Thus, only a single neuron is required to relay the signal from the brainstem cranial nerve motor nuclei or the spinal cord anterior horn for muscle contraction. In the autonomic nervous system two neurons are required to effect a contraction of smooth muscle or cardiac muscle, or to elicit secretion from the cell of a gland. The nerve cell body of the first neuron is in the **brainstem** of the **central nervous system** (**CNS**), whereas the nerve cell body of the second neuron is in a **ganglion** in the peripheral nervous system. Because the second nerve cell body resides in an autonomic ganglion, that nerve cell is referred to as the **post-ganglionic neuron** (**post-synaptic neuron**), whereas the first neuron in this two-neuron chain, is known as the **pre-ganglionic neuron** (**pre-synaptic neuron**). The axon of the pre-ganglionic neuron, the **pre-ganglionic fiber**, is myelinated and it synapses with the post-ganglionic neuron's cell body. The axon of the post-ganglionic neuron, the **post-ganglionic fiber**, is typically not myelinated (although approximately 1% of them are lightly myelinated). It should be noted **hypothalamospinal neurons** have a direct access to second-order neurons in the lateral horn of T1 whose axons synapse with third-order neurons of the superior cervical ganglion, whose sympathetic fibers synapse in the superior tarsal muscles of the eyelids and the dilator muscles of the pupils.

It should be recalled that synapses do not occur in the spinal ganglia, whereas synapses always occur in autonomic ganglia. Generally speaking, it may be stated that the length of pre-ganglionic and post-ganglionic fibers differs in that in the sympathetic nervous system the pre-ganglionic fibers are short and the post-ganglionic fibers are long. In the parasympathetic nervous system the opposite is usually true, in that

Note that the clinical case at the beginning of the chapter refers to a patient who was a long-time, heavy smoker who was experiencing generalized weakness and fatigue as well as dry mouth, reduced tear formation, and impotence. Additionally, the patient experienced orthostatic hypotension.

1 Does the relationship between dry mouth and dry eyes involve the somatic or autonomic nervous systems?

2 What is orthostatic hypotension?

3 Which neurotransmitter is present in skeletal muscle, the lacrimal gland, and the salivary glands?

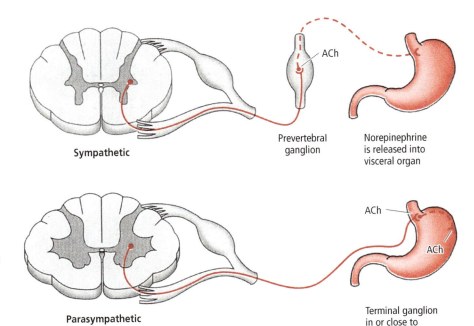

Figure 10.1 • Schematic diagram demonstrating the difference between the sympathetic nervous system (above) as it arises from spinal cord levels T1 to L1, 2 and the spinal component of the parasympathetic nervous system (below) as it originates from the sacral spinal cord. ACh, acetylcholine.

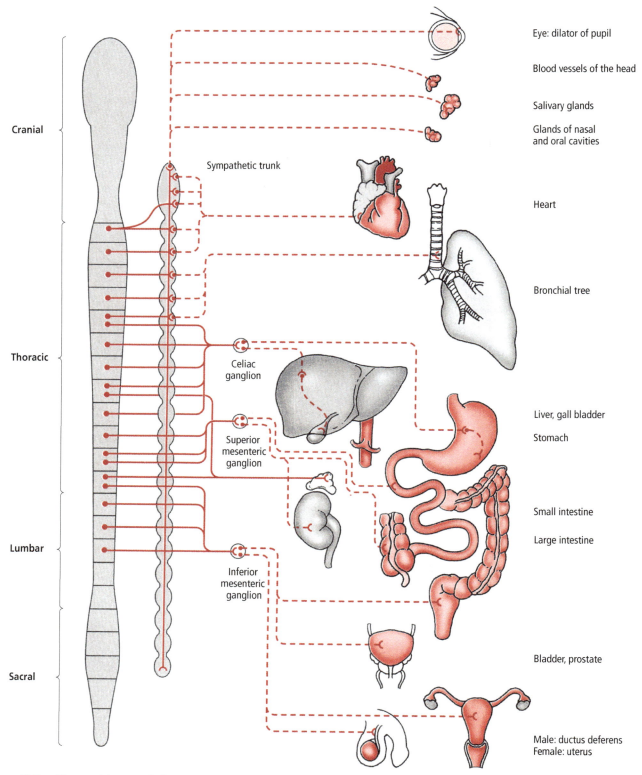

Figure 10.2 • Diagram of the sympathetic nervous system.

the pre-ganglionic parasympathetic fibers are long and the post-ganglionic parasympathetic fibers are usually short.

Formerly it was believed that the **neurotransmitter substance** at the parasympathetic pre-ganglionic *and* post-ganglionic synapse, as well as at the sympathetic preganglionic synapse is **acetylcholine**, whereas at the sympathetic post-ganglionic synapse it is **noradrenaline** (**norepinephrine**) and that the only exception to this rule was that the post-ganglionic sympathetic fibers serving the **eccrine sweat glands** release acetylcholine. In the past 25 years or so it has been demonstrated that additional neurotransmitter substances are released by autonomic neurons; therefore, acetylcholine and noradrenaline are said to be the **principal neurotransmitter substances** and the others, mainly ATP, neuropeptide Y (NPY), vasoactive intestinal peptide (VIP), and nitric oxide (NO) that are also released are said to be **co-transmitters**.

The fibers of the **sensory** (or afferent) **limb** of the sympathetic and parasympathetic nervous systems travel with blood vessels, cranial nerves, and visceral motor fibers. They frequently *pass through but do not synapse* in the autonomic ganglia. Instead, their cell bodies (unipolar/pseudounipolar) are located in the **sensory ganglia** of the cranial and spinal nerves.

Central Connections of the Autonomic Nervous System

Although the implication is that the autonomic nervous system is self-contained and is composed only of pre-ganglionic and post-ganglionic neurons and their processes, as well as the system's sensory components whose cell bodies reside in the sensory ganglia of spinal and some cranial nerves, that implication is far from the truth. Much of the functioning of the autonomic nervous system is controlled at higher levels in the brain and brainstem. Although these higher levels are arranged in four distinct regions of the CNS, namely the telencephalon and diencephalon, mesencephalon, metencephalon and myelencephalon, and spinal levels, they communicate with each other to be able to exercise continuous and immediate control over the essential functions of the autonomic nervous system, namely homeostasis of the body, the overseeing of visceral functions, and the management of response to the challenges presented by the external and internal situations. Table 10.1 details their distribution, their associated regions, and their functions.

SYMPATHETIC NERVOUS SYSTEM

Almost 90% of the nerve cell bodies of the pre-ganglionic neurons of the **sympathetic nervous system** are located in the **lateral cell column** of spinal cord levels T1 to L2, 3 (Fig. 10.2) the other 10% are housed in two additional locations, the **central autonomic nucleus** and the **spinal intercalated nucleus**. Since the axons of these neurons exit the vertebral canal in the anterior root of the spinal nerve at the levels of their origin, the sympathetic nervous system is also known as the **thoracolumbar outflow**. These pre-ganglionic nerve cell bodies receive

Table 10.1 ● Central components of the autonomic nervous system and their functions.

Component	Subcomponent	Function
Telencephalon and diencephalon	Insular cortex	Incorporates information from the viscera as well as temperature and pain sensations
	Anterior cingulate cortex	Integrates sympathetic and parasympathetic functions
	Amygdaloid body	Integrates emotional components of stress especially those of fear
	Hypothalamus	Integrates nervous and endocrine responses required to maintain homeostasis and to respond to changes in the external and internal milieu
Mesencephalon	Periaqueductal gray	Controls responses of the cardiovascular system to pain and stress; controls respiration; and controls urinating reflex
Metencephalon and myelencephalon	Peribrachial complex and pons	Conveys information from the spinal cord to the thalamus, hypothalamus, and amygdaloid body. Manages the functions of the cardiovascular, gastrointestinal, and respiratory systems; assists in coordination of control of urinating reflex; as well coordinates the functions of the lower gastrointestinal system and sex organs
	Nucleus of the solitary tract	Receives information from taste sensations, gastrointestinal tract sensations, as well as information concerning blood pressure, the cardiovascular system, chemoreception, and the respiratory system; assists in controlling reflexes associated with the previously listed systems
	Medulla oblongata and caudal raphe	Assists in the control of blood pressure, cardiac, and pulmonary reflexes. Functions in the response to hypovolemia (decreased plasma volume) and hypotension (decreased blood pressure) by causing the release of vasopressin (antidiuretic hormone); has a major function in regulating temperature, response to pain, and automatic breathing
Spinal levels	Intermediolateral (lateral) cell column (thoracic spinal cord T1 to lumbar 2 or 3 levels)	Segmental reflexes involving the sympathetic nervous system. Controls blood flow distribution during digestion, physical activity, and rest; regulates temperature and cardiovascular pressures
	Sacral parasympathetic nucleus (sacral spinal cord 2–4 levels)	Segmental reflexes involving the parasympathetic nervous system. Controls defecation, urination, and sexual activities

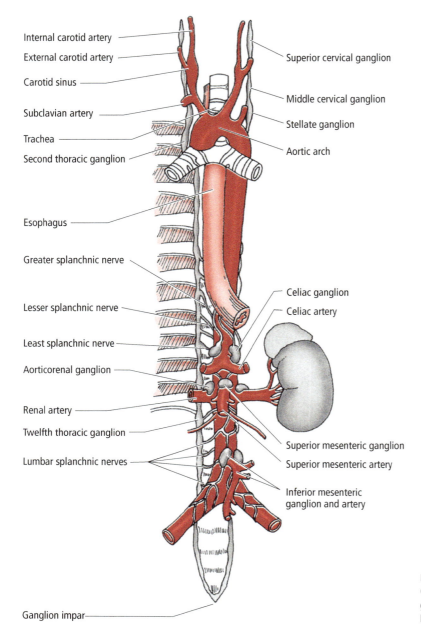

Figure 10.3 • Diagram of the right and left sympathetic trunk. Observe that the two sides fuse inferiorly to form the single ganglion impar. Note that the descending aorta partially hides the left sympathetic trunk.

numerous projections from supraspinal levels that assist in the coordination of the responses that are provided by the sympathetic nervous system. It is important to note that at each thoracic level there are as many as 5,000 pre-ganglionic nerve cell bodies and that each pre-ganglionic neuron probably contacts as many as 20 post-ganglionic neurons. The cell bodies of the post-ganglionic sympathetic neurons reside in the **sympathetic trunk** (**paravertebral ganglia** located on either side of the vertebral column), the **pelvic ganglia**, the **preaortic ganglia**, and the small **aorticorenal ganglia** (Fig. 10.3). The predominant postganglionic neurons are **noradrenergic, multipolar cells**, although many display additional co-transmitter neurotransmitter substances. Some of the functions of the sympathetic nervous system are given in Table 10.2.

Pre-ganglionic fibers and paravertebral ganglia of the sympathetic nervous system

Pre-ganglionic fibers of the sympathetic nervous system arise from nerve cell bodies located mostly in the lateral horn of spinal cord levels T1 to L2, 3

Pre-ganglionic fibers (preganglionic efferent axons) enter the spinal nerve via the anterior rootlets; they subsequently leave the spinal nerve via 14 pairs of small branches, known as **white rami communicantes** ("white" because most of the axons are well myelinated) to reach one of the **sympathetic trunks**. Once pre-ganglionic sympathetic fibers have entered the sympathetic chain ganglion they have one of three options (Figs. 10.4 and 10.5):

Table 10.2 • Some of the functions of the sympathetic nervous system.

Organ/structure innervated	Function
Eccrine sweat glands of skin	Release of sweat
Arrector pili muscle	Contraction to make hair appear longer
Blood vessels of skeletal muscle	Dilation to increase blood flow
Blood vessels of skin/mucous membranes	Vasoconstriction to reduce blood flow
Blood vessels of abdominal viscera	Vasoconstriction
Coronary arteries	No effect
Sinoatrial node of heart	Accelerates heartbeat
Ventricular myocardium	Increases force of contraction
Gastrointestinal tract	Reduces peristalsis; contraction of internal sphincters
Iris	Dilates pupils
Superior tarsal muscles	Contraction of smooth muscles opens upper eyelids
Ductus deferens	Increases peristaltic movements conveying spermatozoa
Bronchial smooth muscle	Relaxation of smooth muscle causes easier breathing
Suprarenal medulla	Releases epinephrine and norepinephrine

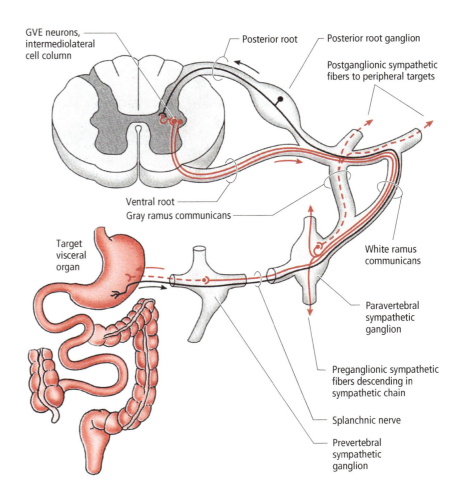

Figure 10.4 • Diagram of the sympathetic nervous system. Solid lines represent pre-ganglionic sympathetic fibers and dashed lines represent post-ganglionic sympathetic fibers. GVE, general visceral efferent.

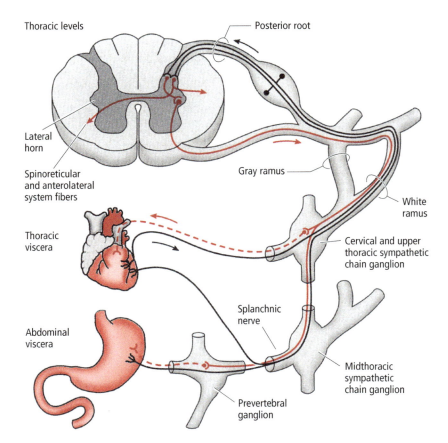

Figure 10.5 ● Diagram demonstrating the difference between the synapse occurring in a sympathetic trunk and that occurring in a prevertebral ganglion. Solid lines indicate pre-ganglionic sympathetic fibers. Dashed lines indicate post-ganglionic sympathetic fibers.

1. To form synapses with post-ganglionic, sympathetic nerve cell bodies in the sympathetic chain ganglion at the level of entry.
2. To proceed superiorly or inferiorly in the sympathetic trunk and synapse in a ganglion above or below the point of entry. This may be as far superiorly as the cervical level or as far inferiorly as the sacral level.
3. To pass through the sympathetic chain ganglion without synapsing in any of the paravertebral ganglia, joining with other pre-ganglionic fibers to form the greater, lesser, or least splanchnic nerves to synapse in one of the prevertebral sympathetic trunks, specifically the pelvic, aorticorenal, or preaortic ganglia.

The paravertebral ganglia in the neck fuse to form three ganglia, the largest of which is the **superior cervical ganglion**, the smaller **middle cervical ganglion**, and the smallest **inferior cervical ganglion**. Quite often the inferior cervical ganglion fuses with the first thoracic sympathetic trunk to form the enlarged **cervicothoracic ganglion**. The caudalmost ganglia of the right and left paravertebral chains fuse with each other in the coccygeal region, thus forming the unpaired **ganglion impar**.

Distribution of the post-ganglionic fibers of the sympathetic nervous system

Post-ganglionic sympathetic fibers may re-enter spinal nerves, may travel with cranial nerves, may travel wrapped around arteries, or may proceed directly to specific organs

The **post-ganglionic fibers** are almost exclusively unmyelinated axons of the post-ganglionic neurons. These axons arise from the sympathetic trunk and are distributed as branches to the spinal nerves or cranial nerves, or they may travel wrapped around arteries, or they may proceed directly to specific organs or to autonomic plexuses (Fig. 10.4). The regions of these unmyelinated fibers that are close to their target cell begin to exhibit **varicosities** (swellings), and the number of these varicosities increases as the distance from the target cells decreases. It has been demonstrated that these varicosities actually move along the post-ganglionic axon, making their way to the presynaptic terminal. These varicosities are packed with mitochondria and neurotransmitter-filled synaptic vesicles that are being delivered to the presynaptic terminal from the post-ganglionic nerve cell body. The possible modes of distribution of sympathetic fibers are as follows:

AUTONOMIC NERVOUS SYSTEM ● ● ● 161

1 Branches to **spinal nerves** re-enter the spinal nerve through connections known as **gray rami communicantes** (gray because the postganglionic axons are almost exclusively unmyelinated). Unlike white rami communicantes, there are 31 pairs of gray rami communicantes, one for each spinal nerve. The post-ganglionic sympathetic fibers are distributed with the cutaneous branches of the spinal nerves, reaching their target cells in sweat glands, smooth muscles of peripheral blood vessels, and arrector pili muscles.

2 Branches to **cranial nerves** may reach the cranial nerves directly, namely, the oculomotor (CN III), the facial (CN VII), glossopharyngeal (CN IX), and vagus (CN X) nerves, or they may follow blood vessels until they are in the vicinity of the cranial nerve that they use to reach their final destination.

3 Branches **wrapped around arteries** travel with these vessels, such as the internal and external carotid arteries and the vertebral artery. There are numerous other arteries that possess postganglionic sympathetic fibers in their tunicae adventitia, and they are discussed at the appropriate place.

4 Branches proceeding directly to **specific organs** may travel on arteries, in nerves, or through plexuses, but they travel on their own, at least for a while. These include post-ganglionic sympathetic fibers destined for the eyes and the heart.

5 Branches that proceed to **autonomic plexuses** possess either post-ganglionic or pre-ganglionic fibers, depending on the plexus. Those that serve the heart, lungs, and pelvis are probably *post-ganglionic*, whereas those that are destined for the abdominal plexuses are *pre-ganglionic* sympathetic fibers destined for the post-ganglionic nerve cell bodies in the walls of the organ to be innervated. The **greater**, **lesser**, and **least splanchnic nerves** are the major nerve fibers that contain pre-ganglionic sympathetic axons destined for the **abdominal plexuses**.

Topographic distribution of the sympathetic nervous system

The sympathetic nervous system may be subdivided into two categories: (1) those associated with the sympathetic chain ganglia; and (2) a collection of autonomic plexuses whose post-ganglionic fibers are distributed to the abdominal and pelvic viscera. The sympathetic chain ganglia, although a single entity, may be viewed as being composed of five discrete regions that correspond to the head, neck, thorax, abdominolumbar region, and pelvis.

Cephalic region of the sympathetic nervous system

The cephalic region of the sympathetic nervous system is composed only of postganglionic sympathetic fibers whose cell bodies are located in the superior cervical ganglia

The cephalic region of the sympathetic nervous system has no ganglia of its own, that is, there are no cephalic ganglia (there are no sympathetic ganglia in the head region); instead, it is composed only of **post-ganglionic sympathetic** fibers whose cell bodies are located in the **superior cervical ganglia**. These fibers are primarily associated with the tunica adventitia of blood vessels passing from the neck into the head and are named according to their vessel of association or by their location, or by the localized region that they serve.

Internal carotid nerves

The internal carotid nerves are composed of post-ganglionic sympathetic fibers derived from the superior cervical ganglion

The **internal carotid nerves** travel with the internal carotid arteries as two separate plexuses, the internal carotid plexus and the cavernous plexus.

The **internal carotid plexus** travels on the lateral aspect of the artery and has the following branches:

• Branches **communicating with cranial nerve V** (trigeminal nerve) and **cranial nerve VI** (abducens nerve).

• **Deep petrosal nerve (sympathetic root of the pterygopalatine ganglion)**: this nerve pierces the cartilage of the foramen lacerum and accompanies the greater petrosal nerve as the two nerves enter the pterygoid canal to form the **nerve of the pterygoid canal** (eponym: nerve of the Vidian canal [named after Vidus Vidius]). This nerve proceeds to the pterygopalatine ganglion in the same named fossa. Some branches derived from the greater petrosal nerve synapse in the pterygopalatine ganglion, but none of the branches that are derived from the deep petrosal nerve synapse (since they are post-ganglionic sympathetic fibers) do; instead, they become distributed by branches of the maxillary division of the trigeminal nerve to smooth muscle of the blood vessels, of the pharynx, palate, and nasal cavity, and the secretory cells of glands.

• **Caroticotympanic nerves**: these are small fibers that are destined to serve regions of the tympanic cavity (middle ear).

The **cavernous plexus** travels on the medial aspect of the internal carotid artery and has the following branches:

• Branches communicating with **cranial nerve IV** (trochlear nerve) and **cranial nerve V₁** (ophthalmic division of the trigeminal nerve).

• Branches destined for the **dilator pupillae muscle** of the iris. These travel with the long ciliary nerve as well as with the short ciliary nerves (without synapsing in the ciliary ganglion) to pierce the eyeball and synapse on the smooth muscle fibers.

• Branches serving the vessels of the hypophysis, brain, meninges, and orbit.

External carotid nerves

The external carotid nerves travel with the external carotid artery and with its branches in the head

The **external carotid nerves**, composed of post-ganglionic sympathetic fibers derived from the superior cervical ganglion, travel with the external carotid artery and its branches in the head. These include the following:

- **Nerve fibers that follow the facial artery**: these travel to and serve the submandibular gland and, using the lingual nerve, reach and serve the sublingual gland.
- Nerve fibers that travel on branches of the facial and superficial temporal arteries serve these arteries as well as the arrector pili muscles and sweat glands of the facial region.
- **Small deep petrosal nerve**: this nerve leaves the middle meningeal artery to pass through (without synapsing in) the otic ganglion. It joins the auriculotemporal nerve and serves the parotid gland.

Cervical region of the sympathetic nervous system

The cervical region of the sympathetic trunk is composed of the **superior**, **middle**, and **inferior sympathetic cervical ganglia** as well as the **sympathetic trunk** that connects them to each other. None of these ganglia receive white rami communicantes because the pre-ganglionic fibers destined for the neck enter the chain ganglia at thoracic levels T1–T4 and proceed superiorly in the cord to reach and synapse on post-ganglionic nerve cell bodies residing in one of these three ganglia. Post-ganglionic fibers leave the ganglia gathered in nerves of various calibers that join the arteries, cervical spinal nerves, and cranial nerves. Branches are derived from each cervical ganglion and are detailed next.

Superior cervical ganglion

The superior cervical ganglion is the largest sympathetic ganglion in the neck

The **superior cervical ganglion**, the largest of the three ganglia (approximately 10–30 mm in length and 5–8 mm in width), is located deep to the carotid sheath, anterior to (and at the level of) the transverse processes of C2 and the longus capitis muscles. It has numerous branches, some of which (internal and external carotid nerves) were detailed in the section above. The additional branches are those that travel with the cranial nerves and cervical spinal nerves, the pharyngeal branches, the intercarotid plexus, and the superior cardiac nerves:

- Branches traveling with **cranial nerves IX, X,** and **XII**.
- Branches traveling with the first four **cervical nerves**.
- The four to six **pharyngeal branches** that form the pharyngeal plexus in unison with cranial nerves IX and X.
- The **intercarotid plexus** that proceeds to the carotid sinus and carotid body to innervate the smooth muscle of the regional blood vessels.
- The **superior cardiac nerves** of the right and left side: these join the cardiac plexus and innervate the cardiac muscle of the heart. They function in concert with the middle and inferior cardiac nerves to accelerate the rate of heartbeat.

Middle cervical ganglion

The middle cervical ganglion is the smallest of the three sympathetic ganglia in the neck

The smallest of the ganglia, the **middle cervical ganglion**, is approximately 6–10 mm in length and 4–6 mm in width, has an inconstant location at the level of vertebrae C3–C7 anterior to the longus colli muscles, and is detected in only about 50% of the individuals examined. The branches arising from the middle cervical ganglion are those that join the cervical spinal nerves, the middle cervical nerve, and the thyroid nerves:

- Gray rami communicantes to the fifth and sixth **cervical nerves** that are distributed with the branches of these cervical nerves.
- **Middle cervical cardiac nerve**, also known as the **great cardiac nerve**: this communicates with the superior cardiac nerve to join the cardiac plexus. It functions, in concert with the superior and inferior cardiac nerves, to accelerate the rate of the heartbeat.
- **Thyroid nerves**: these travel on the tunica adventitia of the inferior thyroid artery to reach and supply the thyroid gland.
- The region of the **sympathetic trunk** connecting the middle and inferior cervical ganglia. This is usually split into an anterior and a posterior portion; the anterior portion passes around the subclavian artery, forming a loop – the **ansa subclavia**.

Inferior cervical ganglion

The inferior cervical ganglion often fuses with the first thoracic sympathetic ganglion, forming the cervicothoracic ganglion (stellate ganglion)

The **inferior cervical ganglion**, usually located at the level of the transverse process of the seventh cervical vertebra, is present in less than 20% of the individuals examined. The branches arising from the inferior cervical ganglion are the ones that join the cervical spinal nerves, the inferior cardiac nerves, and the vertebral nerve:

- Gray rami communicantes to **cervical spinal nerves 6–8**: these are distributed with the branches of these cervical nerves.
- **Inferior cervical cardiac nerves**: these nerves join the cardiac plexus and function, in concert with the superior and middle cardiac nerves, to accelerate the rate of heartbeat.
- **Vertebral nerve**: this nerve travels on the tunica adventitia of the vertebral artery and serves it as well as its intracranial branches with vasomotor function.

The inferior cervical and first thoracic ganglia frequently fuse, and this fused structure is known as the **cervicothoracic** ganglion (**stellate ganglion**). This fused ganglion, approximately 10–37 mm in length and 4–14 mm in width, fulfills all the functions of the inferior cervical ganglion and also gives rise to the gray rami communicantes of spinal nerves T1 and T2. It is usually located anterior to the transverse process of the seventh cervical vertebra, the neck of the first rib, and the first intercostal space.

Thoracic region of the sympathetic nervous system

The thoracic region of the sympathetic nervous system is the only region with white rami communicantes

Theoretically, there are 12 **thoracic sympathetic ganglia** connected to each other by short segments of the sympathetic trunk, but fusion of occasional ganglia with each other reduces their number to approximately 10. **White rami communicantes** are connections between the thoracic spinal nerves and the thoracic sympathetic trunk. They carry myelinated **pre-ganglionic sympathetic fibers** from the lateral column of the spinal cord, which enter each thoracic sympathetic ganglion to synapse with **post-ganglionic sympathetic** nerve **cell bodies** residing in a sympathetic ganglion at any level of the chain ganglia or in a sympathetic ganglion located elsewhere in the body.

Branches arising from the thoracic sympathetic ganglia are the gray rami communicantes, the visceral branches, and the splanchnic nerves.

Gray rami communicantes

The **gray rami communicantes** arise from each thoracic ganglion to unite with their associated thoracic spinal nerve, carrying mostly unmyelinated post-ganglionic sympathetic fibers that will be distributed with the branches of the spinal nerve.

Visceral branches

Visceral branches of the sympathetic nervous system are distributed to specific organs

The **visceral branches** are composed of mostly unmyelinated post-ganglionic sympathetic fibers that are distributed to plexuses on or in the vicinity of the heart, esophagus, lungs, and aorta:

- **Cardiac branches** arise from the second through fifth thoracic ganglia and are destined for the cardiac plexus. They mix with fibers from the superior, middle, and inferior cardiac nerves. They also function in accelerating the rate of the heartbeat.
- **Esophageal branches** arise from many of the thoracic ganglia and serve to modify the function of the enteric nervous system.
- Fibers destined for the **pulmonary plexus** arise from the second through fourth thoracic ganglia and enter the hilum of the lung. They serve the blood vessels of the lung as well as the bronchial musculature.
- The **aortic plexus** is served by fibers arising from the fifth to tenth (or eleventh) thoracic ganglia. These fibers distribute with the branches of the aorta and probably function to modulate the enteric nervous system.

Splanchnic nerves

Splanchnic nerves are pre-ganglionic fibers destined for the celiac ganglion, the medulla of the suprarenal gland, the aorticorenal ganglia, and the renal plexus

The **splanchnic nerves** are composed of mostly **myelinated pre-ganglionic sympathetic fibers** (as well as large myelinated visceral afferent fibers bringing information from the viscera), pre-ganglionic because they have not synapsed in the sympathetic chain ganglia; instead, they passed through the ganglia on their way to the **preaortic ganglia**. There are three splanchnic nerves: the greater, lesser, and least splanchnic nerves.

1 The **greater splanchnic nerves**, formed into a single trunk from pre-ganglionic sympathetic fibers that pass through the fifth to ninth thoracic ganglia, penetrate the crus of the diaphragm to synapse in the **celiac ganglion** (and **splanchnic ganglion**, if present). Some fibers continue through the celiac ganglion without synapsing and continue to the **suprarenal gland medulla** and synapse there on **pheochromocytes** (chromaffin cells).

It is important to recall that the pheochromocytes of the suprarenal medulla, as well as the neurons of sympathetic ganglia, are derived from neural crest cells. Thus, pheochromocytes act as modified post-ganglionic sympathetic neurons and release their secretory product, **epinephrine** and **norepinephrine**, into the capillary beds of the suprarenal medulla. Approximately 85% of the pheochromocytes release epinephrine (as well as a small amount of **enkephalins**) and 15% release norepinephrine. It should be noted that there are at least two important interrelationships between the suprarenal cortex, which is regulated by the hypothalamo-hypophyseal axis and the suprarenal medulla that is controlled by the sympathetic nervous system, namely that (1) both are triggered by physical and/or psychological stresses faced by the individual and that (2) the presence of cortisol, produced by the suprarenal cortex, is necessary for the action of the rate-limiting enzyme, phenylethanolamine-n-methyltransferase, which converts norepinephrine to epinephrine.

2 The **lesser splanchnic nerve**, formed into a single trunk from pre-ganglionic sympathetic fibers that pass through the tenth and eleventh thoracic ganglia, penetrates the crus of the diaphragm, and synapses in the **aorticorenal ganglion**.

3 The **least splanchnic nerve**, formed from pre-ganglionic sympathetic fibers that pass through the twelfth thoracic ganglion, passes through the diaphragm, and synapses in the **renal plexus**.

164 • • • CHAPTER 10

Abdominolumbar region of the sympathetic nervous system

The abdominolumbar portion of the sympathetic nervous system is composed of two to six ganglia and the intervening trunk

The **abdominolumbar sympathetic trunk** is highly variable in that there are two to six ganglia, with an average of four ganglia, connected to one another by cords of varying lengths. The right and left sympathetic trunks in the abdominal cavity are almost hidden by the inferior vena cava and the abdominal aorta, respectively. Branches of the abdominolumbar trunk are composed of gray rami communicantes, lumbar splanchnic nerves, and branches of the celiac ganglion:

- **Gray rami communicantes** in the lumbar area are somewhat longer than those of the thoracic region. They join the spinal nerves to be distributed with them to innervate the smooth muscle of blood vessels as well as the glands and arrector pili of the skin.
- **Lumbar splanchnic nerves**, pre-ganglionic sympathetic fibers that are located caudal to the least splanchnic nerve, synapse in the **inferior mesenteric ganglion**.
- Branches of the **celiac ganglion** are post-ganglionic sympathetic fibers that follow the aorta and its branches and form a large plexus of nerve fibers, the **celiac plexus**.

PARASYMPATHETIC NERVOUS SYSTEM

Topographic distribution of the parasympathetic nervous system

The pre-ganglionic cell bodies of the **parasympathetic nervous system** are located in the parasympathetic nuclei of 4 of the 12 cranial nerves as well as in the lateral horn (parasympathetic cell column) of the sacral spinal cord. Therefore, the parasympathetic nervous system is also referred to as the **craniosacral outflow**. The post-ganglionic nerve cell bodies reside in various ganglia in the head and throughout the body. Generally speaking, unlike in the sympathetic nervous system, acetylcholine is the principal neurotransmitter not only of the pre-ganglionic but also of the post-ganglionic neurons. Additionally, there are the usual co-transmitters, enkephalins for the pre-ganglionic neurons and vasoactive intestinal peptide and/or the neuropeptide Y. Some of the functions of the parasympathetic nervous system are listed in Table 10.3.

Cranial portion of the parasympathetic nervous system

The cranial portion of the parasympathetic nervous system is associated with four cranial nerves: the oculomotor, facial, glossopharyngeal, and vagus nerves

The pre-ganglionic parasympathetic fibers of the cranial portion of the parasympathetic nervous system travel with the branches of four cranial nerves, the oculomotor, facial, glossopharyngeal, and vagus nerves (Figs. 10.6 and 10.7).

Oculomotor nerve (CN III)

The **accessory oculomotor nucleus** (eponym: Edinger–Westphal nucleus) houses **pre-ganglionic parasympathetic** nerve cell bodies whose axons travel in the branches of cranial nerve III. When these pre-ganglionic fibers reach the orbit, they synapse with post-ganglionic parasympathetic nerve

Organ/structure innervated	Function
Iris	Constricts pupils
Ciliary muscles	Contracts to relax suspensory ligaments of the lens (for near vision)
Lacrimal glands	Facilitates flow of tears
Salivary glands	Facilitates flow of serous secretion
Sinoatrial node of heart	Decreases rate of heartbeat
Blood vessels	Usually has little effect
Bronchial smooth muscle	Bronchoconstriction
Glands of conducting portion of respiratory system	Facilitates secretion
Peristalsis of gastrointestinal canal	Stimulates peristalsis
Internal sphincter muscles	Relaxes sphincter muscles (inhibitory function)
Intrinsic glands of the gastrointestinal tract	Facilitates secretion
Pancreas	Facilitates secretion
Gall bladder	Facilitates release of bile
Penis and clitoris	Stimulates erection

Table 10.3 • Functions of the parasympathetic nervous system.

AUTONOMIC NERVOUS SYSTEM • • • 165

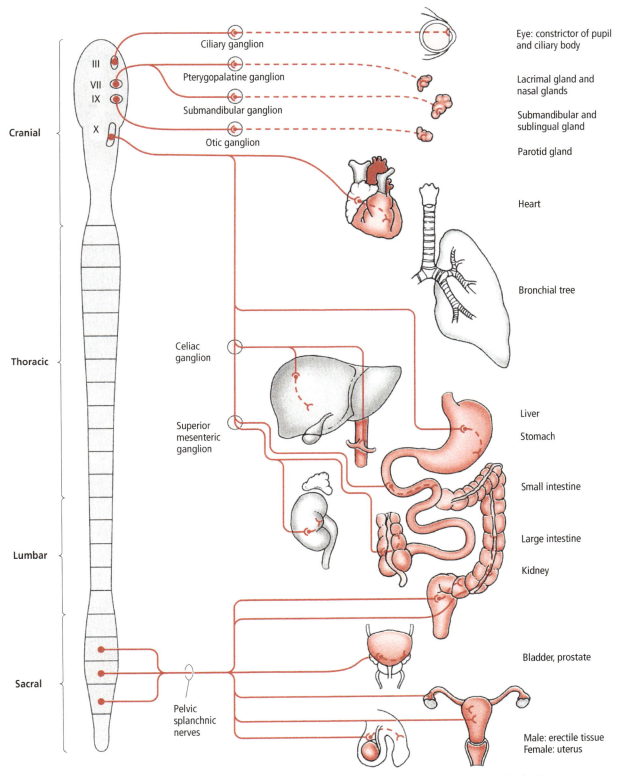

Figure 10.6 • Diagram of the parasympathetic nervous system. Solid lines represent pre-ganglionic parasympathetic fibers and dashed lines represent post-ganglionic parasympathetic fibers.

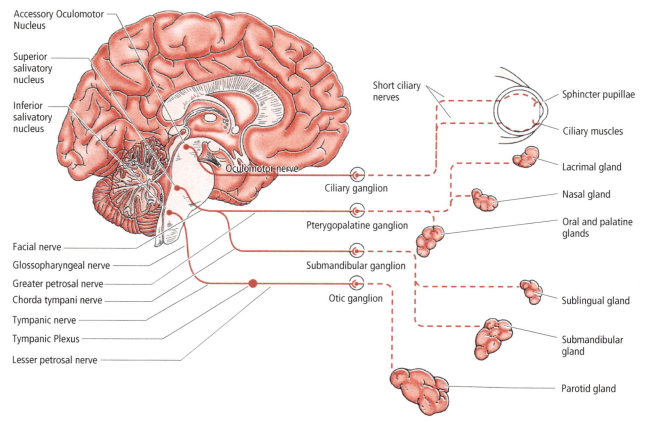

Figure 10.7 • Schematic diagram of the parasympathetic innervation of the head. Solid lines represent preganglionic parasympathetic fibers and dashed lines represent postganglionic parasympathetic fibers.

cell bodies in the **ciliary ganglion** (the parasympathetic ganglion of the oculomotor nerve). The post-ganglionic parasympathetic fibers, known as the **short ciliary nerves**, exit the ciliary ganglion where they pierce the eyeball to innervate the **sphincter pupillae muscle**, which constricts the pupils. They also innervate the muscles of the **ciliary body**, which function in accommodating the lens of the eyeball when focusing on nearby objects.

Facial nerve (CN VII)

The **superior portion of the superior salivatory nucleus** of the brainstem houses the **pre-ganglionic parasympathetic** nerve cell bodies associated with cranial nerve VII.

Pre-ganglionic parasympathetic fibers from the inferior portion of the **superior portion of the superior salivatory nucleus** reach the **pterygopalatine ganglion** (a parasympathetic ganglion of the facial nerve) where they synapse with post-ganglionic parasympathetic nerve cell bodies. Post-ganglionic parasympathetic fibers leave the ganglion and serve the **lacrimal gland**, eliciting the production of tears, as well as **glands in the nasal mucosa**, eliciting the production of mucus.

Pre-ganglionic parasympathetic fibers from the **superior salivatory nucleus** proceed to the **submandibular ganglion** where they synapse with post-ganglionic parasympathetic nerve cell bodies. Post-ganglionic parasympathetic fibers leave the ganglion to innervate the submandibular and sublingual glands, eliciting the flow of saliva.

Glossopharyngeal nerve (CN IX)

The **inferior salivatory nucleus** (nucleus of the glossopharyngeal nerve) of the brainstem houses the **pre-ganglionic parasympathetic nerve cell bodies** associated with cranial nerve IX. Pre-ganglionic parasympathetic fibers from this nucleus reach the **otic ganglion** to synapse with post-ganglionic parasympathetic nerve cell bodies. Their post-ganglionic parasympathetic fibers leave the ganglion to innervate the parotid gland as well as minor salivary glands in the oral mucosa, eliciting the flow of saliva.

Vagus nerve (CN X)

The **nucleus ambiguus** and the **posterior nucleus of the vagus nerve** house the **pre-ganglionic parasympathetic** nerve cell bodies associated with cranial nerve X.

Pre-ganglionic parasympathetic nerve fibers from the **nucleus ambiguus** synapse with post-ganglionic parasympathetic nerve cell bodies located in the **cardiac ganglia** distributed around the great vessels of the heart. Post-ganglionic parasympathetic axons leave the ganglia and innervate the **sinoatrial (SA) node**. Additionally, these fibers also serve occasional atrial and ventricular cardiac muscle fibers. The parasympathetic nervous system decreases the rate of the heartbeat.

Pre-ganglionic parasympathetic nerve fibers from the **posterior nucleus of the vagus** synapse with the following nerve cell bodies:

- Post-ganglionic parasympathetic nerve cell bodies located in ganglia surrounding the **bronchial passages**. Post-ganglionic parasympathetic fibers derived from these ganglia serve to elicit bronchoconstriction.
- Post-ganglionic parasympathetic nerve cell bodies located in the vicinity of the **pancreas**. Post-ganglionic parasympathetic fibers derived from these ganglia serve to elicit the release of pancreatic enzymes and buffer.
- Post-ganglionic parasympathetic nerve cell bodies along the **gastrointestinal tract** extending from the esophagus to the end of the transverse colon. These post-ganglionic parasympathetic nerve cell bodies are housed in the **submucosal plexus** (eponym: Meissner's plexus) and in the **myenteric plexus** (eponym: Auerbach's plexus) located within the wall of the gastrointestinal tract. Moreover, many of the pre-ganglionic parasympathetic axons synapse with the intrinsic nerve cell bodies of the enteric nervous system. Parasympathetic activity facilitates the digestive process and relaxes internal sphincters.

Sacral portion of the parasympathetic nervous system

The pre-ganglionic nerve cell bodies of the sacral portion of the parasympathetic nervous system are housed in the lateral cell column of sacral spinal cord levels 2 through 4

The **lateral cell column** of the second to fourth segments of the sacral spinal cord houses the pre-ganglionic parasympathetic nerve cell bodies composing the **sacral division** of the parasympathetic nervous system. Their pre-ganglionic axons join their corresponding spinal nerves and travel to the **pelvic plexus** and to scattered small **ganglia** in the vicinity of the pelvic organs, where they synapse with the enteric nervous system and with post-ganglionic nerve cell bodies of the intramural parasympathetic ganglia within the wall of the gastrointestinal tract. Post-ganglionic parasympathetic fibers from the pelvic plexus and the additional small ganglia serve the urinary bladder, penis, prostate gland, and seminal vesicles as well as the urinary bladder, uterine tubes, uterus, clitoris, and vagina.

The **urinary bladder**, **prostate gland**, and **seminal vesicles** receive their motor innervation from the small ganglia in their vicinity, except for the internal **sphincter muscle** controlling the **bladder**, which is **inhibited** by the parasympathetic nervous system.

The **uterus** and **vagina** receive *inhibitory* post-ganglionic parasympathetic fibers via the **uterovaginal plexus,**

although in pregnancy the same fibers are believed to have excitatory roles.

The **external genitalia** of individuals with XY chromosomal complement and the **clitoris** of individuals with XX chromosomal complement receive post-ganglionic parasympathetic fibers from the **pelvic plexus** via the **pudendal nerve**. These post-ganglionic parasympathetic fibers release **acetylcholine** and co-release **vasoactive intestinal peptide** and **nitric oxide** at their synapse with their target cells and are responsible for dilation of the cavernous blood sinuses resulting in the erection of the penis and clitoris.

ENTERIC NERVOUS SYSTEM

The intrinsic components of the enteric nervous system are located completely within the wall of the gastrointestinal tract and their neurons are distributed in two sets of ganglia

Submucosal plexus (eponym: Meissner's plexus) and myenteric plexus (eponym: Auerbach's plexus) house the intrinsic components of the enteric nervous system. The several hundred million neurons and even more numerous neuroglial cells of the enteric nervous system originate from neural crest cells, but their region of origin in the neural crest is different from the regions that give rise to the sympathetic and parasympathetic post-ganglionic neurons.

Generally speaking, the **submucosal plexus,** located in the submucosa at its interface with the inner circular layer of the muscularis externa, is responsible for localized phenomena, the contraction of the muscularis mucosae, glandular secretions from the mucosa, and the regulation of localized blood flow. The **myenteric plexus**, located between the inner circular and outer longitudinal layers of the muscularis externa, is responsible for peristalsis and global, rather than local, actions.

The two sets of **plexuses** interact with the sympathetic and parasympathetic nervous systems as well as with each other and function together to ensure that the gastrointestinal tract performs its functions properly.

The **extrinsic components** of the enteric nervous system are the sympathetic and parasympathetic fibers that modulate the functioning of the intrinsic components. The extrinsic components are necessary but not essential for normal functioning of the digestive tract. Parasympathetic innervation facilitates digestion by increasing peristalsis and relaxing internal sphincters. Vigorous activity from the sympathetic innervation slows down peristalsis and increases internal sphincter muscle tonus.

The **intrinsic components** of the enteric nervous system regulate much of the digestive processes by innervating the smooth muscles, glands, as well as the **diffuse neuroendocrine system (DNES) cells** residing in the epithelial lining of the gastrointestinal tract. There are at least 20 different types of DNES cells, and each releases a specific paracrine hormone (although many of these hormones also enter the bloodstream to remote target cells). These hormones act in concert to regulate much of the digestive process. A select list of these cells, their hormones, and their function is presented in Table 10.4.

168 ● ● ● **CHAPTER 10**

Table 10.4 ● Selected hormones secreted by the diffuse neuroendocrine system (DNES) cells of the gastrointestinal canal.

Paracrine hormone	DNES cell	Site of secretion	Function
Cholecystokinin (CCK)	I	Small intestine	Stimulates contraction of gall bladder (with release of bile) and facilitates the release of pancreatic enzymes
Gastric inhibitory peptide (GIP)	K	Small intestine	Inhibits secretion of gastric HCl
Gastrin	G	Pylorus and duodenum	Stimulates gastric secretion of HCl and pepsinogen
Glicentin	GL	Stomach through colon	Stimulates hepatic glycogenolysis
Glucagon	A	Stomach and duodenum	Stimulates hepatic glycogenolysis
Motilin	Mo	Small intestine	Increases gut motility
Neurotensin	N	Small intestine	Inhibits gut motility; stimulates blood flow to the ileum
Secretin	S	Small intestine	Stimulates bicarbonate secretion by the pancreas and biliary tract
Serotonin and substance P	EC	Stomach through colon	Increase gut motility
Somatostatin	D	Pylorus and duodenum	Inhibits nearby DNES cells
Vasoactive intestinal peptide (VIP)	VIP	Stomach through colon	Increases gut motility; stimulates intestinal ion and water secretion

Table 10.5 ● Principal and co-transmitter neurotransmitter substances of neurons of the enteric nervous system.

Myenteric Plexus		
Neuron	**Principal transmitter substance**	**Co-transmitter substance**
Inner circular smooth muscle activator	Acetylcholine	Tachykinin
Inner circular smooth muscle inhibitor	Nitric oxide	ATP, vasoactive intestinal peptide
Outer longitudinal smooth muscle activator	Acetylcholine	Tachykinin
Outer longitudinal smooth muscle inhibitor	Nitric oxide	ATP, vasoactive intestinal peptide
Interneuron (ascending)	Acetylcholine	Not known
Interneuron (descending)	Acetylcholine	ATP (perhaps)
Primary sensory neurons	Tachykinin	Not known
Submucosal Plexus		
Neuron	**Principal transmitter substance**	**Co-transmitter substance**
Noncholinergic vasodilator and secretomotor	Vasoactive intestinal peptide	Not known
Cholinergic vasodilator and secretomotor	Acetylcholine	Not known

In order for the enteric nervous system to be able to function it must have neurons that specialize in gathering information from the various components of the gastrointestinal tract; these neurons are **sensory neurons** and most synapse with another type of neuron, known as an **interneuron**. It is these interneurons that activate either inhibitory or excitatory neurons of the enteric nervous system that synapse with smooth muscles, DNES cells, or intrinsic glands of the gastrointestinal tract. These interneurons communicate with their target cells by releasing neurotransmitter substances into their synaptic clefts. Generally speaking, the principal neurotransmitter substance utilized by the intrinsic neurons of the enteric nervous system is acetylcholine, although some neurons use vasoactive intestinal peptide and tachykinin as well as other members of its family, such as Substance P, neurokinin A and B, neuropeptide K, gamma-hemokinin-1,

and endokinin-A and -B. Additionally, there are several co-transmitters as listed in Table 10.5.

NEUROTRANSMITTERS AND RECEPTORS OF THE AUTONOMIC NERVOUS SYSTEM

As indicated at the beginning of this chapter because the two principal neurotransmitters are acetylcholine and norepinephrine, it is customary to speak of two divisions of the autonomic nervous system, namely **cholinergic** (because they release acetylcholine at their axon terminals) and **adrenergic** (because they release norepinephrine (noradrenaline) at their axon terminals). The principal neurotransmitters of preganglionic and post-ganglionic parasympathetic neurons, pre-ganglionic sympathetic neurons, and those post-ganglionic sympathetic neurons that innervate eccrine sweat glands are all cholinergic. All other post-ganglionic sympathetic neurons are adrenergic. However, co-transmitters are also released, specifically, ATP, nitric oxide, vasoactive intestinal peptide, and neuropeptide Y.

Autonomic receptors

Autonomic receptors are also classified as **cholinergic** or **adrenergic** (noradrenergic), based on their ability to bind and to be activated by acetylcholine or noradrenaline, respectively. Subtypes of both cholinergic and adrenergic receptors have been identified. Additionally, other types of receptors are also present but are not detailed in this discussion of the autonomic nervous system.

Cholinergic receptors

Muscarinic and nicotinic receptors are the two types of cholinergic receptors

There are two types of **cholinergic receptors,** namely, muscarinic receptors and nicotinic receptors. The former are also activated by muscarine, a toxin derived from toadstools, and the latter are also activated by nicotine. It is important to note that nicotine will not activate muscarinic receptors and muscarine will not activate nicotinic receptors, but both types will be activated by acetylcholine.

Muscarinic receptors are located in the membranes of cells synapsing with post-ganglionic parasympathetic fibers and in the membranes of eccrine sweat glands (innervated by post-ganglionic sympathetic fibers).

Nicotinic receptors are located in the membranes of post-ganglionic sympathetic and post-ganglionic parasympathetic nerve cell bodies where they synapse with their pre-ganglionic counterparts. They are also located in the sarcolemma of the neuromuscular junctions of skeletal muscle cells.

Adrenergic receptors

Adrenergic receptors are of two major types: alpha-adrenergic and beta-adrenergic

There are two types of adrenergic receptors, alpha-adrenergic receptors and beta-adrenergic receptors, each with its own subtypes.

Alpha-adrenergic receptors are more easily activated by epinephrine than by norepinephrine and are least activated by isoproterenol. Alpha receptors are located in the plasma membranes of vascular smooth muscle cells, dilator pupillae muscle cells, and cell bodies of enteric neurons controlling internal sphincter muscle contraction.

Beta-adrenergic receptors are more responsive to isoproterenol than to epinephrine and are least activated by norepinephrine. Beta receptors are located in the plasma membranes of cardiac muscle cells, bronchiolar smooth muscle cells, and cell bodies of enteric neurons that facilitate peristaltic activity of the gastrointestinal tract.

PELVIC AUTONOMIC FUNCTIONS

The somatic and autonomic nervous systems interact with each other to perform numerous functions, among them those that control urination, defecation, and erection and ejaculation.

Urine retention and urination

Urine retention and urination require the interaction of the sympathetic, parasympathetic, and somatic motor nervous systems and are controlled by the cerebral cortex and pons

The process of retaining urine in, and evacuating urine from, the urinary bladder is a learned function that requires the interactions of the sympathetic, parasympathetic, and somatic motor nervous systems and is controlled by the cerebral cortex and pons. As the urinary bladder distends, nociceptors and stretch receptors in its wall transmit the information to the posterior horn of the spinal cord. These afferent nerves are **unipolar (pseudounipolar) neurons** whose cell bodies are located in the posterior root ganglia. If the volume of urine in the bladder is low, the skeletal muscles of the **external sphincter** undergo reflex contraction in response to stimulation by axons of somatic motor neurons whose cell bodies are located in the **nucleus of the pudendal nerve** (eponym: Onuf nucleus) in the anterior horn of spinal cord levels S2–S4. Cell bodies of the motor neurons in the nucleus of the pudendal nerve are activated by projections from the **pontine micturition center** and the **cerebral cortex** (Fig. 10.8). Concurrently, the smooth muscle cells of the **internal sphincter** also undergo **reflex contraction** and the **detrusor muscles,** smooth muscles located in the wall of the urinary bladder, undergo **reflex relaxation,** due to the activities of pre-ganglionic sympathetic innervation from the lateral (intermediate) cell column at spinal cord levels T12–L2.

The process of **urination** is facilitated by the micturition centers located in the cerebral cortex and the pontine micturition centers. As the bladder distends and the urine volume reaches a threshold level, sensory information reaches the **ventral posterior lateral (VPL) nucleus of the thalamus** traveling via the posterior column-medial lemniscal pathway, and the anterolateral system. The information is transmitted by tertiary neurons from the thalamus to the **postcentral sulcus** and the individual becomes conscious of the need to urinate. When the individual is ready to void, the retentive reflexes described here become inhibited; thus, the internal and external sphincters relax. Additionally, projections from the pontine micturition centers activate the pre-ganglionic parasympathetic neurons in the lateral horn of spinal cord levels S2–S4, whose post-ganglionic counterparts stimulate the detrusor muscles of the bladder to undergo contraction, thus emptying the urinary bladder.

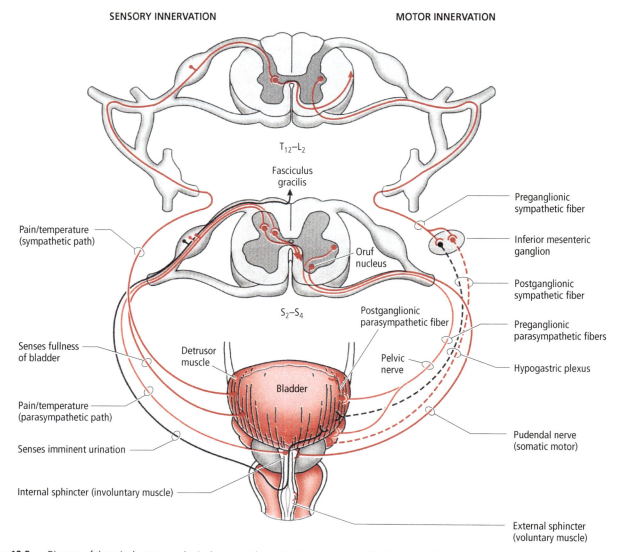

Figure 10.8 • Diagram of the spinal sensory and spinal motor pathways involved in urination. The black solid line indicates the sensation of imminent urination and the black dashed line indicates the inhibition of relaxation of the urinary bladder permitting post-ganglionic sympathetic fibers to cause contraction of the detrusor muscles, thus causing emptying of the bladder.

Defecation

The sigmoid colon distends as the feces continues to accumulate in its lumen causing the urge to defecate

As feces accumulate, the sigmoid colon distends and the individual becomes conscious of the need to **defecate**, but the feces are retained in the colon by the two sphincter muscles, the **internal smooth muscle sphincter** and the **external skeletal muscle sphincter**. The former is supplied by **post-ganglionic sympathetic fibers** from the hypogastric plexus as well as by post-ganglionic parasympathetic fibers located in myenteric plexus (whose pre-ganglionic fibers arise from the sacral spinal cord). The external sphincter is supplied by the somatic nervous system, namely the **inferior rectal nerve**. Additionally, the **puborectalis muscle**, a part of the **levator ani** muscle, participates in retaining the feces in the rectum. The right and left puborectalis muscles originate on the posterior aspect of the pubis and attach to each other by forming a sling around the sigmoid colon pulling it in an anterior direction, forming an anorectal angle, reducing the ability to have a bowel movement. Additionally, when the sigmoid colon becomes distended, the internal sphincter muscle relaxes (due to the parasympathetic nerve supply) and at the same time the external sphincter muscle contracts, preventing the feces from exiting the bowel. While the individual is asleep, this contraction of the external sphincter muscle is a reflex response. When the individual is awake, the contraction of the external sphincter is a voluntary response. When the individual is ready to defecate, the voluntary relaxation of the puborectalis and external sphincter muscles, along with the peristaltic action of the sigmoid colon, and with the intra-abdominal pressure created by contraction of the anterior abdominal wall musculature, the individual is able to expel the feces from the sigmoid colon and the rectum through the anus.

Strong pain killers, such as opioids, reduce colon motility causing constipation and, in worse cases, even impaction which is an emergent condition.

Erection and ejaculation

Erection of the penis and of the clitoris is an autonomic function initiated by tactile or psychogenic stimulation and mediated by parasympathetic innervation derived from sacral spinal cord levels. The parasympathetic fibers are responsible for engorgement of the spongy tissues (cavernous sinuses) of the penis and of the clitoris as well as of the vascular channels of the labia minora, and cause the release of secretions from the greater vestibular glands (eponym: Bartholin glands), lubricating the vulva and vaginal opening. Pre-ganglionic sympathetic fibers from spinal cord levels T10–L2 synapse on post-ganglionic sympathetic nerve cell bodies housed in the inferior mesenteric ganglion. **Post-ganglionic sympathetic fibers** from that ganglion are responsible for rhythmic contractions of the vagina and uterus, and for the release of secretions from the prostate gland and seminal vesicles as well as for **ejaculation**.

CLINICAL CONSIDERATIONS

Congenital megacolon
Congenital megacolon (eponym: Hirschsprung's disease), is an excessive enlargement of the distal part of the colon due to the lack of, or a marked reduction in, the neuronal population in myenteric plexus. Due to the absence of these neurons a region of the colon becomes constricted, the muscles cannot relax, and feces accumulate in the region of the colon proximal to the constriction. With the accumulation of feces the proximal region enlarges, resulting in megacolon.

Oculosympathetic paresis
Oculosympathetic paresis (eponym: Horner syndrome) is a condition resulting from damage to the cervical sympathetic trunk, sympathetic fibers of the carotid plexus, or any region of the spinal cord or brainstem involving the sympathetic pathways. The syndrome is characterized by the following conditions on the affected side: constriction of the pupil (miosis), lack of the ability to sweat (anhidrosis), recession of the eyeball within the orbit (enophthalmos), and drooping of the upper eyelid (ptosis). It is interesting to note that cancer of the apex of the lung (eponym: Pancoast tumor) may involve the cervical sympathetic trunk, thus precipitating oculosympathetic paresis.

Multiple system atrophy
Multiple system atrophy (eponym: Shy–Drager syndrome) patients suffer from orthostatic hypertension and faint frequently because the autonomic nervous system neurons responsible for adjusting blood pressure during postural changes have either degenerated or died. These postural changes can involve the patient standing up from a reclining (or even from a sitting) position.

SYNONYMS AND EPONYMS OF THE AUTONOMIC NERVOUS SYSTEM

Name of structure or term	Synonym(s)/eponym(s)
Myenteric plexus	Auerbach's plexus
Submucosal plexus	Meissner's plexus
Middle cardiac nerve	Great cardiac nerve
Nerve of the pterygoid canal	Nerve of the vidian canal
Post-ganglionic neuron	Post-synaptic neuron
Pre-ganglionic neuron	Pre-synaptic neuron

FOLLOW-UP TO CLINICAL CASE

Serologic testing confirmed that this patient had an autoimmune condition, known as **myasthenic syndrome** (eponym: Lambert–Eaton myasthenic syndrome).

Acetylcholine is released at presynaptic peripheral nerve terminals in response to an action potential. The action potential at the presynaptic terminal opens voltage-gated calcium channels, leading to an influx of calcium, which leads to acetylcholine release. This acetylcholine diffuses across the synapse to reach receptors on the target organ. This is most apparent at the neuromuscular junction, where an action potential along the muscle membrane is induced by the released acetylcholine from the presynaptic nerve terminal. It occurs for both autonomic nerves as well as nerves supplying skeletal muscle, and the effects are seen at both nicotinic and muscarinic receptors.

In myasthenic syndrome, antibodies to the voltage-gated calcium channels are produced. This reduces calcium influx at the presynaptic terminal in response to an action potential, and acetylcholine release is therefore impaired. This produces both skeletal muscle weakness and autonomic dysfunction. Common autonomic symptoms include dry mouth, dry eyes, impotence, heart rate irregularities, constipation, and orthostatic hypotension (can cause dizziness). Both parasympathetic and sympathetic functions can be affected. Most often, however, it is the muscle weakness and fatigue that cause the most disability.

Myasthenic syndrome is a paraneoplastic disorder, meaning it is often associated with cancer. Small-cell lung cancer is the one most often associated. The immune system is induced by an as yet undetermined mechanism to produce antibodies to the voltage-gated calcium channel. Therefore, a cancer work-up should be undertaken in all patients with this syndrome. In particular, this patient needed, at the very least, a chest X-ray and/or chest CT since they had been a heavy smoker.

Autonomic dysfunction can occur with many other conditions affecting acetylcholine release or metabolism. A major culprit is medication with anticholinergic effects (there are many). Botulism, which also reduces acetylcholine release at the presynaptic terminal, causes autonomic dysfunction as well as weakness.

QUESTIONS TO PONDER

1. Why are there only 14 pairs of white rami communicantes, whereas there are 31 pairs of gray rami communicantes?

2. Why is it that pre-ganglionic sympathetic fibers are usually short, whereas pre-ganglionic parasympathetic fibers are usually long?

3. Why can the same neurotransmitter (acetylcholine) be used in both pre-ganglionic sympathetic and parasympathetic synapses, but not at post-ganglionic synapses?

4. Why is it that the short ciliary nerves have both sympathetic post-ganglionic and parasympathetic post-ganglionic fibers, whereas the long ciliary nerves have only post-ganglionic sympathetic fibers?

5. Why is the contraction of the external skeletal muscle sphincter, which controls the emptying of the colon, sometimes a voluntary response and sometimes a reflex response?

CHAPTER 11

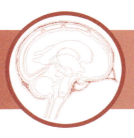  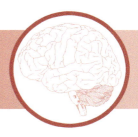

Spinal Reflexes

- CLINICAL CASE
- COMPONENTS OF REFLEXES
- LOWER MOTOR NEURONS
- SKELETAL MUSCLE INNERVATION
- SKELETAL MUSCLE RECEPTORS
- MUSCLE STRETCH REFLEX
- RECIPROCAL INHIBITION
- AUTOGENIC INHIBITION (INVERSE MYOTATIC REFLEX)
- FLEXOR REFLEX (WITHDRAWAL REFLEX, NOCICEPTIVE REFLEX) CROSSED EXTENSOR REFLEX
- MAINTENANCE OF MUSCLE TONE VIA THE GAMMA LOOP
- ALPHA–GAMMA COACTIVATION
- SYNONYMS AND EPONYMS OF THE SPINAL REFLEXES
- FOLLOW-UP TO CLINICAL CASE
- QUESTIONS TO PONDER

CLINICAL CASE

A 72-year-old patient without diabetes mellitus began to experience low back pain 3 months ago, which, 4 weeks after onset began to involve the left anteromedial thigh. Over several weeks the left anterior thigh muscles began to decrease in size and walking became difficult. The patient has fallen several times due to the left knee giving way. Physical examination shows asymmetric flaccid weakness of the left quadriceps femoris and thigh adductor muscles with normal sensation and an absent left patellar tendon reflex. The neuroanatomic issue is to determine where the site of neurologic disease might be: lumbar nerve root vs. lumbar plexus vs. multiple peripheral nerves.

MRI scan of the lower back demonstrates mild degenerative changes in the lumbar spine consistent with age, but no disk herniation, or significant neural foraminal stenosis; after administering gadolinium (contrast agent) the left L3 nerve root enhances (increases in signal intensity) on the T1-weighted imaging sequence. Electromyography shows signs of chronic active denervation (irritability and fibrillation potentials as well as motor unit potential) evidence of a denervating neurogenic process involving the left L3 myotome (left iliopsoas, adductor longus, rectus femoris, and vastus medialis muscles as well as the left mid-lumbar multifidus muscle). A lumbar puncture was performed to collect cerebrospinal fluid, and analysis demonstrates increased CSF protein (>40 mg/dL) and a mild increase in the number of lymphocytes (lymphocytic pleocytosis; >5 cells/mm^3).

A diagnosis of inflammatory left L3 radiculopathy is made.

COMPONENTS OF REFLEXES

A reflex reaction is a simple, stereotyped rapid motor response to a particular sensory stimulus. Reflexes are involuntary, instantaneous, and usually protect the body from potentially damaging stimuli and injury. Reflex activity for the body is confined to the spinal cord and does not involve the brain nor does it reach consciousness. A reflex begins with a stimulus activating a peripheral sensory receptor associated with a unipolar (pseudounipolar) sensory neuron whose cell body is housed in a spinal ganglion (posterior root ganglion). The central process of the sensory neuron projects to the spinal cord where it either synapses with an interneuron or directly with a motor neuron. The motor neuron axon innervates and stimulates the skeletal muscle, causing it to contract, thereby moving the affected body part away from the original stimulus.

A **complete reflex arc** consists of the following components (Fig. 11.1) **Sensory receptors**, **sensory neuron**, **interneuron** (if it is a three-cell reflex arc), **motor neuron**, and **skeletal muscle cells**.

A Textbook of Neuroanatomy, Third Edition. Maria A. Patestas, Amanda J. Meyer, and Leslie P. Gartner.
© 2025 John Wiley & Sons, Inc. Published 2025 by John Wiley & Sons, Inc.
Companion website: www.wiley.com.go/patestas/neuroanatomy3e

CHAPTER 11

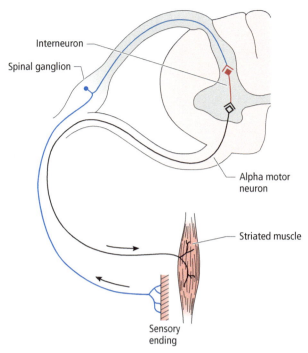

Figure 11.1 • Components of a spinal reflex arc: A receptor, in this case located in the skin. The afferent (sensory) limb is formed by a sensory neuron whose cell body is located in a spinal ganglion. An interneuron is located entirely within the gray matter or the spinal cord. The efferent (motor) limb is formed by a motor neuron whose cell body is located in the anterior horn of the spinal cord. An effector, consisting of a skeletal muscle. Source: Waxman and DeGroot (1995). *Correlative Neuroanatomy*, 22 edn. McGraw Hill, Lange, figure 5.20, p. 63/McGraw-Hill Education.

- The **afferent (sensory) limb** of the reflex arc is formed by the peripheral process of a pseudounipolar (sensory) neuron whose cell body is housed in a spinal ganglion. The peripheral process, which runs in a peripheral nerve, may terminate in the skin, skeletal muscle, tendon, or joint capsule. Its central process enters the spinal cord via the posterior root of a spinal nerve to synapse with an interneuron or a motor neuron in the spinal cord gray matter.
- One or more **interneurons**. Most spinal reflexes include three neurons (sensory neuron, interneuron, and motor neuron) where the interneuron is inserted between the central process of the sensory neuron and the motor neuron. There are some reflexes that include only two neurons, such as the knee jerk reflex, where the central process of the sensory neuron synapses directly with a motor neuron.
- An **efferent (motor) limb** of the reflex arc, which is formed by a motor neuron whose cell body is located in the anterior horn of the spinal cord and its axon runs in a peripheral nerve to terminate in and stimulate a skeletal muscle.

- An **effector**, which is a skeletal muscle. The muscle may be stimulated to contract and produce movement, or it may be inhibited from contracting, depending on the particular reflex.

Damage to any of these components of the reflex arc abolishes the reflex response.

Classification of reflexes

Reflexes are classified according to the number of spinal cord levels involved as well as by their association with the brain. There are three categories of reflexes:

1. **Intrasegmental reflexes**, involve or affect only a single spinal cord level. The central process of the sensory pseudounipolar neuron and the cell body of the motor neuron are located at the same spinal cord level.
2. **Intersegmental reflexes**, involve or affect two or more spinal cord levels. The central process of the pseudounipolar (sensory) neuron enters the spinal cord where it bifurcates into a short ascending and a short descending branch. These branches extend at least one spinal cord level above and or below the level of entry of the central process of the sensory neuron before its branches synapse. This allows synapsing to occur at other spinal cord levels, which contain cell bodies of motor neurons that innervate all the muscles needed for the reflex response. Certain reflex responses require the coordinated action of multiple spinal cord segments in order to influence the necessary muscles and to produce the appropriate muscle response.
3. **Suprasegmental reflexes**, are influenced by neurons residing in the cerebral hemispheres or the brainstem and through their descending projections influence reflex activity occurring at the spinal cord level or levels.

Physicians test a patient's reflexes to check the integrity of the nervous system and to determine the location of neurologic lesions. There are four areas where testing of spinal cord segments and spinal nerves is commonly performed:

1. The biceps brachii reflex (tests spinal cord levels C5 and C6).
2. The triceps brachii reflex (tests spinal cord levels C6–C8).
3. The quadriceps femoris reflex (tests L2–L4 via the "knee jerk reflex").
4. Gastrocnemius reflex (tests S1–S2 via the "ankle jerk reflex," calcaneal tendon reflex).

LOWER MOTOR NEURONS

Lower motor neurons innervate skeletal muscle

The cell bodies of lower motor neurons reside in the brainstem and spinal cord. In the brainstem, they are housed in the cranial nerve motor nuclei, whereas in the spinal cord they reside in the anterior horn. Their axons run in peripheral nerves (cranial and spinal nerves) and synapse with skeletal muscle cells. The lower motor neurons are said to form the final common pathway since they connect the central nervous system and skeletal muscles. Although there are several different types of lower motor neurons, only the alpha and gamma motor neurons that are involved in spinal reflexes are discussed here. The cell bodies of both neuron types are interspersed in the anterior horn of the spinal cord. **Alpha motor neurons**, the largest neurons innervate typical skeletal muscle fibers, referred to as **extrafusal fibers**, which, when stimulated, cause contraction of the gross muscle. **Gamma motor neurons** are smaller neurons and innervate the contractile (polar) portions of the **intrafusal fibers of muscle spindles**. The main function of the gamma motor neurons is to increase the sensitivity of the intrafusal fibers to changes in muscle length and indirectly cause the reflex stimulation of the alpha motor neurons that stimulate contraction of the gross muscle (see next).

SKELETAL MUSCLE INNERVATION

Skeletal muscle consists of extrafusal fibers and muscle spindles

Extrafusal fibers are typical skeletal muscle cells (fibers), which, when stimulated, contract to produce movement. These fibers receive motor innervation from **alpha motor neurons** and sensory innervation from the peripheral processes of **pseudounipolar neurons** whose cell bodies are located in the posterior root ganglia.

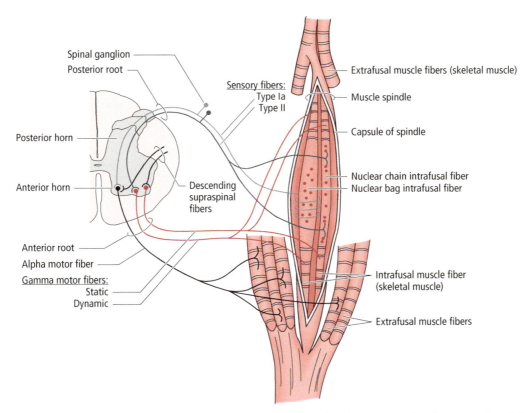

Figure 11.2 ● A muscle spindle covered by a fibrous capsule and enclosing two types of **intrafusal fibers**: *nuclear bag fibers* and *nuclear chain fibers*. A **nuclear bag fiber** has a *non-contractile*, bulging "bag-like" region packed with nuclei, in its central or equatorial aspect, whereas a **nuclear chain fiber** has a *non-contractile* region with nuclei arranged in a linear or "chain-like" manner in its central or equatorial aspect. Both types of intrafusal fibers have skeletal muscle *contractile* portions with myofibrils at both of their two polar ends. The intrafusal (nuclear bag and nuclear chain) fibers of muscle spindles receive both **motor** and **sensory** innervation. The **motor innervation** to the intrafusal fibers is provided by **gamma motor neurons**. The **sensory innervation** to the intrafusal fibers is provided by the peripheral processes of **pseudounipolar neurons** whose cell bodies are located in the spinal ganglia, or in the sensory ganglia of the cranial nerves. There are two types of specialized peripheral sensory endings: (1) **anulospiral (primary) endings (Type Ia fibers)** which are wrapped around the central (non-contractile) portion of both the nuclear bag and nuclear chain intrafusal fibers and (2) **flower spray (secondary) endings (Type II fibers)** that terminate on each side of the Ia terminals, mainly on nuclear chain fibers (and less often on nuclear bag fibers). Source: Haines, DE (2013) *Fundamental Neuroscience for Basic and Clinical Applications*, Elsevier Saunders, figure 24.4a, p. 327. Reproduced with permission of Elsevier.

176 ● ● ● **CHAPTER 11**

SKELETAL MUSCLE RECEPTORS

Skeletal muscle contains two types of receptors specialized in position sense: muscle spindles and tendon organs

Muscle spindles

Muscle spindles are specialized mechanoreceptors (type of proprioceptors, position sense receptors) unique to skeletal muscle that continually *monitor changes in muscle length*. They have a key function during movement and in the maintenance of muscle tone (Fig. 11.2).

Muscle spindles are oriented **parallel** to the longitudinal axes of the extrafusal muscle fibers. Each muscle spindle is covered by a fibrous capsule, and encloses 2–12 **intrafusal fibers**, modified skeletal muscle fibers, each of which in turn is covered by a thin capsule. The connective tissue capsule of the muscle spindles is anchored to the connective tissue septa that extend into the muscle. There are two morphologically and functionally distinct types of intrafusal fibers: *nuclear bag fibers* and *nuclear chain fibers*. Skeletal muscle fibers (extrafusal fibers) contain multiple nuclei. Nuclear bag fibers and nuclear chain fibers are also multinucleated cells, and the organization of their nuclei is their most distinctive characteristic.

- Each **nuclear bag fiber** has a *non-contractile*, bulging "bag-like" region packed with nuclei, in its central or equatorial aspect, and skeletal muscle *contractile* portion with myofibrils at both of its two polar ends. Nuclear bag fibers are classified into *static nuclear bag fibers* and *dynamic nuclear bag fibers* based on their function.

- The **static nuclear bag fibers** respond to *changes in muscle length*.

- The **dynamic nuclear bag fibers** respond principally to the *rate of change in muscle length*.

- Each **nuclear chain fiber** has a *non-contractile* region with nuclei arranged in a linear or "chain-like" manner in its central or equatorial aspect, and skeletal muscle *contractile* portion with myofibrils at both of its polar ends. Nuclear chain fibers detect mostly *changes in muscle length*.

The intrafusal (nuclear bag and nuclear chain) fibers of muscle spindles receive both **motor** and **sensory** innervation. The **motor innervation** to the intrafusal fibers is provided by **gamma motor neurons**. The static nuclear bag fibers are innervated by **static gamma motor neurons**; the dynamic nuclear bag fibers are innervated by **dynamic gamma motor neurons**. They synapse on the *polar (contractile) portions* of the intrafusal fibers. When stimulated, the polar ends contract, pulling away from the central, non-contractile region in the middle. This in turn stretches and disturbs the non-contractile

portion of the intrafusal fibers, causing the sensory fibers wrapped around it to also become stretched and discharge.

The **sensory innervation** to the intrafusal fibers is provided by the peripheral processes (mechanoreceptors) of **pseudounipolar neurons** whose cell bodies reside in the spinal ganglia, or in the sensory ganglia of the cranial nerves (and, in the case of the trigeminal nerve, within its mesencephalic nucleus). There are two types of specialized peripheral sensory endings: (1) **anulospiral (primary) endings (Type Ia fibers)** which are wrapped around the central (non-contractile) portion of both the nuclear bag and nuclear chain intrafusal fibers that become activated at the *onset of muscle stretch or tension* and (2) **flower spray (secondary) endings (Type II fibers)** that terminate on each side of the Ia terminals, mainly on nuclear chain fibers (and less often on nuclear bag fibers); the type II fibers become activated when the muscle is in the process of being *stretched*.

The firing rate of the Type Ia peripheral sensory endings is directly proportional to the extent of intrafusal fiber stretch. The central (non-contractile) portion of the intrafusal fibers can be stretched in two ways: (1) when the gross muscle is passively stretched, as in tapping its tendon (this stretches the extrafusal fibers and, in addition, the muscle spindles containing intrafusal fibers). The central (non-contractile) portion of each intrafusal fiber stretches, which disturbs and stimulates the Ia anulospiral endings, or (2) the central portion stretches when the polar (contractile) ends of the intrafusal fibers contract, in response to gamma motor neuron stimulation, thereby disturbing and stimulating the Ia anulospiral endings.

Tendon organs

Tendon organs are located where skeletal muscle inserts into its tendons

Tendon organs (eponym: Golgi Tendon Organs) are stretch receptors that are unique to skeletal muscle (Fig. 11.3). They are located where the muscle fibers insert into their tendons. Unlike the muscle spindles, which are arranged parallel to the skeletal muscle fibers and contain intrafusal fibers with contractile portions, the tendon organs are arranged in **series**, lack a contractile portion, and do not receive motor innervation. They receive sensory innervation via **Type Ib fibers**. Tendon organs are stimulated during muscle stretch or muscle contraction, and they monitor the *amount of tension* that is placed on a skeletal muscle.

MUSCLE STRETCH REFLEX

The muscle stretch reflex is the simplest of all spinal reflexes. It consists of only two neurons: a sensory neuron and a motor neuron

The **muscle stretch reflex** is also referred to as the *tendon reflex, deep tendon reflex, monosynaptic reflex*, or the *myotatic*

SPINAL REFLEXES • • • 177

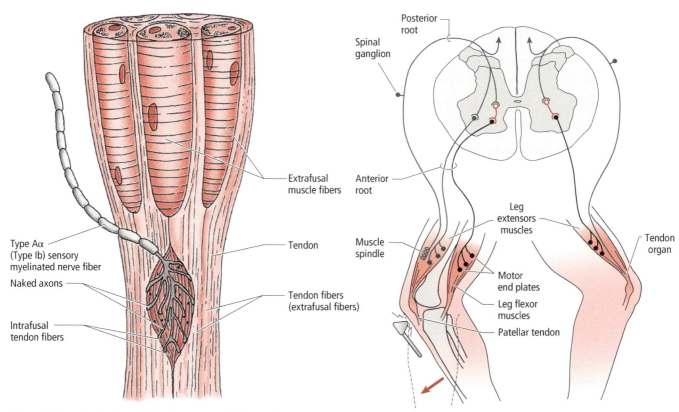

Figure 11.3 • The structure of a tendon organ and its innervation.

Figure 11.4 • Components of the flexor reflex and reciprocal inhibition, left side. Components of autogenic inhibition, right side. Source: Haines, DE (2013) *Fundamental Neuroscience for Basic and Clinical Applications*, Elsevier Saunders, figure 9.9, p. 133/with permission of Elsevier.

reflex (G. *myotatic*, "muscle stretch") (Fig. 11.4). It is referred to as the tendon reflex or deep tendon reflex because it is elicited by tapping a tendon. It is also known as the monosynaptic reflex because there is only a single synaptic contact that occurs in the spinal cord. It is the simplest of all spinal reflexes, consisting only of two neurons: a sensory neuron and a motor neuron. This reflex, routinely tested during a physical examination, is elicited by tapping a large tendon (e.g., tendon of the biceps brachii, triceps brachii, quadriceps femoris, or the calcaneal tendon). The quadriceps stretch reflex (also referred to as the knee jerk reflex or the patellar tendon reflex), its most common example, is described below.

To elicit the knee jerk reflex, the patellar tendon is tapped gently with a reflex hammer (Fig. 11.4). The tap displaces the patella tendon just slightly. This in turn stretches the extrafusal (skeletal muscle) fibers slightly as well as the muscle spindles of the quadriceps femoris muscle.

The primary afferent sensory endings (anulospiral endings, **Group Ia**, heavily myelinated, rapidly conducting fibers) wrapped around the central non-contractile portion of the intrafusal fibers are stretched and stimulated. Thus, the muscle stretch causes the sensory endings on intrafusal fibers to discharge. Nerve impulses are transmitted by the Ia fibers to those **alpha (α) motor neurons** (in lamina IX) of the anterior horn of the spinal cord that innervate the quadriceps femoris muscle. The quadriceps responds by *contracting* slightly, causing extension of the leg at the knee joint. This is an involuntary **reflex-induced contraction** of the quadriceps femoris muscle. The contraction of the gross muscle results in a decrease in the tension on the intrafusal fibers and *silencing* of the sensory Ia endings that innervate the once disturbed muscle spindles. This reflex compensates for the muscle stretch and serves to protect the muscle fibers from overstretching which may cause injury to the muscle. This reflex is a "monosynaptic" reflex because it only involves a single synapse with the alpha motor neuron.

RECIPROCAL INHIBITION

Reciprocal inhibition supports the stretch reflex. As an extensor muscle contracts as part of the extensor reflex, its antagonist, a flexor muscle relaxes.via reciprocal inhibition

Reciprocal inhibition is not only an extension of but also an enhancement of the stretch reflex (Fig. 11.4). In reciprocal inhibition, the receptor is the muscle spindle. One muscle (e.g., the extensor) is stimulated to undergo contraction, while another muscle (e.g., the flexor, its antagonist) is simultaneously inhibited from undergoing contraction and, therefore, relaxes. Two muscles that produce opposite movements at the same joint are said to be antagonists. In this mechanism, a tap on the patellar tendon stretches the extrafusal fibers and muscle spindles of the quadriceps femoris muscle, which ultimately causes the afferent sensory neurons to discharge, and stimulation is relayed to the spinal cord via their Ia fibers. The central process of the afferent neuron bifurcates into the following two branches:

1. One branch forms an **excitatory** synapse directly with an **alpha motor neuron** whose axon leaves the spinal cord to innervate the quadriceps femoris muscle (leg extensor). The quadriceps *contracts* causing extension of the leg at the knee joint to compensate for the muscle *stretch*.
2. The other branch forms an **excitatory** synapse directly with an **inhibitory interneuron,** which in turn inhibits the alpha motor neuron that innervates the antagonist hamstring muscles (leg flexors).

AUTOGENIC INHIBITION (INVERSE MYOTATIC REFLEX)

Tendon organs play a role in the fine-tuning of skeletal muscle contractile force during movement

Autogenic inhibition enhances the stretch reflex (Fig. 11.4). In autogenic inhibition the receptor stimulated is the **tendon organ (TO)**, a high-threshold receptor. It takes considerable tension (stretch) to stimulate this receptor, much more than that which usually stimulates the muscle spindles. The TOs are stretched and stimulated when, for example, the quadriceps femoris muscle is either stretched or undergoes contraction, in either case generating tension in its tendons. During muscle stretch, most of the tension is absorbed by the muscle with only a minimal effect on its tendons. In contrast, during muscle contraction, the TOs are stimulated almost immediately. TOs are specialized receptors that monitor the tension applied to tendons generated mainly by muscle *contraction*. Muscle contraction causes the sensory endings terminating on the TOs to discharge. When the sensory neuron's large diameter, heavily myelinated, fast-conducting **Ib fiber** terminating on the TO is stimulated, this neuron then stimulates an *inhibitory interneuron*, which, in turn inhibits the alpha motor neuron that innervates the quadriceps femoris muscle from firing and, as a result, the quadriceps femoris muscle cannot undergo further contraction. Thus, the TO plays a role in muscle relaxation, whereas stimulation of a muscle spindle leads to muscle contraction.

In the past it was believed that TOs, via autogenic inhibition, limited (over)contraction of a muscle to protect the tendon. More recently, it is thought that the TOs play a role in the fine-tuning of skeletal muscle contractile force during movement.

FLEXOR REFLEX (WITHDRAWAL REFLEX, NOCICEPTIVE REFLEX)

The flexor reflex is elicited by a sufficiently noxious stimulus

Stretch reflexes and autogenic inhibition (as discussed above) are elicited by stimulation of stretch receptors in a muscle or tendon; in contrast, the **flexor reflex** is elicited by stimulation of cutaneous receptors and includes a reflex response across several joints involving an entire limb (Fig. 11.5).

In the **flexor reflex**, the receptor is located in the *cutaneous* (that is, in the skin) or subcutaneous tissue. If the stimulus is painful or injurious the body attempts to remove the affected body part (such as a limb) from the source of injury via this reflex. Lightly myelinated (Aδ) or unmyelinated (C) central fibers of pseudounipolar neurons, whose cell bodies are housed in a spinal ganglion, relay nociceptive (pain) information to the spinal cord *posterolateral tract (eponym: of Lissauer)* where they branch into short ascending

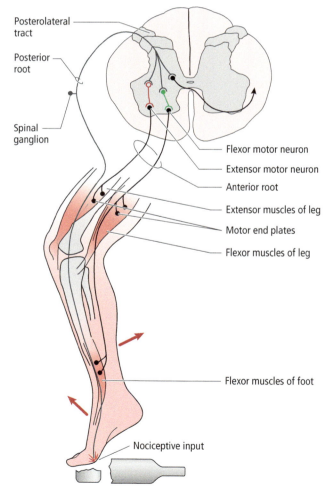

Figure 11.5 ● Components of flexor (withdrawal, nociceptive) reflex. Source: Haines, DE (2013) *Fundamental Neuroscience for Basic and Clinical Applications*, Elsevier Saunders, figure 9.10, p. 133/with permission of Elsevier.

and descending fibers. These terminals synapse in the gray matter where they form excitatory synapses with excitatory and inhibitory interneurons: (1) the excitatory interneurons synapse with the alpha motor neurons that innervate the flexor muscles of the entire limb causing their contraction and withdrawal of the limb; and (2) the inhibitory interneurons synapse with the alpha motor neurons that innervate the extensor muscles (antagonists) of the same limb to inhibit them (causing relaxation of the muscles).

In addition to painful and injurious stimuli, various other types of harmless cutaneous stimulation could potentially elicit the flexor reflex. The reflex response to such excess sensitivity is usually prevented by descending pathways arising from the cerebral hemispheres and brainstem that minimizes unnecessary withdrawal reflex activity. Such descending influences permit the flexor reflex to occur only in response to a sufficiently noxious stimulus. It should be noted, that if a noxious stimulus contacts the anterior aspect of the thigh, it would be the thigh *extensors* that will contract to pull the thigh away from the noxious stimulus. Thus, a better and all-encompassing term for this reflex would be the "**withdrawal**" reflex instead of "flexor" reflex since at times a noxious stimulus will elicit an extensor reflex.

Normally, when the plantar surface of the foot is stimulated, the foot reflexively plantarflexes (toes curl down) mediated by the corticospinal tract. However, in an individual with a corticospinal tract lesion, stimulation of the plantar surface of the foot elicits hyperextension of the foot accompanied by dorsiflexion of the great toe and by the abduction (fanning) of the other four toes. This pathological response is referred to as **Extensor Plantar Response** (Eponym: Babinski sign). In the absence of the corticospinal tract, this pathologic response is mediated by the descending motor pathways arising from the brainstem. In infants under 1 year of age, the Extensor Plantar Response is normal because axons of the corticospinal tract have not yet been myelinated. Once the corticospinal tract becomes myelinated around 1–2 years of age, the Extensor Plantar Response disappears as it becomes suppressed by the myelinated (functional) corticospinal tract. Thus, following a lesion to the corticospinal tract fibers, the Extensor Plantar Response reappears.

CROSSED EXTENSOR REFLEX

The crossed extension reflex occurs to maintain balance and upright posture following contact with a noxious stimulus

In the **crossed extensor reflex** the receptor is in the *cutaneous* or in the subcutaneous tissue. The "crossed extensor reflex" occurs in the contralateral limb that does not come in contact with the unexpected or noxious stimulus (Fig. 11.6). If a person walking barefoot steps on a tack, that limb (the *ipsilateral* limb) will be immediately withdrawn (via the withdrawal reflex) by simultaneous (1) facilitation of the ipsilateral flexors (causing them to contract), and (2) inhibition of the antagonist extensors (causing them to relax). This pathway

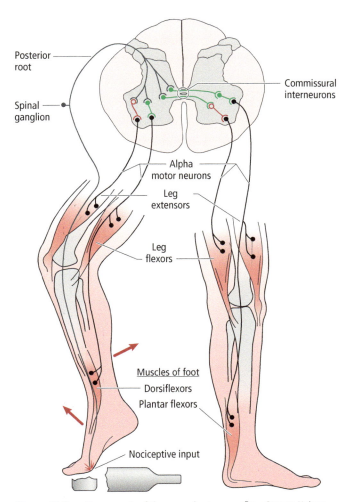

Figure 11.6 • Components of the crossed extensor reflex. Source: Haines, DE (2013) *Fundamental Neuroscience for Basic and Clinical Applications*, Elsevier Saunders, figure 9.11, p. 134/with permission of Elsevier.

and neural circuitry so far is the **withdrawal reflex**. But, if one lower limb is lifted off the floor (via the withdrawal reflex), the weight of the body must shift to the other lower limb in an attempt to maintain balance and prevent falling (and, possibly, causing more injury). This is accomplished by also involving the neural circuitry that controls muscles of the opposite (non-injured) limb in the following manner: The central process (Aδ or C fiber) of the pseudounipolar neuron (carrying nociceptive input) also synapses with and stimulates the **commissural interneurons** whose axons cross the midline in the anterior commissure of the spinal cord and in turn synapse with excitatory and inhibitory interneurons located in the contralateral anterior horn of the spinal cord. The excitatory interneuron synapses with an alpha motor neuron that facilitates the lower limb extensors (antigravity muscles, which enable standing), while the inhibitory interneuron synapses with an alpha motor neuron that inhibits the lower limb flexors.

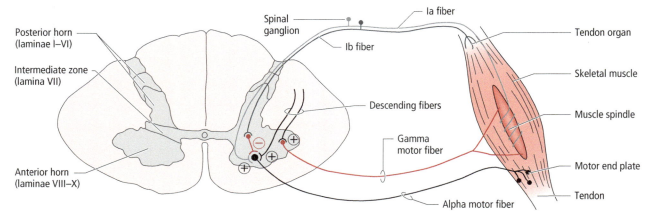

Figure 11.7 ● Components of the gamma loop. Source: Haines, DE (2013) *Fundamental Neuroscience for Basic and Clinical Applications,* Elsevier Saunders, figure 24.5, p. 328/with permission of Elsevier.

MAINTENANCE OF MUSCLE TONE VIA THE GAMMA LOOP

Muscle tone is produced by supraspinal input to the gamma motor neurons

Muscle tone (tonus, residual muscle tension) is muscle tension resulting from a constant unconscious partial contraction of the muscle so that it resists being stretched while at rest. Tonus provides the ability not only to maintain posture but also to react to sudden unexpected stimuli and to execute crucial reflexes. Tonus of muscle fibers is the result of gamma motor neuron and stretch reflex activity. Gamma motor neurons are involved in the maintenance and modulation of muscle tone via an intricate and essential circuit, known as the **gamma loop**, and it occurs in the following fashion (Fig. 11.7): A continuous stream of nerve impulses that are transmitted to a skeletal muscle generates muscle tone. Axons of descending motor tracts (carrying messages from the cerebral cortex and brainstem) terminate on – and synapse with and stimulate – **gamma motor neurons** residing in the anterior horn of the spinal cord. Note that the **gamma motor neurons** are always under **supraspinal control** (that is, controlled by higher levels in the brain) and they *do not* receive input from sensory (either Type Ia or Type II) fibers innervating intrafusal fibers of muscle spindles and, therefore, there are no messages relayed from the muscle spindles back to the gamma motor neurons. As **gamma motor neurons** stimulate the polar ends of intrafusal fibers to contract, the polar ends pull away from the midline of the intrafusal fibers, simultaneously stretching the central (non-contractile) regions of the intrafusal fibers. This stimulates the anulospiral endings of the **Ia sensory fibers** that are wrapped around the central part of the intrafusal fibers. The **Ia sensory fibers** enter the spinal cord, synapse with **alpha motor neurons** (*not* gamma motor neurons) of the spinal cord that innervate **skeletal muscle cells (extrafusal fibers)**, eliciting the stretch reflex. This physiological reflex causes slight contraction of the gross muscle, producing a normal amount of muscle "tone" (tension) so that skeletal muscle is not flaccid or limp. Normally, innervated muscles always have variable amounts of muscle tone; that is not only during movement, but also at rest. Muscle tone is modulated by the descending motor fibers that converge to influence the activity of the gamma motor neurons. Gamma motor neurons are either excited (to increase muscle tone) or inhibited (to decrease or inhibit muscle tone). For example, antigravity muscles (that help a person to stand upright) by counteracting the force of gravity, which is pulling the individual down, have increased muscle tone when a person is standing and decreased muscle tone when the individual is lying down. The gamma loop consists of the following components: (1) **gamma** motor neurons; (2) **intrafusal** muscle fibers; (3) **Ia sensory** fibers; (4) **alpha** motor neurons; and (5) **extrafusal** fibers. Any lesion that may affect the descending projections that converge on the gamma motor neurons or any of the listed components of the gamma loop would affect muscle tone.

If alpha motor neuron axons are severed along their path in the anterior root of a spinal nerve or in a peripheral nerve on their way to innervate a skeletal muscle, the resistance of the skeletal muscle to stretch is greatly diminished. The skeletal muscle exhibits flaccid paralysis and has no muscle tone.

ALPHA–GAMMA COACTIVATION

Alpha–gamma coactivation is a mechanism whereby the muscle spindle's sensitivity to stretch is maintained (during voluntary movement) regardless of muscle length at the onset of muscle stretch

During voluntary movement both **alpha** and **gamma** motor neurons are stimulated *concurrently* by the axons of descending motor tracts (Fig. 11.8). Specifically, when an alpha motor neuron is stimulated to cause contraction of extrafusal muscle fibers, a "copy" of the plan is sent to the gamma motor neuron to cause contraction

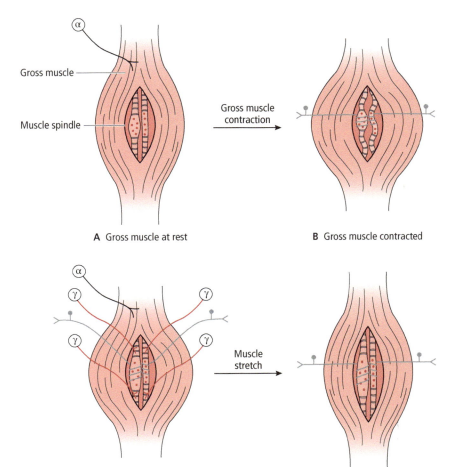

Figure 11.8 ● Components of the alpha–gamma coactivation. (A) Gross muscle at rest. (B) Gross muscle contracted. When the gross muscle contracts, the muscle spindle does not recoil; instead, it remains the same length since its ends are anchored to surrounding connective tissue. Intrafusal fibers sag and loosen and the anulospiral (Ia) endings are silenced. Slight stretching of the gross muscle would cause no response in the Ia fibers, and they would not signal or alert the CNS of the slight muscle stretching. (C) The alpha motor neurons cause contraction of the gross muscle. Gamma motor neurons cause corresponding contraction of the polar (contractile) ends of the intrafusal fibers. This pre-stretches the central (non-contractile) region of the muscle spindles. The intrafusal fibers are always on stretch alert, so when the muscle stretches even slightly (D) the intrafusal muscle fibers (over)stretch and the Ia fibers fire, informing the CNS of even the slightest stretch. Alpha–gamma coactivation maintains spindle sensitivity to muscle stretch regardless of the muscle length at the onset of muscle stretch.

of the polar (contractile) portion of the intrafusal fiber. The descending axons terminate on and synapse with alpha and gamma motor neurons residing in the anterior horn of the spinal cord, stimulating these neurons, simultaneously. This situation is referred to as **alpha–gamma coactivation**. (1) The **alpha motor neurons** cause the *extrafusal fibers* of the gross muscle to contract. Concurrently, (2) the **gamma motor neurons** cause the polar (contractile) ends of the *intrafusal fibers* to contract. Alpha–gamma coactivation is a mechanism which ensures that the sensitivity of the muscle spindles to stretch is maintained (during voluntary movement) regardless of muscle length at the onset of muscle stretch. Although a gross muscle shortens as it contracts, there is no corresponding change in length of the muscle spindles. The muscle spindles cannot shorten because their tapered ends are anchored to the surrounding (non-elastic) collagenous connective tissue. Consequently, as the gross muscle contracts, tension on the central non-contractile region of the intrafusal muscle fibers decreases (they sag), silencing the Ia fibers. During voluntary movement, however, it is the alpha–gamma coactivation that modulates the sensitivity of the intrafusal fibers so that they can retain their ability to serve as muscle stretch receptors by detecting even a slight change in muscle length (resulting from stretch of the gross muscle) regardless of muscle length at the onset of muscle stretch. The gamma motor neurons stimulate and cause contraction of the polar ends of the intrafusal fibers, exerting a pull (tension) on the central (non-contractile) portions of the intrafusal fibers, keeping the central part *taut* so even the slightest additional stress placed on it will stimulate the sensory fibers.

A lesion of a peripheral nerve containing alpha and gamma motor neuron axons results in **flaccid paralysis** (denervated muscle becomes limp and cannot contract), **atrophy** (wasting), **hypotonia** (reduced muscle tone), **hyporeflexia** (diminished reflexes), or **areflexia** (absence of reflexes).

When the descending upper motor neuron axons that modulate the activity of spinal alpha and gamma motor neurons are severed following transection of the spinal cord, the initial result is the diminution of muscle tone. In contrast, the chronic effect of spinal cord transection on muscle tone is an increase in the stretch reflexes (**hyperreflexia**) below the level of the lesion, producing a condition known as **spasticity**.

SYNONYMS AND EPONYMS OF THE SPINAL REFLEXES

Afferent limb of reflex arc	Sensory limb of reflex arc
Efferent limb of reflex arc	Motor limb of reflex arc
Extrafusal fibers	Skeletal muscle fibers
Anulospiral endings	Primary endings
	Type Ia fibers
Flower spray endings	Secondary endings
	Type II fibers
Muscle stretch reflex	Tendon reflex
	Deep tendon reflex
	Monosynaptic reflex
	Myotactic reflex
Autogenic inhibition	Inverse myotactic reflex
Flexor reflex	Withdrawal reflex
	Nociceptive reflex
Babinski sign	Extensor plantar response

FOLLOW-UP TO CLINICAL CASE

To treat the underlying presumed autoimmune inflammation in the left L3 nerve root, immunotherapy is begun with infusions of intravenous methylprednisolone, starting at three treatments each week for the first two weeks. Physical therapy for strengthening and gait training is arranged. Over the ensuing weeks, the pain diminishes and over months, strength and muscle bulk gradually return, although not completely to baseline levels.

QUESTIONS TO PONDER

1. Name the structure of a reflex that may or may not be part of a reflex arc.

2. What is the reason why the tendon organs lack motor innervation?

3. What does the muscle stretch reflex compensate?

4. What is the function of autogenic inhibition?

5. What is unique about the crossed extensor reflex?

6. What controls the gamma motor neurons located in the anterior horn of the spinal cord?

7. Describe the two ways via which the central (non-contractile) portion of the intrafusal fibers can be stretched.

8. Describe the function of the alpha–gamma coactivation.

PART II

Integrative Components of the Nervous System

CHAPTER 12

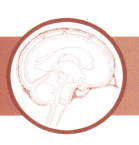

Ascending Sensory Pathways

CLINICAL CASE

SENSORY RECEPTORS

ANTEROLATERAL SYSTEM

TACTILE SENSATION AND PROPRIOCEPTION

SENSORY PATHWAYS TO THE CEREBELLUM

CLINICAL CONSIDERATIONS

MODULATION OF NOCICEPTION

NEUROPLASTICITY

SYNONYMS AND EPONYMS OF THE ASCENDING SENSORY PATHWAYS

FOLLOW-UP TO CLINICAL CASE

QUESTIONS TO PONDER

CLINICAL CASE

A 60-year-old individual reports of falls, imbalance, as well as numbness and tingling in their hands and legs. There is also some lack of coordination of hand use, and they have difficulty manipulating small items such as buttons. The individual is unable to play the piano now since they cannot position their fingers correctly on the piano keys. They think the strength in their arms and legs is adequate, and their vision is normal. Symptoms started with very slight tingling sensations, which they noticed about 5 years ago. The falls and difficulty walking have been present for about 2 months.

Higher-order cognitive functions are intact according to their partner.

On examination, the patient shows normal mental status. Strength seems essentially normal throughout. Sensation, particularly to vibration and joint position, is severely diminished in the distal upper and lower limbs (arms, legs, hands, and feet). Tendon reflexes are normal in the arms, but somewhat brisk in the legs at the knees and ankles. Gait is moderately ataxic, and they have to reach out for support by touching the walls of the hallway at times. Fine movements of the fingers are performed poorly, even though finger and wrist strength seem normal.

A variety of **sensory receptors**, specialized terminals, at the peripheral process of sensory neurons, are scattered throughout the body. Sensory receptors may reside in skin, skeletal muscles, tendons, ligaments, joint capsules, bones, periosteum, arteries, mucous membranes, and other visceral structures. Depending on their location, sensory receptors can become activated by exteroceptive, interoceptive, or proprioceptive input. Exteroceptive input relays sensory information about the body's interaction with the external environment. Interoceptive input relays information about the body's internal state, whereas proprioceptive input conveys information about position sense from the body and its parts. Each type of receptor is specialized in detecting mechanical (mechanoreceptor), chemical (chemoreceptor), nociceptive (L., *nocere*, "to injure," "painful"; nociceptor), thermal (thermoreceptor) or light (photoreceptor) stimuli. This sensory input is then conveyed via the fibers of the cranial or spinal nerves to their respective relay nuclei in the central nervous system (CNS). The sensory information is then further processed as it progresses, via the **ascending sensory systems**

A Textbook of Neuroanatomy, Third Edition. Maria A. Patestas, Amanda J. Meyer, and Leslie P. Gartner.
© 2025 John Wiley & Sons, Inc. Published 2025 by John Wiley & Sons, Inc.
Companion website: www.wiley.com.go/patestas/neuroanatomy3e

186 ● ● ● CHAPTER 12

(**pathways**), to the cerebral cortex or the cerebellum. Sensory information is also relayed to other parts of the CNS where it may function to elicit a reflex response or may be integrated into pattern-generating circuitry.

Each of the ascending sensory pathways begins with peripheral sensory receptors. The brain receives a wide spectrum of sensory inputs (including olfactory, visual, gustatory, auditory, touch, pressure, nociceptive, thermal, proprioceptive, stretch, and chemical) transmitted by a variety of sensory receptors and associated ascending sensory pathways. The brain then converts (translates) all sensory information into a form that enables it to create a representation of the body's external and internal environments. Some information achieves conscious levels, whereas other information does not (this is modulated by the thalamus). The sensory systems, via their widespread and intricate neural connections, play a key role in the integration of sensory, motor, endocrine, and autonomic functions many of which are crucial for the survival of the organism. Furthermore, the sensory systems play a role in protective reflexes, coordination of movement, and ultimately the creation of a myriad of sensory experiences.

People use their fingers to button jackets, search for and find a binder clip in the back of a desk drawer, or find a quarter in their pockets containing various coins without looking, and the visually- impaired can read Braille. Although not looking, they are still able to find a particular object with their fingers.

To facilitate the identification of objects via the sense of touch (without looking), there is a multitude of sensory receptors in people's palms each serving a specific sensory function. The tactile receptors scattered in the skin (including free nerve endings, epithelial tactile discs (eponym: Merkel tactile discs), tactile corpuscles (eponym: Meissner corpuscles), and lamellar corpuscles (eponym: Pacinian corpuscles) collect discriminative sensory information (size, three-dimensional shape, contour, edges, and texture) of objects that contact the palmar surface of the hand and fingers. This exquisitely detailed sensory input is then relayed via a prominent ascending sensory pathway, the posterior column–medial lemniscal (PCML) pathway to the brain where it enters consciousness. This pathway makes the individual aware of the presence of the binder clip in the desk drawer (without looking). Each sensory receptor transmits sensory information from its receptive field. The brain then assembles all of the sensory information and forms a composite three-dimensional image of the item being manipulated by the individual (i.e., binder clip, button). The brain then compares this newly created tactile representation with representations stored in the cerebral cortex to match and identify it. Sometimes the person is unable to identify an unfamiliar object by the sense of touch alone (without looking), because there is no corresponding or similar tactile representation stored in the individual's cerebral cortex.

The PCML pathway not only transmits discriminative (fine, detailed) touch but also relays vibratory sense and proprioception (position sense) from muscle spindles and tendon organs (eponym: Golgi tendon organs) of the limbs to the sensory cortex where the sensory information enters consciousness. The proprioceptors (muscle spindles, tendon organs, and free nerve endings) act to help locate parts of the body without having to look. The proprioceptors transmit electrical signals to the brain informing it as to which muscles are contracted and which muscles are stretched, and if the limbs are still or if they are moving, in which direction, and how fast. Again, the brain collects and integrates all this sensory information to form a representation of the position, orientation, and motion of parts of the body in time and space.

Proprioception from muscle spindles and tendon organs is not only transmitted to the brain via the PCML pathway to consciousness but is also transmitted to the cerebellum via the **ascending sensory pathways to the cerebellum** to *unconscious* levels. The cerebellum uses this information to modulate movement at the unconscious level.

The ascending sensory pathways are classified according to the modalities they carry as well as by their anatomical localization. The two functional categories are the **general somatic afferent (GSA) system**, which transmits sensory information such as touch, pressure, flutter-vibration, pain, temperature, itch, stretch, and position sense from **somatic** structures; and the **general visceral afferent (GVA) system**, which transmits sensory information such as stretch, pressure, pain, and other visceral sensation from **visceral** structures. Anatomically, the ascending sensory systems consist of three distinct pathways: the anterolateral system (ALS), the posterior column–medial lemniscal system (PCML), and the somatosensory pathways to the cerebellum.

The **anterolateral system (ALS)**, which includes the **spinothalamic, spinoreticular, spinomesencephalic, spinotectal, spinohypothalamic,** and **spino-olivary (spinobulbar) fibers,** relays predominantly nociceptive, thermal, and itch sensation, as well as non-discriminative (crude or poorly localized) touch, and pressure sensation. All of these tracts ascend in the anterior portion of the lateral funiculus of the spinal cord, thus their anatomical name, the anterolateral system (Table 12.1). Usually, pain is the body's warning signal following an injury or potential injury and serves as a vital protective mechanism to prevent further injury.

The **posterior column–medial lemniscal system (PCML)** (which includes the **gracile fasciculus, cuneate fasciculus,** and **medial lemniscus**) relays discriminative (fine) tactile sense, flutter-vibratory sense, and proprioception. Proprioception (L. *propria*, "self") is the sensation of the position of the body and its parts in three-dimensional space, a function that has an important role in movement and posture (Table 12.1).

Table 12.1 ● A general description of the anatomical and functional aspects of the ascending sensory pathways.

Anatomical system	Anatomical tracts/fibers	Modalities
Anterolateral (ALS)	Spinothalamic*	Pain, temperature, itch, non-discriminative (crude) touch, and pressure sensation
	Spinoreticular Spinomesencephalic	Pain
	Spinotectal	Pain
	Spino-olivary (spinobulbar)	Pain
	Spinohypothalamic	Pain
Posterior column–medial lemniscal (PCML)*	Gracile fasciculus, Cuneate fasciculus	Discriminative (fine) touch, flutter-vibratory sense, proprioceptive (position) sense
Somatosensory to the cerebellum	Anterior spinocerebellar, Posterior spinocerebellar Rostral spinocerebellar Cuneocerebellar	Proprioceptive information

*Indicates conscious level.

The **somatosensory pathways to the cerebellum**, which include the **anterior**, **posterior**, and **rostral spinocerebellar**, as well as the **cuneocerebellar tracts**, relay proprioceptive information (Table 12.1).

The ascending sensory pathways are the main avenues by which information concerning the body's interaction with the external environment, its internal condition, and the position and movement of its parts, reach the brain. One similarity shared by all three ascending sensory pathways from the body (not including the head or face) is that the first-order neuron cell bodies reside in the **spinal ganglia**. It is interesting to note that **conscious perception** of sensory information from external and internal stimuli is mediated by the **spinothalamic** and **PCML** to the ventral posterior lateral (VPL) nucleus of the **thalamus**, whereas sensations that do *not* reach conscious levels are mediated by the **spinoreticular**, **spinomesencephalic**, **spinotectal**, **spinohypothalamic**, **spino-olivary** (**spinobulbar**), and the **anterior**, **posterior**, and **rostral spinocerebellar**, and **cuneocerebellar tracts**. These fibers terminate in the reticular formation, mesencephalon, hypothalamus, inferior olivary nucleus, and cerebellum, respectively. Conscious perception relies on projections from the thalamus to the cortex igniting neurons in the parietal–frontal network.

Sensory input may ultimately elicit a reflex or other motor response because of the functional integration of the ascending (somatosensory) pathways, the cerebellum, and the somatosensory cortex, as well as the motor cortex and descending (motor) pathways. Furthermore, descending projections from the **somatosensory cortex**, as well as from the **raphe magnus nucleus** and the **dorsolateral pontine reticular formation** to the **somatosensory relay nuclei** of the brainstem and spinal cord, modulate the transmission of incoming sensory impulses to higher brain centers.

The current chapter includes a description of the sensory receptors and the ascending sensory pathways from the body and back of the head, whereas the ascending sensory pathways from the head, transmitted mostly by the trigeminal system, are described in Chapter 17.

SENSORY RECEPTORS

Although sensory receptors vary, they generally all function in a similar fashion

Although sensory receptors vary according to their morphology, their threshold, the velocity of conduction, the modality (the sensation or perception following a stimulus) to which they respond, as well as to their location in the body, generally, they all function similarly.

Sensory receptors are located in somatic and visceral structures. The axon of a pseudounipolar (unipolar) sensory neuron and all of its nerve endings that are associated with sensory receptors that transmit sensory information, form a **sensory unit**. The regions of the body with the highest density of cutaneous tactile receptors include the orofacial region and fingertips. Sensory receptors have a **receptive field**, which is the area or territory where the peripheral terminal of the sensory receptor is located and where it transduces (converts) stimuli into receptor potentials. The stimulus to which a specific sensory receptor responds causes an alteration in the ionic permeability of the nerve endings generating a **receptor potential** (electrical signal) resulting in the formation of **action potentials** that are propagated along the axon to the central terminal. This transformation of the stimulus into an electrical signal is referred to as **sensory transduction**.

Receptors may have small receptive fields or large receptive fields. **Small receptive fields**, located in areas such as the lips and fingertips, have a very large number of receptors per unit area, with each receptor detecting stimuli from and serving a very small surface area of the skin. This type of receptor arrangement results in high discrimination of sensory stimuli applied to those areas. In contrast, **large receptive fields**, located in other regions of the body such as the back, have a very small number of receptors per unit area, with each receptor detecting stimuli from and serving a large surface area of the skin. This pattern of receptor arrangement results in poor discrimination of sensory stimuli.

188 ••• CHAPTER 12

Some receptors that respond quickly and maximally at the onset and termination of the stimulus, but stop responding even if the stimulus continues, are known as **rapidly adapting (phasic) receptors**. These are essential in responding to changes, but they ignore ongoing processes, such as when one wears a wristwatch and ignores the continuous pressure on the skin of the wrist. However, there are other receptors, **slowly adapting (tonic) receptors** that continue to respond to a constant, unchanging stimulus as long as the stimulus is present.

Sensory receptors are classified according to the **source of the stimulus** or according to the **modality** to which they respond. It is important to note that, in general, receptors do not transmit only one specific sensation. A stimulus may be visual, auditory, chemical, mechanical, nociceptive, or thermal in nature.

Classification according to stimulus source

Exteroceptors are close to the body surface and are specialized to detect sensory information from the external environment

Receptors may be classified according to the source of the stimulus and are placed in one of the following three categories: exteroceptors, proprioceptors, or interoceptors. Some receptors may be classified according to the stimulus to which the receptors respond and according to the modality that they perceive (see below).

1 **Exteroceptors** are close to the body surface and are specialized to detect sensory information from the external environment (such as visual, olfactory, gustatory, auditory, and tactile stimuli). Receptors in this class include those that are sensitive to *touch* (light stimulation of the skin surface), *pressure* (stimulation of receptors in the deep layers of the skin, or deeper parts of the body), *temperature*, *pain*, and *flutter-vibration*. Exteroceptors are further classified as teloreceptors or contact receptors:

- **teloreceptors** (G., *tele*, "distant"), include receptors that respond to distant stimuli (such as light or sound), and do not require direct physical contact with the stimulus to be stimulated;
- **contact receptors**, which transmit tactile, pressure, pain, or thermal stimuli, require direct contact of the stimulus with the body.

2 **Proprioceptors** transmit sensory information from the skin, muscles, tendons, and joints about the position

Proprioceptors transmit sensory information from muscles, tendons, and joints about the position of a body part, such as the location of a limb in three-dimensional space

of a body part, such as the location of a limb in three-dimensional space. There is a *static* position sense relating to a stationary position and a *kinesthetic* sense (G., *kinesis*, "movement"), relating to the movement of a body part. The receptors of the vestibular system located in the internal ear, relaying sensory information about the movement and orientation of the head, are also classified as proprioceptors. Although the receptors of the visual system are classified as exteroceptors, the visual system also perceives motion and position of parts of the body, so it contributes to proprioception.

3 **Interoceptors** detect sensory information concerning the status of the body's internal environment. Baroreceptors

Interoceptors detect sensory information concerning the status of the body's internal environment

detect stretching in the wall of the carotid sinus, and blood pressure, as the walls of blood vessels become stretched.

Chemoreceptors, the **carotid body** at the bifurcation of the common carotid artery, and the **central chemoreceptors** in the medulla oblongata of the brain detect the partial pressure of oxygen in the blood (carotid body) and pH and partial pressure of carbon dioxide of the cerebrospinal fluid (medulla oblongata).

Classification according to stimulus modality

Receptors are further classified into the following five categories according to the stimulus to which they are most sensitive, known as the **adequate stimulus**, and according to the modality that they perceive: nociceptors, thermoreceptors, mechanoreceptors, chemoreceptors, and photoreceptors (Table 12.2).

Nociceptors

Nociceptors are rapidly adapting receptors that are sensitive to noxious or painful stimuli

Nociceptors are rapidly adapting receptors that are sensitive to noxious or painful stimuli. They are located at the peripheral terminations of lightly myelinated free nerve endings of **type Aδ fibers**, or unmyelinated **type C fibers**, transmitting pain. Nociceptors are further classified into three types:

1 **Mechanosensitive nociceptors (mechanonociceptors)** (of Aδ fibers) (**Aδ mechanical nociceptors**), which are sensitive to intense mechanical stimulation (such as pinching with pliers) or injury to tissues.

2 **Temperature-sensitive nociceptors (thermonociceptors, thermosensitive nociceptors)** (of Aδ and C fibers), which are sensitive to intense heat and cold.

3 **Polymodal nociceptors** (of C fibers) (**C-polymodal nociceptors**), may be **mechanonociceptors, thermonociceptors**, or **chemonociceptors** depending on their sensitivity to stimuli that are mechanical, thermal, or chemical in nature. Chemonociceptors may be stimulated by endogenous substances released by injured tissues or exogenous substances such as those released by insect bites. Although most nociceptors are sensitive only to one particular type of painful stimulus, some may respond to two or more types of stimuli.

ASCENDING SENSORY PATHWAYS ● ● ● **189**

Table 12.2 ● Sensory receptors.

Sensory	Mediate/respond to	Endings	Location	Associated with	Pathways	Rate of adaptation
Nociceptors						
Pain Temperature	Tissue damage Extreme temperature	Branching free nerve (Aδ, C) endings (lightly-myelinated/ unmyelinated, nonencapsulated)	Epidermis, dermis, cornea, muscle, joint capsules	Aδ (group III) lightly-myelinated fibers, C (group IV) unmyelinated fibers	Anterolateral system	Slow
Thermoreceptors						
Temperature		Branching free nerve (Aδ, C) endings (lightly myelinated/ unmyelinated, nonencapsulated)	Dermis, skeletal muscles	Aδ (group III) lightly myelinated fibers, C (group IV) unmyelinated fibers	Anterolateral system	Slow
Mechanoreceptors						
Free nerve endings	Touch, pressure	Nonencapsulated	Epidermis, dermis, cornea, dental pulp, muscle, tendons, ligaments, joint capsules, bones, mucous membranes	Aδ, C fibers	Anterolateral system	Slow
Epithelial tactile discs (Merkel tactile discs)	Discriminative touch, superficial pressure	Nonencapsulated	Basal epidermis	Aβ (group II) myelinated fibers	PCML pathway	Slow
Tactile corpuscles (Meissner corpuscles)	Two-point discriminative (fine) touch	Encapsulated	Papillae of dermis of hairless skin	Aβ (group II) myelinated fibers	PCML pathway	Rapid
Lamellar corpuscles (Pacinian corpuscles)	Deep pressure and vibratory sensation	Encapsulated	Dermis, hypodermis, interosseous membranes, periosteum, ligaments, external genitalia, joint capsules, peritoneum, pancreas	Aβ (group II) myelinated fibers	PCML pathway	Rapid
Peritrichial (Follicular) nerve endings	Touch, hair movement (bending)	Nonencapsulated	Around hair follicle	Aβ (group II) myelinated fibers	PCML pathway	Rapid
Bulbous corpuscles (Ruffini Endings)	Pressure on, or stretching of, skin	Encapsulated	Joint capsules, dermis, hypodermis	Aβ (group II) myelinated fibers	PCML pathway	Slow
Muscle and tendon mechanoreceptors						
Nuclear bag fibers	Detects onset of muscle stretch, and ongoing muscle stretch		Skeletal muscle	Aα (group Ia) myelinated fibers also secondary afferents, muscle spindle afferents	PCML* pathway and ascending sensory pathways to the cerebellum	Both slow and rapid
Nuclear chain fibers	Activated when muscle stretch is in progress		Skeletal muscle			
Tendon organs (Golgi tendon organs)	Stretching of tendon	Nonencapsulated	Skeletal muscle	Aα (group Ib) myelinated fibers		Slow
Chemoreceptors	Changes in arterial oxygen tension Changes in pH, carbon dioxide concentration in CSF	Chemoreceptor Neurons	Carotid bodies, aortic bodies, Medullary chemosensitive area	Glossopharyngeal nerve, vagus nerve		
Photoreceptors	Light	Bipolar neurons	Retina of eye			

*PCML, posterior column–medial lemniscal.

4 **Pruriceptors** (**itch-sensitive receptors**) (of C fibers) are sensitive to, and respond to, histamine.

Nociception is the reception of noxious sensory information elicited by tissue injury, which is transmitted to the CNS by nociceptors. **Pain** is the subjective perception of discomfort or an agonizing sensation of variable magnitude, evoked by the stimulation of sensory nerve endings.

Thermoreceptors

Thermoreceptors are sensitive to heat or cold

Thermoreceptors are sensitive to heat or cold. These slowly adapting receptors are located in the dermis and skeletal muscles; they are classified into two types, low- and high-threshold receptors:

1 **Low-threshold receptors** are stimulated by temperatures ranging from 15° and 45°C (59 and 113°F). They *do not* elicit a noxious effect/sensation and *do not* produce tissue damage during a brief application of the stimulus.
2 **High-threshold receptors** are stimulated by temperatures lower than 15°C (59°F) *and* by temperatures higher than 45°C (113°F); they elicit a painful sensation. Temperatures below 0°C (32°F) and above 50°C (122°F) can cause considerable tissue damage.

Warm sensation is conducted by free nerve endings of slow-conducting unmyelinated C fibers that respond to *increases* in temperature. Cold sensation is mediated by free nerve endings of slow-conducting thinly myelinated Aδ fibers.

Mechanoreceptors

Mechanoreceptors are activated following physical deformation of the skin, muscles, tendons, ligaments, and joint capsules in which they reside

Mechanoreceptors, which comprise both exteroceptors and proprioceptors, are activated following physical deformation of the areas in which they reside due to touch, pressure, stretch, or vibration of the skin, muscles, tendons, ligaments, and joint capsules. A mechanoreceptor may be classified as **nonencapsulated** or **encapsulated** depending on whether a structural barrier encloses its peripheral nerve ending. Nonencapsulated mechanoreceptors may exist as mechanoreceptive free nerve endings, nonencapsulated endings with accessory structures, or as encapsulated endings covered with a layered capsule.

Nonencapsulated mechanoreceptors

Nonencapsulated mechanoreceptors are slowly adapting; they include free nerve endings and tactile receptors

Free nerve endings (**Bare nerve endings**) (Fig. 12.1) Aδ and C fiber mechanoreceptors are present in the epidermis, dermis,

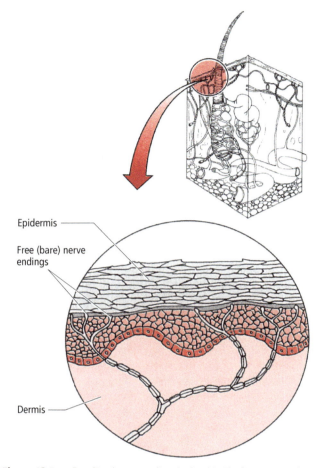

Figure 12.1 • Free (Bare) nerve endings in the skin. The free nerve endings terminating in the epidermis lose their myelin sheath. The terminal portions of the axons of free nerve endings are usually unmyelinated.

cornea, dental pulp, mucous membranes of the oral and nasal cavities, mucosae of the respiratory, gastrointestinal, and urinary tracts, as well as in muscles, tendons, ligaments, joint capsules, and bones. The peripheral nerve terminals of the free nerve endings lack neurolemmocytes (eponym: Schwann cells) and myelin sheaths. They are stimulated by touch, pressure (mechanoreceptors), thermal (thermoreceptors), or painful (nociceptors) stimuli.

Follicular (Peritrichial) nerve endings (Fig. 12.2) are specialized members of this category. They are large-diameter, myelinated, Aβ fibers that coil around a hair follicle below its associated sebaceous gland. They are rapidly adapting receptors that are stimulated only upon bending of the hair; that action stimulates the follicular nerve endings by deforming its mechanically irritable channels and initiating an action potential.

Tactile receptors (Fig. 12.3) consist of disc-shaped, peripheral nerve endings of large-diameter, myelinated, Aβ fibers. Each disc-shaped terminal is associated with a specialized epithelial cell, the **sensory epithelial cell** (eponym: Merkel cell), located in the stratum basale of the epidermis. The sensory epithelial cell and disc-shaped terminal form

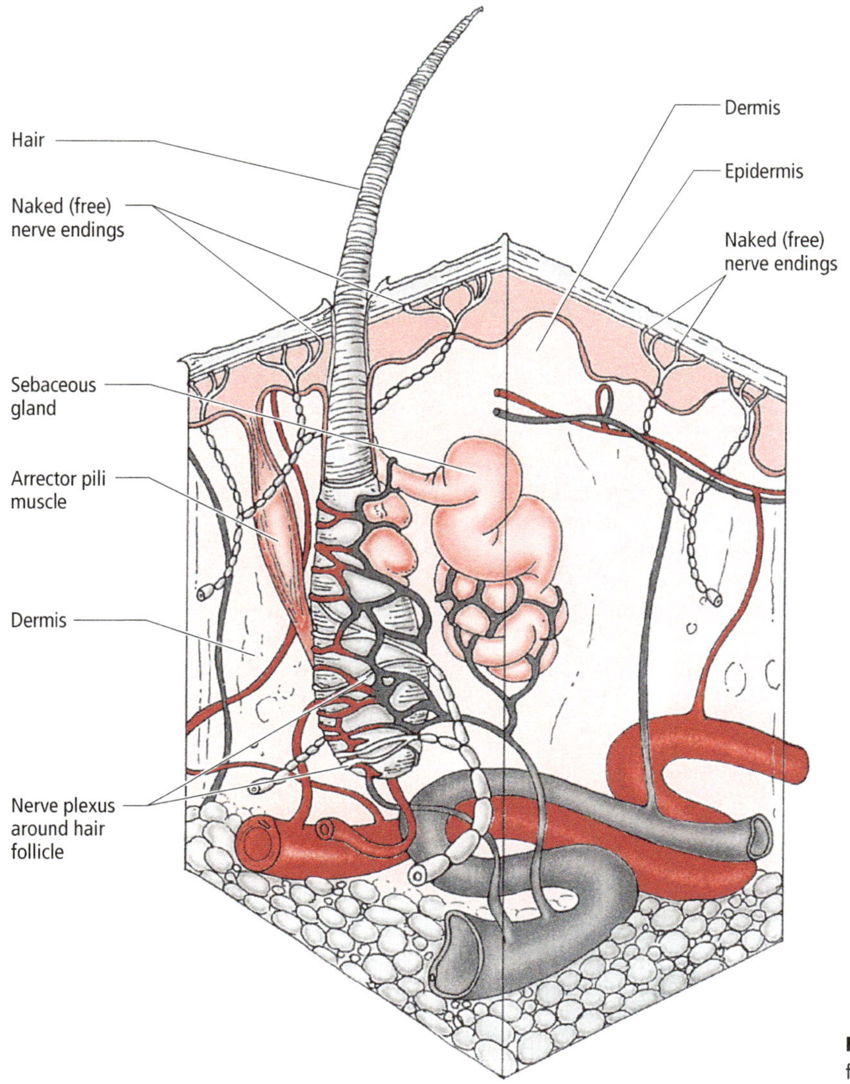

ASCENDING SENSORY PATHWAYS ● ● ● **191**

Figure 12.2 ● Follicular (Peritrichial) nerve endings. These free nerve endings spiral around the base of a hair follicle.

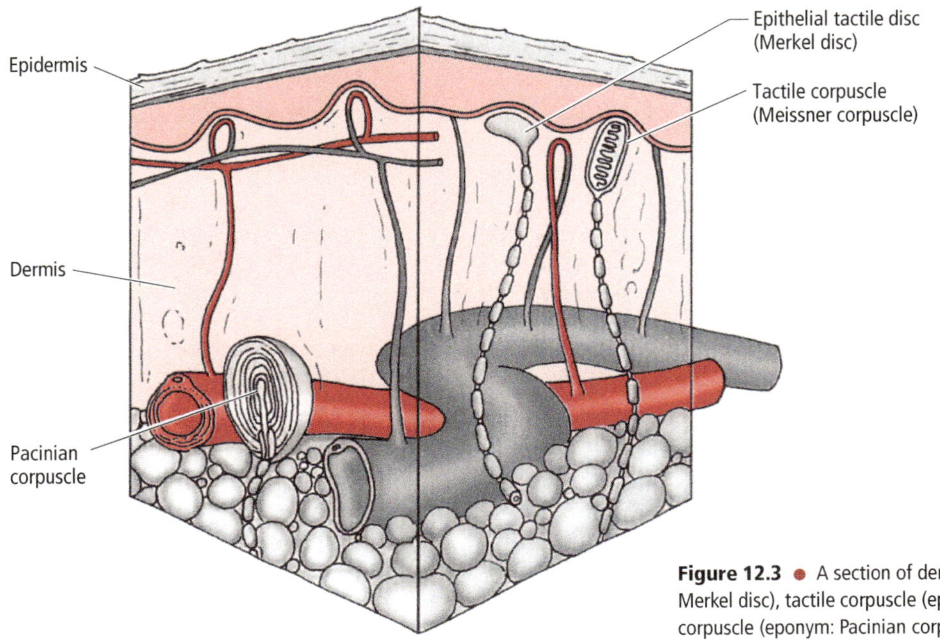

Figure 12.3 ● A section of dermis showing an epithelial tactile disc (eponym: Merkel disc), tactile corpuscle (eponym: Meissner corpuscle), and lamellar corpuscle (eponym: Pacinian corpuscle).

an **epithelial tactile complex** (eponym: Merkel cell-neurite complex). These receptors, frequently referred to as **epithelial tactile discs** (eponym: Merkel discs) (Fig. 12.4), are present mostly in glabrous (hairless), and occasionally in hairy skin. Epithelial tactile discs are slowly adapting mechanoreceptors that respond to discriminative touch stimuli that facilitate the ability to distinguish the texture, shape, and edges of objects.

Encapsulated mechanoreceptors

Encapsulated mechanoreceptors include tactile corpuscles (eponym: Meissner corpuscles), lamellar corpuscles (eponym: Pacinian corpuscles), and bulbous corpuscles (eponym: Ruffini endings).

Tactile corpuscles are present in the dermal papillae of glabrous skin of the lips, forearm, palm, and sole, and in the connective tissue papillae of the tongue

Tactile corpuscles (eponym: Meissner corpuscles) (Fig. 12.5A, B) are present in the dermal papillae of glabrous skin of the lips, forearm, palm, and sole, as well as in the connective tissue papillae of the tongue. These corpuscles consist of the peripheral terminals of Aβ fibers, which are encapsulated by a peanut-shaped structural device consisting of a stack of concentric neurolemmocytes surrounded by a perineural epithelium and then a connective tissue capsule. The nerve terminals interposed between the stack of neurolemmocytes are maximally stimulated by pressure applied to the dermal papillae in which they are situated. They are rapidly

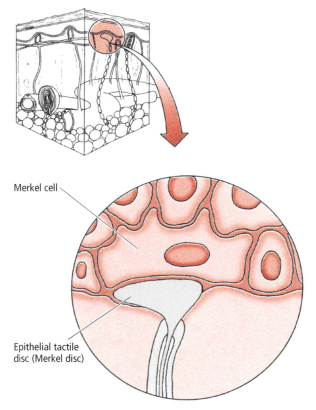

Figure 12.4 • Epithelial tactile discs (eponym: Merkel discs) terminate on the basal surface of the epidermis.

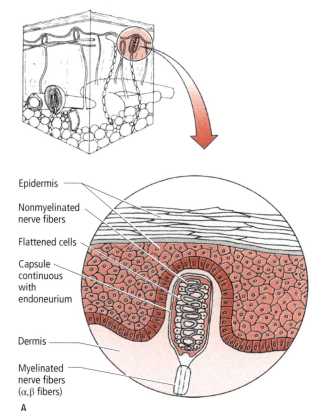
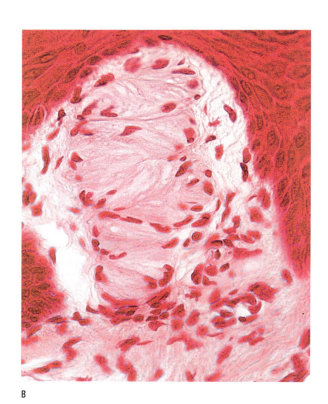

Figure 12.5 • (A) Tactile corpuscles (eponym: Meissner corpuscles) are located in dermal papillae of the skin. (B) Light micrograph of a tactile corpuscle.

adapting and are sensitive to two-point tactile (fine) discrimination; therefore, they are of great importance to the visually impaired by permitting them to be able to read Braille.

Lamellar corpuscles are the largest of the mechanoreceptors

Lamellar corpuscles (eponym: Pacinian corpuscles) (Fig. 12.6A, B), the largest of the mechanoreceptors, are the most rapidly adapting receptors; they resemble an onion in cross-section. Each Pacinian corpuscle consists of Aβ-fiber terminals encapsulated by concentric layers of a perineural epithelium and is enclosed in a connective tissue capsule. Lamellar corpuscles are located deep in the dermis, interosseous membranes, periosteum of long bones, ligaments, external genitalia, joint capsules, as well as in the pancreas and gastrointestinal mesenteries. They are more rapidly adapting than tactile corpuscles (eponym: Meissner corpuscles) and are believed to respond to pressure but mainly to vibratory stimuli, including tickling sensations.

Bulbous corpuscles (eponym: Ruffini endings) are located in the joint capsules, dermis, and underlying hypodermis of hairy and glabrous skin

Bulbous corpuscles (eponym: Ruffini endings) (Fig. 12.7) are located in joint capsules, the dermis, and the underlying hypodermis of hairy and glabrous skin. The unmyelinated

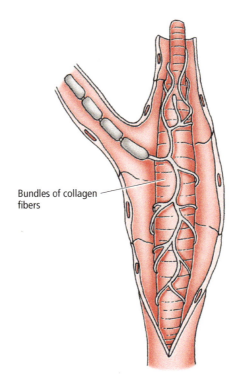

Figure 12.7 • A bulbous corpuscle (Ruffini ending).

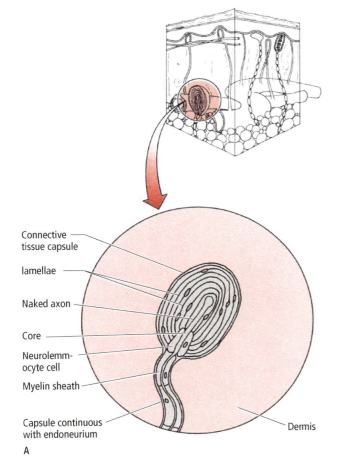

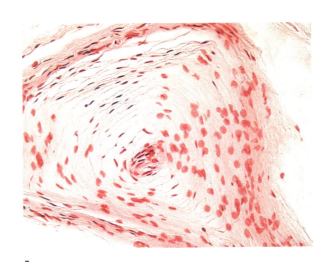

Figure 12.6 • (A) Lamellar corpuscle (eponym: Pacinian corpuscle) are located deep in the dermis of the skin. (B) Light micrograph of a lamellar corpuscle.

194 ● ● ● **CHAPTER 12**

peripheral terminals of Aβ myelinated fibers are slowly adapting. They intertwine around the core of collagen fibers, which is surrounded by a lamellated cellular capsule. Bulbous corpuscles respond to *stretching* of the collagen bundles in the skin or joint capsules (that may occur during movement of a limb), relaying sensory input for **kinesthesia**. Because their morphology resembles that of tendon organs (eponym: Golgi tendon organs), it is believed that they may provide proprioceptive information that aids in maintaining balance and posture.

Chemoreceptors

Chemoreceptors are located within the **carotid bodies** near their fenestrated capillaries and within the **aortic bodies**. The chemoreceptors in the carotid bodies are sensitive to changes in arterial oxygen tension, to prevent hypoxia. When these chemoreceptors are stimulated, they release neurotransmitters that activate the sensory fibers of the glossopharyngeal nerve innervating the carotid bodies which terminate in the nucleus of the solitary tract. This causes a reflex change in breathing rate to correct the blood oxygen levels. Chemoreceptors located in the aortic bodies are minor in the human and send messages to the brainstem via a branch of the vagus nerve. Chemoreceptor neurons located in the **medullary chemosensitive area** of the lateral reticular formation respond to changes in pH and carbon dioxide in the cerebrospinal fluid.

Photoreceptors

Photoreceptors include the rods and cones unique to the retina of the eye that respond to light. They are discussed in detail in Chapter 18.

Muscle spindles and tendon organs (Golgi tendon organs)

Two types of proprioceptors, the neuromuscular (muscle) spindles and the tendon organs (Golgi tendon organs, neurotendinous spindles), are associated with skeletal muscle only

The muscle spindles and tendon organs detect sensory input from the skeletal muscle and transmit it to the spinal cord, where it has an important role in reflex activity and motor control involving the cerebellum. Sensory input from these proprioceptors is also relayed to the cerebral cortex by way of the posterior column–medial lemniscal system (PCML) pathway, which mediates information concerning posture, position sense, as well as movement and orientation of the body and its parts.

Muscle spindles

Structure and function

Skeletal muscle consists of extrafusal and intrafusal fibers

Extrafusal fibers are ordinary skeletal muscle cells constituting the majority of gross muscle, and their stimulation results

in muscle contraction. **Muscle spindles**, composed of small bundles of encapsulated **intrafusal fibers** and their nerve supply, are dispersed throughout the belly of the whole muscle. These are *dynamic stretch receptors* that continuously monitor for changes in muscle length.

Each muscle spindle is composed of 2–12 intrafusal fibers enclosed in a slender capsule, which in turn is surrounded by an outer fusiform connective tissue capsule whose tapered ends are attached to the connective tissue sheath surrounding the extrafusal muscle fibers (Fig. 12.8). The compartment between the inner and outer capsules contains a glycosaminoglycan-rich viscous fluid.

Based on their morphological characteristics, there are two types of intrafusal fibers: nuclear bag fibers and nuclear chain fibers, both of which possess a central, noncontractile region housing multiple nuclei, and a skeletal muscle (myofibril-containing) contractile portion at each end of the central region. The **nuclear bag fibers** are larger, and their multiple nuclei are clustered in the "bag-like" dilated central region of the fiber. The **nuclear chain fibers** are smaller and consist of multiple nuclei arranged sequentially, as in a "chain" of pearls, in the central region of the fiber.

Each intrafusal fiber of a muscle spindle receives sensory innervation via the peripheral processes of unipolar (pseudounipolar) sensory neurons

Each intrafusal fiber of a muscle spindle receives **sensory innervation** via the peripheral processes of unipolar (pseudounipolar) sensory neurons whose cell bodies are housed in spinal ganglia, or in the sensory ganglia of the cranial nerves (and in the case of the trigeminal nerve, within the trigeminal ganglion and its mesencephalic nucleus). Since the large-diameter Aα fibers spiral around the noncontractile region of both the nuclear bag (mainly) and nuclear chain intrafusal fibers, they are known as **anulospiral** or **primary endings**. These endings become activated at the **beginning of muscle stretch** or **tension** but continue to fire at a slower rate while the muscle is stretching. In addition to the anulospiral endings, the intrafusal fibers, mainly the nuclear chain fibers, also receive smaller diameter, Aβ peripheral processes of pseudounipolar neurons. These nerve fibers that terminate on both sides of the anulospiral ending are referred to as **flower spray** or **secondary endings** and although they are not as responsive at the beginning of muscle stretch, they remain active during the muscle stretch (Fig. 12.8).

In addition to sensory innervation, intrafusal fibers also receive motor innervation via gamma motor neurons that innervate the contractile portions of the intrafusal fibers, causing them to contract

In addition to the sensory innervation, intrafusal fibers also receive **motor innervation** via **gamma** motor neurons (**fusimotor neurons**) that innervate the contractile portions of the intrafusal fibers, causing them to undergo contraction. Since the intrafusal fibers are oriented parallel to the longitudinal axis of the extrafusal fibers, when a muscle is *stretched*, the central, noncontractile region of the intrafusal fibers is also stretched, distorting and stimulating the sensory nerve endings coiled around them, causing the nerve endings to

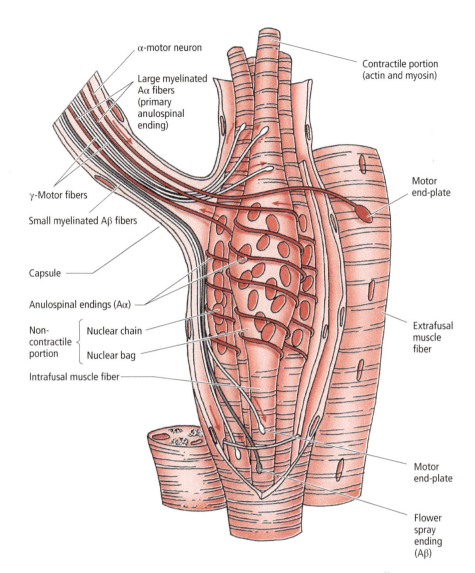

Figure 12.8 ● Right: an extrafusal skeletal muscle fiber. Left: a neuromuscular spindle containing the two types of intrafusal fibers – the nuclear bag fiber with multiple nuclei in the dilated central region and a nuclear chain fiber with a row of nuclei in its central region. Red arrows display the direction of action potential in the nerve fiber.

fire. However, when the muscle *contracts*, tension on the central noncontractile region of the intrafusal fibers decreases (which reduces the rate of firing of the sensory nerve endings coiled around it).

During voluntary muscle activity, simultaneous stimulation of the extrafusal fibers by the alpha motor neurons, and the contractile portions of the intrafusal fibers by the gamma motor neurons, serves to modulate the sensitivity of the intrafusal fibers. That is, the gamma motor neurons cause corresponding contraction of the contractile portions of the intrafusal fibers, which stretch the central noncontractile region of the intrafusal fibers. Thus, the sensitivity of the intrafusal fibers is constantly maintained by continuously re-adapting to the most current status of muscle length. In this fashion the muscle spindles can detect a *change* in muscle length irrespective of muscle length at the onset of muscle activity. It should be noted that even though they contract, the intrafusal fibers, due to their small number and size, do not contribute to any significant extent to the overall contraction of a whole muscle.

Simple stretch (myotactic) reflex

The **simple stretch reflex** is a protective mechanism based on the role of the intrafusal fibers, which function to maintain muscle length caused by external disturbances during muscle stretching (see Fig. 11.4 in Chapter 11, Spinal Reflexes). As a muscle is stretched, the intrafusal fibers of the muscle spindles are also stretched. This in turn stimulates the sensory afferent anulospiral and flower spray endings to transmit this information to those alpha motor neurons of the CNS (spinal cord, or cranial nerve motor nuclei) that innervate the agonist (stretched) muscle as well as to those motor neurons that innervate the antagonist muscle(s). The degree of stretching is proportional to (or related to) the load placed on the muscle. The larger the load, the more strongly the spindles are depolarized, and the more extrafusal muscle fibers are in turn activated. As these alpha motor neurons of the stretched muscle fire, they stimulate the contraction of the required number of extrafusal muscle fibers of the *agonist* muscle. The alpha motor neurons of the *antagonist* muscle(s) are inhibited so the antagonist muscle relaxes. The simple reflex arc involves the firing of only two neurons – an **afferent sensory neuron** and an **efferent motor neuron** – providing dynamic information concerning the changes of the load on the muscle and position of the body region in three-dimensional space.

Tendon organs

Tendon organs (neurotendinous spindles) are fusiform-shaped receptors located at sites where muscle fibers insert into tendons

Unlike muscle spindles, which are oriented parallel to the longitudinal axis of the extrafusal muscle fibers, tendon organs (eponym: Golgi tendon organs) are arranged *in series*. Furthermore, tendon organs do not receive motor innervation since, unlike muscle spindles, they do not have any contractile portions. Tendon organs consist of interlacing **intrafusal** collagen bundles enclosed in a connective tissue capsule (Fig. 12.9). A large-diameter, type Aα (group Ib) slowly adapting sensory fiber, whose cell body is housed in a posterior root sensory ganglion or a cranial nerve sensory ganglion, passes through the capsule and then branches into numerous delicate terminals that are interposed among the intrafusal collagen bundles. The central processes of these Aα (group Ib) afferent neurons enter the spinal cord via the posterior roots of the spinal nerves to terminate and establish synaptic contacts with inhibitory interneurons that, in turn, synapse with alpha motor neurons supplying the contracted agonist muscle.

Combined muscle spindle and tendon organ functions during changes in muscle length

During slight stretching of a relaxed muscle, the muscle spindles are stimulated while the tendon organs remain undisturbed and quiescent; with further stretching both the muscle spindles and the tendon organs are stimulated

As the muscle shortens during concentric muscle contraction, tension is produced in the tendons anchoring that muscle to bone, compressing the nerve fiber terminals interposed among the inelastic intrafusal collagen fibers. This compression activates the sensory terminals in the tendon organs, which transmit this sensory information to the CNS, providing proprioceptive information concerning muscle activity and, by fine-tuning muscle contraction, preventing the placement of excessive forces on the muscle and tendon. In contrast, the noncontractile portions of the muscle spindles are not stretched and are consequently undisturbed. The contractile regions of the muscle spindles, however, undergo corresponding contraction that enables them to detect future changes in muscle length.

During slight stretching of a relaxed muscle, the muscle spindles are stimulated whereas the tendon organs remain

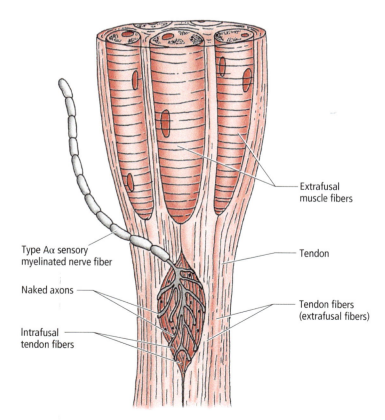

Figure 12.9 ● A tendon organ (eponym: Golgi tendon organ, synonym: neurotendinous spindle).

undisturbed and quiescent. During further stretching of the muscle, which produces tension on the tendons, both the muscle spindles and the tendon organs are stimulated. Thus, tendon organs monitor and check the amount of tension exerted on the muscle (regardless of whether it is tension generated by muscle stretch or contraction), by fine-tuning the muscle's tension on its tendons, whereas muscle spindles check muscle fiber length and rate of change of muscle length.

ANTEROLATERAL SYSTEM

The ALS transmits nociceptive, thermal, itch, and non-discriminatory touch information to higher brain centers, generally by a sequence of three neurons and interneurons

The **anterolateral system (ALS)** transmits *nociceptive, thermal, itch,* and *non-discriminatory (crude) touch* information to higher brain centers (see Table 12.1), generally by a sequence of three neurons and interneurons (Fig. 12.10). The neuron sequence consists of:

1. A **first-order neuron** (pseudounipolar neuron) whose cell body is located in a spinal ganglion (Fig. 12.11). It transmits sensory information from peripheral structures to the spinal laminae in the gray matter of the posterior horn of the spinal cord.
2. A **second-order neuron** whose cell body is located within the spinal laminae of gray matter in the posterior horn of the spinal cord, and whose axon usually decussates and ascends:
    - in the **direct pathway of the ALS (spinothalamic tract)** to synapse in the contralateral ventral posterior lateral nucleus of the thalamus, and sending some collaterals to the reticular formation;
    - in the **indirect pathway of the ALS (spinoreticular tract)** to synapse in the reticular formation, and sending some collaterals to the thalamus; or
    - as **spinomesencephalic, spinotectal, spinohypothalamic,** or **spino-olivary (spinobulbar) fibers** to synapse in several brainstem nuclei and the hypothalamus.

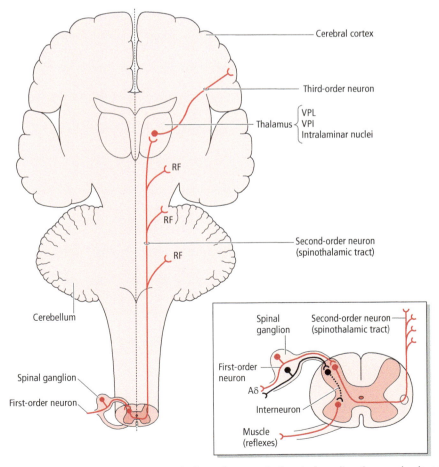

Figure 12.10 ● The direct pathway of the anterolateral system. Note the first-order neuron in the spinal ganglion, the second-order neuron in the posterior horn of the spinal cord, and the third-order neuron in the thalamus. The second-order neuron sends collaterals to the reticular formation (RF) in the brainstem. VPI, ventral posterior inferior; VPL, ventral posterior lateral; VMpo, ventral medial nucleus of the posterior group of the thalamus.

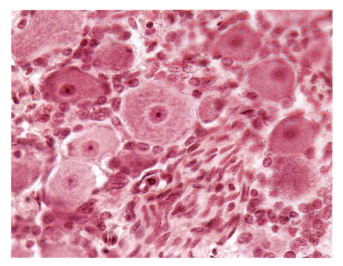

Figure 12.11 • Light micrograph of a spinal ganglion.

3   A **third-order neuron** whose cell body is in the thalamus, and whose axon ascends ipsilaterally to terminate in the somatosensory cortex or other cortical areas.

In some cases, the first-order neuron may synapse with an **interneuron** that resides entirely within the spinal laminae of the posterior horn, and whose axon synapses with the second-order neuron.

Pain pathways from the body

First-order neurons (sensory receptors)

Receptors that transmit nociceptive information consist of high-threshold free nerve endings ramifying near the external surface and internal environment of the organism

Receptors that transmit nociceptive information consist of high-threshold free nerve endings ramifying near the external surface and internal environment of the organism. These are dendritic arborizations of small, pseudounipolar (unipolar), first-order neurons (Fig. 12.12) whose cell bodies are housed in a spinal ganglion. The peripheral processes of these pseudounipolar neurons consist of either (Fig. 12.13):

1   Thinly myelinated **Aδ (fast-conducting) fibers**, which relay sharp, short-term, well-localized pain (such as that

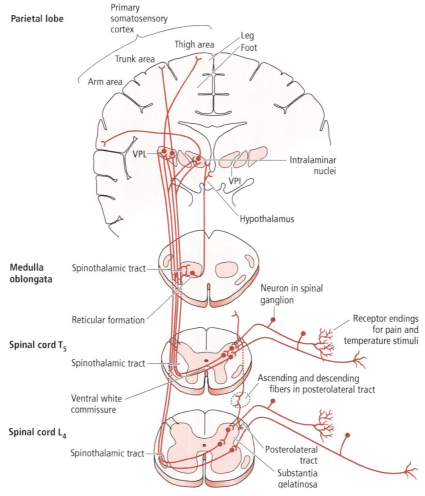

Figure 12.12 • The ascending sensory pathway that transmits pain, temperature, itch, and non-discriminative (crude) touch sensations from the body.
Source: Modified from Gilman, S, Winans Newman, S (1992) *Essentials of Clinical Neuroanatomy and Neurophysiology*. FA Davis, Philadelphia, fig. 19.

ASCENDING SENSORY PATHWAYS

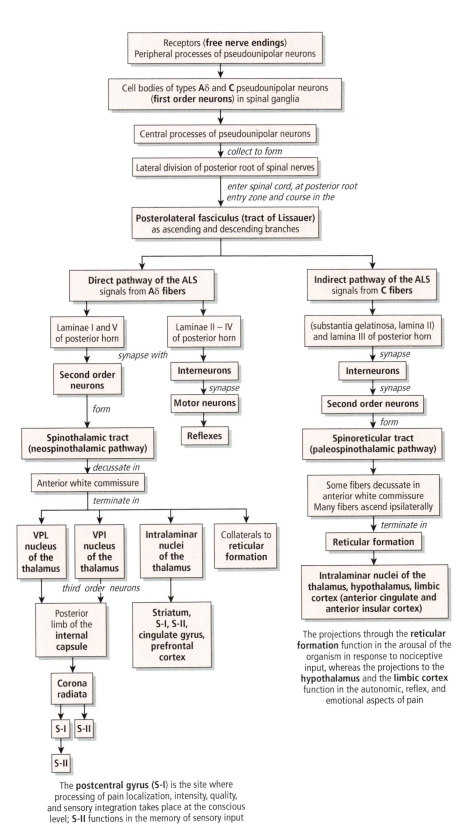

Figure 12.13 ● The spinothalamic (direct) and spinoreticular (indirect) pathways of the anterolateral system (ALS) transmitting pain, temperature, itch and non-discriminative (crude) touch sensation from the body. VPI, ventral posterior inferior; VPL, ventral posterior lateral; VMpo, ventral medial nucleus of the posterior nuclear group of the thalamus.

resulting from a pinprick). These fibers transmit sensations that do not elicit an affective component associated with the experience.
2   Unmyelinated C (**slow-conducting**) **fibers**, which relay dull, persistent, poorly localized pain (such as that resulting from excessive stretching of a tendon). These fibers transmit sensations that elicit an affective, that is an emotional response.

A transient noxious stimulus such as a pinprick elicits a sharp, short-term pain sensation (relayed by Aδ fibers) at first, which is then followed by a dull, persistent pain (relayed by C fibers).

The central processes of these pseudounipolar neurons enter the spinal cord at the posterior root entry zone, via the *lateral division* of the posterior roots of the spinal nerves, and upon entry collectively form the **posterolateral fasciculus** (eponym: tract of Lissauer), which is present at all spinal cord levels. Since the posterolateral fasciculus consists of thinly myelinated and unmyelinated fibers that stain very lightly, this fasciculus presents a pale appearance on histological sections. These central processes bifurcate into short ascending and descending branches within the posterolateral fasciculus. Although stimulation of the peripheral endings of fibers carried by one spinal nerve may enter the spinal cord at a specific spinal level, collaterals of the ascending and descending branches spread the signal to neighboring spinal levels above and below the level of entry.

- The branches of **Aδ fibers**, which are part of the **direct pathway**, either ascend or descend one to three spinal cord levels (Fig. 12.14A, B). The *ascending branches* of the Aδ fibers ascend to terminate in lamina I of the posterior horn where they synapse with second-order projection neurons. The *descending branches* of the Aδ fibers synapse with those interneurons of the spinal cord gray matter that participate in intersegmental spinal reflexes.
- The branches of **C fibers** which are part of the **indirect pathway** either ascend or descend one or two spinal cord levels: The *ascending* and *descending branches* of the C fibers terminate primarily in laminae I and II (substantia gelatinosa) but also terminate in lamina III of the posterior horn where they synapse with interneurons. The interneurons target laminae V–VIII where they synapse with second-order projection neurons.

The spinal cord gray matter is divisible into 10 spinal (eponym: Rexed) laminae. The posterior horn consists of laminae I–VI. Lamina I is the most superficial layer, located at the posterior margin of the posterior horn, whereas lamina VI is located approximately midway between the posterior and anterior horns, at the posterior horn's base. Laminae VII and X (which borders the central canal), collectively referred to as the *intermediate zone* of the spinal cord gray matter, contain interneurons involved in reflexes and other movement. Laminae VIII and IX are in the anterior horn and contain somatotopically arranged

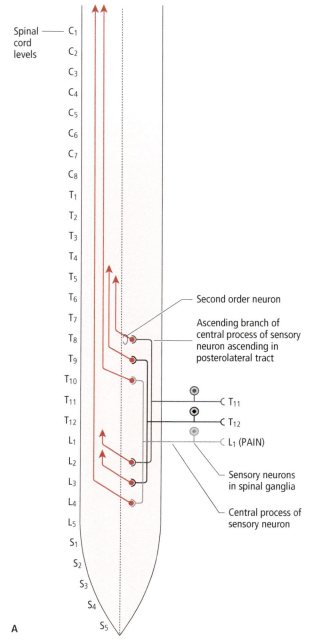

Figure 12.14 ● (A) and (B) Ascending and descending branches of primary sensory endings relaying pain sensation. Cell bodies of sensory neurons are located in spinal ganglia. Peripheral endings terminate in somatic or visceral structures. Central processes enter the spinal cord via the posterior roots of spinal nerves and enter the posterolateral tract (of Lissauer). There the central processes divide into short ascending and descending branches to synapse one to three levels above and descending branches one to three levels below the level of spinal cord entry. The fibers shown ascend and descend three levels to synapse with second-order neurons. Interneurons are not shown. The second-order neurons (appearing in red) give rise to axons that cross the midline of the spinal cord obliquely to reach the opposite side. (B). Simple sketch of the spinothalamic pathway. Central process of first-order neuron bifurcates and its branches ascend and descend two to three levels to synapse with a second-order neuron. Second-order neuron axon crosses in the anterior white commissure (AWC) and joins the ascending spinothalamic tract to the thalamus. Third-order neuron projects from the thalamus to the primary sensory cortex which in turn projects to the sensory association cortex.

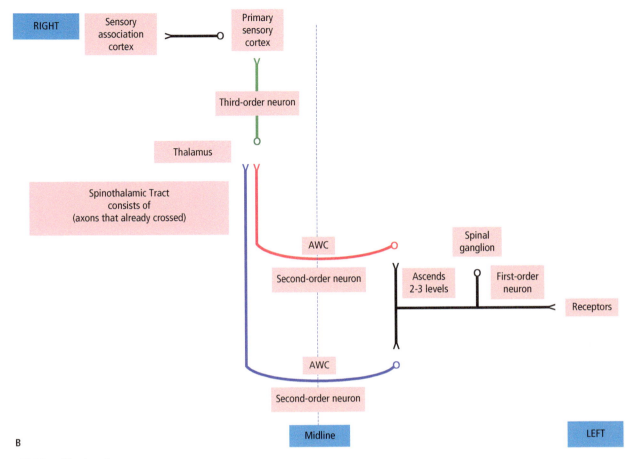

Figure 12.14 • (Continued)

cell bodies of motor neurons whose axons innervate skeletal muscles of the trunk and limbs (see Fig. 5.9 in Chapter 5, Spinal Cord).

The Aδ and C small-diameter fibers have specific laminae of termination where they synapse in the posterior horn. The Aδ and C fiber ascending and descending branches from the posterolateral tract (eponym: Lissauer's tract) terminate in laminae I, II, and V where pain, temperature, and itch sensation processing begins in the CNS. Lamina I (*posteromarginal nucleus*) receives primarily the central processes of Aδ fibers. Lamina I contains not only interneurons that project to other laminae of the spinal gray matter, but also projection neurons (second-order neurons) whose axons ascend to terminate and synapse in the brainstem reticular formation and the thalamus. Lamina II (*substantia gelatinosa*), which consists of an outer (IIo) and an inner (IIi) layer, receives mainly C fiber terminals in both sublayers. The substantia gelatinosa contains a complex neural circuitry that modulates nociception at the spinal cord level. Lamina V receives not only Aδ and C fiber terminals, but also Aα and Aβ fibers, which relay non-nociceptive information. Layer V is involved in the processing of a wide range of sensory nociceptive and non-nociceptive stimuli.

Second-order neurons

The cell bodies of the second-order neurons transmitting nociception reside in the posterior horn of the spinal cord

The cell bodies of the second-order neurons transmitting nociception reside in the posterior horn of the spinal cord (Fig. 12.13). It has been shown that the axons of these second-order neurons course either in the **direct** or **indirect pathways** of the ALS. The **direct pathway** consists of the **spinothalamic tract**. The **indirect pathways** include mainly the **spinoreticular tract**, but also include the smaller **spinomesencephalic, spinotectal, spinobulbar,** and **spinohypothalamic fibers**. Approximately 15% of nociceptive fibers project directly to the thalamus and 85% project to the thalamus via a relay in the reticular formation. Even though the anterolateral system (ALS) is somatotopically organized, the level of its organization is not as well defined as that of the posterior columns.

Direct pathway of the anterolateral system

The spinothalamic tract transmits not only nociceptive input, but also thermal, itch, and non-discriminative (crude) touch input to the contralateral VPL nucleus of the thalamus

Most Aδ fibers participate in the direct pathway of the ALS. Type **Aδ fibers** of first-order neurons synapse primarily with **second-order neurons** in lamina I (posteromarginal nucleus) and lamina V (reticular nucleus) of the spinal cord gray matter. Some first-order neurons synapse with spinal cord **interneurons** that are associated with reflex motor activity. Most of the axons of second-order neurons cross to the contralateral side in the *anterior white commissure of the spinal cord*, forming the **spinothalamic tract**; however, other axons ascend ipsilaterally (Fig. 12.15). As the lateral spinothalamic tract is formed, those axons of second-order neurons that cross the spinal cord are added to it in a lateral-to-medial direction. Thus, axons transmitting sensory information from the sacral levels (innervating lower part of the body) cross in the spinal cord to occupy the posterolateral portion of the lateral spinothalamic tract. As the axons of second-order neurons transmitting sensory information from the cervical levels (innervating upper part of the body) join the lateral spinothalamic tract, they are added medially to occupy its anteromedial portion.

The spinothalamic tract transmits not only nociceptive input, but also thermal, itch, and non-discriminative (crude) touch input to the contralateral **ventral posterior lateral (VPL)** nucleus of the thalamus. It also sends some projections to the **ventral posterior inferior (VPI)**, and the **ventral medial posterior (VMpo) nucleus** of the **thalamus**. Although the spinothalamic tract ends at the thalamus, it also sends collaterals to the reticular formation as it ascends through the brainstem. Since the spinothalamic tract (direct pathway of the ALS: spinal cord → thalamus) is phylogenetically a newer pathway, it is referred to as the **neospinothalamic pathway**. It transmits fast, well-localized, and precise nociceptive input.

Only about 15% of the nociceptive fibers from the spinal cord, ascending in the ALS and carrying nociceptive information, terminate directly in the thalamus via the spinothalamic tract. Although referred to as the "spinothalamic tract," it actually consists of two anatomically distinct tracts: the **lateral spinothalamic tract** (located in the lateral funiculus) and the very small **anterior spinothalamic tract** (located in the anterior funiculus). When these two tracts unite in the brainstem, they form the **spinal lemniscus**. It was believed that the lateral spinothalamic tract transmitted only nociceptive and thermal input, whereas the anterior spinothalamic tract transmitted only non-discriminative (crude) touch; however, it was demonstrated that both the anterior and lateral spinothalamic tracts (as well as the other component fibers of the ALS: spinoreticular, spinomesencephalic, spinotectal, spino-olivary, and spinohypothalamic), transmit **nociceptive**, **thermal**, and **non-discriminative (crude)** tactile signals to higher brain centers.

Indirect pathway of the anterolateral system

Most type C fibers participate in the indirect pathway of the ALS. Type C fibers of first-order neurons terminate on interneurons in laminae I, II (substantia gelatinosa), and III of the posterior horn. Axons of these **interneurons** synapse with **second-order neurons** in laminae VI–VIII. Many of the axons of these second-order neurons ascend ipsilaterally; however, a smaller number of axons sweep to the opposite side in the anterior white commissure of the spinal cord, to form the more prominent **ipsilateral** and smaller **contralateral spinoreticular tracts**. The spinoreticular tracts transmit **nociceptive**, **thermal**, and **nondiscriminatory (crude) touch** signals from the spinal cord to the thalamus indirectly, by forming multiple synapses in the reticular formation prior to their thalamic projections. Since the multisynaptic spinoreticular tract (indirect pathway of the ALS: spinal cord → reticular formation → thalamus) is phylogenetically an older pathway, it is referred to as the **paleospinothalamic pathway** (Figs. 12.13 and 12.16). It transmits slow, poorly localized, dull pain input.

The spinoreticular tract is a bilateral (primarily uncrossed) tract that conveys sensory information to the reticular formation of the medulla oblongata, pons, and midbrain, the region responsible for producing arousal and wakefulness, thus alerting the organism following an injury. Impulses from the reticular formation are then relayed bilaterally to the **intralaminar** and **dorsomedial nuclei of the thalamus**, via *reticulothalamic fibers*. Since these nuclei lack somatotopic

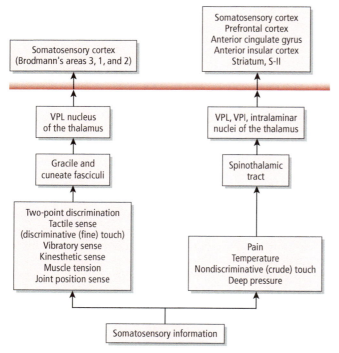

Figure 12.15 • Somatosensory information to consciousness. VPI, ventral posterior inferior; VPL, ventral posterior lateral.

organization, there is only an indistinct localization of sensory signals carried by this pathway. The reticular formation and its rostral continuation into the diencephalon, the intralaminar nuclei of the thalamus, are components of the **reticular-activating system** (**RAS**). The RAS functions in activating the organism's entire nervous system to elicit responses that will enable it to evade painful stimuli. In addition, there are some second-order neurons from the posterior horn that bypass the reticular formation and relay sensory input from C fibers directly to the intralaminar nuclei of the thalamus.

Other component fibers of the anterolateral system

In addition to the spinothalamic and spinoreticular tracts, the ALS also contains spinomesencephalic, spinotectal, spino-olivary (spinobulbar), and spinohypothalamic fibers which are grouped as part of the indirect pathways of the anterolateral system

The **spinomesencephalic fibers** arising from laminae I and V terminate in the periaqueductal gray (PAG) matter and the midbrain raphe nuclei, both of which are believed to give rise to fibers that modulate nociceptive transmission and are thus collectively referred to as the "descending pain-inhibiting system" (see discussion below). Furthermore, some spinomesencephalic fibers terminate in the *parabrachial nucleus*, which sends fibers to the *central nucleus of the amygdala* – a component of the limbic system associated with the processing of emotions. Spinomesencephalic fibers, via their connections to the limbic system, play a role in the emotional component of, and sensitization to, pain.

The **spinotectal fibers** arising from laminae I and V terminate mainly in the deep layers of the *superior colliculus*, but some fibers also terminate in the *anterior pretectum*. The superior colliculi have the reflex function of turning the upper body, head, and eyes in the direction of a painful stimulus via the tectospinal tract (Fig. 12.16).

The **spinohypothalamic fibers** ascend to the hypothalamus where they synapse with neurons that give rise to the *hypothalamospinal tract*. This pathway is associated with the autonomic and reflex responses (i.e., endocrine and cardiovascular responses) to nociception (Fig. 12.16). The hypothalamus also receives information from the spinal cord indirectly via projections to the periaqueductal gray (PAG) matter and the reticular formation.

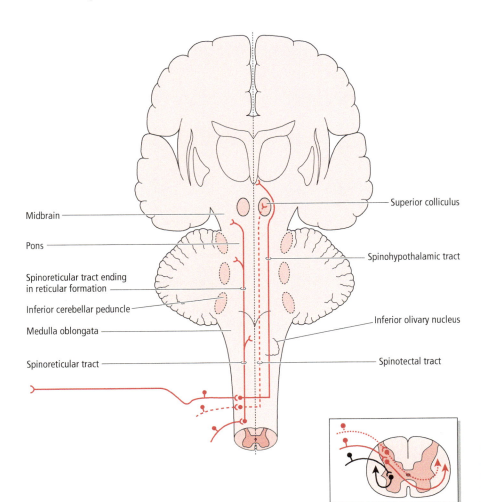

Figure 12.16 • Spinotectal, spinoreticular, and spinohypothalamic tracts of the indirect pathway of the anterolateral system. Note that the spinomesencephalic tract and spino-olivary fibers are not shown.

The **spinobulbar fibers** include the **spino-olivary fibers** which arise from the posterior horn. Spino-olivary fibers (of second-order neurons) decussate at the spinal cord level and ascend to terminate and synapse with third-order neurons in the *inferior olivary nucleus* in the medulla oblongata. Fibers leaving the inferior olivary nucleus decussate to enter the cerebellum via the inferior cerebellar peduncle. The spino-olivary fibers relay proprioceptive sensory input from cutaneous receptors, muscle spindles, and tendon organs. The inferior olivary nucleus, which is involved in motor learning, projects to the cerebellum to terminate in deep cerebellar nuclei and the molecular layer of the cerebellar cortex.

Approximately 85% of the nociceptive fibers from the spinal cord ascending in the ALS terminate in the brainstem reticular formation. From there, the information eventually reaches the thalamus via multiple additional synapses that occur in the brainstem. The reticular formation sends fibers transmitting nociceptive input to the thalamus, to the hypothalamus, which is associated with the autonomic and reflex responses to nociception, and to the limbic system, which mediates the emotional and suffering component of nociception.

Third-order neurons

Cell bodies of third-order neurons of the nociception-relaying pathway are housed in the VPL, the ventral posterior inferior, ventral medial, intralaminar, and dorsomedial thalamic nuclei

The **ventral posterior lateral (VPL) nucleus** gives rise to fibers that course in the **posterior limb of the internal capsule** and in the **corona radiata** to terminate in the **postcentral gyrus** (primary somatosensory cortex, SI) of the parietal lobe of the cerebral cortex. Additionally, the VPL nucleus also sends some direct projections to the secondary somatosensory cortex, SII (Fig. 12.17).

The **ventral posterior inferior (VPI) nucleus** projects mostly to the **secondary somatosensory cortex (SII)**, although some of its fibers terminate in the **primary somatosensory cortex (SI)**.

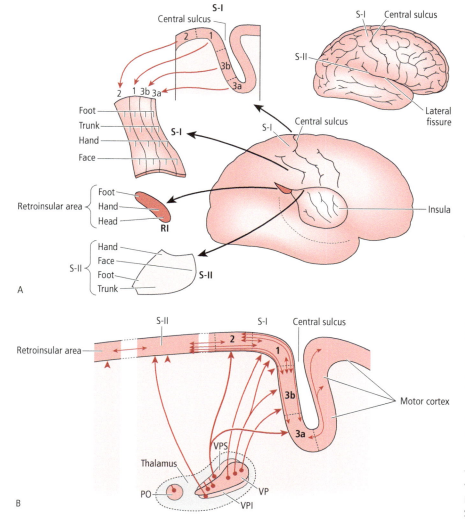

Figure 12.17 • Primary and secondary somatosensory cortex and retroinsular cortex: three major cortical areas receiving somatosensory information from the thalamus. (A) Counterclockwise from top right: lateral view of the brain; a cross-section of the central sulcus (eponym: of Rolando); and the primary somatosensory cortex (postcentral gyrus, S–I) forming the posterior border of the sulcus. Note the components of the primary somatosensory cortex areas 3a, 3b, 1, and 2, and the cortical representation of the foot, trunk, hand, and face. The lateral view of the brain shows the exposed insular cortex, secondary somatosensory cortex (S–II), and retroinsular (RI) cortex. (B) Thalamocortical projections to the primary somatosensory cortex (S–I), secondary somatosensory cortex (S–II), and retroinsular cortex (RI). PO, posterior complex; VP, ventral posterior; VPI, ventral posterior inferior; VPS, ventral posterior superior. Source: Adapted from Burt, AM (1993) *Textbook of Neuroanatomy*. WB Saunders, Philadelphia, figs 10.20, 10.22.

The **ventral medial posterior (VMpo) nucleus** and other nuclei of the posterior thalamic group send fibers to the SII and the retroinsular cortex.

The **intralaminar nuclei** send fibers to the **striatum** (the **caudate nucleus** and the **putamen**), the **SI** and **SII**, as well as to the **anterior cingulate cortex**, the **prefrontal cortex**, **amygdaloid body**, and **hypothalamus**. The intralaminar nuclei are involved in alerting the individual to nociceptive input.

It should be noted that most of the nociception-relaying fibers arriving at the **intralaminar nuclei** transmit nociceptive information relayed there from the reticular formation. Multiple synapses have been formed in the reticular formation prior to synapsing in the intralaminar nuclei, and thus, strictly speaking, the intralaminar relay neurons for this pathway are not the third-order neurons in the sequence, but they function as if they were third-order neurons; therefore, they are treated as such by neuroanatomists.

The **dorsomedial nucleus** projects to the frontal, cingulate, and limbic cortices.

The polysynaptic pathways of the indirect fibers of the ALS involved in the transmission of slow, poorly localized, and dull pain input may be responsible for the persistent pain perceived following lesions to certain thalamic nuclei.

Projections to the somatosensory (somesthetic) cortex

The **primary somatosensory cortex (SI)** consists of the postcentral gyrus of the parietal lobe, which corresponds to eponymous Brodmann's areas 3a, 3b, 1, 2 (Fig. 12.17A). The **secondary somatosensory cortex (SII)** consists of eponymous Brodmann's areas 5 and 7, located in the superior parietal lobule.

Axons of the thalamic third-order neurons terminate in somatotopically corresponding regions of the primary somatosensory cortex. Medial areas of the VPL are represented in the lateral surface of the postcentral gyrus: Regions of the head are represented in the inferior half of the postcentral gyrus near the lateral fissure, whereas those of the upper limb and the trunk are represented in its superior half. The lower limb, represented in lateral areas of the VPL, is represented in the medial surface of the postcentral gyrus (the posterior paracentral gyrus) and the perineum in the paracentral lobule. The body areas with the largest cortical area representation are the head and upper limb (especially the hand), reflecting the great discriminative capability that structures in these regions possess (Fig. 12.18).

In summary, nociceptive signals relayed from the spinal cord directly to the VPL, the VPI, and the ventral medial

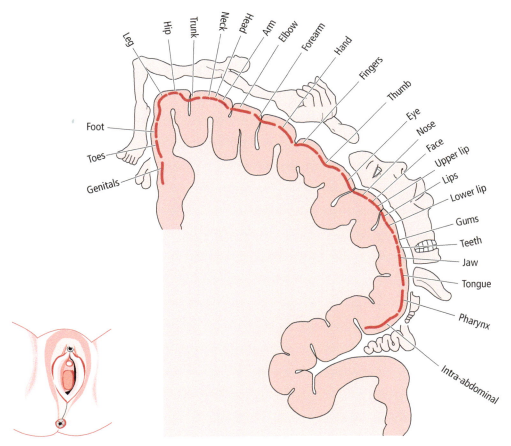

Figure 12.18 • Coronal section through the primary somatosensory cortex (postcentral gyrus) showing the sensory homunculus (L. little person). Note that the amount of cerebral cortex representing each body part is proportional to the extent of its sensory innervation. As an example, the hand although is a small part of the body, receives extensive and exquisite sensory innervation. In contrast, the trunk which is a large part of the body receives less sensory innervation. Source: Adapted from European Journal of Obstetrics and Gynecology.

206 ● ● ● **CHAPTER 12**

(VMpo) nucleus of the thalamus via the spinothalamic tract (neospinothalamic, direct pathway of the ALS) are transmitted to the somatosensory cortex (both to SI and SII). The postcentral gyrus is the site where processing of pain *localization, intensity, quality*, and *sensory integration* takes place at the conscious level. The primary somatosensory cortex sends projections to the secondary somatosensory cortex, which is believed to have an important function in the *memory* of sensory input.

In contrast, nociceptive signals relayed from the spinal cord to the reticular formation via the spinoreticular tracts (paleospinothalamic, indirect pathway of the ALS) are then transmitted to:

- **the intralaminar nuclei of the thalamus** (the cranial extension of the reticular formation into the thalamus), which in turn project to the primary somatosensory cortex;
- **the hypothalamus**; and
- **the limbic system**.

The projections through the reticular formation function in the arousal of the organism in response to nociceptive input, whereas the projections to the hypothalamus and limbic system have an important function in the autonomic, reflex, and emotional responses to a painful experience.

Projections to the cingulate and insular cortices

Cerebral imaging studies such as electroencephalography (EEG), functional magnetic resonance imaging (fMRI), magnetoencephalography (MEG), and positron emission tomography (PET) have demonstrated that nociceptive signals are not only processed at the primary and secondary somatosensory cortices, but also in the **anterior cingulate cortex**, **anterior insular cortex**, and even in the **supplemental motor area** of the motor cortex. The anterior cingulate and anterior insular cortices are connected with the limbic cortex, which plays a role in the emotional aspect of pain. If there is a lesion in the cingulate or insular cortex, patients are consciously aware of pain; however, they do not seem to experience the affective component of pain and do not have an emotional response to it ultimately reflecting a loss of input from the central nucleus of the amygdaloid body.

Visceral pain

A striking characteristic of the brain is that although it receives and processes nociceptive information, the brain itself has no sensation of pain

A striking characteristic of the brain is that although it receives and processes nociceptive information, the brain itself has no sensation of pain. During brain surgery, the patient is often awake and has no pain sensation from the brain tissue itself. The structures that must be anesthetized during brain surgery are the dura mater, the periosteum and bones of the skull, and the extracranial soft tissues.

Visceral pain is characterized as diffuse and poorly localized and is often "referred to" and felt in another somatic structure distant or near, but not at, the source of visceral pain. GVA nociceptive information from visceral structures of the trunk and abdomen is carried mostly by type **C, Aδ**, or **Aβ fibers** (Fig. 12.19A, B). The peripheral terminals of these fibers are associated with lamellar corpuscles (eponym: Pacinian corpuscles) that respond to excessive stretching of the intestinal wall, a lesion in the wall of the gastrointestinal tract, or smooth muscle spasms. The cell bodies of these sensory (pseudounipolar), **first-order neurons** are housed in the spinal ganglia. Their central processes carry the information, via the **posterolateral fasciculus** (eponym: tract of Lissauer) to the posterior horn and lateral gray matter of the spinal cord, where these central processes synapse with second-order neurons as well as with neurons associated with visceral reflex activities.

The axons of the **second-order neurons**, unlike the second-order neurons that are part of the **anterolateral system**, turn in a posterior direction and join the gracile fasciculus (mechanoreceptor pathway) relaying nociceptive signals from visceral structures to the gracile nucleus in the medulla. The third-order neurons of this pathway (in the gracile nucleus) give rise to axons that cross the midline, to ascend as part of the medial lemniscus to the **ventral posterior lateral nucleus (VPL) of the thalamus**. Fibers of a fourth-order neuron of this pathway from the **VPL** project to the **primary somatosensory cortex**.

Visceral pain signals relayed to the primary somatosensory cortex may be associated with referred pain to a somatic structure. It has been shown that in addition to projections to the somatosensory cortex, nociceptive signals are also relayed to the **anterior cingulate** and **anterior insular cortices**, two cortical areas implicated in the processing of visceral pain.

Temperature pathways from the body

1   As stated earlier in this chapter, **low-threshold receptors** are stimulated by temperatures ranging from 15 and 45°C (59 and 113°F) which do not elicit a noxious effect/sensation and do not produce tissue damage during a brief application of the stimulus.
2   **High-threshold receptors** are stimulated by temperatures lower than 15°C (59°F) and higher than 45°C (113°F) and elicit a painful sensation. Temperatures below 0°C (32°F) and above 50°C (122°F) can cause considerable tissue damage.

Warm sensation is conducted by free nerve endings of slow-conducting unmyelinated C fibers that respond to *increases* in temperature. Cold sensation is mediated by free nerve endings of slow-conducting thinly myelinated Aδ fibers (see Fig. 12.13; Table 12.2).

All fibers carrying temperature information to the CNS enter the spinal cord at the posterior root entry zone in the

ASCENDING SENSORY PATHWAYS 207

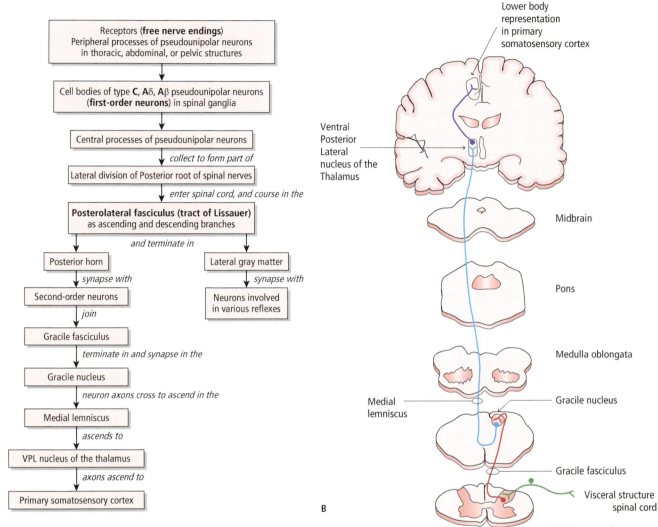

Figure 12.19 ● (A). Flow diagram of the ascending sensory pathway relaying pain sensation from the viscera. VPL ventral posterior lateral. (B). The ascending sensory pathway relaying pain sensation from the viscera.

lateral division of the posterior root of the spinal nerves. These fibers accompany nociceptive fibers and, upon entering the spinal cord, join the **posterolateral fasciculus** (eponym: tract of Lissauer), where they bifurcate into short *ascending* and *descending* fibers. These fibers ascend or descend, respectively, one to three spinal cord levels to synapse in laminae I, II, and III of the posterior horn gray matter, where they relay temperature input to **interneurons** that, in turn, transmit the information to **second-order neurons** in lamina V.

The axons of the second-order neurons, accompanied by nociceptive fibers, decussate in the anterior white commissure of the spinal cord to join the contralateral **anterolateral system**. Some of these fibers ascend to terminate in the brainstem **reticular formation**; however, most terminate primarily in the **VPL nucleus of the thalamus**. Third-order neuron fibers from the thalamus relay thermal sensory information to the **somatosensory cortex**.

TACTILE SENSATION AND PROPRIOCEPTION

The posterior column–medial lemniscal system (PCML), which includes the gracile fasciculus, cuneate fasciculus, and medial lemniscus, relays discriminative (fine) tactile sense, flutter-vibratory sense, and proprioception (position sense) to consciousness. Tactile sensation is divisible into non-discriminative (crude) touch and discriminative (fine) touch

The **posterior column–medial lemniscal system (PCML)**, which includes the **gracile fasciculus**, cuneate **fasciculus**, and **medial lemniscus**, relays discriminative (fine) tactile sense, flutter-vibratory sense, and proprioception (position sense) to consciousness.

Tactile sensation is divisible into non-discriminative (crude) touch and discriminative (fine) touch. **Non-discriminative (crude) touch**, transmitted via the ALS (discussed above), is a sensation that does not include detailed information about the stimulus, as when the skin is gently

CHAPTER 12

stroked with a fine cotton strand. Tactile examination of an object is dependent on **discriminative (fine) touch** sense, which enables one to detect fine detail regarding the *location*, *size*, *shape*, and *texture* of an object even when the eyes are closed. The recognition of the three-dimensional shape of an object by the sense of touch is known as **stereognosis**, whereas the ability to distinguish between two distinct points applied to the skin surface simultaneously is referred to as **two-point discrimination**. Both stereognosis and two-point discrimination rely on fine discriminative tactile sense and are relayed to consciousness via the PCML.

Conscious proprioception may be categorized as static and dynamic proprioception. **Static proprioception (static position sense)** is the awareness of the position of a motionless body part such as a limb, whereas **dynamic proprioception (kinesthetic sense)** is the awareness of movement of a body part, and balance.

Discriminative (fine) touch, pressure, vibratory sense, as well as **proprioceptive** sensory information are transmitted to higher brain centers, reaching consciousness, by three neurons arranged in sequence (Fig. 12.20; Table 12.3).

1 The **first-order neuron** (pseudounipolar neuron), whose cell body is located in a spinal ganglion, has the largest cell body in the ganglion. This neuron transmits sensory information from the periphery to the spinal cord (for reflex activity) and to the **posterior column nuclei** (the **gracile nucleus** and the **cuneate nucleus**) to synapse with a second-order neuron.
2 The axon of the **second-order neuron** decussates and ascends to terminate in the contralateral ventral posterior lateral **thalamus** to synapse with a third-order neuron.
3 The axon of the **third-order neuron** ascends ipsilaterally to terminate somatotopically in the **somatosensory cortex**.

The motor system relies on the sensory information transmitted by the PCML system for the precise execution of movement. In the absence of this information, the execution of voluntary movements is severely impaired.

The specifics of these pathways are discussed in the section below.

Discriminative (fine) touch, proprioception, vibratory sense, and pressure sense from the body

First-order neurons (sensory receptors)

The receptors that transmit discriminative (fine) touch, proprioception, vibratory sense, and pressure sense information to consciousness consist of:

- **free nerve endings** responding to touch, pressure, and proprioception in the skin, muscles, and joint capsules;
- **tactile discs** (eponym: Merkel discs) responding to touch and pressure in the skin;
- **peritrichial (follicular) endings** stimulated by bending of the hair follicles;
- **tactile corpuscles** (eponym: Meissner corpuscles) activated by touch of the skin;
- **lamellar corpuscles** (eponym: Pacinian corpuscles) stimulated by touch, pressure, flutter-vibration, and proprioception in the deep layers of the skin, and in the viscera;
- **bulbous corpuscles** (eponym: Ruffini endings) respond to *stretching* of the collagen bundles in the skin or joint capsules;
- **muscle spindles** stimulated by muscle stretch;
- **tendon organs** (eponym: Golgi tendon organs) stimulated by muscle tension exerted on tendons.

These **first-order pseudounipolar neurons**, whose cell bodies are located in the spinal ganglia, send peripheral processes to somatic or visceral structures. Most of these peripheral processes are medium-size type $A\beta$ and large-size type $A\alpha$ fibers; the remainder are of thin diameter. The peripheral terminals of these neurons either serve as the *sensory receptors* (e.g., free nerve endings, epithelial tactile discs, or peritrichial endings) or are associated with a specialized receptor structure such as tactile corpuscles, lamellar corpuscles, bulbous corpuscles, muscle spindles, or the tendon organs. These structures encapsulate the peripheral sensory terminals, thereby enhancing sensory perception. Upon mechanoreceptors being stimulated by mechanical pressure, the peripheral processes transmit the sensory information to the spinal cord by way of the central processes of the pseudounipolar neurons, which enter the spinal cord at the posterior root entry zone via the *medial division* of the posterior roots of the spinal nerves. Upon entry into the posterior funiculus of the spinal cord, the afferent fibers bifurcate into long ascending and short descending fibers.

The *long ascending* and *short descending* fibers of the central processes of the first-order pseudounipolar neurons give rise to collateral branches that may synapse with several distinct cell groups of the posterior horn interneurons (in laminae III, IV, and V) and with anterior horn motor neurons. These fibers collectively form the **posterior column pathways**, either of the **gracile fasciculus** or of the **cuneate fasciculus**, depending on the level of the spinal cord at which they enter. The fibers ascending in the posterior columns relay sensory information from the **ipsilateral** side of the body and terminate to synapse in their respective nuclei, the **ipsilateral gracile nucleus (GN)** and **ipsilateral cuneate nucleus (CN)** in the caudal medulla.

The fibers in the posterior column–medial lemniscus pathway are somatotopically organized, which becomes evident beginning at the spinal cord level. That is, the orderly and precise arrangement of fibers creates a map of body parts, which is maintained at all levels of the pathway from receptor to somatosensory cortex that permits precise localization of the stimulus.

As the central processes of the pseudounipolar neurons enter the spinal cord at successively higher levels, they are added to the posterior column from medial to lateral. The fibers carrying sensation from the *sacral* dermatomes occupy the most *medial*

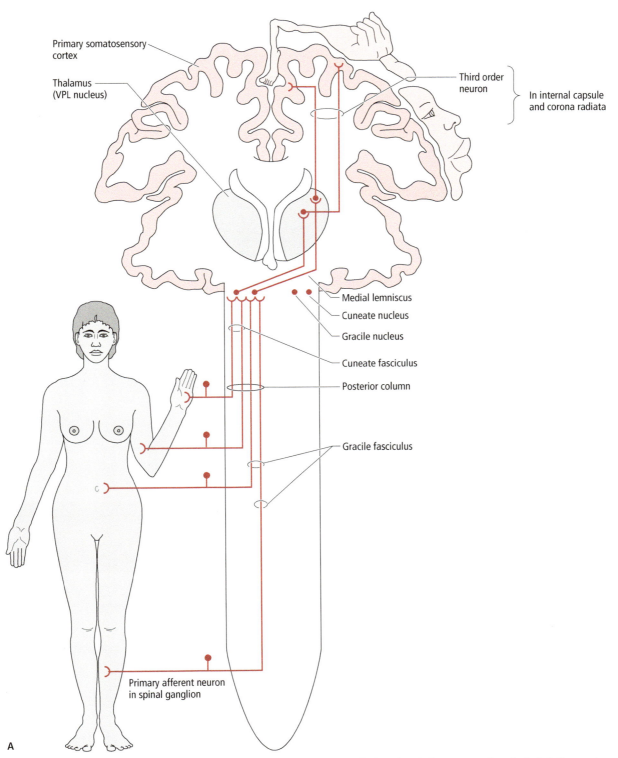

Figure 12.20 ● (A) The posterior column–medial lemniscal pathway relaying discriminative (fine) touch and vibratory sense from the body to the somatosensory cortex. VPL, ventral posterior lateral. (B). Simple sketch of the posterior column-medial lemniscus pathway.

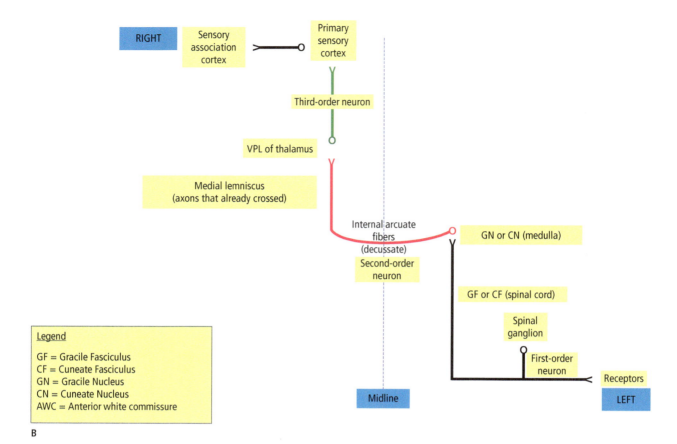

Figure 12.20 • (Continued)

Table 12.3 • Ascending sensory pathways to consciousness.

Sensation	Sensory receptor	Location of cell body of first-order neuron	Location of cell body of second-order neuron (origin of pathway)	Pathway	Decussation
Pain, thermal sense, and itch from the body	Aδ and C fiber endings	Spinal ganglion	Posterior horn	Spinothalamic tract of ALS	Anterior white commissure
Non-discriminative (crude) touch and superficial pressure from the body	Free nerve endings, Epithelial tactile discs (Merkel discs), peritrichial (follicular) nerve endings	Spinal ganglion	Posterior horn	Spinothalamic tract of ALS	Anterior white commissure
Two-point discriminative (fine) touch, vibratory sense, proprioceptive sense from muscles and joints of body	Tactile corpuscles (Meissner corpuscles), Lamellar corpuscles (Pacinian corpuscles),	Spinal ganglion	Gracile nucleus, Cuneate nucleus	First-order fibers: gracile and cuneate fasciculi	
	Muscle spindles, Tendon organs (Golgi tendon organs), bulbous corpuscles (Ruffini endings)			Second-order fibers: medial lemniscus	Medial lemniscal decussation in medulla

ALS, anterolateral system; VPI, ventral posterior inferior; VPL, ventral posterior lateral, VMpo, ventral medial posterior.

aspect of the posterior column, whereas the fibers carrying sensation from the *lumbar*, *thoracic*, and *cervical* dermatomes are added sequentially, but more *laterally*. As the *sacral* fibers ascend in the upper half of the spinal cord, they rotate from the most medial position (next to the spinal cord midline) to the most *posterior* aspect of the posterior column. Furthermore, the fibers that relay *tactile sensation* occupy a more *posterior region* of the posterior columns, whereas the fibers that relay *proprioception* occupy a more *anterior region* of the posterior columns. The fibers that relay tactile sensation terminate in the lower region of the posterior column nuclei, whereas the fibers that relay proprioception ascend further up to terminate in the upper region of the posterior column nuclei.

Ascending fibers relaying visceral pain to the gracile nucleus

Interestingly, about 85–90% of the fibers ascending in the **posterior columns** consist of the central processes of mechanoreceptor neurons (whose cell bodies are housed in the spinal ganglia), the remaining 10–15% of fibers ascending in tandem with those of mechanoreceptors in the gracile fasciculus, emerge from a unique group of cell bodies of second-order neurons that reside *not* in the spinal ganglia (as those of the mechanoreceptors), but instead, in the posterior horn of the spinal cord. These neurons relay *nociceptive* impulses arising from the pelvic viscera and lower part of the gastrointestinal tract. These atypical nociceptive ascending axons do not ascend in the anterolateral system (ALS), but instead join the **postsynaptic posterior column pathway** a minor, auxiliary pathway located within the posterior column. This minor

pathway carries a combination of two types of *second-order neuron* axons: (1) the axons of second-order neurons relaying non-discriminative tactile sensation, which are joined by (2) the axons of second-order neurons relaying nociception from the viscera. Thus, in addition to **mechanoreceptor input**, the gracile fasciculus also carries a minor, but distinct, group of fibers, relaying **visceral nociception** to the gracile nucleus in the medulla oblongata. These visceral nociceptive fibers synapse with a separate group of neurons in the gracile nucleus, and there seems to be no overlap in the synapsing pattern of these pain fibers and the mechanoreceptor fibers. Visceral nociception is then relayed via the medial lemniscus to the ventral posterior lateral nucleus of the thalamus.

Central processes below level T6

The central processes of the first-order pseudounipolar neurons that enter the spinal cord below level T6 include the lower thoracic, lumbar, and sacral levels that bring information from the ipsilateral lower limb and lower half of the trunk

The central processes of the first-order pseudounipolar neurons that enter the spinal cord below the sixth thoracic (T6) level include the lower thoracic, lumbar, and sacral levels. They bring sensory information from the ipsilateral lower limb and lower half of the trunk. The central processes enter the ipsilateral **gracile fasciculus** (**GF**) (L., *gracilis*, "slender") and ascend to the medulla oblongata to terminate in the ipsilateral **gracile nucleus** (**GN**). It should be recalled that the gracile fasciculus is present along the entire length of

Table 12.3 ● *Continued*

Location of pathway in spinal card	Pathway (second order neuron) termination	Location of cell body of third order neuron	Termination of third order neuron	Conscious/ subconscious	Function
Lateral funiculus	VPL, VPI, VMpo, and intralaminar nuclei of the thalamus	VPL, VMpo, and intralaminar nuclei of the thalamus	Postcentral gyrus, cingulate gyrus, prefrontal cortex, retroinsular cortex	Conscious	Relays pain and thermal sensation from the body
Lateral funiculus	VPL nucleus of the thalamus	VPL nucleus of the thalamus	Postcentral gyrus	Conscious	Relays non-discriminative (crude) touch sensation from the body
Posterior funiculus	VPL nucleus of the thalamus	VPL nucleus of the thalamus	Postcentral gyrus	Conscious	Relays two-point discriminative (fine) touch sensation, vibratory sense, proprioceptive sense from muscles and joints of the body, some visceral pain

the spinal cord; it begins to form at sacral cord levels and as it ascends in the spinal cord, fibers are added to it, so it becomes progressively thicker until it reaches the seventh thoracic (T7) level of the spinal cord. From this point on no additional fibers join the GF and, consequently, it remains the same size for the remainder of its ascent in the spinal cord and terminate in the GN in the medulla oblongata. The axon terminals that end in the gracile nucleus are somatotopically organized. Axons transmitting sensory information from sacral levels end in the most medial part of the gracile nucleus.

Central processes Level T6 and above

The central processes of the first-order pseudounipolar neurons that enter the spinal cord at approximately level T6 and above bring information from the upper thoracic and cervical levels; that is, from the upper half of the trunk and upper limb

The central processes of the first-order pseudounipolar neurons that enter the spinal cord at approximately the sixth thoracic (T6) level and above, bring sensory information from the ipsilateral upper thoracic and cervical levels; that is, from the upper half of the trunk and upper limb. These central processes enter the ipsilateral **cuneate fasciculus** (CF) (L., *cuneus*, "wedge") and ascend to the medulla oblongata to synapse with second-order neurons in the ipsilateral **cuneate nucleus** (CN). It should be noted that the cuneate fasciculus begins to form at about the sixth thoracic (T6) spinal cord level and continues to form until the second cervical (C2) spinal cord level. As the CF ascends, more and more fibers are added, so that it becomes progressively thicker until it reaches the second cervical (C2) spinal cord level. Note that the first cervical (C1) spinal nerve does not have sensory fibers, so C1 does not contribute any fibers to the CF. The fibers in the CF terminate in the CN in the medulla oblongata where they are somatotopically organized. Axons transmitting sensory information from thoracic levels end in the most medial part of the cuneate nucleus, whereas axons transmitting sensory information from cervical levels end in the most lateral part of the nucleus cuneatus.

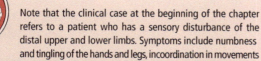

Note that the clinical case at the beginning of the chapter refers to a patient who has a sensory disturbance of the distal upper and lower limbs. Symptoms include numbness and tingling of the hands and legs, incoordination in movements of the hands, imbalance, and difficulty walking. Sensation, particularly to vibration and joint position, is severely impaired in the distal limbs.

1 Which ascending sensory pathway(s) is/are affected in this patient?

2 Where are the nerve cell bodies of the affected pathways located?

3 In which part of the spinal cord are the affected ascending pathways located?

4 How extensive is the damage if symptoms are present in both the upper and lower limbs?

As the ascending fibers of all of the first-order neurons ascend to various spinal cord levels within the gracile fasciculus or the cuneate fasciculus, they give rise to collateral branches that curve away from the posterior columns to enter the spinal cord gray matter. Once there, they form synapses with interneurons and alpha motor neurons, thus participating in intersegmental reflexes of the trunk and limbs. Although the pseudounipolar neurons relaying sensory input from various mechanoreceptors send collateral branches to synapse in the posterior horn, the central processes of pseudounipolar neurons that relay sensory input from the muscle spindles project to the motor neurons in the anterior horn.

The posterior columns on each side of the spinal cord are separated by an intervening connective tissue partition the *median septum*. The gracile fasciculus and cuneate fasciculus are separated from each other by a similar partition, the *posterior intermediate septum*.

Second-order neurons

Axons of first-order neurons synapse with second-order neurons whose cell bodies are located in the gracile and cuneate nuclei within the posterior medulla oblongata

The first-order ascending fibers synapse with second-order neurons whose cell bodies are located in the gracile nucleus (GN) and cuneate nucleus (CN) in the posterior medulla oblongata. The fibers of the second-order neurons emerge from the GN curve and decussate ventral to the fibers that arise from the CN. The fibers of the second-order neurons emerging from the posterior column nuclei collectively form the **internal arcuate fibers** as they curve anteromedially in the caudal medullary tegmentum, interweaving through the reticular formation to reach the opposite side. These fibers ascend as the **medial lemniscus** (L., *lemniscus*, "ribbon") in the caudal part of the medulla, cranial to the prominent pyramidal (motor) decussation, to synapse with third-order neurons in the **ventral posterior lateral (VPL) nucleus of the thalamus**.

The axons of second-order neurons arising from the GN (relaying sensation from the lower limb and lower trunk) occupy the anterior half of the medial lemniscus, whereas the second-order axons arising from the CN (relaying sensation from the upper limb and upper trunk) occupy the posterior half of the medial lemniscus. This fiber arrangement at the level of the medulla oblongata (as evident when viewing a cross section of the medulla oblongata) is depicted as an individual standing erect (with feet anteriorly and head posteriorly). At pontine levels, the medial lemniscus rotates orientation so that the fibers representing the legs are posterolaterally and those representing the head are anteromedially.

The fibers of second-order neurons originating from the cuneate nucleus terminate in the medial aspect of the VPL nucleus of the thalamus, whereas the fibers emerging from the gracile nucleus terminate in the lateral aspect of the VPL nucleus. Due to this pattern of fiber terminations in the thalamus, the upper limb is represented medially and the lower limb laterally. Fibers of the medial lemniscus also terminate in the ventral posterior inferior (VPI), pulvinar, and the lateral posterior group of thalamic nuclei.

Third-order neurons

The VPL nucleus of the thalamus houses the cell bodies of the third-order neurons of the posterior column–medial lemniscal (PCML) pathway

The VPL nucleus of the thalamus houses two distinct populations of neurons, which process sensory input relayed by the PCML pathway: (1) the cell bodies of the third-order neurons of the PCML pathway, and (2) local circuit interneurons which modulate the activity of the thalamocortical cells.

The third-order neurons (thalamocortical cells) give rise to **thalamocortical fibers** that ascend in the **posterior limb of the internal capsule** and the **corona radiata** to terminate in the primary (SI) and secondary (SII) somatosensory cortices. The SI resides in the **postcentral gyrus** (which corresponds to Brodmann areas 3a, 3b, 1, and 2, of the parietal cortex/lobe). Area 1 is responsible for determining the *texture* and area 2 the *size* and *shape* of objects. Areas 2 and 3a receive proprioceptive input. Area 2 receives sensory input derived primarily from tendon organs and joint receptors. Area 3a is stimulated by signals arising primarily from muscle spindles and joint receptors; it functions in proprioception and is believed to participate in motor functions. Areas 3b and 1 receive sensory input derived from cutaneous receptors relaying touch sensation. These two areas also receive pain and temperature input.

Projections to the somatosensory cortex

Brodmann's area 3b receives most of the projections arising from the VPL nucleus of the thalamus and is the site where initial cortical processing of tactile discrimination input occurs

Brodmann's area 3b receives most of the projections arising from the VPL nucleus of the thalamus and is the site where the initial cortical processing of tactile discrimination input takes place. Brodmann area 3b in turn projects to Brodmann areas 1 and 2.

The primary somatosensory cortex consists of six histologically distinct layers of neurons and neuroglia. The thalamus projects its fibers to layer IV of the somatosensory cortex. The thalamocortical fibers relaying sensory information from the VPL nucleus of the thalamus terminate in cortical columns. A cortical column consists of the basic functional unit of the cerebral cortex. Thalamocortical fibers carrying specific types of information from a particular part of the body do not project to a particular layer of the somatic sensory cortex, but instead, project to a cortical column, which includes a block of cortex containing all six layers of the sensory cortex.

The primary somatosensory cortex exports mechanosensory information for further processing to neighboring cortical areas via three pathway streams, the ventral, posterior, and rostral pathways (Fig. 12.21).

The **anterior stream**, which forms the **"what" pathway**, begins at the primary somatosensory cortex. It projects to the **secondary somatosensory cortex** residing on the parietal operculum (L., "lid") and the insular cortex hidden deep in the lateral fissure. The secondary somatosensory cortex in turn projects to the temporal lobe. The "what" pathway functions in object recognition determining what the object is, exclusively via the tactile sense and without the aid of the visual system. If an object such as a spoon is placed in the palms of an individual whose eyes are closed and manipulates the object, this pathway enables the individual to recognize the object as a spoon.

The **posterior stream**, which forms the **"where" pathway**, begins at the primary somatosensory cortex. It projects to the **posterior parietal cortex**, which corresponds to eponymous Brodmann areas 5 and 7. This cortical area is interposed between the primary somatosensory cortex anteriorly, the visual cortex posteriorly, and the auditory cortex inferiorly. This pathway functions not only in the spatial localization of an object, that is, where the object is, relative to the viewer but also functions in the consciousness of body image. Furthermore, due to its location, wedged between

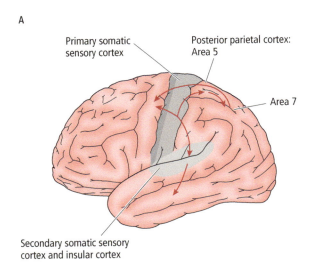
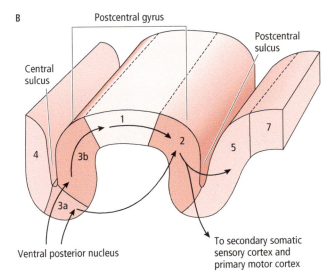

Figure 12.21 ● (A) Lateral surface of the cerebral hemisphere showing the location of the primary and higher-order somatosensory cortex. (B) The postcentral gyrus. Source: Martin (2012). *Neuroanatomy Text and Atlas*, 4th edn, figure 4.12, p. 102/McGraw Hill.

214 ●●● **CHAPTER 12**

the somatosensory, visual, and auditory cortical areas, and the projections it receives from these surrounding cortical areas, the posterior parietal lobe functions in the integration of sensory information to create a multisensory perception and alertness.

In the **frontal stream**, the primary somatosensory cortex and the posterior parietal cortex project to the motor cortical areas of the frontal lobe. This pathway consolidates somatic sensory information, which is used by the motor system to direct movement in the direction of a particular target.

The primary somatosensory cortex projects to the secondary somatosensory cortex, located on the superior border of the lateral fissure. Some third-order neuron fibers from the thalamus terminate directly in the secondary somatosensory cortex.

The PCML pathway relays proprioceptive sensory information to the contralateral somatosensory cortices where it enters consciousness through ignition of the global neuronal network. Sensory information from all the sensory systems converges at the cortical level where it is integrated so that the brain forms a representation of the position or orientation of the body parts in space. Thus, somatotopic information (body map) is assembled as information flows from receptors to nuclei to the sensory cortex.

SENSORY PATHWAYS TO THE CEREBELLUM

Most of the proprioceptive information does not reach conscious levels and instead is transmitted directly to the cerebellum

Only a small portion of the proprioceptive information enters consciousness; instead, most is transmitted directly to the cerebellum as two-neuron pathways (first- and second-order neurons) via the four tracts of the **ascending somatosensory cerebellar pathways**, which process subconscious proprioception from muscles, tendons, and joints, and some cutaneous input from mechanoreceptors. The four tracts are: the posterior (dorsal) spinocerebellar tract, the anterior (ventral) spinocerebellar tract which transmit proprioception mainly from the lower limb, the cuneocerebellar tract, and the rostral spinocerebellar tract which transmit proprioception mainly from the upper limb (Fig. 12.22; Tables 12.1 and 12.4). All four tracts ascend to terminate and synapse in the cerebellar cortex with collaterals to the deep cerebellar nuclei of the spinocerebellum, consisting of the vermis and paravermal zone. These tracts relay sensory input about limb position, muscle length, and tension. The cerebellum uses this afferent information in the modulation of muscle tone, posture maintenance, and the coordination of movement.

Table 12.4 ● Ascending sensory pathways (subconscious).

Sensation	Sensory receptor	Location of cell body of first-order neuron	Location of cell body of second-order neuron (origin of pathway)	Pathway	Decussation of pathway (ipsilateral/contralateral)
Pain and thermal sense from the body	Aδ and C fibers	Spinal ganglion	Posterior horn	Spinoreticular	Anterior white commissure
				Spinotectal	
				Spinomesencephalic	
				Spinohypothalamic	
Proprioceptive sense from muscles and joints of the body, limb position sense	Muscle stretch receptors, tendon organs	Spinal ganglion	Posterior horn (Posterior thoracic nucleus) (Clarke's column) C8–L2, 3	Posterior spinocerebellar	Ipsilateral
		Spinal ganglion	Accessory cuneate nucleus	Cuneocerebellar tract	Ipsilateral
		Spinal ganglion	Posterior horn	Rostral spinocerebellar tract	Ipsilateral
		Spinal ganglion	Posterior horn	Anterior spinocerebellar tract	Decussates in anterior white commissure and decussates again within cerebellum

Posterior spinocerebellar tract

The primary function of the posterior spinocerebellar tract is to relay proprioceptive input from the neuromuscular spindles and tendon organs of the ipsilateral lower trunk and lower limb, to the cerebellum

The primary function of the posterior (dorsal) spinocerebellar tract (Fig. 12.22A; Table 12.5) is to relay proprioceptive input from the neuromuscular spindles and tendon organs of the ipsilateral lower trunk and lower limb to the cerebellum. It should be recalled, however, that it also relays some touch and pressure sensation from the skin of the ipsilateral trunk and lower limb.

First-order neurons (pseudounipolar neurons) whose cell bodies are housed in the spinal ganglia send their peripheral processes to the skin, muscles, tendons, and joints. Here they perceive proprioceptive information, which is then transmitted to the spinal cord by their central processes. These central processes join the *medial division* of the posterior roots of the spinal nerves to synapse at their level of entry in the **posterior thoracic nucleus** (eponym: Clarke's column, synonym: nucleus dorsalis, lamina VII of spinal cord levels C8 to L2, 3, continuous superiorly from the posterior spinocerebellar tract). Sensory information transmitted by spinal nerves entering at the sacral and lower lumbar spinal cord levels is relayed to the caudal extent of the posterior thoracic nucleus (L2, 3) by ascending in the gracile fasciculus.

The posterior thoracic nucleus houses the cell bodies of **second-order neurons** whose axons form the **posterior spinocerebellar tract**, which ascends ipsilaterally in the lateral funiculus of the spinal cord. The heavily myelinated axons of this tract have the largest diameter of all the axons in the central nervous system, and their nerve impulse conduction velocity is unparalleled by any other system. When this tract reaches the brainstem, it joins the restiform body (of the inferior cerebellar peduncle) and then passes (as "mossy fibers") into the vermis of the cerebellum. The posterior spinocerebellar tract relays **proprioceptive** information from the lower limb directly to the cerebellum where this information is processed; it plays an important role in the fine coordination of movements of individual lower limb muscles or synergistic muscles controlling movement of the same joint and in the maintenance of posture.

Table 12.4 ● *Continued*

Location of pathway in spinal card	Pathway (second order neuron) termination	Location of cell body of third order neuron	Termination of third order neuron	Conscious/ subconscious	Function(s)
Lateral funiculus	Reticular formation	Multiple synapses in brainstem reticular formation; fibers terminate in the intralaminar nuclei of the thalamus	Anterior cingulate cortex, anterior insular cortex	Subconscious	Relays pain and thermal sense from the body; input functions in arousal of the organism in response to pain
	Superior colliculus	–	–	Subconscious	Mediates reflex movement of the head, and upper trunk in direction of stimulus
	Periaqueductal gray matter and raphe magnus nucleus	–	–	Subconscious	Involved in pain modulation
	Hypothalamus	–	–	Subconscious	Functions in autonomic, reflex, and emotional aspects of pain
Lateral funiculus	Cerebellar vermis	–	–	Subconscious	Relays proprioceptive information to the cerebellum, and functions in the coordination of movement of the lower limb and posture maintenance
–	Anterior lobe of the cerebellum	–	–	Subconscious	Neck and upper limb equivalent of the posterior spinocerebellar tract
Lateral funiculus	Cerebellum	–	–	Subconscious	Mediates proprioception from the head and upper limb
Lateral funiculus	Cerebellar vermis	–	–	Subconscious	Mediates coordination of muscle activity of the trunk and lower limb

216 ●●● **CHAPTER 12**

Table 12.5 ● Functions of the ascending sensory tracts to the cerebellum (subconscious).

Tract	Location of first-order neuron cell body	Location of second-order neuron cell body	Termination	Function
Posterior spinocerebellar	Spinal ganglion	Posterior horn (posterior thoracic nucleus) (Clarke's column) C8–L2, 3	Ipsilateral cerebellar vermis	Relays proprioceptive input from the ipsilateral trunk and lower limb. Coordination of movements of the lower limb muscles Posture maintenance
Cuneocerebellar	Spinal ganglion	Accessory cuneate nucleus	Ipsilateral anterior lobe of the cerebellum	Relays proprioceptive information from the ipsilateral neck and upper limb Movement of head and upper limb
Anterior spinocerebellar	Spinal ganglion	Posterior horn	Ipsilateral cerebellar vermis	Relays proprioceptive input from the ipsilateral trunk and lower limb Coordination of movements of lower limb muscles Posture maintenance
Rostral spinocerebellar	Spinal ganglion	Posterior horn	Cerebellum	Relays proprioceptive input primarily from the ipsilateral head and upper limb Movement of head and upper limb

Cuneocerebellar tract

Proprioceptive sensory information primarily from the upper limb, but also from the neck and upper half of the trunk enters at spinal cord segments C2–T5

The axons of the **second-order neurons**, whose cell bodies reside in the **external (accessory, lateral) cuneate nucleus**, form the **cuneocerebellar tract**. This tract is referred to as the neck and upper limb counterpart of the posterior spinocerebellar tract. Fibers of the cuneocerebellar tract join the restiform body (of the inferior cerebellar peduncle) and then enter the ipsilateral anterior lobe of the cerebellum. Information about movements in progress carried by the cuneocerebellar tract plays a role in the fine adjustment of movements of the upper limbs.

Proprioceptive sensory information primarily from the upper limb, but also from the neck and upper half of the trunk enters at spinal cord segments C2–T5. The central processes of the pseudounipolar first-order neurons ascend in the cuneate fasciculus and terminate in the external cuneate nucleus located lateral to the cuneate nucleus in the medulla oblongata–the posterior thoracic nucleus homolog for the upper limb (Fig. 12.22A; Table 12.5).

Anterior spinocerebellar tract

The anterior spinocerebellar tract relays proprioceptive information from the muscle spindles and tendon organs mainly of the lower limb that play a role in spinal reflexes and impending movement

The **anterior spinocerebellar tract** relays proprioceptive information from muscle spindles, tendon organs, and cutaneous receptors of the lower limb. It functions in the coordination of movement of the lower limb and maintenance of posture (Fig. 12.22B; Table 12.5).

First-order neurons (pseudounipolar neurons) transmit sensory input to laminae V–VII of the lumbar, sacral, and coccygeal spinal cord levels, where they terminate and synapse with second-order neurons.

The **second-order neurons** are known as *spinal border cells*. They receive messages not only from first-order neurons transmitting proprioception from tendon organs, but also from cutaneous sensory neurons, interneurons, and axon terminals of descending tracts. The axons of these **second-order neurons** form the **anterior spinocerebellar tract**, which decussates in the anterior white commissure of the spinal cord and ascends in the lateral funiculus to the pons. A few fibers ascend ipsilaterally. The anterior spinocerebellar fibers ascend to the rostral pons and then descend to join the *superior cerebellar peduncle* to pass as "mossy fibers" into the cerebellum. These fibers then decussate again to their actual side of origin within the cerebellum to terminate in the vermis of the cerebellum's anterior lobe.

Rostral spinocerebellar tract

Proprioceptive information from the upper limb is transmitted to C7–C8 spinal cord levels

Proprioceptive information from the upper limb is transmitted to C5–C8 spinal cord levels. The central processes of **first-order neurons** synapse with second-order neurons whose cell bodies reside in lamina VII of the posterior horn. The fibers of the **second-order neurons** form the primarily uncrossed **rostral spinocerebellar tract**, the upper limb counterpart of the anterior spinocerebellar tract. These fibers join the restiform body (of the inferior cerebellar peduncle) to enter the cerebellum. Additionally, some fibers pass into the cerebellum via the superior cerebellar peduncle (Fig. 12.22B; Table 12.5). This tract plays a role in the movement of the upper limb.

ASCENDING SENSORY PATHWAYS • • • 217

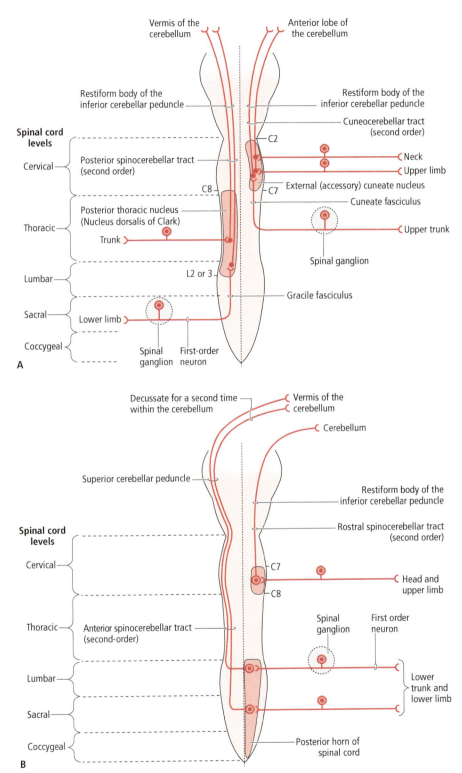

Figure 12.22 • (A) Two of the ascending sensory pathways to the cerebellum: the posterior spinocerebellar tract transmitting sensory information from the lower limb and trunk, and the cuneocerebellar tract transmitting sensory information from the neck, upper limb, and upper trunk to the cerebellum. (B) Two of the ascending sensory pathways to the cerebellum: the anterior spinocerebellar tract transmitting sensory information from the lower trunk and lower limb, and the rostral spinocerebellar tract transmitting sensory information from the head and upper limb to the cerebellum.

CHAPTER 12

CLINICAL CONSIDERATIONS

Lesions involving peripheral nerves
The damage that occurs to a peripheral nerve, and the resulting sensory symptoms, differ depending on whether the damaged nerve carries sensory fibers and which fibers are involved in the lesion

The extent of damage that may occur to a peripheral nerve, and the sensory symptoms that will arise following a lesion, differ depending on whether the damaged nerve carries sensory fibers and which fibers are involved in the lesion. Diminished cutaneous sensation due to damage of the sensory fibers of a particular nerve is usually narrower than the range of distribution of the nerve, due to the overlapping areas of distribution of adjacent nerves.

Posterior root and spinal nerve lesions
The sensory deficits that arise following a lesion to a posterior root or to a spinal nerve are usually revealed in a segmental distribution

The sensory deficits that arise following a lesion to a posterior root or to a spinal nerve are usually revealed in a segmental distribution (Fig. 12.23). Since peripheral nerves branch and extend their innervation into territories of adjacent nerves, a peripheral nerve lesion may include areas supplied by several adjacent spinal cord levels. Due to this innervation overlap, it is difficult to trace the sensory deficit to a single spinal nerve or posterior root. Posterior root irritation results in pain and **paresthesia** (G., "abnormal sensations"), such as tingling, itching, or pricking of the skin.

In chronic pain syndromes, the posterior roots of spinal nerves may be severed by a neurosurgical procedure, known as **posterior rhizotomy** to relieve chronic pain. If the pain persists following this procedure, the spinal ganglia are then removed. It is thought that removing the ganglia eliminates the sensory neurons which may give rise to central processes that instead of entering the spinal cord via their usual path, the posterior roots, they may enter the spinal cord via an alternate route, the anterior roots.

Spinal cord lesions
To be able to identify the site of a spinal cord lesion, one has to be familiar with the anatomical arrangement of the various ascending and descending tracts in the spinal cord. In general, the pathways of the anterolateral system (ALS) relaying **pain**, **temperature**, and **non-discriminative (crude) touch** from the body, ascend in the opposite side of origin, in the anterolateral aspect of the spinal cord. Pathways relaying **discriminative (fine) touch**, **proprioceptive**, and **vibratory** modalities ascend ipsilateral to the side of origin, in the posterior white columns of the spinal cord.

Brown–Séquard syndrome
Although spinal cord injuries are rarely limited to a particular tract, quadrant, or side of the spinal cord, the hemisection of the spinal cord is used for instructive purposes. One such example is the **Brown–Séquard syndrome**, which accounts for 2–4% of all traumatic spinal cord injuries (Fig. 12.24).

When the **spinal cord** is **hemisected** (only the right or left half is severed), all of the tracts (both ascending and descending) coursing through the level of the lesion are severed, and the following will be observed:

1 The lower motor neurons ipsilateral to and at the level of the lesion will be damaged, leading to **ipsilateral lower motor neuron paralysis** at the level of the lesion.

2 Since the corticospinal tract (upper motor neurons) will be severed, the individual will exhibit an **ipsilateral loss of motor function** below the level of the lesion, followed by **spastic paralysis** (see Chapter 11).

For **sensory deficits** at and below the level of the lesion, the following will be observed.

1 In this lesion, the **anterolateral system** (which includes the spinothalamic, spinoreticular, spinomesencephalic, spinotectal, spino-olivary, and spinohypothalamic fibers) has been severed. The fibers that relay pain, temperature, and crude touch sensation to consciousness cross in the spinal cord before joining the spinothalamic tract; thus, the spinothalamic tract located in the anterolateral aspect of the spinal cord is a crossed tract whose fibers carry sensory information from the opposite side of the body. At the spinal cord level, the anterolateral system (ALS) consists of the central processes of both $A\delta$ (fast-conducting fibers, relaying sharp, well-localized pain) and C (slow-conducting fibers, relaying dull, poorly localized pain). Thus, with a lesion at the spinal cord level, there will be:

- a complete loss of **pain and temperature sensation** beginning one to three segments below the level of the lesion, contralaterally (Fig. 12.25A); and
- a diminution of **non-discriminative (crude) touch sensation** beginning one to three segments below the level of the lesion, contralaterally

Although this pathway also relays itch sensation, it is assumed to be impaired although it is not customarily tested.

Note that if a lesion occurs in the brainstem, where the spinothalamic and spinoreticular pain fibers ascend separately and in different locations of the brainstem (the spinothalamic fibers are situated laterally, and the spinoreticular fibers are positioned medially), the deficits will be different. If the lesion selectively damages only the spinothalamic tract pain fibers, but not the spinoreticular pain fibers, it will result in diminution of fast, well-localized pain and thermal sense, leaving the spinoreticular pain fibers undamaged. The intact spinoreticular fibers can still relay pain sensation to the reticular formation and from there to the limbic system which functions in the affective responses to pain.

2 Since the **posterior column pathways (gracile fasciculus and cuneate fasciculus)**, which contain ascending axons that are located ipsilateral to the side of their origin and carry sensory information from the ipsilateral side of the body have been severed, there will be an ipsilateral loss of the following, at and below the level of the spinal cord lesion

- discriminative (fine) touch;
- flutter-vibratory sensation;
- proprioceptive (joint position) sensation: an individual with this lesion will have coordination difficulties referred to as **posterior column sensory ataxia**; if the lesion involves the sensory innervation of the lower limbs, this condition is also referred to as **"stamp and stick ataxia"** because the individual will stamp their feet on the ground to stimulate other sensory receptors in the lower limbs and will also try to feel the ground with their cane to aid with the position

CLINICAL CONSIDERATIONS (*continued*)

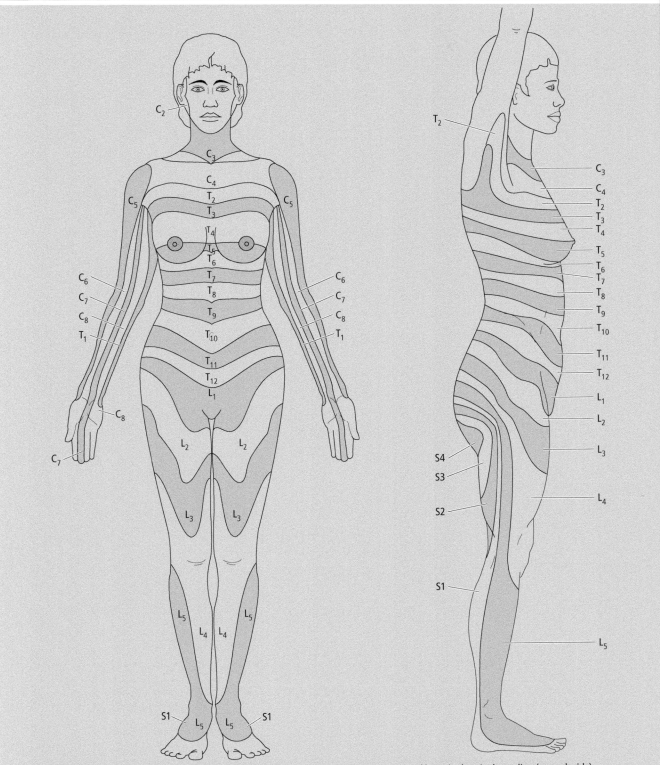

Figure 12.23 • Dermatomes of the skin. Each striped area represents the skin innervated by a single spinal ganglion (on each side).

CLINICAL CONSIDERATIONS (continued)

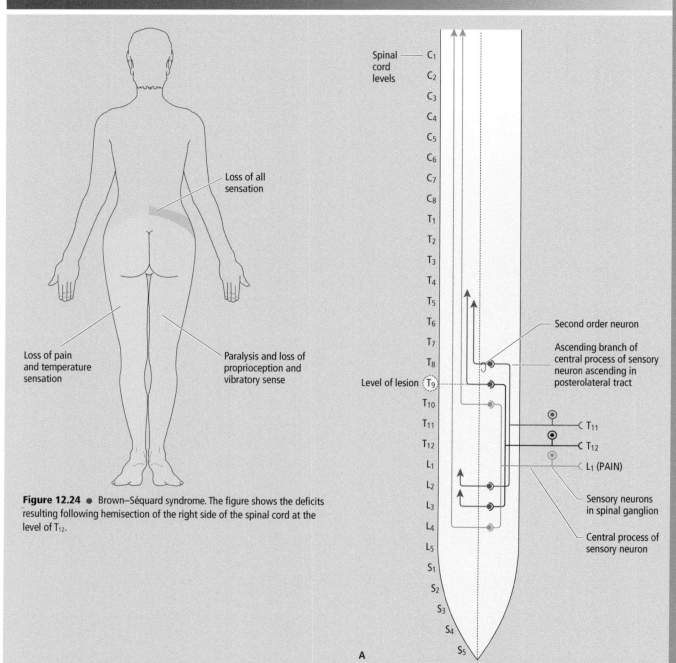

Figure 12.24 • Brown–Séquard syndrome. The figure shows the deficits resulting following hemisection of the right side of the spinal cord at the level of T_{12}.

Figure 12.25 • (A). Nociception from the L_1 spinal nerve territory entering the spinal cord is relayed to second-order neurons located one to three levels above and below the level of spinal cord entry. In this figure, the branches of the central processes of the sensory neurons ascend and descend three spinal cord levels. If a patient has intractable pain at the L_1 sensory territory, a lesion (tractotomy) made a few levels above the level of the pain sensation (at T_9) on the contralateral side will sever the axons of the second-order neurons that synapse with the ascending branches of the L_1 sensory neurons and prevent pain from reaching higher brains centers via this pathway. If the lesion in this case is made below T_9, the axons of the second-order neurons that synapse with the ascending branches of the sensory L_1 neurons will remain intact and relay pain sensation, and the pain will leak to higher brain centers. (B.) A lesion in the spinothalamic tract anywhere along its ascending path (on right side) of the spinal cord, results in diminution or loss of pain & temperature sensation on the contralateral (left) side of the body 2 or more segments below the level of the lesion. Spinothalamic tract is a crossed tract (it carries information from the opposite side of the body).

CLINICAL CONSIDERATIONS (continued)

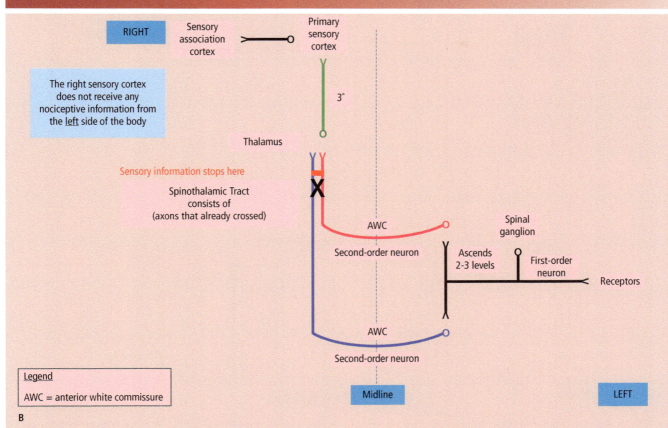

Figure 12.25 ● (Continued)

sense of his lower limbs. The individual will also have difficulty in maintaining their balance when their feet are closely approximated and when their eyes are closed (positive **Romberg sign**); during movement, the brain needs sensory information about the current position and motion of parts of the body (i.e., limbs) so that it can make adjustments to coordinate impending movement. If this feedback is missing, the brain cannot coordinate movement accordingly;

- **stereognosis** (G., stereognosis, "ability to know solids"): an individual with **astereognosis** is unable to identify the shape and form of a known object (such as a fork) following manual examination with the eyes closed, but can identify the object by sight; and
- **two-point discrimination**: the individual is unable to perceive simultaneous stimulation by a blunt instrument at two separate points on the skin as two distinct points of stimulation.

If the spinal cord lesion is at the upper cervical level, both the gracile fasciculus and cuneate fasciculus will be severed. The affected individual will have loss of all the sensory modalities listed above on the ipsilateral side of the body, *at and below* the level of the lesion, which will include loss of sensation from the upper and lower limbs. If the lesion is at the lumbar spinal cord level, sensation from the lower limb will be affected but sparing the upper limb.

Since non-discriminative (crude) touch is carried by more than one ascending sensory pathway, ascending on *both* sides of the spinal cord, some touch sensation remains intact in individuals with Brown–Séquard syndrome. Furthermore, if in a different lesion *only* the **posterior right quadrant** of the spinal cord is damaged, pain and temperature sensation will not be affected, whereas there will be a loss of discriminative (fine) touch, vibratory sense, and proprioceptive sensation, *at and below* the level of the lesion on the ipsilateral (right) side of the body.

Ipsilateral loss of autonomic control below the lesion level would be via lesioning of the descending hypothalamospinal tract. If the sympathetic trunk (adjacent to the vertebral body) was also lesioned and damaged above T1, oculosympathetic paresis (eponym: Horner Syndrome), ptosis, miosis, and anhidrosis would also occur ipsilaterally.

Lesions to the Posterior Column–Medial Lemniscus Pathway
A lesion to the posterior column–medial lemniscal (PCML) pathway below its decussation will result in loss of discriminative touch, proprioception, and flutter-vibratory sense from the ipsilateral side of the body (Fig. 12.26). In contrast, a lesion to the PCML pathway above its decussation will result in loss of the same sensory information from the contralateral side of the body (Fig. 12.27).

CLINICAL CONSIDERATIONS (*continued*)

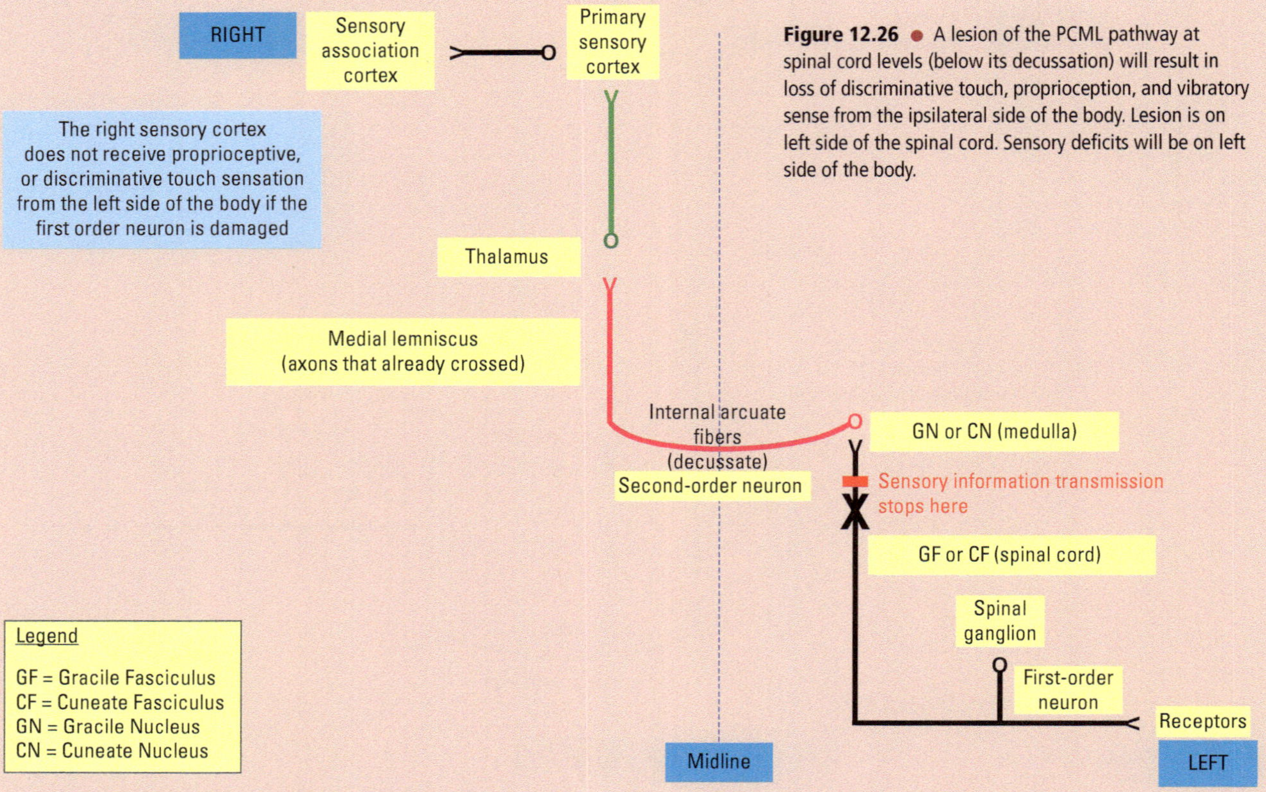

RIGHT | Sensory association cortex | Primary sensory cortex

The right sensory cortex does not receive proprioceptive, or discriminative touch sensation from the left side of the body if the first order neuron is damaged

Thalamus

Medial lemniscus (axons that already crossed)

Internal arcuate fibers (decussate)
Second-order neuron

GN or CN (medulla)

Sensory information transmission stops here

GF or CF (spinal cord)

Spinal ganglion

First-order neuron

Receptors

Figure 12.26 ● A lesion of the PCML pathway at spinal cord levels (below its decussation) will result in loss of discriminative touch, proprioception, and vibratory sense from the ipsilateral side of the body. Lesion is on left side of the spinal cord. Sensory deficits will be on left side of the body.

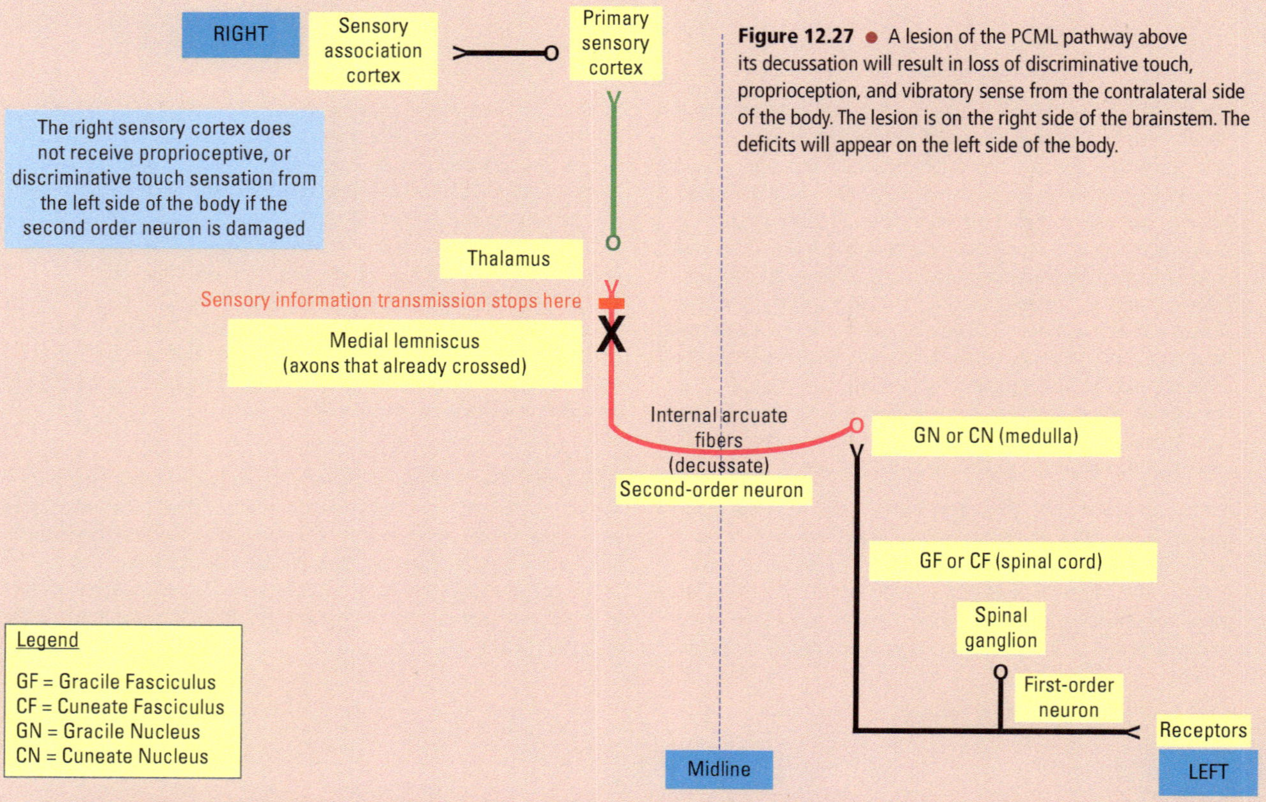

Legend

GF = Gracile Fasciculus
CF = Cuneate Fasciculus
GN = Gracile Nucleus
CN = Cuneate Nucleus

Midline

LEFT

RIGHT | Sensory association cortex | Primary sensory cortex

The right sensory cortex does not receive proprioceptive, or discriminative touch sensation from the left side of the body if the second order neuron is damaged

Thalamus

Sensory information transmission stops here

Medial lemniscus (axons that already crossed)

Internal arcuate fibers (decussate)
Second-order neuron

GN or CN (medulla)

GF or CF (spinal cord)

Spinal ganglion

First-order neuron

Receptors

Figure 12.27 ● A lesion of the PCML pathway above its decussation will result in loss of discriminative touch, proprioception, and vibratory sense from the contralateral side of the body. The lesion is on the right side of the brainstem. The deficits will appear on the left side of the body.

Legend

GF = Gracile Fasciculus
CF = Cuneate Fasciculus
GN = Gracile Nucleus
CN = Cuneate Nucleus

Midline

LEFT

CLINICAL CONSIDERATIONS (continued)

Tabes dorsalis
Tabes dorsalis, a form of tertiary neurosyphilis, is a rare condition that is manifested during the second decade after an individual becomes infected with the microorganism causing syphilis

Tabes dorsalis, a form of tertiary neurosyphilis, is a rare condition that is manifested during the second decade after an individual becomes infected with the microorganism causing syphilis. This condition is characterized by **sensory ataxia** (G., *ataxia*, "without order"), a movement disorder resulting from a sensory impairment characterized by two-point discrimination, vibratory sense, position sense, and kinesthesis. Individuals with this condition find it necessary to look at their lower limbs during walking. Affected individuals also stomp their feet to stimulate any remaining proprioceptive receptors or other receptors. Additionally, they have difficulty standing up straight if their feet are closely approximated when their eyes are closed, or if they are standing in the dark (referred to as **Romberg's sign**). Tabes dorsalis is characterized by degeneration of the spinal ganglia neurons with large-diameter, heavily myelinated fibers coursing in the medial division of the posterior root; thus, sensory information from the mechanoreceptors to the ascending sensory pathways is affected (Fig. 12.28). Additionally, these individuals experience abnormal pain sensations. Note that in contrast, **cerebellar ataxia** is a movement disorder that results from a lesion of the cerebellum or its connections (discussed in Chapter 15).

Friedreich ataxia
Friedreich's ataxia is a hereditary disorder that is manifested prior to, or during, puberty

Friedreich ataxia is a neurodegenerative hereditary disorder that is manifested prior to, or during, puberty. It is the most common inherited ataxia in Europe, the Middle East, India, and North Africa and is rarely seen in other populations. In this condition, the **spinocerebellar tracts** as well as the **posterior column pathways** degenerate and, consequently, produce an increasingly deteriorating ataxia.

Subacute combined degeneration
In subacute combined degeneration both the corticospinal tracts and the posterior column pathways undergo degeneration

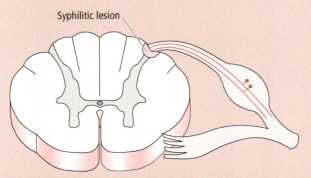

Figure 12.28 • Location of a syphilitic lesion on the thoracic spinal cord.

In **subacute combined degeneration** (also known as funicular myelosis), both the anterior and the posterior columns undergo degeneration. The deficits in the affected individual are characterized by muscle weakness as a result of the degeneration of the corticospinal (motor) tracts, and loss of discriminative (fine) touch, vibratory sense, two-point discrimination, and proprioception as a result of the degeneration of the posterior column pathways. It is caused by vitamin B_{12} deficiency, leading to systemic spongy vacuolation and myelin degeneration of the spinal cord, and is most common in the geriatric population with an incidence of 5–12%.

Syringomyelia
Syringomyelia is a disease in which the central canal of the spinal cord, usually at the lower cervical or upper thoracic spinal cord levels, becomes enlarged

Syringomyelia (G., *syrinx*, "tube") is a disorder in which the central canal of the spinal cord, usually at the lower cervical or upper thoracic spinal cord levels, becomes enlarged. It occurs at an incidence of 8 per 100,000 patients. Although the enlargement of the central canal may extend cranially and/or caudally (Fig. 12.29), the enlarging canal stretches and damages the surrounding spinal cord tissue. The tissue affected first is the anterior white commissure containing crossing fibers (from both sides), followed by the damage to the anterior horn. The deficits are *bilateral*. This results in:

1. **Loss of pain** and **temperature sensation** from the skin of both shoulders and upper extremities due to the destruction of the second-order neuron decussating fibers that relay pain and temperature input from the two sides of the body. If the lesion involves only the cervical levels of the spinal cord, the medial surface of the arm and forearm is not affected since it is innervated by T1 and T2.
2. **Weakness** and **atrophy** of the intrinsic muscles of the hands due to the degeneration of the motor neurons in the anterior horn of the spinal cord. If the disorder progresses to include additional spinal cord levels and more nerve tissue surrounding the increasingly enlarged space, additional deficits will become apparent.

The somatotopic organization of the spinothalamic tract fibers in the spinal cord, orients the fibers relaying pain, temperature, and crude touch from the lower body's lumbar and sacral dermatomes posterolaterally within the tract, whereas the fibers from the thoracic and cervical dermatomes are located anteromedially within the spinothalamic tract. An enlarging lesion in the gray matter of the spinal cord will probably affect the cervical and thoracic fibers first and the fibers from the lower body will be affected later. An unusual clinical occurrence/phenomenon has been observed in which the sensory fibers of the spinothalamic tracts relaying pain, temperature, and crude touch sensation from sacral dermatomes are unaffected due to their posterolateral location/position. This is known as **"sacral sparing."**

Vascular problems of the spinal cord
The spinal cord is supplied mainly by the single anterior spinal artery and the paired posterior spinal arteries, all branches of the vertebral arteries. The blood supply provided by these vessels to the spinal cord cannot adequately perfuse the cords neural tissue below/caudal to the cervical levels. The aorta

CLINICAL CONSIDERATIONS (*continued*)

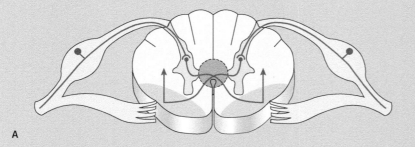

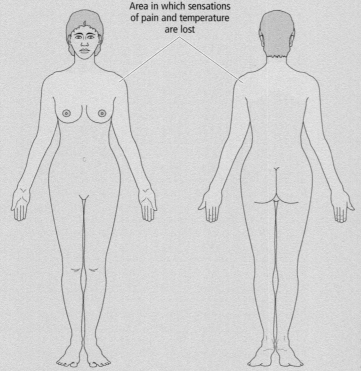

Figure 12.29 ● Syringomyelia. (A) Damage of decussating fibers of the pain and temperature pathway. (B) Skin area in which there is loss of pain and temperature sensation following the development of syringomyelia.

gives rise to multiple posterior intercostal arteries which in turn each give rise to a posterior ramus branch. The posterior ramus branch provides an anterior and a posterior radicular artery which supply the anterior and posterior spinal nerve roots, respectively. The radicular arteries anastomose with the anterior and posterior spinal arteries and thus contribute to and enhance the blood supply to the spinal cord. The anterior spinal artery supplies the anterior 2/3 of the spinal cord, whereas the posterior spinal arteries supply the posterior 1/3 of the spinal cord. Below the cervical level (in the thoracic, lumbar, and sacral regions), the radicular arteries anastomose with the anterior and posterior spinal arteries. In contrast, the cervical region of the spinal cord is supplied only by the anterior and posterior spinal arteries. Thus, without the contribution of the radicular arteries' anastomoses, the spinal cord is most susceptible (to damage from interruption of blood flow in the anterior or posterior spinal arteries) at cervical levels. So, if the anterior or posterior spinal arteries become occluded or compressed in the cervical region, there is no backup from the radicular anastomoses, and the spinal cord neural tissue will be damaged.

Anterior spinal artery syndrome

Occlusion of the anterior spinal artery, either by a thrombus or by compression, will obstruct blood flow to the anterior two-thirds of the spinal cord and result in infarction

Anterior spinal artery syndrome (also known as Anterior Cord Syndrome) is an emergent condition with a 20% mortality rate and a high morbidity rate with half of patients having no resolution of symptoms. Occlusion of the anterior

ASCENDING SENSORY PATHWAYS   225

CLINICAL CONSIDERATIONS (continued)

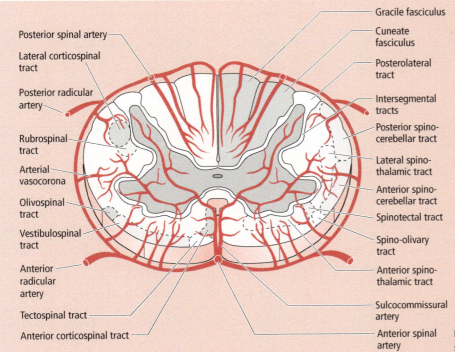

Figure 12.30 • Cross-section of the cervical spinal cord and its arterial supply.

spinal artery, either by a thrombus or compression by a tumor, will obstruct blood flow to the anterior two-thirds of the spinal cord, resulting in infarction (Fig. 12.30). If the anterior spinal artery is occluded or compressed near the artery's origin in the medulla oblongata, structures located anteriorly such as the pyramids and medial lemnisci, will be affected. Since the anterior spinal artery is located in the midline, both pyramids and medial lemnisci will most likely be affected, producing sensory and motor deficits on both sides of the body. Damage of the corticospinal fibers in the pyramids will result in upper motor neuron signs, including bilateral hemiparesis. Damage of the medial lemnisci will result in bilateral loss of discriminative touch, proprioception, and flutter-vibratory sense from the body.

If occlusion or compression of the anterior spinal artery occurs at spinal levels, it will result in damage of the anterior two-thirds of the spinal cord. There will be a bilateral loss of pain and temperature sensation beginning a few levels below the lesion due to damage of the spinothalamic tracts. Damage of the corticospinal tract fibers and the anterior horns of the spinal cord containing lower motor neurons will result in bilateral motor deficits at and below the level of the lesion. Damage between C3 and C5 will compromise breathing (phrenic nerve innervation of the diaphragm) contributing to the high mortality rate of this condition. Damage to the descending pathways controlling sphincter function will cause incontinence.

Since the posterior column pathways (located in the posterior cord) are not affected by occlusion of the anterior spinal artery at spinal levels, discriminatory (fine) touch, vibratory sense, and position sense remain intact.

Posterior spinal artery occlusion or compression
The two posterior spinal arteries supply the posterior one-third of the spinal cord; if either is occluded or compressed, it results in infarction of the posterior funiculi

The two posterior spinal arteries supply the posterior one-third of the spinal cord (Fig. 12.30). If one or both of the posterior spinal arteries is/are occluded or compressed, it results in infarction of one or both of the posterior white columns. If only one of the posterior spinal arteries is occluded, or compressed, it will result in loss of discriminatory (fine) touch and proprioceptive sensation, ipsilaterally. If infarction occurs at caudal medullary levels, near the level of the vessels' origin, the posterior white columns, and their respective nuclei, all processing sensory information from the ipsilateral side of the body will be damaged. Since the spinothalamic tract is located in the anterolateral aspect of the caudal medulla, it is not affected. Posterior spinal artery ischemia is incredibly rare, accounting for 1% of all neurovascular pathologies.

Central or thalamic pain
A lesion of the spinothalamic tract and its nucleus of termination may cause agonizing pain or other unusual sensations

Lesions involving the spinothalamic tract and its nucleus of termination, the VPL nucleus of the thalamus, may initially cause diminished or complete

CLINICAL CONSIDERATIONS (*continued*)

loss of touch, pressure, pain, and temperature sensation, and proprioception from the contralateral side of the body. Spontaneous, inexplicable, agonizing pain and other unusual sensations in the anesthetic parts may follow. This condition is known as central pain (thalamic pain, thalamic syndrome) and is experienced by more than 50% of medullary or thalamic stroke survivors in the first year post-stroke.

Lesions involving the somatosensory cortex

Isolated lesions in the **postcentral gyrus** are uncommon. However, since the postcentral gyrus is supplied by branches of the anterior and also the middle cerebral artery, a vessel that often becomes occluded, this region of the brain may become infarcted. A lesion to the primary somatosensory cortex will result in contralateral loss of:

1 **Two-point discrimination**.
2 **Graphesthesia**, the ability to recognize letters or numbers as they are stroked on the skin.
3 **Stereognosis**, the ability to identify a known object following tactile examination without looking.
4 **Vibratory sense**.
5 **Position sense**.

Furthermore, although the individual has a minimal impairment of **pain, temperature**, and **touch sensation**, they are unable to *localize* the stimulus. Since pain perception is not only processed in the **somatosensory cortex**, but also the **anterior cingulate** and **anterior insular cortices**, pain sensation persists following a lesion to the somatosensory cortex as a result of these additional cortical representations of pain.

In recent years, a bilateral **cingulotomy** (transection of the anterior part of the cingulum bundle) has served as an effective treatment in relieving the emotional, agonizing reaction to pain. An isolated lesion in the **secondary somatosensory cortex** (SII) results in minimal sensory loss, but since SII has an important function in memory of somatosensory information and sensory integration, these functions are impaired.

Referred pain

Pain originating in a visceral structure may be referred to and felt in a somatic structure

The actual origin of visceral pain is imprecisely localized. Although pain may originate deep within a visceral structure, such as the heart, the pain may be "referred to," and felt, in another, distant somatic structure such as the left upper limb. Although several explanations have been proposed for this phenomenon, the following two have the most prominence in the field of neuroscience.

Convergence–projection theory of visceral pain

The **convergence–projection theory** of visceral pain suggests that the central processes of pseudounipolar sensory, GVA neurons supplying visceral structures and the central processes of GSA neurons from a somatic structure, such as the upper limb, enter and terminate at the *same spinal cord level*. Here they converge on and synapse with the same interneurons and/or second-order neurons (viscerosomatic neurons) of the ascending pain pathways in the posterior horn and the intermediate gray matter. Nociceptive information is transmitted by these GSA pathways to higher brain centers.

Concept of referred pain

Second-order GSA projection neurons are continuously being activated by GVA first-order neurons, thereby lowering the threshold of stimulation of the second-order neurons. Consequently, nociceptive sensory information is relayed by the neurons of the GSA pathway to higher brain centers. Thus, GVA nociceptive input is transmitted via the spinoreticular fibers to the reticular formation, the thalamus, and the hypothalamus. The nociceptive signal is subsequently relayed to the region of the somatosensory cortex that normally receives somatic information from other areas, such as the upper limb, and the brain interprets it as if the pain were coming from that somatic structure (upper limb). Therefore, it is the area(s) of the cerebral cortex, in this case the somatosensory cortex, wherein the signals *terminate*, and not the stimulus, the receptor, or the information, that establishes the localization of the sensation.

Phantom limb pain

Individuals who have had a limb amputated may experience pain or tingling sensations that feel as if they were coming from the amputated limb, just as if that limb were still present

A curious phenomenon has been reported by 50–85% of individuals who have had a limb amputated. These individuals experience pain or tingling sensations that feel as if they were coming from the amputated limb, just as if that limb were still present. Although the mechanism of phantom limb pain is not understood, the following two possible explanations are offered.

If a sensory pathway is activated anywhere along its course, nerve impulses are generated that travel to the CNS where they initiate neural activity. This neural activity ultimately "creates" sensations that feel as though they originated in the non-existent limb.

Another possibility is that since there is no touch, pressure, or proprioceptive information transmitted to the CNS from the peripheral processes of the sensory neurons that initially innervated the amputated limb, there are no impulses from touch fibers to attenuate the relaying of nociceptive impulses to the nociceptive pathways, enhancing nociceptive transmission and pain sensation (see the gate control theory of pain, later). Since nociception is not as localized, cortical areas corresponding to the phantom limb will be activated.

MODULATION OF NOCICEPTION

The CNS can prevent and/or suppress the flow of some of the incoming nociceptive signals from peripheral structures

Although the CNS is constantly flooded with sensory information, it can prevent and/or suppress the flow of some of the incoming nociceptive signals at the local circuitry level of the spinal cord posterior horn (and spinal trigeminal nucleus of the brainstem). The CNS can also do this at the level of the descending opioid and nonopioid analgesia-producing pathways that originate in the brainstem and terminate at the relay sensory nuclei of the ascending sensory systems.

Gate control theory of pain

Nociceptive signals from the periphery are filtered by modulation in the substantia gelatinosa of the posterior horn

Rubbing a painful area (activation of touch Aδ/Aβ fibers) reduces the sensation of pain. It has been proposed that pain is filtered by modulation of the sensory nociceptive input to the spinothalamic pain and temperature pathway neurons at the **substantia gelatinosa** (**lamina II**) of the posterior horn gray matter. The theory proposed to explain this phenomenon, known as the **gate control theory of pain**, suggests that the neural circuitry of the substantia gelatinosa (SG) functions as follows (Fig. 12.31).

The activity of nociceptive unmyelinated C fibers and thinly myelinated Aδ fibers terminating in the SG transmits nociceptive impulses by: (1) inhibiting the SG inhibitory interneuron; and (2) simultaneously activating the second-order spinothalamic tract neuron that projects to the thalamus, "keeping the gate open."

The touch Aδ/Aβ myelinated fibers activate the inhibitory interneuron as well as the second-order neuron. However, the inhibitory interneuron, via presynaptic inhibition of the C/Aδ and Aδ/Aβ fibers, prevents impulses from reaching the second-order spinothalamic neuron. The rubbing of a painful region probably stimulates the Aδ/Aβ fibers, which in turn – via the inhibitory interneuron – inhibit the transmission of some of the nociceptive impulses to higher brain centers, providing some relief from pain.

Descending analgesia-producing pathways

Descending analgesia-producing pathways arise from the brainstem and terminate in the spinal cord

Although it has been known for decades that opiates, a group of drugs derived from opium (e.g., morphine), provide powerful relief from pain, their mode of action was not understood. It has been demonstrated that opiates bind to "opiate receptors" on specific nerve cells residing in certain areas of the CNS. The discovery that the nervous system has "opiate receptors," suggested that it must synthesize its own endogenous, opiate-like substances whose binding to these receptors probably modulate afferent nociceptive transmission. Three groups of related endogenous opioid peptides have been identified: **enkephalins** (G. *enkephalin*, "in the head"), **beta-endorphins** ("morphine within"), and **dynorphins** (dynamo + morphine) all of which are known to bind to the same receptors as the opioid drugs.

During a stressful or emotional experience, regions associated with the processing of emotions – namely the telencephalon (frontal cortex), the diencephalon (hypothalamus), and the limbic system (amygdala, insular cortex, anterior cingulate cortex) – project to and stimulate the enkephalin-releasing neurons of the **periaqueductal gray matter** and other nearby regions of the midbrain. The axons of these enkephalin-releasing neurons form excitatory synapses primarily with the serotonin-releasing neurons of the **raphe magnus nucleus** and the **gigantocellular reticular nucleus** of the rostral **medullary reticular formation**. Since these neurons release serotonin, they are said to form the **serotonergic-opioid peptide analgesic system**, which modulates nociception. In addition to the serotonergic-opioid peptide analgesic system, there is a **norepinephrine** (adrenergic, non-opioid) **analgesic system** that arises from the **dorsolateral pontine reticular formation** and terminates in the SG of the spinal cord, and also functions in modulating nociception (see discussion next).

Serotonin-releasing neurons

The axons of the **serotonin-releasing neurons** from the raphe magnus nucleus and the gigantocellular reticular nucleus descend bilaterally in the lateral funiculus of all spinal cord levels to terminate in the SG of the posterior horn. Here, they form excitatory synapses with the inhibitory interneurons, which release the opioid peptides, enkephalin, or dynorphin. These interneurons establish axoaxonic synapses with the central processes of the Aδ and C first-order nociceptive neurons.

The modulation of nociception occurs in the following fashion. Upon stimulation of the peripheral free nerve ending of a pain fiber, the central process of this first-order neuron releases substance P, a neurotransmitter believed to function in the transmission of nociceptive information, which excites the second-order nociceptive relay neurons in the posterior horn of the spinal cord. If the serotonin-releasing neurons excite the inhibitory interneurons of the SG, these interneurons in turn inhibit the central processes of the first-order neurons. Since this inhibition occurs before the impulse reaches the synapse, it is referred to as **presynaptic inhibition** (Fig. 12.32). Thus, these nociceptive incoming impulses are filtered by being suppressed at their first relay

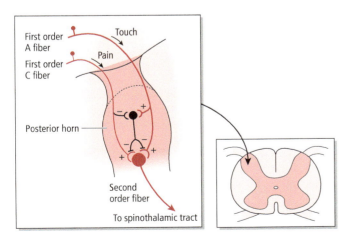

Figure 12.31 • Gate control theory of pain. Stimulation of the C fibers transmitting nociception keeps the gate to higher brain centers open, whereas stimulation of the large-diameter A fiber closes the gate. Source: Modified from Heimart, L (1995) *The Human Brain*. Springer-Verlag, New York, fig. 9.2.

station in the spinal cord, by the SG inhibitory interneurons releasing enkephalin or dynorphin. This inhibition is possible because the central processes of these first-order neurons possess receptors for enkephalin and dynorphin (opioid receptors) in their axolemma.

There is additional evidence suggesting that there are indeed opioid receptors in the axolemma of the central terminals of the first-order nociceptive, substance P-releasing neurons terminating in the SG. This evidence is gathered from reports that naloxone, an opioid antagonist, selectively prevents the blocking of substance P release by the central processes of the first-order neurons.

Norepinephrine-releasing neurons

Another brainstem region, the **dorsolateral pontine reticular formation** (Fig. 12.33), is the site of origin of a norepinephrine (adrenergic, nonopioid) analgesic pathway that descends in the posterolateral funiculus of the spinal cord to synapse in the SG (lamina II). Fibers of this pathway make synaptic contact with inhibitory interneurons of the SG (that may release gamma-aminobutyric acid [GABA]), thus ultimately leading to the inhibition of the second-order projection neurons that reside in lamina V of the posterior horn of the nociceptive pathway. Unlike the descending opioid analgesic pathways, the effects of the adrenergic (norepinephrine-releasing) pathway are not blocked by naloxone.

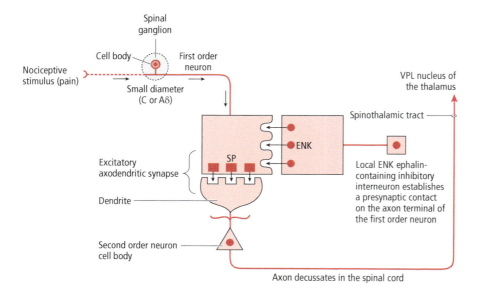

Figure 12.32 ● Opiate-induced suppression of substance P (SP) release in the substantia gelatinosa of the spinal cord posterior horn. Note the substance P release by the central process of the first-order neuron where it synapses with a second-order projection neuron of the spinothalamic tract. The local enkephalin-releasing inhibitory interneuron establishes a presynaptic contact on the axon terminal of the first-order neuron inhibiting the release of substance P. ENK, enkephalin; VPL, ventral posterior lateral.

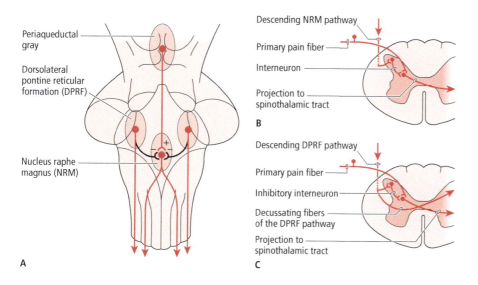

Figure 12.33 ● The descending analgesic pathways. (A) The midbrain periaqueductal gray matter contains enkephalinergic neurons whose axons descend to terminate in the raphe magnus nucleus containing serotoninergic neurons. These neurons in turn descend bilaterally to terminate in the posterior horn of the spinal cord. (B) Descending serotoninergic fibers arising from the raphe magnus nucleus terminate on enkephalinergic and dynorphin-containing interneurons in the substantia gelatinosa of the spinal cord. The interneurons inhibit transmission of nociception via presynaptic inhibition of the first-order afferent neurons. (C) Descending adrenergic fibers arising from the dorsolateral pontine reticular formation (DPRF) terminate bilaterally on inhibitory interneurons of the substantia gelatinosa of the spinal cord. The interneurons inhibit the second-order projection neurons of the spinothalamic tract. Source: Adapted from Burt, AM (1993) *Textbook of Neuroanatomy*. WB Saunders, Philadelphia, fig. 10.25.

NEUROPLASTICITY

Plastic alterations in neural structures may be induced by a noxious insult that may enhance the magnitude of nociception

Classically, the nociceptive system has been believed to be a sensory system that transmits pain signals from peripheral structures to the spinal cord, brainstem, and higher brain centers. Afferent nociceptive input may be modulated at the posterior horn level, filtering some of the pain signals, preventing them from being transmitted to higher brain centers.

It has been shown that not only modulation but also "plastic" alterations may be induced in peripheral nerve terminals, the spinal cord, and the brain in response to a noxious insult. These modifications may enhance the magnitude of nociception and may be a factor in the development of pain that may last only days, or may become persistent and last for months, years, or for the remainder of the individual's life.

SYNONYMS AND EPONYMS OF THE ASCENDING SENSORY PATHWAYS

Name of structure or term	Synonym(s)/eponym(s)	Name of structure or term	Synonym(s)/eponym(s)
Astereognosis	Stereoanesthesia	Lamina VII of the posterior horn of the spinal cord	Posterior thoracic nucleus
Bulbous Corpuscle	Ruffini's end organs Ruffini endings		Clarke column
Cuneocerebellar tract	Cuneatocerebellar tract		
Discriminative tactile sensation	Discriminative touch sensation	Tactile corpuscle	Merkel corpuscle
	Fine touch sensation	Muscle spindle	Neuromuscular spindle
Posterior (dorsal) column nuclei	Nucleus gracilis (NG) and nucleus cuneatus (NC)	Non-discriminative touch sensation	Crude touch sensation
	Gracile nucleus (GN) Cuneate nucleus (CN)	Postcentral gyrus	Primary sensory cortex (SI)
Posterolateral fasciculus	Tract of Lissauer		Primary somatosensory cortex
	Lissauer's tract		Primary somesthetic cortex
			Primary somatic sensory cortex
Dynamic proprioception	Kinesthetic sense		Brodmann's areas 3, 1, and 2
External cuneate nucleus	Accessory cuneate nucleus Lateral cuneate nucleus	Projection neuron	Second-order neuron
Extrafusal muscle fibers	Skeletal muscle fibers (of gross muscle)	Pseudounipolar (unipolar) neuron	Primary sensory neuron
			First-order neuron
Functional component	Functional modality		Afferent neuron
Tendon organ (Golgi tendon organ)	Neurotendinous spindle	Rapidly adapting receptors	Phasic receptors
Hairless skin	Glabrous skin	Secondary somatosensory cortex (SII)	Secondary somesthetic cortex
Interneuron	Internuncial neuron		Brodmann area 43
Intrafusal muscle fibers	Skeletal muscle fibers of muscle spindles	Slowly adapting receptors	Tonic receptors
		Spinothalamic tract	Neospinothalamic pathway
Lamina I of the posterior horn of the spinal cord	Posteromarginal nucleus or zone		Direct pathway of the anterolateral system (ALS)
Lamina II of the posterior horn of the spinal cord	Substantia gelatinosa (SG)	Static proprioception	Static position sense
		Thalamic pain	Thalamic syndrome
Lamina V of the posterior horn of the spinal cord	Reticular nucleus	Thermal receptor	Temperature receptor

FOLLOW-UP TO CLINICAL CASE

This patient has a primarily sensory disturbance that involves the distal extremities bilaterally. Severe sensory disturbance, even in the absence of motor deficits or cerebellar abnormality, often leads to gait ataxia or incoordination of the hands depending on whether the feet and/or hands are involved. Therefore, the cerebellum is not necessarily involved, but it may be a good idea to check it with imaging.

The first thing to determine in this case is whether the sensory dysfunction is from pathology in the CNS or the peripheral nervous system (PNS) (e.g., peripheral neuropathy). The PNS can be evaluated by nerve conductions, which reveal how well an electric impulse is conducted by a peripheral nerve. The CNS and the nerve roots can be evaluated by radiologic imaging, particularly MRI. In the present case, nerve conductions were essentially normal. Other tests indicate normal nerve roots. These tests indicate that the CNS should be evaluated closely. The brisk reflexes in the legs, particularly at the talocrural joints (ankles), also indicate peripheral neuropathy is unlikely. MRI of the brain and spinal cord is unrevealing.

Laboratory tests to check for a metabolic, or possibly a genetic cause of pathology to the sensory pathways in the CNS are indicated.

Laboratory tests indicate that this patient has subacute combined degeneration, secondary to vitamin B_{12} deficiency. Vitamin B_{12} is a cofactor in enzymatic reactions that are critical for DNA synthesis and neurologic function. Deficiency leads to degeneration of white matter in general, but the posterior columns in the spinal cord tend to be affected early and prominently. The reason for this predilection is not clear. Motor fibers of the corticospinal tract in the spinal cord are also affected relatively early and can lead to bilateral leg weakness. Dementia from degeneration in the brain and visual disturbance from optic nerve involvement may also occur.

Pernicious anemia is an autoimmune disease of gastric parietal cells that ultimately leads to decreased absorption of vitamin B_{12} from the small intestine. Treatment with vitamin B_{12} supplementation is very effective if the disorder is caught early.

5. Is there a peripheral nervous system neuropathy?

6. If this patient's disorder remains undiagnosed and untreated, what other systems are likely to be affected?

QUESTIONS TO PONDER

1. Why is it that an individual responds to a fly walking on the wrist of the right hand, but not to the watch on the left wrist?

2. How do muscle spindles detect a change in muscle length (resulting from stretch or contraction) irrespective of muscle length at the onset of muscle activity?

3. In a simple stretch reflex, what indicates the load placed on a muscle?

4. What is the response of the muscle spindles and tendon organs during slight stretching and then during further stretching of a muscle?

5. Why does an individual with a lesion in the primary somatosensory cortex still perceive pain?

6. Why is it that when an individual is experiencing a heart attack they feel pain not only in the chest area, but also in the left shoulder and upper limb?

CHAPTER 13

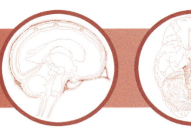
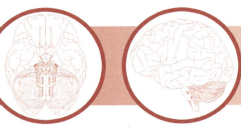

Motor Cortex and Descending Motor Pathways

CLINICAL CASE

CORTICAL AREAS CONTROLLING MOTOR ACTIVITY

DESCENDING MOTOR PATHWAYS

CLINICAL CONSIDERATIONS

SYNONYMS AND EPONYMS OF THE MOTOR CORTEX AND DESCENDING MOTOR PATHWAYS

FOLLOW-UP TO CLINICAL CASE

QUESTIONS TO PONDER

CLINICAL CASE

A 45-year-old patient with a left foot drop, which started 5 months ago. This was very subtle at first but has been getting progressively worse. The patient can now barely lift their left foot up and walks with difficulty. More recently they have noticed some clumsiness of their left hand. Upon further questioning they admit to frequent "twitching" of the muscles in their distal lower limbs (particularly in the leg), forearms, and shoulders. The patient's family thinks they have lost some weight. There is no numbness, tingling, pain, or any other neurologic symptom noted. Family history is negative for any neurologic disorder.

Examination shows severe weakness and flaccidity of left ankle dorsiflexion and mild weakness of the intrinsic left hand muscles and left finger extension. Otherwise, strength seems adequate. There is atrophy noted in the muscles of the left hand and the left dorsiflexor muscles. Fasciculations are noted in several muscle groups of the legs, arms, and back. Deep tendon reflexes are pathologically brisk in the arms and legs bilaterally. An extensor plantar (Babinski) response is present bilaterally. The rest of the neurologic exam was unremarkable.

The desire to move originates somewhere in the **association cortex** in response to something that one sees, hears, smells, tastes, feels, or considers. The association cortex relays this information to the **premotor cortical areas** where impending voluntary movement is *planned* and a design of the plan is generated. Signals of the plan are relayed to the **primary motor cortex** where *execution* of the voluntary movement is initiated. The primary motor cortex and premotor cortical areas house the cell bodies of **upper motor neurons** whose axons descend, along with axons from the **primary somatosensory cortex**, as the **corticonuclear tract** to the brainstem and upper cervical spinal cord, as well as the **corticospinal tract** to the entire length of the spinal cord, sending signals about **voluntary movement** of the head and body, respectively. The upper motor neurons influence lower motor neurons located in the cranial nerve motor nuclei (except the oculomotor, trochlear, and abducens nuclei) and in the anterior horn of the spinal cord. The axons of the lower motor

---

A Textbook of Neuroanatomy, Third Edition. Maria A. Patestas, Amanda J. Meyer, and Leslie P. Gartner.
© 2025 John Wiley & Sons, Inc. Published 2025 by John Wiley & Sons, Inc.
Companion website: www.wiley.com/go/patestas/neuroanatomy3e

neurons form bundles (peripheral nerves, i.e., cranial nerves and spinal nerves) that stimulate skeletal muscle cells and cause them to contract to produce the planned movement.

Motor activity is controlled by intricate interactions of three major regions of the brain: the cerebral cortex, the basal nuclei (basal ganglia), and the cerebellum. The **cerebral cortex**, the ultimate command center of the nervous system, is involved in the planning and execution of complex voluntary motor activities. The **basal nuclei** function in the initiation of movement and modulation of the motor cortex. The **cerebellum** receives information from the cerebral cortex, as well as the visual, auditory, vestibular, and somatosensory systems, which it integrates and utilizes to plan, modify, and coordinate movement. This information enables the cerebellum to play a role in the timing, speed, direction, and precision of every motor activity. The motor cortex relays information to the cerebellum about impending movement. This permits the cerebellum to compare the movement in progress with the movement about to occur. The cerebellum utilizes all of these data to adjust the output of the motor cortex, so that movement is smooth and coordinated. The basal nuclei and the cerebellum exert their influence on the brainstem and spinal cord and, ultimately, on motor activity (at a subconscious level). They do this **indirectly** by regulating the output of the motor and premotor cortex, via the thalamus (basal nuclei and cerebellum → thalamus → motor cortex → brainstem and spinal cord → muscular system → voluntary movement). In contrast, the **primary motor cortex**, and **premotor cortical areas (premotor cortex, supplementary motor cortex/area, frontal eye field,** and **cingulate motor areas)** influence voluntary motor activity via **direct** projections to the brainstem cranial nerve motor nuclei, the reticular formation, and spinal cord, which in turn send commands to the muscular system.

CORTICAL AREAS CONTROLLING MOTOR ACTIVITY

Motor activity is controlled by the primary motor cortex, and premotor cortical areas (premotor cortex, supplementary motor cortex, frontal eye field, and cingulate motor areas)

Motor activity is controlled by projections from the cerebral cortex: the primary motor cortex (MI) and the premotor cortical areas (premotor cortex, supplementary motor cortex, frontal eye field, and cingulate motor areas) of the frontal lobe to the brainstem and spinal cord (Fig. 13.1). These cortical areas are the main origin of descending motor pathways to the

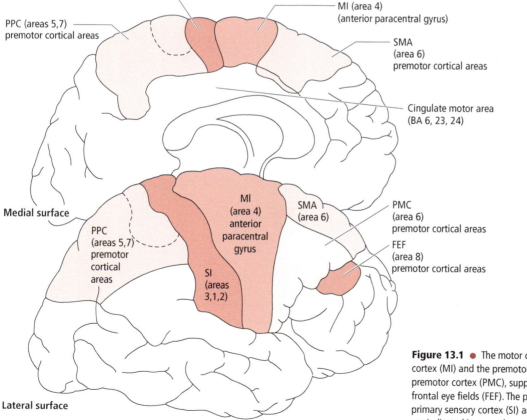

Figure 13.1 ● The motor cortical areas: the primary motor cortex (MI) and the premotor cortical areas consisting of the premotor cortex (PMC), supplementary motor area (SMA), and frontal eye fields (FEF). The posterior parietal cortex (PPC), primary sensory cortex (SI) are also shown. Brodmann areas are indicated in parentheses.

MOTOR CORTEX AND DESCENDING MOTOR PATHWAYS ● ● ● 233

brainstem and spinal cord, although the brainstem also gives rise to descending motor tracts that reach the spinal cord.

Primary motor cortex

The primary motor cortex functions in the execution of distinct, well-defined, voluntary motor activity of the contralateral side of the body

The **primary motor cortex** (Brodmann area 4) resides in the **precentral gyrus** on the lateral surface of the frontal lobe and in the **anterior paracentral gyrus** on the medial surface of the frontal lobe (Fig. 13.1). It has an important function in the execution of distinct, well-defined, voluntary motor activity. The precentral gyri control movements of the contralateral side of the body. In the 1930s Penfield and colleagues mapped the human body on the primary motor cortex somatotopically, as an upside down continuous homunculus (L., "little person") from head to toe (Fig. 13.2a) by applying microstimulations during brain surgery on patients with epilepsy. Although the original classical homunculus was not accurate, it did provide a simple somatotopic map of the body that has been included in neuroanatomy textbooks for many decades. More recent studies using functional magnetic resonance imaging (fMRI) confirmed some, but not all, of the original results. A striking characteristic of the primary motor cortex in humans is that over half of it is associated with the motor activity of the hands, tongue, lips, and larynx, reflecting the manual dexterity and ability for speech that humans possess. The primary motor cortex also influences the motor control of the axial and girdle musculature, and especially the control of the distal muscles controlling movements of the hands and feet. Stimulation of the primary motor cortex produces movement of the body part controlled by the stimulated cortical area.

Some of these fMRI studies have provided functional maps of alterations in the oxygen concentration in cerebral vasculature which accompanies neural activity in the brain. The most recently (2023) published somatotopic map shows not only the proposed motor cortical areas that control movement of specific body parts, but also incorporates three newly detected cortical areas (referred to as intereffector regions) that control integrative, whole-body movements (Fig. 13.2b). The intereffector regions are interposed between the motor cortical sections that control movement of specific body parts and appear to be associated with an external network involved in motor activity as well as pain perception. The new information indicates that the primary motor cortex possesses three separate sections, each representing a different region of the body: lower part of the body, trunk and upper limbs, and head. In each of these motor cortical sections, the distal part of the body region is mapped in the center of the section of the primary motor cortex. Thus, for the section of the primary motor cortex that controls movement of the upper limb, the most distal part, the fingers are mapped in the middle. The parts of the upper limb are arranged as follows: fingers are in the middle and the remaining parts of the upper limb diverge in two directions as follows: shoulder – elbow – wrist – hand – **fingers** – hand – wrist – elbow – shoulder. Thus, the

intereffector regions may control integrative, whole-body movements engaging various body parts, whereas the three separate sections of the primary motor cortex are involved in producing distinct movements of specific body parts.

Neurons in the primary motor cortex are organized into groups; each group sends its axons to one of three regions: the cranial nerve motor nuclei, the reticular formation in the brainstem, or the spinal cord gray matter, where they control the motor activity of a single muscle. The total cortical area that mediates motor activity of a particular body region is proportional to the complexity of the motor activity produced in that body region.

Premotor Cortical Areas

The premotor cortical areas function in the planning of complex motor activity, which is then relayed to the primary motor cortex where the execution of motor activity is initiated

The **premotor cortical areas** reside in the frontal lobe and consist of four regions: the premotor cortex, supplementary motor cortex, frontal eye field, and the cingulate motor areas. In addition to the motor cortical areas located in the frontal lobe, there is an additional cortical area, located in the parietal lobe, the **posterior parietal cortex (PPC)** which is involved in movement (Fig. 13.1). The principal function of the **premotor cortical areas** is the *planning* of complex motor activity, which is then relayed to the primary motor cortex, where the *execution* of that motor activity is implemented. The primary motor cortex then conveys this input mainly to the brainstem or to the spinal cord. Thus, the majority of nerve signals that are generated in the premotor cortical areas cause complex movements produced by groups of muscles performing a task, unlike the discrete muscle contractions elicited by stimulation of the primary motor cortex. Although the premotor cortex and supplementary motor cortex are somatotopically organized, their organization is less precise than that of the primary motor cortex. Stimulation of the premotor cortical areas seldom generates a movement, but instead, interrupts the movement in progress as well as any impending movements by interfering with its planning phase.

The **premotor cortex (PMC)** occupies most of Brodmann area 6, on the lateral aspect of the frontal lobe. It is the recipient of projections from sensory areas of the parietal lobe and, in turn, it projects to the primary motor cortex, the reticular formation (which is the origin of the reticulospinal tracts), and to the spinal cord. The principal function of the PMC is motor control of the axial (paravertebral), and proximal limb (girdle) musculature to effect postural adjustments as background support for impending movements. It also functions in guiding or turning the body and the upper limbs toward the desired direction. Once an intended movement has begun and is in progress, activity in the PMC decreases, reflecting its key function in the planning phase of motor activity.

The **supplementary motor area (cortex) (SMA)** lies in Brodmann area 6. This area has important functions in the programing phase of the patterns and sequences of elaborate

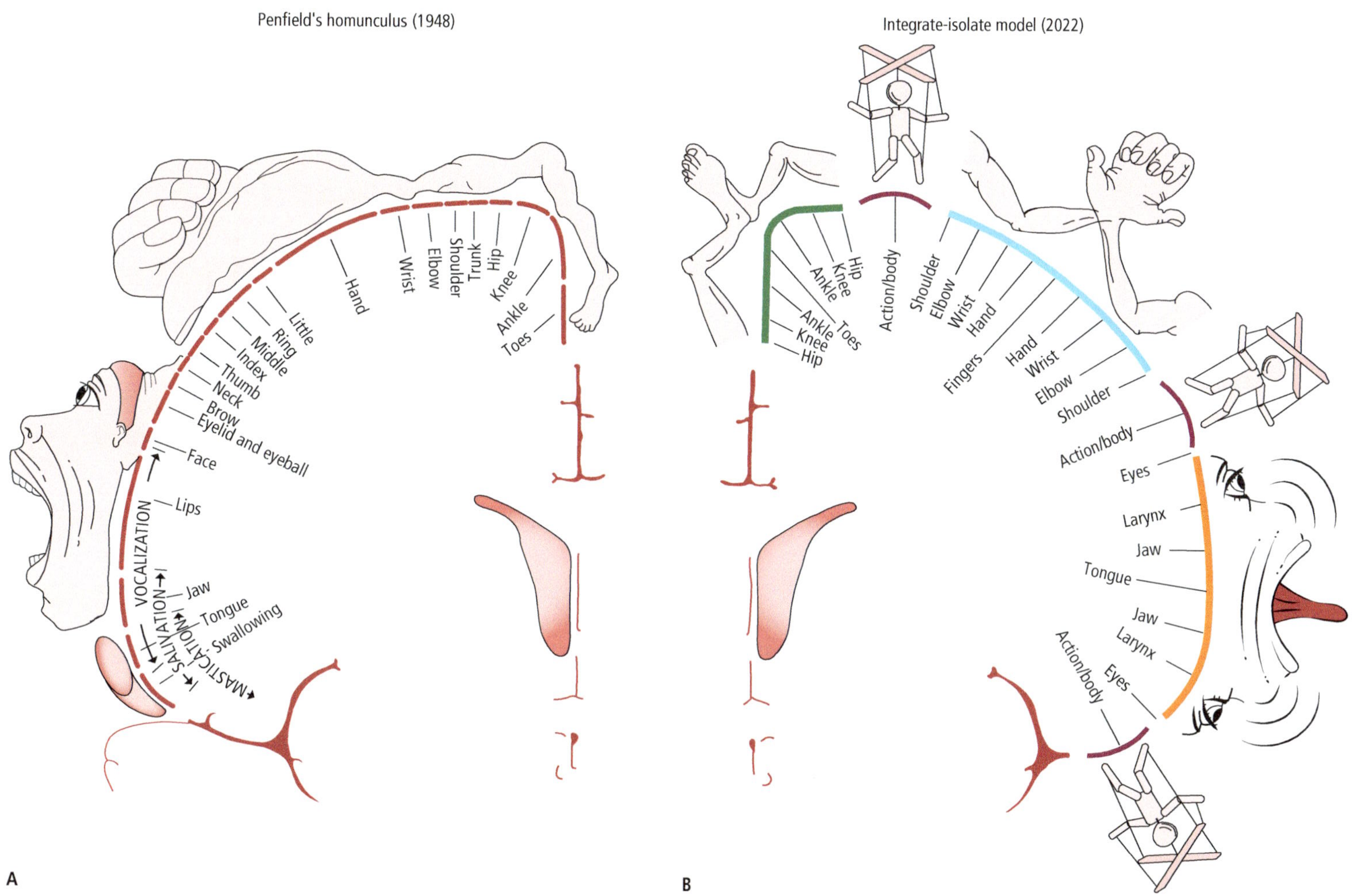

Figure 13.2 ● (A) Coronal section of the brain at the level of the primary motor cortex showing Penfield's classical homunculus. Penfield and colleagues mapped the human body on the primary motor cortex somatotopically, as an upside down continuous homunculus (L., "little person") from head to toe (Fig. 13.2a). (B) The most recently (2023) published somatotopic map shows not only the proposed motor cortical areas that control movement of specific body parts, but also incorporates three newly detected cortical areas (referred to as intereffector regions) that control integrative, whole-body movements (Fig. 13.2b). The new information indicates that the primary motor cortex possesses three separate sections, each representing a different region of the body: lower part of the body, trunk and upper limbs, and head. Source: Gordon et al. (2023)/Springer Nature/CC BY 4.0.

movements, produced by groups of muscles and coordination of movements occurring on the two sides of the body. This cortical area is associated with muscle contractions of the axial (trunk) and proximal limb (girdle) musculature (i.e., muscles controlling movement of the arms and thighs). It also functions in guiding or turning the body or limbs toward the desired (or appropriate) direction.

The **frontal eye field** (**FEF**) occupies Brodmann area 8. This region is located anterior to the premotor cortex on the frontal lobe and functions in the coordination of eye movements, particularly movements mediating voluntary visual tracking of a moving object.

The **cingulate motor cortex** which corresponds to Brodmann areas 6, 23, and 24, is located in the superior and inferior banks of the cingulate sulcus. Similar to the other premotor cortical areas it also projects to the primary motor cortex. Since the cingulate gyrus is part of the limbic system, this motor cortical area may be associated with the motivational or emotional aspects of movement.

The **posterior parietal cortex** resides in the **superior parietal lobule** which corresponds to Brodmann areas 5 and 7. Area 5 receives input from the somatosensory cortex and is involved in tactile discrimination (the ability to perceive a subtle distinction by the sense of touch) and stereognosis (the recognition of the three-dimensional shape of an object by the sense of touch). Area 7 is involved with movements that require visual guidance. Thus, when one reaches for a glass of cold water, both visual guidance (turning the body and aiming the upper limb in the direction of the glass) and tactile sensation (which in this case helps to realize that the glass is slippery and must be grasped firmly) play a role in accomplishing a desired motor task by their projections to the supplementary and premotor cortices.

DESCENDING MOTOR PATHWAYS

The internal pyramidal layer (layer V) of the motor cortex houses the cell bodies of the pyramidal neurons, constituting the main output neurons of the cortical descending motor pathways

The cerebral cortex consists of six histologically distinct layers (Fig. 13.3). The **internal pyramidal layer** (layer V), the most conspicuous layer of the motor cortex, houses the cell bodies of pyramidal cells. These cells constitute the main

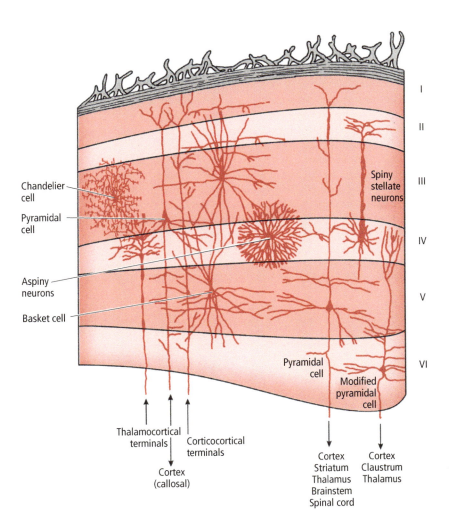

Figure 13.3 • The six layers of the neocortex and the different cell types located in each layer.

CHAPTER 13

output neurons, which contribute to the cortical descending motor pathways, the **corticospinal** and **corticonuclear** (**corticobulbar**) **tracts** that terminate in the spinal cord and brainstem, respectively. The older term corticobulbar (cortico, "cortex"; bulbar, "medulla") referred to all of the cortical projections that terminate in the cranial nerve nuclei located in the brainstem. Since not all cranial nerve nuclei are located in the medulla, this older term was replaced with the newer term "corticonuclear." The corticonuclear tract includes all cortical projections to the cranial nerve motor nuclei located in the pons, medulla oblongata, and upper cervical spinal cord (for the accessory nucleus). Nerve signals arising from the motor cortex elicit skeletal muscle contractions in various parts of the body. All of the pyramidal cell fibers release **glutamate**, an excitatory neurotransmitter, that stimulates excitatory or inhibitory interneurons, or less commonly, lower motor neurons, directly.

Multiple descending motor pathways ultimately exert their influence on muscle activity (Table 13.1). Three of these pathways, the **lateral corticospinal**, the **anterior corticospinal**, and the **corticonuclear tracts**, derive their fibers from the sensorimotor cortex, whereas the remaining four tracts, the **tectospinal, rubrospinal, reticulospinal,** and **vestibulospinal tracts**, derive their fibers from the brainstem. All but one of the descending tracts terminate in the spinal cord; the exception is the corticonuclear tract which terminates in the brainstem and the upper cervical spinal cord.

The motor cortex relays messages to motor neurons and interneurons in the spinal cord via a direct pathway, the **lateral** and **anterior corticospinal tracts**, which control voluntary movements of the body. The motor cortex also relays messages to the spinal cord via multiple other tracts that originate in the motor cortex but descend to connect and synapse in brainstem nuclei which in turn give rise to tracts that descend to synapse in the spinal cord. These pathways are the **corticotectal** and **tectospinal tracts**, the **corticorubral** and **rubrospinal tracts**, the **corticoreticular** and **reticulospinal tracts**, and the **vestibulospinal tracts** (see below).

The anterior corticospinal, tectospinal, rubrospinal, reticulospinal, and vestibulospinal tracts descend to the spinal cord where they synapse with interneurons and lower motor neurons to influence axial and proximal limb (girdle) musculature.

Corticospinal tract

Approximately 85–90% of the axons of the corticospinal tract decussate in the caudal medulla oblongata oblongata, forming the pyramidal decussation

Approximately two-thirds of the **corticospinal tract** fibers (Fig. 13.4) originate from the internal pyramidal cell layer of the frontal cortex; that is, one-third arises from Brodmann area 4, and one-third from area 6 (mostly from the supplementary motor area, with some fibers from the premotor cortex) and the cingulate motor cortex. The other third of the corticospinal tract fibers arises from the parietal cortex; specifically, from the posterior parietal cortex (Brodmann areas 5 and 7) and the somatosensory cortex (Brodmann areas 3, 1, and 2). The corticospinal tract consists of a mix of large-diameter, heavily myelinated, fast-conducting axons and small-diameter, lightly myelinated, and unmyelinated axons. The pyramidal neurons are referred to as the **upper motor neurons** (**UMNs**) of the descending motor pathways. Their axons descend to synapse in the spinal cord gray matter with:

- interneurons in laminae V–VII, which in turn synapse with motor neurons;
- alpha motor neurons, which innervate skeletal muscle fibers; or
- gamma motor neurons, which innervate the contractile portion of intrafusal fibers of muscle spindles (muscle stretch receptors).

Unlike the upper motor neurons, the fibers arising from neurons residing in the somatosensory cortex descend to synapse with second-order sensory neurons in the somatic relay nuclei of the ascending sensory pathways (and thus are not considered to be upper motor neurons). There they influence motor activity by modulating the transmission of sensory information to higher brain centers (see later). Both the cell bodies and axon terminals of the upper motor neurons (arising from the motor cortical areas) and the somatosensory neurons (arising from Brodmann areas 3, 1, 2, 5, and 7) reside entirely within the central nervous system (CNS).

As the fibers of the **upper motor neurons** (from the primary and premotor cortical areas) and the **somatosensory fibers** (from the somatosensory cortex) begin their descent, they traverse the **corona radiata** and then course through the **posterior limb of the internal capsule** near its **genu** (Fig. 13.5). As the fibers continue their descent, they gradually *shift in a posterior direction* to occupy the posterior half (or posterior third) of the **posterior limb of the internal capsule**. When the internal capsule is viewed on a horizontal section of the cerebrum, the corticospinal fibers are somatotopically organized so that the fibers terminating in more rostral levels of the spinal cord (innervating the upper limb) are located more rostrally (anteriorly), whereas the fibers that terminate in lower spinal cord levels (innervating the lower limb) are located more caudally (posteriorly). Axons of upper motor neurons arising from the primary motor cortex descend in the posterior part of the posterior limb of the internal capsule, whereas the axons arising from the premotor cortical areas descend rostral (anterior) to them as they traverse the internal capsule. When the axons reach the midbrain, they are located in the middle third of the **crus cerebri (basis pedunculi)** of the cerebral peduncle. The corticospinal fibers are also somatotopically organized at this level. The axons innervating the upper limb are located medially, next to the

Table 13.1 ● The descending motor pathways.

Pathway	Location of cell bodies	Spinal cord/ brainstem location	Ipsilateral / contralateral	Destination / target	Synapses with	Function
Lateral corticospinal tract	Brodmann areas 4, 6, 5, 7, 3, 1, 2, 23	Lateral funiculus	Pyramidal decussation	All spinal cord levels, primarily cervical and lumbosacral levels	Dorsal horn second-order afferent sensory neurons Lateral intermediate zone interneurons	Sensory modulation Execution of rapid, skilled, voluntary movement, especially those of the hand and foot
					Anterior horn interneurons and lower motor neurons	
				Giant pyramidal (Betz) cell fibers terminate only at lumbosacral levels	Anterior horn lower motor neurons	Execution of rapid, skilled voluntary movement of the lower limb
Anterior corticospinal tract	Brodmann areas 4, 6	Anterior funiculus	Fibers decussate at spinal cord level of termination	Cervical and upper thoracic spinal cord levels	Medial aspect of intermediate zone and anterior horn	Voluntary movement of axial and upper limb (girdle) musculature
Corticonuclear (corticobulbar) tract	Brodmann areas 4, 6, 5, 7, 3, 1, 2, 23	Pons Medulla Cervical spinal cord	Some fibers decussate near termination; others remain ipsilateral	Cranial nerve sensory nuclei	Second-order sensory afferent neurons	Sensory modulation
				Brainstem reticular formation	Interneurons and lower motor neurons	
				Cranial nerve motor nuclei	Interneurons and lower motor neurons	Reflex activity Voluntary movement of muscles of head
Tectospinal tract	Deep layers of the superior colliculus	Anterior funiculus	Decussates at posterior midbrain tegmentum	Cervical spinal cord levels	Medial aspect of intermediate zone and anterior horn interneurons	Coordination of eye and neck (head) movements
Rubrospinal tract	Red nucleus	Lateral funiculus	Decussates at anterior midbrain tegmentum	Upper cervical spinal cord	Lateral aspect of intermediate zone and anterior horn interneurons	Voluntary movement of upper limb muscles
Pontine (medial) reticulospinal tract	Pontine reticular formation	Anterior funiculus	Ipsilateral	Spinal cord	Medial aspect of intermediate zone (interneurons) and anterior horn (lower motor neurons)	Reflex movement of axial (trunk) and limb musculature
Medullary (lateral) reticulospinal tract	Medullary reticular formation	Lateral funiculus	Ipsilateral	Spinal cord	Medial aspect of intermediate zone interneurons	Reflex movement of axial (trunk) and limb musculature
Lateral vestibulospinal tract	Lateral vestibular nucleus	Anterior funiculus	Ipsilateral	Spinal cord	Medial aspect of intermediate zone (interneurons) and anterior horn (lower motor neurons)	Maintenance of posture and balance
Medial vestibulospinal tract	Medial vestibular nucleus	Anterior funiculus	Bilateral, mainly ipsilateral	Cervical spinal cord	Medial aspect of intermediate zone interneurons	Orientation of head

corticonuclear tract, whereas axons that innervate the lower limb are located laterally. The corticospinal tract continues inferiorly through the brainstem where it disperses into multiple bundles in the basal pons and then reassembles to form a distinct bundle known as the **pyramid** (a protuberance on the anterior aspect of the medulla oblongata). Since the corticospinal tract descends in the pyramid, it is sometimes referred to as the **pyramidal tract** which is misleading

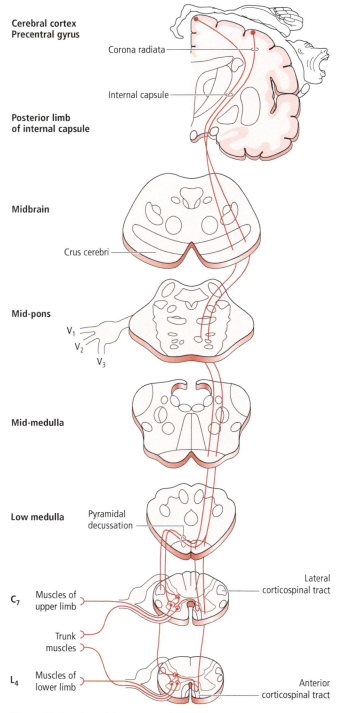

because the pyramid houses not only the descending axons of the corticospinal tract, but also those of the corticonuclear tract and corticoreticular fibers as they descend to their destinations. In the medulla oblongata, the fibers maintain the same somatotopic organization as in the crus cerebri: that is, the fibers that innervate the upper limb are located medially, whereas the fibers that innervate the lower limb are located laterally.

Approximately 85–90% of the axons of the corticospinal tract **decussate** in the caudal medulla oblongata, forming the **pyramidal decussation** (Table 13.1). The axons that emerge from the neurons residing in the lateral surface of the primary motor cortex controlling the upper limb decussate in the anterior aspect of the pyramidal decussation, whereas the axons that emerge from the neurons residing in the medial surface of the primary motor cortex controlling the lower limb decussate in the caudal aspect of the pyramidal decussation. These axons that decussated then descend in the lateral funiculus (L., "cord") of the entire length of the spinal cord, as the **lateral corticospinal tract**, with fibers exiting the tract to synapse in the cervical spinal cord located medially, and the fibers that synapse in the lumbar and sacral levels are located laterally. The lateral corticospinal tract fibers terminate primarily in the cervical, lumbar, and sacral spinal cord levels (Figs. 13.6 and 13.7). The remaining 10–15% of the corticospinal tract axons do not decussate but descend ipsilaterally in the anterior funiculus as the smaller, **anterior corticospinal tract**. Axons of the anterior corticospinal tract terminate mainly in the **anterior (ventral) horn gray matter** of the cervical and upper thoracic spinal cord.

Figure 13.4 ● The origin, course, and termination of the corticospinal tracts. The lateral corticospinal tract synapses with lower motor neurons that innervate the upper and lower limb muscles, whereas the anterior corticospinal tract synapses with the lower motor neurons that innervate the muscles of the neck, shoulder, and trunk, as well as the proximal upper limb (girdle) musculature. Source: Modified from Watson, C (1995) *Basic Human Neuroanatomy: an Introductory Atlas.* Little, Brown & Company, Boston; fig. 25.

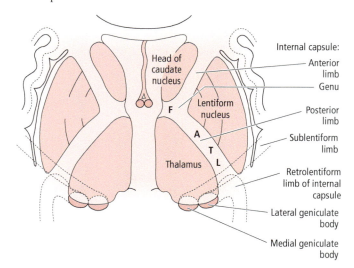

Figure 13.5 ● Horizontal section of the cerebrum showing the anterior limb, genu, and posterior limb of the internal capsule. Note the location of the corticospinal tract axons: A (upper extremity), T (trunk), and L (lower extremity) in the posterior half of the posterior limb of the internal capsule. Fibers carried by the corticonuclear (corticobulbar) tract pass through the genu, F (face). Note that as the corticonuclear fibers descend, they shift posteriorly to the anterior part of the posterior limb of the internal capsule.

MOTOR CORTEX AND DESCENDING MOTOR PATHWAYS • • • 239

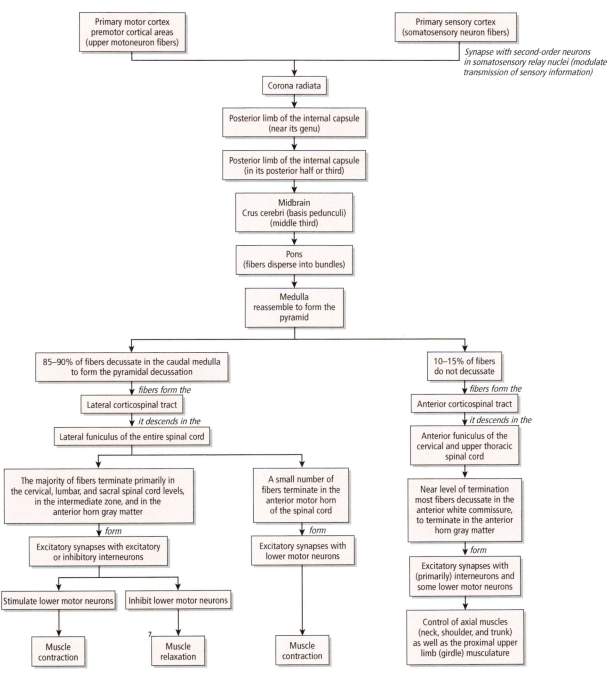

Figure 13.6 • The origin, course, and termination of the corticospinal tract.

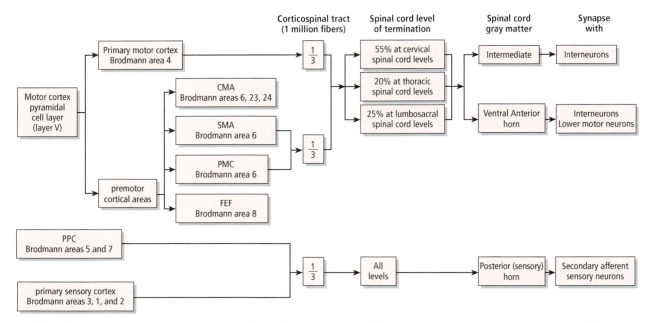

Figure 13.7 • The origin and termination of the corticospinal tract. FEF, frontal eye fields; PPC, posterior parietal cortex; PMC, premotor cortex; SMA, supplementary motor area.

Lateral corticospinal tract

The lateral corticospinal tract mediates the execution of rapid, skilled, voluntary movements of the distal musculature of the upper and lower limbs, especially the intrinsic muscles of the hand

The **lateral corticospinal tract** houses mostly crossed fibers among which is a group of large-diameter myelinated axons that arise from **giant pyramidal cells**, also known as Betz cells, residing in layer Vb of the primary motor cortex.

Most of the lateral corticospinal tract upper motor neurons (whose cell bodies reside in the motor cortex in the frontal lobe) project to the cervical and lumbosacral spinal cord. They end in laminae VII–IX in the **lateral intermediate zone gray matter** (a region of the spinal cord gray matter that is populated by interneurons and is thus similar to the reticular formation of the brainstem, which also contains interneurons), as well as in the **anterior (ventral) horn gray matter** (containing interneurons and lower motor neurons), where they form *excitatory* synapses with **interneurons**. Both the excitatory and inhibitory interneurons synapse with **lower motor neurons**. Excitatory interneurons stimulate the lower motor neurons to cause muscle contraction, whereas the inhibitory interneurons inhibit the lower motor neurons (and may cause muscle to relax). The **first-order afferent fibers** transmitting sensory input from the muscle spindles also establish synaptic contacts with the **inhibitory interneurons** (receiving input from the corticospinal fibers) in the spinal cord, mediating reflex activity. In contrast, a small number of the lateral corticospinal tract upper motor neurons project to the **anterior (motor) horns** of the spinal cord, where they form *excitatory* synapses directly with lower motor neurons. Many of these monosynaptic connections are with lower motor neurons innervating muscles of the hand.

Giant pyramidal (Betz) cells are located exclusively in the primary motor cortex and contribute only ~3% of the one million fibers in the corticospinal tract. Their large-diameter (up to 22 µm) axons descend to synapse only in the lumbar and sacral levels of the spinal cord, where they presumably synapse directly with the lower motor neurons that innervate the musculature of the lower limb. These monosynaptic connections of giant pyramidal cells with the lower motor neurons innervating the muscles of the lower limb are actually fewer in number than the monosynaptic connections to lower motor neurons innervating the hand. Due to their large, myelinated axons, the giant pyramidal cells are capable of fast nerve impulse transmission to the lumbosacral spinal cord. The remaining (~97%) of the corticospinal fibers consist of small-diameter (1–4 µm), slower impulse-conducting fibers.

Although the lateral corticospinal tract contains mostly crossed fibers, it also contains a small group of fibers (approximately 2–3%) that, instead of decussating, descend ipsilaterally. At their termination, the uncrossed fibers synapse with the spinal cord **interneurons** mediating movement of the axial (trunk) and proximal limb (girdle) musculature. The uncrossed fibers are associated with the maintenance of upright posture and general orientation of the limbs.

Approximately, one-third of the lateral corticospinal tract axons (whose cell bodies reside in the parietal lobe) project to laminae IV–VI of the posterior horn gray matter where they function in reflexes and modulate the transmission of sensory information to higher brain centers.

The lateral corticospinal tract mediates the execution of rapid, skilled, voluntary movements of the distal musculature of the upper and lower limbs (especially the intrinsic and extrinsic muscles of the hand and, to some extent, of the foot).

Anterior corticospinal tract

The fibers of the anterior corticospinal tract influence the neurons that innervate the axial and proximal limb (girdle) musculature

The **upper motor neurons** of the **anterior corticospinal tract** do not decussate in the caudal medulla oblongata; instead, they descend in the anterior funiculus of the spinal cord to terminate mainly in the **anterior horn gray matter** of the cervical (mainly) and upper thoracic spinal cord levels. Near their termination, these fibers decussate to the opposite side of the spinal cord via the anterior white commissure to synapse with **interneurons** that in turn synapse with **lower motor neurons**. Other fibers of the anterior corticospinal tract decussate near their termination and synapse directly with lower motor neurons that innervate the axial musculature (such as those of the neck, shoulder, and trunk) as well as the proximal upper limb (girdle) musculature.

Summary of the corticospinal tract

Approximately 55% of the corticospinal tract fibers terminate at cervical spinal cord levels to influence movement of the upper limb musculature, especially that of the hand

Approximately 55% of all of the corticospinal fibers terminate at cervical levels, 20% terminate at thoracic levels, and the remaining 25% terminate at lumbar and sacral levels of the spinal cord (Fig. 13.7). The large percentage of corticospinal fibers terminating at cervical spinal cord levels is indicative of the extensive influence they exert on neurons affecting the motor activity of the upper limb musculature, especially that of the hand.

Corticospinal fibers arising from the somatosensory cortex terminate in the posterior (sensory) horn of the spinal cord, where they synapse with second-order afferent sensory neurons to modulate sensory input (Fig. 13.8). The corticospinal fibers influencing motor activity arise from the motor cortical areas and terminate primarily in the intermediate zone and anterior horn gray matter of the spinal cord, where they synapse with interneurons (which in turn synapse with lower motor neurons); a small number of the corticospinal fibers terminate in the anterior horn of the spinal cord, where they synapse directly with lower motor neurons.

The anterior and lateral corticospinal tracts decrease in size at successively lower spinal cord levels as more and more of their fibers leave the tracts because they have reached their terminations.

The cell bodies of the **lower motor neurons** serving the axial (trunk) and the upper and lower limb musculature reside in the **anterior (motor) horns** of the spinal cord. More specifically, the cell bodies of the lower motor neurons serving the trunk muscles occupy the medial aspect of the anterior horns of the spinal cord; the cell bodies of the lower motor neurons serving the upper and lower limb muscles reside in the lateral aspect of the ventral horns of the spinal cord and are concentrated at cord levels that are the sites of origin of the brachial plexus and lumbosacral plexus, structures that innervate the upper and lower limbs, respectively (Fig. 13.9). The interneurons that may relay messages to the lower motor neurons are located in the intermediate zone. The fibers of the lower motor neurons collect to emerge from the spinal cord as the motor roots of spinal nerves that are a component of the peripheral nervous system.

Although the fibers of the corticospinal tracts terminate principally in the spinal cord, they distribute collaterals to widespread regions of the nervous system in their descent. Collaterals terminate in the ventral nuclei of the thalamus, the basal nuclei, the reticular formation, the red nucleus, the pontine nuclei, the posterior column nuclei, the posterior

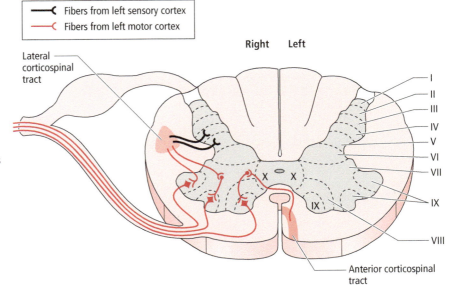

Figure 13.8 • Termination of the corticospinal tracts in the spinal cord. Note that the axons arising from the motor cortex terminate in the ventral horn of the spinal cord where they synapse with interneurons and lower motor neurons that innervate skeletal muscle. In contrast, the axons arising from the sensory cortex terminate in the dorsal horn of the spinal cord where they function in reflexes and modulate the transmission of sensory information to higher brain centers.

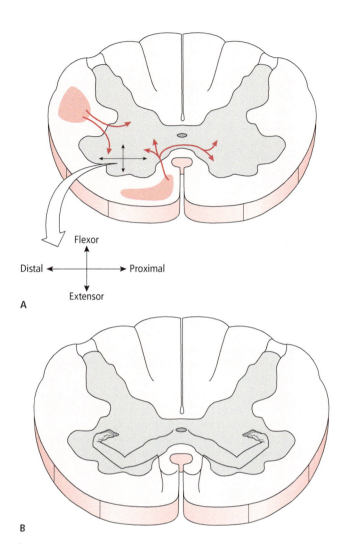

Figure 13.9 • Somatotopic organization of the anterior horns of the cervical spinal cord. (A) Cross-section of the cervical spinal cord showing the location of motor neurons innervating the upper limb and axial musculature. (B) Cross-section of the cervical spinal cord showing the upper limbs and trunk mapped on the ventral horns of the spinal cord.

(sensory) horn of the spinal cord, and the intermediolateral (lateral) cell column (sympathetic) of the spinal cord. Thus, the corticospinal tract probably has more than one function at various levels of the neuronal axis.

Somatosensory fibers influence motor activity

Descending somatosensory fibers synapse with second-order sensory neurons in the somatic sensory relay nuclei of the ascending sensory pathways and the sensory nuclei in the dorsal horn of the spinal cord

The **primary somatosensory area**, located in the postcentral gyrus (Brodmann areas 3, 1, and 2), also projects fibers involved in motor activity to the brainstem and spinal cord. The somatosensory fibers join the upper motor neuron fibers arising from the motor cortex and accompany them in their descent as the corticospinal tracts, to terminate primarily in the **somatic sensory relay nuclei** of the **brainstem** and the **dorsal horn** of the spinal cord. The descending somatosensory fibers, however, do not synapse at their termination with interneurons or motor neurons *that are* receiving synapses from the upper motor neurons. The descending somatosensory fibers synapse with the **second-order sensory neurons** in the somatic sensory relay nuclei of the sensory ascending pathways (the gracile nucleus and the cuneate nucleus of the brainstem), and with the sensory nuclei in the dorsal horn of the spinal cord. Their function in the somatic sensory relay nuclei of the brainstem and spinal cord is to influence motor activity by modulating the sensory information that is relayed to the brainstem and spinal cord from peripheral structures.

Muscle spindle activity

Intrafusal fibers are special stretch receptors dispersed among the extrafusal fibers of a skeletal muscle

Alpha motor neurons innervate the extrafusal fibers of skeletal muscle, whereas gamma motor neurons innervate the intrafusal fibers of muscle spindles. During a normal movement, both the alpha and gamma motor neurons are co-activated. If only the alpha motor neurons innervating a muscle were stimulated, only the extrafusal fibers of the muscle would contract, causing the overall muscle to shorten. Although the intrafusal fibers would passively shorten (because the muscle spindles are attached via connective tissue to the extrafusal fibers), their central non-contractile portion would become slack and be unable to monitor changes in muscle length. Muscle spindles maintain their sensitivity to muscle length, even when the muscle is contracting to a shorter length, via a mechanism known as **alpha-gamma coactivation**. As the alpha motor neurons stimulate the extrafusal muscle fibers to contract, simultaneously the gamma motor neurons stimulate the polar (contractile) portions of the intrafusal fibers to contract. This maintains tension on the central (noncontractile) region of the intrafusal fibers where the sensory endings are located. This alpha–gamma coactivation is necessary to maintain the "stretch sensitivity" of the intrafusal fibers, so that they are ready to detect the slightest stretch at any length or at any state of muscle contraction.

Corticonuclear (corticobulbar) tract

The corticonuclear tract fibers terminate in their target cranial nerve motor nuclei, in the sensory relay nuclei, and in the reticular formation of the brainstem

The corticonuclear tract (Figs. 13.10 and 13.11; Table 13.1) consists of fibers of **upper motor neurons** located in the primary motor cortex (Brodmann area 4), the premotor cortex (Brodmann area 6), supplementary motor area (Brodmann area 6), and the anterior cingulate area (Brodmann area 23). The corticonuclear

MOTOR CORTEX AND DESCENDING MOTOR PATHWAYS ••• 243

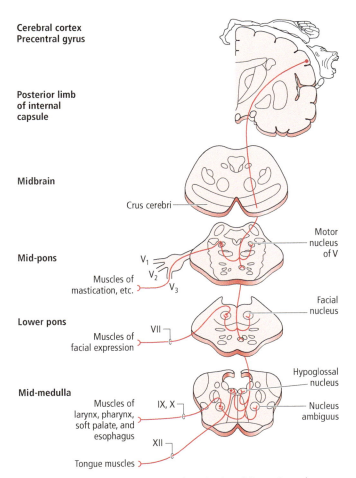

Figure 13.10 ● The origin, course, and termination of the corticonuclear (corticobulbar) tract in the brainstem cranial nerve motor nuclei. Source: Modified from Watson, C (1995) *Basic Human Neuroanatomy*, 5th edn. Little, Brown & Company, Boston; fig. 22.

tract also contains axons arising from the somatosensory cortex (Brodmann areas 3, 1, and 2) and the posterior parietal cortex (Brodmann areas 5, 7). The corticonuclear tract accompanied by the corticospinal tract descend in tandem, traversing the **corona radiata** and the **internal capsule** to reach the **crus cerebri** (**basis pedunculi**) of the midbrain. The corticonuclear tract descends in the **genu** of the internal capsule (anterior to the corticospinal fibers); it, then, leaves the genu of the internal capsule and *shifts posteriorly* into the posterior limb of the internal capsule. The corticonuclear fibers continue their descent into the midbrain where they are located medial to the corticospinal tract fibers in the middle third of the crus cerebri of the cerebral peduncle. The corticonuclear fibers diverge from the corticospinal fibers at various brainstem levels to terminate in their target **cranial nerve somatic motor** or **branchiomotor nuclei**. Similar to the fibers of the corticospinal tracts, some of the corticonuclear fibers synapse directly with motor neurons, but the majority of fibers synapse with **interneurons** housed within the nucleus of termination or with local interneurons of the brainstem

reticular formation. The interneurons in turn synapse with the motor neurons of the cranial nerve motor (or branchiomotor) nuclei. The corticonuclear fibers terminate in the cranial nerve motor nuclei, in the **sensory relay nuclei** (such as the gracile nucleus, the cuneate nucleus, the sensory nuclei of the trigeminal nerve, and the nucleus of the solitary tract), and in the brainstem **reticular formation**. The corticonuclear fibers affect the motor nuclei of the following cranial nerves: trigeminal (CN V), facial (CN VII), glossopharyngeal (CN IX), vagus (CN X), accessory (XI), and hypoglossal (XII). In general, most cranial nerve motor nuclei receive bilateral corticonuclear projections from the motor cortex.

Corticonuclear projections to the motor nuclei of the trigeminal and facial nerves

Most of the corticonuclear fibers terminate in the **reticular formation** at the pontine and medullary levels (Fig. 13.10) to synapse with **interneurons** (which in turn synapse with lower motor neurons housed in the cranial nerve motor nuclei). A number of corticonuclear fibers terminate directly and bilaterally in the motor nuclei of the trigeminal and facial nerves.

At the midpontine levels, some of the corticonuclear fibers diverge from the corticonuclear tract, course into the pontine tegmentum, and terminate bilaterally in the pontine reticular formation and the **motor nucleus of the trigeminal nerve** (Fig. 13.12 A). The remaining corticonuclear fibers project contralaterally to the lower motor neurons that supply motor innervation to the lateral pterygoid muscle and bilaterally to the lower motor neurons that innervate the remaining muscles of mastication.

At the caudal pontine level some of the corticonuclear fibers terminate bilaterally in the **facial nucleus**, responsible for motor innervation of the muscles of facial expression, the platysma, posterior belly of the digastric muscle, stylohyoid, and stapedius muscles. The corticonuclear tract projects bilaterally to the lower motor neurons of the facial nucleus that innervate the muscles of the *upper half* of the face, whereas this same tract projects only contralaterally to the lower motor neurons of the facial nucleus that innervate the muscles of the *lower half* of the face (Fig. 13.12 B).

Corticonuclear projections to the motor nuclei of the glossopharyngeal, vagus, accessory, and hypoglossal nerves

The remaining corticonuclear fibers terminate in the nucleus ambiguus at caudal pontine levels, the hypoglossal nuclei at caudal medullary levels, and the accessory nucleus at upper cervical spinal levels.

At the caudal pontine levels and rostral medullary levels, some corticonuclear fibers diverge from the tract to terminate in the **nucleus ambiguus**, a motor nucleus of the glossopharyngeal, and vagus nerves which supplies motor innervation to the skeletal muscles of the palate, pharynx,

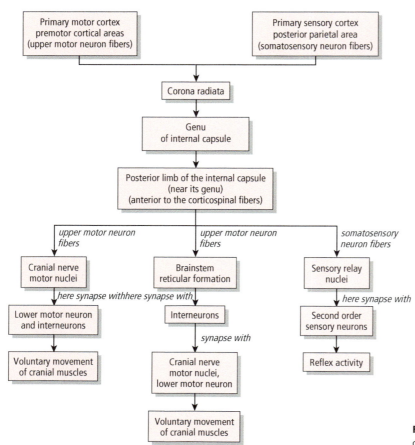

Figure 13.11 ● The origin, course, and termination of the corticonuclear (corticobulbar) tract in the brainstem.

and larynx. The nucleus ambiguus receives bilateral corticonuclear tract projections. A small bundle of corticonuclear fibers, **Pick's bundle**, proceeds inferiorly along with the corticospinal tract to the level of the pyramidal decussation where they cross, then recur, and ascend to terminate in the nucleus ambiguus of the opposite side.

A number of the corticonuclear fibers terminate in the **hypoglossal nuclei** that house the lower motor neurons that innervate the tongue musculature. The corticonuclear fibers project mainly contralaterally on the lower motor neurons that innervate the genioglossus muscle, whereas the remaining lower motor neurons that innervate all other tongue musculature receive bilateral projections (with a contralateral predominance).

At upper cervical spinal cord levels, the remaining corticonuclear fibers terminate primarily in the ipsilateral **accessory nucleus** of the spinal accessory nerve that innervates the sternocleidomastoid and trapezius muscles.

There are no corticonuclear projections to the oculomotor, trochlear, and abducens nuclei

The corticonuclear tract fibers *do not* project to the motor nuclei of the oculomotor, trochlear, and abducent nerves. These nuclei receive motor signals from the cerebral cortex indirectly and do so via a different group of fibers that take a different route. Fibers derived from the **frontal eye field** (Brodmann area 8) and **parietal motor eye fields** descend terminating in the **midbrain reticular formation** and the **paramedian pontine reticular formation (PPRF)**, respectively. They in turn project to the motor nuclei of the oculomotor, trochlear, and abducent nerves that control eye movements. Signals arising from the frontal and parietal motor eye fields result in conjugate eye movements away from the side of origin of the cortical signals.

Corticotectal and tectospinal tracts

The corticotectal tract and the tectospinal tract are involved in the coordination of head movements with eye movements elicited by visual, auditory, and vestibular stimuli

The visual association areas (Brodmann areas 18 and 19) give rise to the fibers of the **corticotectal tract** that descend to terminate in the **oculomotor accessory nuclei** (the interstitial nucleus of Cajal, or the nucleus of Darkschewitch) and the deep layers of the **superior colliculus**.

The neurons of the oculomotor accessory nuclei give rise to fibers that join the medial longitudinal fasciculus (MLF) to terminate in the oculomotor, trochlear, and abducens nuclei, where they influence vertical, rotatory, and smooth pursuit eye movements. The fibers that terminate in the superior colliculus function in the control of eye movements and guiding gaze.

The deep layers of the superior colliculus give rise to the fibers of the **tectospinal tract** (Fig. 13.13; Table 13.1), which decussates at the level of the red nucleus in the midbrain, and then descends to the medulla oblongata, in the medial longitudinal fasciculus. The tectospinal fibers continue their descent in the anterior funiculus of the spinal cord to end at cervical spinal cord levels where they synapse with interneurons. The tectospinal tract is involved in the coordination of head movements with eye movements elicited by visual, auditory, or vestibular stimuli.

Corticorubral, and rubrospinal tracts

The corticorubral tract transmits motor messages from the sensorimotor cortex to the red nucleus. The red nucleus gives rise to the rubrospinal tract. The rubrospinal tract (just like the corticospinal tract) functions in controlling the movement of the hand and digits, by facilitating flexor muscle tone and inhibiting the extensor musculature of the upper limb

The **corticorubral tract**, similar to the corticospinal tract, arises from the sensorimotor cortex and terminates in the ipsilateral **red nucleus** in the midbrain (Fig. 13.14). The cell bodies in the caudal part of the red nucleus give rise to fibers forming the **rubrospinal tract** (which is insignificant in humans). This tract decussates in the anterior midbrain tegmentum and descends in the lateral funiculus of the spinal cord, next to the corticospinal tract. Its fibers terminate in the lateral intermediate zone and anterior horn of the spinal cord where they synapse with interneurons. The red nucleus facilitates the alpha, and gamma motor neurons that innervate the contralateral upper limb flexor muscles, whereas it simultaneously inhibits those of the extensors, specifically the nerve cells that innervate the distal muscles of the upper limbs. This facilitation and inhibition is mediated by the rubrospinal tract terminating in the spinal cord. The rubrospinal tract (along with the corticospinal tract) functions in controlling the movement of the hand and digits, by facilitating flexor muscle tone and inhibiting the extensor musculature of the upper limb.

Corticoreticular fibers and reticulospinal tracts

The corticoreticular fibers transmit motor messages from the premotor and supplementary motor cortex to the pontine and medullary reticular formation nuclei. The reticulospinal tracts influence the motor control of axial (trunk) and proximal limb musculature and are involved in posture maintenance and orientation of the limbs in an intended direction

Corticoreticular fibers, which course with the corticonuclear and the corticospinal tracts to several nuclei dispersed in the brainstem (pontine and medullary) reticular formation, receive bilateral projections from the premotor and supplementary motor cortex. The

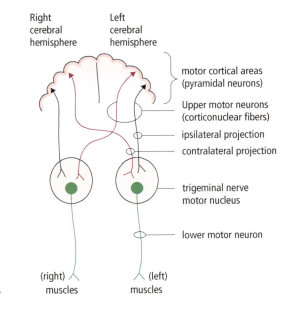

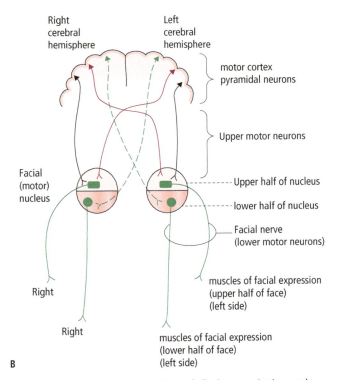

Figure 13.12 ● (A) Corticonuclear (corticobulbar) tract projections to the trigeminal motor nucleus. The trigeminal motor nucleus receives bilateral (ipsilateral and contralateral) projections from the cerebral cortex. Note that the lower motor neurons (LMNs) that innervate the lateral pterygoid muscle receive only contralateral corticonuclear projections. (B) Corticonuclear (corticobulbar) tract projections to the facial motor nucleus. Note that the upper half of the facial motor nucleus receives bilateral corticonuclear projections, whereas the lower half of the facial motor nucleus receives only contralateral projections.

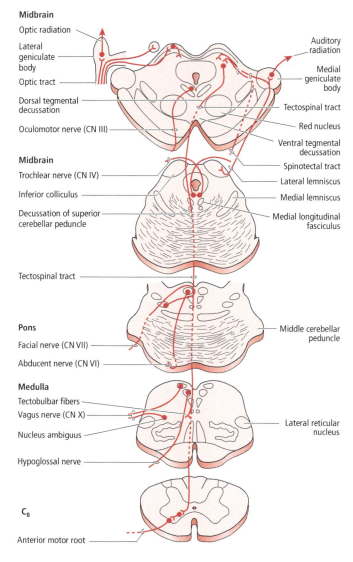

Figure 13.13 ● The origin, course, and termination of the tectospinal tract. Source: Modified from Parent, A (1996) *Carpenter's Human Neuroanatomy*, 9th edn. Williams & Wilkins, Baltimore; fig. 11.18.

multiple sources, primarily, it is the sensory input that directs their function.

The **lateral (medullary) reticulospinal tract** arises from the gigantocellular reticular nucleus and descends bilaterally in the lateral funiculus of the spinal cord. Nerve fibers terminate at all spinal cord levels, synapsing mainly with interneurons in the intermediate zone gray matter but also with alpha and gamma motor neurons of the spinal cord. The medullary reticular fibers have an inhibitory effect on extensors and an excitatory effect on flexors (as in lying down). The lateral reticulospinal tract also relays an autonomic input to the sympathetic and parasympathetic neurons of the spinal cord, which mediate autonomic functions such as pupillary dilation, heart rate modulation, and sweating (Fig. 13.14).

These two tracts influence the motor control of axial (trunk) and proximal limb musculature and are involved in muscle tone, posture maintenance, and orientation of the limbs in an intended direction.

Vestibulospinal tracts

The lateral vestibulospinal tract is involved in the maintenance of posture and balance; the medial vestibulospinal tract mediates head movement while maintaining gaze fixation on an object

The vestibular nuclei receive sensory input related to head movement from the vestibular apparatus of the internal ear via the vestibular division of the vestibulocochlear nerve (CN VIII) and sensory input related to balance from the cerebellum. The **lateral vestibulospinal tract**, arising mainly from the lateral vestibular nucleus (Fig. 13.15), contains ipsilateral fibers that descend in the anterior funiculus to synapse mainly with excitatory interneurons at all spinal cord levels. These interneurons stimulate motor neurons that innervate axial (trunk) and proximal limb extensor muscles, and simultaneously inhibit lower motor neurons that innervate limb flexor muscles. This tract is involved in the maintenance of posture and balance by specifically facilitating motor neurons that innervate the antigravity muscles (extensor muscle tone of the antigravity muscles). This tract also mediates head and neck movements in response to vestibular sensory input.

The **medial vestibulospinal tract**, arising from the medial vestibular nucleus (Fig. 13.15), descends mostly ipsilaterally in the descending medial longitudinal fasciculus (MLF) of the brainstem and then in the anterior funiculus of the cervical and upper half of the thoracic spinal cord. Its fibers synapse with interneurons that synapse with alpha and gamma motor neurons, although some fibers terminate on alpha motor neurons directly. These fibers exert their inhibitory influence on neurons of the cervical spinal cord (maintaining equilibrium elicited by vestibular input) and mediating head movement while maintaining gaze fixation on an object.

The collective function of the lateral and medial vestibulospinal tracts is to stabilize and coordinate the orientation of the head, neck, and body by mediating reflex control of muscle tone, posture, and balance maintenance.

reticular nuclei also receive an input about balance and posture from the cerebellum and from the vestibular nuclei, as well as nociceptive fibers from the spinal cord and brainstem.

Neurons in the pontine reticular nuclei (oral and caudal pontine reticular) give rise to fibers that form the **medial (pontine) reticulospinal tract**, which descends ipsilaterally in the anterior funiculus of the spinal cord. These fibers terminate and synapse with spinal cord interneurons, as well as with alpha and gamma motor neurons at all levels of the spinal cord. (Fig. 13.14). The pontine reticular fibers stimulate extensor muscle (that enable standing) and inhibit flexor muscle movements, enhancing the action of the lateral vestibulospinal tract. Although the nuclei of the medial (pontine) reticulospinal tract receive input from

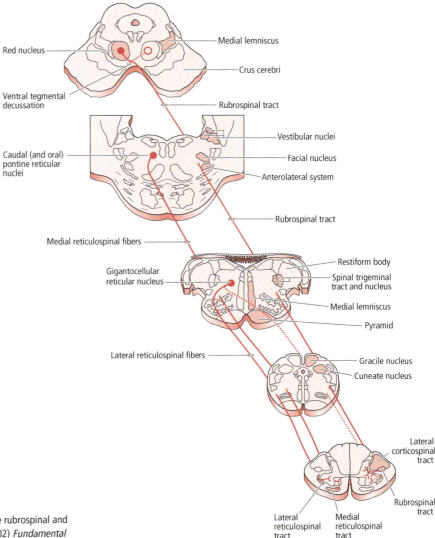

Figure 13.14 ● The origin, course, and termination of the rubrospinal and reticulospinal tracts. Source: Modified from Haines, DE (2002) *Fundamental Neuroscience*. Churchill Livingstone, Philadelphia; fig. 24.9.

Functional classification of the descending motor pathways

The descending motor pathways are classified into three **functional** categories: the ventromedial (anteromedial) group, the lateral group, and the cortical group. In general, the fibers of each of these functional groups synapse in the region of the gray matter in close proximity to their position in the white matter. That is, the fibers of the ventromedial group pathways synapse in the medial aspect and intermediate zone of the spinal cord gray matter (controlling/innervating the axial and proximal limb musculature), whereas the fibers of the lateral group pathways synapse in the lateral aspect and intermediate zone of the spinal cord gray matter (controlling/innervating the upper and lower limb musculature). Furthermore, the lower motor neurons that innervate the flexor muscles occupy a region of gray matter that is posterior to the lower motor neurons innervating the extensor muscles (Figs. 13.9A, B, and 13.16).

The ventromedial group of the descending motor pathways

The ventromedial group of the descending motor pathways controls the axial and proximal limb musculature for balance maintenance and postural adjustment

The **ventromedial group of the descending motor pathways** consists of the anterior corticospinal tract, the medial and lateral vestibulospinal tracts, the medial and lateral reticulospinal tracts, and the tectospinal tract, which are all located in the anterior funiculus of the spinal cord and synapse in the medial aspect of the anterior horn and intermediate zone (Figs. 13.8 and 13.16). These tracts project to the spinal cord where their fibers influence lower motor neurons *bilaterally* either by (1) crossing to the opposite side of the spinal cord, or (2) synapsing with commissural interneurons that cross to the opposite side. Lower motor neurons control the axial and proximal limb musculature of both sides of the body. These tracts function in the simultaneous bilateral control of gross movements of the axial and

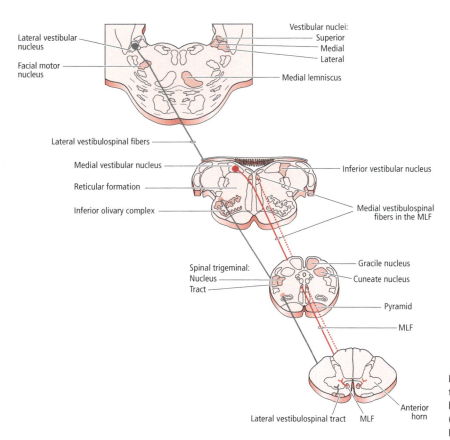

Figure 13.15 ● The origin, course, and termination of the medial and lateral vestibulospinal tracts. MLF, medial longitudinal fasciculus. Source: Modified from Haines, DE (2002) *Fundamental Neuroscience*. Churchill Livingstone, Philadelphia; fig. 24.7.

proximal limb musculature for balance maintenance and during postural adjustments.

The lateral group of descending motor pathways

The lateral group of the descending motor pathways controls the proximal, and especially the distal, musculature of the upper and lower limbs

The **lateral group of the descending motor pathways** consists of the lateral corticospinal and the rubrospinal tracts, located in the lateral funiculus of the spinal cord. The fibers of these two tracts synapse in the lateral aspect and intermediate zone of the anterior horn of the spinal cord (Figs. 13.9A and 13.16). The rubrospinal tract extends only to the upper cervical level of the spinal cord and does not have a very important function in humans. These tracts (mainly the corticospinal tract) function in the control of the distal but, to a certain extent, also of the proximal musculature of the upper and lower limbs, as well as mediating independent digit movements.

The cortical group of the descending motor pathways

The cortical group innervates the distal muscles of the limbs, especially those of the hands

The **cortical group of the descending motor pathways** consists of lateral corticospinal tract fibers that synapse directly

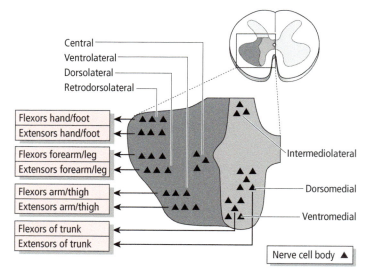

Figure 13.16 ● Somatotopic organization of the ventral horn of the spinal cord. Source: Modified from Fitzgerald, MJT (1996) *Neuroanatomy Basic and Clinical*, 3rd edn. WB Saunders, Philadelphia; fig. 13.1.

with lower motor neurons, particularly the neurons whose fibers innervate the distal muscles of the limbs, such as the intrinsic muscles of the hand. This tract is involved in independent, fractionated movements of the digits.

The lower motor neurons are referred to as the **final common pathway** because they are influenced by the corticospinal, tectospinal, rubrospinal, reticulospinal, vestibulospinal, and reflex neurons. All input is ultimately funneled to the lower motor neurons that innervate the skeletal muscles directly.

Note that the clinical case at the beginning of the chapter refers to a patient who has a motor disturbance in both the upper and lower limbs.

1 Which descending motor pathway has been affected in this patient?

CLINICAL CONSIDERATIONS

Corticospinal and corticonuclear tracts

A lesion of the corticonuclear and/or corticospinal tracts is known as an upper motor neuron lesion

Deficits resulting from an upper motor neuron lesion of the corticonuclear (corticobulbar) tract are described in detail in Chapter 17 on the Cranial Nerves. Since the **corticospinal tract** projects from the cerebral cortex through the corona radiata, the internal capsule, the crus cerebri, the pons, the medulla, and the entire length of the spinal cord, it is vulnerable to lesions from various sources throughout its long descending path. The most common **corticospinal tract lesion** results from a cerebral vascular accident (CVA, a stroke). A stroke can result either from a vascular hemorrhage or a vascular occlusion resulting from a thrombus (loose blood clot or an air bubble) in the brain or the spinal cord.

The **middle cerebral artery** is buried in the lateral fissure (of Sylvius) separating the parietal from the temporal lobe. Its branches that emerge from the lateral fissure supply structures in the frontal, parietal, and temporal lobes. The lateral surface of the primary motor cortex (corresponding to the precentral gyrus, Brodmann area 4) controlling movement of the head, upper limb and trunk, as well as the lateral surface of the premotor and supplementary motor cortices (corresponding to Brodmann area 6) and the frontal eye field (corresponding to Brodmann area 8) receive their arterial blood supply from branches of the middle cerebral artery. The medial extensions of the motor cortices located in the medial surface of the frontal lobe, that is, the primary motor cortex controlling the movement of the lower limb and the medial extension of the premotor and supplementary motor cortices, are supplied by the **anterior cerebral artery**. A vascular infarct resulting from occlusion of the anterior or middle cerebral artery will result in motor deficits in the contralateral side of the body.

A lesion damaging upper motor neurons that is, either their cell bodies (residing in the motor cortex) or their axons anywhere along their descending path in the white matter, results in **upper motor neuron lesion signs**. It is important to note that the motor cortex not only gives rise to the *corticospinal* and *corticonuclear tracts*, but in addition, it gives rise to other descending fibers of the motor system, including the *corticorubral fibers* terminating in the red nucleus, and the *corticoreticular fibers* terminating in the pontine and medullary reticular formation, all of which descend in the internal capsule. It is also important to note that *rubrospinal, reticulospinal,* and *vestibulospinal neurons* are also considered to be upper motor neurons and when damaged produce upper motor neuron lesion signs. Since the upper motor neuron axons of the various descending motor tracts (named previously) *intermingle* to some extent as they descend through different levels (in the internal capsule, the brainstem, or the spinal cord), upper motor neuron lesion signs are usually produced following damage of a mix or combination of descending motor axons due to their intermingling or descending in close proximity to one another. For example, the corticonuclear and corticospinal tracts descend at one point in the genu and posterior limb of the internal capsule, respectively. Further down in the internal capsule both the corticonuclear and corticospinal tracts descend in the posterior limb of the internal capsule. They also descend next to one another at the level of the crus cerebri of the cerebral peduncle in the midbrain. A lesion in the internal capsule or the middle third of the crus cerebri will affect both the corticospinal and corticonuclear tracts, producing upper motor neurons lesion signs (motor deficits) in structures of the head and the body. A lesion involving the medullary pyramid may damage not only corticospinal, corticonuclear, and corticoreticular fibers (which descend in the pyramid), but also reticulospinal fibers due to their close proximity to the pyramid (located posterior to the pyramid). At the level of the spinal cord, a lesion affecting the dorsolateral aspect of the lateral funiculus will damage not only the corticospinal tract but also the rubrospinal tract descending immediately ventral to it, resulting in upper motor neuron lesion signs on the ipsilateral side of the body.

Many of the branches of the middle cerebral artery perforate the cerebral hemisphere to reach deep subcortical structures such as the basal nuclei and the internal capsule. A cerebral vascular accident involving the *lenticulostriate branches* of the middle cerebral artery affecting the internal capsule is referred to as a **capsular stroke**. Since both the corticonuclear and corticospinal tracts course in the internal capsule, a capsular stroke can result in considerable upper motor neuron damage. Initially, a temporary **contralateral paresis** (weakness) or **flaccid paralysis, hypotonia** (decrease in muscle tone or tension), and **hyporeflexia** (decrease in reflexes) occur. Usually, approximately two weeks following the lesion, the individual will regain function of the proximal limb musculature and eventually **spasticity** will appear in the distal limb muscles. **Hypertonia**, characterized by an increase in muscle tone (muscle tension), and **hyperreflexia**, exaggerated tendon reflexes, will appear in the distal limb muscles. Hypertonia can be demonstrated during the passive manipulation of an extremity. The hypertonic extremity will exhibit an increase in resistance to its stretching by the physician. Exaggerated hypertonia/ hypertonicity is referred to as spasticity. The hypertonia is more severe in the antigravity muscles (the upper limb proximal flexors and lower limb extensors). With this type of lesion, fine movements of the distal limbs (hand and foot) will be affected most.

A typical response to the increase in resistance to passive stretch is the **clasp-knife response**. As an extremity (e.g., the forearm) is extended by the physician against the spastic resistance there is an abrupt loss of resistance. The clasp-knife response is believed to be caused by activation of the tendon organs (TOs) as muscles are stretched during the limb manipulation. This in turn stimulates the TOs Ib sensory afferent fibers that relay excitatory messages to inhibitory spinal cord interneurons. These interneurons inhibit the alpha motor neurons that overstimulated the skeletal muscles to create the hypertonia and increased resistance to passive stretch.

Hypertonia, hyperreflexia, and spasticity are not well understood. There are several hypotheses that have offered an explanation of the cause of **spasticity** and related **hypertonia** and **hyperreflexia**.

One hypothesis suggests that an upper motor neuron (UMN) lesion diminishes or eliminates descending inhibitory influences on the dynamic

CLINICAL CONSIDERATIONS (continued)

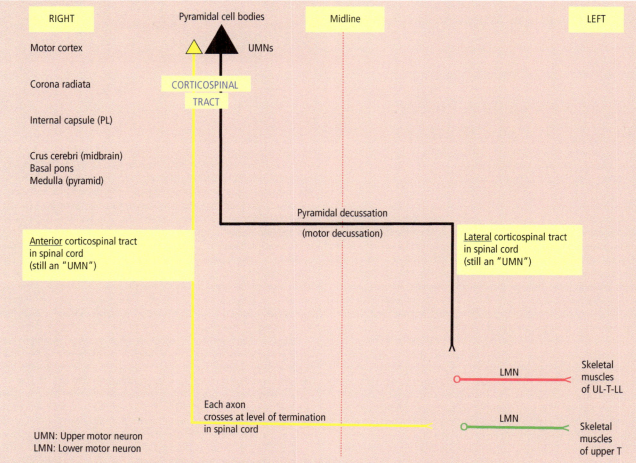

Figure 13.17 • Origin, path, crossing, and termination of the corticospinal tract axons.

gamma motor neurons. This in turn causes overactivity of the gamma motor neurons resulting in contraction of the contractile portions of intrafusal muscle fibers and stretching of the central noncontractile portion of the intrafusal fibers. This overstimulates the type Ia anulospiral afferent fibers from the muscle spindles whose central processes synapse with alpha motor neurons, which are overstimulated causing an abnormal increase in skeletal muscle tone.

Another hypothesis suggests that an UMN lesion may result in diminution of cortical stimulation of spinal cord inhibitory interneurons, resulting in hypertonia and spasticity.

Cortical UMN axons stimulate a recurrent inhibitory interneuron, the recurrent inhibitory interneuron of spinal lamina VII (eponym: Renshaw cell). This interneuron is also stimulated by a collateral branch of a lower motor neuron and in turn inhibits the lower motor neuron. These neural connections inhibit the reflex stimulation of antagonist muscles when the agonists are contracting. An UMN lesion involving cortical neurons would eliminate the inhibitory effect on antagonist muscles. This causes alternating contraction of agonists and antagonist muscles known as **clonus** which accompanies spasticity and hyperreflexia.

A unilateral lesion in the primary motor cortex will affect mainly the fine movements of the contralateral distal limbs (hand and foot)

A *unilateral* lesion in the **primary motor cortex** (Brodmann area 4) will affect mainly the fine movements of the contralateral distal limbs (hand and foot). Lesions confined to the primary motor cortex are rare (Figures 13.17 and 13.18).

A *unilateral* lesion in the **premotor cortex** (PMC, Brodmann area 6) results in the inability to produce voluntary movements in the contralateral side of the body, although there is no apparent weakness, paralysis, or change in muscle tone. Recall that the premotor cortex, along with the supplementary motor cortex, plays a key role in the planning phase of motor activity. Once the plan, namely the sequence of movements is created, it is relayed to the primary motor cortex for execution.

A *unilateral* lesion in the **supplementary motor cortex** (SMC, Brodmann area 6) results in the inability to coordinate different/dissimilar voluntary movements bilaterally. As an example, the affected individual will have no difficulty in coordinating the same movement with both hands concurrently. However, when attempting to perform different movements with each hand concurrently, it will be difficult or impossible to execute. A lesion limited to the supplementary motor cortex does not produce apparent muscle weakness or paralysis and has no effect on muscle tone.

A *unilateral* lesion of the **primary motor cortex** and another motor cortical area will produce more severe motor deficits, typical of upper motor neuron lesion signs: a **contralateral spastic paralysis** with **hypertonia** and **hyperreflexia**.

Lesions of the corticospinal tract and/or the corticonuclear tract in combination with other structures in the midbrain, pons, and medulla produce specific deficits characteristic of various syndromes that are discussed in detail in Chapter 7 on the Brainstem.

CLINICAL CONSIDERATIONS (continued)

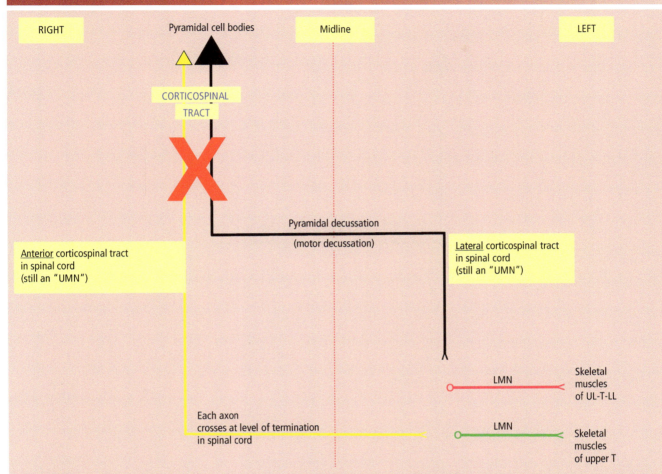

Figure 13.18 • Lesion to the corticospinal tract. Upper motor neuron (UMN) lesion is on the right, above the decussation of the corticospinal tract. Where will the UMN lesion deficits appear? On the left side of the body, contralateral to the side of (and below) the lesion (if the "X" is a clamp and the corticospinal tract is a hose with water, which side will become "dry"? the left side.

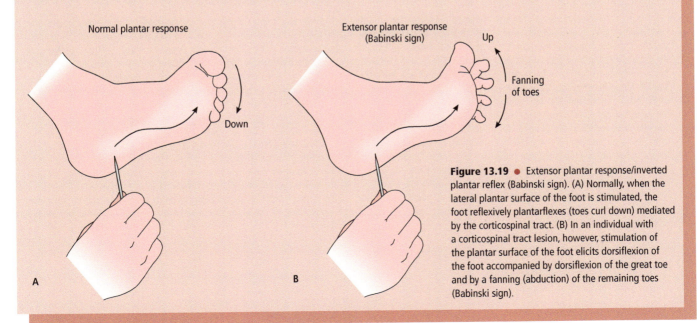

Figure 13.19 • Extensor plantar response/inverted plantar reflex (Babinski sign). (A) Normally, when the lateral plantar surface of the foot is stimulated, the foot reflexively plantarflexes (toes curl down) mediated by the corticospinal tract. (B) In an individual with a corticospinal tract lesion, however, stimulation of the plantar surface of the foot elicits dorsiflexion of the foot accompanied by dorsiflexion of the great toe and by a fanning (abduction) of the remaining toes (Babinski sign).

CLINICAL CONSIDERATIONS (continued)

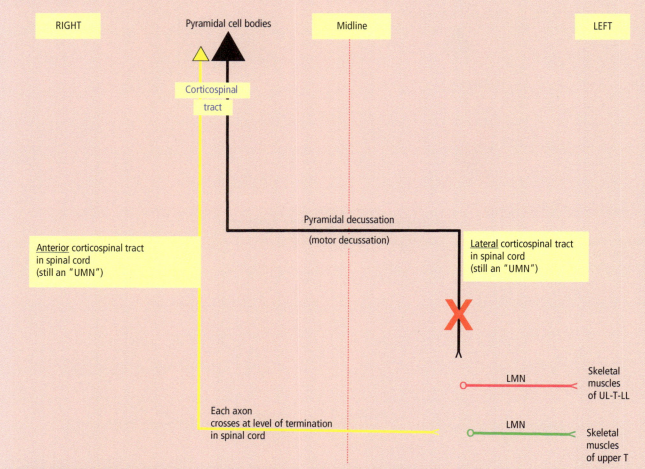

Figure 13.20 ● Lesion to the corticospinal tract. The corticospinal tract consisting of upper motor neuron (UMN) axons is damaged on the left side, below the decussation. Where will the UMN lesion deficits appear? On the left side of the body, ipsilateral to (and below) the lesion.

The most prominent physical deficit resulting from damage of the corticospinal tract fibers (above their decussation) is a **contralateral paralysis (hemiplegia)** of the upper and lower limbs and extensor plantar response/inverted plantar reflex (**Babinski sign**) (Fig. 13.19). If the lesion is below the corticospinal tract decussation, the deficits will appear on the same side as the lesion (Fig. 13.20). Normally, when the lateral plantar surface of the foot is stimulated, the foot reflexively plantarflexes (toes curl down) mediated by the corticospinal tract. In an individual with a corticospinal tract lesion, however, stimulation of the plantar surface of the foot elicits dorsiflexion of the foot accompanied by dorsiflexion of the great toe and by abduction of the remaining toes (Babinski sign). In the absence of the corticospinal tract, this pathologic response is mediated by the descending motor pathways arising from the brainstem. In the newborn, Babinski sign is normal since the lateral corticospinal tract has not yet been myelinated. In the adult, Babinski sign reappears when the lateral corticospinal tract is damaged.

Following trauma to the head involving the brain, scar tissue formation in the brain often results in epileptic seizures. A lesion of the primary motor cortex, with subsequent scar tissue formation, results in **Jacksonian epileptic seizures**. In this type of seizure convulsions may occur only in a certain part of the body, or additional parts of the body may be involved as well, depending on the site of the primary motor cortex that is affected by the lesion.

Lesions in the frontal eye field

Normally, stimulation of the frontal eye field results in deviation of both eyes toward the contralateral side. A unilateral lesion in the frontal eye field (Brodmann area 8) results in both eyes deviating to the side ipsilateral to the lesion. In addition, the affected individual will be unable to turn the eyes contralateral to the lesion. The effects of frontal eye field lesions are not permanent.

Note that the clinical case at the beginning of the chapter refers to a patient who has a motor disturbance in both the upper and lower limbs. Symptoms include muscle weakness, flaccidity, atrophy, and fasciculations, and bilateral, pathologically brisk deep tendon reflexes, and a bilateral Babinski response.

2 Based on this patient's symptoms, which neurons have been selectively affected?

3 What are the symptoms following an upper motor neuron lesion?

4 What are the symptoms following a lower motor neuron lesion?

SYNONYMS AND EPONYMS OF THE MOTOR CORTEX AND DESCENDING MOTOR PATHWAYS

Name of structure or term	Synonym(s)/eponym(s)	Name of structure or term	Synonym(s)/eponym(s)
Alpha (α) motor neuron	Lower motor neuron	Layer V of the cerebral cortex	Internal pyramidal layer
Anterior (ventral) horn of the spinal cord	Motor horn of the spinal cord	Gamma (γ) motor neuron	Lower motor neuron
		Medial reticulospinal tract	Pontine reticulospinal tract
Axial	Trunk	Mesencephalon	Midbrain
Basal nuclei	Basal ganglia	Muscle spindle	Neuromuscular spindle
	Deep cerebral nuclei		
Central sulcus	Central sulcus of Rolando	Oculomotor accessory nuclei	Interstitial nucleus of Cajal and nucleus of Darkschewitch
Cerebral vascular accident (CVA)	Stroke	Posterior (dorsal) horns of the spinal cord	Sensory horns of the spinal cord
Corticonuclear tract	Corticobulbar tract (older term)		
Corticoreticular and reticulospinal tracts	Corticoreticulospinal pathway	Posterior parietal cortex	Brodmann areas 5 and 7
		Premotor area	Premotor cortex
Corticorubral and rubrospinal tracts	Corticorubrospinal pathway		Part of Brodmann area 6
Corticospinal tract	Pyramidal tract (older term)	Primary motor cortex (MI)	Precentral gyrus
Corticotectal and tectospinal tracts	Corticotectospinal pathway		Brodmann area 4
Dorsal (posterior) column nuclei	Nucleus gracilis (NG) and nucleus cuneatus (NC)	Primary somatosensory cortex (SI)	Primary sensory cortex (SI)
	Gracile nucleus and cuneate nucleus		Primary somesthetic cortex
Extensor plantar response	Babinski sign		Brodmann areas 3, 1, and 2
	Inverted plantar reflex		Postcentral gyrus
Extrafusal muscle fibers	Skeletal muscle fibers (of gross muscle)	Principal nucleus of the trigeminal nerve	Main nucleus of the trigeminal nerve
Lower motor neuron	Lower motor neuron		Chief nucleus of the trigeminal nerve
Intermediolateral cell column	Lateral cell column		
	Lateral horn	Proximal limb musculature	Girdle musculature
Interneuron	Internuncial neuron	Red nucleus	Nucleus ruber
Intrafusal muscle fibers	Skeletal muscle fibers of muscle spindles	Somatosensory cortex	Somesthetic cortex
		Supplementary motor area (SMA)	Supplementary motor cortex
Lateral reticulospinal tract	Medullary reticulospinal tract		Part of Brodmann's area 6

FOLLOW-UP TO CLINICAL CASE

Both the history and the clinical exam are very important for the diagnosis in this case. This patient has amyotrophic lateral sclerosis (ALS), commonly known in the USA as Lou Gehrig's disease. This is a neurodegenerative disease affecting only the motor system. Sensory and cognitive functions remain intact throughout the disease. ALS characteristically leads to death of both the upper motor neurons in the motor cortex and also the lower motor neurons in the anterior horns of the spinal cord. It is the most common of the motor neuron diseases – diseases which affect, specifically, the upper and/or lower motor neurons.

The combination of the signs and symptoms of ALS is usually unmistakable. Upper motor neuron loss produces pathologically brisk reflexes, Babinski reflexes (dorsiflexion of the big toe in response to a scratch of the lateral aspect of the bottom of the foot), and sometimes spasticity or "stiffness" of an extremity. Lower motor neuron loss leads to flaccid weakness, atrophy, and fasciculations. Both types of lesions can cause weakness. Head imaging is typically normal.

Electromyography will confirm or refute the presence of motor neuron disease (that which affects the lower motor neurons). This is an electrical test of muscle fiber physiology and gives information about the integrity of the motor unit (a lower motor neuron its axon, and all of the skeletal muscle fibers that it innervates).

ALS is a devastating disease. It is more common in older age groups, but the average person affected by it is middle aged or a bit older. This disease leads to relentless progression of muscle weakness, causing the individual to become wheelchair-bound and eventually incapacitated. Distal muscles in one arm or leg, or bulbar muscles (e.g., speech) are affected first. Any muscle group can be a target, and eventually the muscles of breathing become affected. Death from respiratory failure or another intervening illness, such as pneumonia, typically occurs a few or several years from initial diagnosis. The vast majority of cases of ALS are sporadic. A small percentage of cases are hereditary. The pathophysiology is unknown and there is no effective treatment.

5 What tells you that this patient's symptoms are caused by a neurodegenerative disorder and not a stroke?

QUESTIONS TO PONDER

1. How do the intrafusal fibers of muscle spindles maintain sensitivity to detect the slightest stretch of the gross muscle, even during the contracted state?

2. As you are sitting in class looking at the screen in front of the room, you hear a loud, startling noise in the back of the room, as several large books fall from the bookshelf. The entire class turns in unison toward the back of the room to see the source of the noise. Which of the descending motor pathways is involved in the reflex turning of the eyes, head, and upper trunk, in the direction of the source of the noise?

3. Why does an individual who has suffered a stroke from a lesion confined to the primary motor cortex recover some of the crude motor function in the affected part of the body?

4. A capsular stroke damaged the upper motor neuron fibers descending in the posterior half of the posterior limb of the internal capsule, on the right side of the brain. In which parts of the body do you expect to see motor deficits?

5. Which of the three functional groups of the descending motor pathways are involved in mediating independent, fractionated movements of the digits, especially those of the hand?

CHAPTER 14

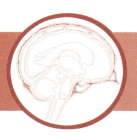

Basal Nuclei

CLINICAL CASE

COMPONENTS OF THE BASAL NUCLEI

NUCLEI ASSOCIATED WITH THE BASAL NUCLEI

INPUT, INTRINSIC, AND OUTPUT NUCLEI OF THE BASAL NUCLEI

CONNECTIONS OF THE BASAL NUCLEI

CIRCUITS CONNECTING THE BASAL NUCLEI, THALAMUS, AND CEREBRAL CORTEX

OTHER CIRCUITS OF THE BASAL NUCLEI

NEUROTRANSMITTERS OF THE BASAL NUCLEI

"DIRECT," "INDIRECT," and "HYPERDIRECT" LOOPS (PATHWAYS) OF THE BASAL NUCLEI

CIRCUITS THAT MODULATE ACTIVITY OF THE BASAL NUCLEI

CLINICAL CONSIDERATIONS

SYNONYMS AND EPONYMS OF THE BASAL NUCLEI

FOLLOW-UP TO CLINICAL CASE

QUESTIONS TO PONDER

CLINICAL CASE

A 39-year-old individual has had abnormal and uncontrollable movements for the past 6 months, as well as forgetfulness, obsessions, and compulsions. Their partner notes some mild personality changes as well. They have become somewhat more irritable and withdrawn. Their movements are spontaneous, unpredictable, and have become almost continuous. Initially, they thought their partner had become "fidgety." Abnormal behaviors as well as extraneous movements have caused some social embarrassment for the pair. The patient's mother died in her fifties at a psychiatric hospital. They have no more information about her. Family history is otherwise unrevealing.

Examination shows obvious chorea consisting of involuntary, nonrhythmic, and persistent jerky movements of the head, neck, and arms that continue during conversation. Memory is mildly diminished. They seem very distractible and have difficulty completing motor tasks. Neurologic exam is otherwise unremarkable. MRI of the brain shows atrophy of the heads of both caudate nuclei.

A Textbook of Neuroanatomy, Third Edition. Maria A. Patestas, Amanda J. Meyer, and Leslie P. Gartner.
© 2025 John Wiley & Sons, Inc. Published 2025 by John Wiley & Sons, Inc.
Companion website: www.wiley.com.go/patestas/neuroanatomy3e

Normal **motor function** is the result of the intricate interaction of the basal nuclei, cerebellum, and cerebral cortex. The **basal nuclei** (frequently called **basal ganglia**) plan and initiate motor activity and modulate cortical output related to motor function. The **cerebellum** functions in the coordination of movement, whereas the **cerebral cortex** is involved in the planning and execution of voluntary movement. The basal nuclei and cerebellum exert their influence on the brainstem and spinal cord, and ultimately on motor activity *indirectly*, by regulating the output of the cerebral cortex via the thalamus. The cerebral cortex then influences the execution of motor activity via *direct* projections to the brainstem nuclei (i.e., the cranial nerve motor nuclei, nuclei of the reticular formation, midbrain tectum, and red nucleus) and spinal cord motor neurons.

The function of the basal nuclei is influenced by input arising not only from **primary sensory** and **sensory association areas** of the cerebral cortex of all five lobes but also from the **thalamus** and **brainstem**. The thalamus, brainstem, and all five lobes of the cerebrum project to the input nuclei of the basal nuclei, mainly to the **caudate nucleus** and **putamen**. These input nuclei then project to the **globus pallidus**, which in turn relays basal nuclear output via the thalamus to the motor and other areas of the **frontal cortex** (Fig. 14.1).

Although more is known about the role of the basal nuclei in motor functions, it has been shown that they have nonmotor functions as well. Therefore, disturbances of the neural connections of the basal nuclei result not only in movement disorders, but also considerable deficits in other functions such as cognition, perception, and emotional behaviors.

COMPONENTS OF THE BASAL NUCLEI

The caudate nucleus and the putamen form the neostriatum; the globus pallidus forms the paleostriatum; the neostriatum and paleostriatum form the corpus striatum

The term "**basal ganglia**" is a misnomer since these structures consist of an assortment of subcortical *nuclei*, rather than *ganglia*. The criterion used by early neuroanatomists for the classification of the basal nuclei was any structure composed of gray matter that is embedded deep within, and close to the basal aspect of, the cerebral hemispheres. Thus, previously, the components of the basal nuclei included the caudate nucleus, putamen, nucleus accumbens, globus pallidus, thalamus, subthalamic nucleus, amygdaloid body, and claustrum. In more recent years, however, neuroanatomists include in the basal nuclei only those deep nuclei of the cerebrum which, when damaged, produce movement disorders. Although the older name basal ganglia is still used by many authors, this textbook adopted the new, approved name, **basal nuclei**.

From a clinical perspective, the components of the basal nuclei involved in motor function are the caudate nucleus, putamen, globus pallidus, and subthalamic nucleus (all of which are embedded deep in the cerebral hemispheres), as well as another nucleus, the substantia nigra, which, although located in the midbrain, is a functional component of the basal nuclei.

The basal nuclei are said to have two divisions, the **dorsal basal nuclei** and the **ventral basal nuclei**.

The **dorsal basal nuclei** consist of the caudate nucleus, the putamen, and the globus pallidus (Table 14.2). Because the **caudate nucleus** and the **putamen** share an embryologic derivation as well as similar morphological characteristics, they are considered to be a single anatomical structure and are collectively referred to as the **neostriatum** (Tables 14.1, 14.2). The **globus pallidus** is a phylogenetically older part of the basal nuclei and is also referred to as the **paleostriatum**. Anatomically, the **neostriatum** and the **paleostriatum** collectively form the **corpus striatum.** The putamen and the globus pallidus form a lens-shaped structure and are collectively referred to as the **lentiform nucleus**.

The **ventral basal nuclei** are composed of the ventral striatum and the ventral pallidum. The **ventral striatum** consists of the **nucleus accumbens** and part of the **olfactory tubercle**, whereas the **ventral pallidum** includes the **innominate substance**.

A salient histochemical characteristic of the neostriatum and particularly of the ventral striatum is the presence of patches, also called striosomes, and a matrix made up of modules called matrisomes. Earliest-born striatal neurons are embedded in striosomes, and the latest-born striatal neurons are embedded in the matrix. **Striosomes** are neurochemically specialized compartments accounting for 15% of the striatum and consist of **spiny** projections neurons (named for the abundant spines exhibited by their dendrites) and neuroglia and have high concentrations of gamma-aminobutyric acid, dopamine, dynorphin, neurotensin, substance P, and

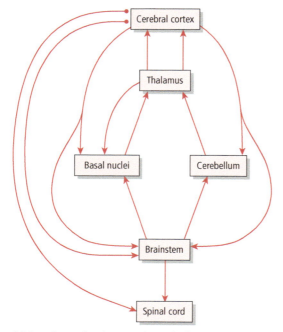

Figure 14.1 • Connections between the cerebral cortex, thalamus, basal nuclei, cerebellum, brainstem, and spinal cord.

Table 14.1 • Classification of the basal nuclei.

Basal nuclei	Striatum (neostriatum)	Ventral striatum	Lentiform nucleus	Corpus striatum
Caudate nucleus	Caudate nucleus	Nucleus accumbens	Putamen	Caudate nucleus
Putamen	Putamen	Part of olfactory tubercle	Globus pallidus	Putamen
Globus pallidus				Globus pallidus
Subthalamic nucleus (of ventral thalamus)				
Substantia nigra (of the midbrain)				

Table 14.2 • Basal nuclei and associated structures.

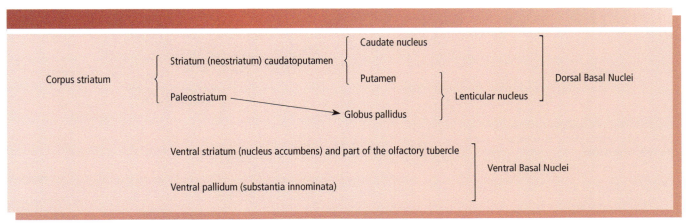

Table 14.3 • Nuclei associated with the basal nuclei.

Nucleus	Location	Function
Thalamic	Diencephalon	
Ventral anterior (VA) Ventral lateral (VL)		Relay signals related to movement from the basal nuclei to the frontal cortex
Mediodorsal (MD)		Relay signals from the basal nuclei to the frontal cortex association areas (which play a role in executive functions) and limbic areas (which play a role in processing of emotions)
Centromedian (CM) Parafascicular (PF)		Relay information from the thalamus (via thalamostriatal fibers) to the caudate nucleus, putamen, and ventral striatum
Amygdaloid body	Temporal lobe	Emotional and motivational aspects of movement
Claustrum	Between the external and extreme capsules in the telencephalon	Multimodal information processing to isolate the sensory information related to the task at hand.

enkephalin. Input to the striosomes comes from the limbic cortex and output goes to the ventral tegmental area of the midbrain, **compact part of the substantia nigra**, and the retrorubral area. **Matrix** accounts for 85% of the neostriatum and consists of matrisomes comprising largely of **aspiny** projection neurons (named for their smooth dendrites) and neuroglia and has high concentrations of acetylcholinesterase, calbindin, the reduced form of nicotinamide–adenine dinucleotide phosphate (NADPH) diaphorase (NO synthase), somatostatin, neuropeptide Y, and cholecystokinin. Input to the matrix comes from sensory, motor, and association areas of the cortex, and output goes primarily to the **reticular part of the substantia nigra**. Different inputs, outputs, and interneuron neuromodulators in striosomes and matrisomes suggest varied functionality and pathways to locally modulate the activity of each set of projection neurons. Ultimately, striosome and matrisome output will modulate striatal projections to the pallidum, and imbalances in this output can contribute to movement disorders and drug addiction.

In addition to the components mentioned so far, the following structures are functionally related to the basal nuclei: the **ventral anterior**, **ventral lateral complex**, **mediodorsal**, and **intralaminar nuclei of the thalamus**, and the **amygdaloid body** of the limbic system (Table 14.3).

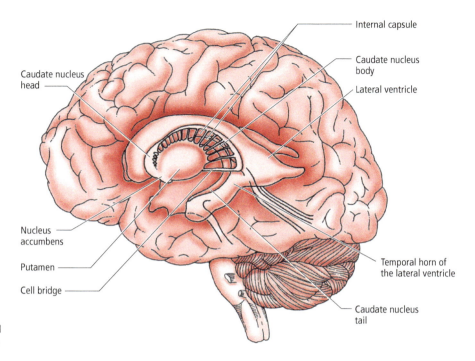

Figure 14.2 ● The caudate nucleus, putamen, and nucleus accumbens and their anatomical relation to the ventricular system.

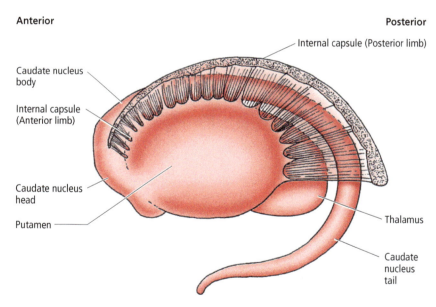

Figure 14.3 ● Lateral view of the corpus striatum and its anatomical relation to the internal capsule.

In this chapter, the **caudate nucleus**, the **lentiform nucleus** with its two components (the **putamen** and **globus pallidus**), the **nucleus accumbens**, the **subthalamic nucleus** of the ventral thalamus, and the **substantia nigra** of the midbrain also are included as components of the basal nuclei because they are functionally connected to the corpus striatum and produce movement disorders when they are damaged. The thalamus, nucleus accumbens, and amygdaloid body do not produce movement disorders when they are damaged; therefore, they are not considered to be part of the basal nuclei.

It is important to note that the **thalamus** in addition to serving as a relay nucleus for *sensory* information to the somatosensory cortex also serves as a relay nucleus for *motor* information that arises from the basal nuclei and from the cerebellum that is destined for the motor cortex.

Caudate nucleus

The caudate nucleus consists of a head, body, and tail, and is located in the walls of the lateral ventricle

The **caudate nucleus** (L., *cauda*, "tail") is a C-shaped structure that is subdivided into a head, body, and tail (Fig. 14.2). Its dilated, somewhat bulbous, rostral end is referred to as the **head of the caudate nucleus**. It is located in the lateral wall of the frontal horn of the lateral ventricle and is continuous rostrally with the nucleus accumbens (Fig. 14.3). It is separated from the

BASAL NUCLEI ••• 259

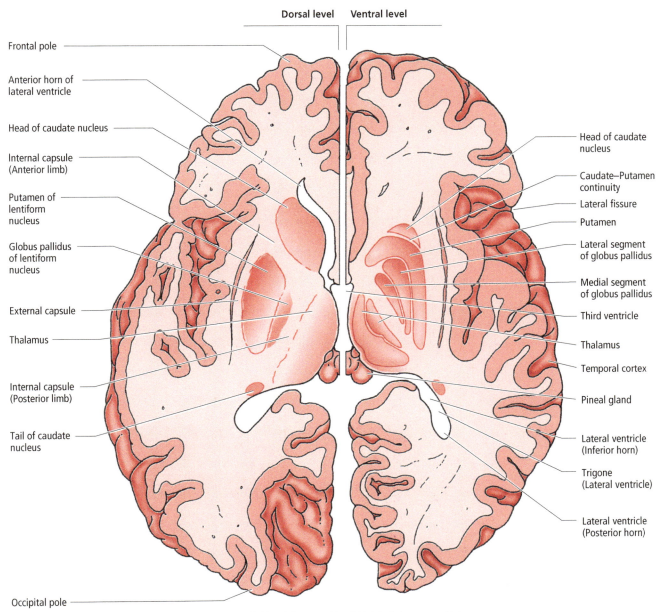

Figure 14.4 • Horizontal sections through the dorsal level (left) and the ventral level (right) of the corpus striatum.

lentiform nucleus by the anterior limb of the internal capsule. Near the interventricular foramen, the head is continuous with the long, curved **body of the caudate nucleus**, which courses posteriorly and lies on the roof of the lateral ventricle. Near the posterior end of the thalamus, the body decreases in diameter and continues as the long, narrow **tail of the caudate nucleus**. The tail gradually bends ventrally and then anteriorly within the roof of the temporal horn of the lateral ventricle passing lateral to the lateral geniculate nucleus of the thalamus, superior to the hippocampus in the floor, and then as far anteriorly as the amygdaloid nucleus.

Lentiform nucleus

The lentiform nucleus is composed of the putamen and the globus pallidus

The biconvex **lentiform nucleus**, located between the insula and the anterior and posterior limbs of the internal capsule, is separated from the caudate nucleus by the anterior limb of the internal capsule and from the thalamus by the posterior limb of the internal capsule. This nucleus is divided by myelinated fibers, the *lateral medullary lamina*, into its two parts: the laterally positioned, convex **putamen** and the medially placed **globus pallidus** (Fig. 14.4).

The **globus pallidus** (**GP**) (L., *Globus*, "globe"; *pallidus*, "pale") is a solid cone that is wedge-shaped on coronal and horizontal sections. Due to its many myelinated fibers, it appears paler in fresh specimens than the putamen. A group of myelinated fibers, the *medial (internal) medullary lamina*, divides the globus pallidus into a medial segment (GPm) and a lateral segment (GPl). The GPm is embryologically related to the diencephalon and the subthalamus (ventral thalamus) and gives rise to the major output from the basal nuclei. GABAergic projections from the ventromedial portion of the

260 ●●● **CHAPTER 14**

subthalamic nucleus (STN) connect it to the GPm, whereas glutaminergic projections from the dorsocaudal portion of the STN connect it to the GPl.

Although, as indicated above, the lentiform nucleus is separated from the caudate nucleus by the internal capsule, the rostral extent of the **putamen** is connected to the head of the caudate nucleus by bridge-like extensions of gray matter. During early developmental stages, the growing axons of neurons in the internal capsule perforated the gray matter of the fused caudate nucleus and putamen (Fig. 14.2). These extensions formed a distinct ventral portion of the striatum, the ventral striatum, which is associated with the limbic system. The intervals between these extensions are traversed by the somewhat vertically oriented, slender but prominent fiber bundles of the *anterior limb of the internal capsule* (see Fig. 14.3). Thus, in fresh brain sections, the region between the caudate nucleus and the lentiform nucleus exhibits alternating stripes of gray matter and white matter. Because of this striped pattern it was named the **corpus striatum** (L., "striped body").

Nucleus accumbens

The nucleus accumbens is associated with the limbic system and processes the emotional aspects of movement

The **nucleus accumbens**, a component of the ventral striatum, resides ventral to the anterior limb of the internal capsule and is connected to the caudate nucleus and the putamen. It receives input information from the limbic system and is involved in the processing of the emotional aspects of movement (see Fig. 14.2).

Subthalamic nucleus

The subthalamic nucleus is a component of the subthalamus

The **subthalamic nucleus** (eponym: nucleus of Luys), a component of the subthalamus, is an oval-shaped, biconvex mass of gray matter that lies lateral to the hypothalamus on the medial aspect of the internal capsule (Fig. 14.5). Its connections with the primary motor cortex, supplementary motor area, associative cortices, and limbic lobes suggest a key role of the subthalamic nucleus in modulating output from the basal nuclei.

Substantia nigra

The substantia nigra, the largest nucleus of the midbrain, consists of two components: the reticular and compact parts

The **substantia nigra** (L., "black substance") extends from the rostral end to the caudal end of the midbrain dorsal to the cerebral peduncle. It is composed of two distinct parts: a ventral, reticular part (SNr), and a dorsal compact part (SNc) (Fig. 14.6).

The reticular part of the substanita nigra (SNr) contains 30% fewer cells than the compact part (SNc) and consists of

(1) neuron terminals arising from the midbrain raphe nuclei, which release serotonin; and (2) striatonigral nerve terminals, which release gamma-aminobutyric acid (GABA). The GABAergic neurons of the SNr are inhibited by the striatonigral axon terminals of neurons whose cell bodies reside in the matrix of the striatum. Although the SNr contains numerous nerve terminals, it is a nuclear structure and contains nerve cell bodies. The SNr is continuous with, and shares many histologic characteristics with, the **medial segment of the globus pallidus** (**GPm**). The neurons in the SNr not only share morphologic, physiologic, and functional characteristics with the neurons of the GPm but also this portion of the substantia nigra receives projections from the striatum and sends its GABAergic neurons to the thalamus. Both the medial segment of the globus pallidus and the reticular part of the substantia nigra give rise to the *output* of the basal nuclei; due to their morphologic and connection similarities and continuity, they are considered to be the same structure.

The compact part of the substantia nigra (SNc) contains approximately 375,000 **neuromelanin**-rich dopaminergic neurons, which give rise to the nigrostriatal dopaminergic pathway that projects to the striatum. Moreover, the dendritic extensions of these dopaminergic neurons extend into the reticular part of the substantia nigra. The dopaminergic nigrostriatal neurons from the compact part are inhibited by the striatonigral axon terminals of neurons whose cell bodies reside in the striatal striosomes.

The SNc and Ventral Tegmental Area (VTA) are derived from the floor plate of the embryonic midbrain in neuromere m1 and share common marker genes CALB1, PITX3, DCC, GIRK2. In addition, the SNc stretches from the isthmus and into the diencephalon and expresses the SOX6 gene. Dopaminergic neurons with cell bodies in the SNc are defined as belonging to the A9 cell group, while those with cell bodies in the VTA belong to the A10 cell group. They also have different connections with the SNc projecting to the striatum for motor function and the VTA projecting to the orbitofrontal and anterior cingulate cortices for limbic function. It has been estimated that each SNc dopaminergic neuron makes over 1 million synaptic connections, compared with VTA dopaminergic neurons which form a little over 120,000 connections, suggesting that dopaminergic neurons in the compact part of the substantia nigra must be highly branched and exceptionally active cells.

NUCLEI ASSOCIATED WITH THE BASAL NUCLEI

Thalamic nuclei

The basal nuclei are interconnected with the ventral anterior, ventral lateral, medial dorsal, and intralaminar nuclei of the thalamus

The **ventral anterior** (VA) and the **ventral lateral (VL) nuclei** of the **lateral thalamic** group (often referred to as the "thalamic motor nuclei") relay movement-related signals from the basal nuclei to the frontal cortex. The medial segment of the globus pallidus is the output nucleus that sends GABAergic projections to the VA/VL. Glutaminergic neurons from the

VA/VL project to layer I of the cerebral cortex in the precentral gyrus of the frontal lobe.

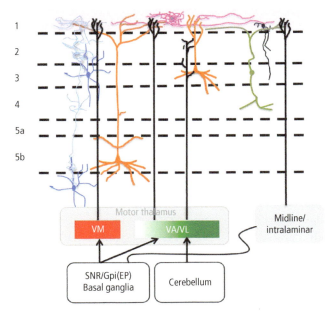

Major projection neurons (black) from the motor nuclei of the thalamus project to layer I of the cerebral cortex. These neurons release neurotransmitters which affect axon terminals in layer I which belong to neurons whose cell bodies are in other layers of the cerebral cortex. Medium-sized pyramidal neurons (orange) whose cell bodies are located in layer III give rise to intracortical projections, and large pyramidal neurons (orange) in layer V give rise to the corticofugal projections associated with motor control. Source: Frontiers Media S.A./https://www.frontiersin.org/articles/10.3389/fncir.2015.00071/full / CC BY 4.0.

The **mediodorsal nucleus** of the **thalamus** relays signals from the basal nuclei to frontal cortex association areas as well as to limbic areas. These areas play a role in cognitive functions and the processing of emotions, respectively.

The **centromedian** and **parafascicular nuclei**, two of the **intralaminar nuclei of the thalamus** representing the rostral extent of the ascending reticular activating system (ARAS), receive inputs from the spinothalamic, trigeminothalamic, and multisynaptic ascending pathways (of the reticular formation) relaying pain sensation. These nuclei convey this information to the somatosensory areas of the cerebral cortex and to the caudate nucleus, putamen, and ventral striatum (see Fig. 14.5; Table 14.3). Based on these connections, these nuclei are believed to function in sensorimotor integration. In addition, as a result of their diffuse cortical connections, they are believed to be involved in the maintenance of arousal of the organism as it relates to pain.

Amygdaloid body

The amygdaloid body is a component of the limbic system and is related to the basal nuclei via neural connections

The **amygdaloid body** is located in the temporal lobe, deep to the uncus. Its centromedial group is thought to be involved in mediating motor movement, visceral responses, and attentional reallocation. Although the amygdaloid body is related to the basal nuclei via its neural connections, it is anatomically and functionally part of the limbic system (Fig. 14.7; Table 14.3).

Claustrum

The **claustrum** consists of a slender layer of gray matter, which is separated medially from the convex surface of the putamen by the *external capsule*, and laterally from the insula by a thin layer of white matter, the *extreme capsule*. It comprises two types of neurons: glutaminergic (85%) and GABAergic (15%). Recent studies suggest that the claustrum modulates cortical excitability and also plays a role in multimodal information processing, isolating sensory information related to the task at hand (Fig. 14.8, Table 14.3).

INPUT, INTRINSIC, AND OUTPUT NUCLEI OF THE BASAL NUCLEI

The basal nuclei are intricately and extensively interconnected with various regions of the central nervous system (CNS). Viewed from the perspective of their connections, the component nuclei of the basal nuclei may be assigned into three categories: **input nuclei**, **intrinsic nuclei**, and **output nuclei** (Table 14.4).

Input nuclei

The input nuclei of the basal nuclei are the caudate nucleus, putamen, and nucleus accumbens

The caudate nucleus, putamen, and the nucleus accumbens are the **input nuclei** of the basal nuclei. They receive prominent glutaminergic *excitatory* input from all five lobes of the cerebral cortex as well as from the thalamus, subthalamic nucleus, substantia nigra, and other brainstem structures. Most of the input reaches the caudate nucleus and the putamen, from where the information is relayed to the intrinsic and output nuclei. The caudate nucleus and the putamen function in the initiation and modulation of gross voluntary movements that are performed at the subconscious level. The caudate nucleus is involved in cognitive functions and plays a role in the control of eye movements, whereas the nucleus accumbens is associated with the processing of the emotional aspects of movement.

Intrinsic nuclei

The intrinsic nuclei of the basal nuclei are the lateral segment of the globus pallidus, the ventral pallidum, the subthalamic nucleus, the compact part of the substantia nigra, and the ventral tegmental area

The **intrinsic nuclei** consist of the lateral segment of the globus pallidus (GPl), the ventral pallidum (innominate substance), the subthalamic nucleus, the compact part of the substantia nigra (SNc),

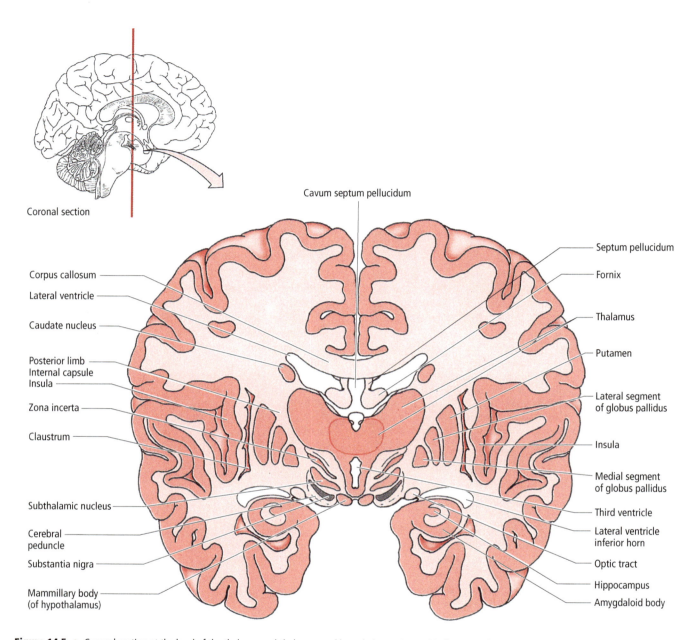

Figure 14.5 ● Coronal section at the level of the thalamus, subthalamus, and hypothalamus. Note, this illustration has come from an individual who had an anatomical variation of a cavum septum pellucidum.

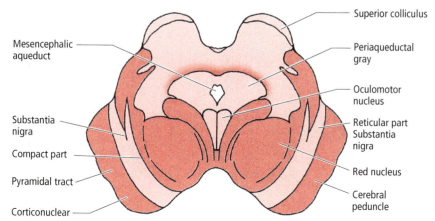

Figure 14.6 ● Transverse section of the rostral midbrain, showing the substantia nigra.

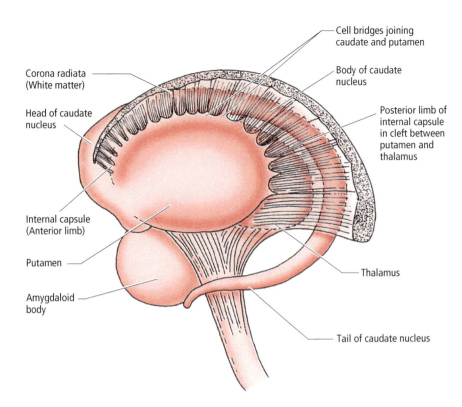

Figure 14.7 • Lateral view of the corpus striatum, amygdaloid body, thalamus, and internal capsule.

and the ventral tegmental area (VTA). These nuclei are not only interconnected by their local circuit projections but are also connected with the input and output nuclei. Although the subthalamic nucleus is a source of *excitatory* signals to the GPl, most local synapses are *inhibitory*.

Both the lateral segment of the globus pallidus and the subthalamic nucleus function in the control of the axial and girdle (proximal limb) musculature. The responsibility of these nuclei is to position and stabilize the trunk and proximal parts of the upper and lower limbs so that the distal limb musculature moving the hands and feet can perform the more discreet movements that are controlled by the motor cortex. This concept is consistent with the observation that neural activity in the globus pallidus precedes that of the motor cortex.

Output nuclei

The output nuclei of the basal nuclei are the medial segment of the globus pallidus (GPm), the reticular part of the substantia nigra (SNr), and the ventral pallidum.

The **output nuclei** consist of the GPm, the SNr, and the ventral pallidum.

Output from the basal nuclei is *inhibitory* (GABAergic), passing from the GPm and SNr mainly to the thalamus, via the pallidothalamic and nigrothalamic fibers, respectively. However, fibers from these nuclei also project to the subthalamic nuclei, the brainstem, and the cerebral cortex. The SNr functions in cognition and control of eye movement.

The basal nuclei exert their influence on motor activity *indirectly* by controlling the premotor and supplementary motor areas of the cerebral cortex, and *not* by direct projections to the brainstem and spinal cord motor neurons. Note that although all five lobes of the cerebral cortex project to the basal nuclei, output from the basal nuclei terminates exclusively in the frontal lobe of the cerebrum.

CONNECTIONS OF THE BASAL NUCLEI

The basal nuclei have the following characteristics:

- they receive **input fibers** from various extrinsic sources;
- they are interconnected by **local (intrinsic) fiber projections**; and
- they project **output fibers** to other areas of the brain that are involved in motor function.

Striatum

Afferent fibers (input)

The caudate nucleus and the putamen are the principal input nuclei of the basal nuclei

The caudate nucleus and the putamen are the principal input nuclei of the basal nuclei. The caudate nucleus is associated primarily with cognitive functions with some association with motor activity, whereas the putamen is associated primarily with motor functions. The caudate nucleus and putamen receive the following **afferent (input) fibers** (Fig. 14.9; Table 14.5):

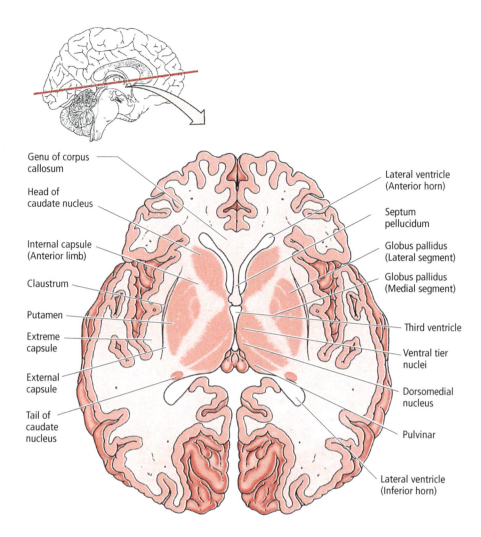

Figure 14.8 • Horizontal section of the brain at the level of the basal nuclei.

Input nuclei	Intrinsic nuclei	Output nuclei
Caudate nucleus	Globus pallidus (lateral segment)	Globus pallidus (medial segment)
Putamen	Subthalamic nucleus	Ventral pallidum
Nucleus accumbens	Compact part of Substantia nigra Ventral pallidum Ventral tegmental area	Reticular part of Substantia nigra

Table 14.4 • Input, intrinsic, and output nuclei of the basal nuclei.

- corticostriatal fibers;
- thalamostriatal fibers;
- nigrostriatal fibers;
- fibers from the ventral tegmental area; and
- fibers from the brainstem pedunculopontine tegmental nucleus (PPN).

Corticostriatal fibers

The corticostriatal fibers form the principal input to the basal nuclei

The **corticostriatal fibers** arise primarily from the motor and somatosensory areas of the cerebral cortex (neocortex). Each cortical part projects fibers that descend in the anterior limb of the internal capsule or in the external capsule to the **caudate nucleus** and the **putamen**, and they release the excitatory neurotransmitters (glutamate or aspartate) at their termination. Fibers arising from the primary sensory (eponym: Brodmann areas 3, 1, and 2), primary motor (eponym: Brodmann area 4), premotor (eponym: Brodmann area 6), and supplementary motor (eponym: Brodmann area 6) cortex terminate principally in the putamen. Fibers arising from the frontal eye field (eponym: Brodmann area 8) and association areas of the cortex terminate principally in the caudate nucleus.

Table 14.5 ● Afferent (input) fibers to the striatum.

Fibers	Origin	Termination	Neurotransmitter (and type)
Corticostriatal	Sensory cortex (Brodmann areas 3, 1, and 2), Primary motor cortex (Brodmann area 4) Premotor cortex (Brodmann area 6) Supplementary motor cortex (Brodmann area 6)	Putamen	Glutamate (excitatory) *or* Aspartate (excitatory)
	Frontal eye field association areas of the cerebral cortex	Caudate nucleus	Glutamate (excitatory) Aspartate (excitatory)
Thalamostriatal	Intralaminar nuclei of the thalamus Centromedian nucleus Parafascicular nucleus Ventral anterior and ventral lateral nuclei of the thalamus	Caudate nucleus and putamen	Glutamate (excitatory)
Nigrostriatal	Substantia nigra Compact part Reticular part	Caudate nucleus and putamen	Dopamine (inhibitory) GABA (inhibitory)
Other	Ventral tegmental area (VTA) Pedunculopontine tegmental nucleus (PPN)	Ventral striatum Caudate nucleus, putamen, pallidum, and subthalamic nucleus	Dopamine (inhibitory) Serotonin (inhibitory)
	Subthalamic nucleus	Caudate nucleus	GABA (inhibitory)

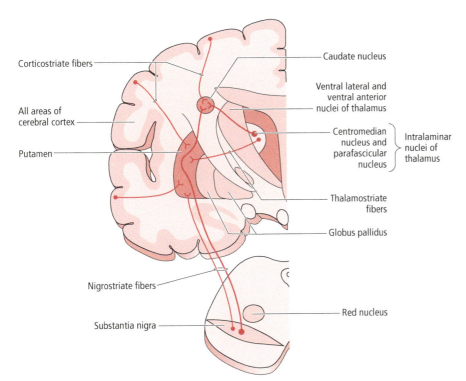

Figure 14.9 ● Afferent (input) projections to the caudate nucleus and putamen. Source: Modified from Watson, C (1995) *Basic Human Neuroanatomy*, 5th ed. Little, Brown & Company, Boston; fig. 39.

Thalamostriatal fibers

The thalamostriatal fibers arise from the intralaminar nuclei of the thalamus and form the second major input to the basal nuclei

The intralaminar (centromedian and parafascicular) nuclei of the thalamus are the cranial continuation of the midbrain reticular formation into the diencephalon and are components of the reticular activating system (RAS). These nuclei receive nociceptive signals via reticulothalamic fibers (of the anterolateral system) and, as components of the RAS, function in alerting the individual and eliciting responses that will enable them to evade painful stimuli.

The **thalamostriatal fibers** are excitatory; they terminate mainly in the **putamen** and the **caudate nucleus**, where they

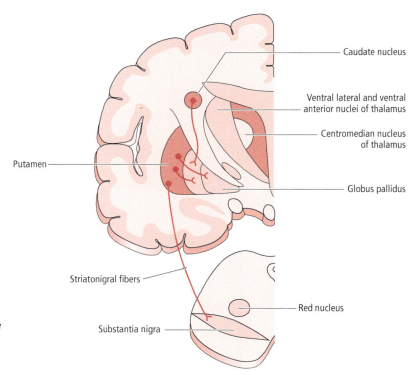

Figure 14.10 • Efferent (output) projections from the caudate nucleus and putamen. Source: Modified from Watson, C (1995) *Basic Human Neuroanatomy*, 5th ed. Little, Brown & Company, Boston; fig. 40.

Table 14.6 • Efferent (output) fibers from the striatum.

Fibers	Origin	Termination	Neurotransmitter (and type)
Striatopallidal	Caudate nucleus and putamen	Globus pallidus 　Internal segment (GPm) 　External segment (GPl)	 GABA and substance P GABA and enkephalin
Striatonigral	Caudate nucleus and putamen	Substantia nigra (reticular part)	GABA, ACh, or substance P

ACh, acetylcholine; GABA, gamma-aminobutyric acid.

release glutamate. In addition, the striatum also receives input fibers from the ventral anterior and ventral lateral nuclei of the thalamus. Projections from the ventral anterior thalamic nucleus project to the caudate nucleus, whereas projections from the ventral lateral thalamic nuclei project to the putamen. Thalamostriatal projections contribute to the smooth initiation and execution of movement sequences.

Nigrostriatal fibers

The **nigrostriatal fibers** arise from the compact part of the substantia nigra and terminate primarily in the **caudate nucleus**, although some fibers terminate in the **putamen**. These predominantly uncrossed fibers are inhibitory or excitatory and release dopamine at their termination. Their influence is determined by the presence of dopamine 1 or dopamine 2 (D1 or D2) receptors.

Fibers from the ventral tegmental area and pedunculopontine tegmental nucleus

The striatum also receives inhibitory dopaminergic nerve fiber terminals from the ventral tegmental area and serotonergic nerve fibers from the pedunculopontine tegmental nucleus

The fibers from the **ventral tegmental area** terminate primarily in the nucleus accumbens of the ventral striatum and the cerebral cortex. These projections function in the initiation of behavioral responses.

The fibers arising from the brainstem **pedunculopontine tegmental nucleus** terminate in the **caudate nucleus**, **putamen**, **pallidum**, and **subthalamic nucleus**. These inhibitory fibers release serotonin at their termination. The pedunculopontine tegmental nucleus may function in the rhythm of movements.

Efferent fibers (output)

Although the caudate nucleus and putamen receive inputs from multiple sources, their output fibers are funneled only to the pallidum and substantia nigra

Although the caudate nucleus and putamen receive inputs from multiple sources such as the cerebral cortex, thalamus, and brainstem, their output fibers are funneled only to the pallidum and substantia nigra.

The **caudate nucleus** and the **putamen** give rise to the following **efferent (output) fibers** (Fig. 14.10; Table 14.6):

- striatopallidal fibers; and
- striatonigral fibers.

Striatopallidal fibers

The striatopallidal fibers terminate in both segments of the globus pallidus

The **striatopallidal fibers** that terminate in the lateral segment of the globus pallidus liberate GABA and enkephalin, whereas those fibers that terminate in the medial segment of the globus pallidus release GABA and substance P.

Striatonigral fibers

The **striatonigral fibers** terminate in the reticular part of the substantia nigra. These fibers release GABA, acetylcholine, or substance P.

Globus pallidus

The medial and lateral segments of the **globus pallidus** receive inputs from the same regions, but their output projections differ.

Afferent fibers (input)

Both segments of the globus pallidus receive the following **afferent (input) fibers** (Fig. 14.11; Table 14.7):

- striatopallidal fibers; and
- subthalamopallidal fibers.

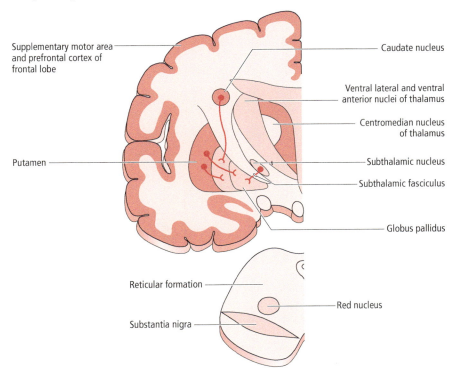

Figure 14.11 • Afferent (input) projections to the globus pallidus. Source: Modified from Watson, C (1995) *Basic Human Neuroanatomy*, 5th ed. Little, Brown & Company, Boston; fig. 41.

Fibers	Origin	Termination	Neurotransmitter (and type)
Striatopallidal	Caudate nucleus and Putamen	Globus pallidus (both segments)	GABA and enkephalin (both are inhibitory) *or* GABA (inhibitory) and substance P (excitatory)
Subthalamopallidal	Subthalamic nucleus	Globus pallidus (mainly medial segment)	Glutamate (excitatory)

GABA, gamma-aminobutyric acid.

Table 14.7 • Afferent (input) fibers to the globus pallidus.

Striatopallidal fibers

The striatopallidal fibers arising from the caudate nucleus and the putamen form the principal input to the globus pallidus

As discussed above, the **striatopallidal fibers** arising from the caudate nucleus and the putamen form the *principal input* to the globus pallidus. They release GABA and enkephalin or GABA and substance P.

Subthalamopallidal fibers

The **subthalamopallidal fibers**, arising from the subthalamic nucleus, form the subthalamic fasciculus. These *excitatory* fibers terminate mainly in the medial segment of the globus pallidus where they release glutamate.

Efferent fibers (output)

Both the medial and lateral segments of the globus pallidus give rise to well-defined output fibers

Even though the medial and lateral segments of the globus pallidus share both morphological and neurochemical characteristics, each gives rise to different, well-defined output fibers (Fig. 14.12).

Output from the lateral segment of the globus pallidus

The **lateral segment of the globus pallidus** gives rise to:

- pallidosubthalamic fibers;
- pallidonigral fibers; and
- other fiber projections that terminate in the striatum, reticular thalamic nucleus, and the medial segment of the globus pallidus (Table 14.8).

Pallidosubthalamic fibers

Pallidosubthalamic fibers arise only from the lateral segment of the globus pallidus

The **pallidosubthalamic fibers** arise only from the lateral segment of the globus pallidus. They join the subthalamic fasciculus, which terminates in the subthalamic nucleus and sends collateral branches to the reticular part of the substantia nigra. These are inhibitory fibers and release GABA at their terminals.

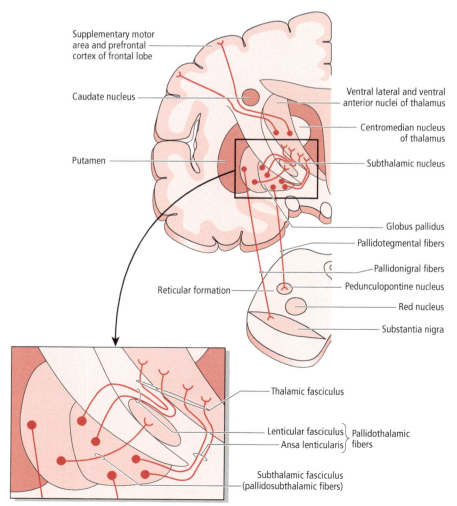

Figure 14.12 ● Efferent (output) projections from the globus pallidus. Source: Modified from Watson, C (1995) *Basic Human Neuroanatomy*, 5th ed. Little, Brown & Company, Boston; fig. 42.

BASAL NUCLEI ••• 269

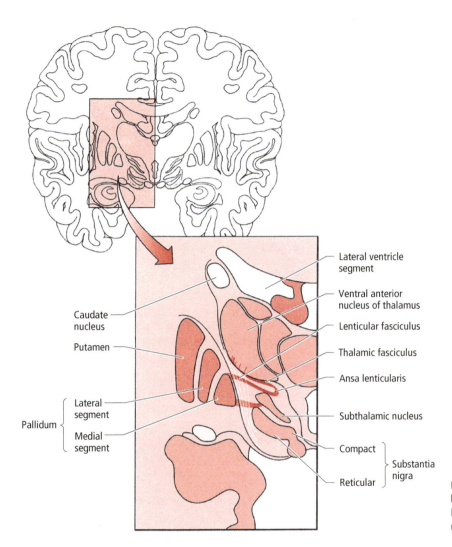

Figure 14.13 • Principal output connections of the basal nuclei arising from the globus pallidus: the ansa lenticularis and lenticular fasciculus. These two fasciculi merge to form the thalamic fasciculus.

Pallidonigral fibers

The pallidonigral fibers arise from the lateral segment of the globus pallidus and terminate in the compact part of the substantia nigra

The **pallidonigral fibers** are inhibitory and release GABA at their termination. In addition, the lateral segment of the globus pallidus also relays signals to the reticular nucleus of the thalamus, which assists in the integration of reciprocal connections between the thalamus and the cerebral cortex.

Output from the medial segment of the globus pallidus

The **medial segment of the globus pallidus**, the main output nucleus of the corpus striatum, gives rise to:

- pallidothalamic fibers (Fig. 14.13);
- pallidotegmental fibers; and
- pallidohabenular fibers (Table 14.8).

Pallidothalamic fibers

The pallidothalamic fibers have two main components: the ansa lenticularis and the lenticular fasciculus

The ventral part of the medial segment of the globus pallidus gives rise to a group of **pallidothalamic fibers** that course ventromedially, and then rostrally, to form a loop, the **ansa lenticularis**, on the medial aspect of the posterior limb of the internal capsule. The ansa lenticularis then courses posteriorly, in the prerubral field (eponym: H field of Forel).

The dorsal part of the medial segment of the globus pallidus gives rise to another group of pallidothalamic fibers that course medially, traversing the posterior limb of the internal capsule. Medial to the subthalamus, these fibers assemble to form the **lenticular fasciculus** (eponym: H_2 field of Forel).

The pallidothalamic fibers coursing in the lenticular fasciculus subsequently join the pallidothalamic fibers of the ansa lenticularis, to form a prominent fiber bundle known

270 ● ● ● **CHAPTER 14**

Fibers	Origin	Termination	Neurotransmitter (and type)
Pallidosubthalamic	Globus pallidus (external segment)	Subthalamic nucleus and collateral branches to the substantia nigra (reticular part)	GABA
Pallidonigral	Globus pallidus (external segment)	Substantia nigra (compact part)	GABA
Pallidothalamic Ansa lenticularis Lenticular fasciculus	Globus pallidus (internal segment)	Thalamic nuclei: ventral anterior (VA) ventral lateral (VL) mediodorsal (MD) centromedian (CM) parafascicular (PF)	
Pallidotegmental	Globus pallidus (internal segment)	Pedunculopontine tegmental nucleus (PPN)	
Pallidohabenular	Globus pallidus (medial segment)	Lateral habenular nucleus	
Other	Globus pallidus (lateral segment)	Striatum Reticular thalamic nucleus Globus pallidus (medial segment)	

Table 14.8 ● Efferent (output) fibers from the globus pallidus.

as the **thalamic fasciculus** (eponym: H_1 field of Forel). This bundle proceeds medially and caudally to the prerubral field rostral to the red nucleus. The fibers then form a loop and course to the thalamus terminating in the following nuclei of the thalamus: ventral anterior, ventral lateral, mediodorsal, centromedian, and parafascicular nuclei. The thalamus in turn projects to the primary motor cortex, the supplementary motor area, and the premotor cortex of the premotor cortical areas, all of which contribute fibers to the corticonuclear, corticospinal, and corticopontocerebellar tracts. The signals arising from the basal nuclei that are relayed to the spinal cord via the contralateral lateral corticospinal tracts are involved in voluntary and skilled limb movements. Recall that the ventral anterior and ventral lateral nuclei of the thalamus are considered to be the "thalamic motor nuclei" since they receive signals related to motor function and send signals to the motor areas of the cortex.

Note that the output fibers arising from the basal nuclei and the fibers arising from the cerebellum that are destined for the ventral anterior and ventral lateral nuclei of the thalamus, do not overlap at their terminations. They terminate in distinct areas of these thalamic nuclei.

Pallidotegmental fibers

The pallidotegmental fibers arise from the medial segment of the globus pallidus and terminate in the pedunculopontine tegmental nucleus (PPN) in the inferior tegmentum of the midbrain

The pallidotegmental fibers arise from the medial segment of the globus pallidus and terminate in the pedunculopontine

tegmental nucleus (PPN) in the inferior tegmentum of the midbrain. The PPN, a nucleus of the reticular formation, is associated with the reticulospinal tract cells, which terminate on lower motor neurons and affect motor function.

Pallidohabenular fibers

Pallidohabenular fibers course in the ansa lenticularis and in the lenticular fasciculus, and then the stria medullaris

The pallidohabenular fibers arise from the medial segment of the globus pallidus, course in the ansa lenticularis and in the lenticular fasciculus within the thalamic fasciculus, and then join the stria medullaris to terminate in the lateral habenular nucleus.

The habenular nucleus, a component of the habenular nuclear complex of the epithalamus, is associated with the limbic system.

CIRCUITS CONNECTING THE BASAL NUCLEI, THALAMUS, AND CEREBRAL CORTEX

The four separate systems of circuits of the basal nuclei are arranged in parallel

The four separate systems of circuits (paths) of the basal nuclei are arranged in parallel, where each circuit relays distinct input from various areas of the cerebral cortex to the basal nuclei where information is integrated and is then relayed to the thalamus. From there the integrated information is conveyed to different areas of the frontal cortex where the execution of motor activity is initiated.

Each **circuit** has "closed-loop" and "open-loop" components. In the "closed-loop" element, information arising from a particular cortical area is relayed through the circuit back to its initial cortical source. In the "open-loop" component, information (arising from cortical areas that have a similar function as the initial cortical area of the closed loop) is conveyed to the cortical source of the closed loop.

Although in the paragraphs above, careful attention was paid to maintaining a difference between "circuits" and "loops," in reality these terms are used synonymously. The four loops described below are circuits, each with closed-loop and open-loop components. Each of the four loops also has a direct and an indirect pathway (see below).

Each system of circuits consists of many distinct parallel circuits that are anatomically and functionally distinct. One circuit within a particular system may be associated with a specific movement of the foot, whereas another circuit in the same system may be associated with the same type of movement of the hand.

The following circuits (loops) have been identified: a motor loop, an oculomotor loop, an association loop, and a limbic loop.

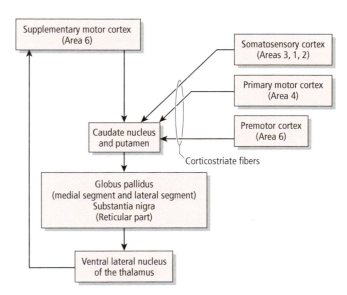

Figure 14.14 • Sensory–motor loops. In the closed loop, information flows from the supplementary motor cortex to the caudate nucleus and putamen, and from there to the globus pallidus and the substantia nigra, continuing to the thalamus and then back to the supplementary motor cortex. In the open loop, input is contributed by the somatosensory cortex, primary motor cortex, and premotor cortex to the closed loop. Source: Adapted from Noback, CR et al. (1996) *The Human Nervous System*, 5th ed. Williams & Wilkins, Baltimore; fig. 24.5.

Motor (sensory-motor) loop

The basal nuclei influence motor activity by projections to the motor cortex

Signals from the basal nuclei are relayed to the motor cortical areas via the motor loop where they influence the upper motor neurons of the corticospinal, corticonuclear, and other descending motor tracts, which in turn affect the lower motor neurons of the brainstem and spinal cord, and ultimately shape motor activity.

The motor areas, as well as various other cortical areas, give rise to a massive number of **corticostriatal fibers**, which terminate in the **caudate nucleus** and **putamen** (Fig. 14.14).

In the **closed loop**, a stream of corticostriatal fibers arising from the **supplementary motor cortex** (eponym: Brodmann area 6) project mostly to the putamen, the "sensory-motor area" of the striatum, as well as to the caudate nucleus. The putamen projects to both the medial and lateral segments of the **globus pallidus** and the reticular part of the **substantia nigra**, all of which in turn project to the **ventral lateral nucleus of the thalamus**. Thalamocortical fibers from the ventral lateral nucleus of the thalamus project back to the **supplementary motor cortex**, thus completing the loop.

In the **open loop**, corticostriatal fibers arising from the **primary motor cortex** (eponym: Brodmann area 4), the **premotor cortex** (eponym: Brodmann area 6), and the **somatosensory cortex** (eponym: Brodmann areas 3, 1, and 2) are conveyed to the putamen. From there the information continues through the motor circuit to terminate in the **supplementary motor cortex**, which is the origin of the closed loop.

In the **motor loop**, signals from the basal nuclei are relayed to the motor cortical areas of the **frontal lobe**. There they exert their influence by planning the sequence of actions to execute learned motor activities whose programs are stored in the putamen. This influence is exerted on the upper motor neurons, which in turn influence (via the corticonuclear and corticospinal tracts) the lower motor neurons of the brainstem and spinal cord, and ultimately motor activity of the face, trunk, and limbs.

Oculomotor (visuomotor) loop

The oculomotor loop has an important function in the control of voluntary saccadic ocular movements

In the **closed loop**, **corticostriatal fibers** arising from the **frontal eye field** (eponym: Brodmann area 8) of the cerebral cortex project to the body of the **caudate nucleus** (Fig. 14.15). The caudate nucleus projects to the medial segment of the **globus pallidus** and the reticular part of the **substantia nigra**. Pallidothalamic and nigrothalamic fibers project to the **ventral anterior** and **dorsomedial nuclei of the thalamus**. Thalamocortical fibers project back to the **frontal eye field**, thus closing the loop.

In the **open loop**, corticostriatal fibers arising from the **prefrontal cortex** (Brodmann's areas 9 and 10) and the **posterior parietal cortex** (Brodmann's area 7) terminate in the body of the caudate nucleus. The signals are relayed through the circuit to terminate in the frontal eye field, which is the origin of the closed loop. The globus pallidus and the substantia nigra also relay this information to the deep layers of the **superior colliculus**.

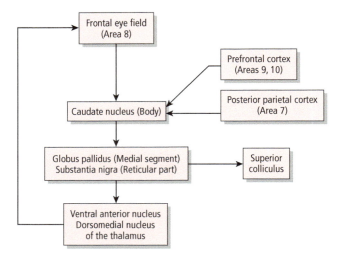

Figure 14.15 • Oculomotor loops. In the closed loop, information flows from the frontal eye field to the caudate nucleus and then to the globus pallidus and the substantia nigra and from there to the thalamus and then back to the frontal eye field. In the open loop, input is contributed by the prefrontal cortex and the posterior parietal cortex to the closed loop. Source: Adapted from Noback, CR et al. (1996) *The Human Nervous System*, 5th ed. Williams & Wilkins, Baltimore; fig. 24.7.

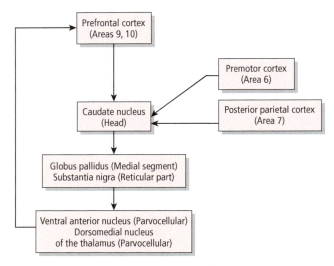

Figure 14.16 • Association loops. In the closed loop, information flows from the prefrontal cortex to the caudate nucleus and then to the globus pallidus and the substantia nigra, and from there to the thalamus and then back to the prefrontal cortex. In the open loop, input is contributed by the premotor and posterior parietal cortex to the closed loop. Source: Adapted from Noback, CR et al. (1996) *The Human Nervous System*, 5th ed. Williams & Wilkins, Baltimore; fig. 24.6.

The **oculomotor loop** has an important function in the control of voluntary saccadic ocular movements, which are rapid jerky movements of the eyes that abruptly change the point of fixation from one object of interest to another.

The association (cognitive) loop functions in the planning of complex motor activity and determining the direction of movement.

Association (cognitive, executive) loop

The association loop functions in the planning of motor activity and determining the direction of movement

In the **association loop**, various association cortical areas of the four lobes project mainly to the ipsilateral head of the **caudate nucleus**, which is the "association area" of the striatum (Fig. 14.16).

In the **closed loop**, **corticostriatal fibers** arising from the **prefrontal cortex** (eponym: Brodmann areas 9 and 10), project to the head of the caudate nucleus, which in turn projects to the medial segment of the **globus pallidus** as well as the reticular part of the **substantia nigra**. They in turn project to the **ventral anterior** and **dorsomedial nuclei of the thalamus**. Thalamocortical fibers terminate in the prefrontal cortex, closing the loop.

In the **open loop**, corticostriatal fibers arising from the **premotor cortex** (eponym: Brodmann area 6) and the **posterior parietal motor area** (eponym: Brodmann area 7) project to the head of the caudate nucleus. The signals are relayed through the circuit to the prefrontal cortex, the origin of the closed loop.

Limbic (motivational) loop

The limbic loop functions in the emotional and motivational aspects of movement manifested as various facial expressions or other body movements

In the **closed loop**, **corticostriatal fibers** arising from the **anterior cingulate gyrus** (eponym: Brodmann area 24) and the **orbitofrontal cortex** (eponym: Brodmann areas 10 and 11) project to the **ventral striatum (nucleus accumbens)** and the head of the **caudate nucleus** (Fig. 14.17). They, in turn, project to the **ventral pallidum**, the medial segment of the **globus pallidus**, and the reticular part of the **substantia nigra**. Fibers arising from these areas project to the **ventral anterior** and **dorsomedial nuclei of the thalamus**. Thalamocortical fibers terminate in the anterior cingulate gyrus and the orbitofrontal cortex, closing the loop.

In the **open loop**, corticostriatal fibers arising from the **medial** and **lateral temporal lobe**, the **hippocampus**, the **amygdaloid body**, and the **entorhinal area** project to the ventral striatum (nucleus accumbens) and to the head of the caudate nucleus. The signals are relayed through the circuit to terminate in the anterior cingulate gyrus and the orbitofrontal cortex, which is the origin of the closed loop.

BASAL NUCLEI

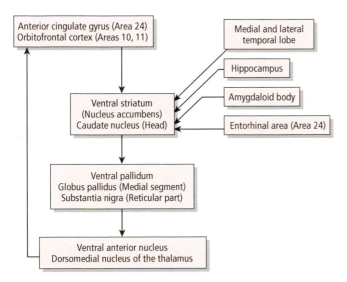

Figure 14.17 ● Limbic loops. In the closed loop, information flows from the anterior cingulate gyrus and orbitofrontal cortex to the ventral striatum and the caudate nucleus and then to the ventral pallidum, globus pallidus, and substantia nigra, and from there to the thalamus and then back to the anterior cingulate gyrus and orbitofrontal cortex. In the open loop, input is contributed by the medial and lateral temporal lobe, hippocampus, amygdaloid body, and entorhinal area to the closed loop. Source: Adapted from Noback, CR et al. (1996) *The Human Nervous System*, 5th ed. Williams & Wilkins, Baltimore; fig. 24.8.

The **limbic loop** functions in the emotional and motivational aspects of movement manifested as various facial expressions or other body movements.

OTHER CIRCUITS OF THE BASAL NUCLEI

Other circuits of the basal nuclei involve the substantia nigra, subthalamic nucleus, and the pedunculopontine tegmental nucleus (Table 14.9).

Substantia nigra

Afferent fibers (input)

The two regions of the **substantia nigra**, the reticular part and the compact part, both receive **afferents (input)** from:

- the striatum, via striatonigral fibers;
- the globus pallidus, via pallidonigral fibers; and
- the subthalamic nucleus.

Efferent fibers (output)

The substantia nigra sends fibers to:

- the ventral lateral, ventral anterior, and mediodorsal nuclei of the thalamus; and
- the striatum.

The former is via inhibitory GABAergic nigrothalamic fibers arising from the reticular part – which also sends collateral branches to the midbrain superior colliculus and tegmental area. The latter sends inhibitory dopaminergic nigrostriatal fibers arising from the SNc.

Subthalamic nucleus

Afferent fibers (input)

The **subthalamic nucleus** receives **afferents (input)** from:

- the lateral segment of the globus pallidus that provides GABAergic fibers, the most prominent subcortical input to the subthalamic nucleus;
- the motor cortex; and
- the pedunculopontine nucleus (PPN), whose fibers are cholinergic.

Nucleus	Afferents	Efferents
Substantia nigra	Striatum Globus pallidus Subthalamic nucleus	Ventral anterior (VA), ventral lateral (VL), and mediodorsal (MD) nuclei of the thalamus Striatum
Subthalamic nucleus	Lateral segment of the globus pallidus Motor cortex Pedunculopontine tegmental nucleus Midbrain raphe nucleus Ventral median nucleus of the thalamus	Medial segment of the globus pallidus Lateral segment of the globus pallidus Reticular part of the substantia nigra Ventral pallidum
Pedunculopontine nucleus	Globus pallidus	Globus pallidus, Substantia nigra

Table 14.9 ● Other circuits of the basal nuclei.

274 ● ● ●   **CHAPTER 14**

In addition, a smaller number of input fibers arise from the midbrain raphe nuclei (serotonergic) and the ventral median nucleus of the thalamus (glutaminergic).

Efferent fibers (output)

The subthalamic nucleus sends projections that terminate in both the medial and lateral segments of the globus pallidus, in the reticular part of the substantia nigra, and in the ventral pallidum. All of these fibers release glutamate and are excitatory.

It is evident, based on these projections, that the globus pallidus and the subthalamic nucleus are interconnected via the subthalamic fasciculus. The subthalamic nucleus modulates the output of the globus pallidus and of the substantia nigra, regions that contribute to the output of the basal nuclei.

Pedunculopontine tegmental nucleus

The **pedunculopontine tegmental nucleus** receives GABA-releasing **input** fibers from the globus pallidus and gives rise to glutaminergic **output** fibers, which terminate in the globus pallidus and the substantia nigra.

NEUROTRANSMITTERS OF THE BASAL NUCLEI

The neurons of the basal nuclei release the following neurotransmitters: GABA, glutamate, dopamine, acetylcholine, and neuropeptides (Table 14.10).

GABA-releasing neurons

GABA-releasing neurons are the principal neurons of the striatum

The principal neurons of the **striatum** are inhibitory neurons that release **gamma-aminobutyric acid (GABA)**, thus making it the principal neurotransmitter of the basal nuclei. In addition to the striatum, the **globus pallidus** and the reticular part of the **substantia nigra** also house GABA-releasing neurons. The fibers of all these GABAergic neurons form the striatopallidal, striatonigral, pallidothalamic, and nigrothalamic pathways (Fig. 14.18).

Glutamate-releasing neurons

Dopamine-releasing neurons are located in the compact part of the substantia nigra and their degeneration is the cause of Parkinson disease

Neurons of various regions of the **cerebral cortex** and the **subthalamic nucleus** give rise to corticostriatal and subthalamopallidal **glutamate-releasing fibers**, which project to the **striatum** and the **globus pallidus**, respectively. The glutamate-releasing neurons are excitatory to the GABAergic and cholinergic neurons of the striatum (Fig. 14.18).

Dopamine-releasing neurons

The neurons of the cerebral cortex and subthalamic nucleus give rise to glutamate-releasing fibers

The axons of dopamine-releasing neurons located in the **compact part** of the **substantia nigra** form the **dopaminergic nigrostriatal pathway**, which terminates in the caudate nucleus and the putamen. These

Table 14.10 ● Neurotransmitters of the basal nuclei.

Neurotransmitter	Excitatory/inhibitory	Region	Fiber pathways	Termination
Glutamate	Excitatory	Cerebral cortex Subthalamic nucleus	Corticostriatal fibers Subthalamopallidal fibers	Striatum Globus pallidus
Gamma-aminobutyric acid (GABA)	Inhibitory	Striatum Globus pallidus Substantia nigra	Striatopallidal fibers Striatonigral fibers Pallidothalamic fibers Nigrothalamic fibers	Globus pallidus Substantia nigra Thalamus Thalamus
Dopamine	Inhibitory	Substantia nigra (compact part)	Nigrostriatal fibers	Putamen where they synapse with GABAergic neurons, which project to the lateral segment of the globus pallidus
	Excitatory	Substantia nigra (compact part)	Nigrostriatal fibers	Putamen where they synapse with GABAergic neurons, which project to the medial segment of the globus pallidus and the reticular part of the substantia nigra
Acetylcholine	Excitatory	Striatum	Intrastriatal	Striatum
Neuropeptides (enkephalin, dynorphin, substance P, somatostatin, neurotensin, neuropeptide Y, cholecystokinin)				

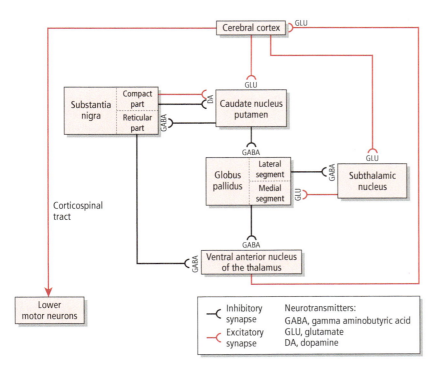

Figure 14.18 • Neural circuitry and neurotransmitters of the basal nuclei. Source: Adapted from Young, PA & Young, PH (1997) *Basic Clinical Neuroanatomy*. Williams & Wilkins, Baltimore; fig. 8.10.

axons have an *inhibitory* effect on the striatal GABAergic neurons that project to the **lateral** segment of the globus pallidus. However, they have an *excitatory* effect on those striatal GABAergic neurons that project to the **medial** segment of the globus pallidus and the reticular part of the substantia nigra. These diametrically opposite effects of dopamine at different terminations are due to differential reactions of the dopamine receptors of the postsynaptic neurons (Fig. 14.18).

Acetylcholine-releasing neurons

The acetylcholine-releasing neurons are local circuit intrastriatal neurons (interneurons)

Acetylcholine-releasing neurons are inhibited by the dopaminergic nigrostriatal neurons, and in turn, are excitatory to the GABAergic output neurons of the striatum (Fig. 14.19).

Neuropeptide-releasing neurons

Neuropeptide-releasing neurons of the basal nuclei may contain the following neurotransmitters: enkephalin, dynorphin, substance P, somatostatin, neurotensin, neuropeptide Y, and cholecystokinin. These neurotransmitters are present concurrently with other neurotransmitters in the same neuron (e.g., both GABA and enkephalin are located in the same neuron) (Fig. 14.19).

"DIRECT," "INDIRECT," AND "HYPERDIRECT" LOOPS (PATHWAYS) OF THE BASAL NUCLEI

Output signals from the basal nuclei are relayed to the thalamus via direct and indirect pathways

Neurons from widespread areas of the cerebral cortex project (via corticostriatal fibers) to the caudate nucleus, putamen, and ventral striatum of the basal nuclei. Output signals arising from the basal nuclei (i.e., in the pallidothalamic and nigrothalamic fibers) are relayed to the thalamus which projects to specific cortical areas of the frontal cortex. The information flows between the basal nuclei and the cerebral

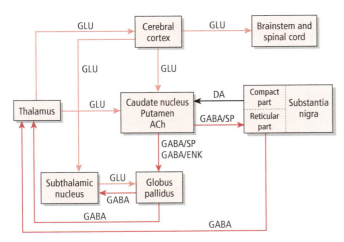

Figure 14.19 • Major pathways of the basal nuclei and their neurotransmitters. ACh, acetylcholine; DA, dopamine; ENK, enkephalin; GABA, gamma-aminobutyric acid; GLU, glutamate and/or aspartate; SP, substance P. Source: Adapted from Fix, JD (1995) *Neuroanatomy*, 2nd ed. Williams & Wilkins, Baltimore; fig. 21.4.

cortex via "**direct pathways**" and "**indirect pathways**," which are wiring patterns depicting the main circuits of the basal nuclei (motor, oculomotor, association, and limbic loops) discussed above. In the following paragraphs, the motor loop is used as an example to demonstrate these direct and indirect pathways.

Direct and indirect pathways

The motor loop of the basal nuclei consists of two distinct pathways that connect it to the cerebral cortex. Through these two pathways, signals from the **input station**, the **putamen** (associated with motor activity), are transmitted to the **output stations**, the **medial segment of the globus pallidus (GPm)** and the **reticular part** of the **substantia nigra (SNr)**. One pathway is referred to as the direct pathway and the other is referred to as the indirect pathway. Although these pathways are in parallel, one has an *excitatory* and the other has an *inhibitory* influence on the cerebral cortex. It is via these pathways that the basal nuclei regulate the output of the motor cortex and thus, indirectly, regulate motor activity.

In the **direct pathway** (Fig. 14.20), **corticostriatal** (**glutaminergic, excitatory**) **fibers** terminate in the **striatum** (mostly in the **putamen**), where they *excite* a group of **GABAergic inhibitory neurons** (also containing substance P). These neurons in turn send their striatopallidal and striatonigral fibers to the GPm and the SNr, respectively, where they have an inhibitory influence on the **GABAergic inhibitory pallidothalamic** and **nigrothalamic neurons**, respectively. The direct loop ultimately negates the inhibitory influence that the globus pallidus has on the thalamus. The inhibition of the inhibitory projection neurons housed in the GPm and the SNr results in *disinhibition* (excitation) of the **thalamocortical excitatory fibers** to the cerebral cortex, and ultimately, *stimulation* of the motor cortex. These signals stimulate the program for the desired movement, causing the motor cortex to initiate and execute the movement.

In the **indirect pathway** (Fig. 14.21), **corticostriatal** (glutaminergic, excitatory) fibers terminate in the striatum (mostly in the putamen), where they *stimulate* another group of GABAergic inhibitory neurons (also containing enkephalin). These neurons in turn send their striatopallidal fibers to the **GPl** where they have an inhibitory influence on the **GABAergic inhibitory pallidosubthalamic neurons**. The subthalamic nucleus houses glutaminergic excitatory neurons that project to the GPm and the SNr, which project GABAergic inhibitory pallidothalamic and nigrothalamic fibers, respectively, to the thalamus, which in turn projects **glutaminergic excitatory thalamocortical fibers** to the cerebral cortex. Usually, the inhibitory pallidosubthalamic

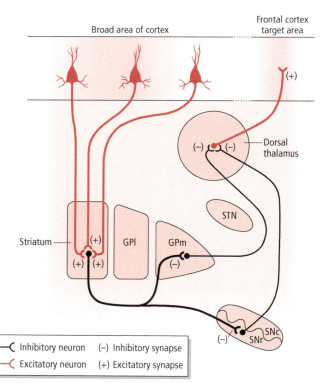

Figure 14.20 ● The direct pathway of the basal nuclei. GPl, globus pallidus (lateral segment); GPm, globus pallidus (medial segment); SNc, substantia nigra (compact part); SNr, substantia nigra (reticular part); STN, subthalamic nucleus. Source: Adapted from Burt, AM (1993) *Textbook of Neuroanatomy*. WB Saunders, Philadelphia; fig. 16.3.

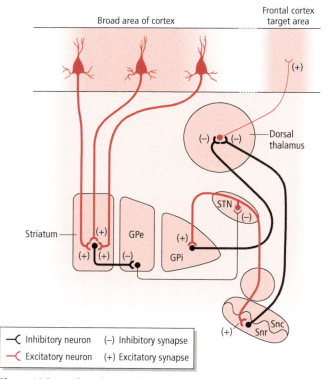

Figure 14.21 ● The indirect pathway of the basal nuclei. GPl, globus pallidus (external segment); GPm, globus pallidus (internal segment); SNc, substantia nigra (compact part); SNr, substantia nigra (reticular part); STN, subthalamic nucleus. Source: Adapted from Burt, AM (1993) *Textbook of Neuroanatomy*. WB Saunders, Philadelphia; fig. 16.4.

neurons cause a reduction of the excitatory influence that the subthalamic nucleus has on the GPm and the SNr. Enhanced stimulation of the indirect loop results in *disinhibition* of the subthalamic nucleus and thus stimulation of the inhibitory GABAergic neurons in the GPm and SNr. Furthermore, corticosubthalamic fibers have an excitatory effect on the subthalamic neurons. The GPm and SNr, via pallidothalamic and nigrothalamic fibers, inhibit the thalamocortical neurons, resulting in the *inhibition* of the motor cortex. The indirect pathway lifts the inhibitory influence of the globus pallidus on the subthalamic nucleus, which in turn stimulates the inhibitory pallidothalamic neurons. Via simultaneous activation of the indirect pathway, another group of thalamic neurons is inhibited, resulting in the inhibition of conflicting movements. The simultaneous activation of the direct and indirect pathways permits the motor cortex to initiate desired, purposeful movements (via the direct pathway) and simultaneously inhibits excessive, purposeless movements (via the indirect pathway). Lesions affecting components of the direct or indirect pathways create imbalances that result in deficits characteristic of malfunction of the basal nuclei.

Hyperdirect pathway

The hyperdirect pathway is a glutaminergic pathway that bypasses the striatum, conveying information from the motor, association, and limbic cortices directly to the STN. This stimulates excitation of the GPm and SNr as illustrated in Fig. 14.21.

CIRCUITS THAT MODULATE ACTIVITY OF THE BASAL NUCLEI

The activity of the basal nuclei is modulated by the dopaminergic nigrostriatal pathway and by the thalamostriatal fibers

The **dopaminergic nigrostriatal pathway** arises from the **substantia nigra compact part** (SNc) and terminates in the **striatum. Dopaminergic fibers** arising from SNc and the retrorubral nucleus form the nigrostriatal pathway whose fibers terminate in the caudate nucleus or the putamen, where they synapse with GABAergic neurons. Dopamine is usually an inhibitory neurotransmitter; however, it has a different effect on different populations of neurons housed in the putamen. It is believed that the putamen contains neurons with up to five different types of dopamine receptors, two of which are D_1 (dopamine-1) and D_2 (dopamine-2) receptors.

The dopaminergic neurons from SNc synapse with striatal **GABAergic–substance P neurons**, which possess D_1 receptors, and project directly from the putamen to the GPm and SNr. These synapses have an *excitatory effect* on the **direct pathway.**

The dopaminergic neurons from the **retrorubral nucleus** synapse with striatal **GABAergic–enkephalin neurons** of the indirect pathway, which possess D_2 receptors. These neurons project from the putamen to the GPl which in turn projects to the subthalamic nucleus, which then projects to the GPm. These synapses have an *inhibitory effect* on the **indirect pathway.**

The **nigrostriatal pathway** strengthens the **corticostriatal fiber input** to the striatum concerning motor activity by stimulation of the direct pathway and by inhibition of the indirect pathway. The cholinergic intrastriatal interneurons, however, counteract the cortical input on the dopaminergic neurons. These **cholinergic neurons** stimulate the **GABAergic–enkephalin striatal output** fibers of the **indirect pathway**, which has an inhibitory influence on the thalamocortical fibers. The net effect on the basal nuclei is **inhibitory**.

The **thalamostriatal fibers** arise from the **intralaminar (centromedian** and **parafascicular) nuclei of the thalamus** and terminate primarily in the striatum, although some fibers terminate in the globus pallidus and the subthalamic nucleus. Projections from the ventral anterior thalamic nucleus project to the caudate nucleus, whereas projections from the ventral lateral thalamic nuclei project to the putamen. Thalamostriatal projections contribute to the smooth initiation and execution of movement sequences.

CLINICAL CONSIDERATIONS

Disorders involving the basal nuclei include hypertonicity and dyskinesias

Hypertonicity

Hypertonicity is an abnormal increase of muscle tone in response to passive stretch. Severe hypertonicity is referred to as **rigidity**. One of the normal functions of the basal nuclei, when stimulated, is to relay signals to the motor cortex and brainstem, which ultimately inhibit (excess) muscle tone. Following a lesion of the basal nuclei, this inhibitory influence is diminished or eliminated and symptoms of hypertonicity are manifested contralateral to the side of the lesion.

Dyskinesias

Dyskinesias may be classified into hyperkinesias or hypokinesias

Dyskinesias include unintentional, disorderly, purposeless movements. The disorders of the basal nuclei may be classified into hyperkinetic disorders or hypokinetic disorders. Dyskinesias "occur at rest," that is, without volitional intent.

Hyperkinetic disorders

The symptoms of hyperkinetic disorders are characterized by abnormal involuntary movements, displayed by affected individuals as tremors, chorea, athetosis, or ballism

The symptoms of hyperkinetic disorders are characterized by abnormal involuntary movements, displayed by affected individuals as tremors, chorea, athetosis, or ballism. Hyperkinetic disorders are caused by a disturbance of the excitatory subthalamopallidal neurons of the indirect pathway.

Tremor is a rhythmic, low-amplitude movement that may be manifested as rhythmic nodding of the head, or in the hands and feet.

Chorea (G., *choros*, "dance") consists of a sequence of rapid, jerky, somewhat agile, and flowing movements involving mainly the hands and feet, the tongue, and facial muscles.

Athetosis (G., *athetos*, "not fixed") is a condition in which the individual displays slow, vermicular ("worm-like") involuntary movements, which usually affect the hands and feet. This condition results from degeneration of the lateral segment of the globus pallidus. Since the globus pallidus becomes silent, the ventrolateral nucleus of the thalamus can send signals spontaneously to the motor cortex.

Ballismus (G., *ballismos*, "jumping about") is the most extreme type of dyskinesia.

Hemiballismus (G., *hemi*, "half;" *ballismos*, "jumping about" or "to throw") is a rare condition most often affecting older individuals, who exhibit involuntary ballistic (violent striking) movements on only one side of the body, that affect only the proximal muscles of a limb. It results following a cerebral vascular lesion of the ganglial branch of the posterior cerebral artery, which involves the contralateral subthalamic nucleus. The subthalamic nucleus functions in the integration of flowing movements produced by different parts of the body. Normally, the subthalamic nucleus projects glutaminergic (excitatory) fibers to the two output nuclei of the basal nuclei, the GPm, and the SNr, which in turn send inhibitory projections to the thalamus. Thalamocortical fibers that are excitatory project to the motor cortex. Projections thus far are ipsilateral; however, the upper motor neurons arising from the cortex decussate in the pyramids, thus affecting the opposite side of the body. A lesion in the subthalamic nucleus would result in disinhibition of the thalamus, which in turn becomes overactive, overstimulating the motor

cortex which in turn results in hyperkinetic symptoms in the opposite side of the body (Fig. 14.22). Although hemiballismus may occur following the destruction of the subthalamic nucleus, it most often occurs following the destruction of the striatal neurons that normally inhibit the lateral segment of the globus pallidus, which in turn exerts an inhibitory influence on the subthalamic nucleus. Whatever the case may be, the inhibitory effect exerted by the medial segment of the globus pallidus on the thalamus is reduced, which in turn overstimulates the motor areas of the cortex. In this condition, the thalamocortical pathway, arising from the ventral lateral nucleus of the thalamus projecting to the supplementary motor area of the premotor cortical areas, is affected. Hemiballismus may be alleviated either with chemical substances, such as dopamine-blocking drugs or GABA-mimetic agents or by surgical removal of the ventral lateral nucleus of the thalamus. Hemiballismus usually disappears several weeks after treatment.

Hyperkinetic disorders are caused by a disturbance of the excitatory subthalamopallidal neurons of the **indirect loop** (Fig. 14.22) (of the motor circuit). This creates an imbalance between the direct pathway (where pallidothalamic neurons are stimulated) and the indirect pathway (where pallidothalamic neurons are inhibited). This causes underinhibition of the thalamic neurons, which ultimately results in excess cortical activity and movement.

Huntington disease

Huntington disease is a rare, genetic disorder; the neurons that degenerate to produce this disease are part of the indirect loop

Huntington disease is a rare, inherited, autosomal dominant genetic disorder, which is associated with a gene defect on chromosome 4. The symptoms of Huntington disease become perceptible in the fourth or fifth decade of life as a result of degeneration of the GABAergic, substance P, and cholinergic-releasing neurons housed in the striatum, and simultaneous degeneration of the cerebral cortex (Fig. 14.23).

At its early stages, the neurons that degenerate are the striatal GABAergic neurons that project to the lateral segment of the globus pallidus, but as the disease progresses, eventually the entire striatal neuron population, including the local circuit ACh neurons also degenerate.

As a consequence of the loss of the GABAergic inhibitory influence, which is normally exerted by the striatum via the striatonigral pathway on the dopaminergic neurons of the substantia nigra (compact part, SNc), the SNc dopaminergic nigrostriatal neurons become overactive. This results in the dopaminergic nigrostriatal inhibitory pathway (over)inhibiting and thus depressing the function of the striatum. This imbalance is responsible for the abnormal movements displayed by affected individuals. The choreiform movements are reduced following treatment with antidopaminergic drugs. As a result of the degeneration (atrophy) of the caudate nuclei that occupy the lateral ventricles, radiographs reveal distended lateral ventricles resulting in **hydrocephalus** *ex vacuo*.

In the early stages of Huntington disease, the cells affected are mainly the medium spiny GABAergic inhibitory neurons that project from the striatum to the lateral segment of the globus pallidus (GPl). This results in a decrease of inhibitory influence on the GPl. Since the GPl normally exerts a tonic inhibitory influence on the subthalamic nucleus, this neural loss will increase the tonic inhibitory influence exerted on the subthalamic nucleus by the globus pallidus. The subthalamic nucleus normally sends excitatory projections to the output stations of the basal nuclei, the GPm and the SNr. The output from the (output stations of the) basal nuclei is normally inhibitory to the thalamus, whose output in turn is excitatory to the motor cortex. This neural

CLINICAL CONSIDERATIONS (continued)

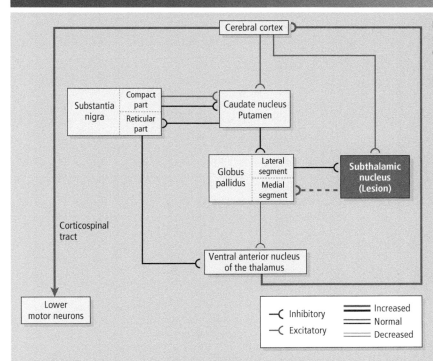

Figure 14.22 • Subthalamic nucleus lesion. Note the neural circuitry modifications resulting in hyperkinetic disorders. Source: Adapted from Young, PA & Young, PH (1997) *Basic Clinical Neuroanatomy*. Williams & Wilkins, Baltimore; fig. 8.11.

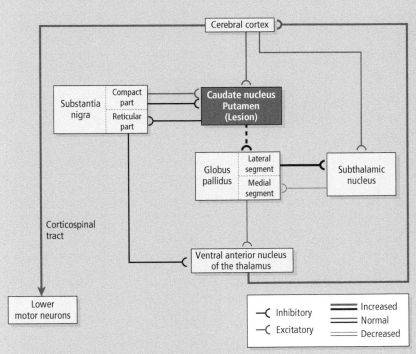

Figure 14.23 • Striatal lesion. Note the neural circuitry modifications resulting in hyperkinetic disorders. Source: Adapted from Young, PA & Young, PH (1997) *Basic Clinical Neuroanatomy*. Williams & Wilkins, Baltimore; fig. 8.12.

loss will result in overinhibition of the subthalamic nucleus, which will in turn under-stimulate the GPm and the SNr. The output nuclei will relay reduced inhibitory signals to the thalamus and brainstem terminations. Inhibition of an inhibitory target results in what is referred to as **disinhibition**, which is excitation. Thus the thalamus and brainstem terminations are disinhibited following the loss of the GABAergic neurons, which results in overstimulation of the motor cortex, and ultimately of the lower motor neurons, producing involuntary movements.

CLINICAL CONSIDERATIONS (continued)

Figure 14.24 ● An individual with Huntington chorea. Symptoms include choreic movements of the limbs and smacking of the lips and tongue. Source: Adapted from Young, PA & Young, PH (1997) *Basic Clinical Neuroanatomy*. Williams & Wilkins, Baltimore; fig. 8.9.

During the early stages of the disease, involuntary, choreiform movements of the limbs and sudden movements of the face appear. As the disease progresses, additional muscles become affected, and the individual is incapable of moving, speaking, or swallowing (Fig. 14.24). These symptoms are accompanied by advancing dementia and cognitive impairment due to degeneration of the caudate nucleus and the cerebral cortex.

Sydenham chorea (St. Vitus' dance)
Sydenham chorea appears as a consequence of childhood rheumatic fever

Sydenham chorea results from a bacterial infection. The infectious bacteria causing rheumatic fever contain antigens that are comparable to the protein receptors located in the cell membrane of the neurons housed in the caudate nucleus and the putamen. The individual infected by this bacterium produces antibodies that bind to the bacterial antigens as well as to the receptors of the striatal neurons. These neurons act as if they were stimulated by neurotransmitter substances and fire, resulting in choreiform movements, involving the face, trunk, and extremities. This condition, however, is temporary and the child will recover completely.

Tardive dyskinesia
Tardive dyskinesia is a condition that affects some individuals who are chronically treated with antipsychotic drugs

In **tardive dyskinesia**, an affected individual displays repetitive movements of the face, mouth, and tongue, as well as choreoathetotic movements of the trunk and limbs.

The antipsychotic drugs (phenothiazines and butyrophenones) block the transmission of dopamine in the brain. These drugs are aimed at the neurons housed in the ventral tegmental area whose axons form the mesolimbic dopaminergic pathway, but cause the dopamine receptors (D_3 receptors) to become hypersensitive. This results in an imbalance of the effect of the nigrostriatal pathway on the motor loop, causing abnormal movements. In recent years newer antipsychotic drugs have become available that do not produce such symptoms.

Hypokinetic disorders
Hypokinetic disorders are characterized by bradykinesia or akinesia and result from neuronal degeneration in the neostriatum

Hypokinetic disorders are characterized by bradykinesia or akinesia. **Bradykinesia** is the slowness of the execution of movement. This is caused by an imbalance created between the direct and indirect pathways and results in inappropriate overstimulation of the antagonist muscles. **Akinesia** is a disorder in which the affected individual is unable to plan or aim a movement toward a desired direction or target. There is an increase in basal nuclear neuronal activity during the planning phase of an impending movement.

Hypokinetic disorders result from neuronal degeneration in the neostriatum, disrupting the **direct pathway** of the basal nuclei. That is the **GABAergic inhibitory neurons** of the neostriatum, which normally send their fibers to the **medial segment of the globus pallidus**, degenerate, eliminating their *inhibitory* influence on the **GABAergic inhibitory pallidothalamic neurons**. Consequently, this permits (frees) the tonically active pallidothalamic neurons to continually (without restraint) over inhibit the **thalamocortical excitatory neurons** whose fibers terminate in the cerebral cortex. In other words, the thalamus is not disinhibited (stimulated). This paucity of thalamic input to the cerebral cortex understimulates the corticospinal and corticonuclear tracts. The components of the **indirect** pathway though are undisturbed, thus creating an imbalance between the direct and indirect pathways. The subthalamopallidal projections (of the indirect pathway) stimulate the pallidothalamic neurons that in turn overinhibit the thalamus. Thus the insufficient thalamic disinhibition (excitation) via the direct pathway is augmented by the overinhibition of the thalamus via the indirect pathway, diminishing normal cortical activity, and increasing abnormal cortical activity. This discrepancy in the funneling of input to the cerebral cortex via the direct and indirect pathways ultimately causes an increase in the stimulation of antagonist muscles.

Parkinson disease
Parkinson disease occurs following degeneration of the dopaminergic neurons of the compact part of the substantia nigra

Parkinson disease (parkinsonism, "paralysis agitans") is the most prevalent disorder associated with the basal nuclei, affecting approximately 1% of individuals over the age of 50. It results from **degeneration** of the **dopaminergic**

CLINICAL CONSIDERATIONS (*continued*)

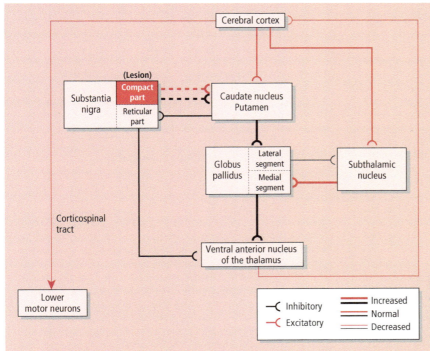

Figure 14.25 ● Decreased dopamine. Note the neural circuitry modifications resulting in hypokinetic disorders. Source: Adapted from Young, PA & Young, PH (1997) *Basic Clinical Neuroanatomy*. Williams & Wilkins, Baltimore; fig. 8.13.

neurons of the compact part of the **substantia nigra** (Fig. 14.25). During the early stages of Parkinson disease, components of the motor loop of the basal nuclear circuits are affected first. Due to the degeneration of these neurons, dopamine is not produced; thus, there is a **reduction of dopamine** levels in the **caudate nucleus** and the **putamen**, where the **nigrostriatal neurons** terminate. Symptoms usually do not appear until about 80% of the dopaminergic neurons of the substantia nigra have degenerated. The reduction of dopamine levels in the striatum results in an increase of acetylcholine release by intrastriatal neurons, and ultimately in the increase of the inhibitory influence exerted by the basal nuclei on the thalamus. An inhibited thalamus understimulates the motor cortex and brainstem terminations, which result in slowness, or loss of movement, characteristic of Parkinson disease.

Although individuals affected by this disease may have a resting tremor, muscle rigidity, hypokinesia, bradykinesia, or impairment of postural reflexes, not all symptoms are necessarily expressed in the same individual (Fig. 14.26). Symptoms usually first appear on one side of the body only, but as the disease progresses, symptoms are present bilaterally. These individuals have no paralysis. However, the akinesia gives one the impression that an individual with Parkinson disease is paralyzed. The akinesia and the tremor generated the older term "paralysis agitans."

Individuals with this disease have a **resting tremor** affecting the hands, which is exhibited as a "pill-rolling tremor," where the index finger rubs on the thumb – similar to the way that pills were made many decades ago. Interestingly, this tremor, caused by the regular alternating contraction of the agonist and antagonist muscles, is not evident when the affected individual is asleep, during voluntary movement, or general anesthesia. In contrast, cerebellar tremor is expressed during movement ("intention tremor") and is absent at rest (see Chapter 15). Also present is muscular **rigidity**, which results from widespread increased muscle tone. Rigidity worsens during the

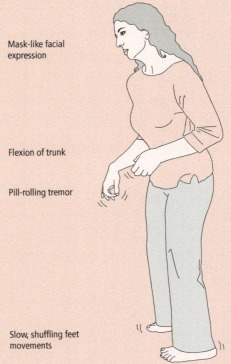

Figure 14.26 ● Individual with Parkinson disease. Symptoms include mask-like facial expression, flexed head, and trunk, arms close to the trunk, pill-rolling tremor of the hands, and slow, shuffling movements of the feet.

CLINICAL CONSIDERATIONS (*continued*)

advanced stages of Parkinson disease. If the rigidity is continuous during a movement, it is referred to as "lead-pipe rigidity," whereas if the rigidity is discontinuous resulting from intermittent muscle relaxations during a movement, it is referred to as "cog-wheel rigidity." The rigidity of all flexors results in the affected individual having a stooped posture.

Parkinson disease is also characterized by **hypokinesia** (G., *hypo*, "few;" kinesia, "movement") and **bradykinesia** (G., *brady*, "slow") in which the affected individual has trouble in both initiating and executing voluntary movements. These are exhibited by a blank, expressionless, mask-like face, slow blinking, slurred speech, and lack of upper limb movements that usually accompany walking. An affected individual has difficulty getting up from a chair, walking, picking up objects, and so on. A shuffling gait may also develop, in which the individual takes short steps. These movements begin with great effort, followed by an increase in speed, although it is difficult to stop a movement in progress. However, muscle strength and deep tendon reflexes are not influenced by this disease. During the later stages of the disease affected individuals show signs of memory loss, and dementia.

During the later stages of the disease, the cognitive, oculomotor, and limbic loops become affected resulting in loss of cognitive function, decrease in eye movements, and dementia.

In humans, melanin begins to appear in the substantia nigra around the fourth or fifth year of childhood. Autopsy of the midbrain of individuals who had Parkinson disease reveals an absence of melanin.

Since dopamine does not cross the blood–brain barrier, patients with Parkinson disease are treated with L-3,4-dihydroxyphenylalanine (levodopa) which can cross this barrier to restore the missing dopamine along with dopamine agonists, an l-DOPA uptake inhibitor (carbidopa), and monoamine oxidase (MAO) inhibitors.

In addition to pharmacologic treatment, the signs and symptoms of Parkinson disease have in recent years been alleviated by surgical procedures that involve lesions of the thalamus (thalamotomy) or the medial segment of the globus pallidus (pallidotomy). An alternative surgical procedure includes stimulation electrodes aimed at target nuclei such as the thalamus, globus pallidus, or the subthalamic nucleus.

Patients receiving pharmacologic or surgical treatment experience relief from some of the symptoms associated with Parkinson disease; however, the progression of the disease cannot be halted.

Parkinson disease (a hypokinetic disorder) and Huntington disease (a hyperkinetic disorder) are two severe disorders of the basal nuclei reflecting two opposite ends of the basal nuclear dysfunction spectrum. If excessive levodopa is administered to an individual with Parkinson disease, the individual will manifest choreic movements. Levodopa administered to a patient with Huntington disease will intensify the choreic movements.

Wilson disease (hepatolenticular degeneration)

Wilson disease, an autosomal recessive genetic disorder, is localized on chromosome 13: gene ATP7B, and is associated with an error in the body's copper metabolism

Wilson disease is seen in young adults (10–25 years of age), and results following degeneration mostly of the putamen and the globus pallidus, although other structures such as the cerebral cortex, thalamus, red nucleus, and cerebellum may also be involved. Degeneration is characterized by the loss of nerve cells and the accumulation of protoplasmic astrocytes. Consequently, the nerve tissue exhibits small spaces, giving it a spongy appearance. Lenticular degeneration is usually accompanied by cirrhosis of the liver.

An individual with this disease manifests symptoms such as tremors, rigidity, and choreiform or athetotic movements, as well as speech disorders and a loss of facial expression. The tremor manifested by an affected individual is not a tremor at rest, as seen in other basal nuclear disease. The tremor here is referred to as **asterixis**, characterized as a "wing-beating" tremor, manifested following the extension of the upper limbs. The tremor may affect the hands at the wrist, or the entire upper limb.

Many individuals with this disease have greenish/golden brown coloring around the corneoscleral junction of their eyes, a condition given the eponymous name of "Kayser–Fleischer rings." These rings are observed in 85% of individuals with Wilson disease who have neurological symptoms.

Individuals affected by this disorder experience relief from symptoms following treatment, which includes decreasing the concentration of copper in the body.

Note that the clinical case at the beginning of the chapter refers to a patient with abnormal, uncontrollable choreiform movements, motor deficits, forgetfulness, and abnormal behaviors. An MRI scan shows atrophy of the head of the caudate nucleus, bilaterally.

1 What neurodegenerative disease produces the unique combination of choreiform movements and progressive deterioration of mental function exhibited by this patient?

2 Which nuclei of the basal nuclei are usually affected in this disease?

3 What do you expect genetic testing results to reveal?

4 In addition to abnormal choreiform movements, what are the most disabling symptoms of Huntington disease?

SYNONYMS AND EPONYMS OF THE BASAL NUCLEI

Name of structure or term	Synonym(s)/eponym(s)	Name of structure or term	Synonym(s)/eponym(s)
Acetylcholine-releasing fibers	Cholinergic fibers	Medial segment of the globus pallidus (GPm)	Internal segment of the globus pallidus
Amygdaloid body	Amygdala		
Basal nuclei	Basal ganglia		Inner segment of the globus pallidus
	Deep cerebral nuclei		
Corticonuclear tract	Corticobulbar tract (older term)		Globus pallidus medial
Corticospinal tract	Pyramidal tract (older term)	Mesencephalon	Midbrain
Dopamine-releasing fibers	Dopaminergic fibers	Paralysis agitans	Parkinson disease
Frontal eye field (FEF)	Brodmann's area 8		Parkinsonism
Gamma-aminobutyric acid-releasing fibers	GABAergic fibers	Premotor cortex	Premotor area Part of Brodmann's area 6
Globus pallidus (GP)	Paleostriatum	Primary motor cortex (MI)	Precentral gyrus
Glutamate-releasing fibers	Glutaminergic fibers		Brodmann's area 4
Hepatolenticular degeneration	Wilson disease	Primary somatosensory cortex (SI)	Primary sensory cortex (SI)
Huntington disease	Huntington chorea (older term)		Primary somesthetic cortex
Interneuron	Internuncial neuron		Postcentral gyrus
Lateral medullary lamina	External medullary lamina		Brodmann's areas 3, 1, and 2
Lateral segment of the globus pallidus (GPl)	External segment of the globus pallidus	Proximal limb musculature	Girdle musculature
	Outer segment of the globus pallidus	Red nucleus	Nucleus ruber
		Striatum	Neostriatum
		Substantia nigra reticular part	Substantia nigra pars reticularis
	Globus pallidus lateral (GPl)	Subthalamic nucleus	Nucleus of Luys
Lentiform nucleus	Lenticular nucleus	Supplementary motor area (SMA)	Part of Brodmann's area 6
Lower motor neuron	Lower motoneuron	Sydenham chorea	St. Vitus' dance
Medial medullary lamina	Internal medullary lamina	Thalamic fasciculus	H1 field of Forel

284 ●●● CHAPTER 14

FOLLOW-UP TO CLINICAL CASE

This patient has Huntington disease, confirmed by genetic testing. This patient has typical symptoms of Huntington disease. Chorea is the most identifiable and specific characteristic, but the most problematic and disabling symptoms are psychiatric disturbances and dementia.

Chorea is a Greek term for "dance." It is characteristic of lesions or pathology in the basal nuclei, especially of the caudate nucleus and putamen. However, the exact localization and pathophysiology of chorea are uncertain. Many lesions involving the caudate and putamen do not produce choreiform or similar movement disorders. Chorea may be generalized, as in Huntington disease, or unilateral, in which case it is called hemichorea. The most common cause of hemichorea is a stroke in the contralateral basal nuclei, although it may be caused by any lesion involving certain parts of the contralateral basal nuclei. Several different conditions, genetic or acquired, can produce chorea. Certain medications such as levodopa, prescribed for the treatment of Parkinson disease, can produce choreiform movements if excess dosages are taken.

Huntington disease is an autosomal dominant neurodegenerative disease. Therefore, an affected parent has a 50% chance of passing the disease to his/her children. It affects the caudate and putamen early and most prominently, but other parts of the basal nuclei and the cerebral cortex are also affected. The disease often presents itself when the affected individual is in his thirties or forties. Usually, by that time the individual has had children already and may have passed the gene to his children. The abnormal gene is on chromosome 4 and contains an expanded trinucleotide sequence. This gene produces the Huntingtin protein. The normal function of this protein, and how the mutated protein causes the death of neurons, is unclear. Genetic testing is available for this disease. Huntington disease causes progressive and relentless deterioration of mental function, which leads to incapacity and death 10–20 years after onset. Chorea and psychosis can be treated by antidopaminergic medication; however, there is no effective treatment for progressive dementia.

QUESTIONS TO PONDER

1. Why are the basal nuclei and the limbic system interconnected?

2. What causes hypertonicity following a lesion of the basal nuclei?

3. One of the causes of hemiballismus is a lesion to the subthalamic nucleus; it is manifested as violent flinging movements of the proximal limb muscles on the opposite side of the body. How do you account for the symptoms appearing contralateral to the side of the lesion?

4. What is the underlying cause of hyperkinetic and hypokinetic disorders?

5. What is likely to occur if excessive levodopa is administered to an individual with Parkinson disease and an individual with Huntington disease?

CHAPTER 15

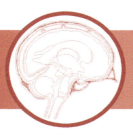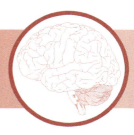

Cerebellum

CLINICAL CASE

MORPHOLOGY OF THE CEREBELLUM

CEREBELLAR PEDUNCLES

DEEP CEREBELLAR NUCLEI

AFFERENTS (INPUT) TO THE CEREBELLUM

EFFERENTS (OUTPUT) FROM THE CEREBELLUM

FUNCTIONAL ORGANIZATION OF THE CEREBELLUM: INTRINSIC CIRCUITRY

CLINICAL CONSIDERATIONS

SYNONYMS AND EPONYMS OF THE CEREBELLUM

FOLLOW-UP TO CLINICAL CASE

QUESTIONS TO PONDER

CLINICAL CASE

A 72-year-old patient, B.K., developed sudden severe vertigo about 5 hours prior to presentation in the emergency room. The vertigo has persisted and remained severe. B.K. became nauseated, vomited a few times, has not been able to walk, and cannot even sit on the side of the bed without falling over. There was initially some double vision, but this has since resolved. The patient reports a "weak" and heavy left arm and of mildly slurred speech. B.K. relays a feeling of being pulled to the left when attempting to stand or sit up.

Physical examination shows mild dysarthria (slurred speech), but language function is normal. Bilateral end gaze nystagmus is present, which is much more prominent toward the left. Facial strength and sensation are normal. Despite the patient's concerns, exam shows strength throughout to be normal. However, there is dysmetria and ataxia involving the left arm, but the right arm is normal. The patient cannot stand or even sit up in bed without assistance. Sensation is normal throughout. Head CT in the emergency room was normal.

Movement is produced by complex interactions of the **basal nuclei**, the **cerebellum**, and the **cerebral cortex** (Table 15.1). To carry out its function, the cerebellum needs to know how the head, trunk, and limbs are oriented at all times. In addition, it needs to know if they are stationary (and in what position) or if they are moving (how fast, in which direction, etc.). Since the cerebellum is part of the motor system, it needs to be constantly informed about the status of each body part with reference to movement so that it can implement its task. How does the cerebellum get this information about what the individual's body parts are doing? The specialized receptors, muscle spindles, and tendon organs (TOs), unique to muscles and tendons, respectively, collect and send sensory information about the position and movement of the body's parts to the cerebellum via the ascending *sensory pathways to the cerebellum* (which are the anterior, posterior, and rostral

---

A Textbook of Neuroanatomy, Third Edition. Maria A. Patestas, Amanda J. Meyer, and Leslie P. Gartner.
© 2025 John Wiley & Sons, Inc. Published 2025 by John Wiley & Sons, Inc.
Companion website: www.wiley.com.go/patestas/neuroanatomy3e

286 ●●● CHAPTER 15

Structure	Function(s)
Cerebral cortex	
Premotor cortex	Planning of voluntary movement
Primary motor cortex	Execution of voluntary movement
Basal nuclei	Modulate premotor and primary motor cortex output
Cerebellum	Coordinates movements
	Maintenance of posture, balance (equilibrium), and muscle tone
	Learning of motor tasks
	Processes memory of skilled motor activity

Table 15.1 ● Role of the cerebral cortex, basal nuclei, and cerebellum in the production of movement.

spinocerebellar tracts and the cuneocerebellar tract). These wide-diameter heavily myelinated tracts carry proprioceptive (position sense) information to the cerebellum at speeds (72–110 m/s) unparalleled by any other ascending pathway. The cerebellum uses this information for its own calculations, but this information does not reach conscious levels. To carry out its function, the cerebellum also uses information from the motor cortex (about impending and desired movements), from the visual system (that helps the cerebellum to aim the movement in the desired direction), from the auditory system (for the auditory guidance of movement), and from the vestibular system (about the orientation and motion of the head, so that it can continuously make corrections to maintain or regain balance). The cerebellum continuously integrates all the information it receives and then projects its output mainly to the thalamus. The thalamus in turn projects to the motor cortex. Via this pathway, the cerebellum selects the muscles that should contract and the muscles that should relax to accomplish the next desired, voluntary movement. In short, the cerebellum coordinates movement by influencing the upper motor neurons in the motor cortex. In this role, the cerebellum acts as if it were the intrinsic puppeteer of the motor system that moves the limbs and other body parts by "pulling on the strings" (of the upper motor neurons) which will ultimately cause movement of specific body parts. There are a large number of upper motor neurons in the motor cortex. The motor cortex executes a movement via the descending motor pathways (consisting of upper motor neurons) by first taking into consideration the cerebellum's input. The cerebellum is part of a silent system that generates a "background" of muscular activity that continuously modulates muscle tone, posture, and balance by influencing the *corticoreticular* and *reticulospinal tracts*, and the *vestibulospinal tracts*. It influences voluntary movement (that is, it acts as the intrinsic puppeteer) produced by the corticonuclear tract (for movement of muscles of the face, mandible, pharynx, larynx, neck, and tongue), the *corticorubral* and *rubrospinal tracts* (for movement of the hand), and the *corticospinal tracts* (for movement of the upper limbs, trunk, and lower limbs). The cerebellum *times* and *steers* a movement as it aims for a target. For example, when the softball batter sees a baseball coming, as they aim to hit the ball with the bat, they also "time" when to swing the bat and in which direction so that they hit the ball. The cerebellum tells the motor cortex which muscles need to contract, and which need to relax, how fast (speed), in which direction (direction), and when (timing), and so on, to produce a particular targeted movement.

Assume that one is about to pick up a container filled with heavy books and as the box is being lifted it turns out that the box is empty. The individual falls backward and staggers because the real situation did not match the assumed situation (instead of an expected heavy box the box was light). Because of the expected heavy weight, the cerebellum told the motor cortex which muscles to contract (to tense the antigravity extensor muscles of the back and the lower limbs, and contract the flexors of the upper limbs) and how much effort to put into lifting the box, confusing the motor system, producing a mighty and forceful (but awkward) movement for lifting an empty box. Consider the following analogies: Think of the motor system as a car and driver. The motor cortex is the engine. The motor cortex plans and executes the movement. The basal nuclei are the ignition system that ignites the motor cortex, to initiate movement. The cerebellum is the driver (or puppeteer) who steers in the right direction, determines the speed (accelerates or decelerates), and times the movement.

What helps people stand upright if the force of gravity is pulling the person down to the ground? What helps people counteract the force of gravity? It's the muscle tone of the antigravity muscles (the extensors of the trunk and lower limbs). The cerebellum modulates muscle tone by sending messages to the muscular system via the *corticoreticular, reticulospinal*, and *vestibulospinal tracts*. These tracts control the gamma motor neurons in their role to produce muscle tone via the gamma loop. Without muscle tone (muscle tension), an individual would not be able to stand upright while gravity is pulling the person down. With soft or flaccid muscles, people would collapse and fall to the ground. It's the cerebellum that helps an individual to counteract the force of gravity (via the tracts indicated previously).

The basal nuclei *initiate* motor activity and *modulate* cortical output related to motor function, the cerebellum functions in the *coordination* of movement, whereas the cerebral cortex is involved in the *planning* and *execution* of voluntary movement. The cerebellum and basal nuclei exert their influence on the brainstem and spinal cord and ultimately on motor activity at the subconscious level, *indirectly*, by regulating the output of the primary motor cortex and premotor cortex. The motor cortex influences the execution of motor activity via *direct* projections to the brainstem nuclei (i.e., cranial nerve motor nuclei, nuclei of the reticular formation, midbrain tectum, and red nucleus) and spinal cord motor neurons.

Input from the motor cortex, and the visual, auditory, vestibular, and somatosensory systems is funneled into the cerebellum where it is integrated, and then utilized to plan and coordinate motor activity; that is, the cerebellum plays a key function in the execution of a motor task as it relates to the timing, speed, direction, and precision of muscular activity, whether of a single muscle or groups of muscles. Additionally, the cerebellum receives information about upcoming motor activity from the motor cortex; thus, it can monitor the commands arising from the motor cortex that are relayed to the lower motor neurons of the brainstem and spinal cord. This capability enables the cerebellum to compare the impending movement with the movement in progress. It also plays an active role in voluntary movement by continuously adjusting the outflow of the motor and premotor cortex, so that movement is modified to be precise, smooth, flowing, and coordinated.

The cerebellum is the conductor behind a specific movement since it not only determines its speed, but with the aid of projections from the visual and other systems, it also orchestrates its course. The cerebellum has important functions in **proprioception, muscle tone, maintenance of posture, balance (equilibrium),** and **coordination of skilled voluntary movements,** as well as in the **learning and memory of motor tasks,** such as playing a musical instrument and riding a bicycle. These functions are all performed at the subconscious level.

MORPHOLOGY OF THE CEREBELLUM

The cerebellum displays alternating, slender, parallel elevations known as folia, and depressions known as sulci, that facilitate a great increase in the surface area of the cerebellar cortex

The **cerebellum** (L., "little brain") accounts for 10–15% of brain volume and is composed of two **cerebellar hemispheres** and the intervening **vermis** (L., "worm"), which is separated from each hemisphere by shallow longitudinally disposed furrows (Fig. 15.1). The surface of the cerebellum displays alternating, slender, parallel elevations (ridges) known as **folia** (plural term, singular: folium) and depressions (grooves), known as **sulci** (plural term, singular: sulcus), which facilitate a great increase in the surface area of the cerebellar cortex (Fig. 15.1C). Despite its smaller volume, the human cerebellum has a surface area equivalent to 78% of the cerebral cortex. The cerebellum is connected to the posterior aspect of the brainstem by three pairs of prominent fiber bundles, the **superior, middle,** and **inferior cerebellar peduncles** (see Fig. 15.8 later). On its anterior surface, near the middle cerebellar peduncle, a small, bulb-like region of each cerebellar hemisphere, known as the **flocculus,** is connected to a region of the vermis known as the **nodulus,** and the two are referred to as the flocculonodular lobe, composing approximately 1% of cerebellar volume.

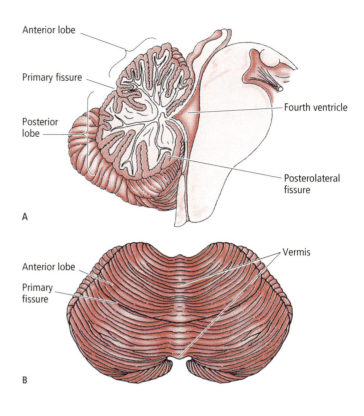

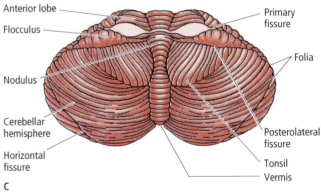

Figure 15.1 • The cerebellum: (A) midsagittal section, medial view, (B) superior view, and (C) inferior view.

Gray matter and white matter

The cerebellum consists of an outer shell of gray cortex and a core of white matter containing deep gray nuclei

Like the cerebrum, the cerebellum consists of gray matter and white matter, organized into an outer shell of gray cortex and a core of white matter containing deep gray nuclei. The **gray matter** is distributed in two regions, as the three-layered outer **cerebellar cortex** and the **deep cerebellar nuclei (cerebellar nuclei)** (Fig 15.2), which are embedded within the white matter. The greater part of the cerebellum is composed of **white matter,** also referred to as the **medullary center** or **medullary white matter,** consisting of myelinated axons and neuroglia. There are three categories of these axons (fibers): intrinsic, afferent, and efferent (Fig. 15.3).

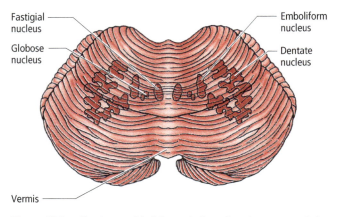

Figure 15.2 • The deep nuclei of the cerebellum, viewed posterosuperiorly.

Folding of the cerebellar cortex

Approximately 85% of the cerebellar cortex lines the banks of the sulci and fissures of the cerebellum and is hidden from the cerebellar surface

As the cerebellum develops, the **cerebellar cortex** undergoes considerable folding resulting in the formation of the cerebellar **folia**, **sulci**, and **fissures**. The cerebellar cortex laminates the underlying white matter of the cerebellum. Each folium consists of a thin layer of gray matter containing nerve cell bodies, as well as a white matter center, composed mostly of myelinated axons and neuroglia. As a result of this folding, approximately 85% of the cerebellar cortex becomes buried (tucked), lining the banks of the sulci and fissures of the cerebellum, whereas the remaining cortex covers the exposed cerebellar surface.

1. The **intrinsic (local, association) fibers** arise from the Purkinje cells of the cerebellar cortex and terminate in the deep cerebellar nuclei.
2. The **afferent (input) fibers** arise from extracerebellar sources and terminate in the deep cerebellar nuclei and/or the cortex.
3. The **efferent (output) fibers** principally originate in the deep cerebellar nuclei (although some arise from the cortex) and exit the cerebellum.

When sectioned in the sagittal plane through the vermis or cerebellar hemispheres, the gray and white matter of the cerebellum are recognizable with the unaided eye. Due to the presence of the deep sulci and the extensive folding of the cerebellar surface, the arborization of the white matter resembles the branching of a tree causing early neuroanatomists to refer to this image as the **arbor vitae** (L., "tree of life") (Fig. 15.4).

Layers of the cerebellar cortex

The number of neurons in the cerebellar cortex far outnumbers that of the much larger cerebral hemispheres

The cerebellar cortex is composed of only three layers (as opposed to the neocortex, which consists of six layers). The number of neurons of the cerebellar cortex (approximately 100 billion) far outnumbers that of the much larger cerebral hemispheres (approximately 23 billion) and new studies, unveiling differences in genetic, morphological, and topographical distributions, are challenging the long-held view that the cerebellar cortex is homogenous, with differences in genetic, morphological, and topographical distributions unveiled.

The cerebellar cortex consists of an outer (superficial) molecular layer that lies deep to the pia mater, a middle (intermediate) Purkinje cell layer, and an inner (deep) granular layer that lies adjacent to the cerebellar white matter (Fig. 15.5). The cerebellar cortex contains (1) **Purkinje**

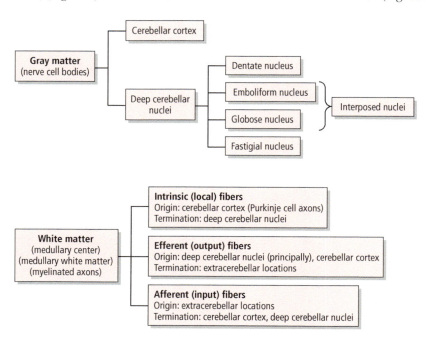

Figure 15.3 • Organization of the gray and white matter of the cerebellum.

CEREBELLUM ••• 289

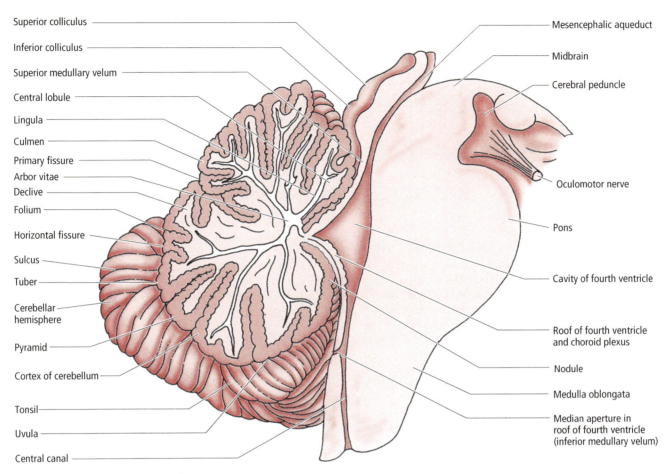

Figure 15.4 ● Midsagittal section through the brainstem and vermis of the cerebellum.

Table 15.2 ● The cerebellar cortex.

Layer	Contents	Neurotransmitter	Synapses
Molecular	Superficial stellate cells	Taurine	Receive excitatory synapses from parallel fibers. Form inhibitory synapses with Purkinje cells
	Basket cells	GABA	Receive excitatory synapses from parallel fibers. Form inhibitory synapses with Purkinje cells
	Purkinje cell dendrites		
	Granule cell parallel fibers		
Purkinje	Purkinje cell bodies (their axons form cerebellar cortical efferent fibers)	GABA	Form inhibitory synapses in deep cerebellar nuclei vestibular nuclei
Granular	Granule cells	Glutamate	Form excitatory synapses with Purkinje, outer stellate, basket, and Golgi cells
	Golgi cells	GABA	Receive excitatory synapses from mossy fibers, climbing fibers, and parallel fibers
		Glycine	Form inhibitory synapses in glomeruli; modulate mossy fiber signal transmission to granule cells
	Unipolar brush cells	Glutamate	Form excitatory synaptic contacts with granule cells and Golgi cells.
	Fusiform horizontal cells	GABA, glycine	Inhibitory interneurons ascending to synapse on stellate, basket, and Golgi cells

projection neurons (with currently nine known subtypes); (2) **excitatory interneurons**: granule cells (five known subtypes) and unipolar brush cells (three known subtypes); (3) **inhibitory interneurons**: fusiform horizontal cells (eponym: Lugaro cells), large Golgi cells, small Golgi cells, basket cells, superficial stellate cell, and deep stellate cells. Note that non-eponymous terms for Purkinje and Golgi cells have not yet been reported.

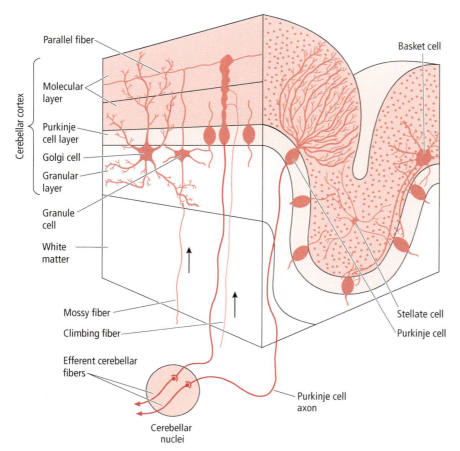

Figure 15.5 ● The cerebellar cortex and its component cell layers.

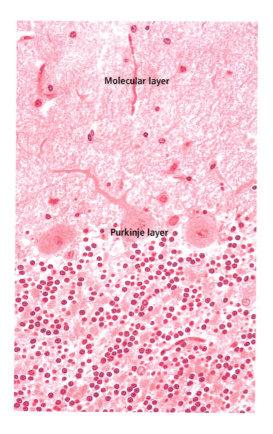

Although the **molecular layer** houses **stellate cells** and the **basket cells**, both of which are inhibitory interneurons, it is mostly occupied by the **dendrites of the Purkinje cells**, and unmyelinated **parallel fibers of the granule cells**, which leaves it relatively unstained using lipoprotein stains such as luxol fast blue.

The **Purkinje cell layer**, the intermediate layer of the cerebellar cortex, lies between the molecular and granular layers. It is occupied by the cell bodies of **Purkinje cells**, which are the only cells whose axons leave the cerebellar cortex, and **Purkinje layer interneurons.**

The **granular layer** is the deepest and thickest layer of the cerebellar cortex, containing 98% of the cerebellar neurons. It houses mostly **granule cells**, some **Golgi cells**, a small number of **unipolar brush cells**, and **fusiform horizontal cells** (eponym: Lugaro cells). The unipolar brush cells reside mostly in the granular layer of the flocculonodular lobe and the vermis.

In addition, this layer also contains synaptic complexes, known as **cerebellar glomeruli**.

The cerebellar cortex receives two special types of fibers: **climbing fibers** and **mossy fibers**. Climbing fibers are the axons of neurons whose cell bodies are in the **inferior olivary complex** (IOC). Mossy fibers include most of the remaining afferents to the cerebellum that arise from widespread areas of the nervous system (see discussion next).

Deep to the cerebellar cortex is the cerebellar white matter, which contains myelinated axons made up of **afferent fibers**, **efferent fibers**, and **intrinsic (local) fibers** of the cerebellum.

Neurons of the cerebellar cortex

The cerebellar cortex has seven types of neurons sorted into three categories: projection neurons, excitatory interneurons, and inhibitory interneurons

The cerebellar cortex has seven distinct cell types grouped into three categories (Fig. 15.5; Table 15.3). The discussion of the cellular components begins with neurons in the deepest layer (the granular layer) and progresses to the most superficial layer (the molecular layer). The interrelationships among them are discussed later in this chapter under "Functional organization of the cerebellum."

Granule cells

Granule cells are small, and they are the main excitatory neurons of the cerebellar cortex

Granule cells possess the smallest neuronal cell bodies of the entire central nervous system. They are the most numerous cells of the cerebellar cortex; in fact, there are so many granule cells that their number exceeds the total neuron population of the cerebral cortex. Although the cell bodies of granule cells are in the granular layer, their axons ascend into the molecular layer where they bifurcate to form a T-shaped configuration. These are known as **parallel fibers** because they are oriented parallel, not only to the cerebellar surface but also to the longitudinal axis of the folium in which they are embedded. Within the molecular layer, these unmyelinated parallel fibers course through the bushy dendrites of

Table 15.3 ● Neurons of the cerebellar cortex.

Cell	Cell body location	Axon location	Dendrite location	Axon synapse type and neurotransmitter	Axon synapses with	Miscellaneous
Granule	Granular layer	Molecular layer where they form parallel fibers		Excitatory (glutamate)	Dendrites of Purkinje cells, and outer stellate, basket, and Golgi cells	Only intrinsic excitatory cells of cerebellar cortex. Axons form parallel fibers
Golgi	Granular layer	Granular layer	All three layers of cortex	Inhibitory (GABA, glycine)	Glomeruli to modulate mossy fiber signal transmission to granule cells	Are inhibitory interneurons. Parallel fibers, mossy fibers, and climbing fibers form excitatory synapses with Golgi cells
Unipolar brush	Granular layer	Granular layer	Granular layer	Excitatory (glutamate)	Granule and Golgi cells	Located mainly in the flocculonodular lobe and vermis, receive input from vestibular afferents
Fusiform horizontal cells	Granular layer	All layers of cortex	Granular layer	Inhibitory (GABA, glycine)	Dendrites of stellate, basket, Golgi cells	Only cerebellar cell excited by serotonin.
Purkinje	Purkinje layer	White matter of cerebellum	Molecular layer; in a single plane; run perpendicular to parallel fibers	Inhibitory (GABA)	Deep cerebellar nuclei (especially dentate); also, some project to vestibular nucleus	Only extracerebellar projection from the cerebellar cortex
Outer stellate	Molecular layer	Purkinje cell layer		Inhibitory (taurine)	Purkinje cells	Are inhibitory interneurons. Parallel fibers form excitatory synapses with outer stellate cells
Basket	Molecular layer	Purkinje and molecular layer		Inhibitory (GABA)	Form basket-like junctions with Purkinje cell body	Are inhibitory interneurons. Parallel fibers of granule cells establish excitatory synapses with basket cells

GABA, gamma aminobutyric acid.

292 ● ● ●   **CHAPTER 15**

Cellular connections of the cerebellar cortex.

a series of Purkinje cells, similar to telephone wires running through branches of a series of trees. A single parallel fiber establishes synapses with hundreds of Purkinje cells, and any one Purkinje cell receives synaptic contacts from thousands of parallel fibers.

Each granule cell, via its parallel fibers, forms **glutamate**-releasing **excitatory synapses** with the dendrites of Purkinje cells, as well as with stellate cells, basket cells, and Golgi cells. Note that the *granule cells are the main intrinsic excitatory nerve cells of the cerebellar cortex.*

Golgi cells

Golgi cells are inhibitory (GABAergic, glycinergic) interneurons

The **Golgi cells** are **inhibitory interneurons**; their cell bodies occupy the granular layer, and their dendritic trees ramify in the molecular layer of the cerebellar cortex. Golgi cells form **glycine** and/or **gamma aminobutyric acid (GABA)**-releasing inhibitory synapses with mossy fibers within the

glomeruli. Glomeruli are synaptic complexes of Golgi cells and their terminals, mossy fibers, and granule cell dendrites ensheathed by glia. They allow neurotransmitters to pool, accumulate, and persist in synaptic clefts, *prolonging their effects*. It should be recalled that parallel fibers of the granule cells, as well as mossy and climbing fibers to the cerebellum, form excitatory synapses with the Golgi cells.

Unipolar brush cells

Unipolar brush cells form a minor group of glutaminergic, excitatory neurons of the cerebellar cortex

The **unipolar brush cells** have an oval-shaped cell body that gives rise to a single dendrite and axon. These cells, located in the granular layer of mainly the flocculonodular lobe and vermis, make synaptic contacts with granule cells and Golgi cells. Although granule cells are the main and most populous excitatory neurons of the cerebellar cortex, the unipolar brush cells form a minor group of **glutaminergic**, excitatory neurons. Due to their preferential topographical location in the flocculonodular lobe and the vermis and to the vestibular afferent input they receive, they may play a role associated with the flocculonodular lobe and the vermis: control of eye movements, the vestibuloocular reflex, and posture.

Fusiform horizontal cells

Fusiform horizontal cells are large GABAergic/glycinergic (inhibitory) interneurons

Fusiform horizontal cells (eponym: Lugaro cells) are large inhibitory interneurons named according to their shape and due to the parallel orientation of their two dendrites originating from opposite ends of their cell body. The sagittal axons of fusiform horizontal cells establish axodendritic synapses with stellate and basket cells in the molecular layer and their horizontal dendrites make dendrodendritic synapses with Golgi cells in the granular layer. In contrast to all other cells of the cerebellar cortex, *input to fusiform horizontal cells comes from serotonin*. Output from fusiform horizontal cells comes via GABA, glycine, or co-transmission of both.

Purkinje cells

The primary target of the Purkinje cell axons is the deep cerebellar nuclei

Purkinje cells are multipolar neurons that have a large cell body and an extensive dendritic tree; in fact, they are one of the *largest* neurons of the central nervous system (CNS). Their flask-shaped cell bodies are in the single-layered Purkinje cell layer, whereas their striking, flattened bushy dendrites reside and ramify in the molecular layer. The Purkinje cell dendrites are aligned in a single plane (similar to the fanned-out feathers of a peacock's tail) and are arranged perpendicular to the parallel fibers and the longitudinal axis of the folium in which they are embedded. The axon of each Purkinje cell descends through the granular layer to reach the underlying white matter core of the cerebellum containing the deep cerebellar nuclei. These axons, which become myelinated upon entering the white matter, terminate principally in the ipsilateral **deep cerebellar nuclei**, particularly the dentate nucleus, where they establish **GABAergic inhibitory synapses**. However, axons of some Purkinje cells (of the flocculonodular lobe) leave the cerebellum, and synapse in the **vestibular nuclei** of the medulla oblongata. This Purkinje cell projection to the vestibular nuclei is the *only* extracerebellar projection *from* the **cerebellar cortex**. Via these projections, the Purkinje cells regulate the activity of the deep cerebellar nuclei and the vestibular nuclei.

Stellate cells

Stellate cells are inhibitory interneurons

Stellate (L., *stella*, "star," "star-shaped") **cells** are **inhibitory interneurons** that occupy the molecular layer of the cerebellar cortex. The stellate cells are believed to form

taurine-releasing inhibitory synapses with the dendrites of the Purkinje cells. Recall that the **parallel fibers** of the granule cells form excitatory synapses with the stellate cells.

Basket cells

Basket cells are inhibitory interneurons

Basket cells are **inhibitory interneurons** and occupy the molecular layer of the cerebellar cortex. Within the molecular layer, axons of numerous basket cells form "basket-like" configurations that cover the body of a Purkinje cell. **GABA** is released at these synapses, inhibiting the Purkinje cells from firing. Recall that parallel fibers of the granule cells establish excitatory synapses with the basket cells.

Zones of the cerebellum

The zones of the cerebellum are composed of cortex, white matter, and the associated deep cerebellar nucleus (or nuclei)

Four pairs of deep cerebellar nuclei, the **fastigial nucleus**, **dentate nucleus**, **emboliform nucleus**, and **globose nucleus** (the last two are frequently referred to as the **interposed nuclei**), are embedded within the cerebellar white matter (see Fig. 15.2). Based on the connections between the regions of the cortex and the deep cerebellar nuclei, the cerebellum is divisible into functional zones, namely the **vermal (median)**, **paravermal (intermediate)**, and **hemispheric (lateral) zone** (Fig. 15.6A). These functional zones are composed of cortex, white matter, and the associated deep cerebellar nucleus (or nuclei), as listed in Table 15.4.

Lobes of the cerebellum

The cerebellum consists of the anterior, posterior, and flocculonodular lobes

As the cerebellum increases in size during development, its cortex is compacted and forms alternating folia and sulci. Three of these sulci are especially deep and are known as the **primary, horizontal, and posterolateral fissures** (Fig. 15.6B). These fissures permit the horizontal division of the cerebellum into three lobes: the **anterior, posterior**, and **flocculonodular lobes**. The flocculonodular lobe is composed of the right and left **flocculi** and the unpaired **nodulus**. The anterior lobe is separated from the posterior lobe by the primary fissure. The posterior lobe is divided into two regions by the horizontal fissure and is separated from the flocculonodular lobe by the posterolateral fissure.

Interestingly, these lobes also correspond to the phylogenetic development of the cerebellum (Table 15.5). The posterior lobe is phylogenetically the newest portion of the cerebellum and is therefore also known as the **neocerebellum** (G., *neo*, "new"). The oldest portion of the cerebellum is the flocculonodular lobe and is also referred to as the **archicerebellum** (G., *archi*, "beginning"). The **paleocerebellum**

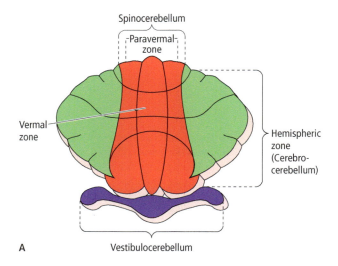

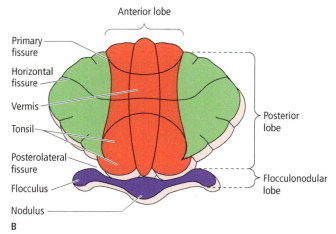

Figure 15.6 • (A) The zones and phylogenetic classification of the cerebellum. (B) The lobes and fissures of the cerebellum.

(G., *paleo*, "ancient") is intermediate in phylogenetic development and corresponds to the anterior lobe.

Flocculonodular lobe (vestibulocerebellum, archicerebellum)

The flocculonodular lobe is reciprocally connected with the vestibular system

The paired **flocculi** (L., *flocculus*, "tuft of wool") and the **nodulus** (of the vermis) together form a small lobe, the **flocculonodular lobe**.

Since this lobe is associated via its connections to the vestibular system, functionally it is referred to as the **vestibulocerebellum** (Fig. 15.6). The vestibulocerebellum is unique since it receives direct projections (first-order neuron fibers) from the vestibular division of the vestibulocochlear nerve (CN VIII). Some of the second-order neurons housed in the vestibular nuclei also project to the vestibulocerebellum which receives, via these projections, a continuous stream of information from the vestibular system about head orientation (position), tilt, and movement. Note that all vestibular projections (first- and second-order neuron fibers) to the vestibulocerebellum are **ipsilateral**. The vestibulocerebellum uses this input information to adjust muscle tone so that it can always maintain posture and balance. It does this in the following manner: Purkinje cells of the vestibulocerebellar cortex and cells of the fastigial nucleus project back to the vestibular nuclei. It is via this projection that the vestibulocerebellum influences the **vestibulospinal** tracts to adjust muscle tone in response to vestibular input. The vestibulospinal tracts terminate in the spinal cord where they influence motor neurons that innervate the axial and proximal limb (girdle) extensor muscles (antigravity muscles) that control posture and balance. The vestibulospinal tracts neutralize the effects of gravity on the body. An example of this is when an

Zone	Cortical region	Cerebellar nucleus
Vermal (median)	Vermis	Fastigial nucleus
Paravermal (intermediate)	Medial region of cerebellar hemispheres	Emboliform, globose nuclei
Lateral (hemispheric)	Lateral region of cerebellar hemispheres	Dentate nucleus

Table 15.4 • The three zones of the cerebellum and associated cortical region and cerebellar nucleus/nuclei.

Table 15.5 • Lobes of the cerebellum.

Lobe	Phylogenetic name	Other name	Components	Function
Anterior	Paleocerebellum	Spinocerebellum	Vermis, paravermal zone, most of anterior lobe	Controls posture, muscle tone, and executes as well as coordinates muscle activity of the trunk and limbs during stereotyped activity. Vermis controls axial and girdle musculature, whereas the intermediate hemisphere controls limb musculature
Posterior	Neocerebellum	Cerebrocerebellum Pontocerebellum	(Lateral) cerebellar hemispheres	Planning, coordination, and execution of rapid, fine, nonstereotyped (skilled) movement
Flocculonodular	Archicerebellum	Vestibulocerebellum	Nodulus and right and left flocculi	Balance and posture maintenance, coordination of head and eye movements

individual loses balance the person immediately stands up straight as the function of the vestibulospinal tracts responds to vestibular input (from head tilt) to stimulate, contract, and firm up the antigravity muscles so that the person regains balance. Since there is a limit to the function of the vestibulospinal tracts, sometimes the individual is unable to regain balance and falls. In summary, the vestibulocerebellum, via its projections to the vestibular nuclei, influences:

1 the **lateral vestibulospinal tract,** which is associated with balance and posture maintenance, and
2 the **medial vestibulospinal tract** (descending to the cervical spinal cord as part of the *descending* **medial longitudinal fasciculus**, MLF) which is associated with the control of neck muscles.

Additionally, due to its projections to, and the influence it exerts on, the vestibular nuclei, it is also responsible for the coordination of head and eye movements (vestibuloocular reflexes, via the *ascending* **medial longitudinal fasciculus**). The vestibular nuclei also project to the motor nuclei of cranial nerves III, IV, and VI via the *ascending* medial longitudinal fasciculus (MLF). Through these cranial nerve connections, ocular movements are coordinated in response to head movement, for example: when gazing at an object, if the head is tilted in one direction, the vestibular receptors detect the head tilt and then send messages via the vestibular nuclei to the motor nuclei that control eye movement. The eyes then automatically roll (via the vestibulo-ocular reflex) in the opposite direction to keep the image on the retina.

Anterior lobe (spinocerebellum, paleocerebellum)

The anterior lobe is reciprocally connected with the spinal cord

The second part of the cerebellum to develop phylogenetically arose later in evolution and is known as the **paleocerebellum**. It is composed of the greater part of the **vermis**, the **paravermal zone** (**paravermis, intermediate hemisphere**), and most of the **anterior lobe**. This part of the cerebellum developed in relation to the brainstem and spinal cord and therefore has reciprocal projections with them. The paleocerebellum is functionally referred to as the **spinocerebellum** (see Fig. 15.6); it has its own somatotopic map, separate from that of the posterior lobe (Fig. 15.7). Sensory (proprioceptive and exteroceptive) input from the muscle spindles and tendon organs of the limbs and trunk is relayed to the spinocerebellum via the **spinocerebellar** and **cuneocerebellar tracts**, and from the head via the **trigeminocerebellar fibers**. These fiber tracts carry proprioceptive information to the cerebellum as a movement that is being performed advances toward completion. This input is then processed by the spinocerebellum, which adjusts if necessary. The spinocerebellum controls posture and muscle tone. It also executes and coordinates muscle activity of the trunk and limbs (particularly of the hands and feet) during stereotyped movements,

such as walking. Stereotyped movements consist of repetitive movements lacking variation. The spinocerebellum is somatotopically organized; that is, the vermis controls axial and proximal girdle musculature (head and trunk), whereas the paravermal hemisphere functions in the control of the distal limb musculature (mainly the hands and feet; Fig. 15.7). The inferior portion of the vermis is associated with gross motor coordination, whereas its superior portion is associated with fine motor coordination. The proprioceptive information that reaches the cerebellum via the spinocerebellar and cuneocerebellar tracts does not reach conscious levels. In contrast, the posterior column–medial lemniscus pathway relays proprioceptive input from the body to the primary somatosensory cortex (which has an accurate map of one's body parts), where the information reaches consciousness so that people are aware of the position and movement of their body parts (e.g., limbs) even when they don't see them. Output from the vermal zone and fastigial nucleus is sent to the thalamus, which in turn projects to the motor cortex to influence the neurons of the **anterior corticospinal tract** that control movement of the axial and proximal limb (girdle) muscles. The vermal zone and fastigial nucleus also influence the neurons of the lateral vestibulospinal tract and reticulospinal tract that control the muscle tone of the extensors, balance, and posture maintenance. Output from the paravermal (intermediate) zone and emboliform/globose nuclei ultimately control limb muscles via the **lateral corticospinal tract**, which controls the motor activity of the distal limb musculature especially the muscles of the hand, and the rubrospinal tract, which complements the corticospinal tract.

Posterior lobe (cerebrocerebellum, pontocerebellum, neocerebellum)

The posterior lobe is reciprocally connected to the cerebral cortex

The most prominent lobe of the cerebellum is the most recent phylogenetically and is thus referred to as the **neocerebellum**. It consists of the **cerebellar hemispheres** but does not include the paravermal zone. Since the neocerebellum has a vast number of reciprocal connections with the cerebral cortex, it is functionally referred to as the **cerebrocerebellum** (or **pontocerebellum**; see Fig. 15.6); in fact, it has its own somatotopic map, separate from that of the anterior lobe (see Fig. 15.7). The cerebellar hemispheres do not exhibit any perceptible anatomical landmark separating the paravermal cortex from the hemispheres. The cerebral cortex projects corticopontine fibers to the ipsilateral pontine nuclei. They in turn project pontocerebellar fibers which cross the midline to join the contralateral middle cerebellar peduncle to enter the contralateral lateral zone cortex and the dentate nucleus. Via this pathway, the cerebellum receives information about voluntary movements that are in progress, or impending voluntary movements. Due to the influence, it exerts on the motor cortex, the cerebrocerebellum is associated with the planning, coordination, and execution of rapid,

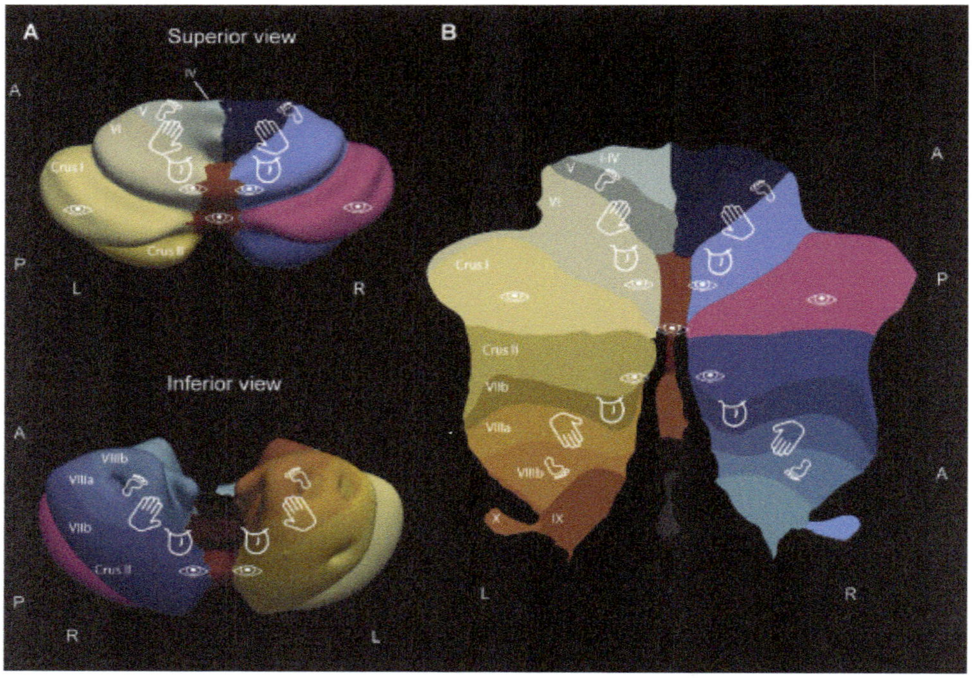

296 ● ● ● **CHAPTER 15**

Figure 15.7 ● Sensory depiction in the cerebellar cortex. Note that visual and auditory information is relayed to the central region of the vermis. Somatosensory information is relayed to the vermal and paravermal zones of the anterior lobe, and the paravermal zone of the posterior lobe.
https://www.sciencedirect.com/science/article/pii/S1053811920301117#:~:text=The%20cerebellum%20is%20known%20to,and%20are%20less%20precisely%20characterized.

fine, nonstereotyped (learned, skilled) movements such as ice skating. Nonstereotyped movement is characterized as nonrepetitive, variable, and of diverse pattern.

CEREBELLAR PEDUNCLES

The cerebellum is attached to the posterior aspect of the brainstem by three pairs of prominent fiber bundles: the **superior**, **middle**, and **inferior cerebellar peduncles** (L., *ped*, "foot") (Fig. 15.8; Table 15.6). These peduncles are major (and the only) communication highways whereby afferent (input) and efferent (output) fibers enter or exit the cerebellum, respectively.

Superior cerebellar peduncle

The superior cerebellar peduncle consists predominantly of efferent pathways arising from the cerebellar cortex and the deep cerebellar nuclei

The **superior cerebellar peduncle** attaches the cerebellum to the caudal midbrain and the pons (Table 15.6). This peduncle consists predominantly of **efferent** pathways arising from the cerebellar cortex and the deep cerebellar nuclei – especially from the dentate, emboliform, and globose nuclei – with some fibers arising from the fastigial nucleus. The efferent pathways have four components:

1 the dentorubrothalamic pathway;
2 the interpositorubrothalamic pathway;
3 the fastigiothalamic tract; and
4 the fastigiovestibular tract.

The fibers of the first two pathways form a bundle of axons known collectively as the **brachium conjuctivum** (L.,

brachium, "arm"; *conjuctivum*, "joining"), which conveys the principal cerebellar efferent (output) pathway.

The superior cerebellar peduncle also carries **afferent** pathways to the cerebellum. The input pathways are:

- the anterior spinocerebellar tract, transmitting proprioceptive input from the lower limb and trunk;
- the tectocerebellar fibers, transmitting visual input from the superior colliculus;
- the rubrocerebellar fibers, transmitting information from the red nucleus;
- the trigeminocerebellar fibers, transmitting sensory information from the head; and
- the caeruleocerebellar fibers from the locus caeruleus that regulate cerebellar activity.

Middle cerebellar peduncle

The massive middle cerebellar peduncle contains afferent fibers from the pons – the pontocerebellar fibers

The middle cerebellar peduncle is the largest of the three peduncles, and it joins the cerebellum to the posterior aspect of the pons. Most fibers arising from the pontine nuclei destined for the cerebellum are the pontocerebellar fibers (transverse fibers of the pons) (Table 15.6). These fibers decussate and then pass into the cerebellum via the contralateral **middle cerebellar peduncle**.

Since signals arising in the cerebral cortex are transmitted via corticopontine fibers to the pontine nuclei, the corticopontine and the pontocerebellar fibers collectively form the **corticopontocerebellar pathway**, which relays **afferents** to the **cerebrocerebellum (neocerebellum)**.

CEREBELLUM

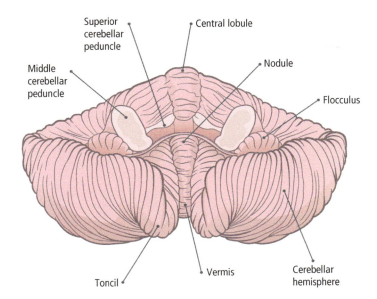

Figure 15.8 • Posterior view of the brainstem and part of the left lobe of the cerebellum. Note the superior, middle, and inferior cerebellar peduncles. Анна Богатырева/Adobe stock.

Inferior cerebellar peduncle

The inferior cerebellar peduncle contains mostly afferent pathways to the cerebellum from the spinal cord and brainstem

The **inferior cerebellar peduncle** connects the cerebellum to the medulla oblongata. It carries mostly **afferent** pathways to the cerebellum from the spinal cord and brainstem (Table 15.6). This peduncle consists of two divisions: the more prominent, laterally positioned, **restiform body**; and the slender, medially located, **juxtarestiform body**.

The **restiform body** includes proprioceptive **afferent fibers** from the spinal cord and brainstem to the cerebellum carried in the following tracts or fibers:

- the posterior spinocerebellar tract;
- the rostral spinocerebellar tract;
- the cuneocerebellar tract;
- the olivocerebellar fibers;
- the trigeminocerebellar fibers; and
- the reticulocerebellar fibers.

Table 15.6 • The cerebellar peduncles.

Cerebellar peduncle	Connects cerebellum to	Afferents (input) fibers	Efferents (output) fibers arise from

Superior (Mostly efferent fibers)	Caudal midbrain and pons	Anterior spinocerebellar tract
		Tectocerebellar fibers
		Rubrocerebellar fibers
		Trigeminocerebellar fibers
		Caeruleocerebellar fibers
			Cerebellar cortex
			Deep cerebellar nuclei
			    Dentorubrothalamic pathway
			    Interpositorubrothalamic pathway
			    Fastigiothalamic tract
			    Fastigiovestibular tract

Middle (afferent fibers)	Pons	Pontocerebellar fibers	
Inferior (Mostly afferent fibers) Restiform body (laterally)	Medulla	Dorsal spinocerebellar tract Rostral spinocerebellar tract Cuneocerebellar tract Olivocerebellar fibers Trigeminocerebellar fibers Reticulocerebellar fibers Vestibulocerebellar fibers	
Juxtarestiformbody (medially)			Cerebellovestibular fibers Cerebelloreticular fibers

The first four tracts/fibers terminate in the **spinocerebellum** (**paleocerebellum**), whereas the olivocerebellar fibers terminate *throughout* the **cerebellar cortex**, but mainly in the **neocerebellum**. The reticulocerebellar fibers, emerging from the nuclei of the pontine and medullary reticular formation, terminate in the **spinocerebellum** and **vestibulocerebellum**.

The **juxtarestiform body** includes mostly **afferent fibers** to the cerebellum, but also some efferent fibers from the cerebellum. The afferent fibers arise from first-order and second-order vestibular neurons and reach the **vestibulocerebellum** and the vermis of the **spinocerebellum** as vestibulocerebellar fibers. The only efferent fibers that leave the cerebellum via the juxtarestiform body are the cerebellovestibular (fastigiovestibular, fastigiobulbar) fibers and the cerebelloreticular fibers.

DEEP CEREBELLAR NUCLEI

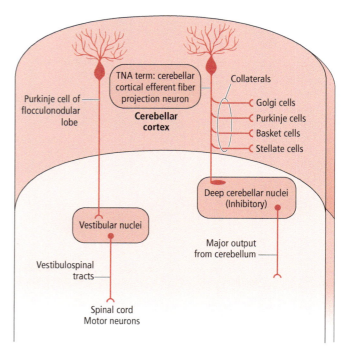

Figure 15.9 • Purkinje cell projections from the cerebellar cortex to the deep cerebellar nuclei and the vestibular nuclei. TNA = Terminologia Neuroanatomica

The cerebellum has four pairs of nuclei embedded deep in its white matter core

As mentioned previously, there are four pairs of intracerebellar nuclei, collectively known as the **deep cerebellar nuclei**. These are embedded deep in the cerebellar medullary white matter, and from lateral to medial, they are the **dentate nucleus**, the **emboliform nucleus**, the **globose nucleus**, and the **fastigial nucleus**.

The deep cerebellar nuclei receive collaterals of cerebellar afferents and cerebellar cortical efferent fibers (Purkinje cell axon terminals) from the overlying cerebellar cortex; they are the main source of cerebellar output

Each of the deep cerebellar nuclei (Table 15.7) receives duplicate input from the collaterals of afferents destined for the corresponding region of the cerebellar cortex, permitting the deep cerebellar nuclei to be informed of input as it is being relayed to the cerebellar cortex. The fastigial nucleus also receives vestibulocerebellar fibers, which include axons of first-order neurons of the vestibular nerve (CN VIII), as well as axons of second-order neurons from the vestibular nuclei. The vestibulocerebellar fibers transmit sensory input directly from the vestibular apparatus of the internal ear. These connections play a role in the maintenance of equilibrium (balance).

Although the deep cerebellar nuclei project to different parts of the brain, each one also has projections to the **inferior olivary complex** (**inferior olive**) as well as to the brainstem **reticular formation**. Furthermore, each deep cerebellar nucleus sends reciprocal projections to the same region of the **cerebellar cortex** that projects to it. Although output from the cerebellum arises from the cerebellar cortex and the deep cerebellar nuclei, the *deep cerebellar nuclei are the main source of cerebellar output*. The deep cerebellar nuclei influence the descending motor pathways and ultimately the motor neurons of the brainstem and spinal cord (Fig. 15.9).

Dentate nucleus

The dentate nucleus is associated with the cerebrocerebellum

The **dentate nucleus** (L., *dens*, "tooth," "tooth-shaped") is the largest, the most laterally positioned, and the most conspicuous of the deep cerebellar nuclei. It appears similar to the inferior olivary complex since it resembles a "crumpled bag." It receives input from the collaterals of pontocerebellar afferents and the **cerebellar cortical efferent fibers** originating from the **cerebrocerebellum** (**pontocerebellum**) (Table 15.7).

The dentate nucleus gives rise to efferent fibers, the **dentatothalamic tract** (**DTT**), that join the **brachium conjuctivum** (eponym: Bundle of Burdach), the major component of the ipsilateral **superior cerebellar peduncle**. These efferent fibers, joined by fibers from the **emboliform and globose nuclei**, decussate in the posterior midbrain, and project to the contralateral **ventral lateral** (**VL**) **nucleus of the thalamus**. The DTT projection is the major output of the dentate nucleus. The VL nucleus of the thalamus gives rise to fibers that ascend to terminate predominantly in the **premotor cortex** and the **primary motor cortex**. One of the outputs of both the premotor and primary motor cortices is the **lateral corticospinal tract**, which has an important function in the control of distal limb muscles. Although the cerebellum receives afferents (input) from both the *motor* and the *sensory* areas of the cerebral cortex about intended motor activity via the **corticopontocerebellar pathway**, the cerebellum projects via a relay in the thalamus to the motor cortical areas, particularly of the primary motor area, where it influences upcoming motor activity.

Additionally, a few fibers arising from the dentate nucleus exit the cerebellum via the same route but project to the parvocellular division of the **red nucleus**, and the **oculomotor nucleus** of the opposite side. The parvocellular

Nucleus	Afferents (input)	Efferents (output) to
Dentate nucleus	Collateral branches of pontocerebellar afferents Purkinje cell axons from cortex of cerebrocerebellum	Ventral lateral (VL) nucleus of the thalamus Red nucleus Oculomotor nucleus Cerebellar cortex Reticular formation Inferior olivary complex
Emboliform and globose nuclei	Collateral branches of cerebellar afferents Purkinje cell axons from cortex of spinocerebellum Anterior spinocerebellar tract	Red nucleus VL nucleus of the thalamus Cerebellar cortex Reticular formation Inferior olivary complex
Fastigial nucleus	Collateral branches of cerebellar afferents Purkinje cell axons from cortex of vestibulocerebellum Vestibulocerebellar fibers First-order neurons of CN VIII Second-order neurons from vestibular nuclei	Lateral vestibular nucleus Medial vestibular nucleus Inferior vestibular nucleus VL nucleus of the thalamus Cerebellar cortex Reticular formation

Table 15.7 ● Connections of the deep cerebellar nuclei.

division of the red nucleus sends fibers via the central tegmental tract to the ipsilateral **inferior olivary complex**, which projects heavily to the cerebellum. The fibers terminating in the oculomotor nucleus synapse with the lower motor neurons that innervate the superior rectus muscle. Since the dentate nucleus neurons, whose axons terminate in the oculomotor nucleus, receive input from the vestibular primary fibers, they may serve as a connection of a vestibulo-oculomotor pathway.

Globose and emboliform nuclei (interposed nuclei)

The globose and emboliform nuclei are associated with the spinocerebellum

Together, the **globose** (L., "globe-shaped") and **emboliform** (G., *emballein*, "to insert") **nuclei** are referred to as the **interposed nuclei**. Afferents to these nuclei include **cerebellar cortical efferent fibers** (Purkinje cell) arising from the **spinocerebellum (paravermal cortex)**, as well as collaterals of input fibers arriving in the cerebellum by way of the restiform body and via the anterior spinocerebellar tract (Table 15.8).

Axons arising from the emboliform and globose nuclei join the brachium conjuctivum, *cross* to the opposite side, and project to the contralateral magnocellular division of the **red nucleus**. Other axons project to the *ipsilateral VL nucleus of the thalamus*. The projection to the red nucleus is the major output of the interposed nuclei. The red nucleus gives rise to the **rubrospinal tract**, whereas the VL nucleus projects to the premotor and primary motor cortexes, which give rise to the *lateral corticospinal tract*. Recall that the rubrospinal and lateral corticospinal tracts together form the **descending lateral systems** and are involved in the control of the **distal musculature of the upper and lower limbs** (see Chapter 13 for more detail).

Fastigial nucleus

The fastigial nucleus is associated with the vestibulocerebellum

The **fastigial nucleus** (L., *fastigius*, "summit" since it is in the roof of the fourth ventricle) receives **cerebellar cortical efferent** (Purkinje cell) fibers arising from the **vestibulocerebellum**, and first-order afferent collaterals arising from the vestibular apparatus and second-order fibers from the vestibular nuclei (Table 15.7).

The fastigial nucleus gives rise to efferent fibers that leave the cerebellum via the **juxtarestiform body** of the **inferior cerebellar peduncle**. These are the fastigiovestibular (cerebellovestibular) fibers and the fastigioreticular fibers.

The **fastigiovestibular fibers** terminate principally in the **lateral and inferior vestibular** nuclei where they influence the vestibulospinal tracts, bilaterally. The **fastigioreticular fibers** terminate in the **pontine and medullary reticular formation** where they influence the reticulospinal tracts bilaterally (but mainly contralaterally). Recall that the vestibulospinal and reticulospinal tracts (collectively referred to as the **descending medial systems**) have an important function in balance and posture maintenance as well as in the motor activity of the **axial and proximal limb musculature** (see Chapter 13 for more detail).

AFFERENTS (INPUT) TO THE CEREBELLUM

Afferent fibers to the cerebellum terminate primarily in the cerebellar cortex, but also send collaterals to the corresponding deep cerebellar nuclei

The **afferent (input) fibers** to the cerebellum outnumber the efferent (output) fibers from the cerebellum by three to one. Afferent fibers pass primarily into the **cerebellar cortex,** and they do so mainly via the inferior and middle cerebellar peduncles, although a number of fibers may enter via the superior cerebellar peduncle.

Afferent signals to the cerebellum arise from three main sources (Fig. 15.10; Table 15.8):

1. The **vestibular system** (vestibular ganglion and nuclei), which terminates in the vestibulocerebellum.
2. The **spinal cord**, which terminates in the spinocerebellum.
3. The **cerebral cortex**, mostly via the corticopontocerebellar pathway but also via the corticoreticulocerebellar and the cortico-olivocerebellar pathways, all of which terminate in the cerebrocerebellum.

In addition to these main sources, the cerebellar cortex also receives afferents from the hypothalamus, superior colliculi, ventral midbrain tegmentum, locus caeruleus, pontine raphe nuclei, trigeminal nerve nuclei, cochlear nuclei, and the deep cerebellar nuclei.

Afferents to the cerebellum arrive as:

- mossy fibers;
- climbing fibers;
- monoaminergic fibers; or
- other fibers.

Mossy fibers

Most of the afferent fibers to the cerebellum are mossy fibers

Mossy fibers include most of the afferents to the cerebellum which terminate in the **cerebellar cortex**. They are **excitatory**, release **glutamate**, and arise from the vestibular system, spinal cord, and cerebral cortex via the brainstem (Table 15.8). On their way to the cerebellar cortex to influence the Purkinje cells, the mossy fibers send collaterals to the **deep cerebellar nuclei**, where they also form **excitatory synapses**.

Mossy fiber afferents from the vestibular system

Afferents from the vestibular system terminate in the vestibulocerebellum as mossy fibers

Sensory information about changes in head position and orientation of the head relative to gravity is transmitted from the vestibular apparatus of the internal ear by the peripheral processes of first-order bipolar neurons whose cell bodies are housed in the ipsilateral **vestibular ganglion**. The central processes of these neurons form the root of the vestibular division of the vestibulocochlear nerve (CN VIII) and enter the brainstem. Most of these central processes terminate in the **vestibular nuclei** to synapse with second-order neurons. However, a small number of these central processes join the **juxtarestiform body** (part of the inferior cerebellar peduncle) and terminate **ipsilaterally** as **mossy** fibers in the **vestibulocerebellum**.

Cell bodies of second-order neurons housed in the **vestibular nuclei** give rise to fibers that form a larger pathway than the fibers of the first-order neurons. They also join the **juxtarestiform body** to terminate **bilaterally** as **mossy** fibers in the **vestibulocerebellum**.

The afferents from the vestibular system that terminate in the vestibulocerebellum (**flocculonodular lobe**) also send collaterals to the **fastigial nucleus**.

Mossy fiber afferents from the spinal cord

Afferents from the spinal cord terminate in the spinocerebellum as mossy fibers

Input from the spinal cord to the cerebellum consists of: (1) the anterior spinocerebellar tract, (2) the posterior spinocerebellar tract, (3) the cuneocerebellar tract, and (4) the rostral spinocerebellar tract. All these tracts relay proprioceptive information (from muscles, tendons, joints, and skin) and terminate in the cortex of the **spinocerebellum** as mossy fibers (see Chapter 12 for more detail).

The **anterior spinocerebellar tract** relays proprioceptive information from the **lower trunk and lower limb**. This tract ascends **contralaterally** in the spinal cord, and its fibers enter the cerebellum via the **superior cerebellar peduncle** and terminate in the cortex of the **vermis**. It is interesting to note that these fibers *cross twice* (once in the spinal cord and then again in the cerebellar white matter), so that they terminate in the cerebellar hemisphere ipsilateral to the origin of the information.

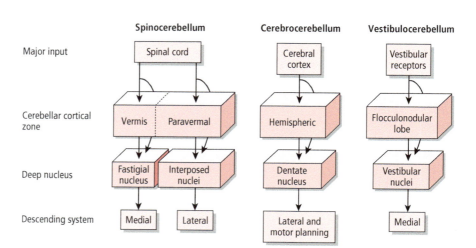

Figure 15.10 • Functional organization of the spinocerebellum, cerebrocerebellum, and vestibulocerebellum. Source: Modified from Martin, JH (1996) *Neuroanatomy, Text and Atlas*, 2nd edn. Appleton & Lange, Connecticut; fig. 10.6.

The **posterior spinocerebellar tract** relays proprioceptive input from the **lower trunk and lower limb**. This tract ascends in the spinal cord **ipsilaterally** and passes through the **inferior cerebellar peduncle** to enter the cortex of the vermis.

The **cuneocerebellar tract** (cuneatocerebellar tract) relays information from the **upper trunk and upper limb** and passes through the **inferior cerebellar peduncle** to terminate in the ipsilateral **paravermal zone** cortex of the **anterior lobe** of the cerebellum.

The **rostral spinocerebellar tract** relays information from the **upper limb**. It passes through the **inferior cerebellar peduncle** to terminate in the cortex of the ipsilateral **vermis**.

The afferents from the spinal cord that terminate in the **spinocerebellum (anterior lobe)** also send collaterals to the **globose** and **emboliform nuclei**.

A striking characteristic of the spinocerebellar tracts is that they relay signals to the cerebellum at very rapid velocities (72–120 m/s). These tracts owe their rapid conduction velocity to their wide-diameter, heavily myelinated axons. This keeps the cerebellum continually informed of the current position and movement of body parts, and it enables it to make rapid adjustments of upcoming motor activity if necessary.

Mossy fiber afferents from the cerebral cortex

Afferent signals from the cerebral cortex are relayed (via the pontine nuclei) to the cerebrocerebellum by mossy fibers

Signals arising from the cerebral cortex are transmitted to the cerebellum via the **corticopontocerebellar pathway** and the **corticoreticulocerebellar pathway**. These pathways terminate in the **cerebrocerebellum (posterior lobe)**.

The **primary motor cortex**, the **premotor cortical areas**, the **somatosensory cortex**, and association cortical areas of the parietal, temporal, and occipital lobes of each cerebral hemisphere give rise to more than 20 million axons that descend in the corona radiata, the internal capsule, and the cerebral peduncle to the pons and spinal cord. Approximately 1 million of these axons form the corticospinal tracts. The other 19 million axons are destined for the pontine nuclei (i.e., as corticopontine fibers of the corticopontocerebellar pathway), the reticular formation, and, via the corticonuclear (corticobulbar) fibers, to the cranial nerve motor nuclei in the brainstem.

Corticopontocerebellar pathway

The corticopontocerebellar pathway is the largest pathway associated with the cerebellum

Most of the 19 million fibers arising from the **cerebral cortex** are the **corticopontine fibers**, and as their name indicates, they terminate *ipsilaterally* in the **pontine nuclei**. Most of the fibers arising from the pontine nuclei, known as **pontocerebellar fibers**, decussate (then referred to as the **transverse fibers of the pons** as they decussate) and

gather to form the **middle cerebellar peduncle** to enter the cerebellum and terminate in the cerebellar cortex as **mossy fibers**. Some of the pontocerebellar fibers project to the ipsilateral cerebellar hemisphere. The pontocerebellar fibers form the *largest* mossy fiber input to the cerebellum. The pontocerebellar fibers send collaterals to the **dentate nucleus** and terminate throughout the cerebellar cortex – except for the flocculonodular lobe – in the following manner: (1) signals originating from the primary motor cortex are relayed via the corticopontocerebellar pathway to the **vermal** and **paravermal zones** of the cerebellum; and (2) signals originating from the premotor, somatosensory, and association cortical areas are relayed via the corticopontocerebellar pathway to the **hemispheric zone** of the cerebellum.

The **corticopontocerebellar pathway** is the largest pathway associated with the cerebellum as evidenced by the massive **middle cerebellar peduncles** connecting the pons to the cerebellum. Based on information relayed to the somatosensory cortex from muscle receptors, this pathway transmits signals that keep the cerebellum informed about a movement in progress. Additionally, based on commands of the primary and premotor cortices relayed to the lower motor neurons of the brainstem and spinal cord, it also keeps the cerebellum informed about upcoming movement. Thus, these signals enable the cerebellum to monitor and adjust motor activity continuously so that muscular movement is precise, smooth, flowing, and coordinated.

Corticoreticulocerebellar pathway

The cerebral cortex alerts the cerebellum to the initiation of a movement via the corticoreticulocerebellar pathway

Various regions of the cerebral cortex give rise to **corticoreticular fibers** that descend to synapse, mostly ipsilaterally, in the nuclei of the pontine and medullary **reticular formation**. **Reticulocerebellar fibers** pass via the **ipsilateral inferior and middle cerebellar peduncles** into the **vermis** of the cerebellum to terminate as **mossy fibers** where they function in feedback systems. Via this pathway, the cerebral cortex alerts the cerebellum of the initiation of a movement. The cerebellum then monitors motor activity and makes any necessary modifications.

Climbing fibers

Fibers arising from the inferior olivary complex (the olivocerebellar fibers) destined for the cerebellum, are referred to as "climbing fibers" when they reach the cerebellar cortex

The cerebral cortex, in addition to projecting to the pons via **corticopontine fibers** (part of the corticopontocerebellar pathway) and to the reticular formation via **corticoreticular fibers** (part of the **corticoreticulocerebellar pathway**), as discussed previously, also projects to the **inferior olivary complex** via **cortico-olivary fibers**.

302 ● ● ● CHAPTER 15

Table 15.8 ● Afferents to the cerebellum.

Types of fibers	Fibers/tract/pathway	Origin	Termination
Mossy fibers	Vestibulocerebellar fibers	Vestibular ganglion Vestibular nuclei	Cortex of vestibulocerebellum and part of vermis
	Spinocerebellar fibers:		
	Anterior spinocerebellar tract	Spinal cord	Cortex of spinocerebellum (vermis)
	Posterior spinocerebellar tract	Spinal cord	Cortex of spinocerebellum (vermis)
	Cuneocerebellar tract	Accessory cuneate nucleus	Cortex of spinocerebellum (paravermal zone)
	Rostral spinocerebellar tract	Spinal cord	Cortex of spinocerebellum
	Corticopontocerebellar pathway:		
	Corticopontine fibers	Primary motor cortex, premotor cortical areas, somesthetic cortex	Pontine nuclei
	Pontocerebellar fibers (transverse fibers of the pons)	Pontine nuclei	Cortex of cerebrocerebellum
	Corticoreticulocerebellar pathway:		
	Corticoreticular fibers	Various areas of the cerebral cortex	Pontine and medullary reticular formation nuclei
	Reticulocerebellar fibers	Pontine and medullary reticular formation nuclei	Cortex of cerebrocerebellum
	Cochleocerebellar fibers	Cochlear nuclei	Cortex of vermis
	Tectocerebellar fibers	Superior colliculus	Cortex of vermis
	Trigeminocerebellar fibers	Trigeminal nuclei	Cortex of cerebellum
	Rubrocerebellar fibers	Red nucleus	
Climbing fibers	Cortico-olivocerebellar pathway:		
	Cortico-olivary fibers	Primary motor cortex, premotor cortical areas, somesthetic cortex	Inferior olivary complex
	Olivocerebellar fibers	Inferior olivary complex	All layers of cerebellar cortex, mainly in cerebrocerebellum
Monoaminergic fibers		Ventral midbrain tegmentum	All layers of the cerebellar cortex (as diffuse fibers)
		Pontine raphe nuclei	Granular and molecular layers of cerebellar cortex
		Locus caeruleus	All layers of cerebellar cortex (as diffuse fibers)
Other fibers		Hypothalamus	All layers of cerebellar cortex
		Deep cerebellar nuclei	Cerebellar cortex

The fibers arising from the pons and the reticular formation destined for the cerebellum (pontocerebellar and reticulocerebellar fibers, respectively), are referred to as mossy fibers when they reach the cerebellar cortex. In contrast, the fibers arising from the **inferior olivary complex** (the **olivocerebellar fibers**), destined for the cerebellum, are referred to as **climbing fibers** when they reach the cerebellar cortex (see Table 15.8).

Cortico-olivocerebellar pathway

The inferior olivary complex, a group of relay nuclei of the cortico-olivocerebellar pathway, is believed to play an important instructing role when one learns to perform a new motor skill

The **cortico-olivocerebellar pathway** originates from the primary motor cortex, the premotor cortical areas, and the somatosensory cortex and follows the same path to the brainstem as does the corticopontocerebellar pathway. The **cortico-olivary fibers**, however, terminate *bilaterally* in the **inferior olivary nuclei**. Input from the spinal cord, the red nucleus (major input), the cerebral cortex (mainly the motor cortex), and the deep cerebellar nuclei is funneled into the inferior olivary complex, which is a cerebellar relay nucleus. The **inferior olivary complex** gives rise to a group of fibers that are destined for the contralateral cerebellar cortex. These are the **olivocerebellar fibers**, the sole efferents from the inferior olivary complex that, as mentioned previously, terminate in the cerebellar cortex as **climbing fibers**. On their way to the cerebellum, the olivocerebellar fibers decussate, course through the **inferior cerebellar peduncle**, enter the cerebellar white matter, send collaterals to the **deep cerebellar nuclei**, and "climb" to the cortical granular layer where they form **aspartate-releasing excitatory synapses**

Table 15.8 ● *Continued*

Collaterals	Neurotransmitter	Excitatory/inhibitory	Function(s)
Deep cerebellar nuclei	Glutamate	Excitatory	Transmit sensory input about changes in head position and orientation
Deep cerebellar nuclei	Glutamate	Excitatory	
	Glutamate	Excitatory	Relay proprioceptive input from trunk and lower limb
	Glutamate	Excitatory	Relay proprioceptive input from lower trunk and lower limb
	Glutamate	Excitatory	Relay proprioceptive input from upper trunk and upper limb
	Glutamate	Excitatory	Relay proprioceptive input from upper limb
Deep cerebellar nuclei	Glutamate	Excitatory	Relays signals that keep cerebellum informed of an actual movement in progress and an impending movement
Deep cerebellar nuclei	Glutamate	Excitatory	Functions in feedback systems
	Glutamate	Excitatory	Relay sensory input about the location and direction of a sound or object in space
	Glutamate	Excitatory	
	Glutamate	Excitatory	Relay sensory input from orofacial structures and stretch receptors of the muscles of mastication
	Aspartate	Excitatory	Functions in feedback systems
Interposed and dentate nuclei	Dopamine	Inhibitory	Modulate cerebellar activity
	Serotonin	Inhibitory	Modulate cerebellar activity
	Norepinephrine (noradrenaline)	Inhibitory	Modulate cerebellar activity
	Histamine		Coordinate somatic and visceral motor functions

with granule cells and Golgi cells. The climbing fibers then continue their course through the Purkinje cell layer, where they terminate in the molecular layer as unmyelinated fibers to form powerful aspartate-releasing excitatory synapses with **Purkinje cell dendrites**.

Monoaminergic fibers

Monoaminergic fibers (dopaminergic, norepinephrine, or serotonergic fibers) projecting to the cerebellum arise from the ventral midbrain tegmentum, the locus caeruleus, and the pontine raphe nuclei

The **ventral midbrain tegmentum** gives rise to **dopaminergic fibers** that terminate in the emboliform, globose, and dentate nuclei as well as in the cerebellar cortex. These

dopaminergic *inhibitory* fibers function in the regulation of cerebellar activity (see Table 15.8).

The **norepinephrine (noradrenergic) fibers** arising from the **locus caeruleus** pass into the cerebellum via the superior and middle cerebellar peduncles to terminate in the cerebellar cortex where they form *inhibitory synapses* with dendrites of Purkinje cells and granule cells. These noradrenergic inhibitory fibers function in the regulation of cerebellar activity.

The fibers arising from the **pontine raphe nuclei** pass into the cerebellum via the middle cerebellar peduncle where they release **serotonin** in the granular and molecular layers of the cerebellar cortex. These fibers *inhibit* Purkinje cells and function to regulate cerebellar activity through fusiform horizontal cells.

304 ● ● ● CHAPTER 15

Afferents from other sources

In addition to the previous afferent projections, the cerebellum also receives afferents from the following sources: (1) the hypothalamus; (2) the cochlear nuclei; (3) the superior colliculi; and (4) the trigeminal nerve nuclei (see Table 15.8).

The histaminergic afferent fibers arising from the **hypothalamus** release **histamine** at their terminals distributed throughout the cerebellar cortex. Histamine **excites** Purkinje cell dendrites in the molecular layer, granule cells in the granular layer of the cerebellar cortex, and the neuronal cell bodies in the deep cerebellar nuclei. These fibers function to coordinate somatic and visceral motor responses and may play a critical role in their integration.

The **cochlear nuclei** (which process auditory information) as well as the **superior colliculi** (which process information associated with the tracking of a moving object) relay sensory input to the dorsal aspect of the vermis regarding the location and shift in position of a sound or object.

The **trigeminal nerve nuclei** also send projections (trigeminocerebellar fibers) relaying sensory input from orofacial structures and the stretch receptors of the muscles of mastication. These fibers pass into the cerebellum via the inferior cerebellar peduncle where they terminate as mossy fibers.

EFFERENTS (OUTPUT) FROM THE CEREBELLUM

Efferent fibers from the cerebellum arise from the Purkinje cells of the cerebellar cortex and the cells of the deep cerebellar nuclei

Efferent fibers from the cerebellum arise from two sources (see Fig. 15.9):

1 **Purkinje cells**, which constitute the ultimate integrating terminal within the cerebellar cortex.
2 Cells of the **deep cerebellar nuclei**. These nuclei house the nerve cells whose *axons form the principal cerebellar output* (see discussion that follows).

Efferent fibers from Purkinje cells

The majority of the Purkinje cell axons terminate locally in the deep cerebellar nuclei. A small extracerebellar projection terminates in the vestibular nuclei

The **Purkinje cell axons** of the cerebellar cortex have two destinations. The only **extracerebellar** destination is as follows: the axons of the Purkinje cells located in the cortex of the **vestibulocerebellum** go past the deep cerebellar nuclei, exit the cerebellum, to end in the **vestibular nuclei.** The majority of Purkinje cell axons, though, terminate **locally** at their main targets (destination) – the **deep cerebellar nuclei.** The Purkinje cell axons always form *inhibitory synapses.*

It is important to note that Purkinje cell axons arising from the cerebellar cortex of the vermis, paravermal zone, or the hemispheric zone overlying the deep cerebellar nuclei project to the corresponding deep cerebellar nucleus (see discussion that follows).

All the Purkinje cells that project to the deep cerebellar nuclei send collaterals to Golgi cells, as well as to other Purkinje cells, basket cells, and stellate cells of the cerebellar cortex.

Output from the flocculonodular lobe to the vestibular nuclei

Purkinje cells of the flocculonodular lobe project to the vestibular nuclei

Purkinje cell axons arising from the cortex of the **flocculonodular lobe (vestibulocerebellum)** form a small extracerebellar projection as they exit the cerebellum via the juxtarestiform body to terminate in the **vestibular nuclei** (see Fig. 15.10). **Vestibulospinal tracts** originate in the vestibular nuclei and terminate on those motor neurons of the spinal cord that control posture and balance.

Note that the flocculonodular lobe projects primarily to the **vestibular nuclei** (which are located outside the cerebellum) and sends only a few Purkinje cell fibers to the **fastigial nucleus.**

Based on the projections of the cerebellum to the vestibular nuclei, and since both the deep cerebellar nuclei and the vestibular nuclei receive *excitatory afferent fibers* from the inferior olivary complex in addition to receiving *inhibitory afferent fibers* from Purkinje cells, the vestibular nuclei are considered by some to be anatomically analogous to the deep cerebellar nuclei. Thus, the vestibular and deep cerebellar nuclei share two common sources of afferent fibers.

Output from the vermal zone and the fastigial nucleus

Purkinje cells of the vermal zone cortex project to the fastigial nucleus

Purkinje cell axons arising from the cortex of the **vermal zone** project to the ipsilateral **fastigial nucleus.** The fastigial nucleus gives rise to fibers that exit the cerebellum via the inferior cerebellar peduncle to terminate bilaterally in the **lateral vestibular nucleus** and in the contralateral brainstem **reticular formation.**

The vestibular nuclei give rise to the **vestibulospinal tracts,** and the reticular nuclei give rise to the **reticulospinal tracts.** Collectively, these tracts form the **medial descending system,** which terminates in the spinal cord and has an important function in the maintenance of posture, balance, and muscle tone of extensor muscles.

Recall that the cortex of the vermis receives input from the spinal cord. The fastigial nucleus also gives rise to a small number of fibers that pass into the superior cerebellar peduncles and ascend bilaterally to the **VL nucleus** of the **thalamus.** The thalamus in turn projects to the **premotor** and **primary motor cortex.** The neurons of the premotor and primary motor cortices, whose axons form the **anterior corticospinal tract,** control motor activity of the axial and proximal limb musculature.

Output from the paravermal zone and the interposed nuclei

Purkinje cells of the paravermal zone cortex project to the interposed nuclei

The **Purkinje cell axons** arising from the cortex of the **paravermal zone** project to the **interposed nuclei**, which are associated with the execution and modification of movements. These nuclei in turn, project, via the superior cerebellar peduncle, to the magnocellular component of the contralateral **red nucleus**. The red nucleus gives rise to the **rubrospinal tract** that terminates in the cervical spinal cord contralaterally and functions in regulating motor activity by facilitating the muscle tone of the ipsilateral upper limb flexors. The interposed nuclei also send some fibers, via the superior cerebellar peduncle, to the **VL nucleus** of the **thalamus**, which projects to the **premotor** and **primary motor cortex**. One of the outputs of these cortices is the **lateral corticospinal tract**. The rubrospinal tract and the lateral corticospinal tract together form the **lateral descending system**, which controls the motor activity of the distal musculature of the limbs, especially the hands.

Output from the hemispheric zone and the dentate nucleus

Purkinje cells of the hemispheric zone cortex project to the dentate nucleus

Purkinje cell axons arising from the cortex of the **hemispheric zone** of the cerebellum project to the **dentate nucleus**, which is associated with the *planning* of movements. Fibers arising from neurons of the dentate nucleus form the **dentorubrothalamic pathway**. These fibers exit the cerebellum by joining the superior cerebellar peduncle, decussate, and terminate to synapse in two places: the parvocellular component of the **red nucleus** and the **VL nucleus** of the **thalamus** (most fibers terminate here). *This projection to the VL nucleus of the thalamus is the principal efferent projection of the cerebellum.* The majority of the fibers in the superior cerebellar peduncle originate from the dentate nucleus.

The **parvocellular component of the red nucleus** gives rise to fibers that join the **central tegmental tract** to terminate in the ipsilateral **inferior olivary complex**.

The **VL nucleus of the thalamus** gives rise to fibers that ascend to terminate in the ipsilateral **primary motor** and **premotor cortex**, which give rise to the **lateral corticospinal tract**. Via its influence on the lateral corticospinal tract, the cerebellum plays an important role in the planning phase, timing, initiation, and precision of discrete movements, specifically the reciprocal contractions of agonist and antagonist muscles of the limbs (especially of the hands).

FUNCTIONAL ORGANIZATION OF THE CEREBELLUM: INTRINSIC CIRCUITRY

As discussed above, the **cerebellar cortex** contains **Purkinje cells** whose functions are *inhibitory*, **inhibitory interneurons** – namely **basket cells, stellate cells, Golgi cells,** fusiform **horizontal cells,** and excitatory interneurons – granule **cells** and the **unipolar brush cells**.

Excitatory inputs to the cerebellar cortex

The climbing and mossy fibers provide the major excitatory input to the cerebellum

Excitatory inputs to the **cerebellar cortex** are derived from both extrinsic fibers and intrinsic fibers. The **extrinsic fibers** are composed of **climbing fibers** and **mossy fibers**. The **intrinsic fibers** are the axons of **granule cells** and the **unipolar brush cells** of the granular cell layer of the cerebellar cortex. An additional source of excitatory synapses comes from the **deep cerebellar nuclei**, which send their axons into the cerebellar cortex as nucleocortical connections.

Climbing fibers

Climbing fibers are the axon terminals of neurons whose cell bodies reside exclusively in the inferior olivary complex

The **inferior olivary complex** gives rise to the **olivocerebellar fibers**. On their way to the cerebellum they decussate, and when they reach the cerebellar white matter they branch, send **excitatory collaterals** to the **deep cerebellar nuclei**, and proceed by "climbing" toward the granular layer of the cerebellar cortex as **climbing fibers**. There, a climbing fiber gives rise to about 10 branches that extend into the molecular layer of the cerebellar cortex where they synapse with Purkinje cells. Each terminal of a climbing fiber forms 300–500 **direct, excitatory** synaptic contacts with the dendrites of an individual **Purkinje cell** (see Fig. 15.5). Each individual Purkinje cell receives synaptic contacts from only a single climbing fiber. *An impulse transmitted from a climbing fiber to the dendrites of a Purkinje cell generates a prolonged, oscillatory-type, powerful depolarization of that Purkinje cell.*

Additionally, the climbing fibers also depolarize the Golgi, stellate, and basket cells. These inhibitory interneurons attenuate Purkinje cell activity, which has a more potent effect on cerebellar cortical outflow than does the limited, intense depolarization of a small population of Purkinje cells that are depolarized directly by a climbing fiber.

Mossy fibers

Most of the afferents to the cerebellum consist of mossy fibers

Most of the afferents to the cerebellum consist of **mossy fibers** (see Fig. 15.5). Mossy fibers relay information to the cerebellum from various sources: (1) the spinal cord; (2) the vestibular ganglion and vestibular nuclei; (3) the pontine nuclei; (4) the reticular nuclei; and (5) the trigeminal nuclei. Afferent fibers enter the cerebellum by passing through all three cerebellar peduncles; they then course through the cerebellar white matter where they branch, and prior to their

306 ●●● CHAPTER 15

termination in the cerebellar cortex, send **excitatory collaterals** to the **deep cerebellar nuclei**. When they reach their destination they terminate in synaptic complexes, known as **glomeruli**, within the granular layer of the cerebellar cortex.

A glomerulus has the following components: a **mossy fiber**, which establishes *excitatory synapses* with the dendrites of a **granule cell**, and a **Golgi cell** and its terminals, which form inhibitory synapses with **granule cell** dendrites. A mossy fiber activates a granule cell within a glomerulus.

It takes a large number of simultaneously stimulated mossy fibers to activate a granule cell. The axon of each granule cell bifurcates in the molecular layer to form parallel fibers, which form **excitatory synapses** with the dendrites of Purkinje cells, basket cells, stellate cells, and Golgi cells (all of which are inhibitory).

A single parallel fiber synapses with thousands of Purkinje cells in succession (as a single electrical wire can pierce through and contact the branches of many trees in succession). The dendritic tree of a single Purkinje cell is traversed by thousands of parallel fibers, which establish synapses with dendritic terminals (as multiple telephone wires can pass among the branches of a single tree). Each parallel fiber establishes only a few synapses with each Purkinje cell. Recall that, in contrast, a climbing fiber forms several hundred direct synaptic contacts with a Purkinje cell. Consequently, as a result of the greater number of synaptic contacts, the **climbing fiber input exerts a stronger excitatory influence on Purkinje cells** than does the mossy fiber input. Thus, it takes a multitude of concurrently stimulated mossy fibers to trigger a nerve impulse in a single Purkinje cell. Interestingly, *mossy fiber stimulation triggers recurring Purkinje cell action potentials, rather than the single, sustained, powerful depolarization generated by climbing fibers*.

It should therefore be noted that a Purkinje cell may be stimulated either by: (1) a **climbing fiber**, directly, resulting in a **very specific output signal**; or (2) **mossy fibers**, indirectly, via granule cell parallel fibers, resulting in a **less specific but tonic output signal**.

Basket cells, stellate cells, and Golgi cells are all inhibitory. The basket cells and the stellate cells form inhibitory synapses with the Purkinje cells. The Golgi cells are stimulated by mossy fibers within the glomeruli and by the parallel fibers that form synapses on their apical dendrites. The stimulated Golgi cells then respond by inhibiting the granule cells back in the glomeruli; that is, they prevent further excitability of the granule cells by upcoming excitatory impulses from the mossy fibers.

Inhibitory inputs to the cerebellar cortex

Inhibitory inputs to the cells of the cerebellar cortex are derived from **Golgi cells**, **fusiform horizontal cells**, **Purkinje cells**, **basket cells**, and **stellate cells.** All four types of cells are **intrinsic** modulator cells. The first three cell types of release **GABA and/or glycine**, whereas the stellate cells release **taurine**.

Inhibitory output from the cerebellar cortex

The single output from the cerebellar cortex is via the axons of Purkinje cells and is inhibitory

There is only a single output from the **cerebellar cortex** – the axons of the **Purkinje cells** (**inhibitory**) called **cerebellar cortical efferent fibers**. These cells transmit integrated information from the cerebellar cortex to the **vestibular nuclei** and to the **deep cerebellar nuclei**.

Output from the deep cerebellar nuclei

Output from the deep cerebellar nuclei is excitatory

The neurons housed in the deep cerebellar nuclei are continually under *excitatory* influences exerted on them by the collaterals of climbing and mossy fibers and *inhibitory* influences exerted on them by Purkinje cell axons. The effect that these excitatory and inhibitory influences have on cerebellar input during motor activity, however, is still uncertain.

These excitatory and inhibitory influences exerted on the cells of the deep cerebellar nuclei are constantly balanced; hence, the intensity of the efferent signals from the cells of the deep cerebellar nuclei remains the same during moderate levels of prolonged excitation. However, during intense motor activity (e.g., when running), the excitatory influence is exerted on the cells of the deep cerebellar nuclei *prior* to the inhibitory influence. Thus, the **deep cerebellar nuclei** first relay an *excitatory* output signal to the motor system to adjust the motor activity in progress and then relay an *inhibitory* output signal to ultimately inhibit the motor activity. Thus, once a movement is initiated, the deep cerebellar nuclear excitatory output modifies the motor activity and, shortly thereafter, it is inhibited to prevent overshooting (missing the mark) of the movement in progress. An increase in the firing rate of the cells of the deep cerebellar nuclei assures that the cerebellum provides an excitatory output signal, whereas a decrease in the firing rate of these cells provides an inhibitory output signal.

Influence of mossy and climbing fibers on cerebellar output

The cerebellum processes information derived from the ipsilateral side of the body and from the contralateral cerebral cortex

The **mossy fibers** form the principal conduit of excitatory signals to the deep cerebellar nuclei by a **direct** projection (mossy fibers → cells of the deep cerebellar nuclei). The mossy fibers simultaneously form an **alternate pathway**, which also ultimately projects to the deep cerebellar nuclei, but first relays in the cerebellar cortex. The mossy fiber stimulates a granule cell that, via its parallel fibers, stimulates Purkinje cells. The Purkinje cells then project primarily to the deep cerebellar nuclei, thus completing the alternate pathway of mossy fiber output to the deep cerebellar nuclei. (Purkinje cells also have a small extracerebellar projection to the vestibular nuclei.)

Stellate cells, basket cells, and Golgi cells modulate Purkinje cell excitation generated by both mossy fibers and climbing fibers. Consequently, Purkinje cells have a strong inhibitory influence on the cells of the deep cerebellar and vestibular nuclei, modulating their output. It is the processing of the afferent input to the cerebellum, and its modulation via Purkinje cells, that ultimately determines the smoothing of movements in progress.

It is important to remember that the cerebellum processes information derived from the *ipsilateral* side of the body and from the *contralateral* cerebral cortex.

CLINICAL CONSIDERATIONS

A lesion confined to one cerebellar hemisphere results in symptoms that appear on the ipsilateral side of the body

It is well known that electrical stimulation of the cerebellum does not produce any type of sensation or movement. This led early neuroanatomists to refer to the cerebellum as a "silent area" of the brain. The cerebellum plays a role in cognitive functions in part via its projections to the frontal lobe and has an important function in motor learning. A lesion involving the cerebellum results in the abnormal execution of voluntary and involuntary motor activities, motor learning, and another cognitive dysfunction. Now that its internal circuitry is better understood, it has helped us understand the source of these effects.

Since the intrinsic circuitry of the cerebellar cortex is identical throughout the cerebellum, it is difficult to correlate a function with a particular region of the cerebellar cortex. Instead, it is the origin and destination of the afferent and efferent fibers that facilitates functional identification in the cerebellum.

Although the cerebellum is subdivided into three functionally distinct zones, each including cortex, white matter, and one or two deep cerebellar nuclei, lesions are rarely confined to one zone only. Lesions involving the deep cerebellar nuclei (the major source of cerebellar output) or the superior cerebellar peduncles (which convey most of the cerebellar output fibers) result in more serious symptoms than lesions that are confined to the cerebellar cortical region. Furthermore, although the cerebellum is involved in motor function, a lesion in the cerebellum results in **abnormal execution** of a motor task and *not* deficits in motor function such as paralysis or paresis. A lesion confined to one cerebellar hemisphere results in symptoms that appear on the body on the **same side as the lesion**. The reason why this occurs is because the cerebellar efferent connections and the descending motor pathways upon which the cerebellum exerts its influence are crossed. Additionally, the main afferent pathways from the spinal cord to the cerebellum, namely the cuneocerebellar and posterior spinocerebellar tracts, are ipsilateral.

The most common symptoms attributed to cerebellar malfunction are **hypotonia** (decrease in muscle tone) and **ataxia** (G., *atactos*, "disorderly," "uncoordinated").

Severe trauma to the cerebellum results in serious symptoms; however, after a variable amount of time, full recovery may follow. It is believed that other parts of the brain may "take over" and compensate for any cerebellar deficits. Persistent conditions, however, such as the prolonged, slow growth of a tumor, cause milder symptoms. In this case, the growing lesion and the damage produced are accompanied by continuous compensation for the cerebellar deficits.

Flocculonodular syndrome

Flocculonodular syndrome has been characterized by the inability to maintain equilibrium

Flocculonodular syndrome has been characterized by the **inability to maintain equilibrium** – without an accompanying ataxia of the limbs, hypotonia, or tremor – and truncal ataxia, unsteady walking, swaying from side to side, and a tendency to fall. The affected individual appears as if inebriated. The individual may also experience **nystagmus**, presented by rhythmic oscillations of the eyes. Since lesions involving the flocculonodular lobe may include midline cerebellar components, the symptoms may be bilateral.

Disorders resulting following a lesion in the vermal and paravermal zones

A lesion involving the vermal and paravermal zones of the cerebellum results in abnormal stance and gait (abnormal standing or walking/running)

Following a lesion in the **vermal** or **paravermal zones**, standing is characterized by the feet positioned far apart, providing a compensatory broad base. While walking, the individual is unable to "walk in tandem." The affected individual also exhibits **titubation**, a condition characterized by a rhythmic tremor (involuntary oscillation) of the trunk or head when sitting or standing, and postural changes of the head, in which the head may be held in a rotated position or tilted toward one side. Due to the lesion's midline proximity, the side in which the head is rotated or tilted may not reflect the side of the lesion.

Cerebellar disease may be manifested in various abnormal movements of the eyes. The most conspicuous abnormal eye movement is spontaneous **nystagmus** that occurs without stimulation of the vestibular apparatus. Nystagmus is characterized by rhythmic oscillation of one or both eyes and is manifested when the eyes are deviated laterally in the horizontal plane toward the side of the lesion.

Disorders resulting following lesions in the hemispheric zone

A lesion affecting one cerebellar hemisphere is exhibited primarily as ataxia and tremor on the ipsilateral side of the body

A lesion involving the **hemispheric zone** may result in numerous disorders, as described later. A lesion affecting one cerebellar hemisphere is exhibited primarily as ataxia and tremor on the ipsilateral side of the body. Damage of the hemispheric zone results in abnormal cerebellar output to the motor cortex via the thalamus, resulting in an abnormal execution of voluntary movement mediated by the corticospinal and corticorubral-rubrospinal tracts.

Ataxia is the lack of coordination of movements. The timing and precision of motor tasks are impaired, resulting in awkward movement.

Ocular movement disorders, of which the most common is **nystagmus**, are most noticeable when the individual deviates his eyes toward the side of the lesion. Ocular movement disorders result as a consequence of asynergic (uncoordinated) activity of the extraocular muscles.

Tremor is characterized by the exhibition of **static** and **kinetic tremors**. **Intention tremor** (a kinetic tremor) is manifested during a voluntary

CLINICAL CONSIDERATIONS (*continued*)

movement and is especially apparent toward the end of a movement, such as when putting glasses on.

Delayed initiation and termination of motor activity. The affected individual has difficulty initiating a motor activity, and once the activity is ongoing, he then has difficulty stopping it.

Abnormal stance and gait are also seen in individuals with hemispheric lesions. The individual stands with feet apart and tends to sway forward, backward, or to the side, resembling a state of intoxication. Ataxia may also be evident when moving the lower limbs.

Decomposition of movement may occur following cerebellar disease. Movements are jerky, awkward, and inaccurate. An individual with such a lesion attempts to compensate for the awkward movements by making gradual movements such as moving one part (arm, forearm, hand, finger) of the body at a time.

Hypotonia (diminished muscle tone) may be exhibited following manipulation of the limbs at a joint. A decrease occurs in resistance to this passive manipulation, resulting in excess movement of the limbs referred to as "loose jointed" or "rag-doll appearance," since the damaged cerebellum does not exert its influence on the stretch reflex. Hypotonia is velocity independent. This condition gradually diminishes over time.

Dysarthria (G., *dys*, "abnormal," "disorderly"; *arthroun*, "to utter distinctly") is characterized by impaired speech that is slow and slurred. It is caused from lack of coordination of the muscles used to produce speech and is referred to as "scanning" speech, because of the excess and equal stress on all syllables.

Dysmetria (G., *dys*, "abnormal," "disorderly"; *metron*, "measure") is a disorder in which the individual is unable to estimate the distance between the moving body part (e.g., hand) and a target ("past pointing"). Overestimating the distance between the body part and object is called **hypermetria**. Underestimating the distance between the body part and object is **hypometria**.

Dysdiadochokinesia (adiadochokinesia) (G., *diadocho*, "alternating"; *kinesis*, "movement") is the inability to carry out rapid alternating movements regularly (e.g., rapid pronation and supination of the forearm).

Dysrhythmokinesis is the inability to maintain rhythm while performing alternating movements at a fast pace.

Impaired check and rebound is a condition that results from loss of cerebellar input to the stretch reflex. When a limb is displaced from a certain position against resistance (e.g., trying to straighten an individual's flexed arm against resistance), its abrupt release does not result in its return to its initial position, but instead it is displaced beyond its previous position. The individual is unable to stop a movement suddenly.

Note that the clinical case in the beginning of the chapter refers to a patient whose symptoms include sudden-onset severe vertigo, nausea, dysarthria, nystagmus, dysmetria, and ataxia in the left arm.

1   A cerebellar stroke is suspected. Which side of the cerebellum has been affected?

2   Which zone or zones of the cerebellum have been affected?

3   Which vessel may have been occluded in this patient?

CEREBELLUM ● ● ● **309**

SYNONYMS AND EPONYMS OF THE CEREBELLUM

Name of structure or term	Synonym(s)/ eponym(s)
Anterior lobe of the cerebellum	Spinocerebellum
	Paleocerebellum
Cuneocerebellar tract	Cuneatocerebellar tract
Deep cerebellar nuclei	Cerebellar nuclei
Basal nuclei	Basal ganglia
Fastigiovestibular fibers	Cerebellovestibular fibers
Flocculonodular lobe	Vestibulocerebellum
	Archicerebellum
Gamma aminobutyric acid releasing fibers	GABAergic fibers
Golgi cells	Inner stellate cells
Hemispheric zone	Lateral zone
Inferior olivary complex	Inferior olive
	Olive
Lower motor neuron	Lower motor neuron
Middle cerebellar peduncle	Brachium pontis
Olivocerebellar fiber terminals	"Climbing fibers"
Paravermal zone	Intermediate zone
Pontocerebellar fibers	Transverse fibers of the pons
Posterior lobe of the cerebellum	Middle lobe of the cerebellum
	Cerebrocerebellum
	Pontocerebellum
	Neocerebellum
Premotor area	Premotor cortex
	Part of Brodmann's area 6
Primary motor cortex (MI)	Precentral gyrus
	Brodmann's area 4
Proprioception	Position sense
Vermal zone	Median zone

FOLLOW-UP TO CLINICAL CASE

This patient had a left cerebellar stroke in the distribution of the left posterior inferior cerebellar artery (PICA). This is a common type of stroke and is therefore important for all physicians to know about, not just neurologists. The inferior and lateral cerebellum, as well as the dorsolateral medulla oblongata, are supplied by the PICA. However, many infarctions involve only part of the PICA territory and may involve only the cerebellum. The PICA arises most commonly from the vertebral artery just proximal to the basilar artery. Thrombosis of the PICA or the vertebral artery, or emboli to the vertebral artery are the most common causes of this stroke syndrome.

The symptoms in the case above are typical. The cerebellum, specifically the cerebellar vermis and surrounding parts of the inferior midline cerebellum, has important connections with the vestibular nuclei. Therefore, vertigo or dizziness is common. Gait is often severely disturbed and gait testing is critical. In fact, sometimes it is the only abnormality in cerebellar disease. Focal cerebellar pathology causes *ipsilateral* ataxia or dysmetria of the limbs, as opposed to the contralateral dysfunction caused by most other brain lesions. Ocular dysfunction is common. Complaints of blurry or double vision, nystagmus toward the side of the lesion, ipsilateral gaze paresis, or other ocular paresis are common. Additional symptoms can occur with involvement of the dorsolateral medulla, according to the structures in that region of the brainstem.

An initial head CT is usually normal unless there is some hemorrhage present. After a day or two it will show the stroke. MRI gives an exquisite view of the cerebellum and brainstem. Often the most distressing and disabling symptom, along with the inability to walk, is vertigo. This can be relieved with antidopamine or similar medication. Symptoms will usually improve or resolve over time (days, weeks, or months).

QUESTIONS TO PONDER

1. Why is it that a cerebellar lesion confined to one cerebellar hemisphere results in symptoms that appear on the same side of the body?

2. Distinguish between the tremor resulting following a lesion in the basal nuclei and a lesion involving the cerebellum.

3. An individual who is inebriated stands with feet apart and waves their arms in different directions to maintain their balance, and tends to sway forward, backward, or to the side. These are also symptoms exhibited by an individual who has a lesion in which of the three cerebellar zones?

4. Based on your knowledge of the intrinsic circuitry of the cerebellum, what is the net output of the cerebellum – inhibitory or excitatory?

5. What causes the hypotonia exhibited (following manipulation of the limbs at a joint) by individuals following a lesion in the hemispheric zone of the cerebellum?

CHAPTER 16

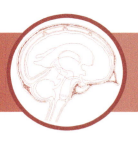

Reticular Formation

- CLINICAL CASE
- MORPHOLOGY OF THE RETICULAR FORMATION
- ZONES OF THE RETICULAR FORMATION
- NUCLEI ASSOCIATED WITH THE RETICULAR FORMATION
- INPUT TO AND OUTPUT FROM THE RETICULAR FORMATION
- FUNCTIONS OF THE RETICULAR FORMATION
- CLINICAL CONSIDERATIONS
- SYNONYMS AND EPONYMS OF THE RETICULAR FORMATION
- FOLLOW-UP TO CLINICAL CASE
- QUESTIONS TO PONDER

CLINICAL CASE

A 58-year-old patient, B.P., presented with sudden onset of complete left-sided paralysis and slurred speech. The head CT in the ER demonstrated evidence of early edematous changes in the distribution of the right middle cerebral artery (MCA). A head CT two days later showed a very large right hemisphere infarction involving the entire right MCA territory. B.P. had been somewhat drowsy but easily aroused during the first few days of hospitalization. The patient also had a right gaze preference, left hemineglect, and severe left hemiplegia. B.P. became more lethargic after the first few days, and then could not be aroused, and was subsequently placed on mechanical ventilation due to respiratory difficulties. Examination in the ICU revealed that the patient was comatose (a sleep-like state from which one cannot be aroused) with a fixed and dilated right pupil. Other brainstem reflexes were intact. There was no movement elicited on the left side, but there was some movement following stimulation on the right. However, there was an **Extensor Plantar Response** (eponym: Babinski reflex) present on the right (indicating dysfunction of the descending motor tracts controlling the right side of the body).

There was subsequent deterioration of the patient's condition, and B.P lost all brainstem reflexes (pupils became unresponsive to light, eye blink to corneal touch was lost, spontaneous respiratory effort ceased, etc.). All limb movement was lost. Brain death was confirmed.

The reticular formation (L., *reticulum*, "little net") consists of various distinct populations of cells embedded in an intricate polysynaptic network of cell processes occupying the central core of the brainstem. From an evolutionary perspective, the reticular formation is phylogenetically an *ancient neural complex* that is closely associated with two other ancient neural systems, the olfactory system, which mediates the visceral sense of smell, and the limbic system, which functions in the visceral and behavioral responses to emotions. The reticular formation and the olfactory and limbic systems are interrelated because of their participation in *visceral functions* and *behavioral responses*. The reticular formation is continually

---

A Textbook of Neuroanatomy, Third Edition. Maria A. Patestas, Amanda J. Meyer, and Leslie P. Gartner.
© 2025 John Wiley & Sons, Inc. Published 2025 by John Wiley & Sons, Inc.
Companion website: www.wiley.com.go/patestas/neuroanatomy3e

informed of activity occurring in almost all areas of the nervous system and responds by influencing a variety of processes: skeletal muscle motor activity; somatic and visceral sensations; autonomic nervous system activity; endocrine functions; and biological rhythms, all via reciprocal connections to the hypothalamus, modulation of behavioral responses, regulating sleep and wakefulness, and adjusting the level of consciousness. It does this via widespread dispersal of neurotransmitters and neuromodulators through the extracellular space rather than being restricted to synaptic transmission between neurons, an activity called *volume transmission*. This diffuse mode of communication enables a large pool of neurons to be targeted, synchronized, and coordinated to integrate and regulate essential brainstem functions.

MORPHOLOGY OF THE RETICULAR FORMATION

Each subdivision of the reticular formation has its own distinct cytoarchitecture, specific connections, and functions

The **reticular formation** is most prominent in the brainstem, where it forms the central core gray matter of the midbrain, pons, and medulla oblongata. The brainstem reticular formation extends from the diencephalon where it merges with the intrathalamic nuclei almost as far inferiorly as the pyramidal decussation in the medulla oblongata. The spinal cord also contains throughout its entire length an analogue of the reticular formation, the **intermediate zone of gray matter**. However, this intermediate zone of gray matter, consisting of **interneurons**, is not as extensive as the brainstem reticular formation.

The reticular formation contains fibers that are oriented in all planes, thus in histologic sections it appears as an interlacing structure that fills the area among various ascending and descending pathways, cranial nerve nuclei, and other gray matter. Although it appears scattered, indistinct, and ambiguous, the reticular formation, in fact, consists of numerous different histologic subdivisions each having its own distinct cytoarchitecture, specific connections, and functions.

Neurons of the reticular formation

The structure of the neurons of the reticular formation facilitates the collection of information from ascending and descending fibers

The **neurons** of the reticular formation possess elaborate dendritic trees whose branches radiate in all directions (Fig. 16.1). Some neurons give rise to an ascending or descending axon, with numerous collateral branches. Other neurons give rise to a primary axon, which bifurcates forming an ascending and a descending branch, both of which then give rise to collateral branches. The dendrites are oriented perpendicular to the long axis of the brainstem. The elaborate radiating dendritic trees, and the axons and their collaterals, facilitate the ability of the neurons of the reticular formation to collect information from ascending and

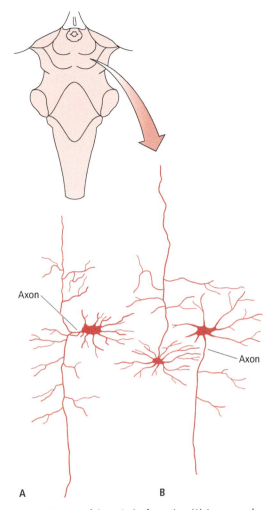

Figure 16.1 • Neurons of the reticular formation. (A) A neuron whose primary axon divides into an ascending and a descending branch. (B) Note the intermingling of neuronal dendrites and axonal collateral branches.

descending fibers arising from various sources, as these fibers traverse the brainstem tegmentum. Information converging on the reticular formation is integrated, and the reticular formation, via its output, influences neuronal activity at all levels of the central nervous system (CNS). As a result of its central location and cytoarchitecture, a single neuron of the reticular formation can exert its influence on various somatic or visceral functions via direct or indirect projections to higher brain centers, local brainstem centers, and the spinal cord. Many neurons of the reticular formation are involved in local multisynaptic and reflex circuits.

Based on its connections (input and output), the reticular formation can be thought of as the CNS "command center" where surveillance of activities occurring in distant sites of the CNS takes place. It is because of its diverse connections that the reticular formation can integrate and modulate various activities and functions of almost all areas controlled by the CNS.

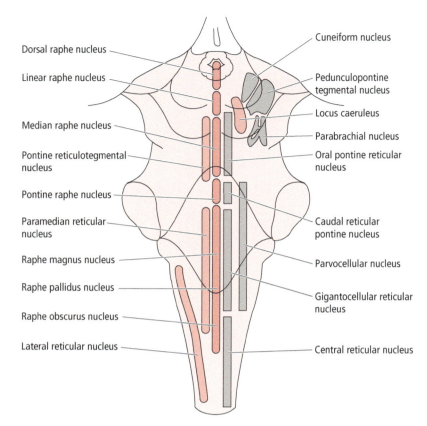

Figure 16.2 ● The nuclei of the reticular formation in the brainstem. Source: Donkelaar HJ, Kachlik D, Tubbs RS 2018/Springer Nature.

Nuclei of the reticular formation

More than 100 nuclei scattered throughout the tegmentum of the midbrain, pons, and medulla oblongata have been identified as being part of the brainstem reticular formation

More than 100 nuclei scattered throughout the tegmentum of the midbrain, pons, and medulla oblongata have been identified as being part of the brainstem reticular formation (Fig. 16.2). Most of the **nuclei** of the reticular formation are not as clearly defined as are other nuclei of the CNS. Although the nuclei of the reticular formation have several diverse functions, they are classified according to the following four general functions:

1. The regulation of the level of consciousness and ultimately cortical alertness.
2. The control of somatic motor movements.
3. The regulation of visceral motor or autonomic functions.
4. The control of sensory transmission.

ZONES OF THE RETICULAR FORMATION

Anatomically, the reticular formation is divided into four longitudinal zones (columns) on the basis of their mediolateral location in the brainstem

The zones of the reticular formation are: the (1) unpaired median zone, the (2) paired paramedian zone, the (3) paired medial zone, and the (4) paired lateral zone (Table 16.1). Some authors consider the median and paramedian zones to be one zone.

Table 16.1 ● Zones of the reticular formation and their component nuclei.

Brainstem location	Median zone	Paramedian zone	Medial zone ("motor" or "efferent" zone)	Lateral zone ("sensory" or "afferent" zone)
Midbrain	Dorsal raphe nucleus Linear raphe nucleus Median raphe nucleus		Cuneiform nucleus Pedunculopontine tegmental nucleus	Parabrachial nucleus
Pons	Median raphe nucleus Pontine raphe nucleus Raphe magnus nucleus	Pontine reticular nucleus Paramedian reticular nucleus	Pedunculopontine tegmental nucleus Oral pontine reticular nucleus Caudal pontine reticular nucleus	Parabrachial nucleus Parvocellular nucleus
Medulla oblongata	Raphe pallidus nucleus Raphe obscurus nucleus	Paramedian reticular nucleus	Gigantocellular reticular nucleus Central reticular nucleus	Parvocellular nucleus

Median zone (also known as the median column, midline raphe, median reticular formation)

The neurons of the median zone that project to higher brain centers are associated with sleep

The neurons that project to the spinal trigeminal nucleus and the posterior horn of the spinal cord modulate or suppress the transmission of nociception

The unpaired **midline raphe** (G., *raphe*, "seam") nuclei consist of intermediate-sized neurons. These nuclei are present along much of the anterior aspect of the brainstem, except in areas of the pons and the medulla oblongata that are occupied by various bundles of decussating fibers such as the pyramidal and lemniscal decussations.

The raphe nuclei are groups of neuronal cell bodies in the midline of the brainstem tegmentum. They are generally split into two groups, a *rostral group* in the midbrain and superior pons (comprising 85% of the serotonergic neurons of the brain), and a *caudal group* extending from the inferior pons to the spinal cord (comprising 15% of the serotonergic neurons of the brain). In the midbrain there are the **linear raphe nucleus**, **dorsal raphe nucleus**, and the **median raphe nucleus** which also extend into the superior pons. The **pontine raphe nucleus** and **raphe magnus nucleus** are also located in the midline of the pons. The **raphe pallidus nucleus** and **raphe obscurus nucleus** reside in the midline of the medulla oblongata (Fig. 16.2).

Serotonergic neurons from the dorsal raphe nucleus, linear raphe, and median raphe project to higher brain centers that are associated with *sleep*, whereas neurons from the raphe magnus, raphe pallidus, and raphe obscurus nuclei that project to the spinal trigeminal nucleus and the posterior horn of the spinal cord modulate the transmission of *nociception*.

Not all neurons in the raphe nuclei are serotonergic. The linear raphe nucleus contains catecholaminergic and substance P-releasing neurons in addition to its serotonergic neurons. Substance P is co-localized in 40% of serotonergic neurons in the dorsal raphe nucleus. The raphe magnus nucleus contains thyrotropin-releasing hormone, substance P, and serotonergic neurons. Many of these nuclei also contain GABAergic and glutaminergic interneurons to modulate the activity of the serotonergic projections from the raphe nuclei.

Paramedian zone (also known as the paramedian column, paramedian reticular nuclei, paramedian reticular nuclear group, paramedian reticular formation)

Via their connections with the cerebral cortex, cerebellum, vestibular nuclei, and spinal cord, the nuclei of the paramedian zone function in feedback systems associated with intricate movements

The **paramedian zone**, located lateral to the midline, receives primarily uncrossed afferent (input) fibers from the cerebral cortex, the fastigial and dentate nuclei of the cerebellum, the vestibular nuclei, and the spinal cord (Fig. 16.2). Efferent (output) fibers from this zone project to the vermis, uvula, and fastigial nucleus of the cerebellum. This nuclear group functions in feedback systems associated with intricate patterns of movement.

Medial zone (also known as the medial column, magnocellular reticular formation, central group of reticular nuclei)

The neurons of the medial zone influence the autonomic nervous system, level of arousal, and motor control of the axial and proximal limb musculature

The large size neurons of the **medial zone** (also referred to as the "motor," "efferent," or "effector" zone) are basal plate derivatives. The nuclei of the medial zone are the **ventral part** of the **central reticular nucleus**, **gigantocellular reticular nucleus**, and the **oral and caudal pontine reticular nuclei**, the **pedunculopontine tegmental nucleus**, and the **cuneiform, intracuneiform,** and **subcuneiform nuclei** (Fig. 16.2).

These medial zone nuclei possess neurons whose axons bifurcate to give rise to long ascending and descending branches, each with collaterals. The *ascending fibers* of these neurons course in the *central tegmental tract* to terminate in the hypothalamus, which controls the autonomic nervous system, and in the intralaminar nuclei of the thalamus, which function in arousal, and in the basal nuclei that influence locomotor behavior. The *descending fibers* extend inferiorly to the spinal cord, with neurons predominantly from the **pontine reticular nucleus** and **oral and caudal pontine reticular nuclei** forming the **medial reticulospinal tract** and neurons primarily from the **gigantocellular reticular nucleus** forming the **lateral reticulospinal tract**. These descending projections terminate chiefly in spinal laminae VII and VIII in trunk and proximal limb segments. They facilitate or suppress extensor and flexor activity to provide task-appropriate postural support in conjunction with the intended movement.

The pedunculopontine tegmental nucleus is located in the mesencephalic locomotor center and comprises largely medium- and large-sized cholinergic neurons. Ascending fibers synapse in the reticular part of the substantia nigra, the subthalamic nucleus, and the medial globus pallidus. Descending fibers synapse in the cerebellum and spinal cord. Lesions of this nucleus induce gait deficits, indicating that the **pedunculopontine tegmental nucleus** plays a central role in the initiation and maintenance of locomotion.

The cuneiform, the intracuneiform, and the subcuneiform nuclei are part of the mesencephalic locomotor center. The cuneiform nucleus contains glutaminergic and GABAergic neurons, and its descending projections exclusively target reticulospinal neurons in the pontomedullary reticular formation.

314 ● ● ● **CHAPTER 16**

Lateral zone (also known as the lateral column, parvocellular reticular formation, lateral nuclear group)

The lateral zone receives sensory information, integrates it, and then relays it to the medial zone. The medial zone then mediates the modulation of sensory afferent input and maintenance of alertness

The **lateral zone**, also referred to as the "sensory" or "afferent" zone, is derived from the alar plate, confined to the rhombencephalon, and consists of a group of nuclei containing small-sized (parvocellular) interneurons, the most numerous type of cells in the reticular formation. These small-sized interneurons have short ascending and descending branches that project mainly locally to the medial zone of the reticular formation. Other interneurons of the lateral zone terminate in the brainstem cranial nerve motor nuclei. The lateral zone receives input from both local and distant regions of the central nervous system.

The lateral zone extends primarily into the pons and medulla oblongata of the brainstem, and is continuous inferiorly with the spinal cord intermediate zone of gray matter, which is also heavily populated by interneurons. The intermediate zone of gray matter should not be confused with the intermediolateral (lateral) horn of the spinal cord gray matter, which contains the cell bodies of preganglionic neurons of the sympathetic nervous system.

The nuclei of the lateral zone are the **parvocellular nucleus**, which extends into the medulla oblongata and pons; the **dorsal subnucleus of the central reticular nucleus**, the **lateral paragigantocellular nucleus**, the **retroambiguus nucleus**, and the **parabrachial nucleus**, which extend into the pons and midbrain.

The **parvocellular nucleus** (L., *parvus*, "small") is located medial to the spinal trigeminal nucleus and ventral to the vestibular nuclei. It receives sensory information from the cerebrum, cranial nerves, cerebellum, and the spinal cord via collateral branches of various somatosensory pathways relaying touch, pressure, pain, temperature, and general proprioception. The lateral zone integrates the sensory information and then relays it to the medial zone. The reticular formation functions in the modulation of sensory afferent input, and in the maintenance of alertness via the **reticulobulbar** and descending **reticulospinal tracts** (all of which arise from the medial zone).

The **parabrachial nucleus** is located in the pons and relays sensory information (visceral discomfort, taste, temperature, pain, itch) to the thalamus, hypothalamus, and amygdaloid body.

NUCLEI ASSOCIATED WITH THE RETICULAR FORMATION

A number of brainstem nuclei are associated with, but are not considered part of, the reticular formation. These nuclei are the **red nucleus** and the **precerebellar reticular nuclei**.

In addition, the **periaqueductal gray matter** of the midbrain also has extensive connections with the reticular formation.

Precerebellar reticular nuclei

The precerebellar reticular nuclei function in the coordination of muscle activity

The **precerebellar reticular nuclei** consist of the **inferior olivary complex, lateral reticular nucleus, perihypoglossal nucleus, arcuate nucleus, pontobulbar nucleus, paramedian tract nuclei,** and **posterior paramedian nucleus**. These nuclei are functionally separate from the other reticular nuclei, send fibers to the cerebellum, and function in the coordination of muscle activity.

Periaqueductal gray matter

The periaqueductal gray matter functions in the processing of autonomic and limbic activities, as well as modulation of nociception

The **periaqueductal gray matter** of the midbrain receives afferent fibers from the cerebral cortex, the hypothalamus, the limbic system, the parabrachial nucleus of the reticular formation, the nucleus of the solitary tract, as well as from the ascending sensory systems. Efferent fibers arising from the periaqueductal gray matter project back to these regions, as well as to the medullary raphe nuclei. Based on its connections, and its association with the reticular formation, it is believed that the periaqueductal gray matter functions in the processing of autonomic and limbic activities as well as the modulation of nociception.

INPUT TO AND OUTPUT FROM THE RETICULAR FORMATION

Afferents (input) to the reticular formation

The reticular formation receives input via collateral branches of the ascending and descending pathways from widespread regions of the CNS. Input sources include the somatosensory, olfactory, visual, auditory, and vestibular systems, the premotor and primary motor cortices, cranial nerve nuclei, the diencephalon, basal nuclei, the cerebellum, and the amygdaloid body.

Afferents from the sensory systems

The reticular formation receives input from all of the sensory systems either via direct or indirect projections:

- The olfactory system projects indirectly via limbic structures.
- The visual system projects indirectly via the superior colliculus.
- The trigeminal system directly relays somatic sensory input from the orofacial region.

- The nucleus of the solitary tract indirectly projects taste sensation (relayed by the facial, glossopharyngeal, and vagus nerves).
- The superior olivary nucleus of the auditory system projects directly.
- The medial vestibular nucleus of the vestibular system projects directly.
- The spinothalamic and spinoreticular tracts project directly.

The ascending sensory systems from the body include two tracts, the spinothalamic and spinoreticular tracts that relay pain and temperature sensation from the spinal cord to higher brain centers. The **spinothalamic tract**, *the "direct" pain pathway*, projects fast, well-localized, discriminative pain, and temperature sensation from the body to the contralateral **ventral posterior lateral (VPL) nucleus** of the **thalamus**. On their way to their target nucleus of the thalamus, as the spinothalamic tract axons ascend through the brainstem, they give rise to collaterals that distribute to the parvocellular nuclei of the reticular formation. The parvocellular nuclei then relay signals to the gigantocellular nuclei of the pontine and medullary reticular formation. The VPL projects to the primary somatosensory cortex where pain and temperature sensation reaches conscious levels. In contrast, the **spinoreticular tract,** *the "indirect" pain pathway relays* slow, poorly localized, pain sensation from the body to the reticular formation, which in turn projects to the **thalamus**, **hypothalamus**, and **limbic system**. The thalamus projects to the cerebral cortex alerting the individual of potential injury. The hypothalamus is involved in the autonomic and endocrine responses to pain, whereas the projections to the limbic system play a role in the affective component of pain.

Afferents from the motor systems

The reticular formation receives input from the primary sensory cortex (eponym: Brodmann areas 3, 1, 2), the primary motor cortex (eponym: Brodmann area 4), and the cerebellum. The primary, sensory, and motor cortices contain the cell bodies of neurons whose axons form the **corticoreticular fibers**. These fibers descend in the corona radiata, and the posterior limb of the internal capsule, to terminate in the *oral and caudal pontine reticular nuclei* located in the pons, and the *gigantocellular nucleus* located in the medulla oblongata. The corticoreticular fibers synapse with the neurons housed in these nuclei to control voluntary and reflex activity of the axial and proximal limb extensor (antigravity) muscles.

The fastigial nucleus of the cerebellum projects to the pontine and medullary reticular formation, via fastigiobulbar fibers, where it plays a role in the control of movement.

Afferents from the autonomic nervous system

Baroreceptor fibers that terminate in the aortic arch and carotid sinus form the afferent limb of the reflex arc that controls blood pressure. The central processes of the GVA neurons enter the brainstem via the glossopharyngeal nerve root, join the **solitary tract**, and terminate in the **nucleus of the solitary tract**. This nucleus relays sensory input to the reticular formation, the brainstem GVE (autonomic) motor nuclei, and the intermediolateral horn (containing preganglionic sympathetic neurons) of the spinal cord gray matter for reflex activity related to the control of arterial vessel lumen diameter and blood pressure.

GVA pseudounipolar neuron cell bodies housed in the inferior ganglion of the glossopharyngeal nerve send their peripheral processes, chemoreceptor fibers, to terminate in the aortic and carotid bodies where they monitor blood carbon dioxide concentration. The central processes of all of the GVA neurons enter the brainstem, course in the solitary tract, and terminate in the nucleus of the solitary tract that projects to the reticular formation.

Afferents from the hypothalamus and limbic system

The hypothalamus sends descending fibers to the pons as well as the periaqueductal gray (PAG) and reticular formation of the midbrain tegmentum. Both the amygdaloid body and prefrontal cortex send projections to the PAG and brainstem tegmentum.

Efferents (output) from the reticular formation

The reticular formation gives rise to output fibers that project to the following areas of the CNS: widespread areas of the cerebral cortex, the diencephalon, basal nuclei, the red nucleus, the substantia nigra, the midbrain tectum, cranial nerve motor nuclei, the cerebellum, nuclei of the autonomic nervous system, and the spinal cord. The efferent projections that are exported to distant targets originate from the median, paramedian and medial zones, whereas projections whose targets are local or near-by originate from the lateral zone of the reticular formation.

FUNCTIONS OF THE RETICULAR FORMATION

The reticular formation, via its vast number of connections to widespread areas of the CNS, is involved in a broad spectrum of activities

The reticular formation receives a continuous flow of somatosensory, auditory, visual, and visceral sensory information through these vast connections and in turn exerts its influence by processing, controlling, and/or modulating the following:

- skeletal muscle motor activity, including muscle tone and reflexes;
- somatic sensation;
- visceral sensation;
- autonomic nervous system activity;
- endocrine functions;

CHAPTER 16

- biological rhythms, via reciprocal connections to the hypothalamus; and
- level of consciousness.

Control of motor activity

Motor activity is controlled by fibers traveling within the following fiber tracts: the corticoreticular fibers, medial reticulospinal tract, and lateral reticulospinal tract (Table 16.2).

Corticoreticular fibers

The corticoreticular fibers terminate bilaterally in the medial zone ("motor," "efferent," or "effector" zone) of the pontine and medullary reticular formation where they modulate the activity of the reticulobulbar and reticulospinal neurons

The **corticoreticular fibers** arise from the premotor cortex and from the supplementary motor area of the premotor cortical areas. In their descent to the brainstem, they accompany the corticonuclear and corticospinal tracts. The corticoreticular fibers terminate bilaterally in the medial zone of the pontine and medullary reticular formation in the nuclei that give rise to the reticulospinal tracts. In addition to receiving corticoreticular fibers from the motor cortex, the medial zone of the pontine and medullary reticular formation also receives fibers arising from other sources involved in motor function such as the basal nuclei, red nucleus, and substantia nigra via the *central tegmental tract*. All of these fibers terminating in the medial zone of the pontine and medullary reticular formation modulate the activity of the reticulobulbar and reticulospinal neurons.

The corticoreticular fibers, their **nuclei** of termination, and the **medial** and **lateral reticulospinal tracts** are referred to as the **corticoreticulospinal pathway** (**system**). The descending fibers of the medial and lateral reticulospinal tracts synapse primarily with spinal cord interneurons, which in turn synapse with lower motor neurons, although some synapse directly with lower motor neurons. The medial and lateral reticulospinal tracts function in locomotion and postural control.

Medial reticulospinal tract

The medial reticulospinal tract stimulates the extensors and inhibits the flexors of the axial and proximal limb musculature

The **medial zone** is the motor (efferent) zone of the **pontine reticular formation**. This zone consists of the **oral pontine reticular nucleus** and **caudal pontine reticular nucleus**. The medial zone of the pontine reticular formation gives rise to the **medial reticulospinal tract** (Fig. 16.3), which descends in the ipsilateral anterior funiculus of the spinal cord. This tract synapses directly and indirectly with alpha and gamma motor neurons in laminae VI–IX at all levels of the spinal cord stimulating the **extensors** and inhibiting the flexors of the axial and proximal limb musculature. The medial reticulospinal tract reinforces the action of the lateral vestibulospinal tract. Furthermore, many medial reticulospinal tract fibers synapse with propriospinal interneurons that function in reflexes involving multiple spinal cord segments. These propriospinal interneurons synapse with lower motor neurons that innervate the axial and proximal limb musculature.

In addition, a few medial reticulospinal tract fibers establish inhibitory synaptic contacts with the termination of the

Table 16.2 ● The reticular formation and associated tracts.

Fibers/tract	Origin	Destination	Function
Corticoreticular fibers	Premotor cortex Supplementary motor area	Medial zone of pontine and medullary reticular formation (oral pontine reticular nucleus, caudal pontine reticular nucleus, and gigantocellular reticular nucleus)	Modulate the activity of the reticulobulbar and reticulospinal neurons
Medial reticulospinal tract	Oral pontine reticular nucleus	Ipsilateral spinal cord to synapse with interneurons and motor neurons	Stimulation of extensor and inhibition of flexor muscles of the trunk and proximal limb
	Caudal pontine reticular nucleus	Ipsilateral spinal cord to synapse with first-order muscle spindle afferents	Prevent transmission of muscle spindle afferent signals, which has an inhibitory effect on stretch reflexes
Lateral reticulospinal tract	Gigantocellular reticular nucleus Ventral part of central reticular nucleus	Spinal cord (primarily ipsilateral) to synapse with interneurons and motor neurons	Inhibition of extensor and stimulation of flexor muscles of the trunk and proximal limb
Central tegmental tract	Basal nuclei Red nucleus Substantia nigra	Medial zone of pontine and medullary reticular formation	Modulates the activity of the reticulobulbar and reticulospinal neurons

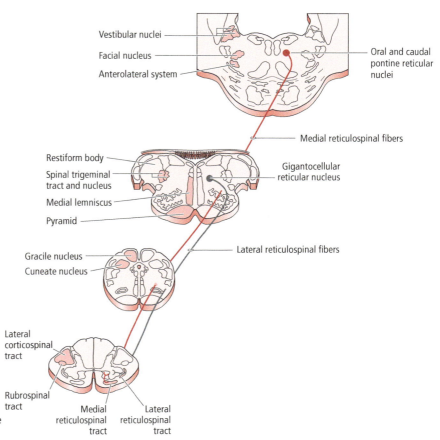

Figure 16.3 ● The origin, course, and termination of the medial and lateral reticulospinal tracts.

central processes of those first-order sensory neurons whose peripheral processes innervate muscle spindles, thus preventing the transmission of afferent input from those muscle spindles. This mechanism has an inhibitory effect on stretch reflexes, with resultant smoothing of voluntary movement.

Lateral reticulospinal tract

The lateral reticulospinal tract fibers have an inhibitory effect on extensors and an excitatory effect on flexors of the axial and proximal limb musculature

The **medial zone** of the rostral **medullary reticular formation** (gigantocellular reticular nucleus and the ventral part of the central reticular nucleus) gives rise to the **lateral reticulospinal tract** (Fig. 16.3). Similar to the medial reticulospinal tract, this tract descends (primarily) in the ipsilateral anterolateral funiculus of the spinal cord. Nerve fibers of this tract terminate at all spinal cord levels, where they synapse primarily with interneurons laminae VI, VII, and VIII. The lateral reticulospinal tract fibers have an opposite effect to that of the medial reticulospinal tract fibers, in that they have an inhibitory effect on extensors and an excitatory effect on **flexors** of the axial and proximal limb musculature.

Additionally, some fibers of the lateral reticulospinal tract synapse with those lower motor neurons of the spinal cord that supply motor innervation to the distal muscles of the upper and lower limbs. These fibers exert primarily an inhibitory influence on the lower motor neurons that innervate axial extensor muscles, and a facilitatory influence on those lower motor neurons that innervate limb flexors. Recall that the lateral corticospinal tract functions in the execution of fine, skilled movements of the hands and feet (see Chapter 13 for more detail).

Note that the descending fibers of the medial and lateral reticulospinal tracts relay signals that arise from the areas mentioned above (premotor cortex, supplementary motor area, basal nuclei, red nucleus, substantia nigra, and cerebellum), as well as signals arising from other areas of the reticular formation. All of these signals are integrated and then influence the activity of the alpha- and gamma-motor neurons innervating skeletal muscle. Furthermore, these tracts also control muscle tone and reflex activity. Since the reticular formation functions in *reciprocal inhibition*, during contraction of flexor muscles, there is simultaneous relaxation of the antagonistic extensor muscles. The medial and lateral reticulospinal tracts, with the contribution of the lateral vestibulospinal tract, function in balance maintenance and in making postural adjustments. Muscle tone, balance maintenance, and postural adjustments form a necessary background upon which voluntary movement is executed.

Control of eye movements

The PPRF of the medial zone of the pontine reticular formation functions in mediating conjugate horizontal eye movements elicited by head movements

The **oral pontine reticular nucleus**, **caudal pontine reticular nucleus**, and **gigantocellular reticular nucleus** receive sensory information from the visual, auditory, and vestibular systems, and function in head and eye coordination.

The **paramedian pontine reticular formation** (**PPRF**) (Fig. 16.4) of the medial zone of the pontine reticular formation is not a nucleus *per se*, but instead is a subset of neurons of the **oral and caudal pontine reticular nuclei** that functions in mediating conjugate *horizontal* eye movements elicited by head movements. It receives afferent fibers from the frontal eye field, the superior colliculus, and the vestibular nuclei, as well as from other regions of the reticular formation. Efferent fibers from the PPRF terminate mainly in the adjacent abducens nucleus of the same side that innervates the ipsilateral lateral rectus muscle, and, via the medial longitudinal fasciculus, to the contralateral oculomotor nucleus to synapse specifically with the motor neurons that supply the contralateral medial rectus muscle.

The **rostral interstitial nucleus of the medial longitudinal fasciculus** located at the cranial end of the midbrain at the level of the superior colliculus contains a collection of nerve cells that function in controlling conjugate *vertical* eye movements (Fig. 16.4A).

Control of pain sensation

The raphe nuclei suppress or modulate the transmission of nociceptive signals from first-order sensory neurons to second-order projection neurons that terminate in higher brain centers

As a result of its location in the brainstem, the reticular formation is well placed to exert its influence on ascending sensory and descending motor pathways, as well as on other pathways.

The fibers of the trigeminothalamic tract transmit nociceptive signals from visceral and somatic structures of the head, whereas the spinothalamic and spinoreticular tracts transmit nociceptive signals from visceral and somatic structures of the body to their target thalamic nuclei. As these fibers traverse the brainstem reticular formation on their way to the thalamus, they give rise to collateral branches that terminate in the lateral zone (the "sensory" or "afferent" zone) of the reticular formation. These signals not only activate the **ascending reticular activating system** (**ARAS**), which relays information to the cerebral cortex alerting the individual, but also relays information to the hypothalamus, the limbic system, the serotonergic raphe magnus nucleus, and the anterior gigantocellular reticular nucleus of the medial zone of the reticular formation. The hypothalamus and the limbic system function in the manifestation of visceral and emotional/behavioral responses to sensory stimuli (such

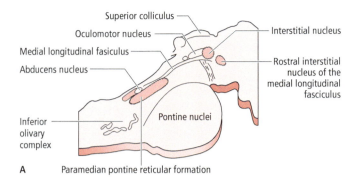
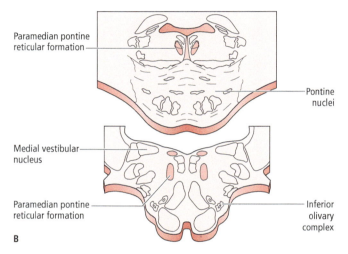

Figure 16.4 ● The location of the horizontal and vertical gaze centers in the brainstem in (A) sagittal and (B) transverse sections.

as pain). The serotonergic neurons of the raphe nuclei give rise to fibers that join the reticulobulbar and the reticulospinal tracts to descend to their target termination, the spinal trigeminal nucleus and the posterior horn of the spinal cord, respectively (Fig. 16.5). The **serotonin** liberated by these descending fibers at their axon terminals stimulates inhibitory enkephalinergic interneurons, each of which forms a presynaptic contact (axoaxonic synapse) on the terminals of the central processes of the first-order nociceptive neurons. By this mechanism, the raphe nuclei suppress or modulate the transmission of nociceptive signals from the first-order neurons to the second-order projection neurons that terminate in higher brain centers, specifically the thalamus. The thalamus relays the nociceptive signals to the primary somatosensory cortex where nociception enters consciousness. For a more detailed description see Chapter 12.

Modulation of the autonomic nervous system

Visceral information arrives at the reticular formation from the cerebral cortex, hypothalamus, and limbic system.

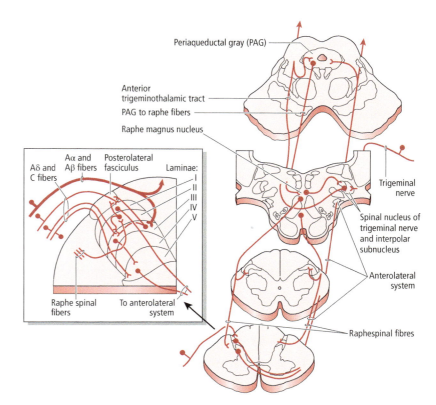

Figure 16.5 ● Descending pathways that modulate the transmission of nociception from the brainstem and spinal cord to higher brain centers. Source: Modified from Haines, DE (2002) *Fundamental Neuroscience*. Churchill Livingstone, Philadelphia; fig. 18.16.

Sensory information from visceral structures that modulate the cardiovascular and respiratory systems is transmitted to the **intermediate reticular formation** via collateral branches derived from the ascending sensory pathways and via the **solitary tract**, which relays baroreceptor (from the carotid sinus, aortic sinus), chemoreceptor (from the carotid body, aortic body), volume receptor (from the **atria of the** heart), and stretch receptor (from the lungs) information from the glossopharyngeal and vagus nerves. The sensory information is processed in vasomotor centers located in the medullary reticular formation, which are believed to be involved in the control of vital functions such as the control of arterial blood pressure (via the rostral and caudal ventrolateral medullary vasomotor neurons) and respiration (via neurons in the rostral and caudal ventral respiratory group, medullary expiratory center, center generating respiratory rhythm). The response from these centers is conveyed via the **reticulobulbar** and **reticulospinal tracts**, to the autonomic nuclei of the brainstem and spinal cord. Adjacent (anterolateral) to the nucleus of the solitary tract, and receiving projections from it, are glutaminergic neurons of the **rostral ventrolateral medulla** that project directly to activate, tonically, sympathetic thoracic preganglionic neurons and, similarly, to mediate all reflexes controlling arterial blood pressure. Concurrently, GABAergic neurons of the **caudal ventrolateral medulla** project directly to inhibit, tonically, the rostral ventrolateral medulla thereby modulating its output. Noradrenergic neurons from the nucleus of the solitary tract and the rostral ventrolateral medulla project to hypothalamic regions including the paraventricular nucleus, supraoptic nucleus, and median preoptic nucleus to regulate blood volume. Descending fibers of the **hypothalamospinal** and **hypothalamomedullary fibers** predominantly originate in the paraventricular nucleus of the hypothalamus. They descend through the mesencephalic periaqueductal gray, adjacent to the mesencephalic reticular formation, and then move to an anterolateral position in the medulla oblongata to synapse in the nucleus of the solitary tract, posterior vagal motor nucleus, nucleus ambiguus, and the anterolateral medulla oblongata. This intricate network underscores the multifaceted interactions and regulatory mechanisms governing autonomic and cardiovascular functions.

The **respiratory rhythm-generating center** is situated anterolaterally in the medulla oblongata, extending from the facial nucleus to the spinal cord. It contains pacemaker neurons that are chemosensitive and respond to various neurotransmitters. Altered levels of carbon dioxide or oxygen can trigger projections from other medullary nuclei which ultimately results in varied breathing patterns such as normal resting pattern, sighing, and gasping. Notably, arterial partial pressure of carbon dioxide is the chief determinant of respiratory drive: a 1 mm Hg increase in the partial pressure of carbon dioxide increases ventilation by 25% in a person who is awake and resting. Hypercapnia (high carbon dioxide levels) or hypoxia (low oxygen levels) can induce raphe neurons to release *serotonin* at the bulbospinal inspiratory premotor synapses onto phrenic motor neurons which induces a long-term (>1 h) increase in ventilation.

Inspiratory behavior is triggered after action potentials in the *pre-inspiratory* neurons in the **parafacial respiratory group** which is rostral to the ventral respiratory group in the medulla oblongata. Inspiratory neurons in the dorsal and ventral respiratory groups fire with increasing frequency for approximately two seconds during inspiration and then are switched off for approximately three seconds during expiration by projections from the pneumotaxic center and pulmonary stretch receptors. They then ramp up firing again to continue the cycle. The **dorsal respiratory group** neurons are predominantly *inspiratory*, are located in the anterolateral portion of the nucleus of the solitary tract, and receive afferent signals from the vagus nerves. They project primarily to the contralateral lower motor neuron cell bodies of the phrenic nerve in the anterior horn of spinal grey at C3, C4, and C5 levels. The **ventral respiratory group** neurons are *inspiratory and expiratory* and are located in the nucleus ambiguus and the retroambiguus nucleus.

Coordination of cranial nerve function

The brainstem reticular formation contains interneurons that are located near the cranial nerve nuclei and participate in the coordination of reflex activity associated with the cranial nerves

Interneurons located in the **lateral reticular formation** project locally to the nearby cranial nerve nuclei. The **ventrolateral reticular formation** contains neurons that participate in the coordination of the visceral functions of the vagus nerve. The vagus nerve innervates the gastrointestinal, respiratory, and cardiovascular systems, and via its connections with the reticular formation plays a role in eliciting reflex responses associated with swallowing, coughing, breathing, and changes in blood pressure.

Interneurons of the lateral reticular formation are also associated with the trigeminal, facial, and hypoglossal nerve nuclei, which innervate structures of the mouth and face. The motor control exerted by these cranial nerves on structures they innervate in the orofacial region is coordinated by the reticular formation. Although motor control of the muscles of mastication – those of the tongue, lips, soft palate, and upper pharynx – is under voluntary control, during chewing the action of all of these structures is automated and is executed without our being aware of it. The branches of the trigeminal nerve transmit general proprioceptive information from the temporomandibular joint, teeth, and muscles of mastication, and general sensation information from the mouth (thermal sensation and information about food consistency) to the sensory nuclei of the trigeminal nerve. These nuclei in turn project to the motor nuclei of the trigeminal and facial nerves and to the reticular formation, which in turn coordinates the movements of orofacial structures during chewing.

One of the most interesting associations of the reticular formation is its participation in the control of the muscles of facial expression during emotional responses. When an individual smiles while posing for a photograph, the imitation smile is voluntarily produced by descending input from the motor cortex (i.e., the corticonuclear tract) to the cranial nerve motor nuclei (i.e., the facial motor nucleus). This is sometimes referred to as a "pyramidal smile" (named after the pyramidal cells of the motor cortex forming the corticonuclear tract). In contrast, a sincere, heartfelt "emotional" smile is elicited by an alternate pathway, circumventing the primary motor cortex. This pathway includes descending motor input from accessory motor areas of the frontal lobe and the basal nuclei that reaches the facial nerve motor nuclei via relays in the reticular formation. This is often referred to as the "Duchenne smile" after the French neurologist and physiologist Duchenne de Boulogne who studied facial expressions. Individuals who have suffered a stroke that damaged the descending corticonuclear tract fibers, paralyzing the muscles of facial expression on one side of the face, are unable to smile voluntarily. However, when they hear pleasant news, a genuine happy facial expression is elicited. This may be hard to believe, but there is a neuroanatomical basis for this phenomenon. Although voluntary control via descending projections from the motor cortex to the lower half of the facial motor nucleus is unilateral, activation of the reticular formation by higher brain centers (i.e., accessory motor areas and basal nuclei) that are involved in eliciting emotional facial expressions is bilateral.

Influence on the endocrine system

The reticular formation also influences the activities of the hypothalamus, modulating the formation or release of hypothalamic-releasing hormones (or release-inhibiting hormones), which in turn influence the activities of the hypophysis.

Control of consciousness

Waking up in the morning, remaining awake all day, and falling asleep at night are all functions of the brainstem reticular formation

On their way to higher brain centers, the ascending sensory pathways (the trigeminothalamic, spinothalamic, and spinoreticular tracts), relaying general and nociceptive sensation, give rise to collateral branches that terminate in the parvocellular nucleus of the lateral zone of the reticular formation. The lateral zone gives rise to fibers that terminate in the medial zone of the reticular formation. The medial zone then projects via the *central tegmental tract* to the hypothalamus and the intralaminar nuclei of the thalamus. A spray of thalamocortical fibers arising from the thalamus project to widespread areas of the cerebral cortex. These fibers are referred to as the **Ascending Reticular Activating System** (ARAS), the arousal system that functions in the sleep–wake cycle. Sensory information must be continually funneled to the cerebral cortex to maintain consciousness. Various levels of *wakefulness* are contingent upon the level of reticular formation activation and the participation of cholinergic, noradrenergic, dopaminergic, and histaminergic neurons. Waking up in the morning, remaining awake all day, and falling asleep at night, are all regulated by functions of the brainstem reticular formation. It is interesting to note that the cerebral cortex requires a continuous stream of sensory input from the ARAS portion of the reticular formation which serves as an arousal system; the cerebral cortex, without this input, is unable to maintain *consciousness* and to function normally. Planning for the day, going about normal daily activities, and interacting with the stimuli in the environment are all mediated by the cerebral cortex.

Nociceptive or other signals that may awaken an individual from sleep are relayed to the cortex via the ARAS. The ARAS stimulates the hypothalamus and limbic system, which are involved in visceral functions and the emotional and behavioral responses to painful, or other stimuli.

Sensory stimuli (visual, auditory, olfactory, or somatic) such as a snake, siren, strong odors, or unexpected contact with an unknown object, that startle or surprise an individual, activate the ARAS and evoke an alerting response. People usually awaken in the morning following the sudden sound of an alarm clock, a loud noise, or perhaps the smell of coffee. The ARAS has to be adequately stimulated to activate or "awaken" the cerebral cortex. On the contrary, lack of or inadequate stimulation of this system, can result in dozing off, as may occur when sitting in a dark, quiet room.

It is interesting to note that, whereas anesthetics block the transmission of sensory signals that are relayed via the ascending sensory pathways through the ARAS to the cerebral cortex, they do not affect the transmission of auditory signals. This may explain a curious phenomenon reported by patients that while they are under general anesthesia they can hear conversations by individuals in the operating room.

Note that the clinical case at the beginning of the chapter refers to a patient who presented with complete unilateral paralysis, slurred speech, ocular deviation, Extensor Plantar Response (eponym: Babinski reflex), respiratory difficulties, and drowsiness, which progressed to a lethargic and comatose state and loss of all brainstem reflexes.

1 Which descending motor tracts have been damaged in this patient to initially cause a complete unilateral paralysis, slurred speech, and Extensor Plantar Response?

2 Which neural system has been damaged to cause lethargy, respiratory difficulties, a comatose state, and loss of all brainstem reflexes?

3 What do unilateral pupillary dilation and fixed pupil indicate?

CLINICAL CONSIDERATIONS

Consciousness can be altered following head trauma, drug intoxication, anesthesia, or metabolic disturbances. A lesion anywhere along the path of the ARAS that affects the transmission of ARAS signals to the cerebral cortex can result in various disorders of consciousness.

Frequently, head trauma can result in increased intracranial pressure. Since the brain is surrounded by the bony skull, intracranial pressure has no outlet, and it compresses the soft brain tissue. An increase in intracranial pressure may push the cerebellar tonsils toward the foramen magnum, which will then apply pressure on or compress the medulla oblongata. This will damage the respiratory centers in the medulla oblongata causing a central apnea (absence of breathing).

The midbrain is quite vulnerable and lesions damaging the reticular formation can cause hypersomnia associated with slow respiration. When the midbrain is compressed, with consequent extensive damage of the ARAS pathways, a comatose state ensues.

An upper spinal cord lesion at T6 or above will damage the descending hypothalamospinal and reticulospinal tracts that terminate in the sympathetic intermediolateral horn of the spinal cord. Normally these tracts synapse with the pre-ganglionic sympathetic neurons residing in the intermediolateral horn, and control autonomic activities. This type of lesion will at first cause a reduction in sympathetic activity such as a lowering of blood pressure, orthostatic hypotension, and slowed heart rate. A gradual sympathetic reflex hyperactivity appears in response to denervation hypersensitivity of the sympathetic neurons. This is exhibited as sympathetic responses such as high blood pressure, sweating, piloerection, urinary retention, and a decrease in peripheral blood flow.

322 ••• CHAPTER 16

SYNONYMS AND EPONYMS OF THE RETICULAR FORMATION

Name of structure or term	Synonym(s)/eponym(s)
Axial musculature	Trunk musculature
Extensor plantar response	Babinski reflex Inverted plantar reflex
Basal nuclei	Basal ganglia Deep cerebral nuclei
Corticoreticulospinal pathway	Corticoreticulospinal system
Corticospinal tract	Pyramidal tract (older term)
Gigantocellular reticular nucleus	Nucleus reticularis gigantocellularis Magnocellular reticular nucleus
Inferior olivary complex	Inferior olive Olive
Intermediolateral horn	Lateral horn
Lateral zone of the reticular formation	Lateral column of the reticular formation Parvocellular reticular formation Lateral nuclear group
Medial zone of the reticular formation	Medial column of the reticular formation Magnocellular reticular formation Central group of the reticular formation
Median zone of the reticular formation	Median column of the reticular formation Midline raphe
Lateral reticulospinal tract	Medullary reticulospinal tract
Linear nucleus	Nucleus raphes dorsalis
Paramedian zone of the reticular formation	Paramedian column of the reticular formation Paramedian reticular nuclei Paramedian reticular nuclear group
Medial reticulospinal tract	Pontine reticulospinal tract
Premotor cortex	Part of Brodmann's area 6
Proximal limb musculature	Girdle musculature
Supplementary motor area	Part of Brodmann's area 6
Trigeminothalamic tracts	Trigeminal lemnisci
Central reticular nucleus	Nucleus reticularis ventralis Central medullary nucleus

FOLLOW-UP TO CLINICAL CASE

This patient had a massive stroke. The above scenario is a typical progression indicative of brain herniation secondary to progressive edema from the stroke. The edema causes swelling of the brain and since the skull is rigid, the swelling follows the route of least resistance downward, and consequently causes compression of other structures. With severe edema there is a shift of the midline structures of the brain to the normal side, opposite the stroke (midline shift), a downward shift of the diencephalon and brainstem, and herniation of the ipsilateral medial temporal lobe through the tentorium cerebelli into the posterior cranial fossa.

Impairment of consciousness occurs with diffuse dysfunction of the cerebral cortex (i.e., metabolic derangements), diencephalic lesions, or upper brainstem lesions. The latter two structures are the locale of the ARAS, which is critical for maintaining consciousness. Lesions involving the ARAS result in coma. In contrast, a focal brain lesion affecting only part of the cerebral cortex and subcortical white matter will cause no or minimal impairment of consciousness. Advancing lethargy with a focal brain lesion, such as stroke or tumor, often indicates progressive brain herniation leading to brainstem function impairment (assuming metabolic problems are ruled out as the cause).

Other signs besides progressive lethargy and coma are also important. Stretching and compression of the ipsilateral oculomotor nerve (CN III) over the tentorial notch leads to ipsilateral pupillary dilation and fixed pupil. Compression or ischemia of the contralateral cerebral peduncle causes ipsilateral weakness or pathologic reflexes (Babinski sign). Other brainstem reflexes such as the corneal blink reflex, vestibulo-ocular or oculocephalic reflexes, gag reflex, and respiratory effort can be lost. Loss of all brainstem reflexes indicates brain death.

QUESTIONS TO PONDER

1. With your knowledge of the ascending sensory pathways and the cranial nerves, describe how nociceptive information is relayed to the reticular formation.

2. What is the general structure of the neurons of the reticular formation and how does their structure facilitate their collection of information from ascending and descending pathways?

3. What is the function of the medial zone of the reticular formation?

4. Why is it that when we are chewing we (fortunately) do not usually bite our tongues?

5. If a lesion in the midbrain damages the ARAS, what is likely to happen?

CHAPTER 17

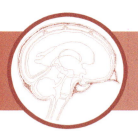  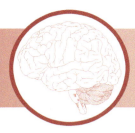

Cranial Nerves

CLINICAL CASE

OLFACTORY NERVE (CN I)

OPTIC NERVE (CN II)

OCULOMOTOR NERVE (CN III)

TROCHLEAR NERVE (CN IV)

TRIGEMINAL NERVE (CN V)

ABDUCENS NERVE (CN VI)

FACIAL NERVE (CN VII)

VESTIBULOCOCHLEAR NERVE (CN VIII)

GLOSSOPHARYNGEAL NERVE (CN IX)

VAGUS NERVE (CN X)

ACCESSORY NERVE (CN XI)

HYPOGLOSSAL NERVE (CN XII)

SYNONYMS AND EPONYMS OF THE CRANIAL NERVES

FOLLOW-UP TO CLINICAL CASE

QUESTIONS TO PONDER

CLINICAL CASE

A 70-year-old patient has excruciating pain in the lower left part of their face. This began one month ago. They describe it as being like a jolt of lightning that radiates from their left ear, down to their mandible, and to the side of their mouth. These jolts of pain occur numerous times each day. Between attacks, their face seems normal. The patient denies any numbness or tingling sensations. There is no hearing abnormality. The pain is triggered by talking, chewing, or touching the lower left part of their face. The patient is unable to eat or brush their teeth, particularly on the left side, since they fear triggering another painful attack. The patient can only drink their meals through a straw and cannot lie in bed on their left side. They had the same symptoms about two years ago. At that time, they were treated with a medication that helped; symptoms subsided, but they stopped taking the medicine. The pain is so distressing that the patient admits to contemplating suicide.

The general and neurologic exam is normal, except that they withdraw and will not let anyone touch the left side of their face.

There are 12 pairs of cranial nerves emerging from the brain and radiating from its surface (Fig. 17.1). They pass through foramina, fissures, or canals of the skull to exit the cranial vault and then to distribute their innervation to their respective structures in the head and neck. One of the cranial nerves, the vagus (L., "wanderer") continues into

---

A Textbook of Neuroanatomy, Third Edition. Maria A. Patestas, Amanda J. Meyer, and Leslie P. Gartner.
© 2025 John Wiley & Sons, Inc. Published 2025 by John Wiley & Sons, Inc.
Companion website: www.wiley.com.go/patestas/neuroanatomy3e

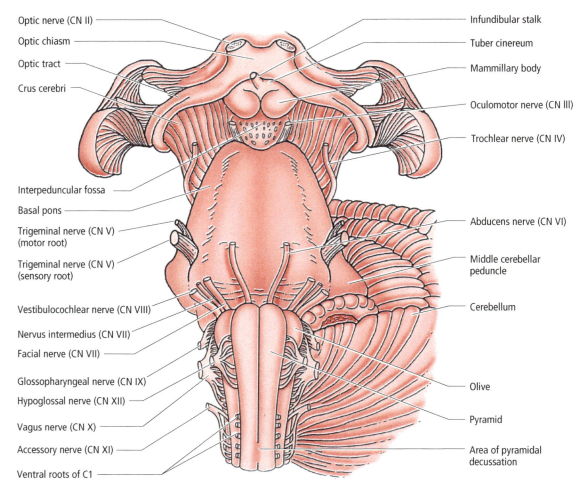

Figure 17.1 • Anterior view of the brainstem showing the cranial nerves.

the trunk where it innervates various thoracic and abdominal structures.

In addition to being named, the cranial nerves are numbered sequentially with Roman numerals in the order in which they arise from the brain, rostrally to caudally. The following list includes their names and corresponding numbers.

I	Olfactory nerve.
II	Optic nerve.
III	Oculomotor nerve.
IV	Trochlear nerve.
V	Trigeminal nerve.
VI	Abducens nerve.
VII	Facial nerve and Nervus Intermedius.
VIII	Vestibulocochlear nerve.
IX	Glossopharyngeal nerve.
X	Vagus nerve.
XI	Accessory nerve.
XII	Hypoglossal nerve.

Although the cranial nerves and their sensory and parasympathetic ganglia (Tables 17.1 and 17.2) form part of the peripheral nervous system, the *optic nerve* is really an outgrowth of the brain that emerges from the prosencephalon (not the brainstem as other cranial nerves) and is therefore not a typical cranial nerve. Another atypical cranial nerve

Table 17.1 • Sensory ganglia of the cranial nerves.

Ganglion	Cranial nerve
Trigeminal (semilunar, Gasserian)	Trigeminal (V)
Geniculate	Nervus Intermedius (VII)
Cochlear (spiral)	Cochlear (VIII)
Vestibular (Scarpa)	Vestibular (VIII)
Superior glossopharyngeal	Glossopharyngeal (IX)
Inferior glossopharyngeal (petrosal)	Glossopharyngeal (IX)
Superior vagal (jugular)	Vagus (X)
Inferior vagal (nodose)	Vagus (X)

Table 17.2 • Parasympathetic ganglia of the cranial nerves.

Ganglion association	Cranial nerve association	Trigeminal	Function(s)
Ciliary	Oculomotor (III)	Ophthalmic division	Constricts pupil; lens accommodation
Pterygopalatine	Nervus Intermedius (VII)	Maxillary division	Lacrimation; nasal gland, and minor salivary gland secretion
Submandibular	Nervus Intermedius (VII)	Mandibular division	Salivation (submandibular and sublingual glands)
Otic	Glossopharyngeal (IX)	Mandibular division	Salivation (parotid gland)
Intramural	Vagus (X)		Gland secretion; peristalsis (thoracic and abdominal organs)

is the *olfactory nerve*. The sensory neurons of the olfactory pathway reside in the olfactory epithelium in the roof of the nasal cavity instead of arising from a ganglion. The axons of these sensory neurons do not collect to form a single bundle of axons, a nerve, but instead, form multiple small collections of axons that traverse the foramina of the cribriform plate of the ethmoid bone to terminate in the overlying olfactory bulb located on the floor of the anterior cranial fossa. A third deviation from the normal cranial nerves is the *accessory nerve* whose spinal portion arises from the cervical spinal cord; thus, there are only nine pairs of cranial nerves that actually emerge from the brainstem.

The main sensory and motor nuclei of the cranial nerves are shown in Fig. 17.2.

In describing the various functional components (**modalities**) of the cranial nerves, the definition of the following terms should be kept in mind: *afferent* is sensory input (**to** the central nervous system); *efferent* is the motor output (**from** the central nervous system) that may be *somatic* to skeletal muscles or *visceral* to smooth muscle in blood vessels, organ walls, and myoepithelial cells of glands, and cardiac muscle, and *special visceral* efferent to skeletal muscles derived from the pharyngeal arches; *general* refers to those components that may be carried by cranial nerves as well as spinal nerves; *special* refers to functional components of the senses (such as smell, vision, taste, hearing, and proprioception) that are carried by cranial nerves only. The following categories describe the functional components carried by the various cranial nerves (see also Table 17.3).

1. **General somatic afferent (GSA).** These fibers carry general sensation (touch, pressure, pain, and temperature)

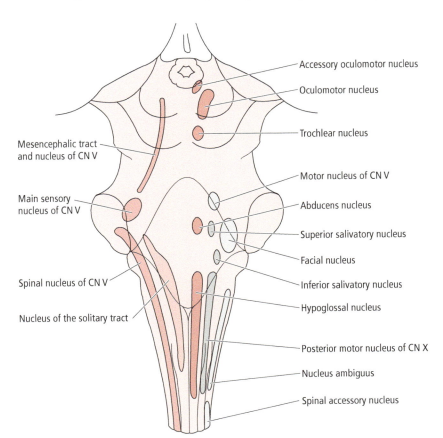

Figure 17.2 • The nuclei of the cranial nerves. The sensory nuclei are illustrated on the left, and the motor nuclei on the right.

326 ● ● ● **CHAPTER 17**

Table 17.3 ● Cranial nerve functional components.

Modality	Cranial nerves	Function(s)
General somatic afferent (GSA)	V, VII, IX, X	General sensation and general proprioception
General somatic efferent (GSE)	III, IV, VI,	Motor supply to extraocular muscles
	XII	Motor supply to the tongue
General visceral afferent (GVA)	VII, IX, X	General sensation from viscera
General visceral efferent (GVE)	III, VII, IX, X	Parasympathetic fibers to viscera
Special somatic afferent (SSA)	II	Special sensory input from the retina
	VIII	Special sensory input from vestibulocochlear apparatus
Special visceral afferent (SVA)	I	Special sense of smell
	VII, IX, X	Special sense of taste
Special visceral efferent (SVE)	V, VII, IX, X	Motor innervation to muscles of pharyngeal arch origin: mandibular, hyoid, 3rd, 4th, and 6th pharyngeal arches
Special visceral efferent (SVE)/ General visceral efferent (GSE)	XI	Motor innervation to the sternocleidomastoid (SCM) and trapezius muscles (based on possible dual origin of the SCM and trapezius muscles)

from mechanoreceptors, nociceptors, chemoreceptors, and thermoreceptors in cutaneous structures and mucous membranes of the head, and **general proprioception** from mechanoreceptors in somatic structures such as muscles, tendons, and joints of the head and neck. The trigeminal, facial, glossopharyngeal, and vagus nerves transmit GSA input to the spinal nucleus of the trigeminal nerve.

2 **General somatic efferent (GSE).** These fibers provide general motor innervation to skeletal muscles derived from embryonic somites. The oculomotor, trochlear, and abducens nerves innervate the extraocular muscles that control eye movements, whereas the hypoglossal nerve supplies motor innervation to the muscles of the tongue (except palatoglossus), mediating movement of the tongue. The accessory nerve innervates the sternocleidomastoid and trapezius muscles. The sternocleidomastoid muscle turns the head to the opposite side, whereas the trapezius muscle elevates and depresses the shoulders.

3 **General visceral afferent (GVA).** General sensation from the viscera is transmitted by the facial, glossopharyngeal, and vagus nerves.

4 **General visceral efferent (GVE).** These fibers provide visceral motor (parasympathetic) innervation to the viscera. Only four cranial nerves transmit parasympathetic fibers: the oculomotor nervus intermedius (facial), glossopharyngeal, and vagus nerves.

5 **Special somatic afferent (SSA).** These fibers carry special sensory input from the eye (retina), for vision, and from the ear (vestibular apparatus for equilibrium and cochlea for hearing). The only nerves transmitting this component are the optic and vestibulocochlear nerves.

6 **Special visceral afferent (SVA).** These are special sensory fibers from the viscera. They convey the special sense of smell transmitted by the olfactory nerve and the special sense of taste transmitted by the nervus intermedius (facial), glossopharyngeal, and vagus nerves.

7 **Special visceral efferent (SVE).** These motor fibers are special because they supply motor innervation to skeletal muscles of pharyngeal arch origin. These fibers are carried by the nerves of the pharyngeal arches, which are the trigeminal, facial, glossopharyngeal, and vagus nerves.

Table 17.4 summarizes the modalities, nuclei, ganglia, and functions of the cranial nerves.

Table 17.4 ● Cranial nerves, their components, distribution, and function.

Cranial nerve	Functional component (modality)	Nucleus	Location of cranial nerve nuclei	Ganglion	Distribution	Function(s)
I Olfactory	SVA	anterior olfactory	Telencephalon	—	Olfactory mucosa	Smell
II Optic	SSA	—	Diencephalon	—	Ganglion cells of the retina	Vision
III Oculomotor	GSE	Oculomotor	Mesencephalon (tegmentum)	—	All extraocular muscles except the lateral rectus and superior oblique	Eye movement

CRANIAL NERVES ● ● ● 327

Table 17.4 ● (*Continued*)

Cranial nerve	Functional component (modality)	Nucleus	Location of cranial nerve nuclei	Ganglion	Distribution	Function(s)
	GVE (parasympathetic)	Accessory oculomotor (eponym: Edinger–Westphal)	Mesencephalon (tegmentum)	Ciliary (parasympathetic)	Sphincter pupillae muscle	Pupillary constriction
					Ciliary muscle	Lens accommodation
IV Trochlear	GSE	Trochlear	Mesencephalon (tegmentum)	—	Superior oblique	Eye movement
V Trigeminal	SVE	Motor nucleus of the trigeminal	Metencephalon (pons)	—	Muscles of mastication: temporalis masseter medial pterygoid lateral pterygoid	Chewing
					Mylohyoid, anterior belly of the digastric	
					Tensor tympani	Tenses tympanic membrane
					Tensor veli palatini	Tenses soft palate
	GSA	Main (chief, principal) nucleus of the trigeminal	Metencephalon (pons)	Trigeminal	Anterosuperior scalp, anterior two-thirds of the dura, cornea, conjunctiva, face, paranasal sinuses, teeth, gingiva, anterior two-thirds of the tongue, TMJ, Periodontal ligament (PDL), nasal cavities	General sensation
		Spinal nucleus of the trigeminal	Metencephalon (pons to C3)			
	GP	Mesencephalic nucleus of the trigeminal	Mesencephalon	—	Muscles of mastication, Periodontal ligament (PDL),	Muscle stretch Pressure sensation
		Trigeminal ganglion			Extra-ocular muscles, TMJ, PDL	Proprioception
VI Abducens	GSE	Abducens	Metencephalon (pons)	—	Lateral rectus	Eye movement
VII Facial	SVE	Facial	Metencephalon (pons)	—	Muscles of facial expression, platysma, posterior belly of the digastric, and stylohyoid	Facial expression
					Stapedius	Tension on stapes
VII Nervus Intermedius	GVE (parasympathetic)	Superior salivatory	Myelencephalon	Pterygopalatine	Lacrimal gland, Glands of the nasal cavity and palate	Lacrimation Mucous secretion
				Submandibular	Myoepithelial cells of the submandibular and sublingual glands	Salivation
	SVA	Solitarius	Myelencephalon	Geniculate	Anterior two-thirds of the tongue	Taste
	GVA	Solitarius	Myelencephalon	Geniculate	Middle ear, nasal cavity, and soft palate	Visceral sensation
	GSA	Spinal nucleus of the trigeminal	Metencephalon (pons)	Geniculate	External auditory meatus and area posterior to ear	General sensation

(*Continued*)

328 ●●● **CHAPTER 17**

Table 17.4 ● *(Continued)*

Cranial nerve	Functional component (modality)	Nucleus	Location of cranial nerve nuclei	Ganglion	Distribution	Function(s)
VIII Vestibulocochlear						
Cochlear	SSA	Posterior and anterior cochlear	Myelencephalon	Spiral	Spiral organ (eponym: organ of Corti) (inner ear)	Hearing
Vestibular	SSA	Vestibular complex	Myelencephalon	Vestibular	Utricle Saccule Semicircular canal ampullae (inner ear)	Equilibrium
IX Glossopharyngeal	SVE	Ambiguus	Myelencephalon	—	Stylopharyngeus	Swallowing
	GVE (parasympathetic)	Inferior salivatory	Myelencephalon	Otic (parasympathetic)	Parotid gland	Salivation
	SVA	Solitarius	Myelencephalon	Inferior ganglion of the glossopharyngeal	Posterior one-third of the tongue and adjacent pharyngeal wall	Taste
	GVA	Solitarius	Myelencephalon	Inferior ganglion of the glossopharyngeal	Middle ear, pharynx, tongue, carotid sinus	Visceral sensation
	GSA	Spinal nucleus of the trigeminal	Myelencephalon	Superior ganglion of the glossopharyngeal	Posterior one-third of the tongue, soft palate, upper pharynx, and auditory tube	General sensation
X Vagus	GVE (parasympathetic)	Dorsal motor nucleus of the vagus	Myelencephalon	Thoracic and abdominal submucosal and myenteric autonomic plexuses	Thoracic and abdominal viscera	Gland secretion, peristalsis
	SVE	Ambiguus	Myelencephalon	—	Muscles of the larynx and pharynx	Phonation
	SVA	Solitarius	Myelencephalon	Inferior (nodose)	Epiglottis	Taste
	GVA	Solitarius	Myelencephalon	Inferior (nodose)	Thoracic and abdominal viscera Carotid body	Visceral sensation
	GSA	Spinal nucleus of the trigeminal	Myelencephalon	Superior (jugular)	Area posterior to the ear, external acoustic meatus, and posterior part of meninges	General sensation
XI Accessory	GSE/SVE	Accessory	spinal cord C1–C5	—	To sternocleidomastoid and trapezius	Head and shoulder movements
XII Hypoglossal	GSE	Hypoglossal	Myelencephalon	—	Muscles of the tongue	Tongue movement

Some authors have suggested renumbering of the traditional cranial nerves based on the concepts that a cranial nerve should have a nucleus within the brainstem, must pass through a cranial foramen and have secondary neurons whose cell bodies are located in the brainstem. Applying this definition still results in 12 cranial nerves as follows: The olfactory and optic nerves do not have a nucleus in the brainstem and are thus eliminated. The oculomotor nerve becomes cranial nerve **I** and the trochlear nerve becomes cranial nerve **II**. The trigeminal nerve is divided into two distinct nerves since it has multiple nuclei. The motor portion

of the trigeminal nerve becomes cranial nerve **III**, the masticatory nerve, whereas the sensory portion of the trigeminal nerve (including the ophthalmic, maxillary and mandibular divisions) becomes cranial nerve **IV**. The abducens nerve becomes cranial nerve V. The facial nerve has separate nuclei and becomes cranial nerve **VI**, whereas the nervus intermedius becomes cranial nerve **VII**. The vestibulocochlear nerve consists of two anatomically distinct nerves with separate nuclei and modalities. The vestibular nerve becomes cranial nerve **VIII**, and the cochlear nerve becomes cranial nerve **IX**. The glossopharyngeal nerve becomes cranial nerve **X**.

The vagus nerve becomes cranial nerve **XI**, but it is divided into two divisions based on the target organs innervated, (1) laryngopalatopharyngeal (formerly the cranial root of cranial nerve **XI**) and (2) thoracoabdominal. The accessory nerve is eliminated since it does not have a nucleus in the brain. The hypoglossal nerve remains unchanged as cranial nerve **XII**. Although this is an interesting concept, this textbook will not adopt these suggested changes until, and unless, there is a general consensus among neuroanatomists concerning the desirability for such a drastic change in terminology.

OLFACTORY NERVE (CN I)

The first-order sensory neurons (olfactory receptor cells) reside in the olfactory epithelium of the superior nasal cavity, and not in a sensory ganglion as is typical of other cranial nerves

The **bipolar olfactory receptor cells** (first-order sensory neurons) of the olfactory apparatus reside not in a sensory ganglion, but in the **olfactory epithelium (neuroepithelium)**, the modified region of the nasal mucosa lining the roof and adjacent upper walls of the right and left nasal cavities (see Fig. 21.1). The axons of these bipolar neurons are **SVA fibers** transmitting *olfactory* sensation. These axons assemble to form 15–20 separate thin bundles of axons, the **olfactory fila** (L., "threads"), which, although collectively form what is considered to be **cranial nerve I**, do not form a single nerve trunk as do the conventional cranial nerves. The olfactory fila traverse the fenestrations of the cribriform plate of the ethmoid bone to terminate in the olfactory bulb where they synapse with **second-order neurons** and **interneurons** (see Chapter 21).

OPTIC NERVE (CN II)

The optic nerve consists of the myelinated axons of the retinal ganglion cells

The **optic nerve** mediates the special sense of *vision* via its **SSA fibers**. Light entering the eye activates cells known as *rods* and *cones*, the photoreceptors of the retina. Electrical signals generated by the rods and cones are transmitted to other cells of the retina that process and integrate sensory input. The **first-order sensory bipolar neurons** of the visual pathway reside in the retina and transmit electrical signals of visual sensory input to the **second-order multipolar ganglion cells**, also residing in the retina. These ganglion cells give rise to unmyelinated axons that converge posteriorly at the optic disc and traverse the lamina cribrosa, a sieve-like perforated area of the sclera, to emerge from the posterior pole of the eye bulb. It is at this point that the axons of the ganglion cells each acquire their myelin sheath and assemble to form the optic nerve. This nerve, an outgrowth of the diencephalon, becomes invested by extensions of the three meninges that cover the brain and then leaves the orbit via the optic canal to enter the middle cranial fossa. Once there, the optic nerves of the right and left sides converge and join each other to form the **optic chiasm** (G., "optic crossing") where partial decussation of the optic nerve fibers of the two sides takes place. All ganglion cell axons arising from the **nasal (medial) half** of each retina *decussate* (through the central region of the chiasm) to the opposite optic tract. All ganglion cell axons arising from the **temporal (lateral) half** of each retina proceed (through the lateral aspect of the chiasm) *without decussating* and join the optic tract of the same side. The ganglion cell axons coursing in each optic tract curve around the cerebral peduncle of the midbrain to terminate and relay visual input in one of the following five regions of the brain: the **lateral geniculate nucleus**, a thalamic relay station for vision; the **superior colliculus**, a mesencephalic nucleus of the visual system associated with somatic reflexes; the **pretectal area**, a midbrain region associated with autonomic reflexes; the **pineal body** to supply it with information concerning the presence or absence of daylight and thus control the circadian rhythm; and the **hypothalamus**, involved in the autonomic visceral responses (such as heart rate, pulse, and breathing rate) that increase in response to visual input that may indicate impending danger (see Figs. 18.5, 18.7, and 18.9).

OCULOMOTOR NERVE (CN III)

The oculomotor nerve provides motor innervation to four of the six extraocular muscles and to the levator palpebrae superioris, as well as parasympathetic innervation to the sphincter pupillae and ciliary muscles of the eye

The **oculomotor nerve** supplies **skeletal motor (somatomotor)** innervation to the superior rectus, medial rectus, inferior rectus, and inferior oblique muscles (which move the bulb of the eye) and the levator palpebrae superioris muscle (which elevates the upper eyelid). It also provides **parasympathetic (visceromotor)** innervation to the ciliary and sphincter pupillae muscles, two intrinsic smooth muscles of the eye.

The triangle-shaped **oculomotor nuclear complex** is located in the rostral midbrain tegmentum. It is partly embedded within, and is situated anterior to, the periaqueductal gray (PAG) matter, adjacent to the midline at the level of the superior colliculus. The oculomotor nucleus consists of several **subnuclei** representing collections of GSE neuron cell bodies, where one of those groups is destined for each of the extraocular muscles and for the levator palpebrae superioris muscle. The *central caudate nucleus*, the cell group innervating the levator palpebrae superioris, is located in the midline, sending motor fibers to this muscle *bilaterally* (to both right and left upper eyelids). The *medial column*, the cell group innervating the superior rectus, sends projections to the *contralateral* side; whereas the cell group innervating the medial rectus, inferior rectus, and inferior oblique sends projections to the *ipsilateral* side. Axons that *cross*, do so *within* the oculomotor nucleus. As the axons that have already crossed exit the oculomotor nucleus, they join the oculomotor nerve root fascicles of the opposite side. Although the oculomotor nerve is unique since it is the only cranial nerve that consists of a *mix* of *crossed* and *uncrossed* fibers, this is of little clinical significance. For example, a lesion that damages the entire (right) *oculomotor nucleus* would result in paralysis of the superior rectus of one (left) eye and paralysis of the remaining extraocular muscles innervated by the oculomotor nerve in the other (right) eye, which is rare. Even if this does occur, the actions of the functional extraocular muscles with intact innervation compensate for the deficits of the paralyzed (left) superior rectus muscle of the same eye. All of the axons that exit the oculomotor nucleus on each side, course anteriorly in the midbrain tegmentum as numerous oculomotor nerve root fascicles, curving medially, passing through the red nucleus and the medial aspect of the substantia nigra, and then gather to form a bundle that exits the brainstem between the two cerebral peduncles at the *interpeduncular fossa*. The *oculomotor nerve that exits* the midbrain contains fibers that have already crossed. Thus, a lesion involving the oculomotor root fascicles sweeping anteriorly in the midbrain tegmentum, or the oculomotor nerve anywhere along its course after it exits the midbrain, would result in deficits only in the *ipsilateral* eye.

The **accessory oculomotor nucleus** (eponym: Edinger–Westphal nucleus), a subnucleus of the oculomotor nuclear complex is located posteriorly, medially, and rostral to the GSE nuclear complex. It contains the cell bodies of two distinct populations of neurons. One group consists of the cell bodies of *preganglionic parasympathetic neurons* whose axons join the oculomotor nerve and course to the orbit where they synapse in the ciliary ganglion. This group of fibers is involved in two distinct parasympathetic functions, pupillary constriction via contraction of the sphincter pupillae, and accommodation of the lens via contraction of the ciliary muscle. The other group of neurons within the accessory oculomotor nucleus are the *centrally projecting* neurons that send their axons to various brainstem nuclei including the gracile, cuneate, spinal trigeminal, parabrachial, and inferior olivary nuclei. Other fibers descend to the spinal cord. These projections play a role in diverse behaviors such as eating, drinking, and stress-related functions.

Several small nuclei that control eye movements reside near the oculomotor nucleus. They are the (1) **elliptic nucleus** (eponym: nucleus of Darkschewitsch) located in the anterolateral margin of the PAG, rostral to the oculomotor nucleus, the (2) **interstitial nucleus** (eponym: interstitial nucleus of Cajal), located in the posteromedial region of the rostral midbrain tegmentum, next to the medial longitudinal fasciculus (MLF), and the (3) **nucleus of the posterior commissure** (a pretectal nucleus). The interstitial nucleus, the elliptic nucleus, and the nucleus of the posterior commissure

play a role in the coordination of head and eye movements, especially in vertical gaze.

As the oculomotor nerve exits the midbrain, it immediately passes between the *posterior cerebral artery* (the terminal branch of the basilar artery) and the *superior cerebellar artery* (a branch of the basilar artery). The proximity of the oculomotor nerves to these vessels makes them susceptible to lesions of vascular origin (e.g., compression of the nerve by calcifications or aneurysms in the adjacent vessel wall). The oculomotor nerve then enters the *interpeduncular cistern* (a dilated region of the subarachnoid space between the cerebral peduncles), passes under the *posterior communicating artery,* and pierces the dura mater to enter the *cavernous sinus*. It courses anteriorly in the lateral wall of the cavernous sinus where it branches into superior and inferior divisions. Each division then exits the cavernous sinus and passes through the superior orbital fissure to enter the orbital cavity. The *superior division* gives rise to branches that innervate the levator palpebrae superioris muscle and the superior rectus muscles. The *inferior division* gives rise to branches that innervate the medial rectus, the inferior rectus, and the inferior oblique muscles. The levator palpebrae superioris muscle elevates the upper eyelid; the superior and inferior rectus muscles elevate and depress the eye, respectively; the medial rectus muscle adducts the eye; and the inferior oblique muscle depresses the eye when it is in the lateral position. The *preganglionic parasympathetic* fibers of the oculomotor nerve terminate in the **ciliary ganglion** where they synapse with postganglionic parasympathetic nerve cell bodies. *Postganglionic parasympathetic* fibers exit the ganglion and reach the sphincter pupillae and ciliary muscles via the **short ciliary nerves** to provide them with parasympathetic innervation. The parasympathetic fibers, when stimulated, cause contraction of the sphincter pupillae muscle, which results in constriction of the pupil. Pupillary constriction reduces the amount of light that reaches the retina. Stimulation of the sympathetic nervous system, which innervates the dilator pupillae muscle via the **long ciliary nerves**, causes pupillary dilation. Ciliary muscle contraction releases the tension on the suspensory ligaments of the lens, changing its thickness to become more convex and accommodating the lens for focusing on nearby objects.

GSA pseudounipolar neurons, whose cell bodies reside within the **trigeminal ganglion** of the trigeminal nerve, send their peripheral processes to terminate in the muscle spindles of the extraocular muscles. These fibers travel via the branches of the ophthalmic division of the trigeminal nerve. GSA (general proprioception) sensory input is transmitted from the muscle spindles via the spindle afferents centrally to the trigeminal nuclear complex, mediating coordinated and synchronized eye movements by reflex and voluntary control of muscles.

CLINICAL CONSIDERATIONS

Unilateral damage to the oculomotor nerve results in deficits in the *ipsilateral* eye. The affected individual cannot elevate the upper eyelid; is unable to move the eye medially or vertically; the pupil is dilated, and the lens is flat, and cannot accommodate for near focus. The following ipsilateral muscles will be paralyzed: the levator palpebrae superioris, resulting in **ptosis** (G., "drooping") of the upper eyelid; the superior and inferior recti, resulting in an inability to move the eye vertically; and the medial rectus, resulting in an inability to move the eye medially. The inferior oblique is also paralyzed. The innervation of the lateral rectus (CN VI) is intact, and since this muscle normally abducts the eye, it pulls the eye laterally due to the unopposed action of its paralyzed antagonist, the medial rectus. The innervation of the superior oblique (CN IV) is also intact and since it normally intorts, depresses, and abducts the eye, it turns the eye "down and out" due to the unopposed actions of the paralyzed, flaccid superior, medial and inferior rectus, and inferior oblique muscles. The lateral deviation of the eye is referred to as **lateral strabismus** (Fig. 17.3). This causes the visual axes of the two eyes to become misaligned as one eye deviates from the midline, resulting in **diplopia** (double vision). Since the affected eye is deviated "down and out," the patient complains of a mixed horizontal and vertical binocular diplopia.

The sphincter pupillae muscle becomes nonfunctional due to the interruption of its parasympathetic innervation. The pupil ipsilateral to the lesion will remain in fixed dilation (**mydriasis**), due to disinhibited sympathetic innervation of dilator pupillae, and does not respond (constrict) to a flash of light. This may be the first clinical sign of intracranial pressure on the GVE fibers that course on the surface/periphery of the oculomotor nerve. The ciliary muscle is also nonfunctional due to interruption of its parasympathetic innervation and cannot accommodate the lens for near vision (i.e., cannot focus on near objects). Consequently, the individual is unable to focus and has blurred vision when looking at near objects with the affected eye.

A lesion involving the oculomotor nerve as it traverses the cavernous sinus, the superior orbital fissure, or the posterior aspect of the orbit, will not present as an isolated oculomotor nerve palsy because other cranial nerves accompany and course in tandem with the oculomotor nerve on their way to the orbit. These nerves include the trochlear nerve, the abducens nerve, and the ophthalmic division of the trigeminal nerve. They may also be affected/damaged since they have a similar course on their way to the orbit, and if all are damaged, a constellation of deficits will be apparent which will help localize the site of the lesion.

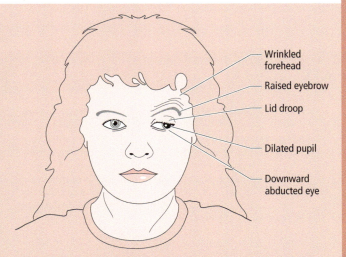

Figure 17.3 • A lesion involving the left oculomotor nerve results in the following symptoms ipsilateral to the side of the lesion: (1) ptosis (drooping of the upper eyelid), (2) downward and outward deviation of the eye, (3) lateral strabismus, (4) pupillary dilation, and (5) loss of accommodation of the lens.

TROCHLEAR NERVE (CN IV)

The trochlear nerve is the thinnest cranial nerve and is the only one whose fibers originate totally from the contralateral nucleus

The **trochlear nerve** provides motor innervation to only one of the extraocular muscles of the eye, the **superior oblique muscle** (a common mnemonic is SO$_4$). Since the superior oblique's tendon threads through the *trochlea* (L., "pulley"), a fibrous loop attached to the anteromedial superior surface of the orbital margin, the nerve innervating it is called the "trochlear nerve."

Normally, contraction of the superior oblique muscle causes the eye to *intort* (rotate medially around an anteroposterior axis), accompanied by simultaneous *depression* (downward movement, about a horizontal axis) and *abduction* (outward movement, about a vertical axis) of the bulb of the eye. This muscle causes the eye to turn inferolaterally ("**down and out**").

The nerve cell bodies of **GSE neurons** reside in the **trochlear nucleus** that lies adjacent to the midline in the tegmentum of the caudal midbrain. Fibers arising from this nucleus initially descend for a short distance in the brainstem and then course posteriorly around the PAG matter toward the tectum. The fibers *decussate* posteriorly before emerging from the brainstem at the junction of the pons and midbrain, just below the inferior colliculus.

The trochlear nerve is unique because it is the only cranial nerve whose fibers originate exclusively from the *contralateral* nucleus. This nerve surfaces on the *posterior* aspect of the brainstem, and it is the *thinnest* of the cranial nerves, consisting approximately of 2,400 axons. As the trochlear nerve emerges from the back of the brainstem, it enters the subarachnoid space and, curving around the cerebral peduncle, reaches the anterior aspect of the midbrain. The trochlear nerve passes between the *posterior cerebral* and the *superior cerebellar arteries*, accompanying but positioned lateral to the oculomotor nerve. The trochlear nerve then proceeds anteriorly, positioned just inferior to the sharp, free edge of the tentorium cerebelli and, piercing the dura mater, it enters and courses within the lateral wall of the cavernous sinus from where it passes into the orbital cavity through the superior orbital fissure. Because the trochlear nerve arises from the back of the brainstem, this cranial nerve has the *longest* intracranial course on its journey to the orbit and is thus highly susceptible to trauma involving rapid deceleration of the head (motor vehicle accidents, boxing). The trochlear nerve does not pass through the common tendinous ring anchoring the recti muscles, so it is at less risk of damage from raised intracranial pressure compared to the oculomotor and abducens nerves.

CLINICAL CONSIDERATIONS

Damage to the trochlear nucleus results in **paralysis** or **paresis** (weakness) of the *contralateral* superior oblique muscle, whereas damage to the **trochlear nerve** results in the same deficits, but in the *ipsilateral* muscle.

When the superior oblique muscle is paralyzed, the individual develops **trochlear nerve palsy**, characterized by the following: the ipsilateral (affected) eye becomes *extorted* about its anteroposterior axis so that the superior surface of the eyeball will approach the lateral wall of the orbit), accompanied by simultaneous *elevation* (upward deviation of the eye known as **hypertropia**) and slight *abduction* (lateral, or outward) deviation of the eye (Figs. 17.4B and 17.5).

The new resting position of the eye is caused by the actions of the unopposed inferior oblique (which **extorts**, **elevates**, and **abducts** the eye).

Paralysis of the superior oblique muscle causes the visual axes of the two eyes to become misaligned and, consequently, the brain receives two non-overlapping images. This is what causes a patient with trochlear nerve palsy to complain of double vision (diplopia). An individual with **trochlear nerve palsy** experiences **vertical diplopia**.

Normally, downward gaze is executed by the joint effort of the superior oblique and inferior rectus muscles. When the superior oblique is paralyzed,

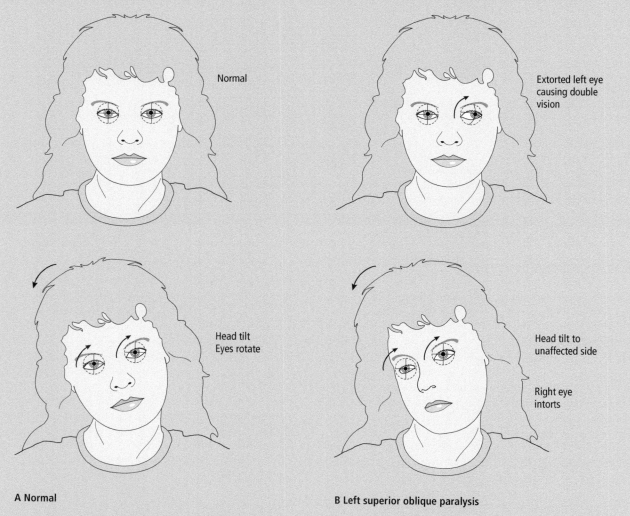

Figure 17.4 ● (A) Normal: When the head is tilted, the eyes automatically rotate in the opposite direction. (B) Left superior oblique paralysis following a lesion to the trochlear nerve: the affected eye becomes extorted with consequent double vision. To minimize the double vision, the individual tilts her head toward the unaffected side, which intorts the normal eye.

CLINICAL CONSIDERATIONS (*continued*)

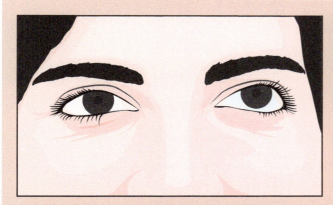

Figure 17.5 • Trochlear nerve paralysis/palsy of the left eye. The affected eye (A) has hypertropia (looks "up") since the superior oblique's ability to depress the eye is impaired, and (B) is extorted (looks "out") since the superior oblique's ability to intort the eye is impaired.

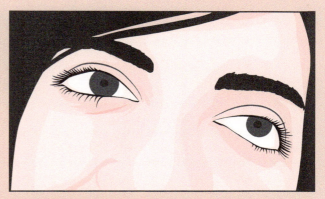

Figure 17.6 • Trochlear nerve paralysis/palsy of the left eye. The affected eye has hypertropia (deviated upward) and is slightly extorted. Hypertropia is due to the impaired ability to depress the affected eye. When the head is tilted toward the side of the palsy (as seen here), the hypertropia and extorsion become worse. That is, the weakness of the superior oblique's depressor and intorsion functions intensify.

Figure 17.7 • Trochlear nerve paralysis/palsy of the left eye. By tilting the head toward the normal side, the normal eye intorts to align it with the affected eye, which is extorted. "Chin tuck" would elevate the normal eye to become closely aligned with the affected (hypertropic) eye.

the patient's ability to depress the affected eye is impaired (i.e., the patient experiences weakness – not total inability – of downward gaze with the affected eye). Consequently, the eye drifts upward caused by the unopposed action of the inferior oblique that extorts, **elevates**, and abducts the eye.

In trochlear nerve palsy, the *hypertropia* of the affected eye (and therefore, the vertical *diplopia*) is most apparent to the individual (and becomes worse) in (1) attempted downward gaze (when descending stairs); in (2) attempted medial gaze (when the affected eye tries to look down and in toward the nose, as in reading a book); and (3) when tilting the head toward the affected side (Fig. 17.6). Because the normal eye becomes depressed, but the affected eye is partially depressed only by the action of the inferior rectus (since the superior oblique is paralyzed and cannot contribute to depressing the eye), it causes the eyes to become misaligned resulting in vertical diplopia. When a patient who has vertical diplopia is descending steps, each (real) step appears to have an overlapping (diplopic, that is an apparently false) step above it. Also, when reading, a sentence appears to have a diplopic, overlapping sentence above the real sentence.

To counteract the diplopia and to restore nearly horizontal proper eye alignment, the affected individual realizes that *the diplopia is minimized by tilting the head toward the normal side* (Figs. 17.4B and 17.7). Normally, tilting of the head to one side elicits a reflex rotation about the anteroposterior axis of the eyes in the opposite direction (Fig. 17.4A), so that the image of an object will remain fixed on the retina.

Tilting of the head to the normal side causes the normal eye to intort and become aligned with the affected extorted eye. This head tilt compensates for the extorsion of the affected eye. Pointing of the chin downward ("chin tuck") rolls the normal eye upward. The chin tuck compensates for the hypertropia of the affected eye. The diplopia is eliminated when the individual closes either eye because only one image reaches the brain.

TRIGEMINAL NERVE (CN V)

The trigeminal nerve, the largest of the cranial nerves, provides the major general sensory innervation to part of the scalp, most of the dura mater, and the orofacial structures; it also supplies motor innervation to the muscles of mastication

The **trigeminal nerve**, the largest cranial nerve, possesses 20,000–35,000 neurons. It provides the major **GSA** innervation (touch, pressure, proprioception, nociception, and thermal sense) to part of the scalp, most of the cranial dura mater, the major intracranial blood vessels, the conjunctiva and cornea of the eye, the face, mucous membranes of the nasal cavities, paranasal sinuses, oral cavity, and hard palate, the temporomandibular joint, mandible, teeth and periodontal ligaments (PDLs), the collagenous connective tissue lining that suspends the teeth in their alveoli. It also provides **SVE** innervation to the muscles of mastication (temporalis, masseter, medial pterygoid, and lateral pterygoid), as well as to the mylohyoid, anterior belly of the digastric, tensor tympani, and tensor veli palatini muscles.

The trigeminal nerve is the only cranial nerve whose sensory root enters and its motor root exits at the pons' lateral aspect at mid-level (see Fig. 17.1). The larger, **sensory root** consists of the central processes (axons) of the pseudounipolar sensory neurons of the trigeminal ganglion. These axons enter the pons to terminate in the trigeminal sensory nuclear complex of the brainstem. The **motor root** is smaller and consists of the SVE axons of motor neurons exiting the pons (Fig. 17.10). The motor root joins the sensory portion of the mandibular division of the trigeminal nerve in the infratemporal fossa, just outside the skull, to form the mandibular trunk. Before the motor root joins the mandibular division, the trigeminal nerve displays a swelling, the **trigeminal ganglion**. The ganglion is enclosed in the **trigeminal cave** (eponym: Meckel's cave), a dural pouch that lies in a bony depression of the petrous portion of the temporal bone on the floor of the middle cranial fossa. Because this is a sensory ganglion, no synapses are occurring there. As the peripheral processes of the pseudounipolar neurons exit the ganglion, they form three divisions (hence "trigeminal," meaning the "three twins"). These divisions traverse the foramina of the skull to exit the cranial vault on their way to reach the structures they innervate. The **ophthalmic division** is purely sensory, exits the superior orbital fissure, and innervates the upper part of the face; the **maxillary division** is also purely sensory (although there may be some exceptions), it exits the foramen rotundum, and innervates the middle part of the face. The **mandibular division** is the largest and is mixed, that is it carries sensory innervation to the lower face and branchiomotor innervation to the muscles listed above and it exits the skull through the foramen ovale.

The trigeminal system

The trigeminal system is formed by the trigeminal nerve, ganglion, nuclei, and its central pathways

The trigeminal system is formed by the trigeminal nerve, ganglion, nuclei, and its central pathways. Similar to the ascending sensory pathways from the body (posterior column–medial lemniscus and the

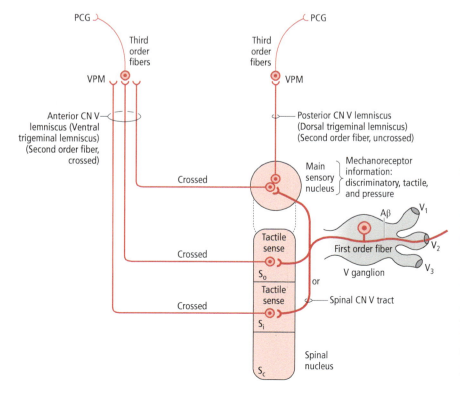

Figure 17.8 ● The trigeminal pathway for touch and pressure. Touch and pressure sensation from the orofacial structures is transmitted to the brainstem trigeminal nuclei, the principal sensory nucleus, and the spinal nucleus via the central processes of first-order pseudounipolar neurons whose cell bodies are located in the trigeminal ganglion. Second-order neurons in these nuclei join the posterior and anterior trigeminothalamic tracts which terminate in the ventral posterior medial (VPM) nucleus of the thalamus. Third-order neurons in the thalamus project to the postcentral gyrus. PCG, postcentral gyrus; S_c, caudal subnucleus; S_i, interpolar subnucleus; S_o, oral subnucleus; V_1, ophthalmic division of the trigeminal nerve; V_2, maxillary division of the trigeminal nerve; V_3, mandibular division of the trigeminal nerve.

CRANIAL NERVES ••• 335

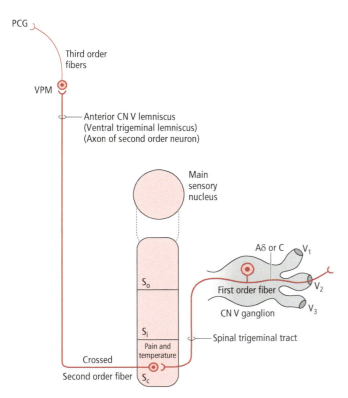

Figure 17.9 • The trigeminal pathway for pain and temperature. Pain and temperature sensation from the orofacial structures is transmitted to the brainstem caudal subnucleus (S_c) of the spinal nucleus of the trigeminal via the central processes of first-order pseudounipolar neurons whose cell bodies are located in the trigeminal ganglion. Second-order neurons from the caudal subnucleus join the anterior trigeminothalamic tract to terminate in the VPM nucleus of the thalamus. Third-order neurons from the VPM terminate in the postcentral gyrus (PCG). For other abbreviations, see Fig. 17.8.

spinothalamic tracts) that relay *touch, pressure, proprioception, nociception, thermal sense*, and *itch* to consciousness, the trigeminal sensory pathways transmit the same type of sensory information to consciousness, but they relay them from structures of the head to the cerebral cortex. Additionally, as the ascending sensory pathways from the body, each of the trigeminal sensory pathways consists of a three-neuron sequence (first-, second-, and third-order neurons) in a somatotopic fashion from the periphery (that is from structures of the head) to the cerebral cortex (Figs. 17.8 and 17.9).

Although most of the cell bodies of the first-order sensory neurons of the trigeminal pathways for *touch, nociception, thermal sense*, and *itch* sensations from the face reside in the trigeminal ganglion, there is an unusual nucleus, the **mesencephalic nucleus of the trigeminal** (discussed below), that also contains pseudounipolar sensory neurons but those neurons transmit sensory input related to *stretch* and *proprioception* from the muscles of mastication and also transmit most *proprioceptive* inputs from the periodontal ligaments surrounding the roots of the teeth. In contrast, the pseudounipolar sensory neurons that transmit sensory input related to *stretch* and *proprioception* from the extraocular muscles, the

TMJ, and from the periodontal ligaments (PDLs) of some teeth, reside in the trigeminal ganglion.

The peripheral processes of the first-order neurons radiating from the trigeminal ganglion gather to form three separate nerves that form the three divisions of the trigeminal nerve. The peripheral endings of these three nerves terminate in sensory receptors of the orofacial region. The same type of sensory receptors located in the skin, skeletal muscles, tendons, mucous membranes, and other visceral structures in the entire body, are also found associated with the peripheral processes of trigeminal sensory neurons that terminate in the region of the head. The sensory receptors at the peripheral endings of the trigeminal sensory neurons transduce tactile, nociceptive, thermal, or itch sensory stimuli to nerve impulses that are relayed to the brainstem. The central processes of trigeminal sensory neurons conveying these impulses enter the pons as part of the nerve's sensory root, join the spinal tract of the trigeminal located laterally in the brainstem, and terminate in the trigeminal sensory nuclei where they establish synaptic contacts with second-order neurons housed in these nuclei. The trigeminal sensory nuclei, except for the mesencephalic nucleus, contain second-order neurons as well as interneurons. The second-order neurons give rise to fibers that may or may not decussate in the brainstem and depending on their nuclear origin, join the **anterior** or **posterior trigeminothalamic tracts**. These tracts ascend to relay trigeminal sensory input to the **ventral posterior medial (VPM) nucleus** of the thalamus, where they synapse with third-order neurons. The third-order neurons then relay sensory information in a somatotopic fashion to the **postcentral gyrus** (primary somatosensory cortex) of the cerebral cortex for conscious awareness of the sensation and for further processing.

Trigeminal nuclei

The trigeminal system includes four nuclei: three *sensory* nuclei, the mesencephalic nucleus of the trigeminal, the principal (main, chief) sensory nucleus of the trigeminal, and the spinal nucleus of the trigeminal and one *motor* nucleus, the motor nucleus of the trigeminal (see Fig. 17.2; Table 17.5). The sensory nuclei form a continuous, longitudinal column of cells that extends from the rostral midbrain rostrally, to the superior two to three cervical spinal cord levels, caudally.

Table 17.5 • The trigeminal nuclei.

Motor nucleus
Sensory nuclei:
• Principal (chief, main) nucleus of the trigeminal
• Mesencephalic nucleus of the trigeminal
• Spinal nucleus of the trigeminal:
Oral subnucleus
Interpolar subnucleus
Caudal subnucleus

336 ● ● ● CHAPTER 17

Sensory nuclei

The sensory nuclei of the trigeminal nerve transmit sensory information from the orofacial structures to the thalamus

As indicated above, the **sensory nuclei** of the trigeminal system consist of a long cylinder of cells, which extends from the mesencephalon to the first few cervical spinal cord levels. Two of these nuclei – the principal sensory nucleus and the spinal nucleus of the trigeminal – receive the first-order afferent terminals of pseudounipolar neurons whose cell bodies are housed in the trigeminal ganglion. These nuclei serve as the **first sensory relay station** of the trigeminal system.

Mesencephalic nucleus

The mesencephalic nucleus contains those sensory pseudounipolar neurons that transmit proprioception from the muscles of mastication and periodontal ligaments (PDLs) to the motor as well as sensory nuclei (principal and spinal) of the trigeminal, and also to the reticular formation and to the cerebellum involved in reflexes; they also act to modulate and coordinate muscle activity during chewing

The **mesencephalic nucleus of the trigeminal** extends from the rostral pons to the rostral midbrain. This nucleus is unique since it is a true "sensory ganglion"; it is not a nucleus because it contains cells that are both structurally and functionally ganglion cells. It is believed that during development, neural crest cells have become embedded within and have become a part of the central nervous system instead of becoming embedded in and becoming a part of the peripheral nervous system. Because this nucleus houses the cell bodies of **sensory (first-order) pseudounipolar neurons** there are *no synapses* in the mesencephalic nucleus. The peripheral large-diameter myelinated processes of these pseudounipolar neurons convey proprioceptive input from the muscle spindles of the muscles of mastication and periodontal ligament to the trigeminal motor nucleus, sensory nuclei (principal and spinal), and extrinsic sites such as the reticular formation, and the cerebellum, which are involved in reflexes, oral stereognosis, and coordination of voluntary muscle activity (during chewing, swallowing, and speaking).

Immediately lateral to the **mesencephalic nucleus** the central and peripheral processes of the mesencephalic pseudounipolar neurons form the **mesencephalic tract** of the trigeminal nerve. The central processes of the mesencephalic pseudounipolar neurons usually branch in the vicinity of the motor nucleus of the trigeminal to send collaterals that terminate in the motor nucleus; some of these form the sensory, *afferent limb* of the **myotatic mandibular "jaw jerk" reflex** (see below). Other neurons of the mesencephalic nucleus which send their peripheral processes to the PDLs of the teeth relay proprioceptive input to the motor nucleus, reticular formation, and cerebellum to modulate and coordinate the activity of the muscles of mastication during the process of chewing.

The **mandibular proprioception** pathway consists of sensory pseudounipolar neurons whose cell bodies are also located in the **mesencephalic nucleus**. The peripheral processes of these pseudounipolar neurons terminate in muscle spindles (low-threshold mechanoreceptors) in the muscles of mastication; the central processes of these pseudounipolar neurons bifurcate to send a branch to the principal sensory nucleus and another branch to the rostral portions of the spinal nucleus. Second-order neurons from both of these nuclei project to the VPM nucleus of the thalamus which in turn projects to the medial surface of the primary somatosensory cortex (eponym: Brodmann area 3a) in the parietal lobe. It is here that the position sense of the mandible enters conscious awareness.

Although proprioceptive information is generally processed by the mesencephalic neurons, their receptors, and their central connections, some proprioception is relayed from the temporomandibular joint (TMJ), extraocular muscle spindles, and some of the PDLs by the sensory pseudounipolar neurons residing in the trigeminal ganglion to the principal sensory and spinal nuclei of the trigeminal.

Principal (main, chief) sensory nucleus

The principal nucleus processes discriminative tactile sensation from the orofacial region

The **principal** (main, chief) **sensory nucleus of the trigeminal nerve** is located in the mid-pons between the superiorly positioned mesencephalic nucleus of the trigeminal and the inferiorly situated spinal nucleus of the trigeminal. Based on the pattern of its afferent projections, this nucleus consists of a posteromedial division and an anterolateral division. Both divisions receive the central processes of first-order neurons whose cell bodies reside in the trigeminal ganglion.

The **posteromedial division** receives the central processes of first-order neurons whose peripheral processes collect sensory input originating in the oral cavity (teeth and surrounding soft tissues). In contrast, the **anterolateral division** receives the central processes of first-order neurons whose peripheral processes relay sensory input originating from a wide range of structures in the trigeminal nerve's entire sensory territory, innervated by all three of its divisions.

Axons of second-order neurons arising from the *posteromedial division* of the principal sensory nucleus form the smaller *ipsilateral* **posterior (dorsal) trigeminothalamic tract**, which ascends to terminate in the VPM nucleus of the thalamus. Most of the second-order neuron axons emerging from the principal sensory nucleus arise from the *anterolateral division* of the principal sensory nucleus and *cross* the midline to join the *contralateral* **anterior (ventral) trigeminothalamic tract**. This tract, which also carries a small number of ascending axons of second-order neurons whose cell bodies reside in the spinal nucleus of the trigeminal relaying tactile, nociceptive, and thermal sensation from the orofacial region, to the VPM nucleus of the thalamus.

Based on its anatomical and functional characteristics, the principal sensory nucleus is homologous to the gracile and cuneate nuclei of the posterior column–medial lemniscus pathway relaying discriminative touch, vibratory sense, and

CRANIAL NERVES ••• 337

proprioception from the **body**. However, the principal sensory nucleus is associated with the transmission of mechanoreceptor information for *discriminatory (fine) tactile, pressure,* and some *proprioceptive sense* from the **head**.

Spinal nucleus

The spinal nucleus is involved in the processing of *mechanical, nociceptive, thermal,* and *itch* sensations from the oral and facial regions

The **spinal nucleus of the trigeminal** is the largest of the three nuclei. It extends from the caudal pontine region (at just below the mid-pons) inferiorly to level C3 of the spinal cord and is continuous inferiorly with the posterior-most laminae (substantia gelatinosa) of the posterior horn of the spinal cord. This nucleus consists of three subnuclei: the rostral-most **oral subnucleus** (subnucleus oralis, pars oralis), the caudal-most **caudal nucleus** (subnucleus caudalis, pars caudalis), and the intermediate **interpolar nucleus** (pars interpolaris, subnucleus interpolaris).

The spinal nucleus of the trigeminal system is involved in the processing of *mechanical, nociceptive, thermal,* and *itch* sensations from the orofacial region. Although this nucleus processes mechanical stimuli, it is mainly concerned with nociception and thermal sensations. It should be noted that *the trigeminal nerve is the only cranial nerve that has a nucleus that processes nociception and thermal sense.* Sensory information is relayed to this nucleus via the central processes of pseudounipolar neurons whose cell bodies reside in the trigeminal ganglion and the mesencephalic nucleus. The two rostral subnuclei are also involved in reflex activity and project to the cerebellum, which functions in the coordination of the muscles that move the mandible. The spinal nucleus contains the cell bodies of second-order neurons whose axons project via the trigeminothalamic tracts to the contralateral **VPM** and the **dorsomedial (DM) nuclei of the thalamus**. The VPM projects to the primary somatosensory cortex residing in the postcentral gyrus, involved in conscious awareness of pain, temperature, and itch sensation. The DM nucleus projects to the anterior cingulate gyrus, which is associated with the affective and motivational components of pain, temperature, and itch sensation.

The **oral subnucleus** (subnucleus oralis) merges with the principal sensory nucleus superiorly and extends to the pontomedullary junction inferiorly. It is associated with the transmission of discriminative (fine) tactile/mechanical sense from the orofacial region.

The **interpolar subnucleus** (subnucleus interpolaris) extends from the rostral extent of the hypoglossal nucleus to the obex and is also associated with the transmission of tactile sense, as well as dental pain.

The **caudal subnucleus** (subnucleus caudalis) extends from the level of the obex (medulla oblongata) to the C3 level of the spinal cord. It is associated with the transmission of nociception and thermal sensations from the orofacial region. Since the caudal subnucleus lies immediately superior to the substantia gelatinosa of the cervical spinal cord levels, it is also referred to as the "**medullary posterior horn**." It is the homolog of the substantia gelatinosa since their neurons have similar cellular morphology, synaptic connections, and functions. Based on its afferent projections, the caudal subnucleus consists of three regions: rostral, intermediate, and caudal. The rostral portion of the caudal subnucleus processes sensory input from the teeth and other structures of the oral cavity. Since the trigeminal nerve is the only cranial nerve that has a nucleus that processes nociception, the central processes of sensory neurons residing in the sensory ganglia of the facial, glossopharyngeal, and vagus nerves converge to terminate in this rostral region within the caudal subnucleus. The intermediate component receives sensory input from the perioral, nasal, and malar (cheek) regions, whereas its caudal portion receives input from the lateral aspect of the face and part of the ear (anterior portion of the external ear and part of the external acoustic meatus).

Motor nucleus

The motor nucleus of the trigeminal nerve contains the cell bodies whose axons form the motor root of the trigeminal nerve, providing motor innervation to the muscles of mastication

The **motor nucleus of the trigeminal** is located at the mid-pontine level, medial to the principal sensory nucleus. It contains interneurons and the cell bodies of SVE **multipolar alpha** and **gamma motor neurons** whose axons form the **motor root** of the trigeminal nerve as they exit the pons. The SVE fibers join the mandibular division of the trigeminal nerve and are distributed to the muscles of mastication as well as to the mylohyoid, anterior belly of the digastric, tensor tympani, and tensor veli palatini muscles.

The trigeminal nerve does not have any parasympathetic nuclei in the CNS or parasympathetic ganglia in the peripheral nervous system. However, it is anatomically associated with the parasympathetic ganglia of other cranial nerves (oculomotor, facial, and glossopharyngeal) and carries their autonomic "hitch-hikers" to their destinations.

Trigeminal tracts

The trigeminal system includes three tracts: the spinal tract of the trigeminal, the anterior trigeminothalamic tract (trigeminal lemniscus), and the posterior trigeminothalamic tract

The **spinal (descending) tract of the trigeminal nerve** consists of ipsilateral first-order afferent fibers of sensory trigeminal ganglion neurons and mediates *tactile, thermal, nociceptive,* and *itch* sensations to the spinal nucleus of the trigeminal from the orofacial region. The spinal tract of the trigeminal also carries first-order sensory axons of the facial, glossopharyngeal, and vagus nerves. The axons of these three cranial nerves terminate in the spinal trigeminal nucleus, conveying GSA sensory input from their respective areas of innervation to be processed by the trigeminal system. The spinal tract descends lateral to the **spinal nucleus of the trigeminal**, its fibers synapsing with neurons at various levels along the rostrocaudal extent of this nucleus. Inferiorly, at upper cervical spinal cord levels, the spinal

338 ●●● CHAPTER 17

tract overlaps the **posterolateral tract** (eponym: fasciculus of Lissauer), its homolog, transmitting similar sensory input, but from the **body**.

Axons of the second-order neurons emerging from the spinal nucleus form the **anterior trigeminothalamic tract**. This tract is analogous to the spinothalamic tract that transmits pain and temperature sensations from the **body**.

The **anterior trigeminothalamic tract** consists mainly of crossed nerve fibers from the ventrolateral division of the principal sensory and spinal nuclei of the trigeminal. This tract relays mechanoreceptor input for *discriminatory tactile* and *pressure sense* (from the principal nucleus) as well as *sharp, well-localized pain* and *temperature* and *non-discriminatory (crude) touch* and *itch* sensation (from the spinal nucleus) to the contralateral **VPM nucleus** of the thalamus.

The **posterior trigeminothalamic tract** carries uncrossed nerve fibers from the posteromedial division of the principal sensory nucleus of the trigeminal, relaying *discriminatory tactile* and *pressure sense* information to the ipsilateral **VPM nucleus** of the thalamus.

The thalamus also receives indirect trigeminal nociceptive (dull, aching pain) input via the reticular formation (reticulothalamic projections).

The **mesencephalic tract** of the trigeminal is formed by the central and peripheral processes of those pseudounipolar neurons whose cell bodies are in the mesencephalic nucleus of the trigeminal.

Trigeminal pathways

Discriminative tactile and pressure sensation

Most of the central processes of pseudounipolar (first-order) neurons enter the pons, bifurcate into short ascending fibers, which synapse in the **principal sensory nucleus**, and into long descending fibers, which terminate and synapse mainly in the **oral subnucleus** and less frequently in the **interpolar subnucleus** of the spinal nucleus of the trigeminal. Some second-order fibers from the principal sensory nucleus cross the midline and join the anterior trigeminal lemniscus to ascend and terminate in the contralateral VPM nucleus of the thalamus. Other second-order fibers from the principal sensory nucleus do not cross; instead, they form the **posterior trigeminal lemniscus** and then ascend and terminate in the ipsilateral VPM nucleus of the thalamus. Descending fibers terminating in the oral subnucleus or interpolar subnucleus synapse with second-order neurons whose fibers cross the midline and ascend in the **anterior trigeminal lemniscus** to the contralateral VPM nucleus of the thalamus. The VPM nucleus of the thalamus houses third-order neurons that give rise to fibers relaying touch and pressure information to the primary somatosensory cortex in the postcentral gyrus of the parietal cortex. The lateral surface of the postcentral gyrus receives a disproportionately large number of projections from the thalamus relaying sensory information from the orofacial region, especially from the perioral regions, which have the highest sensory receptor density.

It should be noted that (1) the central processes of some first-order neurons terminate in the principal sensory nucleus (without bifurcating), and (2) nearly half of the sensory fibers in the trigeminal nerve are Aβ myelinated discriminatory touch fibers.

Nociceptive, thermal, and itch sensation

The caudal subnucleus is involved in the transmission of nociceptive, thermal, and itch sensations from orofacial structures

The remaining sensory fibers in the trigeminal nerve are very similar to the Aδ and C nociceptive and temperature fibers present in the spinal nerves. Aδ fibers are thought to relay sharp pain sensation generated by dental hypersensitivity, whereas C fibers are believed to relay dull aching pain generated from inflammation of the dental pulp. The anterior two-thirds of the cranial meninges receive the peripheral processes of sensory neurons whose cell bodies reside in the trigeminal ganglion. Their central processes descend in the spinal tract of the trigeminal to terminate in the caudal third of the spinal nucleus. Inflammation of the sensory pain fibers may produce pain associated with migraine headaches. As the central processes of pseudounipolar neurons enter the pons, they descend in the **spinal tract of the trigeminal** and most of them synapse in the **caudal subnucleus** of the spinal nucleus of the trigeminal. Nociceptive sensory input relayed in the caudal subnucleus is modified, filtered, and integrated *before* its transmission to higher brain centers.

Interneurons located in the caudal subnucleus project superiorly to the oral and interpolar subnuclei of the spinal nucleus and to the principal sensory nucleus of the trigeminal, where they modulate the synaptic activity and relay of sensory input from all of these nuclei to higher brain centers. Other interneurons residing in the oral and interpolar subnuclei project to the caudal subnucleus where they are believed to influence its neural activity.

Most of the second-order fibers from the caudal subnucleus cross the midline and join the contralateral anterior trigeminal lemniscus, whereas the remainder join the ipsilateral anterior trigeminal lemniscus. All of these second-order fibers ascend to the **VPM nucleus of the thalamus** where they synapse with third-order neurons in that nucleus. The fibers of third-order neurons ascend in the posterior limb of the internal capsule to relay somatosensory information from the trigeminal system to the **postcentral gyrus** of the somatosensory cortex for conscious awareness and further processing.

Electrophysiological observations have indicated that electrical stimulation of the midbrain periaqueductal gray matter, of the medullary raphe nuclei, or of the reticular nuclei, has an inhibitory effect on the nociceptive neurons of the caudal subnucleus.

Substance P, a neuropeptide in the axon terminals of small-diameter first-order neurons, located in the caudal subnucleus, has been associated with the transmission of nociceptive impulses. Opiate receptors have also been found in the caudal subnucleus, which can be blocked by opiate antagonists. These findings indicate that there may be an endogenous opiate analgesic system that could modulate the transmission of nociceptive input from the caudal subnucleus to higher brain centers.

Motor pathway

Fibers of the motor root of the trigeminal nerve innervate the muscles of mastication

Motor neurons housed in the **motor nucleus of the trigeminal** give rise to fibers which, upon exiting the pons, form the **motor root** of the trigeminal nerve (see Fig. 17.10). This short root joins the sensory fibers of the mandibular division of the trigeminal nerve outside the skull, in the infratemporal fossa. Motor fibers are distributed peripherally via the motor branches of the mandibular division, providing motor innervation to the muscles of mastication (temporalis, masseter, medial pterygoid, and lateral pterygoid muscles) as well as to the following muscles: the mylohyoid, anterior belly of the digastric, tensor tympani, and tensor veli palatini.

Mesencephalic neural connections

Pseudounipolar neurons of the mesencephalic nucleus transmit general proprioception input to the sensory and motor nuclei of the trigeminal, and to the reticular formation and to the cerebellum

The peripheral processes of the pseudounipolar neurons housed in the **mesencephalic nucleus of the trigeminal** accompany the motor root of the trigeminal as they both exit the pons. These peripheral processes follow (1) the motor branches of the mandibular division to the muscle spindles of the muscles of

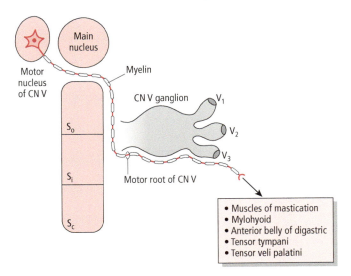

Figure 17.10 • Branchiomotor innervation of the trigeminal nerve. The motor nucleus of the trigeminal nerve contains the motor neurons whose axons assemble to form the motor root of the trigeminal nerve. The motor root exits the pons and joins the mandibular division of the trigeminal nerve and distributes to the muscles of mastication, the mylohyoid, the anterior belly of the digastric, the tensor tympani, and the tensor veli palatini muscles to provide them with motor innervation. S_c, caudal subnucleus; S_i, interpolar subnucleus; S_o, oral subnucleus; V_1, ophthalmic division of the trigeminal nerve; V_2, maxillary division of the trigeminal nerve; V_3, mandibular division of the trigeminal nerve.

mastication; (2) the dental branches of the maxillary and mandibular divisions to the sensory receptors of the periodontal ligaments of the maxillary and mandibular teeth, respectively. The central processes of these pseudounipolar neurons that transmit general proprioceptive input from all the muscles and the periodontal ligaments synapse in the **principal sensory nucleus**, in the rostral subnuclei of the **spinal nucleus,** and in the **motor nucleus** of the trigeminal. Additionally, these central processes also terminate in the reticular formation to mediate reflex responses and in the cerebellum which functions to coordinate movement of the mandible during chewing. The principal sensory and rostral spinal nuclei give rise to axons that ascend in the anterior trigeminothalamic tract to terminate in the contralateral VPM nucleus of the thalamus as part of the **mandibular proprioception pathway**.

The **mandibular proprioception pathway** relays proprioceptive input from the mandible to the primary somatosensory cortex where position sense of the mandible enters conscious awareness. In contrast, the masseteric "**jaw jerk**" **reflex** (see below) is a protective reflex elicited by muscle stretch (of the temporalis and masseter muscles) and causes the reflex muscle contraction (of the same muscles) to compensate for the muscle stretch. This reflex is limited/restricted to the brainstem and does not reach the somatosensory cortex or conscious awareness.

Masseteric ("Jaw jerk") reflex

Both the afferent and efferent limbs of the masseteric reflex are formed by the branches of the trigeminal nerve

The **masseteric reflex** is a monosynaptic, myotactic (G., "muscle stretch") reflex for the masseter and temporalis muscles. Gently tapping a rubber hammer on the operator's finger placed on the patient's mental protuberance (chin) causes the intrafusal muscle fibers within the muscle spindles of the (relaxed) masseter and temporalis muscles to stretch, which stimulates the sensory nerve fibers innervating them. The cell bodies of these sensory pseudounipolar neurons are located in the **mesencephalic nucleus of the trigeminal** (Fig. 17.11). The peripheral processes of these pseudounipolar neurons terminate in the muscle spindles (and are carried by branches of the trigeminal mandibular division) form the *afferent limb* of the reflex arc. The central processes of the mesencephalic pseudounipolar neurons usually branch in the vicinity of the motor nucleus of the trigeminal to send collaterals that terminate in the motor nucleus. The central processes of these neurons synapse not only in the motor nucleus bilaterally, but also in the principal sensory nucleus and the reticular formation as well. The *efferent limb* of this reflex arc is formed by the motor neuron fibers traveling to the masseter and temporalis muscles (bilaterally, via motor branches of the trigeminal mandibular division) to cause them to contract and compensate for the stretch. A bilateral lesion that damages the upper motor neurons that project to the motor nucleus of the trigeminal nerve results in hyperreflexia of the muscles of mastication which is exhibited as a hyperactive masseteric reflex.

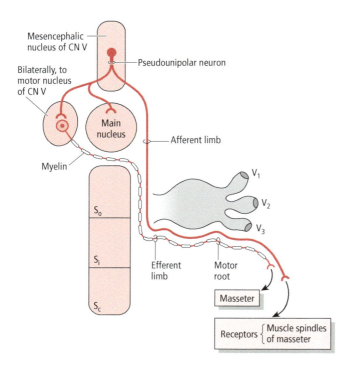

Figure 17.11 • The masseteric reflex. The mesencephalic nucleus of the trigeminal nerve contains the nerve cell bodies of pseudounipolar neurons whose peripheral processes terminate in the muscle spindles of the masseter muscle. Sensory information (about muscle stretch) is carried by the central processes of these neurons to the ipsilateral principal sensory and bilaterally to the motor nucleus of the trigeminal nerve. The motor neurons innervating the masseter muscle cause its contraction. S_c, caudal subnucleus; S_i, interpolar subnucleus; S_o, oral subnucleus; V_1, ophthalmic division of the trigeminal nerve; V_2, maxillary division of the trigeminal nerve; V_3, mandibular division of the trigeminal nerve.

CLINICAL CONSIDERATIONS

Skull fractures involving the foramen ovale may cause a unilateral lesion of the motor fibers to the muscles of mastication, which will result in **flaccid paralysis** or **paresis** with subsequent muscle **atrophy** of the ipsilateral muscles of mastication. This becomes apparent upon muscle palpation when the patient is asked to clench their teeth. Normally, the contraction of one lateral pterygoid muscle swings the mandible to the contralateral (opposite) side. When both lateral pterygoid muscles contract simultaneously, the mandible protrudes. With a lesion that damages the motor root of the trigeminal nerve, when depressing the mandible it deviates *toward* the affected side (weak side) primarily due to the unopposed action of the lateral pterygoid muscle of the unaffected side. This impairs chewing on the lesion side due to muscle paralysis.

Damage to the infratemporal region may also lesion fibers in the deep temporal, buccal, inferior alveolar, lingual and auriculotemporal nerves. A lesion of fibers innervating the tensor tympani muscle results in **hyperacusis** (sounds being heard abnormally loud) and impaired hearing on the ipsilateral side due to flaccidity of the normally tense portion of the tympanic membrane.

Damage to the GSA fibers of the mandibular division will result in loss of sensation from the areas supplied by the branches of this division. Although the trigeminal nerve has an extensive distribution in the head, there is minimal overlapping of the areas innervated by its three divisions, especially in the central region of the face. Lesions in the peripheral branches of the trigeminal nerve can be located by testing for sensory deficits in the areas that are innervated by each of the three trigeminal divisions. If a lesion is located distal to the joining of the autonomic fibers that hitch-hike with the trigeminal branches to the lacrimal, nasal, or salivary glands, then both sensory and autonomic innervation are interrupted.

Herpes zoster (a condition known as shingles) is caused by re-activation of the varicella zoster virus, which has remained latent in the trigeminal ganglion since the initial varicella infection and is characterized by unilateral painful maculopapular vesicles erupting in a single dermatopic area. Of the three trigeminal dermatomes, herpes zoster ophthalmicus is the most commonly affected. Early antiviral treatment can reduce the risk of complications and reduce the duration of post-herpetic neuralgia. Damage to the sensory fibers innervating the cornea (via the ophthalmic division) results in a loss of the corneal reflex when the ipsilateral eye is stimulated (afferent limb damage of the corneal reflex).

Trigeminal neuralgia (trigeminal nerve pain, tic douloureux)

A common clinical concern regarding the trigeminal nerve is **trigeminal neuralgia**. This condition results from idiopathic etiology (unknown cause) and is manifested as intense, sudden onset, and recurrent unilateral pain in the distribution of one of the three divisions of the trigeminal nerve, most commonly the maxillary division. There may be a trigger zone in the distribution of the affected trigeminal division, and if it is stimulated it may trigger an attack that usually lasts for less than a minute. This condition may be treated pharmacologically or surgically. Surgical treatment includes sectioning of the affected trigeminal division as it emerges from the trigeminal ganglion or producing a lesion in the trigeminal ganglion. Although these procedures may alleviate the excruciating pain experienced by patients, they also abolish tactile sensations from the affected area. Sectioning of the descending spinal trigeminal tract proximal to its termination in the caudal subnucleus selectively obliterates the afferents relaying nociception but spares the fibers relaying tactile sensation from the orofacial region.

Note that the clinical case at the beginning of the chapter refers to a patient suffering from intermittent excruciating unilateral pain in the lower half of the left side of their face.

1 Which cranial nerve provides sensory innervation to the lower half of the face?

2 Pain sensation from the lower half of the face is relayed to the brainstem by sensory neurons whose cell bodies are located in which ganglion?

3 In which brainstem nucleus is pain sensation from the lower half of the face relayed to?

4 Name the thalamic nucleus where pain sensation from the lower half of the face is relayed to.

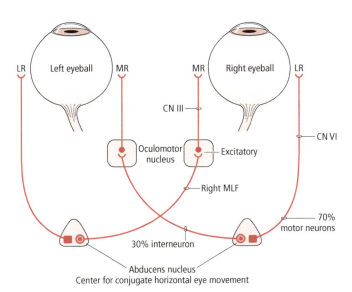

Figure 17.12 ● The connections of the abducens nucleus with the oculomotor nucleus. Note that the abducens nucleus is the center for conjugate horizontal eye movement. It contains two populations of neurons: (1) lower motor neurons whose axons form the abducens nerve that innervates the lateral rectus muscle (LR); and (2) interneurons whose axons cross the midline and join the contralateral MLF to synapse in the oculomotor nucleus with the motor neurons that innervate the medial rectus muscle (MR).

ABDUCENS NERVE (CN VI)

The abducens nerve innervates only the lateral rectus muscle of the eye

The **abducens nerve** supplies motor innervation to the lateral rectus muscle, which abducts the eye (a common mnemonic is LR₆). As the abducens nerve root fascicles exit the **abducens nucleus** they course anterolaterally in the caudal pontine tegmentum to exit the brainstem at the pontomedullary junction. The nerve fibers then pass through the space between the **posterior inferior cerebellar artery** (**PICA**) and the **anterior inferior cerebellar artery** (**AICA**) to enter and ascend in the pontine subarachnoid cistern, next to the **basilar artery**. The nerve fibers continue anteriorly, traversing the cavernous sinus where they lie next to the **internal carotid artery**, and upon leaving the sinus pass via the superior orbital fissure into the orbital cavity, then they pass through the common tendinous ring, to reach and innervate the ipsilateral lateral rectus muscle.

Normally, both eyes move together in unison regardless of the direction which is called conjugate gaze. This is achieved by precise supranuclear coordinated action of all the extraocular muscles of both eyes. The oculomotor, trochlear, and abducens nuclei are interconnected and are controlled by higher brain centers of the cerebral cortex (frontal and parietal motor eye fields) as well as by the brainstem (vestibular nuclei, MLF, and the reticular formation). During horizontal gaze, when looking to one side, the lateral rectus muscle of one side and the medial rectus muscle of the contralateral side contract simultaneously.

Abducens nucleus

The abducens nucleus mediates conjugate horizontal movement of the eyes

The **abducens nucleus** and the internal genu of the facial nerve form an elevation, known as the *facial colliculus* (L., "little hill") in the floor of the fourth ventricle. A group of root fascicles emerging from the abducens nucleus belong to **GSE nerve cell bodies**. The axons course anteriorly in the pontine tegmentum to exit in the anterior aspect of the brainstem at the pontomedullary junction.

The abducens nucleus contains two different populations of neurons (Fig. 17.12). One group (which constitutes approximately 70% of the neuron population) consists of the GSE motoneurons, whose axons form the abducens nerve and project to the ipsilateral lateral rectus muscle. The second group (about 30% of the neurons) consists of internuclear neurons. Their axons emerge from the nucleus, decussate immediately, and join the contralateral MLF to project to the contralateral oculomotor nucleus. There the internuclear neuron terminals synapse with motor neurons that project to and innervate the medial rectus muscle. The MLF interconnects the abducens, trochlear, and oculomotor nuclei so that the two eyes move in unison. The abducens nucleus is the **center for conjugate horizontal eye movement**.

When higher brain centers stimulate the abducens nucleus the following occur simultaneously:

1 Stimulation of the GSE motor neurons of the abducens nucleus causes the *ipsilateral lateral rectus muscle* to contract, causing the eye to *abduct*.
2 Stimulation of the internuclear neurons of the same abducens nucleus that project, via the contralateral MLF, to the contralateral oculomotor nucleus. Here they form excitatory synapses only with the motor neurons projecting to the *contralateral medial rectus muscle* causing it to contract, causing the opposite eye to *adduct*.

Thus, a horizontal gaze to one side is evoked ipsilateral to the stimulated abducens nucleus, which involves the coordinated and simultaneous contraction of the **lateral rectus muscle** *of the ipsilateral eye* and the **medial rectus muscle** *of the contralateral eye*. When the right abducens nucleus is stimulated, the gaze will be to the right. The simultaneous movement of the two eyes to one side (to the right in this case), so they both look in the same direction (to the right) and at the same object, is referred to as *conjugate horizontal eye movement* (eyes move in unison, horizontally). Note that the abducens nucleus (lateral) and the medial longitudinal fasciculus (medial) are next to one another and both are adjacent to the midline. Thus, internuclear neuron axons leaving the *right* abducens nucleus cross immediately to enter the contralateral (*left*) MLF. The left MLF carries the axons to the left oculomotor nucleus. It is the left MLF that connects the right abducens nucleus to the left oculomotor nucleus.

GSA input from the lateral rectus muscle is transmitted centrally to the trigeminal nuclear complex via the processes of pseudounipolar neurons whose cell bodies are believed to reside in the trigeminal ganglion.

CLINICAL CONSIDERATIONS

Abducens nerve lesion
A lesion in the abducens nerve causes paralysis of the ipsilateral lateral rectus muscle, resulting in medial strabismus and horizontal diplopia

A lesion of the abducens nerve (GSE, motor fibers) results in **paralysis** of the ipsilateral lateral rectus muscle that normally *abducts* the eye. Since the paralyzed lateral rectus muscle becomes flaccid and lacks muscle tone, the eye will deviate medially as a result of the unopposed action of its antagonist, the medial rectus muscle (Fig. 17.13). The medial deviation of the affected eye, the new resting position of the eye, is referred to as **medial strabismus (internal strabismus, esotropia)**. The individual can turn the ipsilateral (affected) eye from its medial position to the center (to look straight ahead), but not laterally beyond it. Since the normal eye looks straight ahead and the affected eye is medially deviated (looks toward the nose), the eyes become misaligned, causing double vision, and **horizontal diplopia** (double vision; i.e., a single object is perceived as two separate objects next to each other). The diplopia is greatest when attempting to gaze toward the side of the lesion, and it is minimized by looking toward the unaffected side since the visual axes become parallel. The individual realizes that the diplopia is minimized by turning their head slightly so that the chin is pointing toward the side of the lesion. This would keep the affected eye adducted, and abduct the normal eye, aligning the eyes (eyes become parallel), and minimizing the diplopia.

A bilateral abducens nerve lesion results in the individual becoming "cross-eyed."

When we fix our gaze on an object and then move our head either in the horizontal (or vertical) plane, our eyes reflexively roll away from the direction of the head movement to keep the image (of the object) on the retina. This is a function of the vestibular nuclear projections to the motor nuclei innervating the extraocular muscles, coordinating head and eye movements.

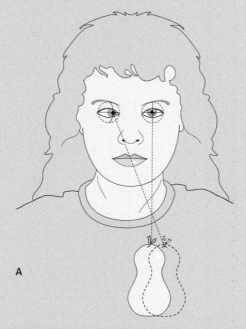
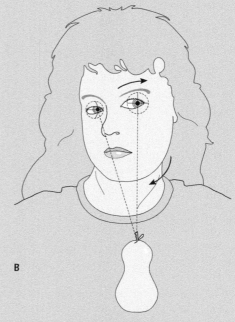

Figure 17.13 ● (A) Medial strabismus of the right eye due to paralysis of the lateral rectus muscle, resulting in diplopia (double vision). (B) To minimize the diplopia, the individual turns her head toward the side of the lesion, which abducts the normal eye.

CLINICAL CONSIDERATIONS (continued)

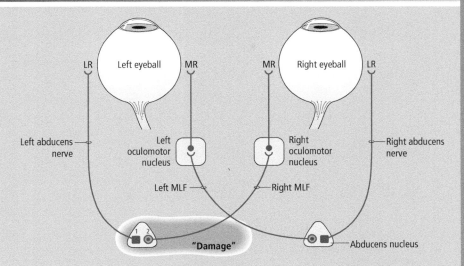

Figure 17.14 ● A lesion of the left abducens nucleus will damage: (1) the lower motor neurons of the abducens nerve, paralyzing the left lateral rectus muscle (LR); and (2) the interneurons that synapse with the lower motor neurons of the oculomotor nucleus that innervate the right medial rectus muscle (MR). The affected individual is unable to gaze to the side of the lesion (left) during conjugate horizontal eye movement. MLF, medial longitudinal fasciculus.

Abducens nucleus lesion

A lesion involving the abducens nucleus or PPRF results in medial strabismus, horizontal diplopia, and ipsilateral lateral gaze paralysis

A lesion involving the abducens nucleus (Fig. 17.14) or the pontine paramedian reticular formation (PPRF) results in the same deficiency as a lesion to the abducens nerve, with the addition of the inability to turn the opposite eye medially as the individual attempts to gaze toward the side of the lesion. This condition, referred to as **lateral gaze paralysis**, occurs because the damaged abducens nucleus no longer provides excitatory input to the opposite oculomotor nucleus neurons that innervate the contralateral medial rectus muscle.

Unilateral medial longitudinal fasciculus (MLF) lesion: internuclear ophthalmoplegia (INO)

A lesion to one MLF results in internuclear ophthalmoplegia (INO)

If the oculomotor, trochlear, and abducens nerves and their nuclei are intact, but there is a **unilateral MLF lesion**, most eye movements are possible. Normally, the eyes move as a pair. However, when there is a lesion to one MLF, the abducens and oculomotor nuclei which are connected by the MLF, become "disconnected," and conjugate horizontal ocular movement will not occur to one side (Fig. 17.15A).

When there is a lesion of the right MLF, and the individual attempts to gaze to the right, the lesion is not apparent, since both eyes can move simultaneously to the right. However, when attempting to gaze to the left, the right eye cannot move inward (medially beyond the midline) but the left eye, which should move outward (laterally) in this lateral gaze, abducts with nystagmus. This deficit is referred to as **internuclear ophthalmoplegia (INO)**. INO is a horizontal eye movement disorder produced by a lesion to the MLF. The function of the MLF comes into play only during conjugate horizontal eye movement. An individual who has a lesion to the MLF cannot gaze away from the side of the lesion. If you ask this same individual to look at a near object (for example a pencil) placed directly in front of their nose, which necessitates that both eyes adduct (converge), they can do so. This indicates that: (1) both oculomotor nerves (which innervate the medial recti) are intact; (2) the cortical projections from the *frontal eye field* and *parietal motor eye fields* to the *midbrain reticular formation* centers that control eye movements are intact; and (3) the projections from the midbrain reticular formation to the motor nucleus of the oculomotor nerve are intact. Therefore, with a lesion to the right MLF, the right eye will have an adduction deficit (only during *conjugate horizontal eye movement*) to the left (when gazing away from the side of the lesion), but *not during convergence of the eyes*. The reason for this is that during conjugate horizontal eye movement the motor neurons that innervate the medial rectus muscle are stimulated by the MLF, but during convergence of the eyes, the motor neurons to the medial rectus are stimulated by the frontal and parietal motor eye fields via the midbrain reticular formation.

"One-and-a-half"

"One-and-a-half" is a rare condition resulting from a lesion in the vicinity of the abducens nucleus, involving the ipsilateral abducens nucleus and decussating MLF fibers arising from the contralateral abducens nucleus

A rare condition referred to as "one-and-a-half" results following a lesion in the vicinity of the abducens nucleus, which involves the entire ipsilateral abducens nucleus as well as the decussating MLF fibers arising from the contralateral abducens nucleus (Fig. 17.16). If a **lesion** is present in the vicinity of the **left** abducens nucleus the following happens:

1 The **GSE motor neurons**, whose axons form the left abducens nerve innervating the left lateral rectus, are damaged. Therefore, the left lateral rectus muscle is paralyzed.
2 The **internuclear neurons** housed in the left abducens nucleus are also damaged. Their crossing fibers (coursing in the right MLF) do not, therefore, form excitatory synapses with the motor neurons of the contralateral oculomotor nucleus that innervate the right medial rectus muscle.
3 The crossing fibers of the internuclear neurons arising from the contralateral (right) abducens nucleus are also damaged; thus, they do not form excitatory synapses with the motor neurons of the left oculomotor nucleus that innervate the left medial rectus.

CLINICAL CONSIDERATIONS (continued)

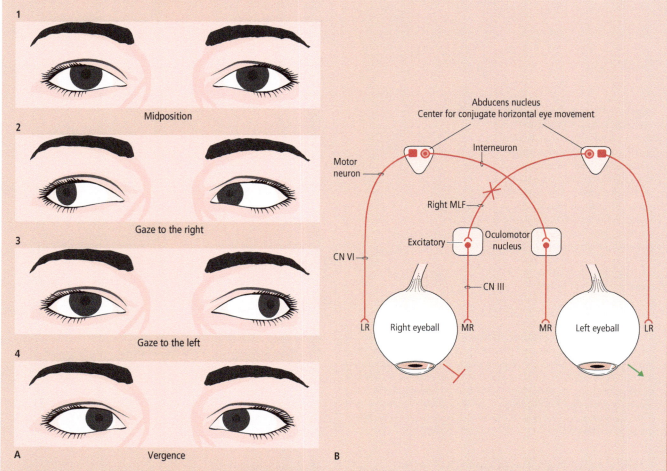

Figure 17.15 ● (A) Lesion to the right MLF. (1) Normal primary position. (2) Normal gaze to the right. (3) Adduction deficit of right eye, due to lesion in ipsilateral MLF. Abduction of left eye with nystagmus. (4) Vergence of the eyes is normal. (B). Lesion to right MLF. When there is a lesion of the right MLF, and the individual attempts to gaze to the right, the symptoms are not apparent, since both eyes can move simultaneously to the right. However, when attempting to gaze to the left, the right eye cannot move inward (medially beyond the midline), but the left eye which should move outward, (laterally), in this lateral gaze, abducts with nystagmus.

Therefore, when attempting to gaze to the left, the left eye will not abduct, and the right eye will not adduct during conjugate horizontal gaze to the left. This is referred to as **ipsilateral lateral (horizontal) gaze palsy** when looking to the side of the lesion. When attempting to gaze to the right, the right eye abducts with nystagmus, whereas the left eye will not be able to adduct during conjugate horizontal gaze to the right. Thus "**one-and-a-half**" is characterized by a lesion involving a combination of "one" abducens nucleus and a "half," the axons of the internuclear neurons arising from the opposite abducens nucleus, ascending in the MLF. That is, the eye ipsilateral to the lesion has no horizontal movement and is medially deviated. The opposite eye can only abduct with nystagmus. It is important to note that the innervation to all the extraocular muscles of both eyes is intact, except one – the left lateral rectus. If you ask this individual to look at a near object placed directly in front of them, both eyes will converge, since both medial recti and their innervation (branches of the oculomotor nerve) are intact.

Thus, this type of lesion becomes apparent only during conjugate horizontal eye movement.

Syndrome of trochlear nerve palsy with internuclear ophthalmoplegia

The trochlear nucleus is located immediately posterior to the MLF, in the caudal midbrain. A lesion in the vicinity of the trochlear nucleus may also affect the nearby MLF. The proximity and combined damage of the two structures make them susceptible to a **syndrome of trochlear nerve palsy with internuclear ophthalmoplegia** (**INO**) (Fig 17.17). The *ascending MLF* contains fibers that interconnect the abducens, trochlear, and oculomotor nuclei as well as vestibular fibers that project to these nuclei. The MLF contains the axons of internuclear neurons (interneurons) whose cell bodies reside in the abducens nucleus in the caudal pons.

CRANIAL NERVES • • • 345

CLINICAL CONSIDERATIONS (continued)

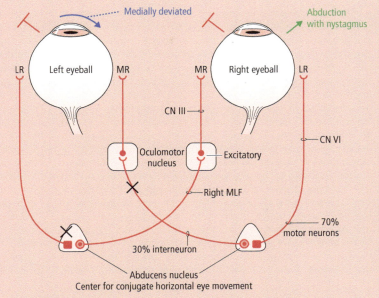

Figure 17.16 • "One-and-a-half" syndrome: Results following a lesion in the vicinity of the abducens nucleus, which involves the entire ipsilateral abducens nucleus as well as the decussating MLF fibers arising from the contralateral abducens nucleus. The individual is unable to gaze to the left and when gazing to the right the right eye abducts with nystagmus.

Figure 17.17 • Trochlear nerve palsy with internuclear ophthalmoplegia. If the (right) trochlear nucleus and the (right) MLF are damaged at the level of the caudal midbrain (when they are next to each other), the superior oblique palsy will show up in the left eye (because the axons leaving the right trochlear nucleus will cross in the midbrain tectum before the nerve emerges from the brainstem as the left trochlear nerve), and the INO will become evident when the individual attempts to gaze to the left (when the right eye is supposed to adduct, but can't, due to the disconnection of the left abducens nucleus and the right oculomotor nucleus).

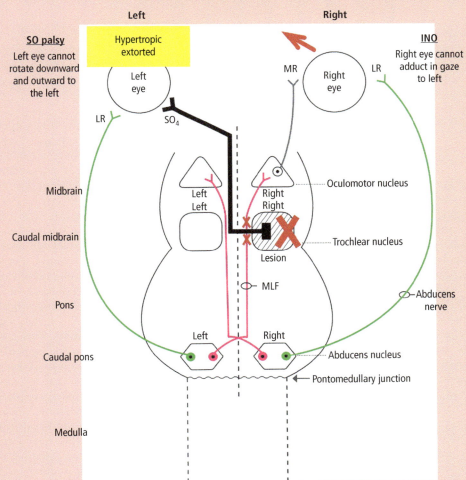

346 ● ● ● **CHAPTER 17**

CLINICAL CONSIDERATIONS (*continued*)

The abducens nucleus receives cortical input from the contralateral frontal eye field (eponym: Brodmann area 8) relaying motor messages that stimulate the neurons of this nucleus. The abducens nucleus contains two distinct populations of neurons: *motor neurons* and *interneurons*. The **motor neurons** give rise to axons that exit the brainstem to form the abducens nerve which courses to the orbit where it provides motor innervation to a single extraocular muscle, the lateral rectus. This muscle mediates abduction (lateral rotation) of the eye. The interneurons connect the abducens nucleus to the contralateral oculomotor nucleus as follows: The **interneurons** give rise to axons that leave the abducens nucleus and immediately cross the midline to join the contralateral (ascending) MLF to ascend to the rostral midbrain where they terminate in the contralateral oculomotor nucleus and synapse specifically with the motor neurons that innervate the medial rectus muscle. The abducens nucleus is the **center for conjugate horizontal eye movement**; thus, simultaneous stimulation of both populations of neurons of this nucleus (via the descending axons of neurons residing in Brodmann area 8) normally mediates (1) abduction of the ipsilateral eye (via the abducens nerve) and (2) adduction of the contralateral eye (via the interneuron axons running in the MLF, and the oculomotor nucleus) to execute lateral gaze ipsilateral to the side

of the stimulated abducens nucleus. As mentioned previously, a lesion in the caudal midbrain, damaging the trochlear nucleus, will most likely also damage the ipsilateral MLF immediately adjacent to it, containing the ascending axons of the interneurons whose cell bodies are housed in the abducens nucleus and whose axons are in route to ascend further up to the rostral midbrain to terminate in the oculomotor nucleus.

The MLF located next to the trochlear nucleus at the level of the caudal midbrain, contains internuclear neuron axons that already crossed down below in the brainstem, at the level of the abducens nucleus which resides in the caudal pons. Thus, if the right MLF (which projects to the right oculomotor nucleus) is damaged at the level of the caudal midbrain, the right medial rectus will not contract (adduct) during conjugate horizontal eye movement in gaze to the left.

If the (right) trochlear nucleus and the (right) MLF are damaged at the level of the caudal midbrain (when they are next to each other), the superior oblique palsy will show up in the left eye (because the axons leaving the right trochlear nucleus will cross in the midbrain tectum before the nerve emerges from the brainstem as the left trochlear nerve), and the INO will become evident when the individual attempts to gaze to the left (when the right eye is supposed to adduct, but can't, due to the disconnection of the left abducens nucleus and the right oculomotor nucleus).

FACIAL NERVE (CN VII)

The facial nerve provides motor innervation to the muscles of facial expression, taste sensation to the anterior two-thirds of the tongue, and parasympathetic innervation to the lacrimal, nasal, sublingual, and submandibular glands

The **facial nerve** (Fig. 17.18) provides **SVE** innervation to the muscles of facial expression, as well as to the platysma, posterior belly of the digastric, the stylohyoid, auricularis, occipitofrontalis, and the stapedius muscles. Nervus intermedius (CN VII) also transmits **taste** sensation from the anterior two-thirds of the tongue, as well as **parasympathetic** (GVE, secretomotor) innervation to the lacrimal, nasal, submandibular, and sublingual glands. Additionally, it provides **general sensation** to the back of the ear, as well as **visceral sensation** from the nasal cavity and the soft palate.

The facial nerve consists of two parts: the **facial nerve proper** and the **nervus intermedius**. The facial nerve proper is the motor root of the facial nerve consisting of the axons of **SVE neurons** whose cell bodies reside in the **facial nucleus**. This nucleus contains subnuclei, each supplying specific muscles or groups of muscles. The nervus intermedius is sometimes referred to as the "sensory root," which is a misnomer since in addition to sensory fibers it also carries parasympathetic fibers. The nervus intermedius consists of the axons of the **GVE parasympathetic neurons**, whose cell bodies reside in the superior salivatory nucleus. It also contains the central processes of first-order, sensory pseudounipolar neurons whose cell bodies are housed in the **geniculate** (L., "bent like a knee") **ganglion**, the only sensory ganglion of the facial nerve. Some of these pseudounipolar neurons

transmit SVA (taste) sensation from the anterior two-thirds of the tongue, others convey GSA sensation from the area posterior to the ear, whereas others carry GVA sensation from the nasal cavity and soft palate.

Both nerve roots (motor root and nervus intermedius) emerge from the brainstem at the cerebellopontine angle. Near their exit from the brainstem, the two roots of the facial nerve accompany one another to the internal acoustic meatus of the petrous portion of the temporal bone and proceed to the facial canal where the nervus intermedius presents a swelling – the **geniculate ganglion**.

The facial nerve gives rise to three of its branches in the facial canal: the *greater petrosal nerve*, the *nerve to the stapedius muscle* (which innervates the stapedius muscle in the middle ear cavity), and the *chorda tympani nerve*. The facial nerve exits the facial canal via the **stylomastoid foramen**, and then immediately gives rise to the *posterior auricular nerve* which divides into the motor branches to the posterior auricular muscle and to the occipitofrontalis muscles. The main trunk of the facial nerve then courses anteromedially to enter the parotid bed where it gives rise to numerous motor branches, which radiate from within the substance of the gland to innervate their respective muscles (muscles of facial expression, platysma, posterior belly of the digastric, and stylohyoid muscles). Unlike other skeletal muscles, the muscles of facial expression, arising from the pharyngeal arches, do not contain muscle spindles (special stretch receptors enclosing intrafusal fibers that receive motor and sensory innervation).

The **superior salivatory nucleus** houses GVE *preganglionic parasympathetic* nerve cell bodies (Figs. 17.18 and 17.19)

CRANIAL NERVES • • • 347

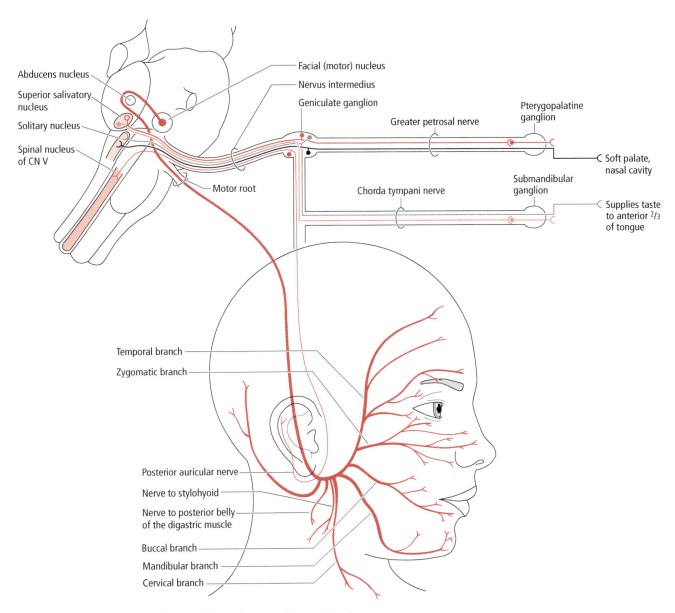

Figure 17.18 • The origin and distribution of the facial nerve and its major branches.

whose axons leave the brainstem via the **nervus intermedius**. These preganglionic fibers are distributed by the greater petrosal and chorda tympani nerves. The fibers in the **greater petrosal nerve** subsequently join the nerve of the pterygoid canal to enter the pterygopalatine fossa where they terminate and synapse in the **pterygopalatine ganglion**, one of the two parasympathetic ganglia of the facial nerve. Postganglionic parasympathetic fibers from this ganglion are distributed to the lacrimal gland and the glands of the nasal and oral cavities to provide them with secretomotor innervation. The **chorda tympani nerve** joins the lingual nerve, a branch of the mandibular division of the trigeminal nerve. The chorda tympani carries *preganglionic parasympathetic* fibers to the **submandibular ganglion** (the second parasympathetic ganglion of the facial nerve), where the fibers synapse with its postganglionic parasympathetic neurons. The postganglionic parasympathetic fibers from this ganglion course to the submandibular and sublingual glands providing them with secretomotor innervation.

The **geniculate ganglion** houses the cell bodies of the SVA neurons, which are responsible for the transmission of **taste** sensation from taste buds on the anterior two-thirds of the tongue (Fig. 17.20). The peripheral processes of these neurons run in the **chorda tympani** and reach the tongue via the lingual nerve of the mandibular division of the trigeminal nerve. The central processes of the SVA neurons enter the brainstem via the **nervus intermedius** to join the ipsilateral **solitary tract** and terminate in the rostral **nucleus of solitary**

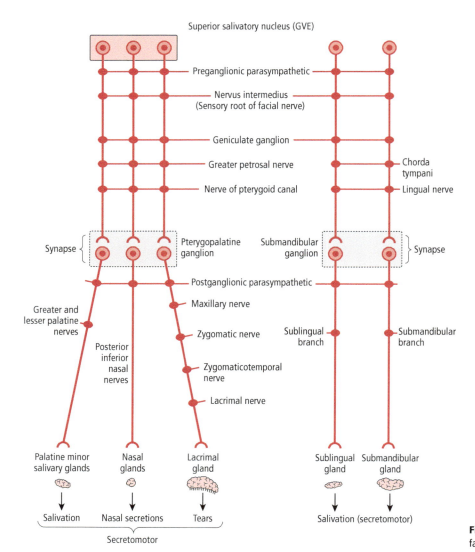

Figure 17.19 • Parasympathetic innervation of the facial nerve. GVE, general visceral efferent.

tract (**solitary nucleus**) in a somatotopic fashion. The rostral nucleus of solitary tract receives gustatory and orosensory primary afferent inputs (mainly from the caudal nucleus of the solitary tract and parabrachial nucleus of the pons) and contributes the most multipolar neurons to the ascending gustatory pathway.

Other pseudounipolar neurons of the geniculate ganglion mediate GVA sensation. Their peripheral processes run in the **greater petrosal nerve** and terminate in the nasal cavity and the soft palate. Their central processes course in the **nervus intermedius**, join the ipsilateral **solitary tract**, and terminate in the rostral **nucleus of the solitary tract**.

Still, other pseudounipolar neurons of the geniculate ganglion are responsible for pain, temperature, and touch sensation (GSA fibers) from the pinna and the external auditory meatus. The peripheral processes of these neurons terminate in the pinna and the external auditory meatus. Their central processes course in the **nervus intermedius** to join the **spinal tract of the trigeminal nerve** and terminate to synapse in the **spinal nucleus of the trigeminal nerve**.

VESTIBULOCOCHLEAR NERVE (CN VIII)

The vestibular division of CN VIII transmits information about position sense and balance, whereas its cochlear division mediates the sense of hearing; the vestibular nerve is the only cranial nerve to send the central processes of some of its first-order sensory neurons to synapse directly in the cerebellum

The **vestibulocochlear nerve** consists of two distinct and separate nerves enclosed within one connective tissue sheath, the **vestibular nerve** (concerned with *position sense* and *balance*) and the **cochlear nerve** (concerned with *hearing*). Both nerves transmit **SSA** information from specialized peripheral ciliated mechanoreceptors, known as "hair cells."

CRANIAL NERVES 349

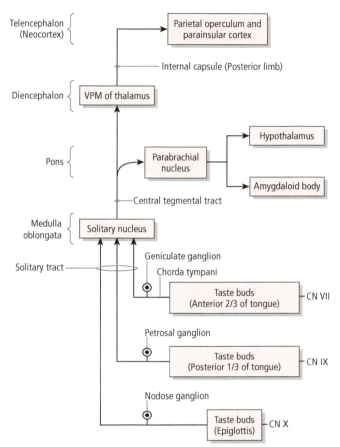

Figure 17.20 • The gustatory pathway. Taste sensation is transmitted by cranial nerves VII (from the anterior two-thirds of the tongue), IX (from the posterior one-third of the tongue), and X (from the epiglottis). The taste sensation is relayed via the solitary tract to the nucleus of the solitary tract (solitary nucleus) in a somatotopic fashion. CNVII fibers project to the rostral nucleus of the solitary tract, and CN IX and CN X fibers project to the caudal nucleus of the solitary tract. The central tegmental tract arising from the nucleus of the solitary tract projects to the parabrachial nucleus and the VPM nucleus of the thalamus, periventricular hypothalamus, and basolateral amygdaloid body. The VPM nucleus of the thalamus projects to the gustatory cortex residing in the parietal operculum and the parainsular cortex. Source: Adapted from Fix, JD (1995) *Neuroanatomy*/Wolters Kluwer Health.

The cell bodies of the **first-order sensory bipolar neurons** of the **vestibular nerve** reside within the **vestibular ganglion** (eponym: Scarpa's ganglion) (see Fig. 20.6). Their peripheral processes terminate in special receptors, the **cristae** in the ampullae of the semicircular ducts and the **maculae** of the utricle and saccule, all housed within the petrous portion of the temporal bone (see Figs. 20.2–20.4). The central processes of these neurons enter the brainstem to terminate not only in the vestibular nuclear complex, where they synapse with **second-order neurons** of the vestibular pathway but synapse also in the cerebellum (see Fig. 20.6). The vestibular nerve is unique since it is the only cranial nerve that sends the central processes of some of its first-order neurons to synapse *directly* in the cerebellum.

The cell bodies of the first-order sensory *bipolar* neurons of the **cochlear nerve** are housed within the **cochlear (spiral) ganglion** (see Fig. 19.3). Their peripheral processes terminate and synapse in the **spiral organ** (eponym: organ of Corti) that house special mechanoreceptors that transduce sound waves into electric impulses. The cochlear ganglion and the spiral organ lie within the cochlea, a snail shell-shaped structure of the internal ear, embedded within the petrous portion of the temporal bone (see Fig. 19.3). The central processes of these neurons accompany the vestibular nerve to synapse in the **cochlear nuclei** in the brainstem with second-order neurons of the auditory pathway (see Fig. 19.4).

GLOSSOPHARYNGEAL NERVE (CN IX)

The glossopharyngeal nerve provides parasympathetic innervation to the parotid gland

The **glossopharyngeal nerve**, one of the smallest cranial nerves, carries five functional components. These are (1) SVA (taste) from the posterior one-third of the tongue and the adjacent pharyngeal wall, (2) GVA sensation from the posterior one-third of the tongue, palatine tonsil, adjacent pharyngeal wall, tympanic cavity, and the carotid sinus (a baroreceptor or blood pressure receptor located near the bifurcation of the common carotid artery), (3) GSA sensation from the external ear, (4) SVE (branchiomotor) innervation to the stylopharyngeus muscle, and (5) GVE parasympathetic innervation to the parotid gland.

CLINICAL CONSIDERATIONS

A lesion to the facial nerve within the facial canal or near its exit from the stylomastoid foramen causes facial (Bell) palsy

Compression or damage of the facial nerve results in ipsilateral paralysis of the muscles of facial expression, the most commonly occurring muscle paralysis in the head. The functional deficits vary and depend on the specific location of the lesion along the course of the facial nerve to its targeted structures.

A unilateral lesion of the facial nerve near its root or in the facial canal (proximal to the geniculate ganglion) and before giving off any of its branches (thus damaging all of its fibers) results in the following conditions *ipsilateral* to the lesion: damage to the SVE fibers, results in a **flaccid paralysis** or **paresis** (impairment) of the muscles of facial expression, the platysma, stylohyoid, and posterior belly of the digastric muscles with subsequent muscle atrophy. The stapedius muscle will also be **paralyzed** and the individual will experience **hyperacusis** (hypersensitivity to sounds, perceiving sounds as abnormally loud). Usually, the stapedius muscle dampens vibrations of the ossicles by leashing stapes, but when it is paralyzed vibrations from the tympanic membrane are transmitted without dampening to the ossicles where stapes can pound on the oval window. This causes turbulence in the cochlear endolymph and is translated by mechanoreceptor inner hair cells as loud noise. Furthermore, damage to the SVA fibers relaying taste results in a **loss of taste** from the anterior two-thirds of the tongue. Damage of the GVE parasympathetic fibers causes **decreased salivary secretion**

CLINICAL CONSIDERATIONS (continued)

from the submandibular and sublingual glands. Since both parotid glands (innervated by a different cranial nerve) and the contralateral sublingual and submandibular glands remain functional, it is difficult to determine from salivary action alone whether there is an interruption of the parasympathetic innervation to the ipsilateral submandibular and sublingual glands. In addition, the efferent limb (contraction of ipsilateral orbicularis oculi) of the corneal blink reflex will be damaged.

Bell palsy may be idiopathic or result following trauma or viral infection of the facial nerve within the facial canal or near its exit from the stylomastoid foramen. This condition is characterized by paresis or paralysis of the muscles of facial expression ipsilateral to the lesion. A viral infection that affects the facial nerve's connective tissue coverings causes inflammation (neuritis) and edema. The swelling compresses the enclosed nerve fibers as they course through the facial canal, a narrow channel, in the petrous portion of the temporal bone and as they exit from the stylomastoid foramen. Compression of the nerve results in ischemia and compromised conduction of the facial nerve fibers.

Bell phenomenon is exhibited by individuals with Bell palsy. As the individual attempts to close the eyes, the eye on the affected side deviates up and out.

A unilateral lesion of the facial nerve proximal to the geniculate ganglion causes loss of tear formation by the ipsilateral lacrimal gland. A condition referred to as **"crocodile tear syndrome"** (lacrimation while eating) may result as follows. As the preganglionic parasympathetic ("salivation") fibers originating from the superior salivatory nucleus are regenerating, they may be unsuccessful at finding their way to their intended destination, the submandibular ganglion, and instead take a wrong route to terminate in the pterygopalatine ganglion. The fibers then establish inappropriate synaptic contacts with postganglionic ("lacrimation") neurons whose fibers project to the lacrimal gland. When the individual is eating, tears are produced by the lacrimal gland ipsilateral to the side of the lesion.

A unilateral lesion damages the facial nerve after it gives off its branches in the facial canal (that is if only its motor fibers are damaged) will result in paralysis of the muscles of facial expression, ipsilateral to the side of the lesion (Fig. 17.21)

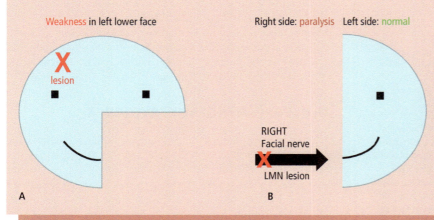

Figure 17.21 ● (A) A lesion that damages the right *upper* motor neuron corticonuclear fibers to the facial nuclei will result in facial weakness on the lower left side of the face. Also see Chapter 13, corticonuclear projections to the facial nucleus and deficits resulting following lesions of corticonuclear tract fibers to the facial nuclei and resulting facial deficits. (B) A lesion that damages the *lower* motor neurons (LMN) of the right facial nerve will result in facial paralysis on the right side of the face. Source: Adapted from Fix, JD (1995) *Neuroanatomy*/Wolters Kluwer Health

---

The glossopharyngeal nerve exits the brainstem as a group of rootlets posterior to the olive in the posterolateral sulcus. These rootlets immediately collect to form the main trunk of the glossopharyngeal nerve, which exits the cranial vault through the jugular foramen where it presents two swellings, its **superior** and **inferior** ganglia (Fig. 17.22) The superior ganglion contains GSA, and the inferior ganglion contains GVA and SVA, cell bodies of **first-order pseudounipolar** neurons.

The **inferior ganglion** of the glossopharyngeal nerve houses the cell bodies of the **SVA** (taste) neurons. Their peripheral processes course with the trunk of the glossopharyngeal nerve to the tongue where they supply the posterior one-third of the tongue and adjacent pharyngeal wall with taste sensation. The central processes of the SVA neurons pass into the brainstem via the glossopharyngeal nerve root, join the **solitary tract**, and terminate in the caudal **nucleus of the solitary tract** (Fig. 17.23A).

GVA first-order nerve cell bodies reside in the **inferior ganglion** of the glossopharyngeal nerve. Their peripheral processes terminate in the mucosa of the posterior one-third of the tongue, lingual tonsil and adjacent pharyngeal wall, tympanic cavity, and the auditory tube (Fig. 17.23B). Unilateral stimulation of the pharyngeal wall elicits a bilateral contraction of the pharyngeal muscles and soft palate (**gag reflex**). The **glossopharyngeal nerve** serves as the afferent (sensory) limb (GVA peripheral fibers whose cell bodies are housed in the inferior ganglion), whereas the **vagus nerve** provides the efferent (motor) limb of the reflex arc. The central processes of the afferent fibers enter the **solitary tract** and synapse in the **nucleus ambiguus**, which sends motor fibers via the vagus nerve to the muscles of the palate (except

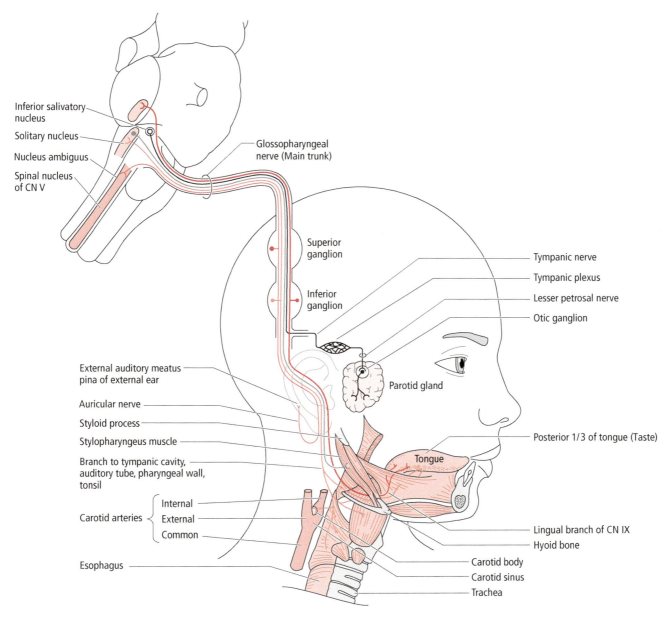

Figure 17.22 ● The origin and distribution of the glossopharyngeal nerve and its major branches.

tensor veli palatini) and pharynx. Interestingly, many individuals in the general population do not have a gag reflex.

Baroreceptor fibers terminate in the carotid sinus, which forms the afferent limb of the reflex arc that controls blood pressure. The central processes of the **GVA** neurons enter the brainstem via the glossopharyngeal nerve root, join the **solitary tract**, and terminate in the caudal nucleus of the solitary tract. The nucleus of the solitary tract relays sensory input to the reticular formation, the brainstem GVE (autonomic) motor nuclei, and the intermediolateral horn (containing preganglionic sympathetic neurons) of the spinal cord for reflex activity related to the control of arterial lumen diameter and, thus, blood pressure.

The glossopharyngeal nerve also provides **GSA** touch, pain, and temperature innervation to the external auditory meatus and to the internal surface of the tympanic membrane (touch, pain, and temperature sensations to the external surface of the tympanic membrane is served by the mandibular division of the trigeminal nerve and auricular branches of vagus nerve). The cell bodies of these sensory neurons are located in the **superior ganglion** of the glossopharyngeal nerve. The central processes of these neurons course in the glossopharyngeal nerve root, enter the brainstem, and join the **spinal tract of the trigeminal nerve** to terminate and synapse in the **spinal nucleus of the trigeminal nerve**. Clinical evidence supports that fibers transmitting nociceptive

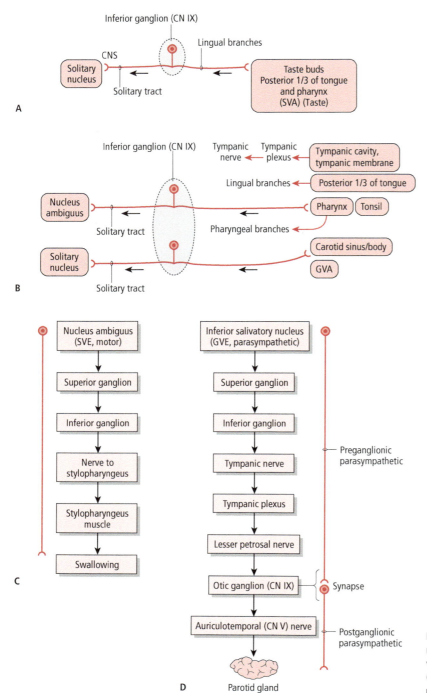

Figure 17.23 • Innervation by the glossopharyngeal nerve: (A) special visceral afferent (SVA; taste); (B) general visceral afferent (GVA); (C) special visceral efferent (SVE; skeletal motor); (D) general visceral efferent (GVE; parasympathetic); and (E) GSA fibers.

sensory input from the pharyngeal wall and posterior one-third of the tongue enter the brainstem and descend in the spinal tract of the trigeminal and terminate in the spinal nucleus of the trigeminal.

The **nucleus ambiguus** contains the **SVE** nerve cell bodies whose axons emerge from the brainstem along with rootlets of the glossopharyngeal nerve, and course with the trunk of the glossopharyngeal nerve (Fig. 17.23C). These axons then leave the glossopharyngeal nerve and become the nerve to the stylopharyngeus muscle, the only muscle innervated by Cranial Nerve IX.

The **inferior salivatory nucleus**, located in the medulla oblongata, houses the **GVE** cell bodies of *preganglionic parasympathetic* neurons whose axons exit the brainstem as part of the glossopharyngeal nerve (Fig. 17.23D). These fibers then leave the trunk of Cranial Nerve IX, join the tympanic nerve,

CLINICAL CONSIDERATIONS

A unilateral lesion to the glossopharyngeal nerve near its exit from the brainstem, damaging all of its fibers, will include the SVA fibers relaying taste sensation and will cause **ipsilateral loss of taste sensation** from the posterior one-third of the tongue. Damage to the GVE parasympathetic fibers will cause a **reduction in salivary secretion** of the ipsilateral parotid gland; and damage to the GVA fibers will result in **diminished visceral sensation** from the pharyngeal mucous membrane, **loss of the gag reflex** (due to damage of the afferent, sensory limb of the reflex arc), and **loss of the carotid sinus reflex**. The SVE fibers innervating the stylopharyngeus muscle (which elevates the pharynx during swallowing), will also be damaged, causing paralysis of this muscle and contributing to **dysphagia**.

and subsequently spread out to form the tympanic plexus, along with fibers from the carotid plexus and the greater petrosal nerve within the tympanic cavity. The *preganglionic parasympathetic* fibers course via the lesser petrosal nerve to the **otic ganglion**, the parasympathetic ganglion of the glossopharyngeal nerve (located in the infratemporal fossa), where they synapse with *postganglionic parasympathetic* neurons whose fibers join the auriculotemporal branch of the trigeminal nerve to reach the parotid gland, providing it with secretomotor innervation.

VAGUS NERVE (CN X)

The vagus nerve has the most extensive distribution in the body, innervating structures in the head, neck, thorax, and abdomen

The **vagus** (L., "wanderer") **nerve** (Fig. 17.24) is a large cranial nerve that has the most extensive distribution in the body. Although it is a cranial nerve, its innervation is not limited to the structures in the head but also extends into the neck, thorax, and abdomen. The vagus, just as the facial and glossopharyngeal nerves, carries five functional components: (1) SVA; (2) GVA; (3) GSA; (4) SVE; and (5) GVE. A group of fine rootlets surface in the medulla oblongata in the posterolateral sulcus, inferior to the glossopharyngeal nerve and superior to the accessory nerve. These rootlets join to form two distinct bundles – a smaller inferior one and a larger superior bundle that collectively form the vagus nerve. The inferior bundle *joins the accessory nerve* and accompanies it for a short distance, but then the two diverge to go their separate ways. The smaller vagal bundle joins the main trunk of the vagus to exit the cranial vault via the jugular foramen. Inferior to the jugular foramen, the vagus nerve displays two swellings, the superior (jugular) and inferior (nodose) ganglia of the vagus nerve. The **superior ganglion** houses the cell bodies of pseudounipolar first-order sensory

neurons carrying **GSA** information from the external auditory meatus and the dura of the posterior cranial fossa. The **inferior ganglion** contains the pseudounipolar first-order nerve cell bodies transmitting **GVA** sensory innervation from the mucosae of the soft palate, pharynx, and larynx. It also houses pseudounipolar first-order nerve cell bodies transmitting **GVA** sensory innervation from the carotid body and a minor **SVA** (taste) sensation component from the epiglottis.

SVA (taste) pseudounipolar neuron cell bodies located in the **inferior ganglion** of the vagus nerve send their peripheral fibers to terminate in the scant taste buds of the epiglottis. Their central processes enter the brainstem along with the other vagal fibers and course in the **solitary tract** to terminate in the caudal **nucleus of the solitary tract** (Fig. 17.25A).

GVA pseudounipolar neuron cell bodies housed in the **inferior ganglion** distribute their peripheral processes in the mucous membranes of the soft palate, and those lining the pharynx, larynx, esophagus, and trachea. Chemoreceptor fibers (GVA also) terminate in the carotid body where they monitor blood carbon dioxide concentration. The central processes of all of the GVA neurons enter the brainstem, course in the **solitary tract,** and terminate in the **nucleus of the solitary tract** (Fig. 17.25A).

GSA pseudounipolar neuron cell bodies conveying pain, temperature, and touch sensation reside in the **superior ganglion** and send their peripheral processes to the external auditory meatus, and external surface of the tympanic membrane. Their central processes enter the brainstem, join the **spinal tract of the trigeminal nerve**, and terminate in the **spinal nucleus of the trigeminal nerve** (Fig. 17.25A).

The cell bodies of the **SVE neurons** are located in the **nucleus ambiguus**. The fibers of these neurons innervate all of the laryngeal and pharyngeal muscles except for the stylopharyngeus, thyrohyoid, and tensor veli palatini muscles (Fig. 17.25B).

The vagus nerve has a very extensive **GVE** distribution. It supplies parasympathetic innervation to the myoepithelial cells of the laryngeal mucous glands and to all of the smooth muscle in blood vessel and organ walls of the thoracic and abdominal viscera. The **posterior motor nucleus of the vagus** houses the nerve cell bodies of *preganglionic parasympathetic* neurons whose fibers accompany the other vagal fibers upon their exit from the brainstem. These fibers proceed in the main trunks (left and right) of the vagus into the thorax where they leave the main trunks and join the autonomic plexuses wrapped around blood vessels and viscera throughout the thoracic and abdominal cavities. The preganglionic fibers terminate and synapse in the terminal parasympathetic ganglia or ganglia *near or within the viscera*. Parasympathetic innervation decreases the heart rate (calms the heart), reduces suprarenal gland secretion, activates peristalsis, and stimulates glandular activity of various organs (Fig. 17.25C).

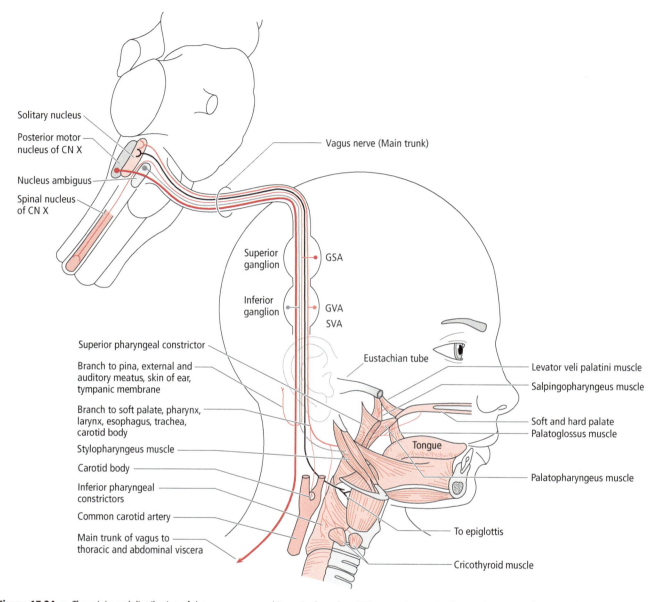

Figure 17.24 ● The origin and distribution of the vagus nerve and its major branches. GSA, general somatic afferent; GVA, general visceral afferent; SVA, special visceral afferent.

CLINICAL CONSIDERATIONS

Unilateral damage of the vagus nerve near its emergence from the brainstem results in several deficiencies on the ipsilateral side. Damage to the SVE fibers will cause **flaccid paralysis** or **weakness** of (1) the pharyngeal muscles and levator veli palatini of the soft palate, resulting in **dysphagia** (difficulty swallowing); (2) the laryngeal muscles, resulting in **dysphonia** (hoarseness) and **dyspnea** (difficulty breathing); and (3) **loss of the gag reflex** (efferent, motor limb). Damage to the GVA fibers will cause **loss of general sensation** from the soft palate, pharynx, larynx, esophagus, and trachea. Damage to the GVE fibers will cause **cardiac arrhythmias** and gastroparesis.

A bilateral lesion of the vagus nerve is incompatible with life, due to the interruption of parasympathetic innervation to the heart.

ACCESSORY NERVE (CN XI)

The accessory nerve supplies motor innervation to the sternocleidomastoid and trapezius muscles.

The **accessory nerve** (Fig. 17.26) supplies motor innervation to the sternocleidomastoid and trapezius muscles. The terminology, classification, and function of the human accessory nerve has been historically inconsistent in the literature due to the varied interpretation of its unusual or unique origin from the CNS, peripheral course, and enigmatic neural components. Terminologia Anatomica (TA) describes the **accessory nerve** as having two roots, a cranial root and a spinal root.

CRANIAL NERVES

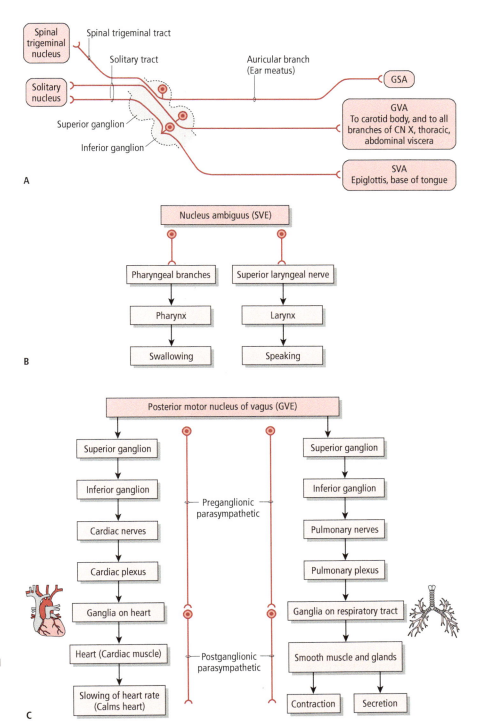

Figure 17.25 • Innervation by the vagus nerve: (A) general somatic afferent (GSA), general visceral afferent (GVA), and special visceral afferent (SVA; taste); (B) special visceral efferent (SVE; skeletal motor); and (C) general visceral efferent (GVE; parasympathetic).

- The **cranial root** (or internal branch) joins the vagus nerve to form the **pharyngeal plexus** to innervate palatal, pharyngeal, and laryngeal muscles derived from the pharyngeal arches.
- The **spinal root** (external branch) also called the **accessory nerve** innervates the sternocleidomastoid and trapezius muscles believed to be derived from somites.

It is now understood that the cranial root of the **accessory nerve** is composed of *aberrant* **vagal fibers** arising from the caudal portion of the **nucleus ambiguus** in the medulla oblongata. These caudal vagal fibers collectively form a distinct root as they emerge from the brainstem, whereas the **accessory nerve** derives its fibers from the **accessory nucleus** residing in the posterolateral aspect of

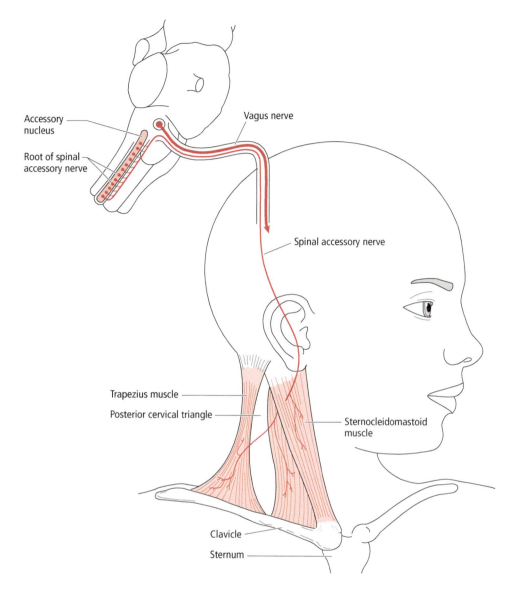

Figure 17.26 ● The origin and distribution of the accessory nerve and its major branches.

the anterior horns of cervical spinal cord levels **C1–C5**. This nucleus is continuous superiorly with the nucleus ambiguus of the medulla oblongata. Delicate rootlets emerging from the surface of the lateral funiculus of the spinal cord (interposed between the posterior and anterior spinal roots) converge and assemble to form the **accessory nerve**. The accessory nerve has been perceived as an irregular or unique spinal nerve (peripheral nerve) since its fibers arise from the spinal cord and then does something unusual. This nerve trunk ascends, enters the cranial vault through the foramen magnum, and proceeds on the lateral aspect of the medulla oblongata to join the aberrant vagal fibers as they emerge from the brainstem. The two groups of fibers accompany one another for a short distance but then diverge to go their separate ways. The aberrant vagal (cranial XI) fibers join the main trunk of the vagus nerve and follow those fibers of the vagus that are destined to supply most of the intrinsic laryngeal muscles. The accessory nerve exits the cranial vault via the jugular foramen. It courses inferiorly to the deep surface of the sternocleidomastoid muscle providing it with motor innervation from C1 to C2 spinal gray anterior horns. It continues its inferior course to the posterior triangle of the neck and then proceeds to the deep aspect of the upper part of the trapezius muscle to supply it with motor innervation from C3 to C5 spinal gray anterior horns. Given its origin, many neuroanatomists no longer consider the accessory nerve to be a true cranial nerve, but instead a unique type of spinal nerve.

Additionally, there are differences of opinion relating to the classification of the functional components (modalities) of the accessory nerve. Some authors consider that this nerve carries **SVE** fibers since neurons of the accessory nucleus develop in a manner characteristic of SVE, not GSE, neurons; whereas others believe that they are **somatomotor**; that is, GSE.

Recent evidence, based on the dual origin of the **sternocleidomastoid** and **trapezius muscles** (from the neural crest and somites), suggests that their innervation, the accessory nerve's modality could be categorized as SVE and GSE.

CLINICAL CONSIDERATIONS

A unilateral lesion confined to the accessory nucleus or the nerve *proximal* to its muscular distribution results in an **ipsilateral flaccid paralysis** and subsequent **atrophy** of the sternocleidomastoid and upper part of the trapezius muscles. An individual with such a lesion is unable to turn their head away from the lesion. Normally, unilateral contraction of the sternocleidomastoid muscle draws the mastoid process inferiorly, bending the head sideways (approximating the ear to the medial end of the clavicle), which is accompanied by an upward turning of the chin toward the opposite side. If the upper part of the trapezius is paralyzed, the upper border of the scapula is rotated laterally and inferiorly with its inferior angle pointing toward the spine. Slight winging of the scapula is noticeable when the arms are positioned lateral to the trunk and become more obvious during abduction of the affected arm. There is also slight drooping of the ipsilateral shoulder, accompanied by a weakening of the shoulder when attempting to raise it and weakness in the abduction of the arm above the horizontal plane. A lesion involving the "cranial root" (**aberrant vagal fibers**) of the accessory nerve results in paralysis of the laryngeal muscles.

However, the results of these investigations need to be confirmed by more extensive studies, leading to more definitive results, before accepting the change in modality categorizations for Cranial Nerve XI.

Due to its unique characteristics, the accessory nerve is viewed as neither a true cranial nerve nor a true spinal nerve, but it should be classified as a new category of peripheral nerve, the transitional nerve.

Recent literature supports that **GSA** proprioceptive fibers are carried by the accessory nerve from the upper cervical spinal cord levels to the structures it innervates but questions the pharyngeal arch origins of the trapezius and sternocleidomastoid muscles.

HYPOGLOSSAL NERVE (CN XII)

The hypoglossal nerve provides motor innervation to the intrinsic muscles of the tongue

The **hypoglossal nerve** (Fig. 17.27) provides motor innervation to the intrinsic muscles of the tongue. The cell bodies of the **GSE** lower motor neurons of the hypoglossal nerve reside in the **hypoglossal**

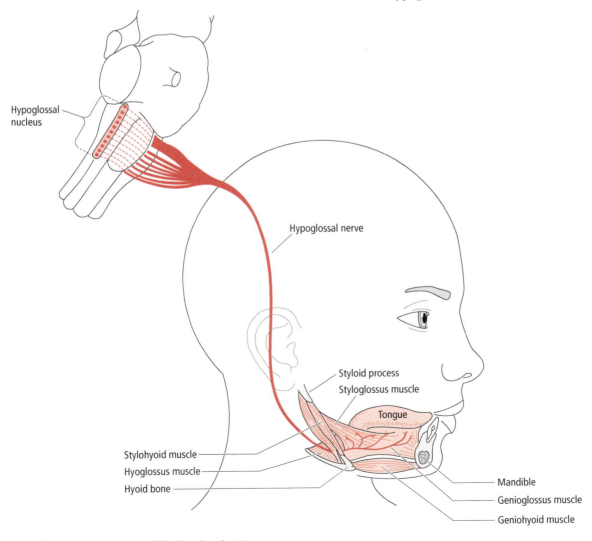

Figure 17.27 • The origin and distribution of the hypoglossal nerve.

358 ••• CHAPTER 17

CLINICAL CONSIDERATIONS

A unilateral lesion of the hypoglossal nerve will cause the tongue to deviate toward the side of the lesion (impaired side)

A lesion in the hypoglossal nucleus or nerve results in **flaccid paralysis** and subsequent **atrophy** of the ipsilateral tongue musculature. **Hemiparalysis** of the tongue causes creasing (wrinkling) of the dorsal surface of the tongue ipsilateral to the lesion. Normally, the simultaneous contraction of the paired genioglossi muscles causes the tongue to protrude. During an examination of the patient, it is important to remember that a unilateral lesion of the hypoglossal nerve will cause the tongue to deviate toward the *side of the lesion* (impaired side) since the functional genioglossus on the intact side is unopposed by the paralyzed, inactive genioglossus on the lesion side.

nucleus, a cell column in the posteromedial aspect of the caudal medulla oblongata. This nucleus, located anterior to the floor of the fourth ventricle near the midline, forms a triangular elevation – the **hypoglossal trigone** – in the floor of the midline of this ventricle. The nerve cell bodies of the hypoglossal nucleus give rise to axons that course antero-laterally in the medullary tegmentum to arise as a series of tiny rootlets on the anterior surface of the medulla oblongata in the anterolateral sulcus separating the pyramid and the olive. These rootlets collect to form the hypoglossal nerve, which exits the cranial vault anteromedially through the hypoglossal foramen. The nerve then courses to the subman-dibular region to serve the *ipsilateral* side of the tongue prior to joining the C1 spinal nerve in the infrahyoid space. The hypoglossal nerve innervates the tongue's intrinsic muscles (transverse, superior and inferior longitudinals, and verti-cal), which alter the shape of the tongue; it also innervates all of the extrinsic muscles of the tongue (styloglossus, hyoglos-sus, and genioglossus), except for the palatoglossus (CN X). The extrinsic muscles of the tongue alter the shape and posi-tion of the tongue. Recent studies indicate that **GSA** fibers terminating in muscle spindles of the tongue musculature transmit proprioceptive sensation to the trigeminal system involved in the reflex activity of mastication. Some investiga-tors believe that the cell bodies of these GSA pseudounipo-lar neurons are located in the mesencephalic nucleus of the trigeminal nerve, whereas others maintain that they belong to the hypoglossal nerve.

SYNONYMS AND EPONYMS OF THE CRANIAL NERVES

Name of structure or term	Synonym(s)/eponym(s)
Accessory oculomotor nucleus	Edinger–Westphal nucleus
Anterior trigeminal lemniscus	Anterior trigeminothalamic tract Ventral trigeminothalamic tract
Caudal Subnucleus of the spinal trigeminal nucleus	Pars caudalis of the spinal trigeminal nucleus
Discriminatory tactile sense	Fine touch sensation
Elliptic nucleus	Nucleus of Darkschewitsch
Frontal eye field (FEF)	Brodmann area 8
Functional components of cranial nerves	Modalities of cranial nerves
General somatic efferent (GSE)	Somatic motor Somatomotor
General visceral efferent (GVE)	Visceral motor
Inferior ganglion of the vagus nerve	Nodose ganglion
Internuclear neurons	Interneurons
Interpolar Subnucleus of the spinal trigeminal nucleus	Pars interpolaris of the spinal trigeminal nucleus
Interstitial nucleus	Interstitial nucleus of Cajal
Main nucleus of the trigeminal	Chief nucleus of the trigeminal Principal nucleus of the trigeminal
Mandibular division of the trigeminal nerve	Mandibular nerve
Maxillary division of the trigeminal nerve	Maxillary nerve
Medial strabismus	Internal strabismus Convergent strabismus Esotropia
Ophthalmic division of the trigeminal nerve	Ophthalmic nerve
Oral Subnucleus of the spinal trigeminal nucleus	Pars oralis of the spinal trigeminal nucleus
Postcentral gyrus	Primary sensory cortex (S-I) Primary somatosensory cortex Primary somesthetic cortex Brodmann areas 3, 1, and 2
Posterolateral tract	Fasciculus of Lissauer
Pseudounipolar neuron	Unipolar neuron
Posterior trigeminal lemniscus	Posterior trigeminothalamic tract Dorsal trigeminal lemniscus
Special visceral efferent (SVE)	Branchiomotor Branchial motor
Sphincter pupillae muscle	Constrictor pupillae muscle
Spinal nucleus of the trigeminal	Descending nucleus of the trigeminal
Spinal tract of the trigeminal	Descending tract of the trigeminal
Spiral ganglion	Cochlear ganglion
Spiral organ	Organ of Corti
Superior ganglion of the vagus nerve	Jugular ganglion
Trigeminal cave	Meckel's cave
Trigeminal neuralgia	Trigeminal nerve pain Tic douloureaux
Vestibular ganglion	Vestibular ganglion of Scarpa
Vestibulocochlear nerve	Acoustic nerve (older term)

FOLLOW-UP TO CLINICAL CASE

This patient has trigeminal neuralgia, also called tic douloureux. This is a purely clinical diagnosis, and tests are usually normal. This is a very common disorder, and this is a typical presentation. The pain is indeed excruciating and should be taken very seriously, as suicide is not an uncommon result!

This condition is most common in the elderly, though it can occur in younger age groups. The common etiology is thought to be from the ephaptic transmission of nerve impulses, or in other words "short-circuiting." The most common cause is from vascular "loops" that develop and surround one of the divisions of the trigeminal nerve at its root, most commonly affecting the ophthalmic and sometimes the maxillary divisions. This causes compression of the nerve, demyelination, and ephaptic transmission. This condition is usually spontaneous and comes "out of the blue." Trigeminal neuralgia in a young person brings up the specter of multiple sclerosis and can be a cause of trigeminal neuralgia. This is thought to result from demyelination of the trigeminal nerve root as it enters the brainstem. Multiple sclerosis is an autoimmune disease that affects only the myelin sheaths of axons in the central nervous system (CNS). Affected individuals display an autoimmune reaction that kills oligodendrocytes, the glial cells that myelinate axons in the CNS. Demyelination of axons follows with the subsequent formation of sclerotic plaques, which damages nerve impulse transmission.

There are effective treatments for this condition. Carbamazepine, an antiseizure medication, is the most effective. Other antiseizure medications have been used. Trigeminal neuralgia from vascular loops can be treated, if refractory to medications, by microvascular decompression surgery. This surgical procedure generated much skepticism when it was first introduced but has produced excellent results for refractory cases and has now become widely accepted.

5 What are some treatment options for trigeminal neuralgia?

QUESTIONS TO PONDER

1. What is the first clinical sign of intracranial pressure on the GVE (parasympathetic fibers) of the oculomotor nerve?

2. What are the functional deficits caused by a lesion to the trochlear nucleus or to the trochlear nerve?

3. What are the functional deficits following a lesion to the right abducens nucleus?

4. What eye movement functional deficits result following a lesion to one medial longitudinal fasciculus?

5. What eye movement deficits result following a lesion in the vicinity of the abducens nucleus?

6. What is the cause of the "crocodile tear syndrome"?

7. Which cranial nerves are likely to be damaged from a growing hypophyseal tumor?

8. Name the cranial nerves that are susceptible to damage from a tumor growing in the vicinity of the cerebellopontine angle.

CHAPTER 18

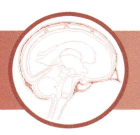  

Visual System

- **CLINICAL CASE**
- **EYE**
- **CENTRAL VISUAL PATHWAYS**
- **VISUAL REFLEXES**
- **CLINICAL CONSIDERATIONS**
- **SYNONYMS AND EPONYMS OF THE VISUAL SYSTEM**
- **FOLLOW-UP TO CLINICAL CASE**
- **QUESTIONS TO PONDER**

CLINICAL CASE

A 32-year-old patient described visual loss in the right eye. They first noticed its onset a week ago, and since then it has become progressively worse over the course of two days. They can no longer see with their right eye. Vision in the left eye was within normal limits. There was also pain in the right eye when they moved their eyes. They had never had eye or visual problems prior to this episode. They otherwise had no other symptoms, their health has always been perfect, and they have never had any neurologic symptoms in the past.

An ophthalmology exam showed visual acuity of 20/20 on the left, while on the right they were unable to read the largest numbers on the number chart but were able to count fingers that were presented to them. Pupillary constriction to light in the left was unremarkable, while reaction on the right was present but seemed less brisk. Fundoscopic exam showed papilledema (swelling) of the right optic disc, while the left was within normal limits. Eye movements were intact, although the patient reported pain on movement of the right eye. The rest of the neurologic exam was unremarkable.

The visual system is the most highly specialized and complex of all sensory systems in the human. It transmits an extraordinary amount of detailed information about color, form, three-dimensional shape, distance, and movement perceived in the environment to highly specialized centers in the brain.

The eyes are intricate sensory organs that mediate the special sense of **vision**. They resemble two autofocusing cameras, each consisting of a lens that not only allows the passage of, but also focuses, light reflected from objects on the retina. The retina contains photoreceptor cells, called rods and cones, that transmit sensory input related to color, light intensity, and form that are reflected from objects to other cells of the retina. This sensory input is further processed, and visual information is conveyed by other retinal cells to higher brain centers, where a final visual image is formed.

EYEBALL

The eyeball (visual organ) is composed of three concentric tissue layers, but not every layer is present throughout the entire wall of the eye. The three layers are the external

---

A Textbook of Neuroanatomy, Third Edition. Maria A. Patestas, Amanda J. Meyer, and Leslie P. Gartner.
© 2025 John Wiley & Sons, Inc. Published 2025 by John Wiley & Sons, Inc.
Companion website: www.wiley.com.go/patestas/neuroanatomy3e

layer: the **sclera**; a vascular intermediate layer: the **choroid**, **ciliary body**, and **iris**; and an internal layer: the **retina** (Fig. 18.1). These three concentric layers enclose the anterior and posterior chambers of the eyeball and the lens.

External layer

The external layer of the eye consists of the sclera and the cornea

The posterior five-sixths of the **external layer** is a tough, opaque, white collagenous connective tissue, the **sclera** (G., *scleros*, "hard"), which forms a protective rigid layer for the eyeball. The anterior one–sixth of the outer layer is modified as the **cornea,** a transparent avascular region that bulges anteriorly. The cornea receives oxygen and nutrients anteriorly via the tear film and posteriorly from aqueous humor in the anterior chamber. The cornea not only permits light to pass through it to enter the eyeball, but it also refracts (bends) it. The region of the sclera, surrounding the cornea, is apparent as the "white of the eyeball." Posteriorly, the sclera permits the optic nerve to pass through its fenestrated region, the **lamina cribrosa.**

Intermediate layer

The intermediate layer of the eye consists of the choroid, ciliary body, and iris

The **intermediate layer** appears dark, due to the presence of melanin pigments in most of its cells. Since it resembles the skin of a dark grape, it is called the **uvea** (L., *uva*, "grape"). The posterior aspect of the intermediate layer, known as the **choroid**, is the vascular coat of the eye. The choroid not only has an important nutritive function for the retina, but its pigmented cells absorb excess light that already passed through the retina, reducing light reflection and glare inside the eyeball. The absorption of light results in the production of heat, which poses a challenge when adjacent to the retina. Therefore, the choroid has the highest blood flow per gram of tissue in the human body. At its anterior margin, the choroid merges with a thickened structure, the **ciliary body** and its processes, at the **ora serrata**, which is the serrated margin between the choroid and the ciliary body. The ciliary body houses the **ciliary muscle**, a smooth muscle that changes the shape of the lens for accommodation (focusing on nearby objects). The ciliary body is continuous anteriorly with the **iris**, a pigmented diaphragm that encircles an opening known as the **pupil**. The

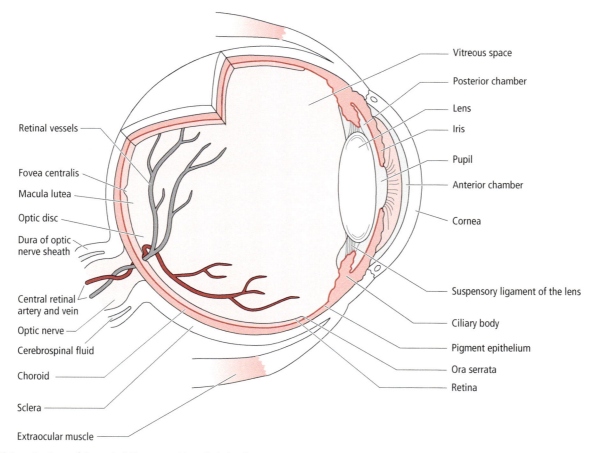

Figure 18.1 ● Anatomy of the eyeball. The external layer includes the cornea anteriorly, which becomes continuous with the sclera that covers the remainder of the eyeball. The intermediate layer consists of the vascular choroid, containing blood vessels and is continuous with the ciliary body and iris anteriorly. The internal layer is formed by the retina consisting of the non-neural pigmented epithelium and the multilayered neural retina. Retinal ganglion cell axons from the retina converge at the posterior aspect of the eyeball to form the optic nerve.

iris overlaps the anterior surface of the lens, and its color is visible through the transparent cornea. Smooth muscle activity, controlled by the autonomic nervous system, in the iris alters the size of the pupil: resulting in the constriction or dilation of the pupil, thereby controlling the amount of light that may enter the eye. The diaphragm-like opening and closing of the iris is controlled by two smooth muscles, the **sphincter pupillae**, innervated by the *parasympathetic nervous system* via the short ciliary nerves whose contraction constricts the pupil, and the **dilatator pupillae**, innervated by the *sympathetic nervous system* via the long ciliary nerves, and whose contraction dilates the pupil.

Internal layer

The internal layer of the eyeball consists of the retina, which is part of the central nervous system

The delicate **internal layer** of the eyeball is the **retina**. Due to its embryological derivation from the optic vesicle, an outgrowth of the diencephalon, the retina is considered to be part of the central nervous system (CNS). The retina, the only part of the CNS that can be observed by the physician during an optical examination, consists of a multilayered sheet of neurons, amacrine, and glial cells. The retina is composed of an external layer, the pigmented epithelium which underlies the choroid, and an internal layer, the retina proper or neural retina (Fig. 18.2). The **pigmented epithelium** consists of a layer of pigmented cells that absorb light that has passed through the other layers of the retina. The **neural retina** houses the layers of nerve cells responsible for the processing and transmission of visual information. The photosensitive portion of the neural retina extends from the ora serrata to the optic disc. The non-photosensitive portion of the retina extends from the ora serrata anteriorly, where it lines the inner surface of the ciliary body and iris.

The axons of the **retinal ganglion cells** forming the **optic nerve** collect into a bundle at the **optic disc** and pierce all the three layers of the eyeball to exit posteromedially (Fig. 18.1). The optic disc (Fig. 18.2C) is insensitive to light (since it lacks photoreceptors) resulting in a **blind spot** in the visual field. Thus, an image that falls on the optic disc of the retina is not perceived by the visual cortex. A curious phenomenon is that individuals are unaware of this interruption of vision since the brain "fills in" the part of the visual field that falls on the blind spot. The **macula lutea** (L., "yellow spot") is a yellow spot of the retina lateral to the optic disc. The **fovea centralis**, a pit in the center of the macula lutea, houses only cones and represents the site of maximal visual acuity (sharpest vision) and color perception in bright light. The number of cones decreases as one moves away from the fovea, but the number of rods increases as one moves closer to the macula.

Retinal layers

The **neural retina** is a multilayered structure composed of 58 distinct cell types within six classes: photoreceptor, horizontal, bipolar, amacrine, retinal ganglion, and non-neuronal cells (including radial glial cells (eponym: Müller cells), protoplasmic astrocytes, microglia, and vascular endothelial cells).

The retina consists of the following 10 layers. Some of the layers are described as inner or outer layers. The outer layers are closer to the external surface (periphery) of the eyeball, whereas the inner layers are closer to the internal chambers of the eyeball. From the point of view of the light entering through the pupil, the light must pass through the entire nine layers of the retina before it hits the tenth layer, pigmented layer of the retina. Viewed from the periphery inward, the layers of the retina are (Fig. 18.2):

1 **Pigmented layer of the retina.** The outermost layer of the retina consists of pigmented cells. This is a non-neural layer that absorbs excess light that has passed through the other retinal layers and is located immediately adjacent to the choroid. This pigmented epithelial layer also has nutritive, barrier, protective, and immunomodulative functions as well as maintaining the physical integrity of rods and cones by phagocytosis of their old membranes (Fig. 18.2A).

2 **Layer of inner and outer segments of the retina.** This layer includes the outer and inner segments of the photoreceptor cells (rods and cones). The discs of the photoreceptors contain visual pigments (rhodopsin and opsins) that undergo chemical changes (following absorption of light) leading to the initiation of local receptor potentials.

3 **Outer Limiting Layer of the retina** (also known as the External Limiting Layer or Outer Limiting Membrane of the retina). Although histologically this layer appears to be a membrane, in actuality, it is not. It is formed by intercellular junctions – the zonulae adherentes – binding the apical aspect of radial glial cells to each other. It has been reported that tight junctions proteins, such as occludins, are also present in these intercellular junctions.

4 **Outer Nuclear Layer of the retina** (also known as the External Nuclear Layer of the retina). This consists of those regions of rods and cones that house their nuclei.

5 **Outer Plexiform Layer of the retina** (also known as the External Plexiform Layer of the retina). This is a synaptic area that contains the terminals of the rods and cones, as well as those of the retinal interneurons (horizontal cells) and bipolar cells. The rods and cones synapse with the horizontal and bipolar cells. The horizontal cells modulate the signal from the photoreceptors to the bipolar cells.

6 **Inner Nuclear Layer of the retina** (also known as the Internal Nuclear Layer of the retina). This consists of the cell bodies of the radial glial cells and the retinal interneurons (horizontal, bipolar, and amacrine cells), which integrate and modulate the activity of the photoreceptors.

7 **Inner Plexiform Layer of the retina** (also known as the Internal Plexiform Layer). This is a synaptic area containing the terminals of the bipolar, amacrine, and retinal ganglion cells. Amacrine cells link the retinal ganglion cells.

8 **Ganglionic Layer of the retina.** This consists of the cell bodies of the retinal ganglion cells (multipolar neurons). These cells are the projection (output) neurons of the retina.

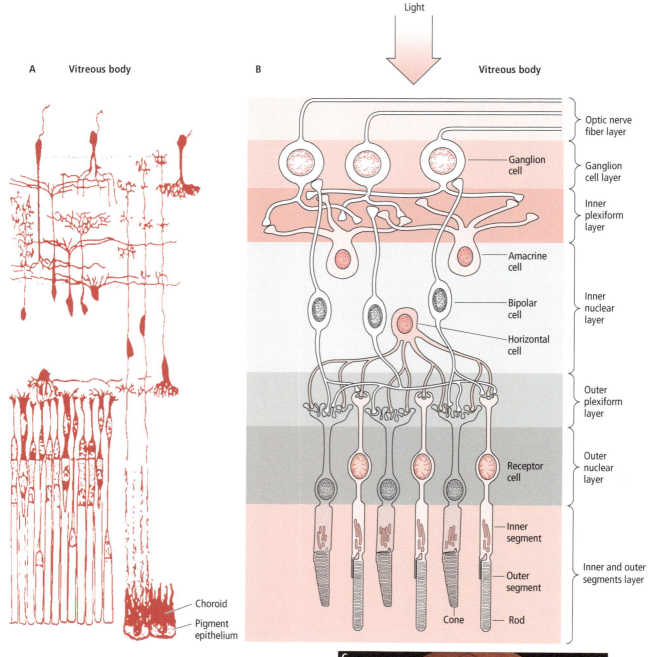

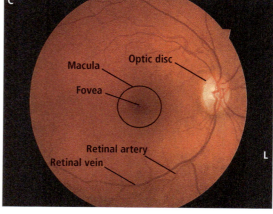

Figure 18.2 ● (A) Layers of the retina including the location of the pigment epithelium and choroid. (B) Layers of the retina and component cells. Note that light must pass from the front of the eyeball through the cornea, anterior chamber, lens, vitreous body, and all of the retinal layers to finally reach the photoreceptor cells, the rods, and cones. Note that layers 1, 3, and 10 are not indicated in this diagram. (C) Retinal blood vessel. Source: Courtesy of Dr. Amanda J. Meyer.

9 **Layer of nerve fibers of the retina** (also known as the optic nerve fiber layer). This is composed of the **unmyelinated axons** of the retinal ganglion cells that gather to exit the eyeball to form a thick bundle, the optic nerve CN II. Once the axons penetrate the sclera posteriorly, they become myelinated and enclosed in meninges. These projection axons terminate primarily in the lateral geniculate nucleus (LGN) of the thalamus.

10 **Inner Liming Layer of the retina** (also known as the Internal Limiting Membrane). This membrane is composed of a basal lamina interposed between the vitreous body and the radial glial cells. The vitreous body is a gelatinous substance filling the vitreous space of the bulb of the eyeball, posterior to the lens. The internal limiting layer separates the neural retina from the vitreous body.

Sensory input from the photoreceptors is transmitted to a synaptic zone stimulating bipolar cells, which in turn transmit the information to retinal ganglion cells. Horizontal and amacrine cells link the conducting neurons laterally within the retina. The convergent circuitry of the rods on the bipolar cells results in a decrease in visual acuity. High acuity would result from a 1:1 ratio of rods to bipolar cells: one tiny portion of the visual field would be mapped to one bipolar cell. Instead, many rods representing a larger portion of the visual field are mapped to one bipolar cell, so that cell represents a larger portion of the visual field. There is little convergence in the macula lutea of the retina, the site of highest visual acuity.

The radial glial cells (eponym: Müller cells) are the most numerous of the non-neuronal cells of the retina and extend almost the entire thickness of the retina from the inner to the outer external limiting layers. Their cell bodies sit in the inner nuclear layer and the radial glial cells contribute to the internal *blood–retinal barrier*. Radial glial cells provide vital support to neurons of the retina including, release of neurotransmitters, transport of metabolites, removal of neuronal waste, scavenging of free radicals, and phagocytosis of dead neurons or pigmented epithelial cells and debris from foreign bodies.

The **cones** (Fig. 18.3B) are activated by *high-intensity light* (**photopic vision**). They mediate sharp vision and color vision. They detect fine detail, have the highest acuity, and are more numerous in the central portion of the retina and less concentrated at the periphery. There are three different types of cones in the human retina (S-, M-, and L-cones), the membrane of each cone being associated with a different pigment and responding to a different wavelength of light: S-cones (short-wavelength) = blue, M-cones (medium-wavelength = green, L-cones (long-wavelength) = red).

Rods are of a single type only (Fig. 18.3A). They are stimulated by *low-intensity light*; thus, they transmit visual input in dim illumination (**scotopic vision**), but they cannot detect

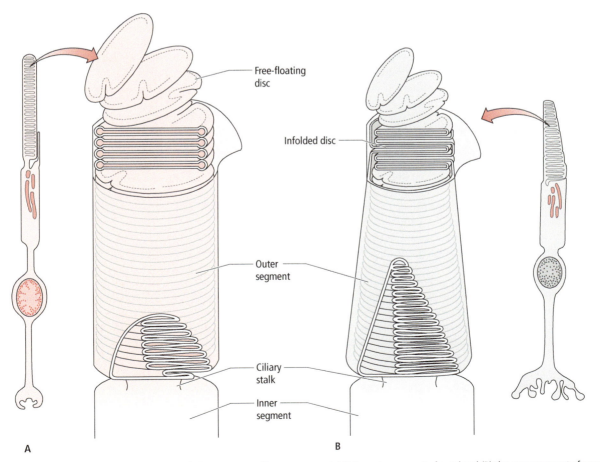

Figure 18.3 • Schematic representation of the retinal photoreceptor cells outer segment: (A) the outer segment of a rod and (B) the outer segment of a cone.

366 ● ● ●   **CHAPTER 18**

colors. They are more concentrated in peripheral regions of the retina. At night, when looking straight at a light located in the distance, it disappears when projected on the fovea (which contains only color-sensitive cones), but when looking away from it, the image falls on more peripheral regions of the retina, where rods are more numerous, and therefore the image becomes visible, but it is seen as various shades of gray.

Molecular biology of the rods and cones

A light-absorbing photopigment is synthesized in the inner segment of both rods and cones; rod membranes are associated with rhodopsin, whereas cone membranes are associated with iodopsin

Both rods and cones are composed of an inner segment, a connecting stalk, and an outer segment. The **inner segment** encloses the cellular organelles and is the site of protein synthesis, such as the protein component of the photopigments. The **connecting stalk** (connecting cilium) is the narrow segment interposed between the inner and outer segments. The photoreceptor cells derive their name from the morphology of their **outer segment** – that of a rod being cylindrical, and that of a cone being conical in shape. The outer segment, the dendritic end of the photoreceptor, is a photosensitive-modified cilium that is located adjacent to the retinal pigmented epithelium. The photoreceptor outer segment consists of a stack of horizontally oriented, membranous, flattened discs derived from the plasma membrane. In rods, the hollow discs located at the apical end of the outer segment are intracellular and are completely separated from the plasma membrane. The discs located closer to the base of the rod outer segment, and the entire outer segment of the cones, consist of shelf-like infoldings of the plasma membrane. They are thus a continuation of the plasma membrane with their interior being continuous with the intercellular space (Fig. 18.3).

A light-absorbing photopigment is synthesized in the inner segments of both rods and cones and is then transported to their respective outer segments (by passing through the connecting stalk), where it is incorporated into disc membranes as transmembrane proteins. The membrane of rods is associated with the pigment **rhodopsin** (G., "visual purple"), whereas the membrane of cones is associated with a pigment that consists of a component chemical of **iodopsin.** Rhodopsin and iodopsin are almost identical chemically, with the exception of their spectral sensitivity difference (thus, only the photochemistry of rhodopsin is discussed next). Both rhodopsin and iodopsin consist of two components: an **opsin**, which is a protein moiety; and a **chromophore** (G., "bearing color"), which absorbs photons. In rods the opsin is known as **scotopsin** (G., *scotos*, "darkness"), whereas in cones the opsins are known as **photopsins** (G., *photos*, "light"). The chromophore associated with rods is **retinal**, a substance derived from *vitamin A*. Normal vision is dependent on a sufficient intake of vitamin A and a severe dietary deficiency of that vitamin results in **night blindness**,

a condition characterized by the inability to see in poor illumination.

Rhodopsin, a transmembrane homologue of G-protein-linked receptor molecules, acts by activating **transducin** (**G$_t$**) a trimeric G-protein. As indicated above, rhodopsin is composed of a protein segment, **opsin** (embedded in the membranes of rod discs), to which a vitamin A derivative, **11-*cis*-retinal**, is covalently attached. The plasma membrane of the rod outer segment contains cGMP-gated sodium channels to which cGMP is bound in the absence of light. The cGMP–sodium channel complex maintains the channel in an open state, so that sodium as well as calcium can enter the outer segment. The constant flow of sodium ions into the outer segment causes a high rate of neurotransmitter release from the synaptic region of the rod, resulting in the inhibition of postsynaptic retinal neurons in the dark.

Interaction of a photon with 11-*cis*-retinal isomerizes it to all-*trans*-retinal, and this change in molecular configuration causes an alteration in the conformation of the opsin to which it is attached. The altered state of the opsin moiety results in the binding of rhodopsin to transducin that, in turn, dissociates and its alpha subunit activates the enzyme **cGMP phosphodiesterase**. Because this enzyme functions to hydrolyze cGMP, thereby decreasing its cytosolic concentration, cGMP is released from the sodium channels causing the closure of these channels. The lack of sodium flow into the outer segment causes a hyperpolarization of the rod with a resultant shutdown of neurotransmitter release from its synaptic region. Since the neurotransmitter substance inhibits the postsynaptic retinal neurons, the absence of the neurotransmitter acts to excite them. Thus, a single photon has the capability of eliciting a visual response.

The closed state of the sodium channel also shuts off the flow of calcium into the outer segment of the rod. This decrease in cytosolic calcium causes the uncoupling of calcium from the cytosolic protein **recoverin**, resulting in its activation. The activated recoverin stimulates guanylyl cyclase to manufacture cGMP that, by binding to the guanylyl cyclase, opens the sodium channels so that the rod can revert to its dark-phase (resting) state.

Lens

The lens is a biconvex, transparent pliable structure, whose shape can be adjusted to focus the cornea image on the retina

Each human eyeball consists of two "lenses," the cornea and the lens (see Fig. 18.1). Light rays entering the eyeball are actually refracted mostly (two–thirds) by the cornea (due to its location at the ocular surface) and not the lens, so that the light converges on the retina. However, due to the rigid organization of the cornea, the refractive power of the human cornea is fixed, leaving the variable focusing (accommodation) to the lens, exclusively. The lens is a biconvex, transparent pliable structure, whose shape can be adjusted to focus the corneal image on the retina. The lens is suspended posterior to the iris by the **ciliary zonule** (also known as the suspensory ligaments of the lens), which are anchored to the ciliary processes of the ciliary body. The lens inverts and reverses the image and focuses it on the fovea centralis of the retina. In order for incoming light to reach the retina, it has to pass through the following refractive media of the eyeball:

1 **The cornea.** This is the chief refracting medium of the eye. It serves to focus a crude image on the retina.
2 **The anterior chamber** of the eye. This is a fluid-filled (aqueous humor) space between the lens and the cornea.
3 **The lens.** The shape (thickness) of the lens can be altered, which can fine focus the corneal image on the retina.
4 **The vitreous body.** This is a transparent, gelatinous substance filling the vitreous space of the bulb of the eyeball, posterior to the lens.
5 **Most layers of the neural retina.** For light to reach the photoreceptors (rods and cones), which are located at the "back" of the retina, light has to go through all of the overlying retinal layers (layers 3–10, see previously) in order to reach them.

CENTRAL VISUAL PATHWAYS

The visual pathway consists of photoreceptors, first-order and second-order neurons residing in the retina, and third-order neurons in the lateral geniculate nucleus of the thalamus

Incoming light rays impinging on the retina pass from its inner layers to its outermost neural layer where they cause the **retinal photoreceptor cells** (modified neurons), the **rods** and **cones**, to become *hyperpolarized*. The photoreceptors then stop releasing their inhibitory neurotransmitters and the **bipolar cells** (**first-order neurons**) are no longer inhibited, and fire. The bipolar cells along with the interneurons, the **horizontal** and **amacrine cells**, process, integrate, and modulate visual input. The bipolar cells relay this sensory input to the **retinal ganglion cells** (**second-order neurons**) of the retina. Therefore, stimulation within the retina proceeds from the rod and cone layer to the bipolar cell layer and finally to the ganglion cell layer. Although light travels through the inner to the outer layers of the retina, visual electrical signals are transmitted in the opposite direction, from the outer to the inner layers of the retina.

There are two main types of **retinal ganglion cells** in the human retina categorized by their size and physiologic function. They are the midget and parasol cells. **Midget ganglion cells** account for more than 80% of retinal ganglion cells in humans, reside mainly in the central region of the retina, receive visual information primarily from cones, and are sensitive to color stimuli. **Midget ganglion cells** are also called **P1-types** because their axons project to the parvocellular layers (containing small cells) of the lateral geniculate nucleus of the thalamus. The **parasol cell** (large **M-type** or small **P2-type**) has a large cell

Although light travels from the inner to the outer layers of the retina, electrical signals relaying visual sensations are transmitted in the opposite direction, namely from the outer to the inner layers of the retina

body, a wide diameter fast-conducting axon, and an extensive dendritic receptive field that permits the cell to collect visual information from a wide region of the retina. Parasol cells account for approximately 10% of retinal ganglion cells and are located mostly in the peripheral region of the retina, receive visual information primarily from rods, and have only a minor role in color perception. Parasol cells respond to movement; they are also referred to as **M-type** since their axons project to the magnocellular layers (containing large cells) of the lateral geniculate nucleus of the thalamus. Unlike the parasol cell, the second type of retinal ganglion cell, the **midget ganglion cell** (P1-type) has a small cell body, a slower-conducting axon, and a smaller dendritic receptive field, which receives visual information from only a small patch of the retina.

There is a third, but minor type of ganglion cell, the **melanopsin-containing retinal ganglion cell**, composing less than 3% of the entire ganglion cell population. These cells, similar to the parasol cells, have a large cell body, and a broad, far-reaching dendritic receptive field. The salient characteristic that makes these cells unique is their response to blue light (even in individuals who were born blind) but are insensitive to the other visual information to which parasol and midget cells respond. The melanopsin-containing ganglion cells project to the suprachiasmatic nucleus (SCN) of the hypothalamus informing it of the presence of daylight and assisting it in establishing the circadian cycle. Neurons from the SCN project to the pineal body which manufactures melatonin, but only at night. These retinal ganglion cells also project to the pretectal nucleus of the midbrain, a component of the visual system that participates in the pupillary light reflex. The other 7% of retinal ganglion cells are composed of seven types of genetically distinct cells.

The axons of **retinal ganglion cells** (midget and parasol cells) are nonmyelinated (see Fig. 18.2B). They course on the *inner surface* of the retina (separated from the vitreous body only by the internal limiting membrane), converge posteriorly at the **optic disc**, and cross a sieve-like perforated area of the sclera, the **lamina cribrosa,** to emerge from the posterior pole of the bulb of the eyeball. At this point, all these axons become myelinated by oligodendrocytes, and they assemble to form a large bundle, the **optic nerve** (**CN II**). The optic nerve is composed of approximately one million retinal ganglion cell axons. This nerve, an outgrowth of the diencephalon, is not a true peripheral nerve, but instead is a tract of the brain. The optic nerve is actually a tract of the central nervous system whose axons are myelinated by oligodendrocytes and consequently are susceptible to demyelinating diseases, such as multiple sclerosis. The optic nerve becomes invested by extensions of the meningeal membranes and the subarachnoid space (containing cerebrospinal fluid) that cover the brain, making it susceptible to compression during periods of high blood pressure. Persistent untreated high blood pressure can cause hypertensive retinopathy.

An increase in intracranial pressure will compress the axons of the optic nerve which will interfere with axonal flow and cause swelling of the optic nerve. The swelling will manifest as **papilledema** at the proximal end of the nerve as it emerges from the posterior pole of the eyeball. The lesion to the optic nerve may result in visual deficits ranging from partial to total loss of vision in the affected eye.

The optic nerve, enveloped in a meningeal cover, exits the orbital cavity by passing through the **optic canal** of the sphenoid bone, and then enters the middle cranial fossa. The optic nerves of the right and left sides converge and join to form an intersection of fibers, the **optic chiasm** (G., "optic cross") on the optic groove on the body of the sphenoid bone in the middle cranial fossa (Fig. 18.4). It is in the optic chiasm where partial decussation of the optic nerve fibers (axons) of the two sides occurs. All retinal ganglion cell axons arising from the lateral (temporal) half of each retina course in the lateral aspect of the optic chiasm without decussating, to join the **optic tract** (*no longer the optic nerve*) of the same side. All retinal ganglion cell axons arising from the nasal (medial) half of each retina decussate (through the central region of the chiasm), and enter the optic tract of the opposite side, to join the temporal fibers. Thus, each optic tract consists of retinal ganglion cell axons arising from both eyeballs (the ipsilateral temporal half and the contralateral nasal half of the retina). These fibers preserve a retinotopic organization in the optic tract, which curves around the cerebral peduncle to terminate and relay their visual information (via retinogeniculate projection) primarily in the anterior aspect of the lateral geniculate nucleus (**LGN**) of the thalamus. It is the LGN that processes visual input for visual perception. The optic tract also ends and relays visual information in: (1) the **superior colliculus,** a midbrain nucleus of the visual system having an important function in saccadic (rapid) eye movements and somatic motor reflexes; (2) the **pretectal area,** which mediates autonomic reflexes such as the control of pupillary constriction and lens accommodation (see discussion on visual reflexes below); (3) the suprachiasmatic nucleus of the **hypothalamus,** which has an important function in circadian rhythms (day–night) and the reproductive cycle; and (4) the **reticular formation** involved in arousal of the organism in response to visual stimuli. The optic tract projects fibers to the superior colliculus and to the pretectal area via the **brachium of the superior colliculus** (Fig. 18.5).

An overview of the central visual pathway is given in Fig. 18.6.

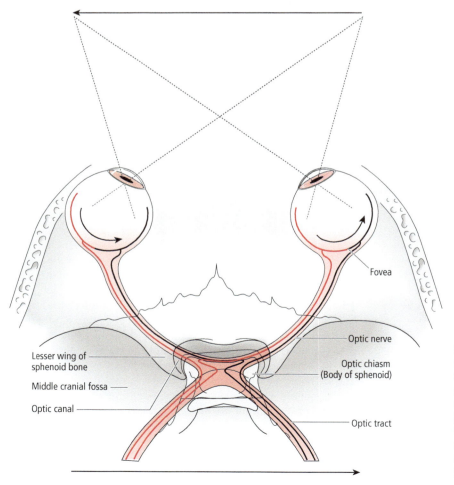

Figure 18.4 • Schematic diagram of the eyeballs, optic nerves, optic chiasm, and optic tracts. Each optic nerve contains visual information from the ipsilateral eyeball. Fibers from the temporal half of each retina course along the lateral aspect of the optic chiasm to join the ipsilateral optic tract. In contrast, fibers from the nasal half of each retina course in the central region of the optic chiasm where they cross to join the contralateral optic tract. Thus, each optic tract carries information from both eyeballs.

VISUAL SYSTEM • • • 369

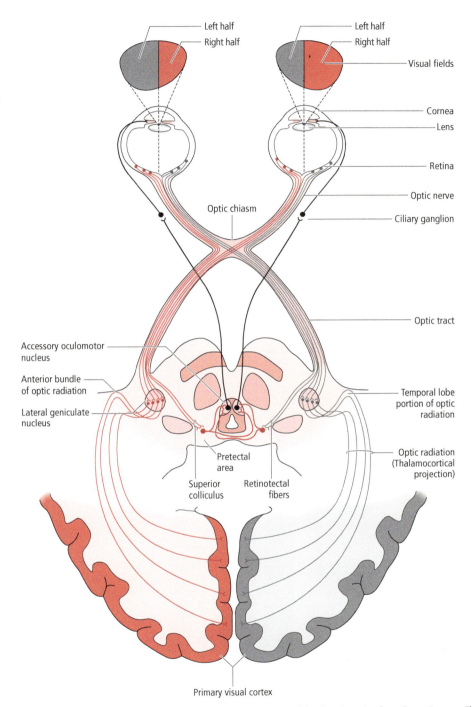

Figure 18.5 • The visual pathways. Retinal ganglion cell axons (axons of second-order neurons) leaving the retina form the optic nerve. The retinal ganglion cell axons arising from the temporal half of each retina pass along the lateral aspect of the optic chiasm to join the ipsilateral optic tract. The retinal ganglion cell axons arising from the nasal half of each retina cross at the optic chiasm to join the contralateral optic tract. Retinal ganglion cell axons terminate in the lateral geniculate nucleus of the thalamus and the superior colliculus and pretectal area of the midbrain. Third-order neurons of the lateral geniculate nucleus project via the optic radiation (geniculocalcarine tract, thalamocortical projections) to the striate area (synonym: primary visual cortex V1; eponym: Brodmann's area 17) located in the banks of the calcarine sulcus on the medial surface of the occipital lobe.

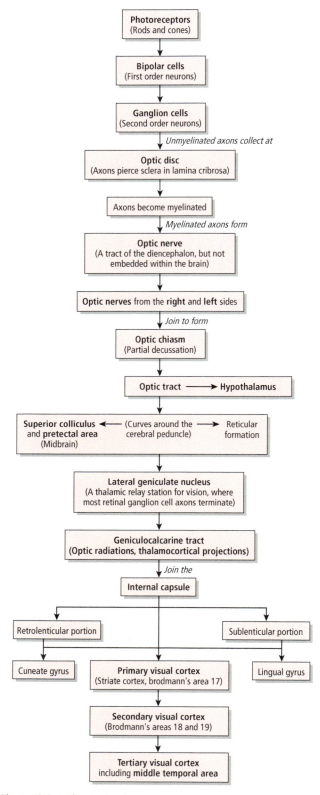

Figure 18.6 • The visual pathway.

Lateral geniculate nucleus (LGN)

The LGN houses the cell bodies of third-order neurons of the visual pathway and serves as a thalamic relay station that processes and regulates the flow of visual information to the primary visual cortex

The **lateral geniculate nucleus** (Fig. 18.7) is a relay station of the posterior thalamus whose principal function is to process and regulate the flow of retinotopically organized visual information and relay it to the **primary visual cortex**. The LGN receives retinal ganglion cell (second-order neuron) axons running in the optic tract at its anterior aspect and houses the cell bodies of **third-order neurons** of the visual pathway. The axons of the third-order neurons leave the nucleus from its posterior and lateral aspects to form the thalamocortical projection (optic radiation, geniculocalcarine fibers) that relays visual input to the primary visual cortex via the retrolentiform (retrolenticular) and sublentiform (sublenticular) parts of the internal capsule and via the corona radiata. The LGN is a laminated structure consisting of six distinct layers, numbered 1–6, which are readily identifiable in an axial section. Although each LGN receives information from the *contralateral visual hemifield*, each of its layers receives input from only one eyeball (the ipsilateral or contralateral retina). Layers 1, 4, and 6 receive retinal ganglion cell axons arising from the nasal half of the contralateral retina, whereas layers 2, 3, and 5 receive retinal ganglion cell axons arising from the temporal half of the ipsilateral retina. Layers 1 and 2 consist of large neurons and are therefore referred to as the **magnocellular (M) layers**; they receive information originating mainly from rods via the **M retinal ganglion cells (parasol cells)** residing in the peripheral region of the retina that are sensitive to *movement* and *contrast* but are have a minor role in color perception. The magnocellular layers are part of the visual pathway associated with the *location* and *movement* of a stimulus in the visual field (are part of the **"where" pathway** (see below). Damage to the magnocellular layers diminishes the perception of fast-moving visual stimuli; however, color perception and high acuity vision are unaffected. Layers 3–6 consist of small neurons and are referred to as the **parvocellular (P) layers**; they receive information originating mainly from cones via **midget (P) retinal ganglion cells** residing in the central region of the retina, responding to *color* and *form* (high-acuity) of stationary stimuli. The parvocellular layers are part of the visual pathway associated with the *color* and *form* of a stimulus in the visual field (are part of the **"what" pathway** (see below). A lesion affecting the parvocellular layers results in loss of color vision and diminishes high-acuity vision, leaving the function of the "where" pathway unaffected. Interposed between these principal six layers, are thin layers housing the smallest cells, known as the **koniocellular neurons** (G. "konis," dust). They are strongly sensitive to color (and complement the parvocellular cells) and detect colors that contribute to shape- or form-discrimination.

VISUAL SYSTEM   371

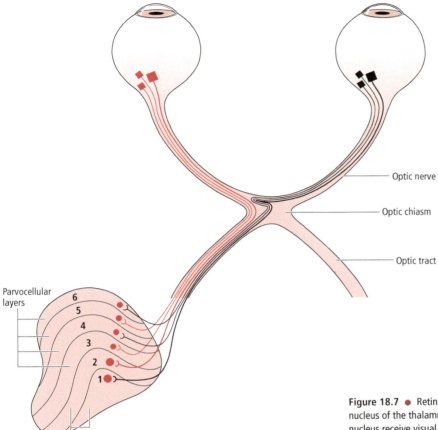

Figure 18.7 • Retinal ganglion cell projections to the lateral geniculate nucleus of the thalamus. Note that layers 1, 4, and 6 of the lateral geniculate nucleus receive visual information from the contralateral retina, whereas layers 2, 3, and 5 receive visual information from the ipsilateral retina.
Source: Modified from Nolte, J (1999) *The Human Brain, An Introduction to Its Functional Anatomy*, 5th edn. Mosby, St Louis, Missouri; fig. 17.25C.

The macular area (highest visual acuity area) of the retina has a greater representation in the LGN than the peripheral (paramacular) areas of the retina. Furthermore, the lateral half of the LGN receives visual information from the inferior retinal quadrants (superior visual field quadrants), whereas it's medial half receives visual information from the superior retinal quadrants (inferior visual field quadrants).

In addition to the retinal ganglion cell axon terminals (second-order neurons) arriving at the LGN (**retinogeniculate projections**), the LGN also receives a greater number of afferents from the visual cortical areas (**corticogeniculate projections**). These fibers are associated with stimulating visual attention of the individual and reciprocally regulating the transmission of sensory input from the LGN to the striate area (primary visual cortex).

Superior colliculus

The superior colliculus functions in the control of reflex movements that orient the eyeballs, head, and neck in response to visual, auditory, and somatic stimuli

The **superior colliculi** (L., *colliculus*, "little hill") (Fig. 18.5) are two prominent elevations projecting from the posterior surface of the midbrain. They are part of the quadrigeminal plate, the two superior and the two inferior colliculi. The superior colliculi are laminated structures, each consisting of seven layers; each superior colliculus is a relay nucleus of the midbrain that receives sensory input (afferents) from:

- The **visual system** to **layers 1–3** (from the retina via retinotectal fibers, and from the visual cortex via corticotectal fibers);
- The **auditory system** to **layers 4–7** (principally from the inferior colliculus); and
- The **somatosensory system** to **layers 4–7** (via the spinotectal tract).

Following the integration of all the sensory input, appropriate reflex responses are produced by the neurons of the superior colliculi. The superior colliculus has important functions in the control of **reflex movements** that **orient the eyeballs**, **head**, and **neck** in response to **visual**, **auditory**, and **somatic stimuli** (tracking moving objects) via its outputs (efferent projections). Efferents from the superior colliculus include projections to the reticular formation, the inferior colliculus, the LGN and the pulvinar of the thalamus, the oculomotor, trochlear, and abducens nuclei via the medial longitudinal fasciculus (MLF), the pontine nuclei and the cerebellum via the tectopontocerebellar tract, and the cervical levels of the spinal cord via the crossed tectospinal tract.

Optic radiation (geniculocalcarine tract)

Axons of third-order neurons originating from the LGN form the optic radiation, which terminate in the primary visual cortex

The fibers of the **optic radiation** follow the contour of the lateral wall of the lateral ventricle (Fig. 18.8), (traveling through the temporal, parietal, and occipital lobes), joining the retrolentiform (retrolenticular) and sublentiform (sublenticular) components of the **internal capsule**, the **corona radiata**, and then pass posteriorly to the **striate area** (primary visual cortex), located on the medial and posterior surfaces of the **occipital lobe**. Each one of the axons of the optic radiation carries visual input only from one retina.

The optic radiation fibers maintain retinotopic organization and form an upper, an intermediate, and a lower division. The **upper division** consists of fibers conveying information from the **inferior retinal quadrants (superior visual field)**. These fibers loop anteriorly and inferiorly around the inferior horn of the lateral ventricle in the temporal lobe then

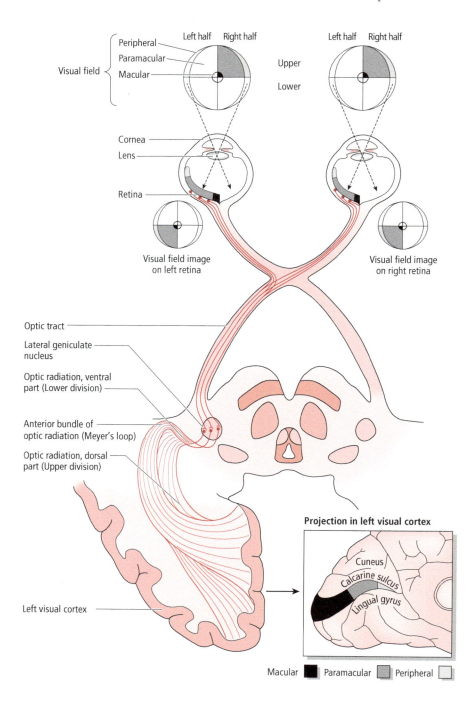

Figure 18.8 ● Visual field representation in the retina and striate area (primary visual cortex).

posteriorly via the sublentiform part of the internal capsule to the striate area on inferior bank of the calcarine fissure of the primary visual cortex (Fig 18.8). The **intermediate division** of the optic radiation consists of fibers relaying visual information to the striate area of the primary visual cortex from the **macular region of the retina** (Table 18.1).

Striate area (primary visual cortex)

The striate area is the site where visual information is processed, and its significance determined

The **primary visual cortex** resides mainly in the medial surfaces of the occipital lobes on the superior and inferior banks of the **calcarine fissure**, corresponding to the eponymously named Brodmann's area 17 although it extends into the posteromedial surface. The primary visual cortex is also referred to as the **striate cortex/area** due to its striped characteristic appearance exhibiting a noticeable band of myelinated fibers, the **"stripe of Gennari."** Other cortical areas that surround the primary visual cortex and process visual information are collectively referred to as extrastriate cortex/areas (eponym: Brodmann areas 18 and 19).

Visual information must travel from the retina, via the visual pathway all the way to the striate area located in the posterior region of the brain, to reach consciousness. The visual cortex consists of a number of functional areas, each specializing in the processing of a specific aspect of vision. The neurons of the visual cortex respond to different visual stimuli transmitted by neurons conveying *color*, *motion*, *three-dimensional vision*, or a combination of various visual stimuli to construct a composite conscious visual image of the external world. If one particular area of the visual cortex processing a specific type of visual information is damaged, that visual component may be lost whereas the other components will remain intact. A lesion involving/affecting the area of the visual cortex that detects movement may cause the affected individual to perceive the world as a series of still snapshots.

The LGN projects visual information to the **striate area of the occipital lobe** (Fig. 18.9), the first cortical stop and site of initial cortical processing of visual information.

The superior bank of the calcarine fissure receives visual input from the **superior retinal quadrants** transmitted by the upper division of the optic radiations. The inferior bank of the calcarine fissure receives visual input from the **inferior retinal quadrants** transmitted by the lower division of the optic radiations (Table 18.1).

The striate area (primary visual cortex) receives visual input from the macula of the retina in its caudal third, and from the paramacular (and more peripheral) fields of the retina in its rostral portions. The macular representation in the striate area is much larger than that of other areas of the retina, reflecting the high visual acuity of the macula. This is similar to the situation of the primary somatosensory cortex that processes sensory information from the highly sensitive face or hand. The striate area neurons project to the **extrastriate areas** (V2, V3, V4, and V5) for further processing. Depending on the retinal region of origin and nature of the visual information, it is subsequently relayed to various higher-order cortical areas, each of which specializes in the processing of specific aspects of vision (see Fig. 18.10).

The "where" and "what" pathways

Visual information leaving the retina is relayed to the visual cortex via two anatomically distinct pathways, the posterior stream and the anterior stream (see Fig. 18.11).

The **magnocellular pathway**, also referred to as the **"where" pathway**, carries retinal information that is ultimately assembled by the visual cortical areas to enable the brain to analyze *movement* and to determine the *spatial localization* of an object in the visual field.

In the **"where"** pathway, visual input is relayed from **rods** to the large parasol retinal ganglion cells that are located mostly in the peripheral region of the retina and respond to *moving* stimuli. **Parasol retinal ganglion cells** relay their information to the **magnocellular layers** (layers 1 and 2) of the lateral geniculate nucleus of the thalamus. From there, the input is transmitted to the **striate area** (primary visual cortex, V1, Brodmann area 17), which, in turn, projects the information to the **extrastriate** areas (V2 and V3 subregions of Brodmann area 18). Subsequently, information is relayed to the **medial temporal area** (V5) which processes the spatial details of visual input related to *movement* (direction and speed of a moving object) and projects to the **posterior parietal cortex** (corresponding to Brodmann area 7a). This pathway projects to cortical areas that analyze both the **movement** and **spatial localization** of objects in the visual field in relation to the viewer. An example demonstrating the function of this pathway is as follows: when one is driving and a large ball appears in the path of the car, the driver must quickly swerve to avoid hitting the ball. This is an unconscious pathway. Damage to this pathway may result in not being able to determine the spatial relationship of two objects that are in front of the viewer (i.e., which of two objects is closer to the viewer). Additionally, an affected individual is unable to determine the location of an object, such as a book, in relation to themselves.

The parvocellular pathway located more anteriorly, is the **anterior stream** of processing, also referred to as the **"what"**

Table 18.1 ● Topographic relationships of projections from the retina to the lateral geniculate nucleus (LGN) of the thalamus to the banks of the calcarine sulcus.

Visual field quadrants	Retinal field quadrants	LGN	Optic radiations	Calcarine banks
Upper	Lower	Lateral half	Inferior part	Inferior
Lower	Upper	Medial half	Superior part	Superior

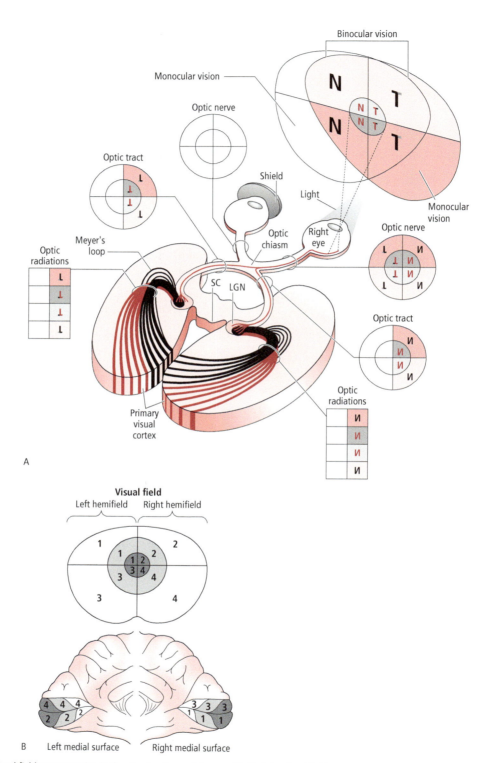

Figure 18.9 ● (A) Visual field representation in the visual pathway: T, temporal half of the visual field; N, nasal half of the visual field; SC, superior colliculus; LGN, lateral geniculate nucleus. In the right retina and right optic nerve, the image from the visual field is backwards and upside down. Note that the left eyeball has a shield over it; thus, no visual information is perceived by that eyeball. At the level of the optic chiasm, the retinal ganglion cell axons coursing in the medial half of the right optic nerve decussate to the opposite side to course in the medial half of the left optic tract, to terminate in the left lateral geniculate nucleus of the thalamus. In contrast, the retinal ganglion cell axons coursing in the lateral half of the right optic nerve remain ipsilaterally and course in the lateral half of the right optic tract, to terminate in the right lateral geniculate nucleus of the thalamus. Visual information is relayed to the striate area in the occipital lobe in a banded pattern. The alternating clear bands represent the areas occupied by the axons of the optic radiations relaying visual information from the left eyeball. (B) Visual field representation in the striate area. The peripheral area of the visual field is represented in the anterior region of the striate area on the medial surface of the occipital lobe. In contrast, the central or macular area of the visual field is represented in the posteriormost region of the striate area. Source: Modified from Fitzgerald, MJT, Folan-Curran, J (2002) *Clinical Neuroanatomy and Related Neuroscience*. WB Saunders, New York; fig. 25.10.

VISUAL SYSTEM ••• 375

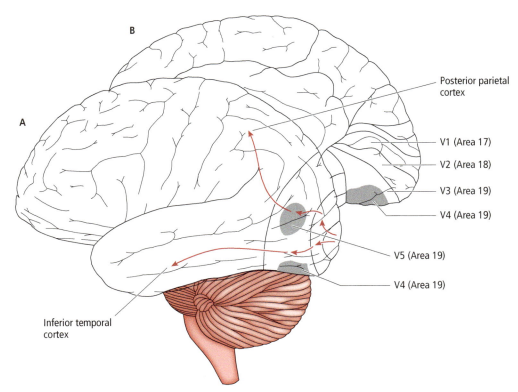

Figure 18.10 • The visual cortical areas and their major connections. (A). Medial view of the brain showing location of V1, V2, V3, and V4. Striate area (primary visual cortex; V1, Brodmann area 17); extrastriate areas (visual association cortical areas; V2, Brodmann area 18; V3, V4, and V5, Brodmann area 19). (B). Visual information flows through two distinct pathways, a posterior pathway and an anterior pathway. In the posterior pathway visual information is processed by the visual cortical areas and the posterior parietal cortex for spatial localization and movement of visual stimuli. In the anterior pathway visual information is processed by the visual cortical areas and the inferior temporal cortex for form and color of visual stimuli. Source: From Martin (2012), figure 7.15, p. 171. *Neuroanatomy Text and Atlas, Fourth Edition.*

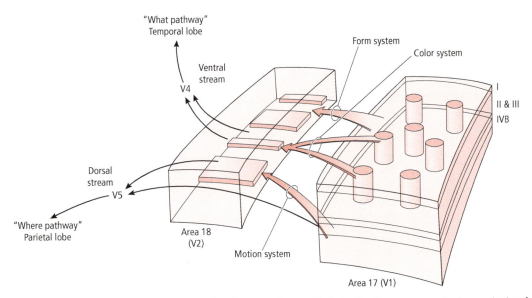

Figure 18.11 • The "what" pathway and "where" pathway from the primary visual cortex. The "what" pathway processes the form and color of visual stimuli. The "where" pathway processes motion of visual stimuli. Source: From Martin (2012), figure 7.16, p. 171. *Neuroanatomy Text and Atlas, Fourth Edition.*

pathway because it carries retinal information that is ultimately assembled by the visual cortical areas to enable the brain to determine the particular details of an object or an individual's face in the visual field.

In the **"what"** pathway, visual input is relayed from **cones** to the small retinal ganglion cells, known as **midget cells**, which are located mostly in the central region of the retina and respond to the fine details (such as texture and shape) and color of stationary stimuli. The midget cells (P cells) relay the information to the **parvocellular layers** (layers 3–6) of the lateral geniculate nucleus of the thalamus. From there, the input is transmitted to the **striate area** (primary visual cortex, V1, area 17) that in turn projects to the **extrastriate areas** (including V2 of Brodmann area 18). Subsequently, information is relayed to the extrastriate area V4 which processes color vision and fine details and is conveyed from the occipital lobe to the **inferior temporal cortex**, (corresponding to Brodmann area 37). Here, visual information is compared to stored information (memories of those specific visual information) to determine its function and significance. Visual information traveling from the retina to the visual cortex via this pathway is processed at each successive relay center specifically for its fine, intricate details of objects in the visual field. The cortical areas integrate all the characteristics to form a composite perception, which is utilized to identify the object or individual (Fig. 18.10).

A lesion in the "what" pathway results in visual *agnosia* (G. *agnostos* "*unknown*") in which the individual sees an object, but is unable to identify it.

The primary visual cortex in the occipital lobe of one hemisphere projects to the primary visual cortex of the opposite hemisphere, thereby combining the visual images from the two retinas into a composite visual perception. The primary visual cortex projects to the brainstem cranial nerve motor nuclei controlling ocular movement in order to support focusing of images on the retina and to direct eyeballs to objects of interest.

In summary: although the "where" and "what" pathways begin in different regions of the retina (where some of the integration of visual information takes place), and proceed to different layers of the lateral geniculate nucleus, the neurons of the optic radiation terminate in the striate area of the occipital lobe where initial cortical processing takes place. From there, most of the fibers of the two pathways terminate in the extrastriate areas (V2 subregion with some ending in the V3 subregion of Brodmann area 18). It is the projections leaving the extrastriate areas (Brodmann area 18 V2 and V3) that, by going their separate ways, terminate in different cortical association areas specializing in the processing of the unique aspects of visual perception carried by the two pathways. Similar to pieces of a puzzle, visual information components including the *shape*, *texture*, *color*, *depth*, and *motion* of an object are assembled and combined so that the brain can form a seamless pattern, a complete visual image. Some of the visual information that reach the frontal lobe reach consciousness. If one is watching a boat sailing in the ocean, the "what" pathway sees the color and other details of the boat, whereas the "where" pathway sees how fast the boat is moving. The tertiary visual areas function in identifying an object as well as in determining its location and color; the middle temporal area has an important function in detecting moving objects.

Columnar organization of the primary visual cortex

The human neocortex consists of a six-layered sheet of neurons and neuroglial cells that processes sensory information of the different ascending sensory systems with similar anatomical and functional organizations. The cortical cell layers are numbered I–VI, with layer I being the most superficial and layer VI the deepest layer located next to the subcortical white matter. The ascending sensory systems send a relay to the thalamus, which, in turn projects to layer IV of the sensory cortical areas. Similar to the other sensory cortical areas, the primary visual cortex also consists of six layers, and the lateral geniculate nucleus, the thalamic relay nucleus for vision, projects to layer IV of the primary visual cortex (striate area).

Neurons with similar functional properties residing in the different cortical layers (I–VI) are oriented longitudinally, above and below one another, to form cellular columns, that are oriented perpendicular to the horizontally oriented cortical layers. The primary visual cortex consists mainly of three

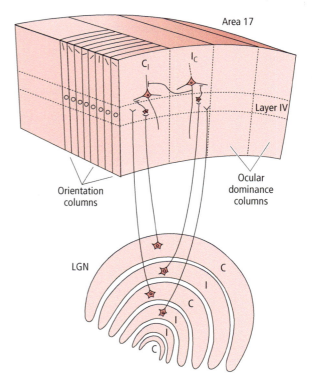

Figure 18.12 ● The lateral geniculate nucleus (LGN) and its projections to the ocular dominance columns and orientation columns of the visual cortex. Source: Haines (2013), figure 20.24, p. 284. Reproduced with permission of Elsevier.

different columns of cellular organization (see Fig. 18.12). They are: (1) **Ocular dominance columns**, whose neurons receive visual information from the ipsilateral or contralateral eyeball. Ocular dominance columns separate information originating from the two retinas, (2) **orientation columns** contain neurons that detect mainly visual stimuli with comparable spatial orientations, and (3) **color columns**, which consist of vertically oriented collections of nerve cells in cortical layers II and III that detect the color of visual stimuli. Visual input from a specific region of the retina of each eyeball is relayed to a pair of ocular dominance columns residing next to one another in the visual cortex. Horizontal communication between cells in these columns plays a role in depth perception.

Visual input projected from the LGN to the striate area (primary visual cortex) is organized on the basis of whether the visual input was derived from the ipsilateral or the contralateral eye. One might expect that the visual input terminates in different cortical layers; however, that is not the case. Visual input from each eye follows a different pattern of termination, relaying to different alternating bands of cortex, referred to as ocular dominance columns.

VISUAL REFLEXES

Pupillary light reflex

The pupillary light reflex, by enabling the eye to adapt to varying light intensities, protects the eye and also facilitates vision

Inherent in the structure of the visual system is its ability to adapt to environmental changes such as varying light intensity, to protect the eye and to facilitate vision. The **pupillary light reflex**, an autonomic response, is mediated independent of cortical input.

The iris contains two smooth muscles, the sphincter pupillae and dilator pupillae, which control the diameter of the pupil. When the circularly arranged fibers of the *sphincter pupillae* contract the diameter of the pupil is decreased, therefore reducing the amount of light that falls on the retina. The fibers of the *dilator pupillae* muscle are arranged radially and, when they contract, the pupillary aperture dilates, allowing more light to enter the eyeball to reach the retina. The size of the pupillary aperture depends on two factors: (1) the tone of the autonomic nervous system – if the activity of the *sympathetic* nervous system dominates, the pupillary aperture *dilates*; if the activity of the *parasympathetic* nervous system dominates, the pupillary aperture *constricts*; and (2) the amount of light that reaches the retina – varying the light intensity results in a corresponding alteration of pupillary size. Normally, bright light causes the pupils to constrict, whereas dim light causes the pupils to dilate. Various pharmacologic agents and illicit drugs, such as anticholinergics and LSD, can cause dilation of the pupil. In contrast, cholinergic agonists, opioids, and barbiturates can cause constriction of the pupils to pin prick size.

The pupillary light reflex of an individual may be tested by illuminating one eyeball with a small flashlight (while preventing the light shining into the other eyeball) and observing that normally, simultaneous pupillary constriction results in both. The reaction in the illuminated eyeball is the **direct pupillary light reflex**, whereas the reaction in the nonilluminated eyeball is the **consensual pupillary light reflex**.

The pupillary light reflex pathway is summarized in Fig. 18.13.

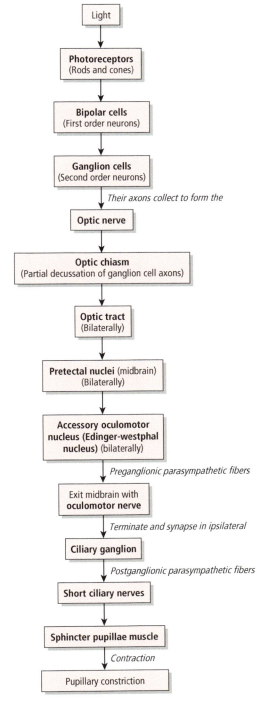

Figure 18.13 • The pupillary light reflex pathway.

Afferent limb of the pupillary light reflex

The information that bright light is stimulating the photosensitive **retina** is transmitted via the **optic nerve**, **optic chiasm**, **optic tract**, and the **brachium of the superior colliculus** (without synapsing at the LGN) to the **pretectal area** located in the cranial aspect of the midbrain (Fig. 18.14). Neurons of the pretectal area send bilateral projections to the **accessory oculomotor nuclei** (eponym: Edinger–Westphal nuclei) of the oculomotor nuclear complex activating the efferent limb of the pupillary light reflex (see below), providing instructions to the sphincter pupillae muscles of both eyeballs to contract, thereby reducing the diameter of the pupils and decreasing the amount of light impinging on the retinae. This is a fast reflex which protects the retina of each eyeball.

Efferent limb of the pupillary light reflex

Each accessory oculomotor nucleus (containing preganglionic parasympathetic neurons) (Fig. 18.14) projects via the **oculomotor nerve** to the ipsilateral **ciliary ganglion** (a parasympathetic ganglion of the oculomotor nerve) synapsing with postganglionic parasympathetic neurons housed in this ganglion. The postganglionic parasympathetic fibers project via the **short ciliary** branches of the **trigeminal nerve** to the ipsilateral **sphincter**

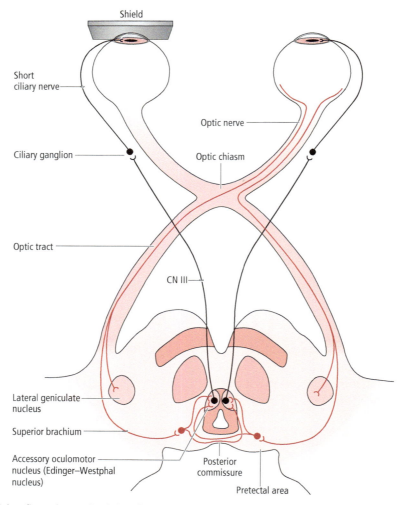

Figure 18.14 ● The pupillary light reflex pathway. When light is flashed into one eyeball, normally, both pupils constrict simultaneously as follows. The information is transmitted from the illuminated eyeball via the optic nerve and then the optic tract to the pretectal area in the midbrain. The pretectal area projects to the **accessory oculomotor nucleus** (eponym: Edinger–Westphal nucleus) bilaterally – connecting it to the parasympathetic neurons of both sides, thus initiating a bilateral pupillary response. The accessory oculomotor nucleus contains the cell bodies of preganglionic parasympathetic neurons whose axons join the oculomotor nerve of its respective side, to terminate in the ciliary ganglion. In the ciliary ganglion, the preganglionic parasympathetic terminals synapse with the postganglionic parasympathetic neurons whose axons project via the short ciliary nerve of the trigeminal nerve to the sphincter pupillae muscle, causing the pupil to constrict.

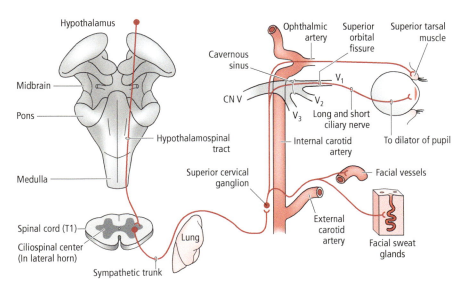

Figure 18.15 • The pupillary dilation pathway. Axons arising from the hypothalamus form the hypothalamospinal tract, which descends ipsilaterally in the lateral funiculus of the spinal cord to terminate in the ciliospinal center of the intermediolateral cell column at the T1 level. Preganglionic sympathetic neurons project their axons arising from the ciliospinal center to the superior cervical ganglion where they synapse with postganglionic sympathetic neurons. The postganglionic neurons give rise to axons that form a perivascular plexus, following vessels to the orbit where they terminate in the dilator pupillae muscle, causing pupillary dilation. Source: Modified from Fix, JD (1995) *Neuroanatomy*. Williams & Wilkins, Media; fig. 17.4.

pupillae muscle, causing it to contract. This results in bilateral pupillary constriction, with a consequent reduction in the amount of light reaching the retina of both eyeballs.

Pupillary dilation reflex

Pupillary dilation occurs when sympathetic activity is dominant, such as when the individual is experiencing pain, fear, rage, or illicit drugs

Pupillary dilation, mediated by the sympathetic division of the autonomic nervous system, occurs

1. When sympathetic activity is dominant, such as when the individual is experiencing pain, fear, rage, or illicit drugs with no change in the amount of light entering the eyeball and it
2. Also occurs slowly in dim illumination to increase the pupillary aperture to allow more light into the eye and to impinge on the retina.

Neurons of the posterior aspect of the **hypothalamus** project to the **ciliospinal center** of the spinal cord located in the intermediolateral cell column of cord levels (C8) T1–T2 (Fig. 18.15). Here, their axons synapse with preganglionic sympathetic neurons. Preganglionic sympathetic fibers exit the spinal cord, enter the sympathetic trunk, and ascend to the **superior cervical ganglion** where they synapse with postganglionic sympathetic neurons. Postganglionic sympathetic fibers form the internal carotid arterial perivascular plexus and follow vessel branches to the orbit where they join the **long ciliary** branches of the **trigeminal nerve** and terminate in the **dilator pupillae muscle**. Sympathetic innervation of this muscle causes it to contract, increasing the pupillary diameter (Fig. 18.16).

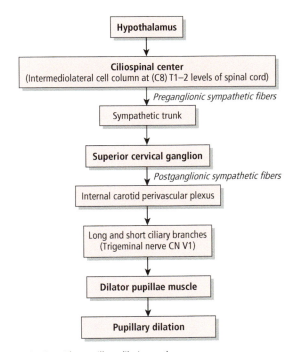

Figure 18.16 • The pupillary dilation pathway.

Convergence accommodation reflex

The convergence accommodation reflex alters the thickness of the lens, which facilitates the projection of a focused image on the retina

As visual attention is consciously switched from a faraway object to one nearby, the thickness of the lens of each eyeball changes from a flatter to a more rounded, biconvex shape in order to be able to project a focused image on the retina. This is accomplished by an initial conscious fixation on the near object, followed by the subconscious convergence accommodation reflex, consisting of three reflex changes (Fig. 18.17).

1 **Convergence**. As attention is shifted from a far to a near object, both eyeballs converge simultaneously. This is mediated by bilateral contraction of the medial recti muscles, which are innervated by the **medial rectus subnucleus** of the oculomotor nerve (CN III). This convergence permits the visual image to be projected and focused on the foveae of both retinas. If this is not accomplished, the individual will experience diplopia (G. *diploos* "double"; double vision). Unlike other reflexes, this reflex involves the cerebral cortex.
2 **Accommodation**. Parasympathetic stimulation causes contraction of the ciliary body of the iris thus releasing the tension it exerts on the ciliary zonules attached to the lens. Since the lens is no longer being stretched thin, it shortens and thickens (lens **accommodation**), creating a higher refractive index, and focuses the image on the retina.
3 **Pupillary constriction**. Pupillary constriction enhances the outline of the image formed on the retina. This is a distinct and independent process from that which occurs in the pupillary light reflex.

Visual input is transmitted from the retina to the striate area (primary visual cortex) via the visual pathway. The primary visual cortex transmits the information to the visual association cortex (Brodmann area 19) where reflex activity is initiated. Fibers from this area form the **corticotectal tract** (the afferent limb of reflex), which projects bilaterally, to the superior colliculus and/or the pretectal area. These areas then project to the **interoculomotor nucleus** of the oculomotor nuclear complex, that, in turn, project to the **accessory oculomotor** nucleus and the **medial rectus subnucleus** (also of the oculomotor nuclear complex). Motor fibers from the medial rectus subnucleus and preganglionic parasympathetic fibers

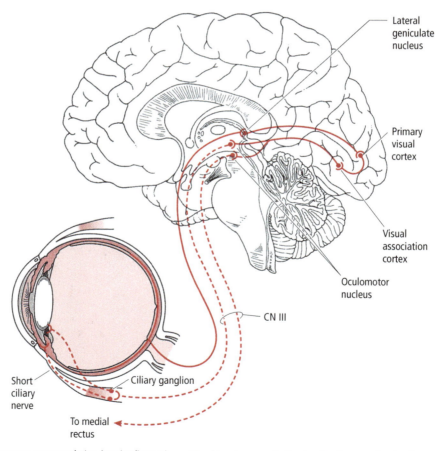

Figure 18.17 ● The convergence accommodation (near) reflex pathway. Visual input passes via the visual pathway to the visual association cortex. It is believed that the visual association cortex projects bilaterally, to the superior colliculus and/or the pretectal area, both of which in turn project to the **interoculomotor nucleus** (eponym: nucleus of Perlia) of the oculomotor nuclear complex. Both interoculomotor nuclei project to the **accessory oculomotor nucleus** (eponym: Edinger–Westphal nucleus) and the **medial rectus subnucleus** of the oculomotor nucleus. Each oculomotor nucleus stimulates the contraction of the ipsilateral medial rectus muscle, causing adduction of the eyeball. The accessory oculomotor nucleus projects preganglionic parasympathetic fibers to the ciliary ganglion where they synapse with postganglionic parasympathetic neurons that innervate the sphincter pupillae (pupillary constriction) and the ciliary muscle (lens accommodation) causing the lens to thicken for near vision. Note that the accommodation reflex, a bilateral response, is as described above.

VISUAL SYSTEM • • • 381

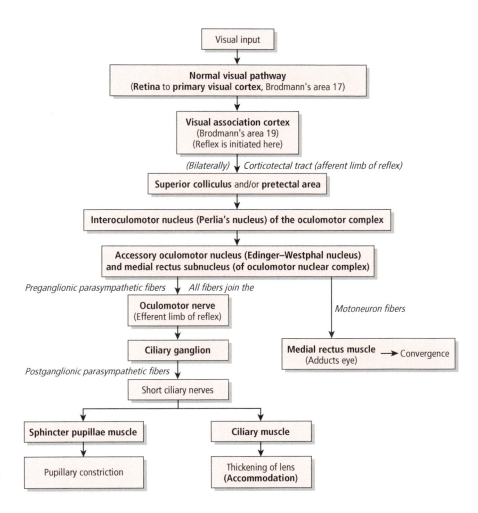

Figure 18.18 • The convergence accommodation (near) reflex pathway.

from the accessory oculomotor nucleus (Edinger–Westphal nucleus) join the **oculomotor nerve** (the efferent limb of reflex) to distribute to the medial rectus muscle and the ciliary ganglion, respectively. The medial recti muscles adduct the eyeballs (convergence). Postganglionic parasympathetic fibers arising from the ciliary ganglion join the short ciliary nerves of the ophthalmic division of the trigeminal nerve to distribute to the sphincter pupillae muscle and to the ciliary muscle. Contraction of the sphincter pupillae muscle results in pupillary constriction, whereas contraction of the ciliary body results in thickening of the lens (accommodation) for near focus (Fig. 18.18).

Corneal blink reflex

When a foreign object contacts the cornea, the corneal blink reflex elicits a forceful blinking of both eyelids to protect the eyeballs from possible injury

The **corneal blink reflex** consists of (Fig. 18.19):

- **Receptors** (at the peripheral terminals of pseudounipolar neurons in the ophthalmic division of the trigeminal nerve);
- An **afferent limb** (peripheral processes of pseudounipolar neurons in the ophthalmic division of the trigeminal nerve);
- An **integrator** (spinal nucleus and main sensory nucleus of the trigeminal nerve);
- An **efferent limb** (facial nerve branches to the orbicularis oculi muscle); and
- An **effector** (orbicularis oculi muscle).

When a fine strand of hair or any other foreign object is brushed against the cornea unilaterally, it evokes an immediate, forceful blinking of both eyes simultaneously. **General somatic afferent** (**GSA**) sensation from the cornea is transmitted by the receptors at the peripheral processes of **first-order pseudounipolar** neurons whose cell bodies are housed in the **trigeminal ganglion**. These peripheral processes course in the ophthalmic division of the trigeminal nerve (afferent limb). The central processes of most of these neurons enter the pons, join the ipsilateral **spinal tract of the trigeminal nerve**, and descend to terminate in the rostral two-thirds of the ipsilateral **spinal nucleus of the trigeminal nerve** where they synapse with **second-order neurons** (interneurons). The remainder of the first-order neurons

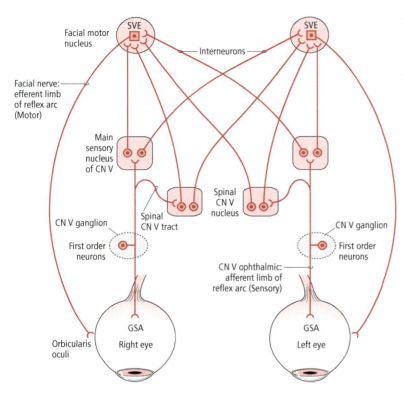

Figure 18.19 • The corneal blink reflex. When a wisp of cotton is gently brushed against the cornea of one eyeball, both eyelids blink. Touch sensation is transmitted from the cornea via the pseudounipolar neurons of the trigeminal ganglion to the ipsilateral spinal nucleus and main sensory nucleus of the trigeminal nerve. Both the spinal and main sensory nuclei of the trigeminal nerve send bilateral projections to the facial motor nuclei where they synapse with the motor neurons that innervate the orbicularis oculi muscles, causing simultaneous blinking of both eyelids. GSA, general somatic afferent; SVE, special visceral efferent.

terminate in the ipsilateral **main sensory nucleus of the trigeminal nerve**. Second-order neurons from the spinal trigeminal nucleus as well as from the main sensory nucleus project bilaterally to the facial nucleus where they synapse with motor neurons. Fibers of the motor neurons course in the **facial nerve** to terminate in and innervate the orbicularis oculi muscle, causing bilateral blinking (Fig. 18.20).

Visual fields

The **visual field** is that part of the environment that is perceived by the individual with both eyelids open and in primary position (looking straight ahead). In the center there is a **binocular zone/field** that is seen by both eyeballs and on each side of it, like parentheses, there is a crescent-shaped right and left **monocular zone** that is perceived by the ipsilateral retina (Fig. 18.9).

The **visual field** is divided by imaginary lines into nasal (medial) and temporal (lateral) halves, which, in turn, are further subdivided into upper (superior) and lower (inferior) quadrants (Fig. 18.21A). Since light rays travel in straight lines, the temporal half of the visual field (hemifield) is projected onto the nasal half of the retina (hemiretina), whereas the nasal half of the visual field (hemifield) is projected onto the temporal half of the retina (hemiretina). The image within the visual field quadrants is reversed and inverted by the lens when it is projected on the respective retina.

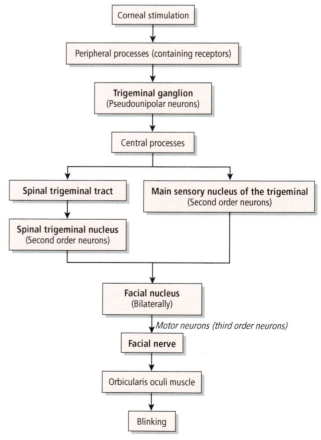

Figure 18.20 • The corneal blink reflex.

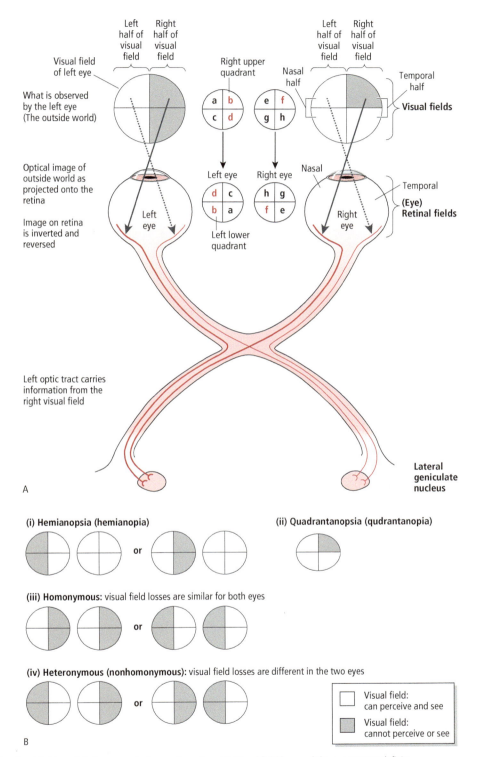

Figure 18.21 ● (A) The visual fields and their representation on the retina. (B) Visual field loss and the consequent deficits.

CLINICAL CONSIDERATIONS

Visual deficits

The naming of visual deficits is based on the visual field loss and not the actual impaired or damaged region of the retina or other visual system component

The naming of visual deficits is based on the *visual field loss* and not the actual impaired or damaged region of the retina or other visual system component (Table 18.2).

Anopsia (anopia) (G., "loss of vision") refers to visual field loss. *Hemianopsia* (G., *hemi*, "half") is the loss of only one-half (temporal or nasal) of the visual field of one or both retinas (Fig. 18.21B). **Quadrantanopsia** (L., *quadr*, "one–fourth") is the loss of one–fourth (upper temporal or lower temporal; upper nasal or lower nasal) of the visual field of one or both retinas.

A deficit may be homonymous (G., *homo*, "same") or heteronymous (G., *hetero*, "different") (Fig. 18.21B). In a **homonymous** deficit, the visual field losses for the two retinas are the same. That is, there is a right or left visual field loss for both retinas. In a **heteronymous** deficit, the visual field losses for the two retinas are not alike. That is, there is a right half visual field loss in one retina, and a left half visual field loss in the contralateral retina, or vice versa.

A *unilateral* lesion of the **optic nerve** results in **permanent blindness** in the ipsilateral (corresponding) retina (A in Fig. 18.22).

Damage to the fibers at the optic chiasm results in heteronymous deficits. *Unilateral* damage to the **temporal (non-crossing) fibers of the optic chiasm** (arising from the temporal half of the ipsilateral retina) results in an **ipsilateral nasal hemianopsia** (B1 in Fig. 18.22). The non-crossing fibers are often damaged by an aneurysm and/or calcification of the internal carotid artery due to its close proximity to the fibers. *Bilateral* damage to the **temporal fibers of the optic chiasm** (arising from the temporal half of each retina) results in a **binasal heteronymous hemianopsia** (B2 in Fig. 18.22).

A lesion that damages the **crossed fibers** (arising from the nasal half of each retina) occupying the central region of the **optic chiasm** results in a bitemporal heteronymous (non-homonymous) hemianopsia ("tunnel vision") (C in Fig. 18.22). The central fibers may be damaged due to compression by a tumor of the hypophysis (pituitary gland).

Damage to the fibers posterior to the optic chiasm results in homonymous deficits.

A *unilateral* lesion caudal to the optic chiasm, affecting the **optic tract** fibers or the **lateral geniculate nucleus**, results in a **contralateral homonymous hemianopsia** (D in Fig. 18.22). In this case the right optic tract is damaged; thus, objects in the left half of the visual field of each eyeball are not perceived. The visual field loss of each eyeball is the same (homonymous).

A *unilateral* lesion from occlusion of the middle cerebral artery affecting the **lower division of the optic radiations**, results in a **contralateral upper homonymous quadrantanopsia** or **"pie in the sky visual disorder"** (E in Fig. 18.22). A lesion affecting the fibers of the **upper division of the geniculocalcarine tract (optic radiations)** results in a **contralateral lower homonymous quadrantanopsia** or **"pie in the floor visual disorder"** (F in Fig. 18.22).

A *unilateral* lesion that damages the **upper and lower divisions of the optic radiation** will result in a **contralateral homonymous hemianopsia** (G in Fig. 18.22).

A *unilateral* **vascular** lesion that damages the striate area (primary visual cortex) may result in **contralateral homonymous hemianopsia with macular sparing** (H in Fig. 18.22). The cortical area that receives visual information from the macula (located in the posterior pole of the primary visual cortex) receives blood supply from the *posterior cerebral artery* (PCA) with collateral blood supply contributed by the *middle cerebral artery* (MCA).

Macular sparing occurs because a large area of the primary visual cortex is allocated for processing visual information from the macula (a rather small area of the retina, with high visual acuity) that often some central vision is spared due to the collateral circulation. Thrombosis of the *posterior cerebral artery* often results in damage of the primary visual cortex. This in turn results in a contralateral visual field defect that at times spares the part of the

---

Table 18.2 ● Visual deficit(s) resulting following a lesion at various locations in the visual pathway. Lesions damaging the fibers of the optic chiasm result in *heteronymous* defects. A lesion damaging structures of the visual pathway located posterior to the optic chiasm produces *homonymous* visual defects.

Structure of visual pathway	Lesion	Deficit
Optic nerve	Unilateral	Permanent blindness of ipsilateral retina
Optic chiasm		
Temporal fibers (noncrossing fibers)	Unilateral	Ipsilateral nasal hemianopsia
	Bilateral	Binasal heteronymous hemianopsia
Nasal fibers (crossing fibers)	Bilateral	Bitemporal heteronymous hemianopsia ("tunnel vision")
Optic tract	Unilateral	Contralateral homonymous hemianopsia
Lateral geniculate nucleus	Unilateral	Contralateral homonymous hemianopsia
Lower division of optic radiations (eponym: Meyer's loop)	Unilateral	Contralateral upper homonymous quadrantanopsia
Upper division of optic radiations	Unilateral	Contralateral lower homonymous quadrantanopsia
Both upper and lower divisions of optic radiations	Unilateral	Contralateral homonymous hemianopsia
Primary visual cortex (due to vascular lesion resulting from posterior cerebral artery thrombosis)	Unilateral	Contralateral homonymous hemianopsia with macular sparing

CLINICAL CONSIDERATIONS (continued)

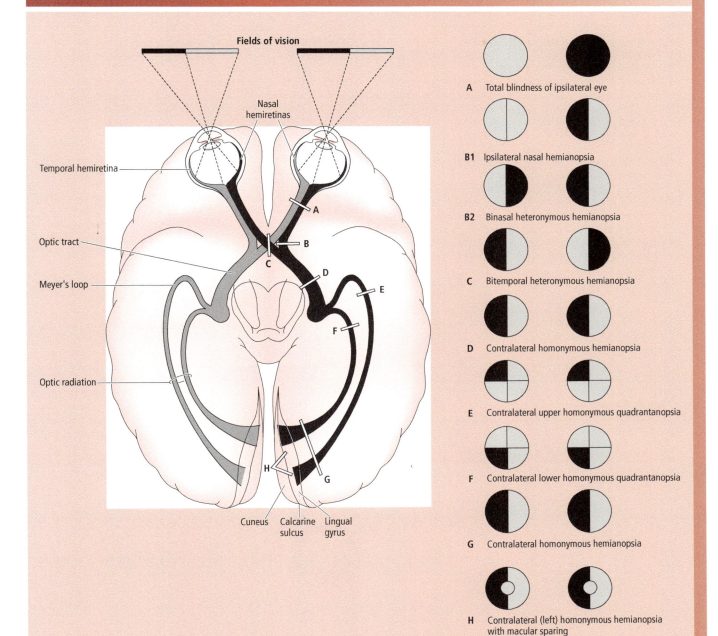

Figure 18.22 • Lesions at various points along the visual pathway and consequent visual field deficits: (A) lesion of the optic nerve: total blindness in the ipsilateral eye; (B1) unilateral lesion of the temporal (noncrossing) fibers of the optic chiasm: ipsilateral (right) nasal hemianopsia; (B2) bilateral lesion of the temporal (noncrossing) fibers of the optic chiasm: binasal heteronymous hemianopsia; (C) lesion of the crossing fibers of the optic chiasm: bitemporal heteronymous hemianopsia; (D) lesion of the optic tract: contralateral (left) homonymous hemianopsia; (E) lesion of Meyer's loop (lower division of the optic radiations): contralateral (left) upper homonymous quadrantanopsia; (F) lesion of the upper division of the optic radiations: contralateral (left) lower homonymous quadrantanopsia; (G) lesion of the upper and lower divisions of the optic radiations: contralateral (left) homonymous hemianopsia; (H) a vascular lesion of the primary visual cortex resulting from posterior cerebral artery thrombosis: contralateral (left) homonymous hemianopsia with macular sparing.

CLINICAL CONSIDERATIONS (*continued*)

cortex that processes visual information from the macula (central vision). Two factors support macular sparing: (1) The cortical area that processes information from the macula is supplied by the *posterior cerebral artery* with collateral blood supply provided by the *middle cerebral artery*. If the posterior cerebral artery becomes occluded, the middle cerebral artery can nourish the macular cortical region. (2) The region of the visual cortex that processes information from the macula is quite extensive. It is rare that a single infarction damages the entire cortical region that represents the macula.

Lesion in the afferent or efferent limb of the pupillary light reflex pathway
If one of the optic nerves (afferent limb of the pupillary light reflex arc) is damaged, and light is flashed into the eyeball ipsilateral to the damaged nerve, both direct and consensual pupillary responses will not be present since sensory input is not transmitted centrally. When the light is flashed into the eyeball on the side of the undamaged optic nerve, both the direct and the consensual pupillary responses will be elicited, if both oculomotor nerves are functional.

If the optic nerve is functional bilaterally, but there is unilateral damage to the oculomotor nerve (efferent limb of the pupillary light reflex arc), light flashed into either eye is perceived by the individual and sensory input is transmitted centrally. When light is flashed into the eyeball ipsilateral to the oculomotor nerve lesion there is only a consensual pupillary response. When light is flashed into the eyeball contralateral to the oculomotor nerve lesion, there is only a direct pupillary response. If both oculomotor nerves are damaged and parasympathetic innervation to the sphincter pupillae muscles is interrupted bilaterally, then neither pupil will exhibit the direct or consensual responses.

An individual may be blind but present a normal pupillary light reflex (see Fig. 18.14). This indicates that the site of the lesion producing the blindness involves structures caudal (posterior) to the divergence of the optic tract fibers (destined for the accessory oculomotor nucleus and/or the pretectal area, which is associated with the pupillary reflex). The lesion may involve the LGN, or the optic radiations, or the primary visual cortex. If the individual is completely blind, a bilateral lesion of the visual cortex is more probable since a bilateral lesion of the optic radiations is incompatible with life as trauma or disease of a size to affect both would involve extensive damage to surrounding structures.

Lesion in the pupillary dilation pathway
A lesion that damages the sympathetic (preganglionic or postganglionic) fibers of this pathway results in **oculosympathetic paresis** (eponym: Horner syndrome) ipsilateral to the lesion. One of the characteristics of this syndrome is a persistent **miosis** (pupillary constriction) ipsilateral to the lesion due to the unopposed action of the parasympathetic nervous system. Although an individual with oculosympathetic paresis exhibits loss of the pupillary dilation reflex, the pupillary constriction reflex is intact, both ipsilateral and contralateral to the lesion.

Lesion in the afferent or efferent limb of the corneal blink reflex pathway
If there is a lesion in the ophthalmic nerve (afferent limb) of the corneal blink reflex arc, corneal sensation is lost on the ipsilateral side, and the corneal blink reflex is not elicited on either side upon stimulation of the cornea ipsilateral to the lesion. If the cornea ipsilateral to the intact ophthalmic nerve is stimulated, it evokes bilateral simultaneous blinking.

If the ophthalmic nerve (afferent limb) of the corneal blink reflex arc is intact bilaterally, but there is a unilateral lesion of the facial nerve (efferent limb), both corneas can sense stimulation when either one is stimulated. However, only the eyelid ipsilateral to the intact facial nerve will blink.

Note that the clinical case at the beginning of the chapter refers to a patient with visual loss, papilledema of the optic disc, an abnormal pupillary light reflex, and pain in the right eye.

1 What forms the optic disc?

2 At what point do the retinal ganglion cell axons become myelinated?

3 What component of the pupillary light reflex arc is affected by the papilledema in this patient?

SYNONYMS AND EPONYMS OF THE VISUAL SYSTEM

Name of structure or term	Synonym(s)/eponym(s)
Hemianopsia	Hemianopia
Heteronymous	Non-homonymous
Intermediolateral horn of the spinal cord	Lateral horn of the spinal cord
	Lateral column of the spinal cord
Miosis	Pupillary constriction
Mydriasis	Pupillary dilation
Optic radiation	Geniculocalcarine tract
	Geniculostriate tract
Primary visual cortex	Striate area
	Brodmann area 17
Sphincter pupillae muscle	Constrictor pupillae muscle
Striate area	Band of Gennari
	Primary visual cortex

FOLLOW-UP TO CLINICAL CASE

This patient has optic neuritis. This refers to inflammation of the optic nerve arising from any cause. However, it most commonly occurs in multiple sclerosis (MS) or another demyelinating disease. Not all patients with optic neuritis have or will ever develop MS, but most do. Therefore, optic neuritis typically refers to demyelination of the optic nerves. MS affects only the central nervous system. Although almost all of the cranial nerves are components of the peripheral nervous system, the optic nerves are unusual since they are actually tracts and thus components of the central nervous system.

This case is a typical presentation of optic neuritis, that is, visual loss affecting one eyeball (occasionally both) and sometimes with pain on eye movement and papilledema. Papilledema refers to swelling of the optic disc as seen on ophthalmoscopic exam. This may or may not be present, depending on the part of the optic nerve that is affected. If visual acuity is severely impaired then the pupil of the affected eyeball will not constrict to light properly, that is, less than the unaffected eyeball. This can result in a "Marcus Gunn pupil," as determined in the swinging light test. The examiner shines a penlight into each eyeball in succession. When the light is directed in the normal eye, both pupils constrict. When it is swung to shine into the affected eyeball, the pupil paradoxically dilates (this is enhanced by the fact that it had been constricted just prior). This is referred to as an afferent pupillary defect in the affected eyeball.

Optic neuritis is treated by intravenous corticosteroids over three to five days. The majority of cases resolve completely, or nearly so, over the course of weeks or months. An MRI of the optic nerves may reveal the inflammation. However, an MRI is used mainly in the detection of any other brain lesion that may be indicative of MS. Visually evoked potentials can also be used to detect subclinical abnormality of the optic nerves to confirm the presence of MS in suspected cases.

QUESTIONS TO PONDER

1. What is the pathway of the light entering the eyeball and that of the visual electrical signals through the layers of the retina?

2. A physician was testing the pupillary light reflex of a patient. When the right eyeball was illuminated, both direct and consensual pupillary light reflexes were elicited. However, when the left eyeball was illuminated, neither pupil responded. Where might the lesion be?

3. A physician was testing the pupillary light reflex of a patient. When the right eye was illuminated, only the consensual pupillary light reflex was elicited. When the left eye was illuminated, only the direct papillary light reflex was elicited. Where might the lesion be?

4. A physician was testing the corneal blink reflex in a patient. When brushing a wisp of cotton across the patient's right cornea, only the right eyelid blinked. When the cotton was brushed across the patient's left cornea, again only the right eyelid blinked. Where may the lesion be?

5. Following a thorough examination of a patient's visual system, the physician concluded that the patient had a bitemporal heteronymous hemianopsia. Where in the visual pathway is the lesion? What might cause this type of visual disorder?

6. What causes oculosympathetic paresis (Horner syndrome)?

CHAPTER 19

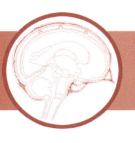  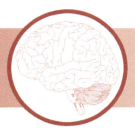

Auditory System

CLINICAL CASE

EAR

AUDITORY TRANSMISSION

CENTRAL AUDITORY PATHWAYS

CLINICAL CONSIDERATIONS

SYNONYMS AND EPONYMS OF THE AUDITORY SYSTEM

FOLLOW-UP TO CLINICAL CASE

QUESTIONS TO PONDER

CLINICAL CASE

A 49-year-old patient complained of right-sided hearing loss, which began a year ago. This had slowly and progressively become worse. Hearing from the left ear was normal. They first noticed this when it became difficult to hear with the phone held to their right ear. They have also noted vague and persistent imbalance and dizziness over the past couple of months. There is some ringing in their right ear. The patient had no other symptoms.

Examination revealed mild hearing loss and difficulty distinguishing different sounds on the right. The appearance of the external auditory canal and tympanic membrane is unremarkable. Gait examination revealed mild ataxia when the patient was attempting to walk in a straight line. The rest of the neurologic exam was within normal limits.

Auditory information is transmitted via the **auditory pathway** from the internal ear to the auditory cortex in the temporal lobe. This pathway that mediates the special sense of hearing begins with the first-order neurons of the **cochlear ganglion** in the cochlea (G., *kokhlias*, "snail shell"), a bony structure located in the internal ear within the petrous portion of the temporal bone (Fig. 19.1). The peripheral processes of these neurons innervate a specialized receptor organ, the **spiral organ (eponym: organ of Corti)** housed within the cochlea. The central processes (axons) of these neurons form the **cochlear nerve** (the cochlear division of the vestibulocochlear nerve, CN VIII) which transmits auditory information to the cochlear nuclei in the brainstem. From there, the pathway continues as a sequence of **multisynaptic** neural connections through the medulla oblongata, pons, midbrain, diencephalon, and the cerebral cortex. In its ascent, the auditory information is processed with respect to frequency (pitch), amplitude (loudness), and location of the sound. Auditory information originating in one ear ultimately is transmitted to both cerebral hemispheres. When auditory information arrives at the primary auditory cortex it enters consciousness (a sound is heard). The primary auditory cortex transmits the information to the secondary auditory cortex where it is interpreted (determines meaning of sound), and via connections to the posterior portion of the superior temporal gyrus (eponym: Wernicke's area), functions in the comprehension of language. Secondary auditory cortical areas project to higher-order auditory areas via the "what" pathway and the "where" pathway. Via the "what" pathway, one determines who is speaking or what the source of the sound is. Via the "where" pathway, the location of a sound is determined relative to the listener's ears. Originally, the

---

A Textbook of Neuroanatomy, Third Edition. Maria A. Patestas, Amanda J. Meyer, and Leslie P. Gartner.
© 2025 John Wiley & Sons, Inc. Published 2025 by John Wiley & Sons, Inc.
Companion website: www.wiley.com.go/patestas/neuroanatomy3e

principal function of the internal ear was the maintenance of equilibrium (balance), but the phylogenetically older vestibular apparatus (for balance) developed a special receptor for sound perception.

EAR

The ear is divisible into three parts: the **external ear**, the **middle ear**, and the **internal ear**, each performing a necessary function for the perception of sound (Fig. 19.1; Table 19.1).

External ear

The components of the external ear facilitate the reception of sound waves by funneling them toward the tympanic membrane

The **external ear** is comprised of the cartilaginous **auricle (pinna)** and the **external auditory meatus (canal)**, both of which serve to facilitate the reception of sound waves conducted by the air by funneling them toward the **tympanic membrane** located at the internal extent of the meatus. The contour of the helical-shaped auricle alters the frequency spectrum of sound depending on the position of the listener in reference to the location of the sound source. The tympanic membrane resembles a resilient trampoline that oscillates in response to incoming sound waves. Moreover, it forms a partition and seals off the external auditory meatus from the tympanic (middle ear) cavity.

Middle ear

The middle ear functions in the transmission and transduction of the vibrations of the tympanic membrane to the internal ear (i.e., from an air to a fluid medium)

The **middle ear**, also referred to as the **tympanic cavity**, consists of an irregular-shaped *air-filled* chamber, embedded in the petrous portion of the temporal bone. It contains several important elements:

1. Three small articulating bones: the auditory ossicles (the **malleus**, the **incus**, and the **stapes**).
2. Two miniature skeletal muscles: the **tensor tympani muscle**, which attaches to the malleus, and the **stapedius muscle**, which attaches to the stapes.
3. Two membrane-covered foramina in the bone: the **vestibular (oval) window** and the **cochlear (round) window**.
4. Two additional openings: a. the opening of the **auditory tube** (pharyngotympanic tube eponym: Eustachian tube), which permits communication between the middle ear and nasopharynx, and b. a posterior opening, known as the **aditus to mastoid antrum**.
5. The **chorda tympani nerve**: a branch of the facial nerve (CN VII), which, although it passes through the middle ear and name reflects it, has no function there.

The middle ear functions in the *transmission* and *transduction* of the vibrations of the tympanic membrane to the internal ear (from an air to bone to a fluid medium) (Fig. 19.2).

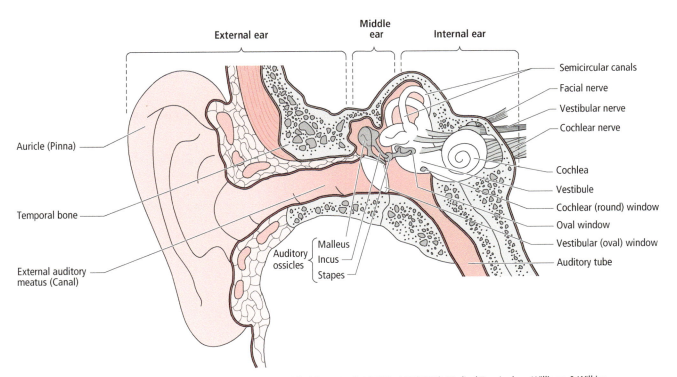

Figure 19.1 ● The external, middle, and internal ear. Source: Modified from Canfield Willis, MC (1996) *Medical Terminology*. Williams & Wilkins, Baltimore; plate 26.

Table 19.1 ● Components of the outer, middle, and internal ear.

Part of ear	Components	Function
External ear	Auricle	Alters the frequency spectrum of sound received
	External auditory meatus	Funnels sound waves toward the tympanic membrane
	Tympanic membrane	Separates external ear from middle ear
		Transmits sound wave vibrations to the ossicles
Middle ear	Ossicles	Conduct and amplify sound waves
	-Malleus	Conduct oscillations from the tympanic membrane to the membrane of the
	-Incus	vestibular window
	-Stapes	
	Muscles	Dampen intensity of tympanic membrane vibrations
	-Tensor tympani	Tenses tympanic membrane by pulling on malleus
	-Stapedius	Pulls on and stabilizes the stapes
	Windows	Permit the conversion of air waves into fluid waves of the internal ear
	-Vestibular (oval) window	
	-Cochlear (round) window	
	Openings	
	-Auditory tube	Equalizes atmospheric pressure between middle and external ear
	(pharyngotympanic tube, eponym: Eustachian tube)	
	-Opening to the mastoid air Cells	Unknown
	Nerve	
	-Chorda tympani	No function in ear, nerve just passes by
Internal ear	Bony labyrinth	Contains perilymph
	-Vestibule	Encloses the saccule and the utricle
	-Semicircular canals	Enclose the (membranous) semicircular ducts
	-Cochlea	Encloses the cochlear duct
	-Scala vestibuli	Waves propagated in its perilymph agitate the vestibular membrane
	-Scala tympani	Its perilymph propagates waves toward the cochlear window
	Membranous labyrinth	Contains endolymph
	-Three semicircular ducts	Contain the receptors for head movement
	-Saccule	Contains the receptors for head movement
	-Utricle	Contains the receptors for head movement
	-Endolymphatic sac	Connects the scala vestibuli and scala tympani to the subarachnoid space
	-Endolymphatic duct	Contains endolymph and the spiral organ (the receptor for hearing)
	-Cochlear duct	Contains endolymph whose vibrations are converted into electrical impulses by mechanical stimulation of hair cells of the spiral organ

As discussed below, both the **vestibular window** and the **cochlear window** serve necessary functions in these processes.

The **ossicles** (malleus, incus, and stapes) are linked in series by synovial joints and are suspended within the tympanic cavity by tiny ligaments. Since they extend from the tympanic membrane (via malleus) to the membrane of the vestibular window (via stapes), they conduct oscillations of the tympanic membrane to the membrane of the vestibular window. The intensity of these vibrations is dampened by the **tensor tympani** and the **stapedius muscles**. When the tensor tympani contracts, it tenses the tympanic membrane by pulling on malleus, therefore reducing the intensity of the vibrations transmitted from this membrane to the ossicles. When the stapedius muscle contracts, it pulls on and stabilizes the stapes, therefore reducing the vibrations that are transmitted by the footplate of the stapes to the vestibular window of the membranous labyrinth

of the internal ear, and eventually to the auditory receptors (hair cells). By dampening the vibrations, they serve a protective reflex function for the auditory apparatus by reducing the intensity with which the stapes impinges on the membrane of the vestibular window safeguarding it from being damaged.

The auditory tube, also referred to as the pharyngotympanic tube (eponym: Eustachian tube), connects the middle ear (tympanic cavity) to the nasopharynx. It serves to equalize the atmospheric pressure between the middle ear and external ear. At higher elevations, the atmospheric pressure in the external auditory meatus is lower than that of the middle ear cavity and causes the tympanic membrane to curve outwards (toward the external auditory meatus) stimulating pain receptors of the tympanic membrane. The external surface of the tympanic membrane receives sensory innervation from the mandibular branch of trigeminal (CN V3) and the auricular

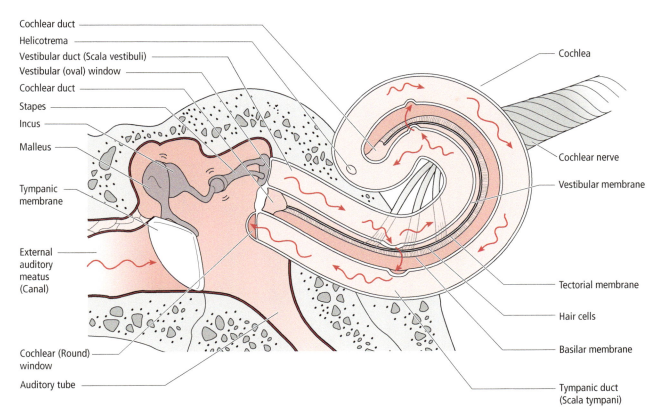

Figure 19.2 ● The transmission of sound waves from the external to the internal ear. Sound waves strike the tympanic membrane, causing it to vibrate. The vibrations of the tympanic membrane are transmitted to the ossicles in the middle ear, which, in turn transmit the vibrations from the foot of stapes to the membrane of the vestibular window. Oscillations of the vestibular window's membrane are then sequentially transmitted to the perilymph of the vestibular duct and the endolymph of the cochlear duct, and then to the basilar membrane. Sound is detected by the bending of stereocilia of hair cells of the spiral organ due to movement of the basilar and tectorial membranes. Vibrations are conveyed to the perilymph in the tympanic duct, which causes the elastic membrane covering the cochlear window to release pressure waves into the middle ear cavity, where they dissipate.

branch of vagus (CN X) nerves, while the internal surface of the tympanic membrane receives sensory innervation from the glossopharyngeal nerve (CN IX). Since the walls of the auditory tube are normally collapsed, pressure differences can be relieved by the action of swallowing, chewing, yawning, blowing one's nose while holding the nares closed, or coughing, actions that serve to open the auditory tube via the contraction of tensor veli palatini muscles, allowing equalization of pressure on the two sides of the tympanic membrane. When pressure differences cannot be relieved, such as during a middle ear infection, pain can be referred to other areas innervated by the same nerves including the mandible (for CN V3), the external acoustic meatus (for CN X), and the oropharynx (for CN IX). The auditory tube also allows any fluid accumulation in the middle ear to drain into the nasopharynx.

Internal ear

The internal ear encloses the cochlea, which contains the spiral organ (the receptor for hearing) and the vestibular apparatus, which contains the receptors associated with maintenance of equilibrium

The *fluid-filled* **internal ear** is embedded in the petrous portion of the temporal bone. It consists of the **bony (osseous) labyrinth** (G., "maze") and the **membranous labyrinth** forming an interconnected system of fluid-filled channels and chambers. The bony labyrinth is divided into three regions: the **vestibule**, the three **semicircular canals**, and the **cochlea** (Fig. 19.3). The vestibule and the semicircular canals are associated with the maintenance of equilibrium and are discussed in Chapter 20. The membranous labyrinth fits inside and is supported by the bony labyrinth and follows its general contour.

Cochlea

The cochlea's interior is divided into three parallel, fluid-filled, compartments by two membranes, the vestibular (eponym: Reissner) membrane and the basilar membrane

The **cochlea** is a spiral, bony shell, resembling a snail shell (Figs. 19.1–19.3), which winds 2.5–2.75 turns about its osseous axis, the **modiolus**. Bubble-like hollow areas dispersed within the modiolus house groups of nerve cell bodies of the bipolar sensory neurons that collectively form the **cochlear (spiral) ganglion**. The peripheral processes of these bipolar neurons terminate in the basal aspect of the hair cells of the **spiral organ**, the cochlear receptor for hearing, whereas the central processes of these neurons form the root of the **cochlear nerve**.

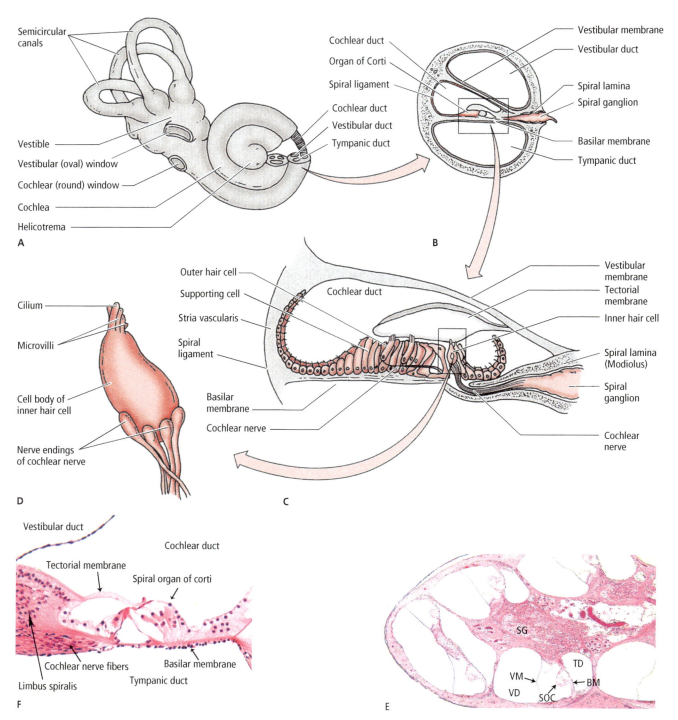

Figure 19.3 ● Components of the internal ear. (A) The vestibulocochlear apparatus. (B) Cross-section through the three ducts (scalae) of the cochlea: the vestibular duct (scala vestibuli), cochlear duct (scala media), tympanic duct (scala tympani). (C) The spiral organ resting on the basilar membrane. (D) An inner hair cell synapsing with nerve endings from the cochlear nerve. (E) Light micrograph of the cochlea. SG, Spiral Ganglion; BM, Basilar Membrane; VM, Vestibular Membrane; TD, Tympanic Duct; VD, Vestibular Duct; SOC, Spiral Organ of Corti. (F) Light micrograph of the spiral organ.

The cochlea's interior is divided into three parallel, fluid-filled, compartments by two membranes, the vestibular (eponym: Reissner) membrane and the basilar membrane. Each compartment is referred to as a scala (L., "staircase") since each spirals about the modiolus like a winding staircase. They are: the (bony) **vestibular duct** (scala vestibuli); the (membranous) **cochlear duct** (scala media), which is wedged between the vestibular and the basilar membranes (as a blind-ending duct ending at the apex of the cochlea); and the (bony) **tympanic duct** (scala tympani) (Fig. 19.3B,E).

The interior of the vestibular and tympanic ducts is lined by a simple squamous epithelium. The tympanic and vestibular ducts are filled with **perilymph**, a fluid that has a similar ionic composition as extracellular fluid and cerebrospinal fluid that has a high concentration of sodium compared with potassium levels. The perilymph is mixture of plasma from the labyrinthine vessels and cerebrospinal fluid entering from the subarachnoid space through the cochlear aqueduct (perilymphatic duct). The cochlear duct is filled with a different fluid, **endolymph**, which is similar to intracellular fluid and has a high potassium concentration and low sodium concentration. Endolymph is produced from perilymph by selective membranous ionic transport.

The perilymph-filled, bony, vestibular duct of the cochlea is in direct communication with the perilymph-filled, bony, vestibule of the vestibular apparatus. The **vestibular and tympanic ducts**, both of which are filled with **perilymph**, are joined at the **helicotrema** (G., *helicon*, "helix"; trema, "hole"), a small connecting aperture located at the apex of the cochlea, which permits perilymph between the two structures. These ducts are connected to the **subarachnoid space** via the **cochlear aqueduct**, and the perilymph flows into the subarachnoid space thus contributing to the cerebrospinal fluid. It should also be noted that the vestibular duct communicates with the **vestibular window** and the tympanic duct communicates with the **cochlear window**.

The **cochlear duct**, also referred to as the scala media, contains endolymph. On cross-section of the cochlea, the triangular-shaped cochlear duct is wedged between the vestibular and tympanic ducts with its apex attached to the spiral lamina of the **modiolus**. The **vestibular membrane** forms the *roof* of the cochlear duct and serves as the barrier between the cochlear duct and the vestibular duct. This thin, somewhat elastic membrane consists of two layers of simple squamous epithelium, separated from each other by a basement membrane. The cells of the simple squamous epithelium contacting the vestibular duct form tight junctions with each other thus establishing an impermeable barrier that prevents the perilymph from entering the cochlear duct and the endolymph from entering the vestibular duct. The layer of squamous epithelial cells that face the cochlear duct also form tight junctions with each other; moreover, they also possess numerous microvilli, suggesting that they function in ion transport. Because the vestibular membrane is somewhat elastic in nature, it is believed to play a role in the transmission of vibrations from the perilymph of the vestibular duct to the endolymph of the cochlear duct. The **basilar membrane** that separates the cochlear duct from the tympanic duct forms the *floor* of the cochlear duct. The basilar membrane is an elastic structure exhibiting a gradual increase in width and a decrease in stiffness from the vestibular window (at the cochlear base) to the helicotrema (at the cochlear apex). The gradual decrease in stiffness permits this membrane to be sensitive to *high-frequency* vibrations near the **cochlear base**, and *low-frequency* vibrations near its **apex**, at the helicotrema. The inner surface of the bony labyrinth bordering the cochlear duct is the **stria vascularis**, which produces endolymph. Endolymph is not derived directly from the blood but is instead derived from perilymph via selective ion transport carried out by those epithelial cells of the vestibular membrane that possess microvilli. Endolymph is continuously formed, percolates through the membranous labyrinth, and is reabsorbed.

The receptors associated with the sense of hearing form a structure, the spiral organ, which rests like a carpet along the floor of the cochlear duct on the surface of the basilar membrane. The spiral organ consists of **supporting epithelial cells** and neuroepithelial receptor **"cochlear hair cells"** (Fig. 19.3F). Each hair cell displays numerous **stereocilia** (long microvilli) and a single **kinocilium** projecting from its apical cell surface. The free ends of the stereocilia are embedded in the **tectorial membrane**, an acellular, gelatinous substance, joined to the **osseous spiral lamina** (a bony shelf protruding from the modiolus). The hair cells are the mechanoreceptors of the auditory system that act as transducers, that is, they serve to convert mechanical energy into electrical impulses that can be relayed to the brainstem. Hair cells synapse with the peripheral ends (dendritic processes) of the bipolar neurons whose cell bodies are housed in the **cochlear ganglion**. The central processes of these bipolar neurons form the **cochlear nerve**, which joins the vestibular nerve to form the **vestibulocochlear nerve (CN VIII)**. This nerve courses through the internal acoustic meatus to enter the cranial cavity and then enters the brainstem at the pontomedullary angle.

AUDITORY TRANSMISSION

As sound wave vibrations are transmitted from the tympanic membrane to the membrane of the vestibular window, they are amplified approximately 20-fold

The sound waves, which the **auricle** and the **external auditory meatus** channel to the **tympanic membrane**, cause it to vibrate (see Fig. 19.2). Oscillations of the tympanic membrane cause the three **ossicles** to vibrate in sequence, conducting the vibration to the **vestibular window**. The articulating ossicles form a lever system that not only mediate vibrations from the tympanic membrane to the membrane of the vestibular window, but also – a fact of paramount importance – *amplify* these vibrations as they are converted from **air waves** of the external ear into **fluid waves** (perilymph) of the internal ear. Since liquids are more resistant to the conduction of sound waves, the incoming sound waves hitting the tympanic membrane must be amplified by the ossicles to ensure adequate vibration of the perilymph. As the tympanic membrane is much larger than the vestibular window membrane, and the sound waves reaching the vestibular window membrane are amplified by the ossicles, the compound effect is that the vibrations reaching the vestibular window have been amplified approximately 20-fold from the original vibrations at the tympanic membrane. The rapid oscillations of the **stapes footplate** cause the membrane covering the vestibular window to vibrate, which in turn agitates the perilymph of the vestibular duct, causing fluid displacement and the formation of pressure waves. The pressure waves are propagated from the base of the cochlea, through the vestibular duct, along the helical path to the apex of the cochlea. Although perilymph waves in the **vestibular duct** also flow via the **helicotrema** to the **tympanic duct**, the effect of these waves is probably minor and causes little stimulation of the spiral organ.

Perilymph waves in the vestibular duct agitate the **vestibular membrane**, which begins to oscillate. The oscillating vestibular membrane generates waves in the endolymph of the **cochlear duct**, which in turn causes the **basilar membrane** to oscillate. Basilar membrane oscillations are initiated at the base of the cochlea and advance along the basilar membrane as a traveling wave toward the apex of the cochlea. Wave formation in the vestibular and tympanic ducts is possible only because of the flexibility of the elastic membrane covering the **cochlear window**, which vibrates and allows pressure formed by the waves to be released into the tympanic cavity, as the cochlear window membrane protrude towards it. Without the cochlear window to compensate for the pressure exerted on the vestibular window, waves could not be propagated in the tympanic duct since fluid completely enclosed in a bony compartment is resistant to compression.

The bases of the hair receptor cells rest on the basilar membrane, whereas the tips of their stereocilia and singular kinocilium extend into the overlying **tectorial membrane**. It is the basilar membrane oscillations that cause a shearing force and deformation of the hair cells' stereocilia and kinocilia, which are inserted into the less mobile, overlying tectorial membrane (Fig. 19.3C). When this deformation pushes stereocilia toward the kinocilium, mechanical ion channels open resulting in an influx of potassium ions which causes depolarization of the cell membrane. Membrane depolarization opens voltage-gated calcium channels leading to an influx of calcium ions which triggers the release of neurotransmitters. The inner and outer hair cells transduce sound waves into neural impulses which are transmitted to the dendritic processes of the bipolar type I and type II auditory sensory neurons, respectively. Sensory stimulation is then relayed by the cochlear nerve to the cochlear nuclei in the brainstem (Fig. 19.4).

The pathway of auditory stimulation is summarized in Fig. 19.5.

CENTRAL AUDITORY PATHWAYS

The auditory pathway transmits sensory input by projecting bilaterally in the brain, that is, auditory signals from each ear ultimately reach both cerebral hemispheres

Auditory input travels in the **auditory pathway**, which begins with the **first-order neurons** housed in the **cochlear ganglion** and continues as a chain of multisynaptic relays into the medulla oblongata, pons, midbrain, diencephalon (thalamus), and finally the cerebral cortex. The entire ascending auditory pathway is tonotopically organized from the cochlea to the auditory cortex, which is analogous to the somatotopic organization of the other sensory systems. Due to the complexity of the auditory pathways, only the main pathways are discussed here. The auditory pathway transmits sensory input related to *frequency* (pitch), *amplitude* (loudness), and *location* (direction) of a sound.

The auditory pathway relays sensory input by projecting bilaterally in the brain; that is, auditory input from each ear (similar to visual input from each eye) ultimately reaches both cerebral hemispheres. The only site of unilateral (ipsilateral) projection in the auditory pathway is the first relay, since each cochlear nerve projects to its ipsilateral cochlear nuclei.

Thus, these four structures: (1) the cochlea (in the internal ear), (2) the first-order bipolar neurons of the cochlear ganglion whose central processes form the cochlear nerve and synapse in the anterior and posterior cochlear nuclei, (3) the anterior and posterior cochlear nuclei, and (4) the second-order neurons whose cell bodies are housed in the cochlear nuclei are all involved in the transmission of auditory information from the ipsilateral ear. Most subsequent relays of the auditory pathway are bilateral.

The cochlear ganglion is unusual since its nerve cell bodies are not aggregated into a single ganglion. Instead, the cell bodies form numerous swellings that are embedded in multiple small (bubble-like) spaces within the modiolus of the cochlea. The dendritic terminals of the peripheral processes of these neurons terminate in the spiral organ and

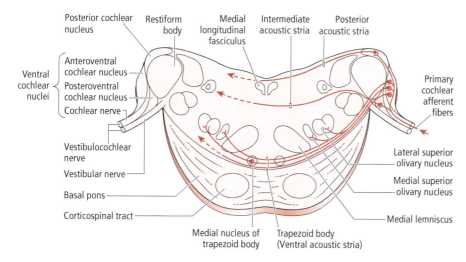

Figure 19.4 ● The main efferent projections from the cochlear nuclei. Source: Modified from Burt, AM (1993) *Textbook of Neuroanatomy*. WB Saunders, Philadelphia; fig. 12.15.

AUDITORY SYSTEM   395

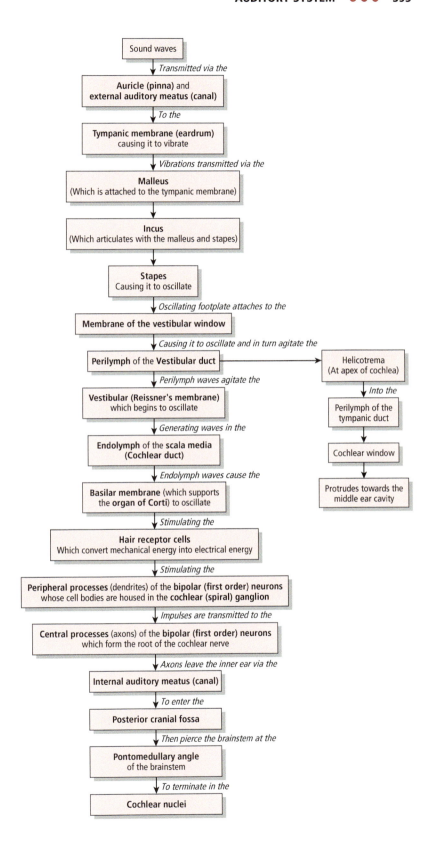

Figure 19.5 ● The pathway of auditory stimulation from the external auditory meatus to the cochlear nuclei in the brainstem.

synapse with its specialized "hair cell" receptors. The central processes of these neurons collect to form the **cochlear nerve** (the cochlear division of the vestibulocochlear nerve, CN VIII). In the internal ear, the cochlear nerve joins the vestibular nerve to form the vestibulocochlear nerve. From the internal ear, the vestibulocochlear nerve then passes through the internal acoustic meatus to enter the posterior cranial fossa of the cranial cavity. As it enters the brainstem at the pontomedullary angle, the vestibulocochlear nerve again separates into its two components, the vestibular nerve and the cochlear nerve. The vestibular nerve terminates in the vestibular nuclei, whereas the **cochlear nerve** terminates in the **cochlear nuclei**. The vestibular and cochlear nuclei extend into the pons and the medulla oblongata. The central processes of the first-order neurons of the cochlear nerve enter the anterior cochlear nucleus where they bifurcate, some branches terminating in the posterior cochlear nucleus, and the remaining branches terminating in the anterior cochlear nucleus. Thus, the **anterior cochlear nucleus** and the (smaller) **posterior cochlear nucleus** are the first relay stations of auditory input. Each of these cochlear nuclei subsequently gives rise to separate axon bundles, each of which has a distinctive pattern of termination and function. The **posterior cochlear nucleus** functions in *vertical sound localization* (localizes a sound source based on whether it is located above or below the level of the listener's ears), whereas the **anterior cochlear nucleus** functions in *horizontal sound localization* (localizes a sound source based on whether it is located to one side or the other or from the front or back), see below.

The cochlear nuclei are the origin of distinct, parallel, ascending auditory pathways that relay auditory input through a sequence of nuclei where it is analyzed at multiple levels before it reaches the auditory cortex for further processing.

The **cochlear nuclei** house the cell bodies of the **second-order neurons** of the auditory pathway. The second-order neurons give rise to axons whose projections are as follows:

Second-order fibers arising from the cochlear nuclei either project ipsilaterally from the anterior cochlear nucleus to the medial and lateral superior olivary nuclei, or decussate forming three different separate pathways: the **posterior, intermediate**, and **anterior acoustic striae** (see Fig. 19.4).

(1) Most of the decussating second-order fibers arising from the **anterior cochlear nucleus** form a readily evident group of fibers, the **trapezoid body**, also known as the **anterior acoustic striae** in the anterior pontine tegmentum. The fibers in the trapezoid body project to and terminate *contralaterally* in one of the following:

- the **medial nucleus of the trapezoid body**, which in turn projects to the **lateral superior olivary nucleus**;
- the **medial superior olivary nucleus**; or
- the **posterior nucleus of the lateral lemniscus** and the **inferior colliculus** (by ascending in the contralateral lateral lemniscus) (Fig. 19.6).

The superior olivary complex (medial and lateral superior olivary nuclei) projects via the ipsilateral and contralateral lateral lemniscus to the inferior colliculus for *binaural processing*. This pattern of bilateral projections reflects the nuclei's significance in the *horizontal localization of sounds*.

(2) **Second-order fibers** arising from the **posterior cochlear nucleus** forming the **intermediate acoustic striae** anterior to the medial longitudinal fasciculus. These fibers subsequently join the ipsilateral and contralateral lateral lemniscus to ascend to, and terminate in, the **anterior nucleus of the lateral lemniscus** and the **inferior colliculus**, bilaterally (Fig. 19.7).

(3) **Second-order fibers** arising from the **posterior cochlear nucleus** forming the **posterior acoustic striae**, which decussate on the anterior aspect of the floor of the fourth ventricle (in the posterior pontine tegmentum). These fibers join the contralateral lateral lemniscus to ascend to, and terminate in, the **inferior colliculus**, completely *bypassing* the superior olivary nuclear complex. It is by the use of this pathway that fibers arising from the posterior cochlear nucleus transmit *monaural* auditory signals via the **lateral lemniscus** directly to the inferior colliculus.

Superior olivary complex

The superior olivary nuclei are the first nuclei of the auditory pathway that process auditory information from both ears and determine the direction from which a sound is coming, as well as its intensity

The **superior olivary nuclei** form a nuclear complex housing **third-order neurons**, located in the pontine tegmentum at the level of the facial nucleus. Starting with the third-order neurons all the way to the auditory cortex, they carry mainly binaural auditory information; thus, if one of the central auditory pathways is damaged, there is a "back up" projection on the other side resulting in minor hearing impairments, mainly in the contralateral ear. The nuclei of this complex are the **medial superior olivary nucleus**, the **lateral superior olivary nucleus**, and the **nucleus of the trapezoid body**, all of which receive second-order fiber terminals from the cochlear nuclei (as discussed above) and have an important function in *sound localization in the horizontal axis*. The anterior cochlear nucleus projects to the superior olivary complex which is the origin of the pathway for *horizontal localization of sounds*. The superior olivary complex processes auditory information from both ears and determines the direction and location that a sound is coming from in the following manner. The **medial superior olivary nucleus** processes auditory input by comparing the *time difference* for a sound to reach each ear (interaural time differences), whereas the **lateral superior olivary nucleus** processes auditory input by comparing the *intensity* (volume) of a sound arriving at each ear (interaural intensity differences). Thus, a sound arising from a source, such as a telephone located to one's right, arrives at the right ear sooner than it does at the left ear, and the sound is slightly louder in the right ear than it is in the left ear. Alternately, if a sound arises

AUDITORY SYSTEM ••• 397

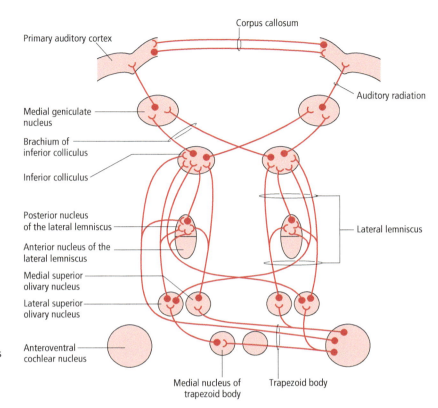

Figure 19.6 • The principal ascending auditory pathways emerging from the anterior cochlear nucleus. Source: Modified from Burt, AM (1993) *Textbook of Neuroanatomy*. WB Saunders, Philadelphia; fig. 12.16.

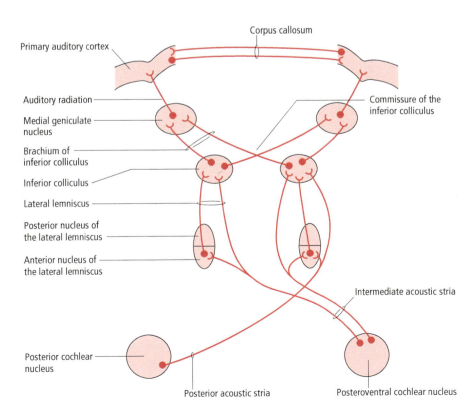

Figure 19.7 • The principal ascending auditory pathways emerging from the posterior cochlear nuclei. Source: Modified from Burt, AM (1993) *Textbook of Neuroanatomy*. WB Saunders, Philadelphia; fig. 12.17.

398 ●●● CHAPTER 19

from a source that is located at an equal distance from both ears, then the sound arrives at both ears at the same time. The neurons in the lateral superior olivary nucleus are *stimulated* by impulses transmitted by second-order fibers arising from the ipsilateral anterior cochlear nucleus and are *inhibited* by impulses transmitted by second-order fibers arising from the contralateral nucleus. The **posterior cochlear nucleus** projects via the **dorsal acoustic striae** to the contralateral lateral lemniscus relaying auditory information about the elevation (height) of the sound source directly to the inferior colliculus, completely *bypassing* the superior olivary complex.

Lateral lemniscus

The lateral lemniscus is the main ascending pathway of the auditory system in the brainstem, conveying auditory input from both ears

The lateral lemniscus is the main ascending pathway of the auditory system in the brainstem, conveying auditory input from both ears. In summary, the lateral lemniscus (L., "ribbon") contains the following (Fig. 19.6):

1 Second-order fibers emerging from the contralateral **anterior cochlear nucleus** (which bypass the superior olivary complex) to terminate in the **posterior nucleus of the lateral lemniscus** and the **inferior colliculus**.
2 Second-order fibers arising from the ipsilateral and contralateral **posterior cochlear nucleus** that terminate in the **anterior nucleus of the lateral lemniscus** and in the **inferior colliculus**.
3 Second-order fibers arising from the contralateral posterior cochlear nucleus that terminate in the anterior nucleus of the lateral lemniscus and in the inferior colliculus. These second-order fibers form a prominent component of the lateral lemniscus.
4 Third-order fibers originating from the **superior olivary complex** (the fibers arising from the medial superior olivary nucleus join the ipsilateral lateral lemniscus, whereas those that arise from the lateral superior olivary nucleus join the ipsilateral and contralateral lateral lemniscus) that terminate in the **posterior nucleus of the lateral lemniscus** and in the inferior colliculus. The third-order fibers arising from the superior olivary complex to the contralateral inferior colliculus also form a prominent component of the lateral lemniscus.
5 Fourth-order fibers arising from the posterior and anterior nuclei of the lateral lemniscus that project to the ipsilateral inferior colliculus.

Nuclei of the lateral lemniscus

The diffuse anterior and posterior nuclei of the lateral lemniscus are located within the substance of the lateral lemniscus

The **anterior nucleus of the lateral lemniscus** contains neurons that process auditory input from only *one ear,* whereas the **posterior nucleus of the lateral lemniscus** houses neurons that process auditory input from *both ears.*

Inferior colliculus

The inferior colliculus is associated with sound localization

The **inferior colliculi** (L., colliculus, "little hill"), located in the roof of the caudal midbrain, are the mesencephalic relay nuclei of the auditory system. They receive afferents ascending in the lateral lemniscus from the cochlear nuclei, the superior olivary complex, and from the nuclei of the lateral lemniscus. The inferior colliculi also receive afferents from the contralateral inferior colliculus via the **commissure of the inferior colliculus,** and from the auditory cortex.

Each inferior colliculus consists of three components: the **central nucleus, external nucleus,** and the **posterior cortex.**

Fiber bundles arising from the posterior cochlear nucleus, the superior olivary complex and the nucleus of the lateral lemniscus ascend in the lateral lemniscus to the midbrain where they are funneled into the **central nucleus of the inferior colliculus,** the main destination of the lateral lemniscus. The fibers arising from the posterior and anterior cochlear nuclei relay auditory messages about *horizontal* and *vertical* sound source localization, to the tonotopically organized central nucleus. Tonotopical organization is mapped on the multiple laminae that constitute the central nucleus, with each lamina containing neurons that are stimulated by similar tonal frequencies. The central nucleus then relays auditory information to the medial geniculate nucleus of the thalamus. The external nucleus of the inferior colliculus plays a critical role in the processing, integration, and relay of auditory information, contributing to functions like sound localization, sound processing, and reflex responses to auditory stimuli. The posterior cortex of the inferior colliculus is specialized for the processing of spatial auditory information, allowing for the localization and discrimination of sound sources in the surrounding environment.

The inferior colliculus gives rise to a prominent fiber bundle, the **brachium of the inferior colliculus** (L., brachium, "arm"), an arm-like structure whose fibers, carrying auditory output, end in the ipsilateral **medial geniculate nucleus,** a thalamic relay station of the auditory system. The inferior colliculus also projects to the contralateral medial geniculate nucleus and the superior colliculus (which is involved in visual reflexes) to mediate audiovisual reflex activity such as turning the head and eyes in the direction of a startling noise. The inferior colliculus is associated with sound localization.

Medial geniculate nucleus

The medial geniculate nucleus processes auditory input related to sound intensity (loudness) and frequency (pitch), and transmits it to the auditory cortex

The **medial geniculate nucleus** is the thalamic relay nucleus of the auditory system. It consists of three divisions: **medial, posterior,** and **ventral (anterior).** The **ventral division** (referred to as the medial geniculate nucleus) is its main component; it is laminated and tonotopically organized. The ventral division is the thalamic relay target of the ascending auditory fibers arising from the central nucleus of the inferior colliculus. Its key function is to process auditory information. The remaining two divisions (medial and posterior) of the medial geniculate nucleus are targets of all three component nuclei of the inferior colliculus and, in addition, receive afferents from the somatic sensory and visual systems. These projections may be involved in arousal, attention, and wakefulness.

Fibers arising in the **medial geniculate nucleus** form the **auditory radiations** that join the sublentiform portion of the posterior limb of the internal capsule to terminate in the primary auditory cortex. The medial geniculate nucleus processes auditory input related to *sound intensity* (loudness) and *frequency* (pitch) and transmits it to the auditory cortex.

Auditory cortex

The primary auditory cortex, corresponding to the eponymous Brodmann area 41, is located deep in the lateral fissure, in the transverse gyri

The auditory cortex consists of **primary**, **secondary**, and **higher-order cortical areas**. They are concentrically arranged with the primary cortex forming the core, the secondary auditory cortex forming an intermediate, and the higher-order cortical areas forming the outer component of the auditory cortex (Fig. 19.8).

The **primary auditory cortex** corresponding to Brodmann area 41 is located deep in the lateral fissure, on the superior surface of the superior temporal gyrus of the temporal lobe.

The superior surface of the superior temporal gyrus has one to several gyri oriented transversely in relation to the gyri located on the lateral surface of the temporal lobe, thus their name, the **transverse gyri**. The anterior gyrus (eponym: gyrus of Heschl) is the site of the primary auditory cortex (Fig. 19.8).

The **primary auditory cortex** receives the auditory radiations from the medial geniculate nucleus of the thalamus. This cortical region has a tonotopic representation of frequencies; that is, neurons responding to low frequencies reside laterally, whereas neurons responding to high frequencies reside medially, in its caudal extent. The primary auditory cortex is arranged into two-dimensional, alternating, vertically oriented columns of neurons (Fig. 19.9). One dimension of the auditory cortex is composed of **frequency columns**. The cells in each frequency column respond to an auditory stimulus of a particular frequency. As mentioned above, cells responding to low frequencies reside in the frequency columns located on the lateral aspect of the transverse superior temporal gyrus, whereas frequency columns containing cells responding to gradually higher frequencies are arranged in sequence toward the medial extent of the primary auditory cortex. The other dimension of the auditory cortex is composed of alternating **binaural columns**. There are two types of binaural columns: summation columns and suppression columns. The neurons residing in the **summation columns** respond to an auditory stimulus that stimulates *both ears* simultaneously. In contrast, the neurons in the **suppression columns** respond maximally to an auditory stimulus that stimulates *only one ear* but respond minimally when the auditory stimulus stimulates both ears.

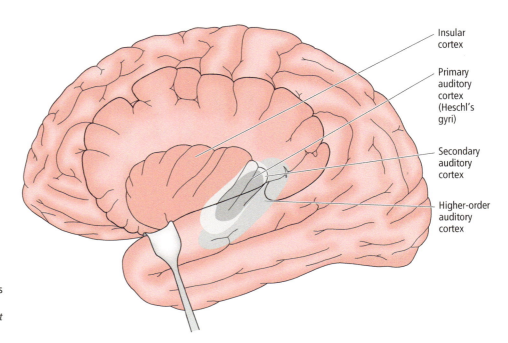

Figure 19.8 ● Auditory cortex consists of the primary, secondary, and higher-order cortical areas in the superior temporal gyrus of the temporal lobe. Source: From Martin 2012, figure 8.8, p. 193. *Neuroanatomy Text and Atlas, Fourth Edition.*

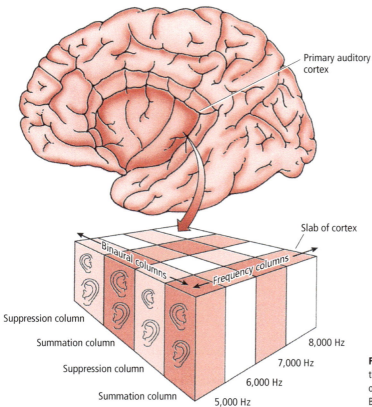

Figure 19.9 ● Lateral view of the cerebral hemisphere showing the location of the primary auditory cortex and an illustration of its organization. Source: Modified from Matthews, G (2001) *Neurobiology*. Blackwell Publishing, Oxford; fig. 17.14.

The primary auditory cortex of each side sends projections to the contralateral side via the corpus callosum. The **primary auditory cortex** has an important function in the conscious perception of sound, the detection of alteration in pattern as well as the localization of a sound (direction and distance of sound source).

The **secondary auditory cortical areas**, although numerous, are not well defined but are believed to correspond to the eponymous Brodmann area 42. They surround the primary auditory cortex and form reciprocal connections with the primary auditory cortex. The secondary auditory cortical areas play an important function in the interpretation of sounds and, via connections with the posterior aspect of the superior temporal gyrus (eponym: Wernicke's area), function in the comprehension of language. The secondary auditory cortical areas project to the higher-order auditory areas.

Auditory information flows from the primary to the secondary auditory cortex and, from there, to the higher-order auditory cortical areas. The auditory cortex is known to be the source of two distinct, prominent auditory pathways one of which is involved in the localization of sound sources (the **"where" pathway**), and the other that is involved in the processing of complex sounds and language (the **"what" pathway**) (Fig. 19.10).

The **"where" pathway** begins in the primary auditory cortex, which relays auditory input to the inferior portion of the secondary auditory cortex, which in turn projects to the inferior portion of the higher-order auditory cortex in the superior temporal gyrus. The higher-order auditory cortex then exports messages to the (1) **posterior parietal cortex** in the parietal lobe, which is interposed between somatosensory and visual cortical areas, and to the (2) **posterolateral prefrontal cortex** and **premotor cortex** in the frontal lobe. The posterior parietal cortex receives input from the auditory cortex as well as somatosensory and visual inputs from neighboring cortical areas. The **posterior parietal cortex**, via multisensory integration of all inputs, assembles a seamless composite image of the body's surroundings that enables the brain to determine the body's orientation in reference not only to its surroundings but also the location of the stimuli.

The **posterolateral prefrontal cortex** and **premotor cortex** of the frontal lobe function in the *planning* phase of movement. The posterior parietal cortex informs these areas in the frontal lobe of the location of a sound ("where") relative to the location of the body, which helps the brain determine the *direction* of desired movement based on the location of the sound. These frontal areas then relay the plan of impending movement to the primary motor cortex for execution.

AUDITORY SYSTEM   401

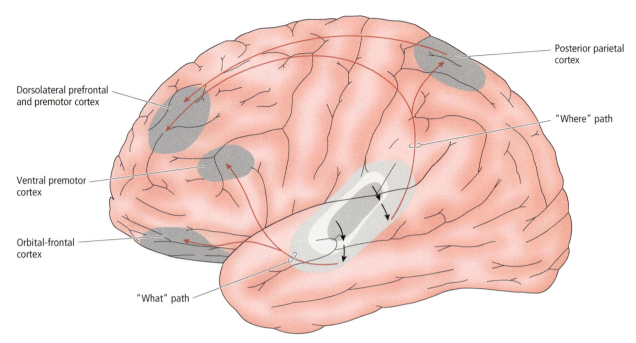

Figure 19.10 ● Distinct "what" and "where" pathways emerging from the auditory cortex are exported to the frontal and parietal lobes, respectively. Source: From Martin 2012, figure 8.10, p. 194. *Neuroanatomy Text and Atlas, Fourth Edition.*

The **"what" pathway** emerges from the primary auditory cortex and continues with connections to the rostral portions of the secondary cortex and then to the higher-order auditory cortex. From there, output is sent to the frontal lobe including the **anterior premotor cortex** (eponym: Broca's area) and **orbital–frontal cortical areas**. These projections to the frontal lobe not only function in the production of speech, but also in determining who is speaking as well as the source of sounds.

Descending auditory projections

The olivocochlear bundle has an inhibitory effect on cochlear nerve activity, modulating and sharpening auditory transmission

The **superior olivary nucleus**, one of the nuclei of the olivary complex, gives rise to the **olivocochlear bundle** (an efferent pathway) whose fibers descend both ipsilaterally and contralaterally to terminate in the spiral organ, where they synapse with receptor hair cells (Fig. 19.11). This tract has an inhibitory effect on cochlear nerve activity, modulating and sharpening auditory transmission. Thus, the cochlea is innervated by the peripheral (afferent) processes of the first-order bipolar neurons of the cochlear nerve, as well as by the olivocochlear (efferent) central fibers arising from the brainstem.

Sound attenuation reflex

A loud noise heard in one ear causes a reflex contraction of both the tensor tympani and stapedius muscles of both ears

A loud noise heard in one ear causes a reflex contraction of the tensor tympani muscles and the stapedius muscles of both ears, and, therefore, dampen, bilaterally, the transmission of excessively loud and/or startling sound wave signals from the external to the internal ear. The sound waves entering one ear follow the normal auditory pathway via the **cochlear nerve** to the ipsilateral **anterior cochlear nucleus**. Each anterior cochlear nucleus projects bilaterally to the **superior olivary nuclei**. They in turn project bilaterally to the **facial nuclei** and the **motor nuclei of the trigeminal nerve**. Each facial nucleus projects to the ipsilateral stapedius muscle, and each motor nucleus of the trigeminal nerve projects to the ipsilateral tensor tympani muscle (Fig. 19.12).

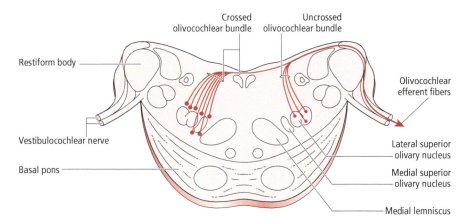

Figure 19.11 ● The olivocochlear pathway arises from the superior olivary nuclei and terminates in the spiral organ where it inhibits the transmission of auditory information to higher brain centers. Source: Modified from Burt, AM (1993) *Textbook of Neuroanatomy*. WB Saunders, Philadelphia; fig. 12.18.

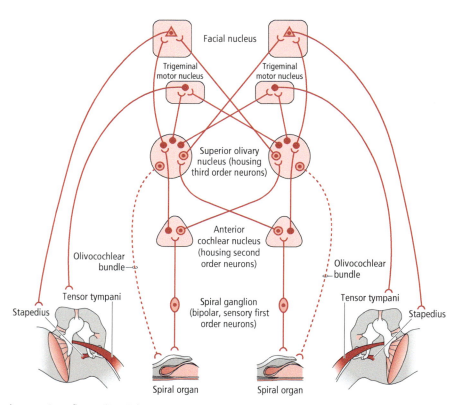

Figure 19.12 ● The sound attenuation reflex. Auditory information is transmitted from the spiral organ via the central processes of bipolar sensory neurons to the ipsilateral anterior cochlear nucleus. Subsequent connections are bilateral. Information is then transmitted sequentially to the superior olivary nucleus which in turn projects to the trigeminal motor and facial nuclei. The trigeminal motor nucleus projects to the tensor tympani and the facial nucleus projects to the stapedius muscle, causing their contraction and dampening of auditory input.

CLINICAL CONSIDERATIONS

Hearing defects can result from many causes. A unilateral lesion in the facial nerve (CN VII) proximal to giving rise to the branch to the stapedius muscle results in **hyperacusis**, characterized by abnormally acute hearing in the affected ear. This condition often occurs in individuals with idiopathic facial paralysis (eponym: Bell palsy).

Hearing defects may be categorized into conduction deafness and sensorineural (nerve or perceptive) deafness.

Conduction deafness usually involves the auditory apparatus of the external and/or middle ear and is caused by a defect in the *mechanical transmission* of sound from the air-filled external and/or middle ear to the fluid-filled internal ear, which interferes with sound conduction. Conduction deafness may result from:

1 **Cerumen** (wax) **buildup** in the external auditory canal.
2 **Perforation of the tympanic membrane**.
3 **Otitis media**, inflammation of the middle ear, otitis media, which is accompanied by fluid accumulation in the middle ear cavity (which, if not treated, may progress to meningitis or brain abscesses).
4 **Otosclerosis** (G., *oto*, "ear"; sklerosis, "hardening"), which is the most common pathologic condition resulting in conduction deafness in adults. It is a consequence of new bone deposition of the labyrinthine spongy bone bordering the vestibular window with subsequent fixation of the stapes on the vestibular window. Otosclerosis can be surgically corrected by releasing the stapes. A hearing aid often improves conduction deafness.

Sensorineural (nerve) deafness results following a lesion involving the cochlea, the cochlear nerve, or the central auditory pathways. A lesion involving the cochlea, the cochlear nerve, or the cochlear nuclei results in an ipsilateral hearing deficit. Injury to the hair receptor cells may be caused from extensive exposure to loud noise, antibiotics, or other drugs such as aspirin, or various prenatal infections, such as rubella, cytomegalovirus, or syphilis.

The most prevalent sensorineural hearing deficit occurring in the elderly is **presbycusis** (G., *presbus*, "old man"; *cusis*, "hearing"). This condition results from the gradual degeneration of the spiral organ.

Acoustic neuroma (schwannoma or neurilemoma) is a tumor arising from the neurolemmocytes (eponym: Schwann cells) covering the vestibulocochlear nerve (CN VIII). This tumor appears in the internal auditory meatus, or close to the cerebellopontine angle after the nerve leaves the brainstem in the posterior cranial fossa. This tumor causes nerve destruction resulting in **total loss of hearing** (deafness) in the ipsilateral ear and **tinnitus** ("ringing" sensation in the ear) and/or vestibular symptoms.

Lesions involving the central auditory pathways (anywhere rostral to the cochlear nuclei) result in **minor hearing impairments** mainly in the **contralateral** side. This is due to the *bilateral* projections of the auditory pathway. The individual will have difficulty in localizing sounds on the contralateral side if the lesion involves the ascending auditory pathways rostral to the level of the pons, or if the primary or secondary auditory cortex is involved.

Genetic mutations are the underlying cause of some cases (<1%) of congenital bilateral deafness. The **Jervell–Lange-Nielsen syndrome** is a recessively inherited genetic disorder associated with bilateral deafness. This syndrome results from the formation of defective ion channel complexes in the stria vascularis lining the lateral wall of the cochlear duct. The cells of the stria vascularis regulate the ionic composition of endolymph in the cochlear duct, which is an important component in the formation of nerve impulses in the auditory system.

Note that the clinical case at the beginning of the chapter refers to a patient whose symptoms include hearing loss and ringing in the right ear, imbalance, dizziness, and ataxia.

1 What can cause hearing loss?

SYNONYMS AND EPONYMS OF THE AUDITORY SYSTEM

Name of structure or term	Synonym(s)/eponym(s)
Auditory tube	Eustachian tube
	Pharyngeal tube
Auricle	Pinna
Bony labyrinth	Osseous labyrinth
Cochlear ganglion	Spiral ganglion
Cochlear duct	Scala media
External auditory meatus	External auditory canal
Internal auditory meatus	Internal acoustic canal
Lateral fissure	Sylvian fissure
Mesencephalon	Midbrain
Primary auditory cortex	Transverse temporal gyrus of Heschl
	Brodmann area 41
Sensorineural deafness	Nerve deafness
	Perceptive deafness
Spiral organ	Organ of Corti
Superior olivary nuclei	Superior olivary complex
Trapezoid body	Ventral acoustic striae
Tympanic duct	Scala tympani
Tympanic membrane	Eardrum
Vestibular duct	Scala vestibuli
Vestibular membrane	Reissner membrane

FOLLOW-UP TO CLINICAL CASE

Hearing defects may result from conduction deafness or sensorineural deafness. Conduction deafness usually involves the auditory apparatus of the external and/or middle ear and is caused by a defect in the mechanical transmission of sound from the air-filled external and/or middle ear to the fluid-filled internal ear, which interferes with sound conduction. Sensorineural (nerve) deafness results following a lesion involving the cochlea, the cochlear nerve, or the central auditory pathways.

This case required further testing since clinical evaluation alone would not determine the cause of deafness. The other noted symptoms were vague or nonspecific, although the dizziness and imbalance could have possibly indicated vestibular or cerebellar dysfunction. This patient was referred to an otolaryngologist for a full examination and imaging of the base of the skull, mastoid, internal ear, and auditory canals. MRI would be best, but CT may be adequate. This patient's MRI (with contrast) showed an acoustic neuroma of the right eighth cranial nerve, clinically known as a schwannoma.

An acoustic neuroma is a relatively common, benign tumor of neurolemmocytes (eponym: Schwann cells; which provide myelin for all peripheral nerves, including cranial nerves). Schwannomas commonly arise from cranial nerve VIII although they can affect any nerve, particularly the cranial nerves. Acoustic neuromas are bulbous or fusiform tumors that are located in the cerebellopontine angle. The most common and initial symptom of an acoustic neuroma is typically unilateral hearing loss. Dizziness (even occasionally vertigo) and ataxia can arise from dysfunction of the vestibular portion of cranial nerve VIII or from cerebellar dysfunction secondary to compression. Other symptoms, such as facial weakness, can arise from compression of the adjacent facial nerve.

Acoustic neuromas are most often unilateral, with one notable exception. Neurofibromatosis type II leads to multiple tumors, including bilateral acoustic neuromas. A cerebellopontine angle meningioma can be confused with an acoustic neuroma. The mainstay of treatment is surgical excision. Hearing can be restored only some of the time.

2 The patient's MRI showed the presence of an acoustic neuroma on the right side. How does this tumor cause the patient's symptoms of hearing loss, imbalance, dizziness, and ataxia?

3 What other cranial nerve emerges from the brainstem near cranial nerve VIII that may also be compressed by an acoustic neuroma and cause symptoms?

QUESTIONS TO PONDER

1. What is the function of the middle ear (tympanic cavity)?

2. What is the function of the auditory (pharyngotympanic eponym: Eustachian) tube?

3. What is the tonotopic organization of the basilar membrane?

4. What structures are involved in the transduction of the sound waves from the external ear to the internal ear?

5. What is the function of the olivocochlear bundle?

6. Which cranial nerves are involved in the auditory attenuation reflex?

CHAPTER 20

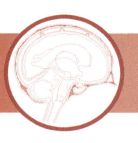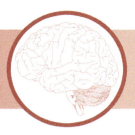

Vestibular System

CLINICAL CASE

VESTIBULAR APPARATUS

VESTIBULAR NERVE (CN VIII)

CENTRAL PATHWAYS OF THE VESTIBULAR SYSTEM

CONTROL OF OCULAR MOVEMENTS

VESTIBULAR NYSTAGMUS

CALORIC NYSTAGMUS

SYNONYMS AND EPONYMS OF THE VESTIBULAR SYSTEM

FOLLOW-UP TO CLINICAL CASE

QUESTIONS TO PONDER

CLINICAL CASE

A 65-year-old patient presented to the office with two weeks of dizziness. They described the dizziness as a whirling sensation that made them feel as though their head was spinning. Each spell lasted for about 10–20 seconds. The patient had several spells per day and noted that turning in bed and looking upward exacerbated this condition. Since these spells were so incapacitating, they were afraid to move. The patient felt normal in between these spells and did not have any ringing in the ears or visual, hearing, or speech problems. Walking and balance were normal, except during the dizzy spells. There has not been any trauma, history of similar dizziness, or recent infection.

Kinesthesis (G., *kinein*, "to move;" esthesis, "sensation"), the sense of movement of the body as a whole or of its parts and the sense of the body's position in three-dimensional space with respect to objects in the immediate environment, is mediated by a somatosensory system. The position and motion of different body parts are detected and monitored by special receptors located in the muscles, tendons, and joints. The kinesthetic sense assists in the maintenance of posture and the control of a myriad of voluntary motor activities.

In conjunction with the kinesthetic sense is the sense of equilibrium (balance), mediated by the **vestibular system**. The vestibular system, a proprioceptive (somatosensory) system, mediates the special functions of muscle tone, posture maintenance, equilibrium, and coordination of head and ocular movements. Also, it transmits sensory information to the brain about the position of the body and its spatial orientation.

The vestibular system is equipped with two groups of receptors. One group detects *angular acceleration* (rotational movement) of the head, as in turning the head from side to side. The other group of receptors detects *spatial orientation* of the head in space, relative to gravity and *linear acceleration* or *deceleration* forces.

Sensory input from the *visual*, *vestibular*, and *proprioceptive* systems is integrated, at the subconscious level, by the brain, especially the cerebellum, to generate motor responses

---

A Textbook of Neuroanatomy, Third Edition. Maria A. Patestas, Amanda J. Meyer, and Leslie P. Gartner.
© 2025 John Wiley & Sons, Inc. Published 2025 by John Wiley & Sons, Inc.
Companion website: www.wiley.com.go/patestas/neuroanatomy3e

that maintain equilibrium, posture, muscle tone, and reflex ocular movements.

Muscle spindles, tendon organs (eponym: Golgi tendon organs) and free nerve endings relay *conscious* proprioceptive sensory information via the **posterior column-medial lemniscus pathway** to the somatosensory cortex. These same types of receptors relay *unconscious* proprioceptive information from muscles, tendons, and joint capsules via the **ascending sensory pathways to the cerebellum**, which informs the cerebellum of the current orientation of the body. The sensory cortex and the cerebellum, using the information supplied by these pathways, construct a map of the body's orientation and motion in real time in the coordination of impending and current movement.

The vestibular apparatus relays sensory information about the orientation and motion of the head to the vestibular nuclei and the cerebellum. The vestibular nuclei project fibers via the **medial longitudinal fasciculus (MLF)** to the oculomotor, trochlear, and abducens motor nuclei that innervate the extraocular muscles to coordinate ocular movements in response to head movements. The vestibular nuclei also project fibers to thalamic nuclei, which, in turn project to multiple vestibular sensory cortical areas for *conscious* awareness of the current orientation and motion of the head. Working in the background, the intercommunication between the vestibular nuclei and cerebellum acts to adjust, continuously, muscle tone, posture, and balance via two descending tracts, the lateral vestibulospinal tract and the medial vestibulospinal tract.

When a specific movement causes a loss of balance, the tilting of the head and leaning of the body to one side is detected by the proprioceptors in the head and in the body. The sensory receptors of the head include the vestibular apparatus embedded in the internal ear, whereas receptors of the body include the muscle spindles, tendon organs, and free nerve endings of muscles, tendons, and joints. This collective sensory input about the orientation of the head and limbs not only reaches consciousness via the ascending sensory pathways from the body and head to the somatosensory cortex, but also reaches other structures of the CNS, including the reticular formation, the vestibular nuclei, the cerebellum, and spinal cord, which function, at the unconscious level, to trigger automatic, reflex responses that modulate muscle tone of appropriate muscles to correct the imbalance by adjusting and regaining an upright posture. Ultimately, the neural interconnections between the vestibular system and the cerebellum exert an excitatory influence on the motor neurons that innervate the extensor (anti-gravity) muscles to regain balance and to maintain upright posture. The **lateral vestibulospinal tract** is one of the tracts used by the unconscious balance mechanism working in the background to regain an upright posture.

VESTIBULAR APPARATUS

The sensory apparatus of the vestibular system resides in the internal ear, embedded in the petrous portion of the temporal bone, and is composed of a bony (osseous) labyrinth (G., "maze") and a membranous labyrinth. Sensory information about the orientation and motion of the **head** (*not* the of body), is detected by the receptors in the **vestibular (sensory) apparatus** (Figs. 20.1 and 20.2)

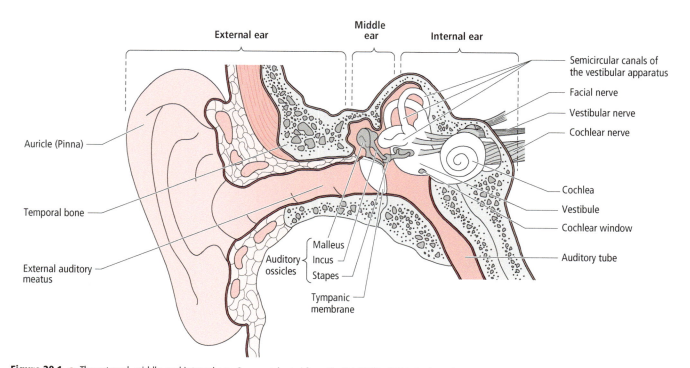

Figure 20.1 ● The external, middle, and internal ear. Source: Adapted from Canfield Willis, MC (1996) *Medical Terminology*/Marjorie Canfield Willis.

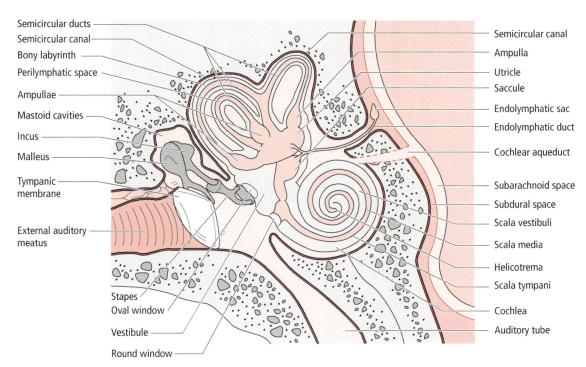

Figure 20.2 • Schematic illustration showing the vestibulocochlear apparatus (the 3 semicircular canals and cochlea) of the inner ear embedded in the petrous portion of the temporal bone.

and is relayed to the **vestibular nuclei** in the brainstem and to the **cerebellum** by the **vestibular nerve** (the vestibular division of the vestibulocochlear nerve, CN VIII). The vestibular system, in concert with the cerebellum, then mediates the appropriate postural adjustments.

The membranous components housed within the bony (osseous) labyrinth are associated with equilibrium

> The membranous components housed in the bony labyrinth that are associated with equilibrium include the three semicircular ducts and the utricle and saccule, housed in the vestibule

The **bony labyrinth** of the internal ear (Fig. 20.2) includes the **cochlea**, three bony **semicircular canals**, and the **vestibule**. The cochlea is a special receptor that mediates hearing and has no function in equilibrium (and is discussed in Chapter 19). The semicircular canals and the vestibule, however, are associated with equilibrium.

Semicircular canals

There are three **semicircular canals** in the internal ear that collectively form the housing of the kinetic labyrinth. They are the **anterior** (superior), **lateral** (horizontal), and **posterior** (inferior) canals. The semicircular canals are oriented along three planes of angular or rotational movement, at nearly right angles (perpendicular) to each other. Both ends of the C-shaped canals attach to, and open into, the bony vestibule (Fig. 20.2). The **anterior canal** (a vertically-oriented semicircular canal) of each side is positioned *anterolateral* to the median plane (about 45° from the coronal plane). It is oriented parallel to the **posterior canal** (also a vertically oriented semicircular canal), on the contralateral side, which is positioned *posterolateral* to the median plane (about 45° from the coronal plane). These two canals function as a pair. The two **lateral canals** of the right and left sides are oriented in the same (horizontal) plane, and they too function as a pair.

Pilots and astronauts refer to the three planes of rotational movement as pitch, roll, and yaw. **Pitch** refers to forward and backward tilting of the head (flexion/extension of the head at the atlanto-occipital joints), as when nodding the head to indicate "yes." **Roll** refers to tilting the head sideways from one shoulder to the other shoulder (lateral flexion of the head at the cervical joints), and **yaw** refers to the movement of turning the head from right to left (or left to right), (left and right rotation of the head at the atlantoaxial and atlanto-odontoid joints) as when indicating "no." The lateral semicircular canals are oriented in the horizontal yaw plane, whereas the anterior and posterior semicircular canals are oriented in the half pitch/half roll planes.

When the head is in its normal upright position, the anterior and posterior canals are oriented almost in the vertical plane, whereas the lateral canals are oriented almost in the horizontal plane. During a certain movement that affects a particular functional pair of canals (i.e., the two lateral canals, or the right anterior and left posterior, or the right posterior and left anterior), the receptors in one canal of the pair are stimulated and excite the afferent axons of the vestibulocochlear nerve, whereas the receptors in its complementary canal on the contralateral side, are inhibited and diminish neural activity.

Vestibule

The **vestibule** is a component of the bony labyrinth, receiving the open ends of the three semicircular canals. It is a perilymph-filled space that is continuous with the perilymphatic space of the semicircular canals. The vestibule houses two membranous sacs, the **utricle** and **saccule**, forming the static labyrinth. The utricle and saccule each contain an elliptical-shaped sensory receptor, the **macula** (L., "spot") (see below).

Membranous labyrinth associated with equilibrium

The components of the membranous labyrinth associated with equilibrium are the three semicircular ducts, the utricle, and the saccule. They all contain endolymph and are bathed in perilymph

The **membranous labyrinth** is housed within the bony labyrinth and reflects its general shape and contour. The space separating the bony labyrinth from the membranous labyrinth is the **perilymphatic space**, filled with a fluid referred to as **perilymph**, which resembles extracellular fluid (has low potassium and high sodium concentrations). The membranous labyrinth consists of a series of **membranous sacs** and **ducts** filled with another viscous fluid, referred to as **endolymph** (has high potassium and low sodium concentrations), which resembles intracellular fluid. The components of the membranous labyrinth associated with equilibrium are the **three** (membranous) **semicircular ducts** (which are enclosed in their respective bony semicircular canals) and the (membranous) **utricle** and **saccule**, two saclike structures that are enclosed within the bony vestibule. The region of the membranous labyrinth containing the receptor for hearing is the cochlear duct, which is enclosed in the bony cochlea (discussed in Chapter 19). Endolymph percolates through the three membranous semicircular ducts, the utricle, the saccule, and the cochlear duct. A connection, the **ductus reuniens** permits communication between the lumen of the cochlear duct and the lumen of the components of the vestibular membranous labyrinth. Both ends of each of the semicircular ducts are attached to, and empty into, the utricle.

Each semicircular duct has a single dilated segment, an **ampulla** (L., "dilation"), near one of its ends. The ampullae of the three semicircular ducts, as well as the utricle and saccule, contain in their interior the **sensory receptors of the vestibular system**. The receptors in the ampullae of the semicircular ducts are the **ampullary crests**, whereas the receptors in the utricle and the saccule are the **macula of utricle** and the **macula of saccule**, respectively (Figs. 20.3 and 20.4). Thus, there are a total of five receptor sites (the three crests and the two maculae) in each vestibular sensory apparatus. The ampullary crests of the semicircular canals rest in a patch of neuroepithelium at the base of each ampulla. The macula of utricle rests in a strip of neuroepithelium on the base of the utricle, whereas the macula of saccule is oriented vertically (in sagittal plane) on the medial wall of the saccule.

Note that the clinical case at the beginning of the chapter refers to a patient whose symptoms include intermittent spells of dizziness. The patient was diagnosed with benign positional vertigo (BPV) that may be caused by free floating debris or particles in the endolymph of the membranous labyrinth in the internal ear.

1 What structures make up the membranous labyrinth?

2 What is the sensory structure that is enclosed within, and is unique to, the utricle and the saccule?

Vestibular receptors

The vestibular receptors (ampullary crest, macula of utricle, and macula of saccule) consist of a distinct area of neuroepithelium containing mechanoreceptor "hair cells"

The **vestibular receptors** consist of a distinct area of neuroepithelium inside the membranous labyrinth, composed of mechanoreceptor "hair cells" and support cells that fill the spaces among the hair cells (see Figs. 20.3 and 20.4). The hair cells display many stiff **stereocilia** (specialized, elongated microvilli, and not true cilia), known as "hairs," and a single immotile **kinocilium**. These hair cells, similar to the hair cells in the cochlea that are involved in hearing, are *transducers* that transform the mechanical stimulation they experience into electrical signals that are transmitted to, and are then conveyed by, the vestibular nerve to the brainstem and cerebellum. Both afferent and efferent nerve endings terminate on the basal and lateral aspects of the hair cells. The neuroepithelia of the **ampullary crests**, **macula of utricle**, and **macula of saccule** are covered with a **gelatinous glycoprotein membrane**. The stereocilia and kinocilia extend into and are embedded in this membrane. Movement of the hair cells with respect to the membrane results in mechanical deformation of the stereocilia and kinocilia which is then transformed into an electrical potential. The ampullary crests respond to angular acceleration of the head, whereas the macula of utricle and macula of saccule respond to linear acceleration and reorientation of the head.

Statoconia

In the utricle and saccule, the gelatinous glycoprotein membranes overlying their macula have statoconia (otoconia or otoliths) on their free surface

The gelatinous glycoprotein membranes overlying their macula in the utricle and the saccule, have on their free surface, **statoconia,** also

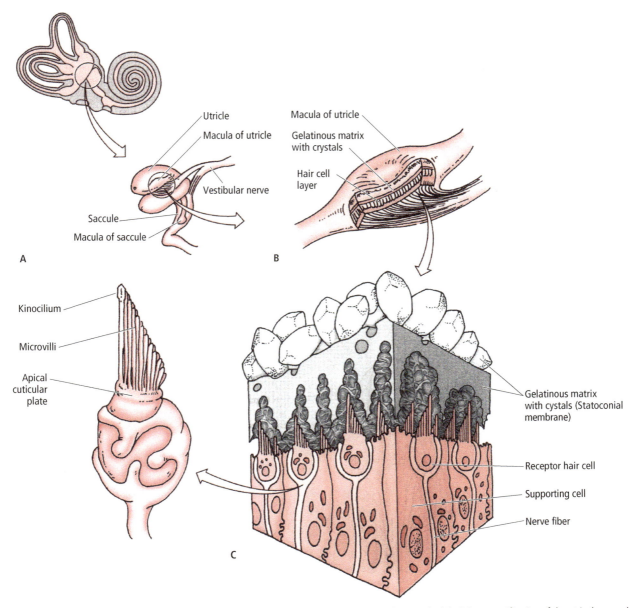

Figure 20.3 ● The macula. (A) Location of the macula within the utricle and saccule. (B) The utricular macula. (C) Higher magnification of the utricular macula illustrating the statoconia on the statoconial membrane, the receptor cells, and the peripheral fiber terminals of the vestibular nerve synapsing with the receptor hair cells. (D) Higher magnification of a receptor hair cell showing its single kinocilium and numerous microvilli.

referred to as otoconia (G., "ear dust") or otoliths (G., "ear stones") – crystals consisting of calcium carbonate (see Fig. 20.3). Since the gelatinous glycoprotein membrane of the utricle and the saccule has statoconia, it is referred to as the **statoconial membrane**. The specific gravity of the statoconia is greater than that of the endolymph that surrounds them, consequently, they respond to a greater aspect to the force of gravity. Gravity applies a continuous linear acceleration on the head, and thus deflects the gelatinous membrane, which in turn stimulates the hair cells.

The glycoprotein membrane of the **ampullary crest** is dome-shaped and is called the **cupula**, which is devoid of crystals (see Fig. 20.4). The cupula increases resistance to the flow of endolymph and unlike the statoconial membranes is *not* influenced to as great a degree by gravitational forces **as it is influenced by endolymphatic flow**. The cupula is a mechanical structure that narrows the lumen of the canal and so increases resistance to flow. The three ampullary crests enclosed in the semicircular duct ampullae (one crest in each ampulla) detect angular acceleration or deceleration (rotational movement) of the head (as in turning or tilting the head). The differing orientations of the three semicircular ducts enable an individual to perceive motion in all planes. Head rotation in a particular plane will result in endolymphatic flow in the functional pair of canals oriented closest to the plane of rotation and is almost perpendicular to the axis

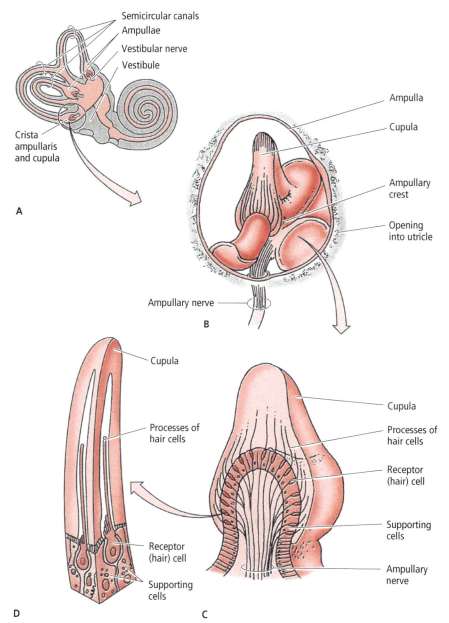

Figure 20.4 • Three-dimensional view of the semicircular canals. (A) Location of the ampullary crest within the ampullae of the three semicircular canals. (B) An enlarged ampullary crest. (C) Higher magnification of a sectioned ampullary crest showing the cupula, receptor hair cells, and peripheral processes of the ampullary nerve (branch of CN VIII) synapsing with the receptor hair cells (D).

of rotation. Head rotation in a plane that is not parallel to that of any one canal, will activate more than one canal.

The receptors, the macula of utricle and macula of saccule within the utricle and the saccule, respectively, detect spatial orientation of the head in space relative to gravity and to linear acceleration or deceleration forces. The receptors in the utricle and in the saccule are oriented *perpendicular* to each other and, therefore, can detect motion in two different planes. When the head is upright, the **macula of utricle** is oriented in the *horizontal plane* (on the base of the utricle) and can be stimulated (activated) by linear forces (acceleration or deceleration) in the horizontal plane (as occur in a car, when it accelerates or decelerates). The **macula of saccule** is oriented in the *vertical plane* (on the medial wall of the saccule) and can be stimulated by linear forces (acceleration or deceleration) in the vertical plane (as occurs in an ascending or descending elevator). Note that the macula of utricle and the macula of saccule perceive head orientation when the body itself is at rest but is exposed to external forces.

3 What is associated with the macula that may create debris that can float in the endolymph of the saccule or utricle?

Mechanisms of action of the vestibular apparatus

When the body is at rest, and the head is tilted, the statoconia (resting on the gelatinous membrane of the macula of utricle and the macula of saccule) respond to gravitational forces, causing the membrane to slant, which in turn deflects the "hairs" (of the hair cells) embedded in it. The hair deflection causes opening of ion channels of, and consequent depolarization of the mechanoreceptor hair cells, which in turn causes depolarization and firing of the vestibular nerve dendrites synapsing with them.

In angular (rotational) movement of the head (Fig. 20.5), the bony semicircular canals and the enclosed membranous semicircular ducts rotate at the same velocity as the head. However, the viscous endolymph filling the ducts reacts slower thus "falling behind" due to inertia (that is, the endolymph does not move as fast as the ducts and canals that house them). This resistance causes a relative motion difference between the faster moving cupula (which is attached to the interior of the semicircular ducts) and the slower endolymphatic flow. This motion difference results in the apparent flow of the endolymph in the direction opposite to that of the cupula, pressing against and tilting it. This action deflects the stereocilia embedded in the cupula. The stereocilia that are deflected away from the kinocilium (caused by the endolymphatic flow away from the kinocilium), closes ion channels and elicits hyperpolarization (inhibition) of the hair cell. Whereas the stereocilia that are deflected toward the kinocilium (by endolymphatic flow toward the kinocilium) stretch the hair cell membrane, opening ion channels, and eliciting depolarization (excitation) of the hair cell, and, therefore, initiate a nerve impulse that is relayed to the vestibular nerve afferent peripheral nerve endings. Flow of endolymph *in the direction of the ampulla* of the horizontal canal results in deflection of the stereocilia toward the kinocilium which **stimulates** the **hair cell**. In the anterior and posterior semicircular canals, flow of endolymph in the direction of the ampulla has the opposite effect, that is, the **hair cells are inhibited**. If rotation is sustained, the endolymph of the semicircular canals eventually moves at the same rate as the cupula. When this occurs, the cupula is no longer pushed and bent by the endolymph, and the stereocilia and kinocilia are no longer deflected. Since the hair cells are no longer depolarized, the **semicircular canal receptors are not activated by sustained rotation**. Following about 20 seconds of continued rotation, however, the cupula slowly recoils back to its resting position. When the head stops rotating, so do the ducts and the canals. However, the endolymphatic flow persists within the membranous duct due to its momentum, pressing on, and bending the cupula from its resting position in the opposite direction (in the same direction as the head had been rotating), causing the stereocilia to bend away from the kinocilium, thus inhibiting the hair cell from firing completely. A few seconds later, endolymphatic flow ceases and within 20 seconds the cupula once again recoils back to its resting position, permitting the hair cell to emit impulses at a tonic level. Therefore, the semicircular canal conveys a positive signal at the onset of head rotation only if movement is in the correct direction and conveys a negative signal when head rotation ceases.

It is important to note that the function of the equilibratory senses mediated by the vestibular system is to *perceive differences in the relative movement* between the endolymph and the receptor statoconial membrane (as occur during the onset and cessation of rotation), rather than to merely detect motion (when sustained rotation is in progress).

Slow, slight rotation sends signals to the brain that indicate movement. Fast, intense rotation (as well as motion sickness that occurs in a car or boat) overwhelms the receptors, sending a flood of signals to the brain, with consequent dizziness and/or nausea. When an individual is aboard a ship in a room without a window and cannot see outside, their vestibular system sends information to the brain that they are moving (since the ship is moving), however the individual cannot "see" that they are actually moving. The inconsistency of the sensory information (or sensory mismatch) that is sent to the brain by the vestibular and visual systems causes them to feel **seasick**. Going out on the deck to *see* movement, confirms the vestibular input to the brain, and alleviates the symptoms of nausea.

Astronauts can experience space sickness when exposed to an environment without gravity. **Space sickness**, an unusual phenomenon affecting an astronaut's sense of equilibrium, is believed by some scientists, to be caused by a lack of vestibular sensory input due to the zero-gravity environment of space. Others believe that space sickness is similar to car or sea sickness and is caused by a sensory mismatch as described earlier.

VESTIBULAR NERVE (CN VIII)

> The cell bodies of the first-order bipolar neurons of the vestibular pathway reside in the vestibular ganglion (eponym: Scarpa ganglion), a sensory ganglion of the vestibular nerve

The cell bodies of the first-order **bipolar neurons** of the vestibular pathway reside in the vestibular ganglion, a sensory ganglion of the vestibular nerve. The peripheral processes of these bipolar neurons (Fig. 20.6) terminate in the **ampullary crests** of the semicircular ducts, the **macula of utricle**, and the **macula of saccule**, where they form synapses with the mechanoreceptor epithelial "**hair cells.**" The hair cells are transducers; that is, they transform mechanical stimulation into electrical stimulation (action potentials), which is relayed to the dendritic terminals of the first-order bipolar neurons. The central processes of the bipolar neurons gather to form the root of the **vestibular nerve** (the vestibular division of the vestibulocochlear nerve, **CN VIII**). The vestibular nerve transmits sensory information about the orientation and motion (linear and angular acceleration) of the head to the brainstem. In the internal ear, the vestibular

412 ●●● CHAPTER 20

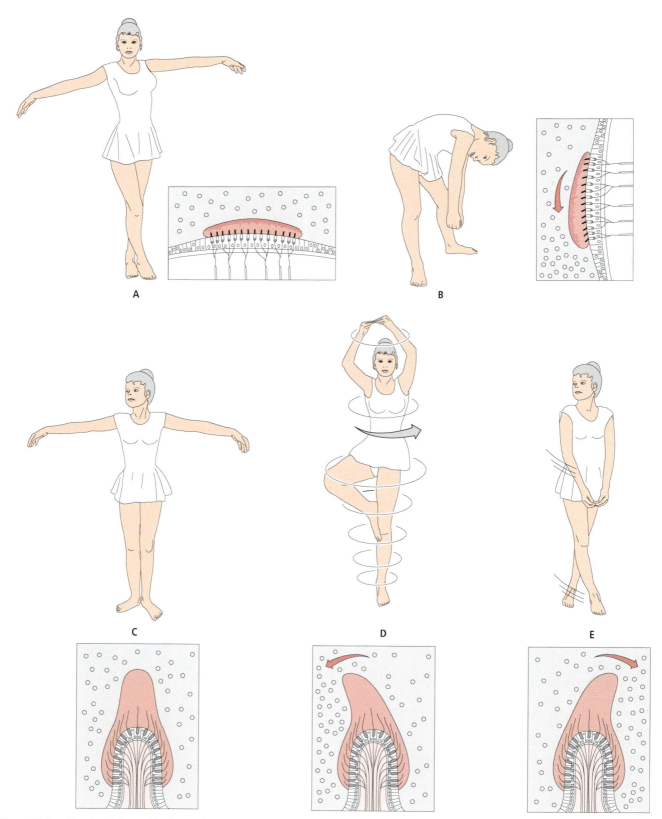

Figure 20.5 ● Function of the semicircular canals in balance maintenance. (A, B) The maculae detect spatial orientation of the head relative to gravity and linear acceleration or deceleration forces. (C–E) The crests respond to endolymphatic flow. They detect angular acceleration or deceleration (rotational movement) of the head (as in turning or tilting the head). As a person at rest (C) begins to rotate (D), the ampullary crest is tilted by the flow of the endolymph opposite to the direction of the rotation. When the person stops rotating (E), the ampullary crest is tilted by the flow of the endolymph in the same direction as the rotation.

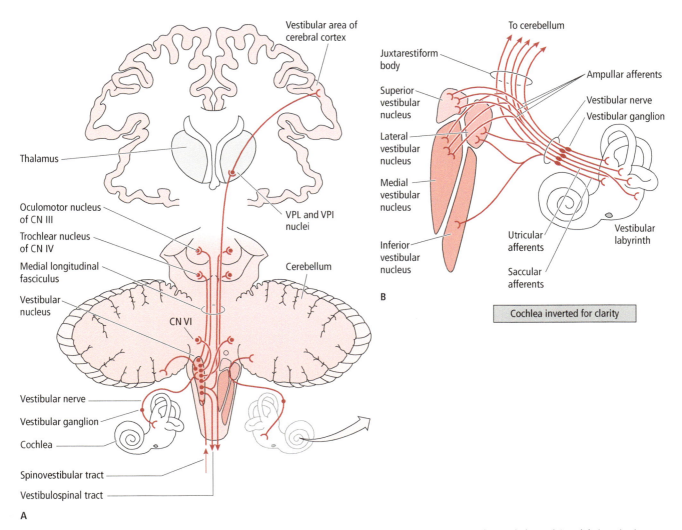

Figure 20.6 ● (A) The vestibulocochlear apparatus, termination of the central processes of the vestibular nerve in the vestibular nuclei, and their projections to the oculomotor, trochlear, and abducens nuclei. VPI, ventral posterior inferior; VPL, ventral posterior lateral. (B) The termination of the central processes of the first-order afferent neurons of the vestibular ganglion in the brainstem vestibular nuclei and the cerebellum.

nerve joins the cochlear nerve, to form the **vestibulocochlear nerve**. Although the two nerves accompany one another wrapped in a common connective tissue sheath, they are anatomically and functionally two distinct nerves. From the internal ear, the vestibulocochlear nerve passes through the internal auditory meatus to enter the posterior cranial cavity. As the vestibulocochlear nerve enters the brainstem at the pontomedullary angle, it separates into its two component nerves. The fibers of the cochlear nerve synapse in the cochlear nuclei in the pons and medulla oblongata (as discussed in Chapter 19). The vestibular nerve primary afferents are distributed to the vestibular nuclei and the cerebellum as follows:

Most of the vestibular nerve fibers enter the **vestibular nuclei** located in the pons and superior medulla oblongata, where the central processes of the vestibular nerve bifurcate into ascending and descending terminals. Some first-order vestibular fibers (the primary vestibulocerebellar fibers), however, do not terminate in the vestibular nuclei, but take an alternate route by going around them, joining the **juxtarestiform body** in the inferior cerebellar peduncle and terminating *directly* in the **ipsilateral flocculonodular lobe of the cerebellum**. This direct termination of the central processes of the first-order bipolar neurons in the cerebellum is unique to the vestibular system.

CENTRAL PATHWAYS OF THE VESTIBULAR SYSTEM

Sensory input from the vestibular system is integrated to produce coordinated movement of the head, eyeballs, and body, and to maintain equilibrium and muscle tone

Unlike other sensory systems, the projections of the vestibular system are mainly reflexive in nature, in that they initiate reflex reactions (Fig. 20.7). Sensory input is

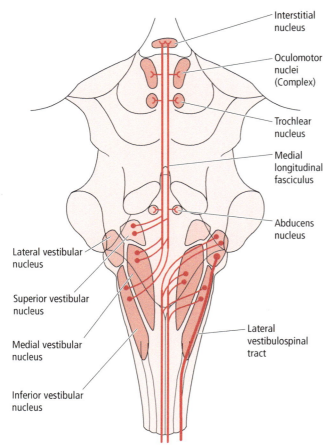

Figure 20.7 ● The vestibular nuclei and their ascending and descending projections.

integrated to produce coordinated movement of the head, eyeballs, and body, and to maintain equilibrium and muscle tone. To accomplish and to mediate these functions, the vestibular system projects mainly to the following structures:

1. The **motor nuclei** of the oculomotor, trochlear, and abducens nerves that innervate the extraocular muscles and thus control reflex ocular movements in response to head movements.
2. The **cerebellum**, which integrates sensory input from various systems and coordinates head and body movement.
3. The **reticular formation** at the pontine and medullary levels, where reflex motor activity is initiated.
4. The **spinal cord**, where postural adjustments can be made.

Vestibular nuclear complex

The vestibular nuclear complex is composed of four vestibular nuclei, and the central processes of the first-order bipolar neurons of the vestibular ganglion that terminate there

The **vestibular nuclear complex** is composed of four vestibular nuclei (Fig. 20.6), and the central processes of the first-order bipolar neurons of the vestibular ganglion that terminate there. The four vestibular nuclei are:

1. The **superior** (eponym: Bechterew) vestibular nucleus.
2. The **medial** (eponym: Schwalbe) vestibular nucleus.
3. The **lateral** (eponym: Deiter) vestibular nucleus.
4. The **inferior** (spinal, descending) vestibular nucleus.

The vestibular nuclei are located in the superior medulla oblongata and inferior pons in close proximity to the cerebellum, lateral to the fourth ventricle. The superior nucleus is located entirely in the pons, whereas the other three nuclei extend into both the caudal pons and the rostral medulla oblongata. These nuclei contain the cell bodies of second-order neurons whose fibers carry outflow signals to various destinations.

The **superior** and **medial vestibular nuclei** receive the first-order neuron terminals relaying sensory input from the ampullary crests of the **semicircular canals**. Following the reception of this sensory input, these nuclei then relay it via two structures:

1. The ascending **MLF** projecting to the extraocular muscle nuclei to elicit compensatory ocular movements triggered by movements of the head.
2. The **medial vestibulospinal tract** (descending MLF) projecting to the cervical spinal cord to elicit appropriate head movements.

The **lateral vestibular nucleus** receives vestibular sensory input mainly from the macula of the **utricle** but may also receive input from the **saccule** and semicircular ducts. This nucleus projects, via the *lateral vestibulospinal tract*, to motor neurons or interneurons at all spinal cord levels to make postural adjustments.

The **inferior vestibular nucleus** receives vestibular sensory input from the **semicircular canals** as well as the **utricle**. Most of the first-order vestibular fibers terminate in this nucleus, which then projects to the reticular formation and the cerebellum.

Termination of the vestibular nerve first-order afferent fibers

The first-order afferent fibers of the vestibular nerve project to, and terminate directly in, either the vestibular nuclear complex or the flocculonodular lobe of the cerebellum

The **vestibular nuclei** and the **cerebellum** serve as the **first relay centers of the vestibular pathway**. The vestibular nerve is unique since it is the only cranial nerve that sends some of its first-order afferent fibers (the primary vestibullocerebellar fibers) relaying sensory input *directly* from peripheral receptors to the cerebellar cortex (of the flocculus, nodulus, and uvula).

VESTIBULAR SYSTEM   415

Afferent projections (input) to the vestibular nuclei

Most of the afferents arriving at the vestibular nuclei arise from the cerebellum

The vestibular nuclei receive afferent (input) projections from the following sources

1 The central processes of the bipolar first-order neurons of the **vestibular nerve**, which transmit sensory input from the vestibular receptor apparatus in the internal ear (as discussed above; see Fig. 20.6).
2 The **vestibulocerebellum** (the flocculus and nodulus) and part of the uvula (of the cerebellum), which project their cerebellar corticovestibular fibers to the ipsilateral vestibular nuclei via the juxtarestiform body (Fig. 20.8).
3 The **spinocerebellum** (the vermis of the anterior lobe of the cerebellum), which projects its fibers to the ipsilateral vestibular nuclei via the juxtarestiform body.
4 The **fastigial nucleus** of the cerebellum, which projects bilaterally to the vestibular nuclei via the fastigiovestibular tract.
5 The reciprocal connections from the **contralateral vestibular nuclei**.
6 The **spinal cord** (via spinovestibular fibers).
7 The **pretectal nuclei**.

The spinal cord (all levels) sends proprioceptive information to the vestibular nuclei via the **spinovestibular tract** regarding the orientation status of the body so that postural adjustments can be made.

Efferent projections (output) from the vestibular nuclei

The vestibular nuclei project **second-order fibers** to the following areas (see Fig. 20.7):

1 The **oculomotor, trochlear,** and **abducens motor nuclei** as well as the **accessory oculomotor nuclei**, the **interstitial nucleus** (eponym: of Cajal), and the **elliptic nucleus** (eponym: nucleus of Darkschewitsch) via the **MLF** for the coordination of head and ocular movement.
2 The **contralateral vestibular nuclei**, most of which are reciprocal inhibitory connections (commissural vestibulovestibular fibers).

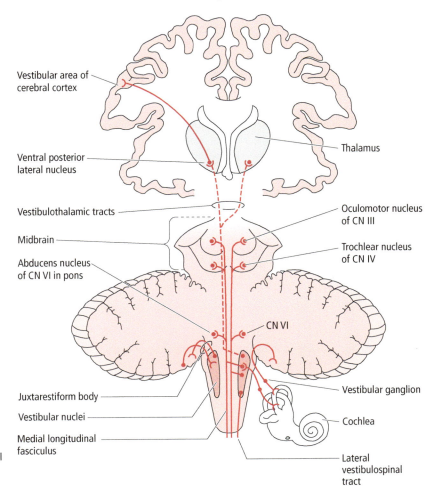

Figure 20.8 ● The main central projections of the vestibular system. Output fibers from the vestibular nuclei join the ascending MLF to terminate in the motor nuclei innervating the extraocular muscles where they function in vestibulo-ocular reflexes. Output fibers from the vestibular nuclei joining the descending MLF and the lateral vestibulospinal tracts terminate in the motor horn of the spinal cord where they function in postural reflexes.

416 ● ● ● **CHAPTER 20**

3 The **inferior olivary nucleus** (via the vestibulo-olivary tract), which in turn projects to the cerebellar vermis via the restiform body.

4 The **ipsilateral flocculus** and **nodulus** (both of which are associated with equilibrium and receive sensory information from the semicircular ducts) and the uvula (they also receive secondary vestibulocerebellar fibers from the vestibular nuclei).

5 The **uvula** (of the cerebellum).

6 The **fastigial and dentate nuclei** of the cerebellum (bilaterally), via the juxtarestiform body (note that the juxtarestiform body consists of two-directional traffic; i.e., it carries both afferent [cerebellovestibular] and efferent [vestibulocerebellar] fibers connecting the vestibular nuclei and the cerebellum).

7 **Spinal cord** anterior horn alpha- and gamma-motor neurons and interneurons.

8 The **vestibular labyrinth**, where descending efferent fibers modulate afferent input to the vestibular nuclei.

9 **Thalamus**

10 **Visceral nuclei of the brainstem**, the solitary, vagal, and parabrachial nuclei, to influence autonomic cardiovascular adjustments (blood pressure control) during changes in posture

Ascending projections

The superior and medial vestibular nuclei give rise to fibers that join the MLF and terminate in the cranial nerve nuclei which innervate the extraocular muscles

The **superior vestibular nucleus** fibers join the ipsilateral and contralateral MLF and project to the oculomotor nuclear complex as well as the trochlear nuclei bilaterally. Some of the fibers arising from the **medial vestibular nucleus** join the ipsilateral and the contralateral MLF to terminate in the oculomotor nuclear complex and the abducens nucleus bilaterally, whereas other medial vestibular nuclear fibers join only the contralateral MLF and terminate in the contralateral trochlear nucleus and interstitial nucleus.

Fibers arising from the **inferior vestibular nucleus** join the ipsilateral and contralateral MLF to terminate in the oculomotor and trochlear nuclei bilaterally.

All of these ascending (output) connections function in the *coordination of ocular movements*. Since the MLF also carries descending fibers from the medial vestibular nucleus to the spinal cord (see below), it functions in the coordination of head, eyeball, and neck movements.

Descending projections

The vestibular nuclei regulate the activity of spinal cord alpha- and gamma-motor neurons by their lateral and medial vestibulospinal tracts to control the anti-gravity muscles and neck muscles

The **lateral vestibulospinal tract** originates primarily from the lateral and inferior vestibular nuclei. This tract descends ipsilaterally and terminates at all levels of the spinal cord although it projects more fibers at the cervical and lumbar levels, where it synapses with both interneurons and with alpha- and gamma-motor neurons in the anterior horn of the spinal cord. On its way to the spinal cord, the lateral vestibulospinal tract gives rise to collaterals that synapse in the reticular formation of the medulla oblongata. This tract also gives rise to collaterals that synapse at different spinal cord levels where they function to modulate posture by coordinating the activities of different muscle groups. Functionally, the lateral vestibulospinal tract regulates the activity of spinal cord motor neurons to *maintain posture and balance*. It forms excitatory synapses with the alpha- and gamma-**motor neurons** whose fibers innervate extensor (anti-gravity) muscles in the neck, trunk, and lower limb, as well as inhibitory synapses via interneurons to the alpha- and gamma-motor neurons that innervate flexor muscles (causing their relaxation). The lateral vestibulospinal tract neurons control the axial and proximal limb musculature.

The **medial vestibulospinal tract** originates primarily from the medial vestibular nucleus. This tract descends bilaterally in the descending MLF and terminates in the upper cervical spinal cord where it, too, forms synapses with interneurons and alpha- and gamma-**motor neurons**, innervating neck flexors and extensors. Functionally, since the medial vestibulospinal tract terminates in the cervical spinal cord, it serves to stabilize head position/orientation and *coordinate head and ocular movements during gaze*.

The **inferior vestibular nucleus** gives rise to fibers that terminate in the inferior olivary nucleus, the reticular formation in the medulla oblongata, the cerebellum, and in the cervical spinal cord via the MLF.

All output vestibular projections distribute collaterals in the medullary, pontine, and mesencephalic reticular formation. These relays play an important function in the reflexive control and coordination of head and ocular movements.

Vestibular thalamic and cortical projections

The vestibular nuclei project second-order fibers to the thalamus bilaterally either by joining the auditory fibers of the lateral lemniscus or by ascending through the reticular formation near the MLF. The thalamus in turn relays vestibular input to the vestibular cortex

Unlike other sensory systems that form ascending tracts (i.e., spinothalamic, medial lemniscus) that terminate in the thalamus, second-order vestibular fibers do not form a distinct ascending vestibular tract. The vestibular nuclei project second-order fibers to the thalamus bilaterally either by joining the auditory fibers of the lateral lemniscus or by ascending through the reticular formation near the MLF. The thalamus in turn relays this information to the vestibular cortex including the eponymous Brodmann area 3a, the retroinsular cortex and the posterior parietal lobe. The **superior** and **lateral vestibular nuclei** give rise to second-order fibers that ascend bilaterally to terminate in the **ventral posterior lateral** (**VPL**) and ventral posterior inferior (VPI) nuclei of the thalamus. The thalamus gives rise to third-order fibers that terminate in the eponymous **Brodmann area 3a** of the primary somatosensory cortex in the parietal lobe, located next to the primary motor area (eponym: Brodmann area 4). The eponymous Brodmann area 3a is believed to be the site of integration of sensory input from the vestibular and other proprioceptive systems (see Fig. 20.8) that send sensory information from the vestibular apparatus and from muscle spindles from the neck muscles respectively, to form a composite image that informs the brain about the orientation and motion of the head. The associations between eponymous Brodmann areas 3a and 4 serve in the regulation of motor activity. Other second-order fibers arising mainly from the superior, medial, and inferior vestibular nuclei ascend to terminate in thalamic nuclei, which, in turn project to the **retroinsular cortex** and the **posterior parietal lobe**. These projections are involved in the conscious awareness of the orientation of the head and body parts in the context of the external environment.

CONTROL OF OCULAR MOVEMENTS

The majority of ocular movements are mediated by reflex activity involving several neural systems including the vestibular system that are interconnected by complex pathways

Ocular movements are categorized into conjugate and disconjugate (disjunctive) movements. **Conjugate eyeball movements**, in which both eyeballs turn toward the same direction simultaneously (i.e., to the right or to the left), include: (1) **smooth pursuit** eyeball movements, which are involuntary, slow, and smooth, and occur during the tracking of a moving object (note that smooth pursuit can only be elicited if there is an actual moving stimulus to track); (2) **optokinetic** eyeball movements, which are tracking eyeball movements and are triggered by motion of the visual field; (3) **saccadic** eyeball movements, which are rapid and abrupt, voluntary or involuntary, and change the point of visual fixation; and (4) **vestibular** eyeball movements, which are involuntary eyeball movements triggered by head movements (i.e., the vestibulo-ocular reflex). **Disconjugate eyeball movements** consist of **vergence** eyeball movements in which both eyeballs turn in opposite directions simultaneously. Vergence eyeball movements are involuntary and may be: (1) **convergent** (both eyeballs are adducted, i.e., rotate toward the midline, when focusing on an object close to one's face); or (2) **divergent** (i.e., eyeballs rotate away from the midline toward their normal position when one is looking at an object as it moves away from one's face).

The motor nuclei (oculomotor, trochlear, and abducens nuclei, containing lower motor neurons) innervating the extraocular muscles receive inputs from various sources, which indirectly control voluntary or involuntary ocular movements. There are four principal projection sources to these nuclei.

1 **Upper motor neurons** from the **frontal eye field** controlling *voluntary ocular movements*.

2 The **vestibular system**, which detects head motion and *coordinates head* and *ocular movements* via: (i) the pontine reticular formation lateral gaze center (also referred to as the **paramedian pontine reticular formation**) that controls *conjugate horizontal ocular movements*; and (ii) a **center for vertical gaze** in the periaqueductal gray of the mesencephalon, which controls *conjugate vertical ocular movements*, to keep the eyeballs fixed on a stationary object as the head is moving.

3 Projections from the **visual cortex** to the rostral midbrain (vertical gaze center and lateral gaze center), that assists in the *tracking of a moving object* by moving the eyeballs to maintain the image of a moving object on the retina.

4 Projections from the *auditory system* that play a role in causing a sudden (reflex) turning of the eyeballs in the direction of a startling sound (via the tectospinal tract).

Vestibular control of eye movements

As the head moves, reflex activity mediated by vestibular neural circuits connects the vestibular system with the cranial nerve nuclei to elicit compensatory ocular movements, so that an object can remain in the field of vision

The extraocular muscles that control movements of the eyeballs receive innervation from **motor neurons** whose cell bodies are housed in the oculomotor, trochlear, and abducens nuclei. The superior oblique muscle is innervated by the trochlear nerve, the lateral rectus muscle is innervated by the abducens nerve, whereas the remainder of the extraocular muscles, the superior and inferior rectus, the medial rectus, and the inferior oblique, are innervated by the oculomotor nerve. To produce smooth, coordinated ocular movements as one extraocular muscle contracts, its antagonists relax.

As the head moves, the eyeballs may gaze at a motionless object by reflex activity mediated by vestibular neural circuits that connect the vestibular system (which detects orientation and movement of the head in space) with the cranial nerve nuclei (innervating the appropriate extraocular muscles). This elicits compensatory ocular movements, so that the object can remain in the field of vision.

Conjugate horizontal ocular movements

Vestibulo-ocular reflex

During movement of the head in the horizontal plane (right to left or vice versa), information is transmitted from the canal-specific vestibular system via reflex circuit connections to the abducens and oculomotor nuclei, which innervate extraocular muscles controlling horizontal ocular movement

The vestibular nuclei have an essential role in the vestibulo-ocular reflex. When the head is turned (e.g., to the right), it induces subtle endolymphatic flow in both horizontal semicircular ducts. The endolymph flows to the left since its inertia causes it to "fall behind" relative to the movement of the head. The endolymphatic flow causes deflection of the cupula in the right horizontal canal ampulla, bending the "hair cell" sensory hairs toward the kinocilium, and increasing the neural activity of the right ampullary crest of the right horizontal canal. There is a simultaneous decrease in neural activity of the ampullary crest of the left horizontal canal. The depolarized hair cells (in the right horizontal canal ampulla) transmit sensory information to the peripheral (dendritic) terminals of the first-order, vestibular ganglion neurons. The central processes of these neurons terminate in the ipsilateral (in this example, right) vestibular nuclei where they synapse with neurons whose axons join the MLF and terminate in the contralateral (left) **abducens nucleus** (Fig. 20.9). In the abducens nucleus, the axon terminals synapse with and *stimulate* the motor neurons that project directly to, and innervate, the contralateral (left) **lateral rectus muscle** as well as another group of neurons that cross the midline, join the contralateral (right) **MLF** and terminate in the contralateral (right) **oculomotor nucleus** where they synapse with the motor neurons that innervate the right **medial rectus muscle**. It is in this manner that the muscles that would cause conjugate horizontal ocular movement in the opposite direction to that of the head movement are stimulated.

The second-order vestibular fibers that terminate in the ipsilateral (right) abducens nucleus *inhibit* the motor neurons that project directly to, and innervate, the ipsilateral (right) lateral rectus muscle as well as another group of neurons that cross the midline, join the contralateral MLF, and terminate in the contralateral (left) oculomotor nucleus where they synapse with the motor neurons that innervate the contralateral (left) medial rectus muscle. Thus, the muscles that would cause conjugate (simultaneous) horizontal ocular movement in the same direction as the head movement are inhibited.

The outcome of this mechanism is the **reflex turning of both eyeballs to the left** to maintain an image on the retina as the head is turned to the right.

In summary, when an individual fixes their gaze on ("stares at") an object and then begins to turn their head, for example to the right, their eyes will reflexively (via MLF vestibular connections to the extraocular nuclei) "turn" to the left (as the head is moving), which is opposite to the direction of head movement. This turning of the eyes (dictated by vestibular reflex connections) compensates for the shifting of head position to *maintain visual fixation* on an object. Otherwise, without this vestibular reflex mechanism coupling head and eye movements, as the erect head moves from a certain position (to the right or to the left) the eyes would remain still, perceiving a changing visual field in the duration of head movement. The vestibulocerebellum regulates

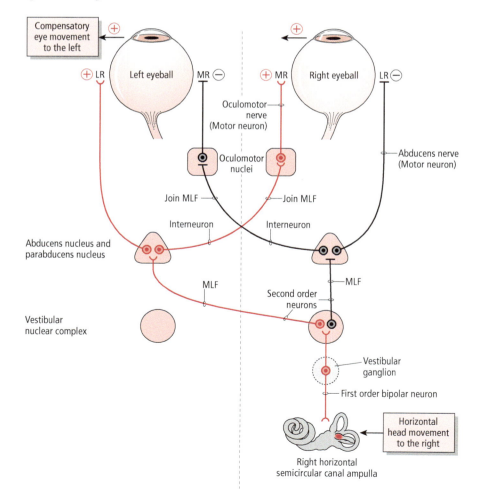

Figure 20.9 • Central connections mediating compensatory horizontal eye movement in response to horizontal head movement. LR, lateral rectus muscle; MLF, medial longitudinal fasciculus; MR, medial rectus muscle.

vestibulo-ocular reflex activity to produce compensatory ocular movement for the head movement and can cancel it so that the eyes can move from a target.

Conjugate vertical ocular movement

During vertical head movement, information is transmitted from the vestibular system to the trochlear and oculomotor nuclei, which innervate extraocular muscles controlling vertical eye movement

During movement of the head in the vertical (sagittal) plane (flexion or extension), information is transmitted from the canal-specific vestibular system via reflex circuit connections to the trochlear and oculomotor nuclei, which innervate extraocular muscles controlling vertical ocular movement. If the head is erect and the eyeballs are fixed on an object, as the head is tilted forward (flexion) or backward (extension), a reflex will be automatically activated. This causes a "reflex turning" of the eyeballs upward or downward, respectively, to maintain visual fixation on the object.

Conjugate upward ocular movement

When the head tilts forward, the eyes reflexively turn upward

When tilting the head forward (flexion), the eyes will reflexively turn upward. This is caused by endolymphatic flow in the anterior (superior) vertical semicircular duct and the subsequent deflection of its cupula. Deflection of the cupula stimulates the hair cells, which transmit sensory input to the ipsilateral vestibular nuclei. The first-order fibers synapse with both inhibitory and excitatory second-order vestibular neurons (Fig. 20.10).

Second-order inhibitory vestibular neurons join the ipsilateral MLF and ascend to, terminate in, and inhibit: (1) the ipsilateral trochlear nucleus, whose neurons give rise to fibers that decussate and then exit the brainstem to innervate the **contralateral superior oblique muscle**; and (2) the oculomotor nucleus, which sends fibers to innervate the **ipsilateral inferior rectus muscle**.

Second-order excitatory vestibular neurons give rise to fibers that decussate, join the contralateral MLF, and ascend

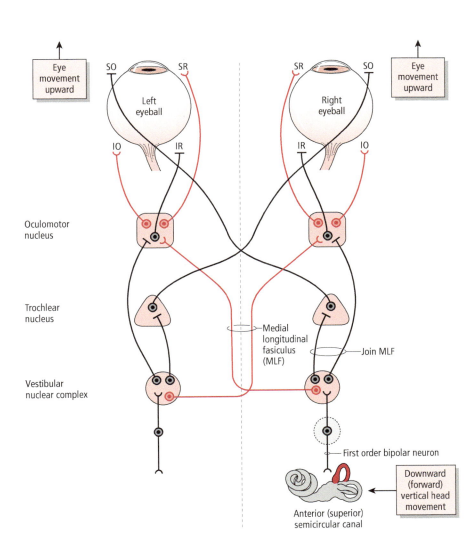

Figure 20.10 • Central connections mediating compensatory vertical (upward) eye movement in response to downward (forward) movement of the head. IO, inferior oblique muscle; IR, inferior rectus muscle; SO, superior oblique muscle; SR, superior rectus muscle.

to terminate in the contralateral oculomotor nucleus where they synapse with neurons that innervate the **contralateral superior rectus** and **inferior oblique muscles**.

In summary, as the head tilts *forward*, the eyeballs *turn upward* by the simultaneous inhibition of the neurons that innervate the superior oblique and inferior rectus muscles of both eyes, and the stimulation of the neurons that innervate the left and right superior rectus and inferior oblique muscles.

Conjugate downward ocular movement

When the head tilts backward, the eyes reflexively turn downward

When tilting the head backward, the eyeballs will reflexively turn downward. This is caused by endolymphatic flow in the posterior (inferior) vertical semicircular duct and the subsequent deflection of its cupula. Deflection of the cupula stimulates the hair cells, which transmit sensory input to the ipsilateral vestibular nuclei. The first-order neurons synapse with both inhibitory and excitatory second-order vestibular neurons (Fig. 20.11).

Second-order inhibitory vestibular neurons join the ipsilateral MLF and ascend to terminate in the oculomotor nucleus where they synapse with neurons that inhibit the **ipsilateral superior rectus** and **inferior oblique muscles**.

Second-order excitatory vestibular neurons give rise to fibers that decussate, join the contralateral MLF, and terminate in: (1) the trochlear nucleus, where they synapse with neurons that give rise to fibers that cross the midline and exit in the trochlear nerve, to innervate the **superior oblique muscle**; and (2) the oculomotor nucleus, where they synapse with neurons that innervate the **inferior rectus muscle**.

In summary, when the head tilts *backward (extension)*, the eyeballs *turn downward* by the simultaneous inhibition of the superior rectus and inferior oblique muscles of both eyeballs and the stimulation of the superior oblique and inferior rectus muscles of both eyeballs.

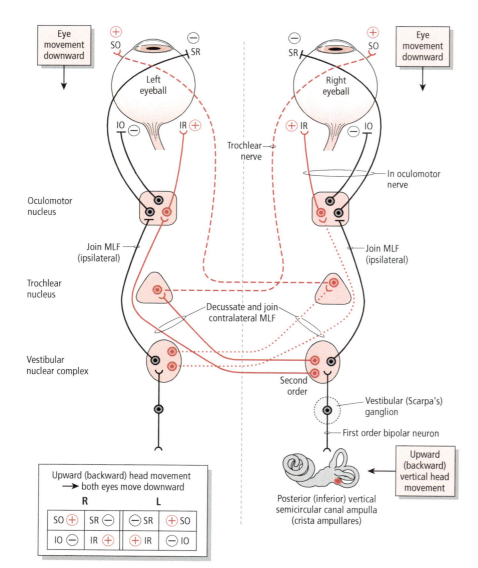

Figure 20.11 • Central connections mediating compensatory vertical (downward) eyeball movement in response to upward (backward) movement of the head. IO, inferior oblique muscle; IR, inferior rectus muscle; MLF, medial longitudinal fasciculus; SO, superior oblique muscle; SR, superior rectus muscle.

VESTIBULAR NYSTAGMUS

Horizontal nystagmus is a normal ocular movement when it is induced by activation of the vestibular apparatus; spontaneous nystagmus is indicative of a lesion affecting the vestibular nuclei

Horizontal nystagmus is a normal ocular movement when it is induced by activation of the vestibular apparatus. The distinguishing characteristic of nystagmus is a rhythmic, back and forth movement of the eyeballs, with a slow component in one direction, and a fast component (rapid return) in the other direction following asymmetric stimulation of the vestibular sensory apparatus of the two sides.

Nystagmus induced following stimulation of the visual system by a moving object is referred to as **optokinetic nystagmus**. As an example, imagine sitting on the right side of a moving train, looking out the window (and while keeping your head still), looking at the series of electrical poles as they emerge one by one, on the sidewalk. Maintaining focus on one pole at a time, while the train is moving forward, the eyes will follow it slowly toward the back of the train (this ocular movement is referred to as the *slow phase* of nystagmus, to the right), and then, when the pole is no longer visible, the eyeballs will flick back quickly, to the front end of the train (this ocular movement is referred to as the *fast phase* of nystagmus, to the left), to establish focus on another pole. Since the nystagmus is induced by the stimulation of the visual system by a moving object, in this case the "moving" electrical poles, it is referred to as optokinetic nystagmus.

Spontaneous nystagmus is pathologic and is indicative of a lesion affecting the vestibular nuclei. A lesion in the vestibular apparatus, nerve, nuclei, or pathways, or the vestibulocerebellum, may produce nystagmus.

In addition to horizontal nystagmus, there are also vertical and rotatory nystagmus. **Vertical nystagmus** may result following damage involving the superior vestibular nucleus in the pons. Extensive damage in the pontomedullary junction affecting the entire vestibular nuclear complex results in **rotatory nystagmus**.

To test the functional integrity of the vestibular system, two clinical tests – the rotation test and the caloric test – may be performed.

In the **rotation test**, the individual sits in a "Bárány chair" (a device used for aerospace physiology training) with their head held upright and tilted at 30° anteriorly. In this position the horizontal semicircular canals are oriented parallel to the plane of rotation. The individual is then rotated about 10 complete rotations to either the right or left, maintaining a continuous angular acceleration of the head. The rotation is then suddenly ceased. If, for example, the individual (with normal vestibular apparatus) is rotated to the right, the rotation will stimulate the right horizontal semicircular duct and then the stimulation will be transmitted to the right vestibular nuclei. The endolymphatic flow will tilt the cupula of the right horizontal semicircular canal to the left, due to the relative motion difference of the cupula and the endolymph. As rotation continues, the endolymph will "catch up" and flow at the same speed as the duct and head,

which will no longer activate the cupula. Sudden stopping of the rotation causes the endolymph, due to its momentum, to continue flowing in the direction of rotation, tilting the cupula to the left, which will inhibit the hair cells. During rotation, the eyeballs will move slowly to the left (slow component) to maintain visual fixation, and then will flick back (fast component) to the center (right) (Fig. 20.12). The nystagmus is in the direction of the spin, to the right. As the rotation is suddenly stopped, the direction of nystagmus is in the opposite direction (to the left). This is known as **post-rotatory nystagmus**. Normally, post-rotatory nystagmus lasts for about half a minute.

In addition to these reactions, the individual will lean or fall to the right (the same direction as the rotation) and experience vertigo (subjective sensation of rotation) to the left (opposite to the direction of rotation). If the subject is asked to point at a target, the subject will exhibit **past pointing**, in which the subject's arm points in the direction of the rotation. Since nystagmus is named after the fast component, in this case it is to the right.

CALORIC NYSTAGMUS

The caloric test may be utilized to stimulate each vestibular apparatus, or specific semicircular canals, in conscious or unconscious individuals

The **caloric test** may be used to stimulate each vestibular apparatus, or specific semicircular canals, in conscious or unconscious individuals. The caloric test allows for the examination of the vestibular apparatus of only one side at a time, whereas in a rotational test both sides are stimulated simultaneously.

For the caloric test, the **horizontal canal** has to be oriented in the vertical plane. To stimulate the horizontal canal, the subject sits upright with their head inclined about 60° posteriorly. The external auditory meatus is subsequently irrigated with cold or warm water. The cold or warm temperature of the water produces currents in the fluid within the vestibular apparatus. With cold water, the endolymph becomes denser which causes a sinking of the fluid, whereas with warm water, the endolymph rises.

In a subject with a normal vestibulo-ocular system, when the left ear is irrigated with cold water, flushing causes nystagmus (fast component) toward the opposite side (to the right), with accompanying "past pointing" and falling on the same side as the stimulation. When the same ear is irrigated with warm water, the nystagmus (fast component) is to the left (i.e., the subject exhibits the opposite response). A mnemonic for this is "COWS": cool opposite, warm, same).

If the subject is comatose, nystagmus is not observable. If the brainstem is undamaged, cold irrigation will cause both eyeballs to turn to the side of the cold irrigation. If the subject has a lesion affecting the MLF bilaterally, cold irrigation will cause the ipsilateral eyeball to turn toward the side of the cold irrigation. If there is a lesion in the caudal brainstem, affecting the vestibular nuclear complex, there will be no deviation of the eyeballs.

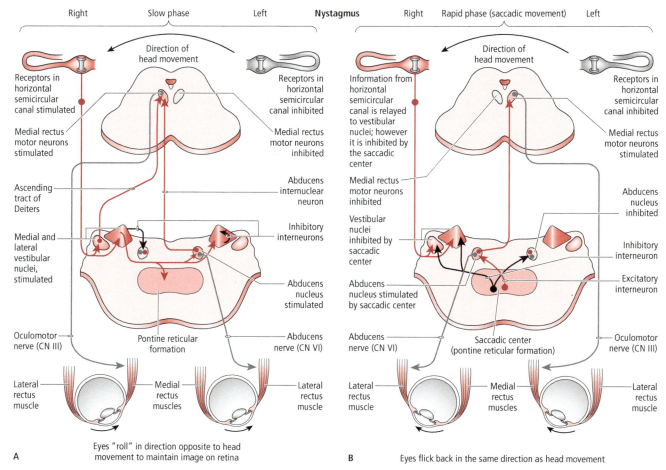

Figure 20.12 • Vestibular nystagmus: (A) slow phase and (B) rapid phase. Source: Modified from Netter, FH (1983) *The Ciba Collection of Medical Illustrations*, Vol. 1 *The Nervous System*, Part I *Anatomy and Physiology*. CIBA, New Jersey; plate 31.

Positional alcohol nystagmus and rotatory vertigo

The normal function of the ampullary crest (the receptors of the semicircular canals), is to detect angular acceleration of the head, but they are insensitive to gravity and linear acceleration of the head (those functions belong to the utricle and saccule). Normally, the density of the gelatinous cupula of the ampullary crest and that of the surrounding endolymph that bathes them is the same and since the cupula lacks otoconia it is insensitive to gravitational force. However, during and following acute alcohol intoxication, a condition in which the density of the cupula and the surrounding endolymph differ, may result in aberrant sensations. The alcohol diffuses from the local capillaries into the cupula. In this case, the density of the cupula becomes lighter thus the relative densities of the cupula and endolymph differ. This causes the semicircular ducts to become sensitive to gravity resulting in the "illusion" of movement when the head is oriented in certain positions and causes nystagmus. This condition is referred to as positional dependent vertigo and nystagmus. The nystagmus ceases when the alcohol level of the cupula and the endolymph equilibrate. Then, as the blood alcohol concentration decreases, the alcohol diffuses out of the cupula increasing its density relative to the endolymph resulting in positionally dependent nystagmus in the opposite direction that persists for several hours.

SYNONYMS AND EPONYMS OF THE VESTIBULAR SYSTEM

Name of structure or term	Synonym(s)/ eponym(s)
Anterior semicircular canal	Superior semicircular canal
Ampullary crest	Crista ampullaris
Bony labyrinth	Osseous labyrinth
Cranial nerve VIII	Vestibulocochlear nerve
Disjunctive ocular movements	Disconjugate ocular movements
Frontal eye field (FEF)	Brodmann area 8
Inferior vestibular nucleus	Spinal vestibular nucleus
	Descending vestibular nucleus
Lateral semicircular canal	Horizontal semicircular canal
Lateral vestibular nucleus	Deiter vestibular nucleus
Medial vestibular nucleus	Schwalbe vestibular nucleus
Posterior semicircular canal	Inferior semicircular canal
Saccule	Sacculus
Saccule and utricle	Statoconial organs
Statoconia	Otoconia, otoliths
Superior vestibular nucleus	Bechterew vestibular nucleus
Utricle	Utriculus
Vestibular ganglion	Scarpa ganglion
Vestibular nuclei	Vestibular nuclear complex

FOLLOW-UP TO CLINICAL CASE

This patient has **benign positional vertigo**. This is the most common identifiable vertigo, and it occurs most frequently in older individuals, although it can occur at any age. It is common and probably underdiagnosed. It is thought to be caused from free floating debris or particles in the endolymph of one of the semicircular canals, most commonly the posterior semicircular canals possibly because they are located near the utricle. Free floating particles are also derived from calcium carbonate crystals or statoconia. These crystals are normally associated with the macula in the utricle or the saccule and can at times become "dislodged and float in the endolymph." The cause of this pathology is unknown, but it is sometimes associated with trauma. This condition most often resolves on its own eventually, but can recur.

Medications are often ineffective in the treatment of this condition.

The evaluation of dizziness requires that the patient describes exactly what they mean by dizziness. Dizziness can result from vertigo, which is an illusion of movement when there is actually no movement occurring. This is the same sensation produced when one is spun around and then suddenly stopped. The term dizziness is often used to refer to a vague sensation of imbalance, disequilibrium, or light headedness. This is often nonspecific and indeed non-neurologic. In contrast, vertigo is quite specific and usually indicates a problem either in the internal ear, the brainstem (specifically the posterolateral medulla oblongata), and/or the cerebellum. If it is an internal ear problem, it is often associated with hearing loss and/or tinnitus.

QUESTIONS TO PONDER

1. Which sensory systems work with the vestibular system to maintain equilibrium, posture, muscle tone, and reflex ocular movements?

2. Give the location of the receptors of the vestibular apparatus.

3. Differentiate between the structure and function of the statoconial membrane of the macula of utricle, the macula of saccule, and the ampullary crest.

4. What is the effect of endolymphatic flow in the direction of the ampullae of the horizontal, anterior, and posterior semicircular canals?

5. What is unique about the vestibular first-order bipolar neuron central projections?

6. What is the source of most of the afferents arriving at the vestibular nuclei?

7. What is the function of the vestibulo-ocular reflex (VOR)?

CHAPTER 21

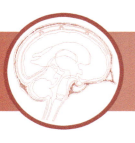

Olfactory System

- CLINICAL CASE
- OLFACTORY RECEPTOR NEURONS
- OLFACTORY TRANSDUCTION
- OLFACTORY NERVE (CN I)
- CENTRAL CONNECTIONS OF THE OLFACTORY SYSTEM
- BLOOD SUPPLY AND DRAINAGE
- CLINICAL CONSIDERATIONS
- SYNONYMS AND EPONYMS OF THE OLFACTORY SYSTEM
- FOLLOW-UP TO CLINICAL CASE
- QUESTIONS TO PONDER

CLINICAL CASE

A 32-year-old carpenter, L.P., fell backward off a ladder while at work about a month ago. The patient's coworkers said that the back of L.P.'s head hit the ground, but the patient does not remember this. L.P. apparently lost consciousness for approximately 10 minutes and was sent to the Emergency Room where a head CT (including the sinuses) was performed and was normal. There was no apparent skull fracture. L.P. recovered alertness and awareness and insisted on going home. Symptoms of headache and dizziness subsided after a few days. However, the patient continued to experience a persistent loss of sense of smell and taste since the accident. L.P. says that food tastes bland and coffee, freshly cut wood, and perfume have only a slight odor.

Examination shows the patient's sense of smell was greatly diminished when tested with several types of odiferous substances, in both nostrils. Taste sensation was normal for sweet, salty, bitter, umami, and sour. Examination of the nose and sinuses was normal. The remaining general physical and neurologic exam was normal.

The **olfactory system** mediating the special sense of smell is phylogenetically one of the most primitive sensory systems. Although humans do not rely to a great extent on their olfactory sense, the survival of many other animals is greatly dependent on olfaction. This is reflected in the relative size difference of the olfactory systems – that of humans being rather rudimentary compared with that of many other animals. In humans the olfactory epithelium consists of about 10 cm^2, compared to about 170 cm^2 in dogs; dogs also have more than 100 times more receptors in each square centimeter than humans. The olfactory system is not only implicated in the perception and transmission of *olfactory sensation*, but is also believed to affect other neural systems associated with producing or influencing *emotional*, *behavioral*, and *reflex responses* such as reproductive and nurturing behaviors, and autonomic visceral functions such as salivation, gastric secretion, and nausea.

The olfactory system has a direct projection to the limbic system which is associated with the processing of learning, memory, emotions, pleasure, and drives. Thus, a scent may

---

A Textbook of Neuroanatomy, Third Edition. Maria A. Patestas, Amanda J. Meyer, and Leslie P. Gartner.
© 2025 John Wiley & Sons, Inc. Published 2025 by John Wiley & Sons, Inc.
Companion website: www.wiley.com.go/patestas/neuroanatomy3e

have an emotional label, based on prior experiences individuals had with a particular scent. Therefore, the olfactory system may be considered to be a direct channel to an individual's past. A scent may be enough to relive an experience from years ago and to feel all the emotions originally associated with it. Scents experienced during the early years can trigger vivid memories when re-encountered in later life.

The nostrils are located just above the mouth so that the olfactory system can serve as an alarm system for potentially noxious substances that one may inhale or is about to ingest. The association of a scent to an unpleasant experience may also have, from an evolutionary perspective, a protective function for future avoidance of a noxious substance.

The neuroanatomical connections of the olfactory system are unique among the sensory systems. Although the other sensory systems that relay touch, pain, temperature, vision, hearing, and taste sensation project to the thalamus *prior* to projecting to the cerebral cortical areas specialized for processing sensory information for each of the senses, the olfactory system projects *directly* to the olfactory and limbic cortices, *bypassing* the thalamus. In contrast, the other sensory systems project to the limbic structures after their relay to the sensory cortex, for further processing that is essential for learning, memory, and emotions. Higher-order processing, abstract thinking, and decision-making associated with visual and auditory information take place at the cortical level. In contrast, the olfactory system projects to the limbic structures evading any prior cortical filtering of the information, eliciting an immediate emotional response and a decision about a particular situation made at the limbic system level. The limbic system connections to all the sensory systems, including olfaction, evoke memories that may initiate an emotional response which in turn results in a behavioral response ranging from a smile to taking the proper course of action to avoid a noxious stimulus. Overall, the wiring of the olfactory system to the limbic system is direct, eliciting an immediate and powerful emotional response.

The ancestral cerebrum consisted of a pair of swellings, which received olfactory input from the olfactory bulbs and whose principal function was to mediate olfactory sensation. During evolution, the ancestral brain underwent modifications resulting in the enlargement and increased complexity of the cerebrum. These gradual changes were accompanied by the emergence and vast expansion of the neocortex with a consequent movement of the paleocortex to a more internal position at the base of the brain.

OLFACTORY RECEPTOR NEURONS

Olfactory receptor neurons reside in the olfactory epithelium and are the only neurons that are in direct contact with the environment

Olfactory receptor neurons, the first-order bipolar neurons of the olfactory pathway, are in the olfactory epithelium (and not in a ganglion) and are the only neurons that are in direct contact with the external environment.

The nasal cavities are lined by a special membrane, the nasal mucosa, which is modified at the roof and the adjacent superior walls of these cavities into a patch of specialized thick **olfactory epithelium** (neuroepithelium) and underlying connective tissue (Fig. 21.1). This pseudostratified ciliated columnar epithelium includes several cell types, namely basal cells, immature receptor cells, olfactory receptor neurons, and supporting epithelial cells. The **basal cells** are stem cells that proliferate to form **immature receptor cells** that differentiate to become **olfactory receptor neurons**. Between 10 and 25 cilia, each approximately five micrometers long, project from the **olfactory knob**, the distal, expanded, end of the olfactory receptor neuron (Fig. 21.1). The cell bodies and axons of olfactory receptor neurons are surrounded by the **supporting epithelial cells** (synonym: sustentacular cells) which structurally and electrically isolate olfactory receptor neurons. The supporting epithelial cells are glial-like cells that provide both physical and physiological support to all other cells of this epithelium. These supporting cells metabolize toxins and phagocytose dead cells as part of their neuroprotective role. Supporting epithelial cells are targeted by SARS-CoV-2 variants due to several factors including a glycosyltransferase coded for by an allele at the UGT2A1 locus, and high levels of ACE2 and TMPRSS2 expression. The olfactory receptor neurons are **first-order afferent bipolar neurons** that have a lifespan of about 1–2 months. Cells that degenerate either desquamate or are phagocytosed by supporting epithelial cells. These neurons are unique since they are the only nerve cells that come in direct contact with the environment external to the body and are thus more susceptible to damage. **Olfactory-ensheathing cells** (OEC) are a specialized type of glial cells that encircle the axons of olfactory neurons not only in the lamina propria of the nasal mucosa, but also along the passage of the axons through the foramina of the cribriform plate of the ethmoid bone, and into the olfactory bulb. Regenerating olfactory neurons are guided to their targets within the olfactory bulbs by OECs that secrete bioactive factors such as nerve growth factor, basic fibroblast growth factor, and glial-cell-line-derived neurotrophic factor. When compared with astrocytes and neurolemmocytes, OECs express higher levels of innate immune factors, and have been found to phagocytose apoptotic or necrotic olfactory receptor neurons and myelin debris. In these ways, the OECs provide a microenvironment favorable for nerve regeneration. It used to be believed that the olfactory receptor neurons were the only nerve cells to retain the capability to regenerate throughout the lifetime of the organism. However, research in the past decade or so showed that adult neurogenesis also occurs in the dentate gyrus of the limbic system as well as in other regions of the central nervous system.

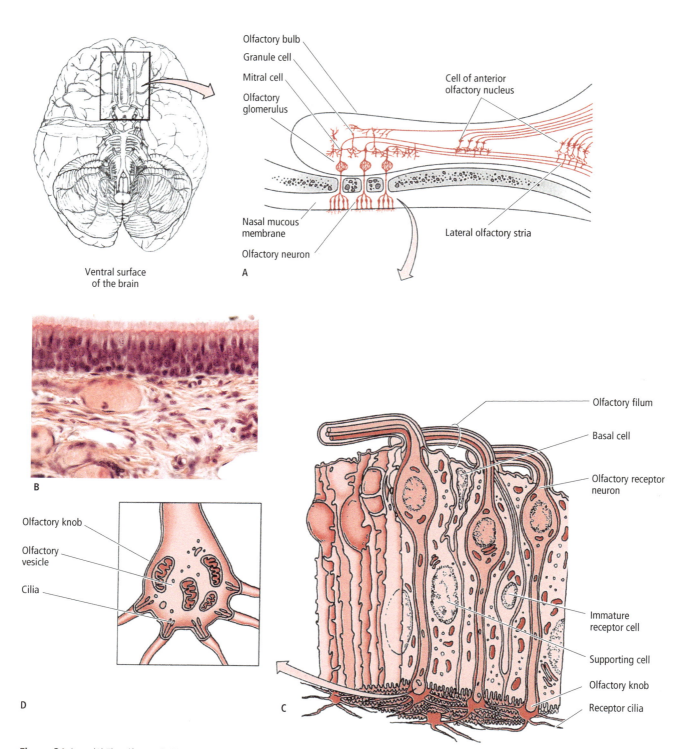

Figure 21.1 ● (A) The olfactory bulb and tract. The olfactory receptor neuron axons pass through the cribriform plate of the ethmoid bone to enter the cranial vault where they terminate and synapse in the olfactory bulb. (B) Note the olfactory receptor neurons – the first-order neurons of the olfactory pathway. (C) Enlargement of an (D) Olfactory knob.

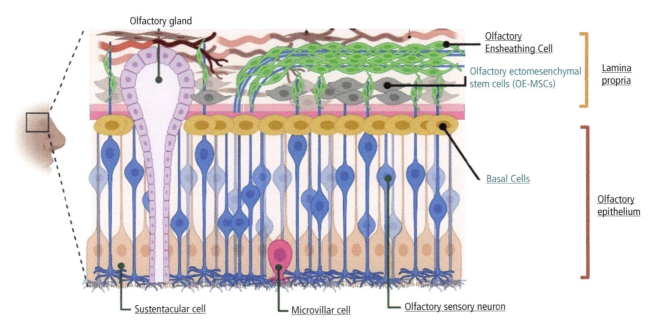

Source: Adapted from: https://www.mdpi.com/ijms/ijms-22-06311/article_deploy/html/images/ijms-22-06311-g001.png

The connective tissue (lamina propria) underlying the olfactory epithelium contains **olfactory glands** (eponym: Bowman glands), whose ducts carry aqueous secretion that forms a moist blanket on the epithelial surface. The remainder of the nasal mucosa produces a mucous secretion. The apical surface of the olfactory receptor neurons possesses modified cilia, which are embedded in the mucous secretions. Odorous substances entering the nasal cavities dissolve in the moist layer of mucus and interact with receptor molecules embedded in the plasmalemma of the ciliary membrane, stimulating the **chemosensitive cilia** of the olfactory receptor neurons. Most of the flavor that we "taste" in foods is actually the aroma that passes from the back of the mouth and the nostrils, reaching the superior recesses of the nasal cavity lined with olfactory epithelium.

OLFACTORY TRANSDUCTION

Odorous substances interact with and bind to odorant-receptor protein molecules embedded in the plasmalemma of the chemosensitive cilia, where transduction takes place

Odorous substances entering the nasal cavities dissolve in the moist mucus layer overlying the olfactory epithelium. They may reach the olfactory epithelium via either retronasal or orthonasal respiration. Retronasal delivery of odorants occurs when food or drink is introduced into the oral cavity and odorants disperse into the oropharynx and up into the nasopharynx. Orthonasal delivery of odorants occurs when odorants directly enter the nasal cavity through the nares, by way of nasal inhalation or sniffing (>18 L/min).

These substances then interact with **odorant-binding proteins** (**receptors**) of the chemosensitive cilia of the olfactory receptor neurons (trans-membrane G-protein-coupled receptors that function in the transduction of information) immersed in the mucus layer. Each olfactory receptor cell contains about a dozen structurally identical receptors, all of which are coded for by a single gene. Each olfactory receptor cell contains only one type of receptor; however, this same type of receptor may also be exhibited by thousands of other olfactory receptor neurons distributed throughout the olfactory epithelium. A particular type of receptor may bind multiple odorous substances. This suggests that each olfactory receptor cell is responsive to numerous odorous substances. In addition, the route of delivery (orthonasal versus retronasal) affects the central structures activated. In the case of the odor of chocolate the insula/operculum, thalamus, hippocampus, amygdaloid body, and posterolateral orbitofrontal cortex are activated more strongly via orthonasal olfaction, and the perigenual cingulate and medial orbitofrontal cortex are more strongly activated via retronasal olfaction.

In humans, there are about 400 distinct olfactory receptors that constitute the largest family of G-protein-coupled receptors. These receptors function in the detection of a wide range of odorous molecules. The binding of an odorous substance to the odorant-receptor protein molecules stimulates the initiation of a second messenger pathway associated with an olfactory-specific G-protein, which subsequently stimulates adenyl cyclase, which generates cAMP. The increase in cAMP levels within the olfactory receptor cell unlocks a cyclic nucleotide-gated cation channel embedded in the ciliary membrane permitting sodium and calcium cations to enter

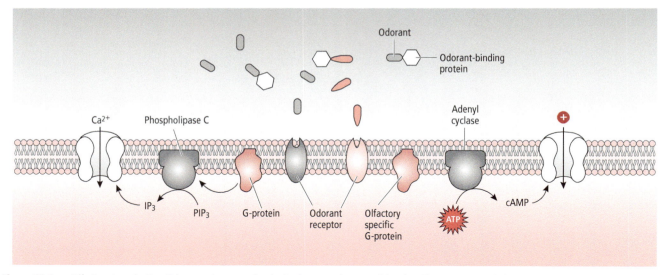

Figure 21.2 ● Olfactory transduction. Odorous substances dissolve in the mucus layer overlying the olfactory neuroepithelium and then bind to odorant-binding proteins that move them through the mucus. The odorous substances subsequently bind to the receptors in the olfactory epithelium, which in turn activates a second messenger pathway. cAMP = cyclic adenosine monophosphate; IP3 = inositol 1,4,5-triphosphate; PIP3 = phosphatidylinositol-3,4,5-triphosphate.

the cell (Fig. 21.2). This stimulates the **chemosensitive cilia** of the olfactory receptor neurons, and a wave of depolarization (generator potential) is propagated from the dendrites to the cell body of the olfactory receptor neuron. A high enough depolarization triggers a nerve impulse that is propagated along the axon of the olfactory receptor neuron, centrally to the olfactory bulb.

OLFACTORY NERVE (CN I)

Olfactory receptor neurons give rise to unmyelinated axons, the slowest impulse-conducting axons of the CNS

Each olfactory receptor neuron gives rise to an unmyelinated centrally directed axon. They are the slowest impulse-conducting axons of the central nervous system (CNS). The axons of these bipolar cells do not assemble to form a single nerve trunk enclosed in a connective tissue sleeve; instead, they form 15–20 separate bundles (fascicles) – the **olfactory fila** (L., "threads") (Fig. 21.1C). The olfactory fila course superiorly, traversing the sieve-like dura-lined perforations of the cribriform plate of the ethmoid bone of the skull, becoming invested in a sheath of olfactory-ensheathing cells and arachnoid mater, to terminate, in a topographic fashion, in the ipsilateral **olfactory bulb**. Olfactory receptor neurons from the anteroinferior portion of the middle nasal conchae project to the posterolateral olfactory bulb, and those from the superior nasal conchae terminate in the anterolateral olfactory bulb. Olfactory receptor neurons from the superior nasal septum project to the medial side of the ipsilateral olfactory bulb with the anterior fibers terminating in the posteromedial olfactory bulb and the posterior fibers synapsing on the anteromedial olfactory bulb. The olfactory receptor neurons synapse with second-order neurons. Collectively, all the olfactory fila from each nasal cavity form the **olfactory nerve (CN I)** on both right and left sides. The olfactory nerve fibers carry only **special visceral afferent (SVA)** modality, denoting the special sense of olfaction. The olfactory receptor neurons are unique not only because they serve the functions of **sensory receptors** (chemoreceptors that become activated by chemical stimuli in their surroundings), **transducers** (generate graded potentials), and **first-order neurons** (relaying olfactory sensation centrally to the olfactory bulbs), but also because they reside in a peripheral structure, the olfactory epithelium, rather than being housed within a sensory ganglion.

In addition to the sympathetic and parasympathetic innervation supplying the blood vessels and glands of the connective tissue, the olfactory mucosa also receives general sensory afferent innervation from the trigeminal nerve. The free nerve endings of the peripheral trigeminal afferents ramify in the olfactory epithelium and the underlying lamina propria. Trigeminal fibers not only mediate pain and temperature sensations from the nasal mucosa but are also believed to transmit noxious odor perception to subconscious levels, thus eliciting the autonomic secretomotor reflex in response to the presence of noxious chemicals (such as ammonia).

Note that the clinical case at the beginning of the chapter refers to a patient with a persistent loss of the sense of smell and taste following head trauma.

1 Which component of the olfactory system is most susceptible to injury during head trauma such as that sustained by the patient in this case?

CENTRAL CONNECTIONS OF THE OLFACTORY SYSTEM

Olfactory bulb

The olfactory bulbs, outgrowths of the telencephalon, receive the central processes (axon terminals) of the olfactory receptor neurons

The **olfactory bulbs** are two oval-shaped, flattened structures located on the basal aspect of the frontal lobes, which are lodged in two fenestrated depressions formed by the cribriform plate of the ethmoid bone in the anterior cranial fossa. The two olfactory bulbs receive the central processes (axon terminals) of the first-order olfactory receptor neurons (see Fig. 21.1A) whose terminations are organized and mapped in an orderly manner. This is the **first relay station** of the olfactory system where synaptic integration occurs (Fig. 21.3). Since the olfactory bulbs are rostral dilations of the olfactory tracts, which are outgrowths of the telencephalon rather than typical cranial nerves, this first relay station of the olfactory system bypasses the diencephalon and is actually located in the cerebral cortex (this situation is contrary to other sensory systems that normally project to regions of the diencephalon such as the thalamus, *prior to* the cortical projection).

The olfactory bulbs are laminated structures consisting of seven layers surrounding a core of white matter. There are two layers of the **olfactory nerve layer** (external layer which is neurotrophin receptor p75 (p75NTR) positive, and the internal layer that is p75NTR negative), which is the most superficial, the **glomerular layer**, the **external plexiform layer**, the **mitral cell layer**, the **internal plexiform layer**, and the deepest layer **granular layer,** containing granule cells in direct contact with the white matter. This laminated cellular architecture is more clearly defined in animals other than humans, reflecting the exquisite olfactory system that these animals possess. The core of white matter of the olfactory bulb consists of *afferent* fiber terminals (centrifugal afferents), which have reached the bulb from other brain centers to modulate neural activity within the bulb, and *efferent* axons leaving the olfactory bulb via the **olfactory tract** to reach other cortical areas.

The axons of olfactory receptor neurons are highly convergent as they terminate in the glomeruli of the olfactory bulbs. **Glomeruli** (Fig. 21.3) are spherical synaptic complexes where olfactory receptor neuron axons establish synapses with **inhibitory interneurons (periglomerular neurons, superficial granule neurons, intermediate granule neurons, and the deep granule neurons)** and **excitatory interneurons (juxtaglomerular neurons)** and the **main projection second-order neurons** of the olfactory bulb, the **mitral** and **tufted neurons**. Each olfactory receptor neuron has an input to several dendritic processes (residing in a glomerulus) of a single mitral or tufted neuron. Neural activity involving combinations of glomeruli is responsible for the perception of a "single" odor. That is, if several sets of dendrites are stimulated, the "odor" perceived is different than if only one of those sets of dendrites is stimulated. The olfactory interneurons modulate the transmission of the sensory afferent information from the olfactory receptor neurons that is communicated

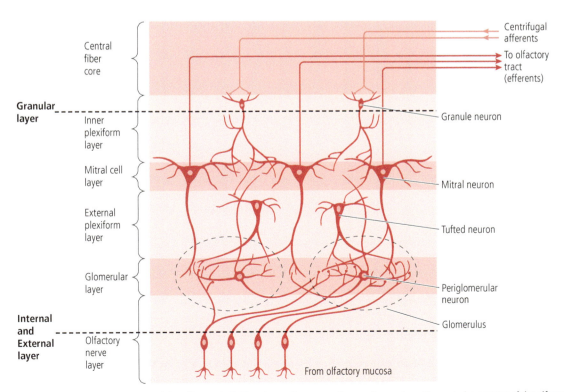

Figure 21.3 • The layers and neuronal circuitry of the olfactory bulb. The glomeruli are synaptic complexes where the central processes of the olfactory receptor neurons synapse with the neurons in the olfactory bulb.

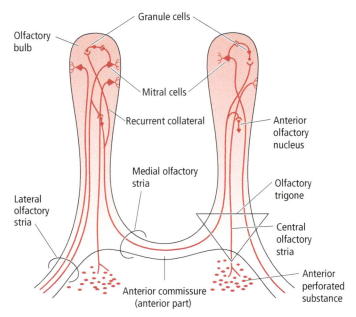

Figure 21.4 ● The olfactory bulb, anterior olfactory nucleus, and olfactory tract, and its division into the medial and lateral olfactory striae.

to the mitral and tufted neurons, thereby facilitating olfactory discrimination. These interneurons also control mitral and tufted neuronal output by regulating the stream of signals that reach other brain centers from the olfactory bulb. The olfactory bulbs also receive **centrifugal fibers** providing biofeedback input from neurons residing in other sites of the CNS. Centrifugal afferent fibers are projected to the olfactory bulb from the **locus caeruleus**, the **medial and dorsal midbrain raphe nuclei**, the **medial septal nucleus**, and the **nuclei of the diagonal band**. Each olfactory bulb also receives sensory input from the olfactory bulb of the opposite side, which is first processed in the contralateral **anterior olfactory nucleus** (Fig. 21.4). This input is believed to adjust the capacity of the second-order neurons and interneurons to respond to incoming signals.

Olfactory tract

The olfactory tract carries afferent fibers to the olfactory bulb and efferent fibers from both the olfactory bulb (mostly axons of the mitral and tufted neurons) and the anterior olfactory nucleus

Each olfactory tract, a cord-like structure, relays olfactory sensory information to the cortex. It arises from the olfactory bulb, courses posteriorly, and diverges to form the **medial, lateral,** and **central olfactory striae (tracts)** (Fig. 21.4). The three striae and a small region rostral to the anterior perforated substance collectively form a triangular region, the **olfactory trigone**. The anterior perforated substance may be involved in the emotional and motivational aspects of olfaction.

The olfactory projection consists of a sequence of only two neurons from receptor to cortex: the olfactory receptor neurons and the projection neurons residing in the olfactory bulb.

The principal fibers carried by the **olfactory tract** are the *efferent* axons of second-order output neurons of the bulb – the mitral and tufted neurons. Others include the *afferent* centrifugal fibers.

Mitral and tufted neurons occupying the olfactory bulb send collaterals to a cortical structure, the ipsilateral **anterior olfactory nucleus** located in the vicinity of the junction between the olfactory bulb and tract (Fig. 21.4). Efferent fibers arising from neurons of this nucleus accompany the efferent axons of the mitral and tufted neurons coursing in the olfactory tract. The fibers arising from the anterior olfactory nucleus project to both olfactory bulbs; some of these fibers project to the nearby ipsilateral olfactory bulb, whereas others join the medial olfactory stria and cross via the **anterior commissure** to terminate in the *contralateral* olfactory bulb, anterior olfactory nucleus, and **medial septal nuclei**. From the medial septal nuclei, the fibers split into two bundles: 1. the medullary stria which projects to the habenular nuclei and then onto the superior and inferior salivatory nuclei, and 2. the olfactory–hypothalamic–tegmental bundle which projects to the posterior nucleus of vagus nerve to stimulate increased peristalsis and gastric secretion. The majority of the fibers coursing in the medial olfactory stria are derived from neurons housed in the anterior olfactory nucleus.

The central olfactory stria also carries fibers from mitral and tufted neurons, which terminate in the ipsilateral **anterior perforated substance** and the **medial septal nuclei**. These nuclei form connections with the limbic system and are therefore implicated in the *emotional* and *visceral response to odors* (to the great benefit of companies manufacturing perfumes).

The largest is the lateral olfactory stria, consisting primarily of the axons of mitral neurons, is the **main central pathway of the olfactory system**. Its second-order fibers project mostly to **paleocortical structures** including the parahippocampal gyrus, entorhinal cortex, and the amygdaloid body.

Olfactory cortex

The primary olfactory cortex projects to the thalamus, which in turn projects to the orbitofrontal cortex

The **lateral olfactory stria**, consisting of second-order fibers, projects mostly to **paleocortical structures**. The paleocortical structures that are targets of the lateral olfactory stria consist of parts of the **primary olfactory cortex** (the lateral olfactory gyrus and anterior half of the uncus, collectively referred to as the **"piriform cortex"**) and the **periamygdaloid cortex**, which forms the rest of the olfactory component of the uncus and the **amygdaloid body** (Fig. 21.5). *Some* fibers terminate in the **lateral olfactory gyrus (nucleus of the lateral olfactory stria)**. However, *most* of the fibers terminate in the anterior half of the **uncus** forming part of the **primary olfactory cortex** located in the anteromedial aspect of the temporal lobe. The lateral olfactory stria also sends collaterals to the anterior olfactory nucleus and to subcortical limbic structures. The primary olfactory cortex (consisting of paleocortex)

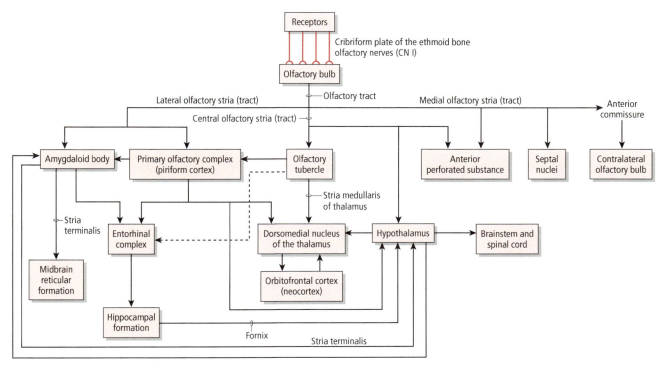

Figure 21.5 ● The connections of the olfactory system. The olfactory receptor cell axons pass through the cribriform plate of the ethmoid bone to terminate in the olfactory bulb. Each olfactory bulb is continuous posteriorly with the olfactory tract, which divides into the medial, central, and lateral olfactory striae. From there, olfactory information is transmitted to the amygdaloid body, the primary olfactory cortex, the olfactory tubercle, the anterior perforated substance, the septal nuclei, and the contralateral olfactory bulb via the anterior commissure. Source: Modified from Fix, JD (1995) *Neuroanatomy*, 2nd edn. Williams & Wilkins, Baltimore; fig. 20.1; and Burt, AM (1993) *Textbook of Neuroanatomy*. WB Saunders, Philadelphia; fig. 13.13.

projects to the **dorsomedial nucleus of the thalamus**, which in turn projects mainly to the orbital gyri of the **orbitofrontal cortex** of the frontal lobe (the **neocortical** region of the olfactory system). However, the primary olfactory cortex also sends *direct* projections to the neocortex and is the only sensory system that *bypasses the thalamus* in projecting to the telencephalon. The olfactory system's neocortical projections terminating in the orbitofrontal cortex have an important function in the *conscious awareness of odors*. Not only does the orbitofrontal cortex receive projections from the olfactory system, but it also receives projections from gustatory cortical areas. The olfactory and gustatory sensory integration that occurs in the orbitofrontal cortex is believed to produce what is experienced as flavor. The neocortex functions as a correlating and coordinating center for the vast spectrum of sensory and motor input it receives from all sensory and motor systems (Fig. 21.5).

The primary olfactory cortex also projects to the **hypothalamus**, the amygdaloid body, and the **olfactory association cortex** (secondary olfactory cortex, entorhinal cortex) located next to the primary olfactory cortex. The entorhinal cortex also receives *direct* projections from the olfactory tract. Functionally, the entorhinal cortex is closely associated with the hippocampus, affecting memory and emotional behavior. Projections to the hypothalamus are associated with sexual and maternal behavior.

BLOOD SUPPLY AND DRAINAGE

The olfactory bulb and tract receive arterial blood from the olfactory artery which is a branch of the A2 segment of the anterior cerebral artery. Contributions are also made by the medial frontobasal artery, and the anterior and posterior ethmoidal arteries. Venous drainage is highly variable and is suggested to be via the olfactory veins into the frontopolar vein or the origin of the superior sagittal sinus.

432 ••• CHAPTER 21

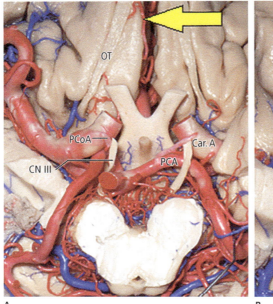
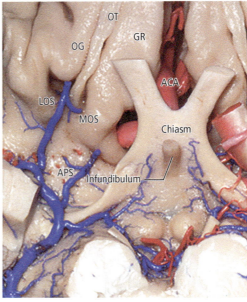

Vascular relationships of the olfactory nerve. A. Inferior view of the basal surface of the frontal lobes, and the mesencephalon, where the oculomotor nerves pass below the posterior cerebral arteries. B. Enlarged view of the olfactory nerve, anterior perforated substance, and surrounding structures. Abbreviations: ACA, anterior cerebral artery; APS, anterior perforated substance; CN III, cranial nerve III; LOS, lateral olfactory stria; MOS, medial olfactory stria; OT, olfactory tract; OG, orbital gyri; Pit. Stalk, pituitary stalk; GR, gyrus rectus. Source: Color figure can be viewed at wileyonlinelibrary.com López-Elizalde et al. (2017). Clinical Anatomy, 31(1): 109–117./John Wiley & Sons.

CLINICAL CONSIDERATIONS

Anosmia, the complete loss of the sense of smell, or **hyposmia**, the partial loss of the sense of smell, may result following injury to the olfactory apparatus, ipsilateral to the lesion. These conditions may occur subsequent to head trauma, especially if there is fracture or damage to the cribriform plate of the ethmoid bone. If the olfactory bulb becomes dislodged from its normal position, the delicate first-order afferent fibers traversing the fenestrations of the cribriform plate of the ethmoid bone may be injured or torn.

Other prevalent conditions such as **rhinitis** or **sinusitis**, resulting from irritation and/or swelling of the nasal mucous membrane, may temporarily disrupt olfaction. Moreover, exposure to various contaminants may adversely affect olfaction. Extensive damage may result in permanent loss of the sense of smell. Nasal congestion resulting from adenovirus (common cold) infection may also affect the sense of smell, whereas SARS-CoV-2 interrupts olfactory signal transduction to cause anosmia. A complete loss of olfaction is usually accompanied by a partial or complete loss of gustation (taste), as commonly experienced during nasal congestion when the individual is unable to taste food.

Individuals with lesions in the orbitofrontal cortex – the neocortex necessary for conscious awareness of odors – are able to detect the presence of odors but are unable to distinguish between different odorous substances.

A viral or bacterial infection in the ears, paranasal sinuses, or throat may spread to the upper recesses of the nasal cavities introducing infectious agents into the cranial vault. The meningeal membranes may subsequently become infected spreading the infection to the subarachnoid space and cerebrospinal fluid, resulting in **meningitis**, a serious and life-threatening inflammation of the brain and its surrounding membranes, the meninges. The upper recesses of the nasal cavity communicate with the cranial vault via perforations of the cribriform plate of the ethmoid bone, which can serve as a conduit for the spread of infection, especially following head trauma involving the cribriform plate.

Individuals diagnosed with Alzheimer disease, Parkinson disease, or Huntington disease also experience a deterioration in olfactory function because of degenerating olfactory pathways. Individuals suffering from epilepsy may experience **parosmia (dysosmia)**, a distorted perception of odors present in their environment or **olfactory hallucination (phantosmia)**, the perception of odors that are not really present in their environment, prior to an epileptic seizure.

SYNONYMS AND EPONYMS OF THE OLFACTORY SYSTEM

Name of structure or term	Synonym(s)/eponym(s)
Amygdaloid nucleus	Amygdala
Interneurons	Internuncial neurons
Lateral olfactory gyrus	Nucleus of the lateral olfactory stria
Lateral olfactory stria	Lateral olfactory tract
Medial olfactory stria	Medial olfactory tract
Olfactory association cortex	Secondary olfactory cortex
	Entorhinal cortex
Olfactory epithelium	Olfactory neuroepithelium
Parosmia	Dysosmia
Plasmalemma	Cell membrane

FOLLOW-UP TO CLINICAL CASE

In this case the etiology of the patient's olfactory dysfunction is obvious: head trauma, which is a very common cause of olfactory dysfunction. Its cause is rarely of neurologic origin. The most common etiologies include SARS-CoV-2 (COVID-19 disease), upper respiratory infections, sinus disease, or any other nasal obstruction or congestion. Medications such as some antibiotics (amoxicillin, azithromycin, ciprofloxacin), blood pressure medications (amlodipine, enalapril), statins (atorvastatin, lovastatin, pravastatin), and thyroid medication (levothyroxine) can induce loss of, or alteration of, the sense of smell. Neurodegenerative diseases such as Alzheimer disease, Parkinson disease, or Huntington disease are also associated with loss of olfactory sensation.

Most cases of olfactory loss caused by head trauma are thought to arise from shearing injury to the olfactory nerve fibers as they pass through the cribriform plate of the ethmoid bone disconnecting the olfactory nerve fibers from the overlying olfactory bulb. However, it can also arise from injury to the olfactory bulbs, olfactory tracts, or the frontal or temporal lobes. An individual is often unaware of unilateral loss of olfactory sensation, but it becomes noticeable when the lesion is bilateral. Olfactory dysfunction (loss of/or alteration of the sense of smell) occurs in approximately 5–10% of patients with head trauma. There is some correlation between the severity of head trauma and the presence of, or the degree of, olfactory disturbance.

Patients with a loss of the sense of smell often complain of loss of, or alteration in, taste as well. However, loss of taste sensation to sweet, sour, salty, umami, and bitter is much rarer than olfactory dysfunction. Trigeminal nerve endings in the nasal membranes detect irritants. Therefore, to only test olfactory function, non-irritating substances such as coffee or vanilla should be used. Ammonia or vinegar are irritants and can be detected through the trigeminal nerve endings.

Recovery of normal olfactory function in cases caused by head trauma is rare. There is no effective treatment.

2 Besides head trauma what are some other causes that may lead to loss of the sense of smell?

3 What is the likelihood of the recovery of normal olfactory function following head trauma?

QUESTIONS TO PONDER

1. What makes the olfactory receptor neurons unique?

2. Where are the odorant-binding proteins located?

3. Where are the odorant-receptor molecules located?

4. How many types of receptors are in the ciliary membrane of a single olfactory receptor cell?

5. Describe the events following the binding of an odorous substance to an odorant-receptor protein.

CHAPTER 22

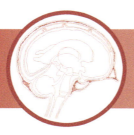  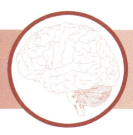

Limbic System

- CLINICAL CASE
- LIMBIC LOBE
- BRAINSTEM CENTERS ASSOCIATED WITH LIMBIC SYSTEM FUNCTION
- PATHWAYS OF THE LIMBIC SYSTEM
- LIMBIC ASSOCIATION CORTEX
- LIMBIC SYSTEM INPUT TO THE ENDOCRINE, AUTONOMIC, AND SOMATIC MOTOR SYSTEMS
- CLINICAL CONSIDERATIONS
- SYNONYMS AND EPONYMS OF THE LIMBIC SYSTEM
- FOLLOW-UP TO CLINICAL CASE
- QUESTIONS TO PONDER

CLINICAL CASE

A 38-year-old, M.S., reported to their primary care physician of two separate attacks during which they experienced fear and anxiety. These occurred suddenly without warning or precipitation by any circumstance. These episodes were brief, lasting a minute or so each. M.S. noted increasing levels of anxiety and irritability even in between these attacks. Friends and family were concerned about increasingly "odd" behavior and unprovoked outbursts of anger from M.S. There were brief episodes during which M.S. became incoherent. There was concern about panic attacks or even a more severe psychiatric condition such as schizophrenia.

The patient had another attack that was witnessed by a friend. There was a sudden feeling of fear and anxiety as in the prior attacks, accompanied by agitation and crude verbalizations. This quickly progressed to loss of responsiveness and convulsive activity that lasted for less than a minute. After this episode M.S. had a neurologic evaluation and testing.

The **limbic system** consists of a diverse group of structures including phylogenetically ancient cortical areas, a group of associated subcortical nuclei, as well as associated pathways that interconnect regions of the telencephalon, diencephalon, and brainstem. A major function of the limbic system is to modulate the hypothalamus.

As a result of its complex nature and widespread connections, the areas considered to constitute the limbic system are sometimes inconsistent among authors; however, the structures most commonly included in the limbic system are the *cingulate*, *subcallosal*, *parahippocampal*, and *dentate gyri*, the *isthmus of the cingulate gyrus*, the *uncus*, the *hippocampus proper*, the *subiculum*, the *amygdaloid body*, the *septal area*, the ventral tegmental area of the midbrain, and some nuclei of the thalamus and hypothalamus (Table 22.1). The two main structures of the limbic system are the **hippocampal formation** (which includes the **hippocampus proper**, the **dentate gyrus**, and the **subiculum**) and the **amygdaloid body**. In humans, the hippocampal formation is involved in the consolidation of short-term memory into long-term memory, learning, and the regulation of aggressive behavior, whereas the amygdaloid body regulates emotional expression via modulation of the hypothalamus. An emotion (L., *emovere*, "to disturb") is an intense subjective sensation – including anger, distress, excitement, fear, happiness, hate, love, or sadness – that may be felt toward another individual or an object. The behavioral expression of emotions is usually accompanied by physiological (visceral and/or somatic) responses, mediated by the limbic system and the hypothalamus, via their influence on the endocrine, autonomic, and somatic motor systems.

A Textbook of Neuroanatomy, Third Edition. Maria A. Patestas, Amanda J. Meyer, and Leslie P. Gartner.
© 2025 John Wiley & Sons, Inc. Published 2025 by John Wiley & Sons, Inc.
Companion website: www.wiley.com.go/patestas/neuroanatomy3e

LIMBIC SYSTEM ●●● **435**

Table 22.1 ● Components of the limbic system.

Cortex	Limbic lobe
	Cingulate gyrus
	Isthmus of cingulate gyrus
	Subcallosal area
	Parahippocampal gyrus
	Uncus
	Hippocampal formation
	Dentate gyrus
	Hippocampus proper (mythonym: Ammon's horn, cornu ammonis)
	Subiculum (subicular cortex)
Nuclei	Amygdaloid body (nucleus, complex)
	Mesencephalic ventral tegmental area
	Nucleus accumbens
	Septal nuclei (nuclear complex)
	Some thalamic nuclei
	Some hypothalamic nuclei
Pathways	Alveus
	Fimbria
	Fornix
	Perforant pathway
	Cingulum
	Septohippocampal tract
	Ventral amygdalfugal pathway
	Mammillothalamic fasciculus
	Mammillointerpeduncular tract
	Mammillotegmental tract
	Stria terminalis
	Stria medullaris of prethalamus

Associated pathways such as the alveus, fimbria of the fornix, fornix, perforant pathway, cingulum, septohippocampal tract, ventral amygdalofugal tract, mammillothalamic fasciculus, mammillointerpeduncular tract, mammillotegmental fasciculus, stria terminalis, and stria medullaris of prethalamus are also included in the limbic system (Table 22.1).

The limbic system, which is unique to mammals, functions in: (1) *species preservation*, which includes reproduction and associated instinctive behavior; (2) *self-preservation*, which includes feeding behavior and aggression; and (3) the *expression of fear*, *motivation*, and other emotions, as well as *memory* and *learning*. The limbic system not only exerts its influence on the expression of emotions, but also via its extensive connections with the hypothalamus, the region of the brain that integrates the functions of the endocrine and the autonomic nervous systems. This influences the corresponding visceral responses (i.e., increase in heart rate and breathing, and sweating), which accompany the expression of emotion.

The limbic system not only generates the emotional background for intellect, but it also serves to balance emotional and cognitive mechanisms. The limbic system, also referred to as the "visceral brain," processes information about a certain situation and then, via its visceral control, produces visceral responses that may help the individual to deal with the situation (i.e., fight or flight, or survival). What a person "knows" or "thinks" consciously, is mediated by the neocortex.

Since the limbic system is involved in diverse functions and consists of intricate interconnections with its component parts, it is often difficult to associate a particular component of the limbic system with a particular function.

LIMBIC LOBE

The limbic lobe consists of the subcallosal area, cingulate gyrus, isthmus of the cingulate gyrus parahippocampal gyrus, the uncus as well as the hippocampal formation, collectively forming a cortical perimeter around the corpus callosum

The **limbic lobe** (L., *limbus*, "border") (Fig. 22.1) was characterized by Broca in 1878 as a composite of cortical structures ("*le grand lobe limbique*") that formed a transitional border intervening between the centrally positioned diencephalon and the overlying telencephalon. The limbic lobe consists of the **subcallosal area**, **cingulate gyrus**, **isthmus of the cingulate gyrus, parahippocampal gyrus**, the **uncus,** as well as the **hippocampal formation**, all of which collectively form a cortical perimeter around the corpus callosum. The subcallosal area and parahippocampal gyrus consist of the **neocortex** (G., *neo*, "new"); the cingulate gyrus consists of the **mesocortex** (G., *mesos*, "middle," "intermediate"); and the hippocampal formation consists of the **archicortex** (G., *archein*, "beginning," "original," "primitive") (Table 22.2). The neocortex is phylogenetically new and consists of six layers, whereas the archicortex is the ancient cortex and comprises three layers. The mesocortex is a transitional cortex positioned between the archicortex and neocortex. The limbic lobe is considered to be a component of the more extensive limbic *system*.

Subcallosal, cingulate, and parahippocampal gyri

The **subcallosal gyrus** consists of gray matter, which underlies the basal surface of the anterior extent of the corpus callosum (Fig. 22.1). It curves superiorly around the genu of the corpus callosum and then extends posteriorly, superior to the corpus callosum, as the **cingulate gyrus**. At its posterior inferior extent, the cingulate gyrus in turn continues as the **parahippocampal gyrus**.

Hippocampal formation

The hippocampal formation consists of three regions: the hippocampus proper, the dentate gyrus, and the subiculum

The **hippocampal formation** consists of three regions, the **hippocampus proper** (also referred to as the hippocampus, mythonym: Ammon's horn, or cornu ammonis), the **dentate gyrus**, and the **subiculum** (subicular cortex) (Fig. 22.1; Table 22.3). Structures associated with the hippocampal formation, via neuroanatomical connections, include the **entorhinal cortex**, the **indusium griseum, fasciola cinerea,** and the **septal area** (a primitive precommissural

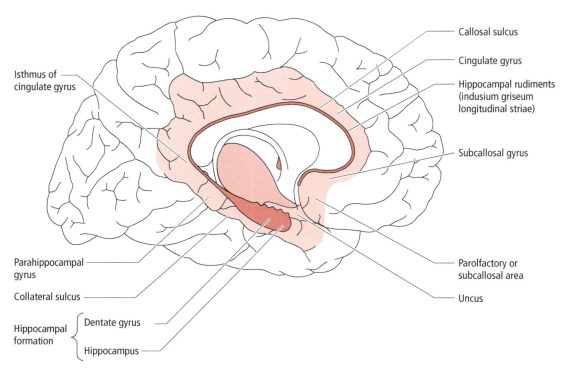

Figure 22.1 ● The cortical structures of the limbic lobe: the subcallosal, cingulate, and parahippocampal gyri and the hippocampal formation.

Table 22.2 ● Cortical components of the limbic lobe.

Part of cortex	Components
Allocortex	Archicortex
	Hippocampal formation
	Hippocampus proper, dentate gyrus, subiculum
	Paleocortex
	Piriform cortex of the parahippocampal gyrus
Juxtallocortex	Mesocortex
	Cingulate gyrus
Isocortex	Neocortex
	Subcallosal area
	Parahippocampal gyrus

area). The hippocampal formation extends from the amygdaloid body anteriorly to the splenium of the corpus callosum posteriorly. On cross (coronal) section, its intrinsic structure is most striking in its middle third.

The hippocampal formation consists of the archicortex, which, during the early stages of its development, scrolls inferiorly and medially on the floor of the lateral ventricle (Fig. 22.2). The function of the hippocampal formation in recent memory and learning is well established.

Hippocampus proper

The hippocampus proper is a phylogenetically ancient cortical structure that forms a comma-shaped prominence on the floor and medial wall of the temporal (inferior) horn of the lateral ventricle

The **hippocampus proper** (G., *hippocampus*, "seahorse") is a phylogenetically ancient cortical structure that forms a comma-shaped prominence on the floor and medial wall of the temporal (inferior) horn of the lateral ventricle. The hippocampus is an infolding of the cortex of the human brain that forms during prenatal development and becomes embedded within the parahippocampal gyrus of the temporal lobe. The ventricular surface of the hippocampus is coated with an ependymal layer. At its anterior extent, the hippocampus displays a swelling with several grooves resembling a paw and is thus referred to as the **pes hippocampus** (L., *pes*, "foot"). The hippocampus extends from the amygdaloid body anteriorly and then tapers as it courses posteriorly to the inferior surface of the splenium of the corpus callosum.

On coronal section (Fig. 22.3A), the hippocampus resembles a seahorse as a result of its arched form, thus its name. Due to its C-shaped outline in coronal section, the hippocampus also resembles a ram's horn, and its mythonym is Ammon's horn (cornu ammonis or CA) after an Egyptian deity with a ram's horns on its head.

Table 22.3 ● The hippocampal formation.

Component	Layers	Cell bodies	Dendrites	Axon terminals
Hippocampus proper	Molecular (deepest)		Pyramidal cell apical dendrites	Granule cells of dentate gyrus ("mossy fibers") Perforant pathway Septohippocampal tract Pyramidal cell axon collateral branches
	Pyramidal (middle)	Pyramidal cells		
	Polymorphic (superficial)	Interneurons	Pyramidal cell basal dendrites	Pyramidal cell axon collateral branches
Dentate gyrus	Molecular (outer)	Some	Apical dendrites of granule cells	Perforant pathway Alveus Collateral branches of granule cells
	Granule cell (middle) Polymorphic (deepest)	Granule cells Interneurons		
Subiculum	Three-layered cortex next to hippocampus Six-layered cortex next to parahippocampal gyrus			

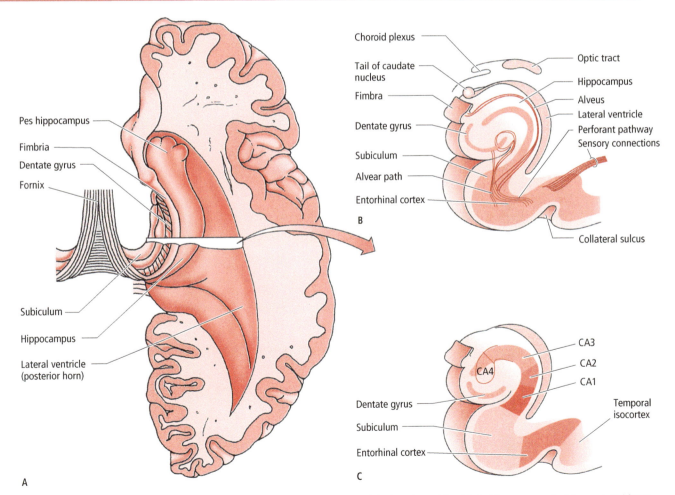

Figure 22.2 ● (A) Dissection of the right cerebral hemisphere illustrating the interior of the lateral ventricle with the hippocampus, dentate gyrus, and fornix. (B) Coronal section exposing the fiber pathways. (C) Coronal section exposing the three sectors of the hippocampus proper.

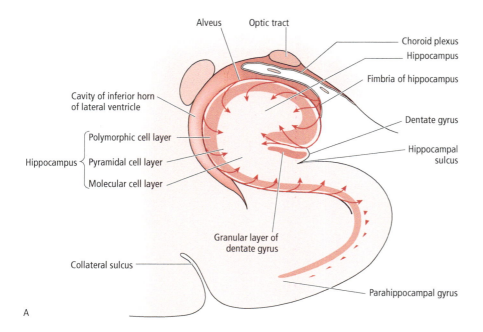

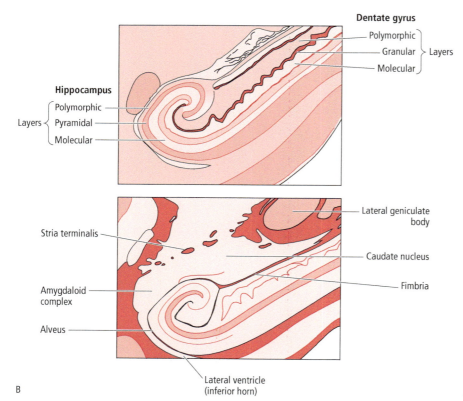

Figure 22.3 ● (A) Coronal section through the hippocampus and its associated structures. Source: Modified from Snell, RS (1997) *Clinical Neuroanatomy for Medical Students*, 4th edn. Lippincott-Raven, New York; fig. 16.4. (B) Coronal sections of the hippocampal formation illustrating the association of the hippocampus and dentate gyrus to the inferior horn of the lateral ventricle, and the caudate nucleus. The respective layers of the hippocampus and the dentate gyrus are illustrated in the middle image

Layers of the hippocampus proper

The hippocampal archicortex consists of three layers

When viewed in transverse (coronal) section, the human hippocampus (consisting of archicortex) displays four zones: CA1, CA2, CA3, and CA4 (Fig. 22.2C). Area CA1 lies next to the subiculum, whereas CA4 lies close to the dentate gyrus. Histological examination displays three layers of the hippocampal archicortex: the **molecular**, **pyramidal**, and **polymorphic layers** (Fig. 22.3A; Table 22.3).

Molecular layer

The molecular layer is the deepest layer of the hippocampal archicortex and is located in the core of the hippocampal formation; it merges with the molecular layer of the dentate gyrus and neocortex

The **molecular layer** (the deepest layer), which is located in the core of the hippocampal formation, consists of apical dendritic trees and axon terminals. The apical dendrites belong to hippocampal pyramidal cells (see below). The axon terminals belong to granule cells whose cell bodies reside in the granule cell layer of the dentate gyrus, the perforant pathway arising in the entorhinal cortex, the septohippocampal tract arising from the septal area, and the pyramidal cell axon collateral branches, arising from the pyramidal cell layer of the hippocampus. This layer merges with the molecular layer of the dentate gyrus and the neocortex.

Pyramidal layer

The pyramidal layer is the middle, most prominent, layer of the hippocampal archicortex; it merges with the internal pyramidal layer of the neocortex

The **pyramidal layer** (the middle layer) is the most prominent layer of the hippocampal formation and consists of **pyramidal neurons**. Their dendrites ramify in the molecular layer, whereas their axons pass in the opposite direction, across the polymorphic layer, and then course in the alveus, fimbria, and the fornix. Pyramidal cell axon collateral branches, referred to eponymously as Hilar Schaffer collaterals, cross the polymorphic and pyramidal layers to reach the molecular layer where they form synaptic contacts with the dendrites of other pyramidal neurons. The pyramidal layer of the hippocampal formation merges with the internal pyramidal layer of the neocortex.

Polymorphic layer

The polymorphic layer of the hippocampal archicortex is the superficial layer, consisting of interneurons

The **polymorphic layer** (**stratum oriens**) (the superficial layer) shares structural characteristics with the deepest layer of the neocortex. This layer lies deep to the alveus and consists of interneurons, as well as pyramidal cell dendrites and axon collateral branches.

Dentate gyrus

The dentate gyrus has a tooth-like configuration (hence its name), created by numerous blood vessels that pierce its ventricular surface and that of the hippocampus

The **dentate gyrus** (L., *dentate*, "tooth-shaped") (see Fig. 22.3; Table 22.3) is a notched band of cortex that is interposed between the superior aspect of the parahippocampal gyrus and the fimbria. This gyrus received its name from its tooth-like configuration created by numerous blood vessels that pierce the ventricular surface of the hippocampus and then that of the dentate gyrus, which imparts a notched appearance. Posteriorly, the dentate gyrus courses along with the fimbria near the splenium of the corpus callosum where it extends to the indusium griseum. Anteriorly, the dentate gyrus extends to the uncus of the parahippocampal gyrus.

Layers of the dentate gyrus

Similar to the hippocampus, the dentate gyrus also consists of three layers of archicortex. They are an outer **molecular layer**, a middle **granule cell layer** (corresponding to the pyramidal layer of the hippocampus), and a deep **polymorphic layer** (see Fig. 22.3B).

Molecular layer

The **molecular layer** of the dentate gyrus archicortex consists mainly of a small population of nerve cell bodies and granule cell dendrites.

Granule cell layer

The granule cell layer of the dentate gyrus archicortex contains the cell bodies of granule cells whose axons form the output of the dentate gyrus

The **granule cell layer** consists of the cell bodies of the **granule cells**, the most predominant cell type of the dentate gyrus. The dendritic arborizations of the granule cells ramify in the molecular layer of the dentate gyrus where they form synapses with the terminals of the perforant pathway. The axons of the granule cells (the output neurons of the dentate gyrus) give rise to collateral branches, then exit the dentate gyrus, pass into the molecular layer of the hippocampus as **mossy fibers**, and terminate on the dendrites of the pyramidal cells of the hippocampus. *The hippocampus is the sole target of the dentate gyrus output.* A large number of the collateral branches project to the molecular layer of the dentate gyrus.

An area of the dentate gyrus known as the **hilus** merges with the polymorphic layer of the CA3 zone of the hippocampus. The hilus consists of interneurons and granule cell axons.

A cross-section of the hippocampal formation exhibits a perceptible double interlocking C. This salient characteristic of the hippocampal formation results from the interlocking of the prominent **pyramidal layer** of the hippocampus proper and the granule cell layer of the dentate gyrus.

Polymorphic layer

The **polymorphic layer** of the dentate gyrus archicortex consists of interneurons.

Subiculum

The subiculum is a thin band of cortex interposed between the hippocampus proper and the parahippocampal gyrus of the temporal lobe

The **subiculum** is a transitional zone that displays a three-layered archicortex adjacent to the hippocampus, but progressively becomes a more elaborate six-layered neocortex as it approaches the parahippocampal gyrus. The subiculum receives information relayed by the hippocampal pyramidal cells. It gives rise to fibers that join the fornix and terminate in the mammillary nuclei of the hypothalamus, and the anterior nuclear group of the thalamus.

Structures associated with the hippocampal formation

Entorhinal cortex

The entorhinal cortex gives rise to the most prominent input to the dentate gyrus

The **entorhinal cortex** (eponym: Brodmann area 28) is located anterior to the amygdaloid body and to the anterior half of the hippocampal formation. The entorhinal cortex exhibits distinguishing characteristics in two of its layers (layers II and IV) that are visible at light microscopy level. Layer II contains the cell bodies of large multipolar neurons that assemble to form multiple "islands" throughout this layer. Since this layer is close to the cortical surface, the cell islands protrude from the surface of the entorhinal cortex, appearing as tiny swellings that are visible with the naked eye and resemble the texture of lemon peel. Unlike layer II, layer IV of the entorhinal cortex includes a dense fiber plexus that forms a distinct boundary between layer III, containing small pyramidal cells, and layer V, containing large pyramidal cells.

Widespread areas of the association cortex relay input to the entorhinal cortex, and it is via the entorhinal cortex that these cortical association areas have access to the hippocampus and play a role in memory consolidation. The entorhinal cortex gives rise to the most prominent *input* to the **dentate gyrus** of the hippocampal formation via the **perforant pathway** arising from the lateral part of the entorhinal cortex and the **alvear pathway** arising from the medial part of the entorhinal cortex. These pathways are part of the intrinsic circuitry of the hippocampal formation.

Indusium griseum

The subcallosal gyrus gives way to the indusium griseum, a slender stripe of gray matter that overlies and follows the contour of the superior surface of the corpus callosum in the midline

The **indusium griseum** (L. *indusium* "tunic," L. *griseum* "grey") represents a portion of the hippocampal formation that remained attached to the corpus callosum during its development, as the rest of the hippocampal formation moved laterally, inferiorly, and then medially to its final position in the temporal lobe (Fig. 22.4). Dorsally, it curves around the splenium of the corpus callosum and then merges with the dentate and parahippocampal gyri by way

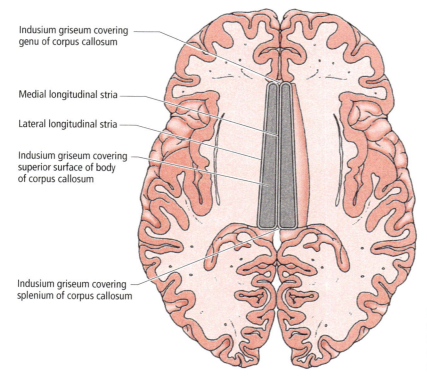

Figure 22.4 • Dissection of the superior surface of the cerebrum to reveal the superior surface of the corpus callosum. Source: Modified from Snell, RS (1997) *Clinical Neuroanatomy for Medical Students*, 4th edn. Lippincott–Raven, New York; fig. 16.4.

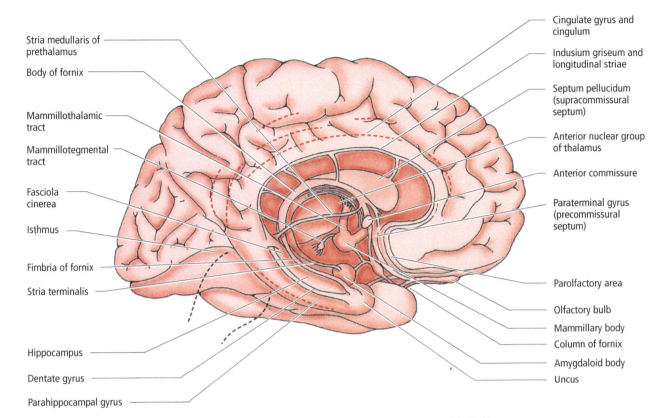

Figure 22.5 • Dissection of the medial surface of the cerebral hemisphere showing the main components of the limbic system.

of the fasciola cinerea. Two pairs of rudimentary stripes of indusium griseum white matter, the **medial** and **lateral longitudinal striae**, arise and terminate in the hippocampal formation, and course bilaterally along the superior surface of the corpus callosum, lateral to the indusium griseum.

Fasciola cinerea

The **fasciola cinerea** (L., *fasciola*, "small bandage"; *cinereus*, "ashen-hued") is a transitional zone of intervening cortex that joins the dentate gyrus with the indusium griseum (Fig. 22.5).

Septal area

The **septal area** is located in the telencephalon, in close proximity to the genu of the corpus callosum. This area, which is well developed in humans, includes the **septal nuclear complex** whose nuclei are gathered rostral to the anterior commissure. The septal nuclei are classified into two main groups, the **medial** and the **lateral septal nuclei**. They are connected with the hippocampal formation via the fornix, and with the hypothalamus via the medial forebrain bundle. Furthermore, the septal nuclei send fibers via the stria medullaris of prethalamus to the habenular nuclei. Human studies have demonstrated that the septal nuclei play a role in aggressive behavior, sexual behavior, modulation of autonomic functions, as well as with cognitive and memory functions.

Afferents (input) to the hippocampus proper

Afferent fibers to the hippocampus proper arise from the following sources (Table 22.4): the indusium griseum, cingulate gyrus, septal nuclei, the entorhinal cortex, the dentate gyrus, the parahippocampal gyrus, and the contralateral hippocampus.

Table 22.4 • Connections of the hippocampus.

Input sources to the hippocampus	Output targets of the hippocampus
Indusium griseum	Septal nuclei
Cingulate gyrus	Lateral preoptic area of the hypothalamus
Septal nuclei	Anterior part of the hypothalamus
Entorhinal cortex	Mammillary body
Dentate gyrus	Anterior nuclear group of the thalamus
Parahippocampal gyrus	Midbrain tegmentum
Contralateral hippocampus	Habenular nuclei

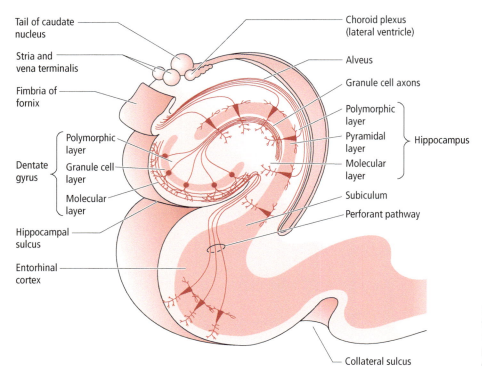

Figure 22.6 ● A transverse section through the temporal lobe at the level of the hippocampus and dentate gyrus. The main pathways connecting the hippocampus, dentate gyrus, and entorhinal cortex are shown.

Efferents (output) from the hippocampus proper

Efferent fibers of the hippocampus proper arise mostly from the pyramidal cells of the hippocampal pyramidal cell layer

Efferent fibers of the hippocampus proper (Table 22.4) arise mostly from the pyramidal neurons of its pyramidal cell layer. Their axons course in the alveus, fimbria, and fornix, which distributes them to various nuclei. Note that pyramidal cells are also the output neurons of the neocortex of the five lobes. Axons coursing anterior to the anterior commissure terminate in the septal nuclei, the lateral preoptic area of the hypothalamus, and the anterior part of the hypothalamus. Axons coursing posterior to the anterior commissure are distributed to the mammillary body of the hypothalamus, the anterior nuclear group of the thalamus, and the tegmentum of the mesencephalon. Some hippocampal fibers terminate in the habenular nuclei of the epithalamus via the habenulo-interpeduncular tract (fasciculus retroflexus). A second pathway, the **hippocampo–entorhino–neocortical pathway**, emerges from the hippocampus proper and follows the perforant path back to deeper layers of the entorhinal cortex and from there to the neocortex (Fig. 22.6). This pathway is involved in the consolidation of episodic memories (associated with events, e.g., going to a dental appointment last month).

Intrinsic circuitry of the hippocampal formation

Widespread areas of the neocortex, particularly the cingulate gyrus, relay signals via the cingulum to the entorhinal cortex

The components of the hippocampal formation (the hippocampus proper, dentate gyrus, and subiculum) serve as successive processing stations in a series of local projections within the hippocampal formation. This local circuit consists of a pathway that arises from the entorhinal cortex.

Although the intrinsic circuitry of the hippocampal formation consists primarily of a closed loop of signal transmission, collateral branches and feedback connections are also present along this pathway, which serve to modify the relay of signals within the hippocampal formation (Figs. 22.6 and 22.7).

Axons arising from the **entorhinal cortex** (of the parahippocampal gyrus) form the **perforant** and **alvear pathways** (also known as the **lateral** and **medial perforant pathways**, respectively), which end in the molecular layer of the **dentate gyrus**, where their fibers synapse with the dendrites of its **granule cells**. Furthermore, these axons also send collaterals to the **CA1** and **CA3** zones of the hippocampus and the **subiculum**. *The entorhinal cortex provides the major input to the hippocampal formation.*

The granule cells of the dentate gyrus give rise to axons referred to as **mossy fibers**, which project not only to the CA3 zone of the hippocampal formation where they synapse with **pyramidal cells**, but also to its own **polymorphic layer**. The dentate gyrus polymorphic-layer neurons give rise to fibers that synapse with the dendrites and cell bodies of the granule cells, modifying the information relayed by the granule cells. Note that the *dentate gyrus projects only locally* and does not project outside the hippocampal formation.

The pyramidal cells of the CA3 zone give rise to axons that join the **fimbria**. Prior to joining the fimbria, these axons

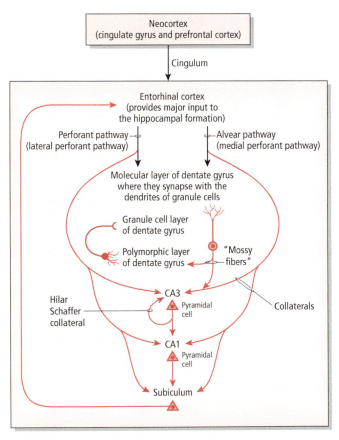

Figure 22.7 • The intrinsic circuitry of the hippocampal formation.

emit collaterals (eponym: Hilar Schaffer collaterals) that project within CA3, and to the CA1 zone of the hippocampal formation (to synapse with the pyramidal cells there), as well as to the **septal nuclei**.

The pyramidal cells of the CA1 zone send their axons to the subiculum (subicular cortex) which, in turn projects to the deeper layers of the entorhinal cortex, completing a closed circuit. The cells in the deeper layers of the entorhinal cortex give rise to axons that project to the superficial layers of the entorhinal cortex from which the perforant pathway arises. The entorhinal cortex also gives rise to a prominent projection to **neocortical association areas**.

Widespread areas of the cerebral cortex project to the entorhinal cortex, which in turn relays this input to specific areas of the hippocampal formation. Association (intrinsic) connections link different areas of the hippocampal formation, whereas commissural fibers link the hippocampal formation of the two sides.

Papez circuit

The **Papez circuit** was proposed by James Papez in the 1930s as the neuroanatomical basis of emotion. Although it is known today that the connections of the limbic system are much more complex, this circuit served as a basis for more modern research. It was proposed that emotional expression is consolidated in the hippocampus, perceived in the cingulate gyrus and expressed via the mammillary bodies. The behavioral expression of emotions is usually accompanied by physiological (visceral and/or somatic) responses. It is by its connections to the hypothalamus that the limbic system influences the autonomic nervous system to produce the corresponding visceral responses (e.g., increase in heart rate and breathing, and sweating), which supplement the expression of emotion. It is now known that the Papez circuit is also implicated in cognitive processes, such as mnemonic functions and spatial short-term memory. The Papez circuit consists of the following anatomical structures: the hippocampus, fornix, mammillary body, anterior nucleus of the thalamus, mammillothalamic fasciculus, and cingulate gyrus.

The **subiculum** of the **hippocampal formation** sends fibers via the **fornix** to the medial mammillary nucleus of the **mammillary body**, and other nuclei of the hypothalamus. The **medial mammillary nucleus** sends the majority of its output fibers via the **mammillothalamic fasciculus** to the **anterior nuclear group of the thalamus**, which in turn projects **thalamocortical fibers** to the **cingulate gyrus**. The cingulate gyrus projects back via the **cingulum** to the hippocampal formation, completing the circuit. This circuit, though, is not a closed one, since each relay station has numerous connections with other areas of the nervous system. The subiculum also gives rise to fibers that join the fornix to terminate in the amygdaloid body. This connection may be involved in the consolidation of memories associated with emotions.

Function of the hippocampal formation in memory

Learning is the manner by which knowledge is acquired, and memory is the process by which knowledge is stored and is retrievable in the future. Memory is classified very broadly into **immediate**, **short-term**, and **long-term memory**. Newly learned information is initially stored into immediate or short-term memory, lasting for seconds and minutes, respectively. Recently learned information, if reinforced, may be conveyed and stored into long-term memory for a prolonged period of time, a function which is mediated by the hippocampal formation. Individuals who have lesions in the hippocampal formation are unable to transfer immediate and short-term memory into long-term memory. Learning and memory processes may be interrupted by trauma, cerebral anoxia, Alzheimer disease, the use of alcohol and illicit drugs, and amnestic confabulatory syndrome (eponym: Korsakoff syndrome) involving the hippocampus. Although a well-known function of the hippocampal formation is the consolidation of short-term memory into long-term memory, it is also associated with the body's reaction to stress and emotions. The anterior region of the hippocampal formation is believed to function in processing of the body's reaction to stress/anxiety and emotions, whereas the posterior region functions in cognition, explicit memory, and spatial memory.

Memory is classified into **declarative (explicit) memory** and **nondeclarative (implicit) memory**. Declarative memory is further classified into **semantic memory** (which encodes facts, individuals, and objects/items), and **episodic memory** (which is the ability to look back and place facts and events that have a spatial and temporal background, such as remembering our high school graduation, and also looking into the future to visualize a vacation in Europe). In contrast, implicit memory mediates the learning of skills such as running or swimming, which once mastered can be performed without any conscious effort (their execution becomes automatic). This type of memory is associated with the basal nuclei and the cerebellum which store the programs of motor activity for future access.

A lesion that damages the hippocampal formation results in impairment of declarative memory, whereas with a lesion that damages the medial temporal lobe, implicit memory (skills such as riding a bicycle) is preserved.

Recent studies indicate that the mechanism by which short-term memory is assembled/converted into long-term memory may be **long-term potentiation**, a form of neuronal plasticity. Long-term potentiation is a physiological process that occurs at neuronal synaptic contacts, as follows. When a neuron transmits a succession of nerve impulses to another neuron, the summation of excitatory postsynaptic potentials increases the probability that the target neuron will be depolarized by this particular synapse and by additional synaptic contacts established on the target neuron. Depolarization of the target neuron may be due to a greater chance that neurotransmitter will be released by the presynaptic terminal and/or an enhanced response of the postsynaptic terminal. Long-term potentiation in the hippocampal formation has been established at the axon terminals of the perforant pathway (the main cortical input pathway to the hippocampal formation terminating in the **dentate gyrus**), and at the axon terminals of the **CA3 zone** pyramidal cells forming synaptic contacts with **CA1 zone** cells. Recall that the perforant pathway consists of cortical association fibers that emerge from the entorhinal cortex, and migrate to the molecular layers of the hippocampus and the dentate gyrus, where they synapse with pyramidal and granule cells, respectively. Long-term potentiation may be the physiological process via which certain types of memory are formed.

Amygdaloid body

The amygdaloid body is divided into three groups of subnuclei, the basolateral, olfactory, and centromedial groups

The **amygdaloid body** (G., *amygdalon*, "almond") (Fig. 22.8) is an almond-shaped group of nuclei located deep to the uncus of the parahippocampal gyrus anterior to the hippocampus and the temporal horn of the lateral ventricle. The amygdaloid body is divided into three groups of subnuclei (Table 22.5): the large basolateral group, and the smaller olfactory and centromedial groups. The basolateral and olfactory group subnuclei receive *afferent* fibers, whereas the centromedial group subnuclei give rise to *efferent* fibers. Each subnuclear group has different connections and functions.

The amygdaloid body receives sensory input from widespread areas of the nervous system including somatosensory, visual, auditory, and visceral (such as olfactory) stimuli.

The **olfactory nuclei** of the **amygdaloid body** are the oldest phylogenetically. These nuclei are the termination of olfactory fibers arising from the olfactory bulb via the lateral olfactory stria and the olfactory cortex. These nuclei also receive fibers from the thalamus, hypothalamus, and the brainstem. Since these nuclei project to the ventromedial nucleus of the hypothalamus, they are believed to play a role in eating behaviors.

The **basolateral nuclear group** is the newest phylogenetically. It is reciprocally connected to the cortical sensory

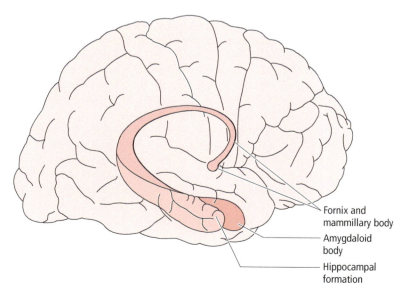

Figure 22.8 ● The relationship of the amygdaloid body and hippocampal formation. Also illustrated are the fornix, the principal efferent pathway of the hippocampal formation, and its termination, the mammillary body of the hypothalamus.

Nuclear Group	Connections	Functions
Olfactory	Afferents from olfactory bulb, olfactory cortex, thalamus, hypothalamus, and brainstem Efferents to the ventromedial nucleus of the hypothalamus	Behavior associated with hunger and eating
Basolateral	Reciprocally connected to the sensory association cortical areas Afferents from parahippocampal gyrus and insular cortex Efferents to the medial dorsal nucleus of the thalamus, basal nucleus (of Meynart), and ventral striatum	Behaviors associated with eating and drinking Autonomic and somatic reflex activity Behavioral reaction to stressful situations
Centromedial	Reciprocally connected to the visceral sensory and autonomic nuclei in the brainstem	Respiratory and cardiovascular responses

Table 22.5 ● Nuclear groups of the amygdaloid complex.

association areas of four lobes (the visual, auditory, and somatosensory systems) and projects to the medial dorsal nucleus of the thalamus, the basal nucleus (eponym: basal nucleus of Meynart), and the ventral striatum. Fibers from the parahippocampal gyrus and insular cortex also terminate here. This nuclear group influences the various functions of the hypothalamus associated with feeding and drinking behavior, autonomic and somatic reflexes, and responses to stress.

The basolateral and olfactory nuclear groups project to the **central nucleus** of the centromedial group of the amygdaloid body. This nucleus has reciprocal connections with the visceral sensory and visceral motor nuclei located in the brainstem. The central nucleus gives rise to the ventral amygdalofugal pathway, which terminates in the autonomic nervous system nuclei of the brainstem. These include the **posterior nucleus of the vagus**, as well as nuclei of the **reticular formation** involved in respiratory and cardiovascular functions. In addition, the central nuclei control the function of the ANS via their projections to the **lateral hypothalamus**.

Afferent projections to the amygdaloid body

The amygdaloid body receives *afferent* (input) fibers from the following sources (Table 22.6): olfactory bulb, orbitofrontal cortex, cingulate gyrus, basal forebrain, medial thalamus, hypothalamus, and brainstem.

Efferent projections from the amygdaloid body

The principal function of the limbic system is to modulate hypothalamic activity, and it does so via two main pathways, the stria terminalis and the ventral amygdalofugal (amygdaloid) pathway

One of the principal functions of the limbic system is to modulate hypothalamic activity; this modulation is mediated via two main output pathways arising from the amygdaloid body, which distribute fibers mainly to the

hypothalamus, but also to sources of amygdaloid input. These *efferent* pathways are: the **stria terminalis** (which conveys fibers primarily from the centromedial nuclear group to the septal nuclei, hypothalamus, nucleus accumbens, caudate nucleus and putamen) and the **ventral amygdalofugal (amygdaloid) pathway** (which also conveys input fibers to the amygdaloid body) (Table 22.6). The latter pathway is less distinct and distributes fibers from the central nucleus to the brainstem as well as fibers from the basolateral nuclear group that terminate in the thalamus, ventral striatum, and basal nucleus. The signals arising from the striatum are ultimately relayed to the medial dorsal nucleus of the thalamus, which in turn projects to the prefrontal and orbitofrontal cortices. The limbic system projections to the basal nuclei are associated with underlying emotional aspects that influence movement.

Functions of the amygdaloid body

One of the key roles of the amygdaloid body is the control of autonomic nervous system (ANS) function in relation to previous experiences

One of the key roles of the amygdaloid body is the control of autonomic nervous system (ANS) function in relation to previous experiences. ANS modulation is mediated indirectly by reciprocal projections between the amygdaloid body and the hypothalamus (centromedial nuclear group), which in turn coordinates ANS activity, and directly via the descending pathways that arise from the amygdaloid body and terminate in the brainstem autonomic centers (central amygdaloid nucleus).

The mode of amygdaloid control on ANS function differs from that of the hypothalamus (without amygdaloid input). Hypothalamic control of ANS activity is reflexive in nature, continuously monitoring the body's internal environment via baroreceptors and osmoreceptors, and making appropriate adjustments, whereas control of ANS activity by the amygdaloid body is mediated by instinct. The amygdaloid

446 ••• CHAPTER 22

body utilizes prior experiences to modulate ANS activity. The basolateral nuclear group has abundant reciprocal projections with the sensory association areas of all five lobes and is associated with the processing of input related to past experiences. In contrast, the olfactory nuclear group projects to the hypothalamus, whereas the centromedial nuclear group projects to the brainstem. The amygdaloid body, through its connections with the hypothalamus, influences the visceral and somatic aspects that correspond to the behavioral expression of emotion.

BRAINSTEM CENTERS ASSOCIATED WITH LIMBIC SYSTEM FUNCTION

Hypothalamus

The hypothalamus mediates ANS (visceral) responses that accompany the expression of emotions

Various emotional states initiate intricate psychological, endocrine, and autonomic responses. The limbic system sends numerous projections to the hypothalamic nuclei, especially to the mammillary and preoptic nuclei (Table 22.7). The hypothalamus in turn mediates ANS (visceral) responses that accompany the expression of emotions.

Thalamus

The amygdaloid body and hypothalamus project to the "limbic nuclei" of the thalamus: the anterior nuclear group, and the lateral dorsal and medial dorsal nuclei of the thalamus

The amygdaloid body and hypothalamus project to the "limbic nuclei" of the thalamus, which include the **anterior nuclear group** and the **lateral dorsal** and **medial dorsal nuclei** of the thalamus (Table 22.7). These nuclei then relay this information to the limbic lobe. Furthermore, the hippocampal formation receives *afferent* fibers from the anterior, lateral dorsal, lateral posterior, and intralaminar nuclei of the thalamus.

Habenular nuclei

The limbic system projects to the reticular formation of the mesencephalon via a relay in the habenular nuclei

The limbic system projects to the reticular formation of the mesencephalon via a relay in the **habenular nuclei**. These nuclei project to the **interpeduncular** and **raphe nuclei** of the mesencephalon via the **habenulointerpeduncular tract**. Other habenular efferent fibers terminate in the **hypothalamus**, the **ventral tegmental area**, and the **substantia nigra**.

Ventral tegmental area

The **ventral tegmental area** of the midbrain modulates the processing of memory via *dopaminergic* fibers that terminate in cortical areas related to the limbic system.

Locus caeruleus and dorsal raphe

The **locus caeruleus** sends *noradrenergic* fibers whereas the **dorsal raphe** sends *serotonergic* fibers to the hippocampal formation, to the limbic lobe, and to the amygdaloid body, where they modulate the processing of memory.

PATHWAYS OF THE LIMBIC SYSTEM

A number of pathways connect the various components of, and areas associated with, the limbic system (Tables 22.1 and 22.8). They are the alveus, fimbria, fornix, perforant pathway, cingulum, septohippocampal tract, ventral amygdalofugal pathway, mammillothalamic fasciculus, mammillointerpeduncular tract, mammillotegmental tract, stria terminalis, stria medullaris of prethalamus, anterior commissure, diagonal band, and habenulointerpeduncular tract.

Table 22.6 ● Connections of the amygdaloid body.

Afferent projections to the amygdaloid body	Efferent projections from the amygdaloid body
Olfactory bulb	Hypothalamus
Orbitofrontal cortex	Septal nuclei
Cingulate gyrus	Ventral striatum
Basal forebrain	Basal nucleus
Medial thalamus	
Hypothalamus	
Brainstem	

Table 22.7 ● Brainstem centers associated with limbic system function.

Structure	Function

Hypothalamic nuclei
  Mammillary nucleus
  Preoptic nucleus
	Control autonomic nervous system responses associated with the expression of emotions

Thalamic nuclei
  Anterior
  Lateral dorsal
  Medial dorsal
	Process information from the hypothalamus and amygdaloid body, and relay information to the limbic lobe, prefrontal, and temporal areas

Habenular nuclei	Serve as relay centers for information arising from the limbic system destined for the midbrain reticular formation
Ventral tegmental area	Regulates processing of memory
Locus caeruleus	Regulates processing of memory
Dorsal raphe	Regulates processing of memory

Table 22.8 ● Pathways of the limbic system.

Pathway	Origin	Termination
Alveus	Pyramidal cells of the hippocampal formation	
Fimbria	Continuation of the alveus	
Fornix	Continuation of the fimbria	Hypothalamus
		Septal area
		Basal forebrain
		Hypothalamus
		Nucleus accumbens
		Thalamus
Perforant	Entorhinal cortex	Hippocampus proper
		Dentate gyrus
Cingulum	Cingulate gyrus	Neocortex
Septohippocampal tract	Medial septal nucleus	Hippocampus proper
	Nuclei of the diagonal band	Dentate gyrus
		Subiculum
		Entorhinal cortex
Ventral amygdalofugal	Amygdaloid body	Hypothalamus
Mammillothalamic fasciculus	Mammillary body	Anterior nuclear group of the thalamus
Mammillointerpeduncular tract	Mammillary body	Interpeduncular nucleus
Mammillotegmental tract	Mammillary body	Midbrain tegmentum
Stria terminalis	Amygdaloid body	Septal area
		Preoptic area of the hypothalamus
		Bed nucleus of the stria terminalis
Stria medullaris (prethalamus)	Habenular nucleus	Septal nuclei
		Anterior hypothalamus
Anterior commissure	Olfactory bulb	Contralateral olfactory bulb and
	Parahippocampal gyrus	parahippocampal gyrus
Diagonal band	Parolfactory area	Periamygdaloid area
Habenulointerpeduncular tract	Habenular nucleus	Interpeduncular nucleus of the midbrain

Alveus, fimbria, and fornix

The fornix is the main output pathway of the hippocampal formation relaying information to the hypothalamus and the septal nuclei

A fine layer of white matter, the **alveus** (L., "trough," "canal"), is located deep to the ependymal layer overlying the ventricular surface of the hippocampal formation. The alveus is composed of a two-way bundle of myelinated axons, most of which belong to the pyramidal cells of the hippocampal formation. These axons gather to become the **fimbria** (L., "fringe"), which extends to the subsplenial area inferior to the corpus callosum. At the subsplenial area each hippocampus (right and left) proceeds as the **crus of the fornix** (L., *crus*, "leg"; *fornix*, "arch") (Fig. 22.9). The fornix consists of two crura, one on each side, which are joined by commissural fibers forming the **commissure of the fornix (hippocampal commissure)**.

Anterior to the commissure of the fornix, the two crura approach one another and unite in the midline to become the **body of the fornix** which courses anteriorly following the inferior edge of the septum pellucidum. At the level of the interventricular foramen and the anterior commissure, the fornix becomes arched, proceeding inferiorly and posteriorly diverging again into the **columns of the fornix**. As each column turns inferiorly on each side, it dives into the lateral wall of the third ventricle and then enters the mammillary body of the hypothalamus. In the human, the fornix consists of approximately 1.2 million axons.

At the level of the anterior commissure, the fornix separates into a precommissural part that lies anterior to the anterior commissure and a postcommissural part that lies posterior to the anterior commissure. The **precommissural fibers** take their origin primarily from the pyramidal cells of the hippocampus (although some fibers originate from the

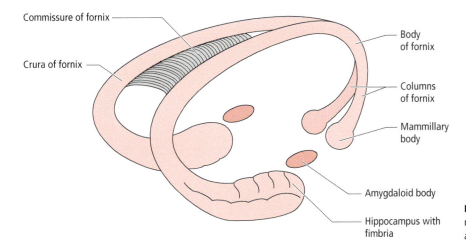

Figure 22.9 ● Three-dimensional view of the relationships between the hippocampus, the fornix, and the mammillary bodies of the hypothalamus.

subiculum and entorhinal cortex) and project to the septal area, basal forebrain, hypothalamus and nucleus accumbens. The **postcommissural fibers** emerge from the subiculum and then enter the hypothalamus where the majority of the fibers project primarily in the medial mammillary nucleus. Some fibers also terminate in the thalamus and other nuclei of the hypothalamus.

The **fornix** is the *main output* pathway of the hippocampal formation relaying information to the hypothalamus and the septal nuclei. Although most of the axons in the fornix are *efferents* to the ipsilateral hypothalamus and the septal area, some of the fibers are *afferents* to the hippocampus from various brain centers. In addition, the fornix carries commissural fibers to the habenular nuclei, and others connecting the hippocampal formation of the two sides.

Through its connections with the hypothalamus (which coordinates visceral and endocrine functions) and with the temporal lobe, the hippocampus can influence emotional behavior and memory, respectively. Neurosurgeons are very wary of the fornix and are sure to say, "preserve the fornix"!

Perforant pathway

The perforant pathway is the main cortical input pathway to the hippocampal formation

The **perforant pathway** (see Figs. 22.6 and 22.7) is the main cortical input pathway to the hippocampal formation. It consists of cortical association fibers that emerge from the entorhinal cortex, and migrate to the molecular layers of the hippocampus and the dentate gyrus, where they synapse with pyramidal and granule cells, respectively.

Cingulum

The **cingulum** consists of cortical association fibers located on the deep aspect of the cingulate gyrus. It transmits information that interconnects the surrounding neocortex.

Septohippocampal tract

The **septohippocampal tract** carries fibers that arise from the medial septal nucleus and the nuclei of the diagonal band. These fibers join the fornix to terminate in the hippocampus, dentate gyrus, subiculum, and entorhinal cortex next to the hippocampal formation.

Ventral amygdalofugal pathway

The **ventral amygdalofugal pathway** is the main projection pathway from the amygdaloid body to the hypothalamus, which it reaches by traversing the internal capsule. This pathway originates from the basolateral nucleus and central nucleus of the amygdaloid body and connects to the nucleus accumbens, basal forebrain, medial dorsal nucleus of the thalamus, and lateral hypothalamus. These connections of the ventral amygdalofugal tract mediate emotions of conscious perception of emotion.

Mammillothalamic fasciculus

The **mammillothalamic fasciculus** is the principal output pathway arising from the hypothalamic mammillary nucleus. It consists of heavily myelinated axons that terminate in the anterior nuclear group of the thalamus. The mammillothalamic tract fasciculus plays a key role in temporal and contextual memory.

Mammillointerpeduncular tract

The **mammillointerpeduncular tract** arises from the mammillary body and terminates in the interpeduncular nucleus. The interpeduncular tract is involved in sleep regulation and pain sensitivity.

Mammillotegmental tract

The **mammillotegmental tract** arises from the mammillary body and terminates in the midbrain tegmentum's anterior and posterior tegmental nuclei and posterior gray matter areas.

Stria terminalis

Fibers of the **stria terminalis** take their origin mainly from the amygdaloid body and terminate in the septal area, the medial preoptic area of the hypothalamus, and the bed nucleus of the stria terminalis. The bed nucleus of the stria terminalis plays an important role in controlling autonomic, neuroendocrine, and behavioral responses.

Stria medullaris of prethalamus

The **stria medullaris of prethalamus** consists of fibers that interconnect the habenular nucleus with the septal nuclei and the anterior hypothalamus.

Cerebral Anterior commissure

The cerebral **anterior commissure** carries decussating fibers transmitting olfactory information between the olfactory bulbs, as well as information between the parahippocampal gyri.

Diagonal band

The **diagonal band** (eponym: diagonal band of Broca) consists of fibers that arise from the paraolfactory area and terminate in the periamygdaloid area in the temporal lobe.

Habenulointerpeduncular tract (tractus retroflexus)

The **habenulointerpeduncular tract** arises from the habenular nucleus and terminates in the interpeduncular nucleus of the midbrain.

LIMBIC ASSOCIATION CORTEX

The limbic association cortex consists of the cortex on the medial surface of the frontal, parietal, and temporal lobes and the ventral (orbital) surface of the frontal lobe

The structures of the limbic association cortex include the cingulate, parahippocampal, orbitofrontal, and medial frontal gyri and the anterior pole of the temporal lobe. The cingulate gyrus includes three anatomical regions: anterior, mid-cingulate, and posterior, each of which is associated with a different function. The anterior region, via its connections to other limbic structures, including the amygdaloid body, the orbitofrontal and insular cortices, is involved in the processing of emotions associated with physical and emotional pain. The mid-cingulate region is associated with motor function and may play a role in the production of movement elicited by emotions. The posterior region is involved in memory and sensory processing. The orbitofrontal and orbital gyri are involved in higher-order thinking, judgment, and reward. The anterior pole of the temporal lobe shares reciprocal connections with the orbitofrontal cortex, the hypothalamus, and the amygdaloid body; it is involved in personality characteristics.

LIMBIC SYSTEM INPUT TO THE ENDOCRINE, AUTONOMIC, AND SOMATIC MOTOR SYSTEMS

The limbic system, via its extensive and widespread projections, influences the endocrine, autonomic and somatic motor systems to produce physiological responses in tissues and organs throughout the body that are appropriate for, and accompany, a particular emotional situation

The limbic system influences the endocrine system via a pathway that begins in the limbic association cortex and which projects to the olfactory and centromedial nuclear groups of the amygdaloid body. The amygdaloid body modulates hypothalamic activity via the stria terminalis, consisting of prominent amygdaloid output fibers some of which project to the ventromedial (VM) nucleus of the hypothalamus. The VM nucleus in turn relays signals mainly to the parvocellular neurons of the arcuate nucleus of the hypothalamic periventricular zone. Then, via the tuberoinfundibular tract, consisting of the axons of neurosecretory neurons, the hypothalamus controls the secretory activity of the hormone producing cells in the anterior lobe of the hypophysis (pituitary gland). In response to limbic signals, the hypophysis (via its hormone secretions) controls the function of various tissues and organs throughout the body.

To prepare the body to deal with a challenging emotional or a physically demanding situation (e.g., the fight-or-flight-or-freeze-or-fornicate response), the limbic system also influences the function of the autonomic nervous system via its direct and indirect projections to the brainstem and spinal cord autonomic nuclei. In this pathway, the limbic association cortex projects to the centromedial nuclear group of the amygdaloid body and to neurons located in the middle and lateral zones of the hypothalamus. The amygdaloid body in turn projects to the same nuclei of the hypothalamus. The amygdaloid body and the paraventricular nucleus and the lateral hypothalamus give rise to fibers that descend to synapse directly with the pre-ganglionic neurons located in the autonomic nuclei of the brainstem and spinal cord. Thus the amygdaloid body can switch on autonomic function both via direct projections to the autonomic centers and indirectly by projecting first to the hypothalamus. Post-ganglionic neurons project to the various tissues and organs where they produce the autonomic responses that accompany an emotional response.

The limbic system also activates the somatic motor system via its influence on the reticular formation to prepare the body for strenuous motor activity required for fighting or escaping a dangerous situation. This pathway begins with the limbic association cortex which projects to the hippocampus proper, amygdaloid body, and septal nuclei. They in turn project to the lateral zone of the hypothalamus which projects to the reticular formation. It is the neurons of the reticular formation that form the reticulospinal tracts; it is these tracts that descend to the spinal cord to influence motor neurons.

The limbic system is connected to the basal nuclei via the limbic loop (see Fig. 14.17 in Chapter 14, Basal Nuclei). This loop functions in the emotional and motivational aspects of movement, manifested as various facial expressions or other body movements. As an example, in an emergency, intense emotion can spark a sudden and quick movement.

CLINICAL CONSIDERATIONS

Injury to the temporal lobes results in the inability to form new memories

Bilateral lesions of the hippocampus may be produced during a forceful impact of the anterior poles of the temporal lobes on the greater wings of the sphenoid bone. Injury to the temporal lobes results in the inability to form new memories.

Studies have shown that the pyramidal neurons in area CA1 of the hippocampal formation (referred to as the Sommer sector) are particularly susceptible to cell injury and death if blood oxygen concentration drops to low levels. Cell degeneration within CA1 occurs within 3–4 minutes as a consequence of the decrease in oxygen supply that occurs following a cardiac arrest to individuals who are resuscitated. Injury to the pyramidal cells of the hippocampal formation results in memory loss.

Bilateral lesion or removal of the medial temporal lobe, including the amygdaloid body and most of the hippocampal formation, has been performed in the past on individuals who have epileptic seizures. The hippocampus has an exceptionally low threshold for epileptic seizures, which may be transmitted to other parts of the limbic lobe and to the neocortex. Although this surgical treatment reduces the occurrence of seizures in these individuals, they are unable to learn anything or to form new memories, a condition referred to as **anterograde amnesia**. In addition, removal of the hippocampi results in **retrograde amnesia**, a condition in which there is a deficit in more recent memories, but not of those in the distant past. These findings have led to the conclusion that the hippocampal formation is necessary in the formation of long-term memories. It is believed that the hippocampus generates the drive and mediates the translation of short-term into long-term memories, which are stored in the neocortical association areas (visual association cortex, auditory association cortex, somatosensory association cortex, etc.).

Limbic system syndromes

Alzheimer disease

Alzheimer disease is a degenerative disorder of the brain and is the most common form of progressive dementia in the elderly

Alzheimer disease is a degenerative disorder of the brain and is the most common form of progressive dementia in the elderly. Individuals with this disease are unable to form new memories.

This disease is caused by pathologic alterations, including neurofibrillary tangles, neuritic plaques, and neuronal degeneration, which appear initially in the pyramidal cell islands (of layer II) of the entorhinal cortex. From there, degeneration spreads to the CA1 zone of the hippocampus proper, and then back to the deeper layers of the entorhinal cortex. Consequently, this neuronal degeneration hinders the normal flow of information through the hippocampal formation. Confusion and deficits in executive function occur following further spread of neurofibrillary tangles to the temporal pole and prefrontal cortex. Subicular pathology occurs roughly at the same time that neurofibrillary tangles invade the temporal neocortex.

Amnestic confabulatory syndrome (Eponym: Korsakoff syndrome)

Amnestic confabulatory syndrome is a disorder that most often results from thiamin nutritional deficiency in chronic alcoholism

Amnestic confabulatory syndrome is a memory disorder that most often results from thiamin (vitamin B1) nutritional deficiency in chronic alcoholism. Affected individuals exhibit amnesia (memory loss) of both anterograde and retrograde memory. Anterograde memory comprises memory loss that follows after hippocampal formation injury, whereas retrograde memory loss consists of memory loss of events that occurred prior to the lesion. Patients with Korsakoff syndrome have loss of both recent memory and events that occurred long ago. To compensate for this when they converse, they make up fictitious information or events to fill in the gaps of memory loss. Morphological changes are reported in the hippocampal formation, the columns of the fornix, and the mammillary bodies of the hypothalamus. However, the area that exhibits the most drastic modification is the medial dorsal nucleus of the thalamus.

Klüver–Bucy syndrome

Experiments performed on monkeys with bilateral temporal lobe lesions, which included the amygdaloid body as well as the surrounding cortical areas that project to the amygdaloid body, produced what is referred to as the **Klüver–Bucy syndrome**. This syndrome, however, is rare in humans, and the following description refers to the original experimental findings. The most striking behavioral characteristics of this syndrome are: lack of fear and anger in previously wild animals, docility, changes in feeding behavior, sexual abnormalities, excessive oral curiosity in examining objects, and visual agnosia (the inability to recognize objects).

Note that the clinical case at the beginning of the chapter refers to a patient whose symptoms included anxiety attacks, irritability, odd behavior, loss of responsiveness, and convulsive activity.

1 Which part of the brain has a low threshold for epileptic seizures?

SYNONYMS AND EPONYMS OF THE LIMBIC SYSTEM

Name of structure or term	Synonym(s)/eponym(s)
Amnestic confabulatory syndrome	Korsakoff syndrome
	Korsakoff psychosis
Amygdalofugal pathway	Amygdaloid pathway
Amygdaloid body	Amygdala
Archicortex	Archipallium
	Allocortex
Commissure of the fornix	Hippocampal commissure
Habenulointerpeduncular tract	Tractus retroflexus
Hippocampus proper	Hippocampus
	Ammon's horn
	Cornu ammonis
Mammillothalamic fasciculus	Mammillary fasciculus
Mesocortex	Juxtallocortex
Neocortex	Neopallium
	Isocortex
	Homogenetic cortex
Paleocortex	Paleopallium
	Periallocortex
Stria medullaris of prethalamus	Stria medullaris thalami
Subiculum	Subicular cortex
Telencephalon	Forebrain

FOLLOW-UP TO CLINICAL CASE

This patient initially had symptoms of a psychiatric disturbance and seemed to display some elements of psychosis. However, M.S. then had another episode that began like the previous attacks but then clearly became a seizure. The knowledge that M.S. had a seizure shifts the whole scheme of diagnostic possibilities. A seizure work-up was performed, including an electroencephalogram (EEG) and magnetic resonance imaging (MRI) of the brain.

The EEG showed epileptiform activity coming from the left anterior temporal region. The MRI revealed an astrocytoma in the left anteromedial temporal lobe. The patient underwent resection of the tumor. There were subsequently no more seizures and many, but not all, of M.S.s other psychological symptoms improved.

Psychological symptoms are very common in epilepsy. They often occur as an aura preceding the onset of more characteristic manifestations of seizure. However, the aura in this case is actually the beginning of the seizure. Fear, anxiety, anger, disorientation, bizarre behavior, altered perception, a feeling of *déjà vu*, and agitation are some of the more common manifestations. Psychological or emotional symptoms are associated with temporal lobe seizures, especially involving the anteromedial temporal lobe, and sometimes frontal lobe seizures. In fact, the most epileptogenic regions of the brain are the temporal and frontal lobes, especially the limbic regions. Thus, epilepsy with emotional or psychological manifestations is occasionally confused with psychiatric disease.

Patients with epilepsy, especially temporal lobe epilepsy, sometimes display bizarre behaviors or personality, or have psychological disturbances even between the seizures. This can occur in patients with lesions, such as tumors, in specific regions of the brain or even in patients without such lesions. This leads to the hypothesis that the temporal lobe, specifically the medial temporal lobe, may be structurally or physiologically abnormal even if no observable lesion is detected there.

Very-fine-resolution MRI imaging of the temporal lobes, using coronal images, often shows medial temporal sclerosis in those with chronic temporal lobe epilepsy. This shows up as subtle scarring and loss of volume of the parahippocampal gyri. There is a loss of neurons and gliosis in the hippocampus and to a variable degree in the surrounding structures.

2 Why are psychological symptoms common in epilepsy?

3 What does fine-resolution MRI imaging of the temporal lobe of patients with chronic temporal lobe epilepsy show?

QUESTIONS TO PONDER

1. Distinguish between the limbic lobe and the limbic system.

2. What are the components of the hippocampal formation?

3. What causes the dentate gyrus to appear "tooth-shaped"?

4. Where do the cell bodies of the output neurons of the dentate gyrus reside?

5. What is the source of the most prominent input to the dentate gyrus?

6. What is the source of the efferent fibers of the hippocampus proper?

7. What is the function of the hypothalamus in the expression of emotions?

8. What is the main output pathway of the hippocampal formation?

9. What is the main cortical input pathway to the hippocampal formation?

10. Describe the cause and symptoms of Alzheimer disease.

CHAPTER 23

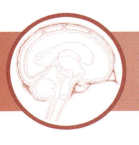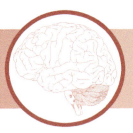

Hypothalamus

- CLINICAL CASE
- BORDERS
- HYPOTHALAMIC ZONES AND COMPONENT NUCLEI
- HYPOTHALAMIC REGIONS (AREAS) AND COMPONENT NUCLEI
- CONNECTIONS OF THE HYPOTHALAMUS
- PATHWAYS OF THE HYPOTHALAMUS
- FUNCTIONS OF THE HYPOTHALAMUS
- HYPOTHALAMOHYPOPHYSEAL CONNECTIONS
- CLINICAL CONSIDERATIONS
- SYNONYMS AND EPONYMS OF THE HYPOTHALAMUS
- FOLLOW-UP TO CLINICAL CASE
- QUESTIONS TO PONDER

CLINICAL CASE

A 23-year-old patient, F. G., describes daytime sleepiness and poor sleep at night. Sometimes F.G. has an uncontrollable urge to sleep during the daytime. When F.G. feels sleepy and then takes 5–15-minute naps, they feel somewhat refreshed upon awakening. These episodes occur a few times per week. F.G. has also noticed that occasionally their body feels somewhat limp and their head drops. These episodes last only a few seconds and seem to be prompted by laughing, getting very angry, or excited. A few times F.G. has noticed brief episodes of complete paralysis while lying in bed at night prior to falling asleep. Sometimes F.G. has strange and occasionally frightening visual hallucinations before drifting off to sleep at night. F.G. usually wakes up a few times each night. F.G.'s most concerning symptom is the daytime sleepiness, which interferes with their ability to concentrate. These symptoms began several years ago. Family history is negative. F.G. is not taking any medications or drugs.

The general physical and neurologic exam is normal. MRI of the brain is normal.

The **hypothalamus**, a subdivision of the diencephalon, weighs approximately 4 g, and forms less than 1% of the brain's total weight. Although this small region of the brain may seem inconsequential, it is, on the contrary, the center of numerous, and various, very important functions associated with the survival of the organism and continuation of the species.

The hypothalamic functions that are crucial to the survival of the organism include the control of appetite, fluid balance, electrolyte balance, glucose concentration, metabolism, sleeping, and body temperature regulation. In addition, hypothalamic function is associated with the continuation of the species, since it plays a role in gonadal function. The hypothalamus mediates these functions by integrating the

---

A Textbook of Neuroanatomy, Third Edition. Maria A. Patestas, Amanda J. Meyer, and Leslie P. Gartner.
© 2025 John Wiley & Sons, Inc. Published 2025 by John Wiley & Sons, Inc.
Companion website: www.wiley.com.go/patestas/neuroanatomy3e

tasks of the endocrine, autonomic (visceral motor), somatic motor, and limbic systems.

The hypothalamus maintains **homeostasis**, a state of constant internal environment (physiological equilibrium) and responds to both neural and non-neural stimulations. The hypothalamus consists of a group of nuclei each playing a role in one or more of the above-mentioned functions. Some nuclei consist of neurons that receive and respond to neural input from widespread areas of the nervous system. Other nuclei consist of neurons that respond to non-neural input, such as fluctuations in the temperature, osmotic pressure, and hormone levels of the circulating blood. During stressful situations, however, this state of equilibrium may be disturbed. The body possesses receptors that detect changes in various physiological systems and the hypothalamus, through its connections with the endocrine, autonomic, and limbic systems, mediates corrective mechanisms that compensate for the imbalance, and aid the body in restoring physiological equilibrium. Furthermore, through its extensive interconnections with the limbic system, the hypothalamus plays an important role in memory and emotional behavior, and mediates the appropriate visceral responses, such as changes in heart rate and blood pressure, accompanying behavioral expression. For example, an individual who is feeling anxious, experiences an increase in heart rate and breathing.

The various hypothalamic nuclei integrate the functions of the autonomic, endocrine, and limbic systems. The allocation of a particular function to a particular nucleus is a complex task since some of the nuclei have more than a single function and because the functions of some of the nuclei are not well understood. Furthermore, a particular region of the hypothalamus may contain nuclei that function in concert to produce a particular effect.

BORDERS

The hypothalamus is located anterior and inferior to the thalamus, surrounding the narrow anterior portion of the third ventricle

The following structures form the borders of the hypothalamus (Figs. 23.1 and 23.2): *anteriorly*, from superior to inferior, the cerebral anterior commissure, lamina terminalis, and optic chiasm; *posteriorly*, the interpeduncular fossa; *superiorly*, the hypothalamic sulcus; *inferiorly*, the tuber cinereum (L., "gray swelling"); *medially*, the third ventricle; and *laterally*, the innominate substance, subthalamic nucleus, and internal capsule.

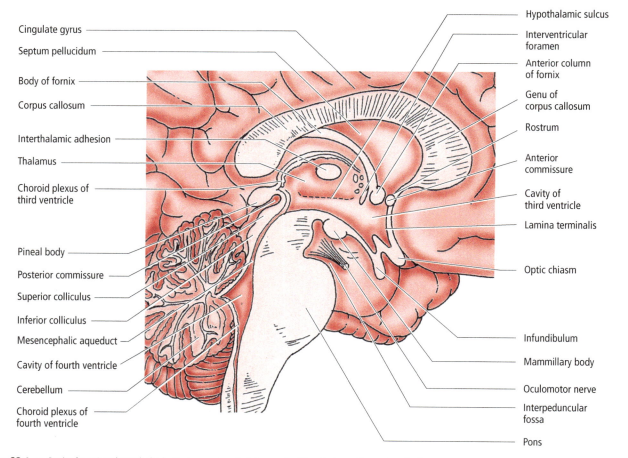

Figure 23.1 ● Sagittal section through the brainstem and part of the cerebral hemispheres illustrating the hypothalamus and its neighboring structures, the diencephalon and midbrain. The hypophysis is absent from this illustration as it is retained in the skull.

HYPOTHALAMUS

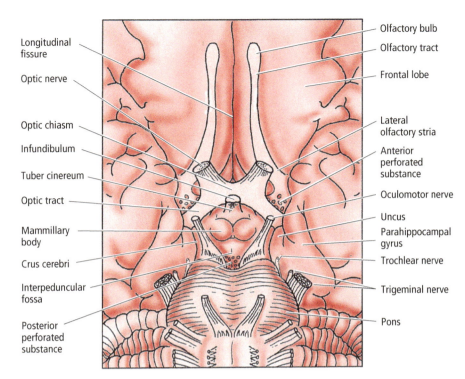

Figure 23.2 ● Schematic ventral view of the brain illustrating the hypothalamus and surrounding structures.

HYPOTHALAMIC ZONES AND COMPONENT NUCLEI

The hypothalamus is divided *sagittally* from medial to lateral into three zones: the **periventricular**, **medial**, and **lateral zones** for descriptive purposes (Table 23.1). Two prominent, heavily myelinated fiber bundles associated with the hypothalamus – the **mammillothalamic fasciculus** and the column of the **fornix** – form a conspicuous boundary between its medial and lateral zones (Fig. 23.3).

Periventricular zone

The periventricular zone is located next to the midline and consists of numerous small nuclei

The **periventricular zone** is located next to the midline, immediately deep to the ependymal layer that lines the third ventricle and consists of numerous, rather small, and difficult to delineate nuclei that line the wall of the third ventricle along the anterior–posterior extent of the hypothalamus. The nuclei of this zone are the **anterior periventricular**, **suprachiasmatic**, and **arcuate**

Table 23.1 ● Hypothalamic zones, regions, and component nuclei.

	Anterior region		Intermediate region	Posterior region
Zone	**Preoptic region**	**Supraoptic (chiasmatic) region**	**Infundibular (tuberal) region**	**Mammillary region**
Periventricular	Preoptic nucleus Anterior periventricular nuclei	Suprachiasmatic nucleus Anterior periventricular nuclei	Arcuate nucleus	
Medial	Medial preoptic nucleus	Anterior hypothalamic nucleus	Dorsomedial hypothalamic nucleus	Mammillary nuclei: Dorsal premammillary nucleus Ventral premammillary nucleus Tuberomammillary nucleus Lateral mammillary nucleus Medial mammillary nucleus Supramammillary nucleus
		Paraventricular nucleus Supraoptic nucleus	Ventromedial nucleus of hypothalamus	Posterior hypothalamic nucleus
Lateral	Lateral preoptic nucleus	Lateral hypothalamic nucleus	Lateral tuberal nuclei Lateral hypothalamic nucleus	Lateral hypothalamic nucleus

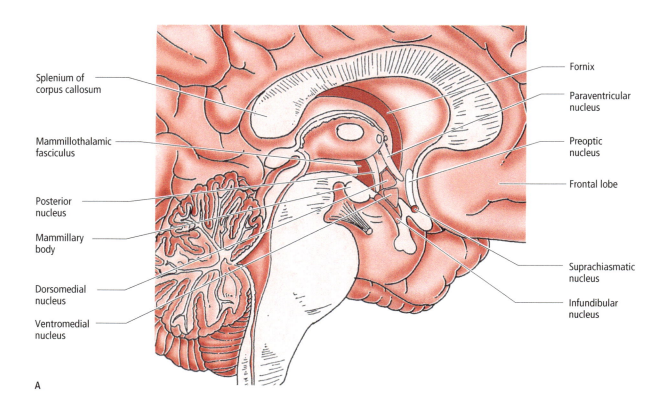

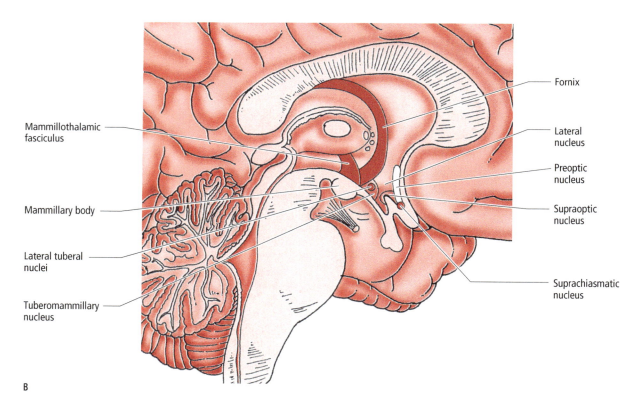

Figure 23.3 ● Schematic paramedian sagittal section of the brainstem illustrating the hypothalamic nuclei. (A) Nuclei of the medial zone of the hypothalamus positioned medial to the fornix and mammillothalamic fasciculus. (B) Nuclei of the lateral zone of the hypothalamus positioned lateral to the fornix and the mammillothalamic fasciculus.

HYPOTHALAMUS ● ● ● 457

nuclei (Figs. 23.4B and 23.5). The anterior periventricular and arcuate nuclei control the functions of the **adenohypophysis** (anterior pituitary gland), which synthesizes and releases various endocrine hormones into the bloodstream.

The suprachiasmatic nucleus functions in the control of circadian rhythms and is referred to as the "master clock" of the body.

Medial zone

The medial zone is the intermediate zone and includes numerous, well-defined nuclei that are involved in various functions

The nuclei located in this zone, listed in an approximate anterior to posterior sequence, are the **medial preoptic, anterior hypothalamic, paraventricular, supraoptic, dorsomedial hypothalamic, ventromedial hypothalamic, premammillary** (dorsal, ventral), **tuberomammillary, supramammillary, mammillary** (medial, lateral), and **posterior hypothalamic nuclei** (Figs. 23.4–23.7). The nuclei in the medial zone control the functions of the autonomic nervous system and the secretory activity of the **neurohypophysis** (posterior pituitary gland).

Lateral zone

The lateral zone receives input from the limbic system and plays an important role in the behavioral expression of emotions

The lateral zone contains the **lateral preoptic nucleus** anteriorly, the **lateral tuberal nuclei** in the tuberal region, and the **lateral hypothalamic nucleus**, which extends throughout the anterior-posterior extent of the lateral zone (Figs. 23.4–23.7). In addition, this area also contains the longitudinally oriented fibers of the *medial forebrain bundle*. This zone contains nuclei that receive information from the limbic system and mammillary bodies and then relay this input to other hypothalamic nuclei and the brainstem. This zone plays an important role in the behavioral expression of emotions.

HYPOTHALAMIC REGIONS (AREAS) AND COMPONENT NUCLEI

The hypothalamus is divided by the optic chiasm, tuber cinereum, and mammillary bodies into four regions in an anterior-posterior sequence

In general, the hypothalamic nuclei are often difficult to identify since their boundaries are not sharply delineated from adjacent nuclei, as they are in other parts of the central nervous system (CNS). In order to define their location more precisely, the hypothalamus is, in addition to being divided sagittally into three mediolateral zones, further divided coronally into four regions (areas) in an anterior-posterior sequence. These four regions, which are located near three prominent anatomical structures visible on the ventral surface of the brain (the optic chiasm, tuber cinereum, and mammillary bodies), are the **preoptic region, supraoptic (chiasmatic or anterior) region, tuberal (infundibular or middle) region**, and **mammillary (posterior) region**. Each of these regions contains nuclei as described next; however, some nuclei may extend into the adjacent zones and regions (see Table 23.1).

Preoptic region

The preoptic region controls the release of gonadotropic hormones from the adenohypophysis

The **preoptic region**, although located at least in part anterior to the lamina terminalis in the telencephalon, has been accepted as the anterior telencephalic portion of the hypothalamus (which is a diencephalic structure). It consists of gray matter located in the anterior most extent of the third ventricle, both anterior and posterior to the plane of the lamina terminalis. Due to the similarities of its connections and functions to those of the anterior part of the medial region of the hypothalamus, the preoptic area is grouped with the anterior hypothalamus.

The preoptic area extends into the periventricular, medial, and lateral zones. It contains the **preoptic** and **anterior periventricular nuclei** in the periventricular zone, the **medial preoptic nucleus** in the medial zone, and the **lateral preoptic nucleus** in the lateral zone. The preoptic area contains a heterogenous assembly of neurons which have a wide variety of functions, including regulating gonadal function, sleep, and thermoregulation.

The preoptic region controls the release of gonadal hormones from the adenohypophysis. There is sexual dimorphism evident in this region, reflecting the different function it has in biological males and females. The medial preoptic nucleus is more prominent in the male. In the female, the adenohypophysis release of gonadotropin hormones is cyclic that ultimately regulates the menstrual cycle, whereas in the male, the adenohypophysis releases gonadotropins continually. The medial preoptic nucleus synthesizes and releases gonadotropin-releasing hormone (Gn-RH) which is delivered to the adenohypophysis via the tuberoinfundibular tract. Gn-RH stimulates the release of follicle-stimulating hormone (FSH) and luteinizing hormone (LH). The function of the lateral preoptic nucleus is not well understood, but its connections to the ventral pallidum may indicate a role in the control of locomotion.

The preoptic area, important for the promotion of sleep, projects GABAergic neurons to arousal centers in the brain including the tuberomammillary nucleus, raphe nucleus, and locus caeruleus. Neuromodulatory transmission from these arousal centers is silenced during REM sleep, due to projections from the preoptic areas. Lesions of the preoptic area lead to insomnia.

The preoptic area also plays a key role in the integration of peripheral and central temperature information. Warm thermal information is conveyed from the spinal ganglion to the lateral parabrachial nucleus in the pons and then to warm-sensitive neurons in the preoptic area. Heat-loss mechanisms

458 ••• CHAPTER 23

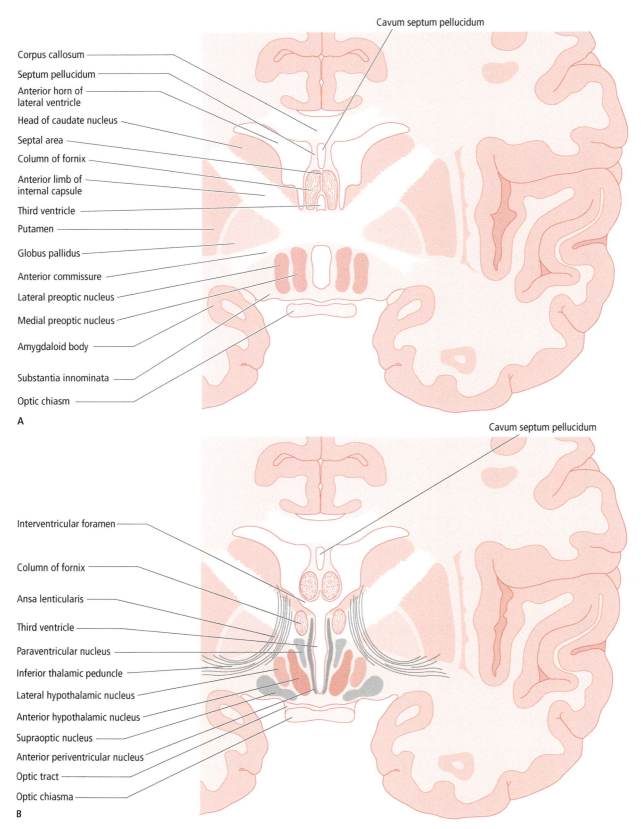

Figure 23.4 • Coronal sections through the hypothalamus at the level of: (A) the preoptic region and (B) the supraoptic region. These figures are based on an individual with a cavum septum pellucidum.

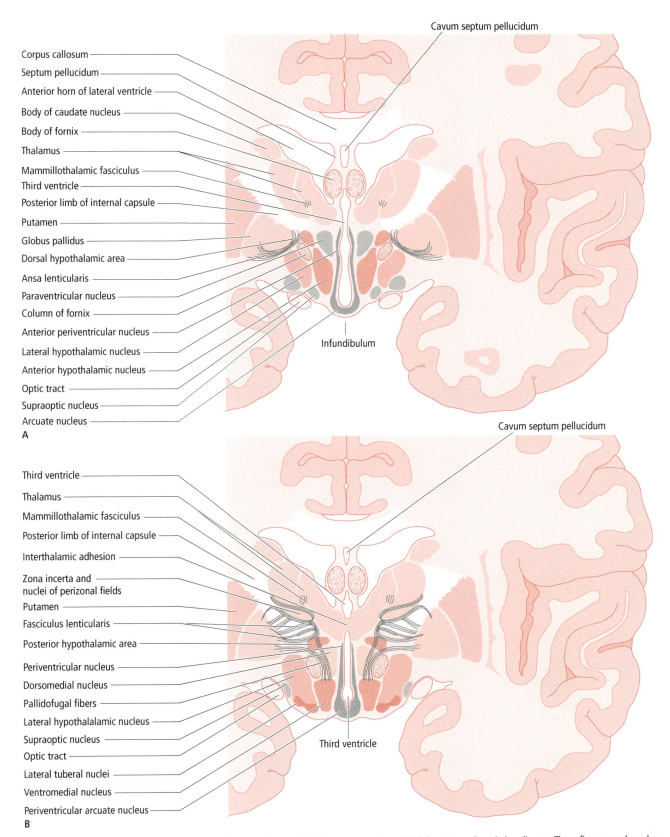

Figure 23.5 ● Coronal sections through the hypothalamus at the level of: (A) the supraoptic and (B) the intermediate (tuberal) areas. These figures are based on an individual with a cavum septum pellucidum.

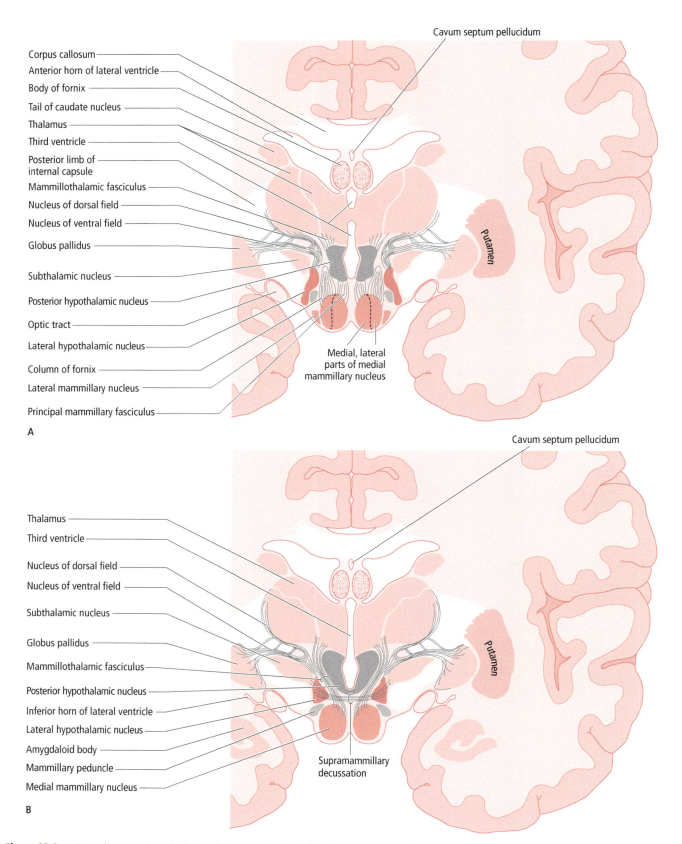

Figure 23.6 ● Coronal sections through the hypothalamus at the level of: (A) the anterior mammillary region and (B) the posterior mammillary region. These figures are based on an individual with a cavum septum pellucidum.

HYPOTHALAMUS

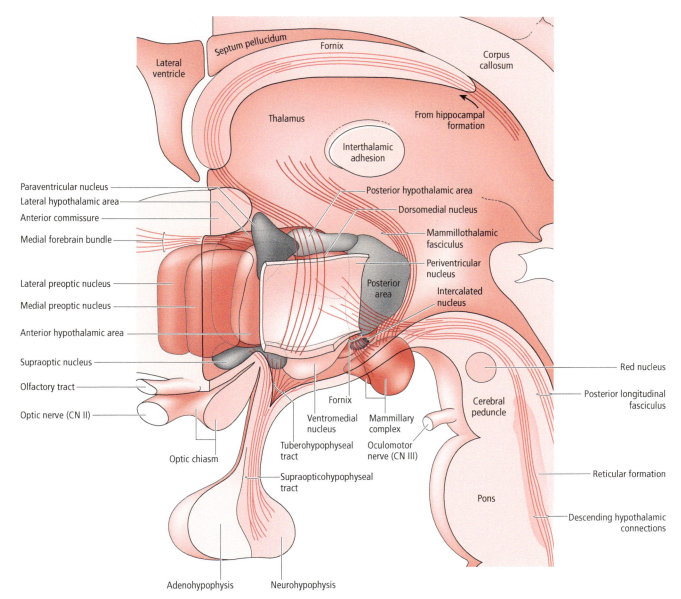

Figure 23.7 ● Schematic diagram of the hypothalamus illustrating its component nuclei. Source: Modified from Netter, NH (1983) *The CIBA Collection of Medical Illustrations*. Vol. 1, part 1. CIBA, New Jersey; p. 207, plate 55.

such as sweating, increased respiration, and inhibition of thermogenesis in brown adipose tissue are triggered in the preoptic area.

Supraoptic (anterior, chiasmatic) region

The supraoptic region contains nuclei that function in the control of circadian rhythms, temperature regulation, and control of water balance

The **supraoptic region,** located superior to the optic chiasm, is continuous anteriorly with the preoptic area and contains six nuclei. They are, from medial to lateral, the suprachiasmatic and anterior periventricular nuclei in the periventricular zone, the anterior hypothalamic, paraventricular, and supraoptic nuclei in the medial zone, and the lateral hypothalamic nucleus in the lateral zone.

The **suprachiasmatic nucleus** is a small nucleus located next to the midline, superior to the optic chiasm, adjacent to the third ventricle. This nucleus receives visual information via axons arising from the retina (retinohypothalamic fibers) as they course through the optic chiasm. The neurons in this nucleus respond to the information from the photoreceptors, conveyed by the ganglionic layer of the retina, that respond to the onset and offset of light. The suprachiasmatic nucleus functions in the *control of circadian rhythms* and is referred to as the "master clock" of the body.

462 ••• **CHAPTER 23**

The **anterior hypothalamic nucleus** (see Fig. 23.4B) is located superior to the supraoptic nucleus and is involved in *temperature regulation*.

The **paraventricular** and **supraoptic nuclei** (see Figs. 23.4B and 23.5A) have the most abundant blood supply in the brain. Some of the neurons of these two conspicuous nuclei synthesize and release the hormone called vasopressin or antidiuretic hormone (ADH), whereas others synthesize and secrete the hormone oxytocin. These hormones are transported via axons arising from neurons housed in these hypothalamic nuclei down to the neurohypophysis. These unmyelinated axons gather to form the **supraopticohypophyseal tract**, which courses in the infundibular stalk and then terminates in the neurohypophysis. The paraventricular nucleus *controls water balance* by functioning in the conservation of water.

In addition to the magnocellular cell group, which projects to the neurohypophysis, the paraventricular nucleus also houses other groups of neurons. One of these groups of cells, the parvocellular group, projects to the median eminence, whereas another group of cells projects to the reticular formation and the autonomic centers of the brainstem and spinal cord.

Tuberal (intermediate, middle) region

The tuberal region contains nuclei that are involved in the control of adenohypophyseal hormone release, caloric intake, and appetite

The **tuberal region** is located posterior to the tuber cinereum and contains the following main nuclei: the arcuate nucleus in the periventricular zone, the dorsomedial and ventromedial nuclei in the medial zone, and the lateral hypothalamic nucleus in the lateral zone.

The **arcuate nucleus** is located in the periventricular zone, in the tuber cinereum, arching under the anterior portion of the third ventricle. The cells of this nucleus produce *hypothalamic-releasing hormones*, and their axons form the **tuberohypophyseal** tract that carries these hormones to the **infundibulum**, where they are delivered into the **hypophyseal portal system** from which the hormones are released into the general circulation.

The **dorsomedial hypothalamic nucleus**, when stimulated, produces aggressive behavior in animals and is believed to be involved in emotional behavior.

The **ventromedial nucleus of the hypothalamus** is a "satiety center," which upon stimulation results in satiety. This nucleus contains cells that are sensitive to blood glucose levels and which are believed to monitor *caloric intake*. Stimulation of the ventromedial nucleus influences metabolic processes by facilitating the release of glucagon and elevates glycogenolysis, gluconeogenesis, and lipolysis. This nucleus receives visceral afferents from the **nucleus of the solitary tract**.

The **lateral hypothalamic nucleus** is involved in the *control of appetite* and is referred to as a "feeding center." When stimulated, it induces eating, release of insulin, and results in a decrease in glycogenolysis, gluconeogenesis, and lipolysis.

Mammillary (posterior) region

The mammillary nuclei are associated with emotions, whereas the posterior hypothalamic nucleus serves to regulate heat-gaining behaviors

The **mammillary region** includes the prominent mammillary nuclei and the posterior hypothalamic nucleus. The three **mammillary nuclei** (**medial**, **tubero-(intermediate)**, and **lateral**) and the **posterior hypothalamic nucleus** collectively form the **mammillary region**. The mammillary bodies appear as two conspicuous elevations on the inferoposterior part of the hypothalamus. Unlike the other hypothalamic nuclei, which are associated with functions of the endocrine and/or autonomic nervous systems, the mammillary nuclei are a major target projection area of the **hippocampal formation** via the **fornix**, relaying input related to *emotions*. In addition, information is relayed to the mammillary nuclei via the **mammillary peduncle** from the **dorsal and ventral tegmental nuclei** as well as from the **raphe nuclei**. The mammillary nucleus relays input to the anterior nucleus of the thalamus via the prominent **mammillothalamic fasciculus.** The thalamus in turn projects to the **cingulate gyrus** of the limbic cortex.

The mammillary bodies are also implicated in the conversion of short-term memory into *long-term memory*. A lesion damaging the mammillary nuclei has no effect on the recall of events that occurred months or even years prior to the lesion, but the affected individual has difficulty remembering events and information that occurred after the occurrence of the lesion.

Neurons projecting from the mammillary region are also involved in pain modulation. Histaminergic tuberomammillary neurons, the sole source of histamine in the brain, project to the dorsal raphe, periaqueductal gray, and posterior horn spinal cord to alter pain perception. They also are also involved in wakefulness with histamine binding to H1 receptors triggering cortical arousal, whereas histamine binding to H3 receptors leads to drowsiness. In fact, patients with excessive daytime sleepiness (known as narcolepsy) are treated with inverse agonists which have a higher affinity for uncoupled H3 receptors. Therefore, the **tuberomammillary neurons** are important in *arousal and analgesia*.

The **posterior hypothalamic nucleus** contains neurons that are sensitive to a *decrease* in blood temperature. This nucleus serves as a "thermostat" and *regulates body temperature* by conserving heat and stimulating heat production. Heat is conserved by vasoconstriction of cutaneous vessels, whereas heat is produced by an increase in thyroid activity, activation of thermogenesis of brown adipose tissue, and unconscious rapid repetitive muscular contractions (as in shivering).

CONNECTIONS OF THE HYPOTHALAMUS

The hypothalamus is connected with widespread regions of the nervous system. It receives input related to emotion from the limbic system, as well as input from sensory and motor nuclei from the brainstem and spinal cord. The hypothalamus exerts its influence via its outputs on the endocrine and autonomic nervous systems.

Afferents (input) to the hypothalamus

Most neural input to the hypothalamus arises from the limbic system. Non-neural (i.e., hormonal, chemical) input is carried to the hypothalamus via the vascular system

The hypothalamus receives **afferent fibers** (neural input) from various structures, although most neural input arises from the limbic system. Other fibers arise from visceral structures and the brainstem, specifically from the following sources: the orbitofrontal cortex, basal forebrain, septal area and septal nuclei, ventral striatum, midline and medial dorsal nucleus of the thalamus, hippocampus, amygdaloid body, retina, dorsal and ventral tegmental nuclei, raphe nuclei, locus caeruleus, spinal cord, and reticular formation.

In addition to the neural input, the hypothalamus also receives non-neural (hormonal, chemical) input from the vascular system. Hypothalamic neurons serve as receptors that are sensitive to rapid changes in the temperature, osmotic pressure, and hormone levels of the circulating blood. These receptors are referred to as circumventricular organs (CVOs). Although there are six circumventricular organs, only two are associated with the hypothalamus. They consist of hypothalamic neurons that are embedded in the walls of the third ventricle and are sensitive to alterations in the chemical composition of the cerebrospinal fluid. One of these structures is associated with the lamina terminalis and is referred to as the **vascular organ of the lamina terminalis**, a chemosensory structure that detects and responds to peptides and macromolecules present in the bloodstream. The other CVO is the **subfornical organ** located inferior to the fornix, near the interventricular foramen. The subfornical organ serves in the regulation of body fluids. The CVOs consist of networks of fenestrated capillaries; thus, they permit substances present in the bloodstream to pass through the capillary endothelium and enter the extracellular space surrounding the receptor cells. These vessels are also covered by specialized ependymal cells, called **tanycytes**, which only provide a partial barrier between the cerebrospinal fluid and the CVOs. Therefore, the CVOs are strategically placed and composed to sample blood and cerebrospinal fluid and to activate hypothalamic efferents to affect behavior.

Efferents (output) from the hypothalamus

In addition to the neural output, the hypothalamus also provides output to the adenohypophysis and neurohypophysis

Most hypothalamic output projections terminate in the sources of hypothalamic input. The hypothalamus projects **efferent fibers** (neural output) to the following structures: the septal area and septal nuclei, medial dorsal nucleus of the thalamus, anterior nucleus of the thalamus, hippocampus, amygdaloid body, midbrain reticular formation, and brainstem and spinal cord autonomic nuclei.

In addition to the neural output, the hypothalamus also provides output to the adenohypophysis and neurohypophysis. The hypothalamus exerts its influence on the endocrine system via these two projections to the hypophysis.

Hypothalamic influence is exerted on the endocrine system directly, via the magnocellular neurosecretory projection fibers of the **supraopticohypophyseal tract**, which terminates in the neurohypophysis. There, these neurons release hormones that pass into the capillaries of the neurohypophysis and then enter the bloodstream.

The hypothalamus also exerts its influence on the endocrine system indirectly, via the parvocellular neurosecretory projections which deliver releasing hormones to the local capillary networks (**primary capillary plexus**) of the **hypophyseal portal system**, which carries hormones synthesized by the hypothalamus to the **secondary capillary plexus** of the adenohypophysis where they leave the vascular system to bind to receptors of their target cells.

PATHWAYS OF THE HYPOTHALAMUS

The pathways connecting the hypothalamus with other regions of the nervous system are categorized into afferent or efferent pathways based on the direction of most of the information that they carry. However, *none* of them carry exclusively afferent or efferent fibers since they all carry information that is *bidirectional* (Tables 23.2–23.4).

Afferent pathways to the hypothalamus

See Table 23.2.

Fornix

The fornix is the most prominent neural input to the hypothalamus

The **fornix** is a myelinated tract that arises from the hippocampal formation of the limbic system and conveys the *most prominent neural input to the hypothalamus*. It distributes fibers to the septal and preoptic nuclei, the anterior areas of the hypothalamus, and then proceeds through the hypothalamus to terminate mostly in the medial nucleus of the mammillary body posteriorly.

Mammillary peduncle

The **mammillary peduncle** carries prominent sensory input from the dorsal and ventral tegmental nuclei of the mesencephalon as well as from the raphe nuclei to the lateral mammillary nucleus.

Stria terminalis

The **stria terminalis** transmits information from the olfactory portion of the amygdaloid body to the septal nuclei, medial preoptic area, and the anterior nucleus of the hypothalamus (of the medial hypothalamic zone). This pathway transmits olfactory information, which is believed to affect sexual behavior.

464 ● ● ● **CHAPTER 23**

Table 23.2 ● Afferent pathways to the hypothalamus.

Tract/pathway	Origin	Termination	Function
Fornix	Hippocampal formation	Preoptic and anterior areas of the hypothalamus	
		Medial mammillary nucleus Septal nuclei	
Mammillary peduncle	Dorsal and ventral tegmental nuclei Raphe nuclei	Lateral mammillary nucleus	Relays sensory input from sensory pathways
Stria terminalis	Amygdaloid body	Medial preoptic area Anterior hypothalamic nucleus Septal nuclei	Olfactory information, which influences sexual behavior
Thalamohypothalamic tract	Medial dorsal and midline nuclei of the thalamus	Lateral preoptic area of the hypothalamus	
Ventral amygdalofugal tract	Basolateral nucleus of the amygdaloid body	Lateral hypothalamic nucleus	Influences autonomic nervous system activities
Retinosuprachiasmatic tract	Retina	Suprachiasmatic nucleus	Control of circadian rhythms
Spinohypothalamic tract	Spinal cord	Autonomic control centers of the hypothalamus	Neuroendocrine and cardiovascular responses
Fibers from the reticular formation	Reticular formation	Autonomic control centers of the hypothalamus	Relay nociceptive input to the hypothalamus

Thalamohypothalamic tract

The **thalamohypothalamic tract** carries fibers from the medial dorsal and midline nuclei of the thalamus to the lateral preoptic area of the hypothalamus.

Ventral amygdalofugal tract

The **ventral amygdalofugal tract** relays amygdaloid signals from the basolateral nucleus to the lateral hypothalamic nucleus where they influence autonomic nervous system activities.

Retinosuprachiasmatic tract

The **retinosuprachiasmatic tract** consists of a bundle of fibers arising from the retina that terminates in the suprachiasmatic nucleus of the hypothalamus. It is involved in the control of circadian rhythms.

Spinohypothalamic fibers

The spinohypothalamic fibers relay nociceptive input from the spinal cord to the autonomic control centers of the hypothalamus

The **spinohypothalamic fibers**, a component of the anterolateral system of the ascending sensory pathways, consist of axons arising from the spinal cord that relay nociceptive input to the autonomic control centers of the hypothalamus. There

they synapse with neurons that give rise to the **hypothalamospinal tract**, which is associated with the autonomic and reflex responses (i.e., endocrine and cardiovascular responses) to nociception.

Fibers from the reticular formation

The ascending reticular activating system (ARAS) relays nociceptive input to the hypothalamus, which mediates the autonomic and reflex responses to nociception

Most of the nociceptive fibers from the spinal cord ascending in the anterolateral system of the ascending sensory pathways, and the fibers of the trigeminothalamic tract, send collateral branches to the reticular formation as they ascend through the brainstem. These collateral branches synapse in the reticular formation and activate the **ascending reticular activating system (ARAS)**, which relays information to the cerebral cortex alerting the individual, but also relays information to the hypothalamus, limbic system, and serotonergic raphe nucleus magnus. The hypothalamus mediates the autonomic and reflex responses to nociception, the limbic system mediates the emotional/behavioral responses to nociception, and the raphe nucleus is involved in descending control of pain.

Efferent pathways from the hypothalamus

See Table 23.3.

HYPOTHALAMUS ● ● ● 465

Table 23.3 ● Efferent pathways of the hypothalamus.

Tract/pathway	Origin	Termination	Function
Principal mammillary fasciculus	Mammillary nuclei	Anterior nucleus of the thalamus	
Mammillothalamic fasciculus	Medial and lateral mammillary nuclei		
Mammillotegmental fasciculus	Lateral mammillary nuclei	Dorsal and ventral tegmental nuclei	Influences ANS activity
Mammillointerpeduncular tract	Mammillary nucleus	Interpeduncular nucleus	
Hypothalamothalamic fibers	Lateral preoptic area	Dorsomedial nucleus of the thalamus	
Supraopticohypophyseal tract	Supraoptic and paraventricular nuclei	Neurohypophysis	Releases hormones that pass into the capillaries of the neurohypophysis and then enter the bloodstream
Periventricular bundle	Periventricular zone nuclei	Frontal cortex	Stimulate ANS responses
		Medulla oblongata	
		Spinal cord	
Tuberohypophyseal (tuberoinfundibular, infundibular) tract	Arcuate and anterior periventricular nuclei	Infundibular stalk	Regulates the synthesis and release of adenohypophyseal hormones
Posterior longitudinal fasciculus (PLF)	Hypothalamus	Brainstem and spinal cord ANS nuclei and reticular formation	Control of skeletal muscles involved in chewing, swallowing, and shivering
Hypothalamospinal tract	Paraventricular, dorsomedial, ventromedial, and posterior nuclei	Brainstem	Influences ANS activity
		Spinal cord lateral cell column	
		Sympathetic and parasympathetic (sacral) nuclei	
ANS, autonomic nervous system.			

Principal mammillary fasciculus

The **principal mammillary fasciculus** arises from the mammillary body and then quickly bifurcates to give rise to the **mammillothalamic** and **mammillotegmental fasciculi.**

Mammillothalamic fasciculus

The mammillothalamic fasciculus (eponym: tract of Vicq d'Azyr) is a myelinated bundle of fibers that arises from the medial and lateral mammillary nuclei and terminates in the anterior nucleus of the thalamus.

Mammillotegmental fasciculus

The mammillotegmental fasciculus is a component of the medial forebrain bundle (MFB). It carries mammillothalamic fasciculus axon collaterals from the lateral mammillary nuclei to the midbrain dorsal and ventral tegmental nuclei in the reticular formation located in the periaqueductal gray (PAG) matter of the caudal midbrain. The PAG neurons then relay signals to the nucleus of the solitary tract and to the posterior nucleus of the vagus. The mammillotegmental fasciculus ultimately affects the ANS nuclei of the brainstem.

Mammillointerpeduncular tract

The **mammillointerpeduncular tract** carries fibers from the mammillary nucleus that terminate in the interpeduncular nucleus. The interpeduncular nucleus plays a role in the sleep–wake cycle.

Hypothalamothalamic fibers

Originate from the lateral preoptic area and terminate in the medial dorsal nucleus of the thalamus.

Supraopticohypophyseal tract

The supraopticohypophyseal tract carries oxytocin or vasopressin to the neurohypophysis

The **supraopticohypophyseal tract** is a short tract consisting of axons that carry oxytocin or vasopressin (ADH), synthesized in the cell bodies of hypothalamic neurons residing in the supraoptic and paraventricular nuclei, to the neurohypophysis.

Oxytocin and vasopressin are synthesized in different populations of hypothalamic neurosecretory neurons. The secretory substances are transported via these axons to the neurohypophysis. Upon stimulation, the hypothalamic neurosecretory cells transmit impulses to their terminals, which

466 ● ● ● **CHAPTER 23**

results in the release of the hormones into widespread capillaries in the neurohypophysis. These capillaries drain the hypophyseal veins which ultimately empty into the dural venous sinuses and leave the cranium via the internal jugular vein.

Periventricular bundle

The **periventricular bundle** consists of fibers that connect the periventricular zone nuclei with the frontal cortex, the posterior nucleus of the vagus in the medulla oblongata, the lateral cell column of the spinal cord from T1 to L2, and the sacral parasympathetic nucleus of the spinal cord from S2 to S4. The functions of the projections to the brainstem and spinal cord are autonomic system responses.

Tuberohypophyseal (tuberoinfundibular, infundibular) tract

The tuberohypophyseal tract carries fibers from the arcuate and anterior periventricular nuclei to the infundibular stalk and to the adenohypophysis

The tuberohypophyseal tract carries fibers from the arcuate and anterior periventricular nuclei to the infundibular stalk and to the adenohypophysis. The axon terminals of the **tuberohypophyseal tract** discharge "**releasing hormones**" or "**release-inhibiting hormones**" into the hypophyseal portal system, which delivers them to the adenohypophysis where they regulate the synthesis and release of hormones. The tuberohypophyseal tract and the hypophyseal portal system serve as a connection between the hypothalamus and adenohypophysis.

Posterior longitudinal fasciculus

The posterior longitudinal fasciculus carries fibers from the hypothalamus to the brainstem and spinal cord autonomic nuclei and to the reticular formation

The **posterior longitudinal fasciculus** carries fibers that connect various regions of the CNS. Among these fibers are some that arise from the medial zone of the hypothalamus and descend to terminate in the brainstem PAG, which in turn projects to the nucleus of the solitary tract and to the posterior nucleus of vagus. The posterior longitudinal fasciculus ultimately affects the ANS nuclei of the brainstem. In addition, this pathway carries signals from the hypothalamus to the reticular formation, which are then relayed via the reticulobulbar tract to the brainstem lower motor neurons, and via the reticulospinal tract to the spinal cord lower motor neurons. By way of the posterior longitudinal fasciculus, the hypothalamus ultimately influences chewing, swallowing, and shivering.

Hypothalamospinal tract

The hypothalamospinal tract is a direct route via which the hypothalamus influences pre-ganglionic sympathetic and parasympathetic neurons residing in the spinal cord

The **hypothalamospinal tract** carries fibers arising mainly from the paraventricular nucleus of the hypothalamus.

Additional fibers originate from the dorsomedial, ventromedial, and posterior nuclei. All of the fibers descend to terminate in the brainstem and spinal cord to synapse directly with pre-ganglionic sympathetic (lateral cell column) and parasympathetic (sacral) nuclei neurons. These direct projections to the spinal cord are small in number; the majority of the descending fibers from the hypothalamus that are destined to influence the autonomic centers are believed to synapse at multiple relay centers.

Bidirectional pathways

(Table 23.4)

Medial forebrain bundle

The medial forebrain bundle contains fibers that are believed to play a role in motivation and sense of smell

The **medial forebrain bundle** (MFB) is not really a distinct bundle as its name implies, but instead consists of a vast array of axons with various origins and terminations. It includes both afferent and efferent fibers to and from the hypothalamus. The afferent fibers to the hypothalamus arise primarily from the septal area. Other fibers arise from the basal forebrain and primary olfactory cortex, as well as from the brainstem raphe nuclei, ventral tegmental area, and locus caeruleus. Efferent fibers arising from the hypothalamus and running in the medial forebrain bundle terminate in the septal area, septal nuclei, and brainstem reticular formation and autonomic nuclei, as well as in the autonomic nuclei of the spinal cord. These fibers are believed to play a role in *motivation* and *sense of smell*. The medial forebrain bundle is more prominent in other animals with a keen sense of smell.

Stria medullaris of prethalamus

The **stria medullaris of prethalamus** (stria medullaris thalami) carries afferent and efferent fibers to and from the hypothalamus, connecting its supraoptic nucleus and pre-optic area with the habenula. The **habenula** is visible in a midsagittal section of the brain, in the medial surface of the thalamus near its caudal end. It appears as a small swelling anterior to the pineal gland, above the posterior commissure. The habenula plays a pivotal role in controlling motor and cognitive behaviors by influencing the activity of dopaminergic and serotonergic neurons.

FUNCTIONS OF THE HYPOTHALAMUS

The hypothalamus functions in the regulation of body temperature, food intake, fluid intake, and control of the autonomic nervous system (Table 23.5). It also plays a role in emotional expression, memory, and aggression.

Tract/pathway	Origin	Termination	Function
Medial forebrain bundle (MFB)	Afferents Septal area Basal forebrain Primary olfactory cortex Brainstem raphe nuclei Ventral tegmental area Locus caeruleus	Hypothalamus	Motivation and sense of smell
	Efferents Hypothalamus	Septal area Septal nuclei Brainstem reticular formation Autonomic nervous system nuclei	
Stria medullaris of prethalamus	Afferents Habenula	Hypothalamus Supraoptic nucleus Preoptic area	
	Efferents Hypothalamus Supraoptic nucleus Preoptic area	Habenula	

Table 23.4 ● Bidirectional pathways of the hypothalamus.

Nucleus	Function(s)
Preoptic	Controls the release of reproductive hormones from the adenohypophysis Promotes heat-loss behaviors Promotes sleep
Suprachiasmatic	Regulates circadian rhythms; "master clock"
Arcuate	Produces hypothalamic-releasing and -inhibiting hormones
Anterior Periventricular	Produces hypothalamic-releasing and -inhibiting hormones
Medial preoptic	Regulates the release of gonadal hormones from the adenohypophysis Promotes sleep
Anterior	Regulates parasympathetic nervous system activity Promotes heat-loss behaviors
Dorsomedial	Stimulation of this nucleus causes savage behavior in animals
Ventromedial	Involved in eating behavior; "satiety center"
Mammillary	Processes information related to emotional expression
Posterior	Regulates sympathetic nervous system activity Promotes heat-gain behaviors Promotes wakefulness
Lateral preoptic	Possible role in movement
Lateral	Regulates sympathetic nervous system Involved in eating behavior; "feeding center"
Paraventricular	Produces vasopressin/ADH[*] and oxytocin
Supraoptic	Produces vasopressin/ADH[*] and oxytocin

* ADH, antidiuretic hormone.

Table 23.5 ● Hypothalamic nuclei and their functions.

Regulation of body temperature

The hypothalamus serves as a "thermostat" and plays a central and vital role in body temperature regulation

Temperature-sensitive receptors (thermosensitive free nerve endings) located in the skin detect information about the environmental temperature. Other peripheral temperature-sensitive receptors (similar to those in the skin but located deep in the body) detect information about the body's core temperature. Both environmental and body core (visceral) temperature input is relayed to higher brain centers, including the hypothalamus, via the ascending sensory pathways.

In addition to the peripheral temperature-sensitive receptors, the **preoptic region** and **anterior hypothalamic nuclei** house two types of temperature-sensitive (thermosensitive) neurons that respond to very slight temperature changes (less than 0.1°C) of the blood. One group of these neurons becomes stimulated by increases in blood temperature, whereas another group of neurons becomes stimulated by decreases in blood temperature.

The hypothalamus is a well-vascularized brain center. As the circulating blood percolates in the capillary bed surrounding the thermosensitive preoptic and anterior hypothalamic neurons, these neurons become stimulated by warming or cooling of the blood. It is believed that the preoptic and anterior hypothalamic nuclei initiate *heat loss* mechanisms, whereas the **posterior hypothalamic nuclei** activate *heat gain (conservation and production)* mechanisms (Table 23.5). Thus, the hypothalamus serves as a "thermostat" that mediates mechanisms that enable the body to return to, and maintain, a normal core body temperature.

Temperature regulation involves hypothalamic integration of autonomic, endocrine, somatic motor, and limbic system activities. For example, in response to an increase in body temperature, the **anterior hypothalamus** triggers the appropriate body responses that will result in a decrease in body temperature. These responses include autonomic activities such as sweating and cutaneous vasodilation, and a decrease in somatic motor function. Furthermore, behavioral changes such as seeking a cooler environment or taking off a layer of clothes to reduce body temperature may be appropriate. The neurons that control the parasympathetic nervous system are located in the anterior nucleus of the hypothalamus to mediate the parasympathetic responses associated with decreasing body temperature.

In response to a decrease in body temperature, the **posterior hypothalamus** triggers the appropriate body responses that will produce an increase in body temperature. These responses include: (1) piloerection (formation of "goose bumps") and the vasoconstriction of cutaneous vessels, both of which are mediated by the autonomic nervous system to conserve body heat; (2) an increase in metabolism (via the release of thyroid-stimulating hormone (TSH) by the adenohypophysis), mediated by the endocrine system to increase body heat; (3) shivering, mediated by the somatic motor system to produce heat; and (4) behavioral (motivational) responses, such as turning on a heater, putting on warmer clothes, and drinking something warm, mediated by the limbic system to increase body temperature. Frontal lobe connections to the limbic system mediate behavioral changes to optimize conditions for homeostasis. The neurons that control the sympathetic nervous system are located in the posterior hypothalamic nucleus. These neurons send their axons to the lateral horn of the spinal cord to mediate the sympathetic responses associated with increasing body temperature.

Although body temperature may be regulated by the hypothalamus and associated physiological processes, the temperature range in which these processes are effective is rather *narrow*. For example, the hypothalamus cannot keep the body at a normal temperature range in an extremely cold environment.

Regulation of food intake and the sleep–wake cycle

The hypothalamus contains a "feeding center" and a "satiety center" and has a very important function in food intake regulation

In the past, the **lateral** and **ventromedial nuclei of the hypothalamus** were believed to be associated with eating behavior. Stimulation of the lateral hypothalamic nucleus in experimental animals induced eating, and it was referred to as a "hunger" or "feeding center." Stimulation of the ventromedial nucleus inhibited eating and was referred to as a "satiety center." It is now known that things are not as simple as were once believed. In addition to the lateral and ventromedial nuclei, the **paraventricular nucleus** also plays a role in the regulation of food intake (Table 23.5), possibly through its projections to the brainstem reticular formation, as well as to the brainstem and spinal cord autonomic nuclei. Food intake is regulated by a combination of central hormonal mechanisms and peripheral hormonal and neural mechanisms.

The lateral hypothalamic nucleus contains a group of neurons that synthesize **orexins** (G., *orexis*, "appetite") (also referred to as **hypocretins**), which are hypothalamic neuropeptides that not only regulate food intake but also the sleep–wake cycle. Fasting causes an *increase* in orexin synthesis, whereas administration of exogenous orexin induces eating. Orexin is also involved in the waking and sleep cycle, specifically, keeping an individual awake throughout the day. At night, hypothalamic orexin production decreases, contributing to sleep pressure.

Note that the clinical case at the beginning of the chapter refers to a patient whose symptoms include daytime sleepiness, an occasional limp feeling, complete paralysis, and visual hallucinations prior to falling asleep.

1   What is the neuropeptide synthesized by the hypothalamus that not only regulates food intake but also the sleep–wake cycle?

2   What effect do orexins have on food intake and the sleep–wake cycle?

Regulation of fluid intake

The hypothalamus contains neurons that play a role in water intake

A group of cells located lateral to the lateral nucleus of the hypothalamus, known as the **zona incerta**, is believed to play a role in water intake. Stimulation of the zona incerta results in the drinking of large amounts of water (known as polydipsia).

Control of the autonomic nervous system

The hypothalamus is the master controller of autonomic functions

The hypothalamus receives visceral input via the ascending sensory system and input related to emotions from the limbic system. The hypothalamus in turn exerts its influence on both the sympathetic and parasympathetic nervous systems, coordinating their functions in order to maintain homeostasis or to produce the appropriate visceral responses that accompany behavioral expression.

Stimulation of the **anterior** (**preoptic** and **anterior nucleus**) and **medial hypothalamus** controls the activities of the **parasympathetic nervous system** (see Table 23.5). Parasympathetic responses include pupillary constriction, increase in salivary secretion, decrease in heart rate and blood pressure, increase in intestinal peristaltic waves, vasodilation, sweating, piloerection, and contraction of the detrusor muscles of the urinary bladder, resulting in urination.

Stimulation of the **posterior** and **lateral hypothalamus** controls the activities of the **sympathetic nervous system**. Sympathetic responses include pupillary dilation, decrease in salivary secretion, increase in heart rate and blood pressure, decrease of peristaltic waves, vasoconstriction of cutaneous blood vessels, and relaxation of the detrusor muscles of the urinary bladder.

Although fibers destined for the control of autonomic centers (sympathetic and parasympathetic) of the brainstem and spinal cord arise from various hypothalamic nuclei, the principal source of these hypothalamic descending fibers is the parvocellular part of the **paraventricular nucleus**. Interestingly, some of the paraventricular neurons project to both parasympathetic and sympathetic centers in the brainstem and spinal cord. The neurotransmitters released by these descending fibers forming a separate bundle, include glutamate and the two peptides, vasopressin and oxytocin (the same peptides that are transported by the hypothalamohypophyseal tract to the neurohypophysis).

It is important to realize that fibers arising from the anterior and medial hypothalamic nuclei pass posteriorly and laterally on their way to the brainstem and spinal cord. Thus, stimulation of the posterior and lateral hypothalamus may not only stimulate the local neurons that project to the sympathetic centers, but also those fibers, arising from the anterior and medial hypothalamus, that pass by the posterior and lateral hypothalamus, which are destined for the parasympathetic centers.

Fibers arising from the posterior and lateral hypothalamus synapse with the pre-ganglionic sympathetic neurons at the T1 to L2 levels of the spinal cord to elicit the appropriate sympathetic responses. Fibers arising from the anterior and medial hypothalamus synapse with the parasympathetic neurons at the accessory oculomotor nucleus (eponym: Edinger–Westphal nucleus), the superior and inferior salivatory nuclei, the posterior nucleus of the vagus, and the S2–S4 parasympathetic nucleus to elicit the appropriate parasympathetic responses.

Role in emotion, memory, and aggression

Through its extensive interconnections with the limbic system, which processes emotions, the hypothalamus plays an important role in emotional behavior

Emotions trigger the relay of a heavy input from the limbic system to the hypothalamus. The hypothalamus in turn, via its connections with the autonomic nervous system, mediates the appropriate visceral responses (such as changes in heart rate, blood pressure, breathing rate, and sweating) accompanying behavioral expression. For example, a student who is feeling anxious before an examination may experience an increase in heart rate, breathing rate, have sweaty palms, and a dry mouth, all of which are responses mediated by the sympathetic nervous system. The magnitude of the visceromotor responses to emotions depends on the individual, their prior experiences, and their ability to control their emotions and the consequent visceral responses at the neocortical level.

The hippocampus, a region involved in learning and *memory*, gives rise to a prominent fiber bundle, the **fornix**, which, among other targets, projects to the mammillary bodies of the hypothalamus. Hippocampal input to the hypothalamus is involved in the influence of memories on emotion.

Stimulation of the **lateral hypothalamus** not only induces eating and drinking behaviors in animals, but also causes the animal to become more active, and even exhibit **aggressive behavior** or rage. Stimulation of the **ventromedial nuclei** of the hypothalamus and surrounding areas produces satiety and tranquility.

HYPOTHALAMOHYPOPHYSEAL CONNECTIONS

The hypothalamus regulates hypophyseal functions

The hypothalamus regulates hypophyseal functions. Hypothalamic influence on the hypophysis, and ultimately on the endocrine system, travels via two paths: a neural path and a non-neural (vascular) path. The **hypophysis** (also known as the pituitary gland) consists of two lobes, the anterior and posterior lobes. The anterior lobe is named the **adenohypophysis** (G., *adena*, "gland") due to its glandular properties, whereas the posterior lobe is named the **neurohypophysis** due to its neural properties.

Hypothalamic influence on the hypophysis, and ultimately on the endocrine system, travels via two paths: a neural path and a non-neural vascular path. The two paths are formed by two distinct groups of **neuroendocrine cells**. These cells display morphologic and functional characteristics of neurons since they have dendrites, axons, and conduct nerve impulses. Furthermore, since these cells do not release neurotransmitters, but instead synthesize and release neurosecretory hormones (releasing hormones or factors) into the local capillary network, they also have endocrine properties.

In the **neural path**, hypothalamic control is exerted on the neurohypophysis via a neural pathway, the **supraopticohypophyseal tract**. The axon terminals of this pathway release hormones into the extracellular space, which then pass directly into fenestrated capillaries of the neurohypophysis, which in turn deliver their blood into the general circulation (Fig. 23.8).

In the **non-neural path**, hypothalamic control is exerted on the adenohypophysis via a vascular network, the **hypothalamic portal system**. This path involves the release of hypothalamic hormones by the tuberoinfundibular tract into the hypophyseal portal system, a vascular network confined to the hypothalamohypophyseal region, which connects the hypothalamic and hypophyseal vascular capillary beds, and which carries these hormones to the adenohypophysis. The appropriate cells of the adenohypophysis then release hormones into the general circulation (Fig. 23.9).

Neurohypophysis

The hypothalamus contains the cell bodies of neurosecretory cells that display the morphological characteristics of neurons, transmit nerve impulses down their axons, and have endocrine properties

The **supraoptic** and **paraventricular nuclei** of the medial zone of the hypothalamus house the magnocellular (L., large cell bodies) of hypothalamic **neurosecretory cells** (Fig. 23.10). These cells are **neurons** since they have both the morphological characteristics of neurons and upon stimulation transmit nerve impulses down their axons. In addition, these cells also have endocrine properties since they produce hormones (not neurotransmitters) that are released into the bloodstream. Therefore, they are considered to be specialized neurons.

The profuse vascularization of the hypothalamus enables the neurosecretory cells of the supraoptic and paraventricular nuclei to sample the osmolarity of the blood flowing through the hypothalamus. The neurosecretory cells are stimulated even by a slight increase of osmotic pressure (a mere 1% elevation) in the blood which triggers the relay of nerve impulses down their axons.

The neurosecretory cells synthesize two peptide hormones, **antidiuretic hormone** (**ADH**, also known as **vasopressin**) and **oxytocin**. Only one of these hormones is synthesized by any one (large) cell, and both hormones are

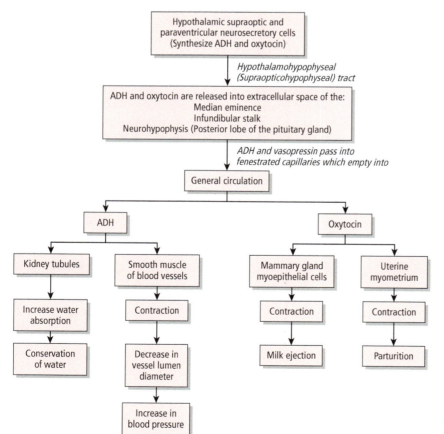

Figure 23.8 • Hypothalamic control of the hypophysis (neural path). ADH, antidiuretic hormone.

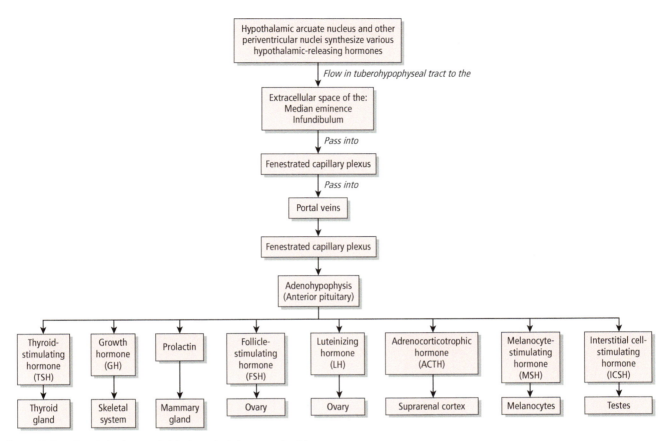

Figure 23.9 • Hypothalamic control of the hypophysis (non-neural path).

synthesized by the (large) cells of both nuclei. The cells of both nuclei synthesize vasopressin and oxytocin, with each neuron containing one hormone or the other, not both. Each of these hormones is synthesized as a prohormone in the cell body of the neurosecretory cells. Each of these hormones then becomes bound to a carrier protein known as a **neurophysin** and is then transported within vesicles (membrane-bound saccules), down the axons of these neurosecretory cells, which gather to form the **supraopticohypophyseal tract**. Although these axons arise from both the supraoptic and paraventricular nuclei, most arise from the supraoptic nucleus and thus the tract is named the supraopticohypophyseal tract. This tract descends only a short distance to terminate in the median eminence, infundibular stalk, and the neurohypophysis. The supraopticohypophyseal tract consists of thin, non-myelinated axons that, in the infundibulum, are covered by astrocytes. In the neurohypophysis, these axons are supported by **pituicytes**, a type of neuroglia, which constitute the major cell type of the posterior lobe. These hormones are either stored in the axon terminals or are released and then pass into the blood circulation. Hormone release is triggered when nerve impulses are transmitted to the axon terminals of the neurosecretory cells of the supraopticohypophyseal tract. The axon terminals are expanded and terminate close to the fenestrated capillary wall. There are two basal laminae that intervene between the axon terminal and the capillary wall. One layer covers the axon terminal, and the other lamina covers the capillary wall. These terminals then liberate vasopressin (ADH), which passes into the fenestrated capillaries located in the neurohypophysis and then flows into the general circulation. Unlike other neurons, those residing in the supraoptic, and paraventricular nuclei do not establish synaptic contacts with other nerve cells. Instead, their function is to detect alterations in the chemical composition of the blood. When these cells are stimulated, they liberate their hormones into the bloodstream.

The vasopressin (ADH) released by these nerve terminals passes into the capillary beds of the median eminence, infundibular stalk, and of the neurohypophysis to gain entrance into the general circulation. When the water content of the body is low, the amount of vasopressin released into the bloodstream is increased. Vasopressin acts on the kidney's distal and collecting tubules causing them to reabsorb water from the tubule lumen, until the water content of the body is restored to normal, resulting in minimal water loss in the urine. When the body's water content is high, only small amounts of vasopressin are released into the bloodstream. In response to the decreased vasopressin blood levels, the kidney reabsorbs less water, resulting a dilute urine and water loss. By modulating the volume of water lost in the urine,

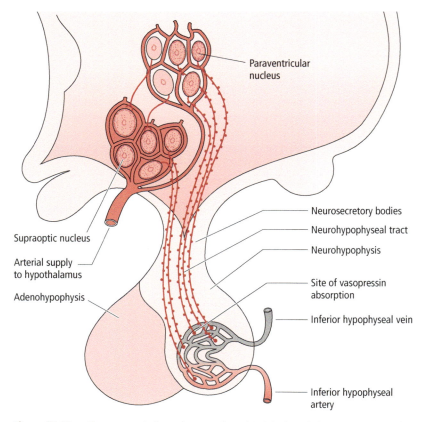 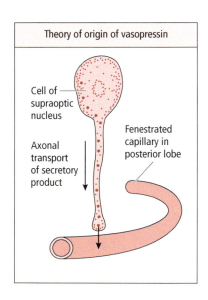

Figure 23.10 ● The paraventricular and supraoptic nuclei of the hypothalamus. Source: Modified from Netter, NH (1983) *The CIBA Collection of Medical Illustrations.* Vol. 1, part 1. CIBA, New Jersey; p. 211, plate 59.

the kidney controls the osmotic pressure in the blood enabling the body to conserve water and regulate body fluid content. Vasopressin also acts on the smooth muscle of the vessel wall, causing it to contract which results in a decrease in lumen diameter (thus its name, vasopressin) and a consequent elevation in blood pressure.

Vasopressin is also released in response to a decrease in blood volume as would result following hemorrhage or in response to a decrease in blood pressure. Pressure receptors are located in the internal carotid artery (carotid sinus), while stretch receptors are located in the walls of the left atrium and pulmonary veins. These receptors relay this sensory input to the nucleus of the solitary tract in the medulla oblongata via the glossopharyngeal nerve. From there, input is relayed via ascending fibers to the magnocellular **supraoptic nucleus**.

The oxytocin released by the nerve terminals, like vasopressin, also passes into the capillary bed of the median eminence, infundibular stalk, and the neurohypophysis, to gain entrance into the general circulation. Oxytocin (G., *oxy*, "rapid"; *tokos*, "birth"; thus, "rapid labor") has two functions: (1) it causes contraction of the uterine myometrium during the final stages of pregnancy, which facilitates parturition, and (2) contraction of the mammary gland myoepithelial cells, resulting in milk ejection from the active mammary gland.

Adenohypophysis

The hypothalamus communicates with the adenohypophysis via the hypophyseal portal system

Unlike the neurohypophysis, the adenohypophysis does not receive axon terminals from hypothalamic neurons. Parvocellular neurons (small cells) located in the **arcuate nucleus** and other **periventricular zone nuclei**, and cells of the paraventricular, suprachiasmatic, tuberal, and medial preoptic nuclei of the hypothalamus give rise to fibers that become incorporated in the **tuberohypophyseal tract**. These cells synthesize various releasing hormones (factors) and release-inhibiting hormones (factors) (Figs. 23.9 and 23.11). The releasing hormones produced by the hypothalamus are:

- **somatotropin-releasing hormone** (**SRH**), which stimulates the release of somatotropin (growth hormone) (produced in the anterior periventricular and arcuate nuclei);
- **prolactin-releasing hormone** (**PRH**), which stimulates the release of prolactin;
- **corticotropin-releasing hormone** (**CRH**), which stimulates the release of adrenocorticotropin; neurons are located in the parvocellular paraventricular nucleus;
- **gonadotropin-releasing hormone** (**GnRH**), which stimulates the release of luteinizing hormone (LH) and follicle-stimulating hormone (FSH);

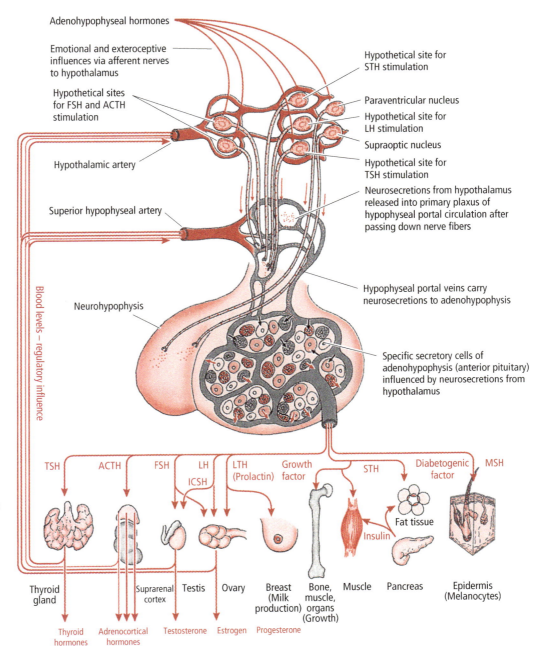

Figure 23.11 ● The interaction between the hypothalamus and adenohypophysis. LRH, lutein-releasing hormone; SRH, somatotropin-releasing hormone; for other abbreviations see Fig. 23.9. Source: Modified from Netter, FH (1983) *The CIBA Collection of Medical Illustrations.* Vol. 1, part 1. CIBA, New Jersey; p. 210, plate 58.

- **thyroid-stimulating hormone-releasing hormone** (thyrotropin-releasing hormone, **TRH**), which stimulates the release of TSH; and neurons are located in the anterior periventricular, ventromedial, and dorsomedial nuclei of the hypothalamus;
- **melanocyte-stimulating hormone-releasing hormone**, which releases melanocyte-stimulating hormone (MSH).

The release-inhibiting hormones produced by the hypothalamus are:

- **somatostatin**, which inhibits growth hormone (GH) and TSH release (produced by neurons in the anterior periventricular nucleus);
- **prolactin-inhibitory factor** (**PIF**, dopamine), which inhibits the release of prolactin; and
- **melanocyte-stimulating hormone inhibitory factor** (**MSHIF**), which inhibits melanocyte-stimulating hormone (MSH) release.

These releasing or inhibiting hormones are packaged into vesicles that flow down the axons of these neurons, which collectively form the **tuberohypophyseal** (**tuberoinfundibular**) **tract**, and are released in the **median eminence** and the hypophyseal (infundibular) stalk. There these releasing and inhibitory hormones pass into the primary capillary plexus that gives rise to portal veins that carry these hormones to the secondary capillary plexus located in the adenohypophysis. The adenohypophysis contains cells that synthesize and, when appropriately stimulated, release into the general circulation the following trophic hormones: growth hormone (GH, somatotropin), prolactin, adrenocorticotropic hormone (ACTH, corticotropin), FSH, LH (interstitial-cell-stimulating hormone [ICSH] in males), TSH (thyrotropin), and MSH. These hormones are released into the general bloodstream and are carried to their target organs (endocrine glands) where they bind to receptors on cells and stimulate (or inhibit) the release of hormones produced by the respective organs. Growth hormone stimulates growth of long bones and the body in general; prolactin stimulates the proliferation of the mammary glands prior to and following birth along with many other physiological functions; adrenocorticotropic hormone stimulates the release of steroid hormones such as cortisol by the suprarenal cortex; the gonadotropic hormone FSH stimulates the development and growth of ovarian follicles and oocytes in biological females and the release of androgen-binding proteins by sustentacular (eponym: Sertoli) cells of the testes in biological males, whereas LH triggers ovulation and the formation of the corpus luteum; ICSH stimulates the formation of testosterone in biological males; thyroid-stimulating hormone stimulates the synthesis and release of the thyroid hormones triiodothyronine (T3) and tetraiodothyronine (T4, thyroxine); MSH stimulates melanocytes to form melanin.

The hypophysis receives its blood supply mainly from two vessels: the superior and inferior hypophyseal arteries (Fig. 23.12). The **superior hypophyseal artery** is a branch of the internal carotid or the posterior communicating artery and supplies the hypophyseal stalk and the adenohypophysis.

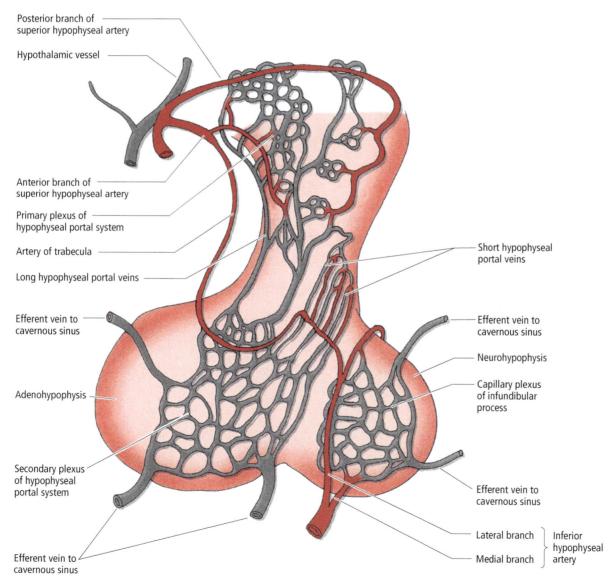

Figure 23.12 • Vascularization of the hypothalamus and the hypophysis. Source: Modified from Netter, FH (1983) *The CIBA Collection of Medical Illustrations.* Vol. 1, part 1. CIBA, New Jersey; p. 209, plate 57.

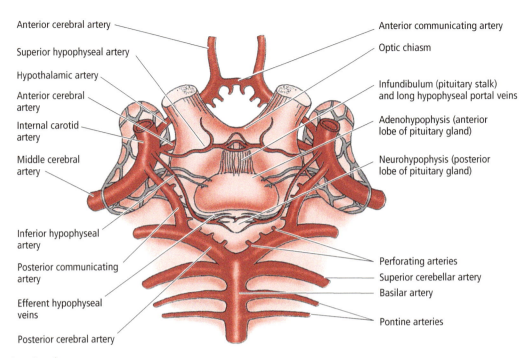

Figure 23.12 ● *(Continued)*

The **inferior hypophyseal artery** is a branch of the internal carotid artery and supplies the neurohypophysis.

Branches of the superior hypophyseal artery drain into a **capillary plexus** in the median eminence and superior part of the infundibular stalk. This plexus in turn drains into venules, which drain into long portal veins that descend in the infundibular stalk to the adenohypophysis where they branch into another capillary plexus. Both of these plexuses consist of fenestrated capillaries. These capillaries then drain into the nearby cavernous sinus.

CLINICAL CONSIDERATIONS

A hypothalamic lesion usually extends into a number of its nuclei, affecting their function and resulting in more than one symptom

The hypothalamus is a small region of the brain and a lesion confined to only a single nucleus is uncommon. Instead, a hypothalamic lesion usually extends into a number of nuclei, affecting their function and resulting in more than one symptom. Hypothalamic lesions may result following skull fractures involving the sphenoid bone, cerebrovascular accidents, or most commonly, hypophyseal (pituitary) tumors. Depending on the nucleus or nuclei damaged, symptoms result from the dysfunction of the autonomic nervous system, water intake, food intake, temperature regulation, and sexual behavior.

Damage to nuclei that control body temperature
Bilateral destruction involving the preoptic and anterior hypothalamic nuclei results in hyperthermia (an elevation in body temperature)

The **preoptic** and **anterior hypothalamic nuclei** are involved in the control of body temperature. In the event that body temperature rises above normal limits, these nuclei initiate heat loss mechanisms in order to restore normal body temperature. Bilateral destruction involving these nuclei results in **hyperthermia**, an elevation in body temperature.

The presence of infectious agents in the body results in the production of pyrogens which may enter the bloodstream and eventually reach the hypothalamus. The pyrogens pass through the fenestrated capillaries of the anterior hypothalamus and contact the hypothalamic receptor cells of the circumventricular organs. In response to this contact, the body's thermostat is changed to a higher temperature point; thus, the body's temperature becomes elevated above normal. This is accomplished by inhibition of heat loss mechanisms, cutaneous vasoconstriction, and shivering. Cutaneous vasoconstriction results in the lowering of the body's surface temperature, which is perceived as a cold sensation, usually accompanying a fever.

Bilateral destruction involving the posterior hypothalamic nuclei results in poikilothermia (the inability to regulate body temperature), regardless of environmental temperature

The **posterior hypothalamic nuclei** are also involved in the control of body temperature. In the event that body temperature drops, they initiate mechanisms that will conserve and produce heat to restore body temperature. Bilateral destruction involving the posterior hypothalamic nuclei results in the inability to regulate body temperature (**poikilothermia**) regardless of the environmental temperature. The reason for this phenomenon is that

CLINICAL CONSIDERATIONS (continued)

when the cells of the posterior nuclei are damaged, the fibers arising from the anterior hypothalamic nuclei are also damaged as they course through the posterior hypothalamic nuclei on their way to their destination.

Damage to nuclei that regulate food intake
A lesion in the lateral nucleus results in hypophagia or aphagia and a drastic drop in body weight. A bilateral lesion in the ventromedial nucleus results in hyperphagia and excess body weight

The **lateral** and **ventromedial nuclei** of the hypothalamus are involved in eating behavior. The lateral hypothalamic nucleus contains a group of neurons that synthesize **orexins** (G., *orexis*, "appetite") (also referred to as hypocretins), which are hypothalamic neuropeptides that regulate food intake and the sleep–wake cycle. Stimulation of the lateral nucleus ("feeding center") induces eating, whereas stimulation of the ventromedial nucleus ("satiety center") inhibits eating. Fasting causes an increase in orexin synthesis, whereas administration of exogenous orexin induces eating. A lesion in the lateral nucleus results in **aphagia** (G., "no eating") and a drastic drop in body weight. A bilateral lesion in the ventromedial nucleus results in **hyperphagia** (G., "overeating") and excess body weight.

Undetectable or low levels of orexin (hypocretin) production are associated with **narcolepsy**, a disorder in which an individual falls asleep suddenly, several times during the day and at inopportune times. Studies show that the hypothalamus of an individual suffering from narcolepsy contains only one-tenth of the normal number of neurons that produce orexin/hypocretin. It is believed that narcolepsy may be caused by deficient orexin/hypocretin neurotransmission in some individuals.

3 What is believed to be the cause of narcolepsy?

Damage to the area that controls water intake
A lesion involving the cell group lateral to the lateral hypothalamic nucleus, known as the **zona incerta**, results in a lack of interest in water intake. This may result in severe dehydration and death.

Damage to nuclei that control the autonomic nervous system
A lesion involving the anterior hypothalamic nucleus will result in the disruption of parasympathetic activities, whereas a lesion involving the posterior hypothalamic nucleus will result in the disruption of both sympathetic and parasympathetic activities

The anterior and posterior hypothalamic areas are associated with the control of the autonomic nervous system. A lesion involving the **anterior hypothalamic nucleus** will result in the disruption of parasympathetic activities, whereas a lesion involving the **posterior hypothalamic nucleus** will result in the disruption of both sympathetic and parasympathetic activities. The reason for this is that although the posterior hypothalamic nucleus is involved in the control of the sympathetic nervous system, fibers arising from the anterior nucleus course through the posterior nucleus on their way to their destination. Thus, a lesion in the posterior nucleus will damage local hypothalamic neurons controlling sympathetic activities and fibers passing through the nucleus controlling parasympathetic activities.

Damage to hypothalamic nuclei involved in aggression
Stimulation of the **lateral hypothalamus** not only induces eating and drinking behaviors in animals, but also causes animals to become more active, and even exhibit **aggressive behavior** or rage. Bilateral lesions in the lateral hypothalamus cause a lack of interest in eating, drinking, and extreme passivity of the animal. Stimulation of the **ventromedial nuclei** of the hypothalamus and surrounding areas produces satiety and tranquility. Bilateral lesions of the ventromedial areas of the hypothalamus induce excessive eating, drinking, excitability, and aggression.

Diabetes insipidus
Diabetes insipidus is a condition that can result following anterior hypothalamic damage of the paraventricular and supraoptic nuclei (but primarily the supraoptic nucleus)

Damage to the anterior hypothalamus may result following the compression of the paraventricular and supraoptic nuclei by a growing tumor at the anterior aspect of the brain. The cells of this nucleus produce vasopressin (ADH) which acts on the kidney tubules to conserve water. As a result of nuclear damage, an inadequate amount of vasopressin results in **diabetes insipidus**. This condition is characterized by **polydipsia** (excess thirst, with the drinking of large volumes of water) and then **polyuria** (the production of large volumes of dilute urine). Diabetes insipidus may also result following damage of the tuberoinfundibular tract. This is a hormonal-type diabetes insipidus, as opposed to the renal type where the renal tubules are incapable of responding to plentiful circulating vasopressin.

Cavum Septum Pellucidum
Figures 23.4–23.6 illustrate an individual with cavum septum pellucidum, a condition resulting from the incomplete fusion of the fetal left and right septum pellucida, thereby creating a potential cavity between them. Typically, the septum pellucida fuse between approximately 24 weeks' gestation to 24 weeks' postnatal life. However, this fusion does not occur in approximately 15% of the population. The cavum septum pellucidum is not lined with ependymal or choroid plexus cells; hence, the internal fluid is a filtrate of cerebrospinal that has crossed the septal laminae and generally remains asymptomatic. Cyst formation within the cavum septum pellucidum is rare (0.04%) but can lead to neurological and psychiatric symptoms by obstruction of the interventricular foramina, compression of the hypothalamo-septal triangle, or chronic deep venous impairment caused by displacement and stretching of the internal cerebral and subependymal veins.

SYNONYMS AND EPONYMS OF THE HYPOTHALAMUS

Name of structure or term	Synonym(s)/eponym(s)
Adenohypophysis	Anterior lobe of pituitary gland
Adrenocorticotropic hormone (ACTH)	Corticotropin
Amygdaloid body	Amygdala
	Amygdaloid nuclear complex
Antidiuretic hormone (ADH)	Vasopressin
Arcuate nucleus	Infundibular nucleus
Growth hormone (GH)	Somatotropin
Hypophyseal	Hypophysial
Hypothalamic-release-inhibiting hormones	Hypothalamic-release-inhibiting factors
Hypothalamic-releasing hormones (HRH)	Hypothalamic-releasing factors
Mammillary region of the hypothalamus	Posterior region of the hypothalamus
Mammillotegmental tract	Mammillotegmental fasciculus
Mammillothalamic fasciculus	Mammillary fasciculus
	Tract of Vicq d'Azyr
Neurohypophysis	Posterior lobe of the pituitary gland
Prolactin inhibitory factor (PIF)	Dopamine
Stria terminalis	Amygdalohypothalamic fibers
Supraoptic region of the hypothalamus	Chiasmatic region of the hypothalamus
Supraopticohypophyseal tract	Hypothalamohypophyseal tract
	Neurohypophyseal tract
Thyroid-stimulating hormone (TSH)	Thyrotropin
Thyroid-stimulating hormone-releasing hormone	Thyrotropin-releasing hormone (TRH)
Trigeminothalamic tract	Trigeminal lemniscus
Tuberal region of the hypophysis	Infundibular region of the hypophysis
	Middle region of the hypophysis
Tuberohypophyseal tract	Tuberoinfundibular tract
	Infundibular tract
Ventral amygdalohypothalamic tract	Amygdalofugal tract

FOLLOW-UP TO CLINICAL CASE

This patient has narcolepsy. Narcolepsy is a condition that causes abnormal sleep patterns. It is almost always an acquired disease and symptoms typically begin in older childhood or young adulthood. The most common symptom is daytime sleepiness, which often leads to "sleep attacks." Nighttime sleep is often of poor quality, so the total sleep time during a 24-hour period is often less than adequate. Some of the most characteristic and notable symptoms arise secondary to the intrusion of REM (rapid eye movement) states into waking hours. REM is a stage of normal sleep in which there is paralysis of all the muscles except those controlling eye movements and respiration. Dreams often occur in REM sleep. This intrusion of REM into waking hours accounts for episodes of cataplexy, sleep paralysis, and hypnagogic hallucinations.

Cataplexy refers to brief episodes of loss of muscle tone, which is brought about by intense emotions such as laughing or anger. This can cause falls or, in cases where it is just partial, an involuntary head flexion or mandibular depression. It can cause extreme social embarrassment contributing to other psychosocial symptoms. Sleep paralysis refers to brief episodes of complete paralysis, which can occur just before sleep onset or upon waking from sleep. Hypnagogic hallucinations are hallucinations, commonly visual and sometimes frightening, which occur before the onset of sleep (hypnagogic) or upon waking from sleep (hypnopompic).

Narcolepsy can be confirmed by a multiple sleep latency test (MSLT). An abnormally short onset of sleep (within 5 minutes) and an abnormally quick onset of the REM stage of sleep confirm the diagnosis of narcolepsy. A regular sleep study should be performed to look for other causes of excessive daytime sleepiness, such as obstructive sleep apnea.

The cause of narcolepsy has recently been determined. The posterior hypothalamus contains neurons that contain orexin (hypocretin). These neurons project widely, including to the cerebral cortex and brainstem. Orexin mediates wakefulness and alertness, and it has been demonstrated that a lack of orexin in animals, or an alteration of the orexin receptors, leads to symptoms reminiscent of narcolepsy. It is thought that there is a selective absence of these orexin containing neurons in the posterior hypothalamus in narcolepsy. The etiology of this condition is unclear.

The sleepiness and sleep attacks are treated with stimulants such as amphetamines, methylphenidate, or caffeine. A newer medication for this, which has less abuse potential and side effects, is modafinil. Cataplexy can be treated with tricyclic antidepressants or similar medicines, which promote the effects of amines such as norepinephrine, dopamine, or serotonin. These medications inhibit REM sleep.

QUESTIONS TO PONDER

1. What are the functions of the periventricular zone nuclei of the hypothalamus?

2. What are the functions of the medial zone nuclei of the hypothalamus?

3. What are the functions of the lateral zone nuclei of the hypothalamus?

4. What is the function of the nuclei in the preoptic region of the hypothalamus?

5. What are the functions of the supraoptic region of the hypothalamus?

6. What are the functions of the tuberal region of the hypothalamus?

7. What are the functions of the mammillary region of the hypothalamus?

8. What is the most prominent source of input to the hypothalamus?

9. What is the function of the circumventricular organs?

HYPOTHALAMUS ● ● ● **479**

10. In addition to neural output, what other structure does the hypothalamus influence?

11. What is the most prominent neural input to the hypothalamus?

12. What is the function of the spinohypothalamic fibers?

13. What is the function of the reticular formation fibers terminating in the hypothalamus?

14. What is the function of the supraopticohypophyseal tract?

15. What is the function of the tuberohypophyseal tract?

16. Summarize the functions of the hypothalamus.

17. A biological female was brought to the doctor because, although she was 18 years old, menstruation had not occurred. Blood tests confirm that the patient had almost no follicle-stimulating hormone or luteinizing hormone. Name the possible problems that could cause this condition.

CHAPTER 24

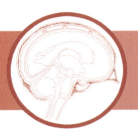  

Thalamus

- CLINICAL CASE
- BORDERS
- ANATOMY
- INTERNAL AND EXTERNAL MEDULLARY LAMINAE
- THALAMIC NUCLEI
- CLINICAL CONSIDERATIONS
- SYNONYMS AND EPONYMS OF THE THALAMUS
- FOLLOW-UP TO CLINICAL CASE
- QUESTIONS TO PONDER

CLINICAL CASE

A 68-year-old hypertensive patient, K.J., had sudden onset of headache and complete loss of sensation from left face, upper limb, and lower limb. This began 10 hours before admission to the emergency room. Patient exhibited no other symptoms.

K.J.'s blood pressure in the ER was 230/120. Other vital signs were relatively normal. Brief examination in the ER confirmed sensory loss on the left side, with normal strength but not well coordinated. K.J. was able to ambulate quite well. An urgent head CT was performed, which was diagnostic.

Information arising from the **basal nuclei**, **cerebellum**, **limbic system**, and **sensory systems** is relayed to the **thalamus**, a major subcortical relay station. Here, the information is processed, integrated, and is then transmitted to specific areas of the ipsilateral cerebral cortex. These cortical areas are in turn connected via reciprocal feedback projections to the thalamic nuclei (subnuclei), making the cerebral cortex the most prominent input source to the thalamus. The thalamus is interconnected not only with the neocortex of all five lobes, but also with the phylogenetically older cortical areas including the paleocortex (piriform cortex) and archicortex (hippocampal formation).

The thalamus functions not only in the further processing and integration of *sensory* and *motor* information, but also serves as the main gateway, via which this and additional information reaches the cerebral cortex. Therefore, the thalamus not only controls the flow of numerous streams of information to the cerebral cortex for additional processing but also, ultimately, regulates cortical activity. Although the thalamus projects mainly to the cerebral cortex, it also provides input to the basal nuclei and the hypothalamus.

BORDERS

The thalamus has an anterior and posterior pole, and four surfaces. Its boundaries are: *anteriorly*, the interventricular foramen (eponym: of Monro); *posteriorly*, the posterior extent of the pulvinar; *medially*, the third ventricle; *laterally*, the posterior limb of the internal capsule; *superiorly*, its free surface, which contributes to the floor of the lateral ventricle; and *inferiorly*, the hypothalamic sulcus on the lateral wall of the third ventricle, separating it from the hypothalamus (Figs. 24.1 and 24.2).

A Textbook of Neuroanatomy, Third Edition. Maria A. Patestas, Amanda J. Meyer, and Leslie P. Gartner.
© 2025 John Wiley & Sons, Inc. Published 2025 by John Wiley & Sons, Inc.
Companion website: www.wiley.com.go/patestas/neuroanatomy3e

THALAMUS ••• 481

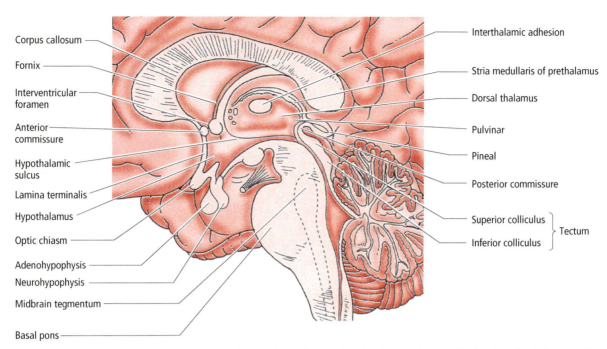

Figure 24.1 ● A midsagittal section through the brainstem and part of the overlying cerebral hemisphere with a medial view. Note the thalamus and its neighboring diencephalic structures and midbrain.

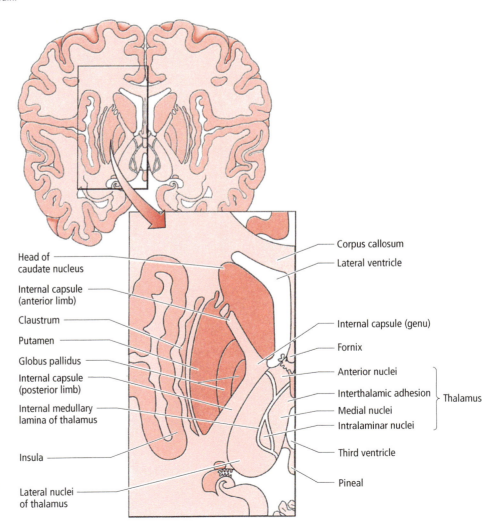

Figure 24.2 ● An axial section through the cerebral hemispheres showing the dorsal thalamus, basal nuclei, and internal capsule in superior view.

ANATOMY

The thalamus is the largest subcortical gray mass of the CNS and is the main constituent of the diencephalon

The **thalamus** (G., "inner chamber") (Fig. 24.2) is an oval-shaped mass of gray matter, embedded deep within the white matter of each cerebral hemisphere. The right and left thalami are separated along most of their medial surface by the narrow, sagittally oriented third ventricle. A short, narrow, cylindrical structure – the **interthalamic adhesion**, which traverses the third ventricle – forms a bridge between the left and right thalami in 25–85% of the brains studied. Although the interthalamic adhesion connects the right and left sides of some brains, it is not considered to be a true "commissure" since this structure consists of gray matter only (but is devoid of neuronal cell bodies) and has no decussating fibers; its function is unknown.

The **stria medullaris of prethalamus** and the **stria terminalis**, two thin strands of white matter, lie on the superior surface of the thalamus (Fig. 24.3). The stria medullaris of prethalamus courses on the superomedial margin of the thalamus, separating its superior from its medial surface. It is a slender bundle of two-way fibers connecting the septal nuclei and hypothalamus with the habenular nucleus of the epithalamus. The stria terminalis carries fibers primarily from the amygdaloid body, which terminate in the hypothalamus, septal area, and the bed nucleus of the stria terminalis. It courses on the superolateral surface of the thalamus, separating its superior surface from its lateral surface and from the body of the caudate nucleus.

The lateral aspect of the superior surface of the thalamus contributes to the floor of the lateral ventricle, and the medial aspect of the superior surface of the thalamus displays a groove formed by the fornix.

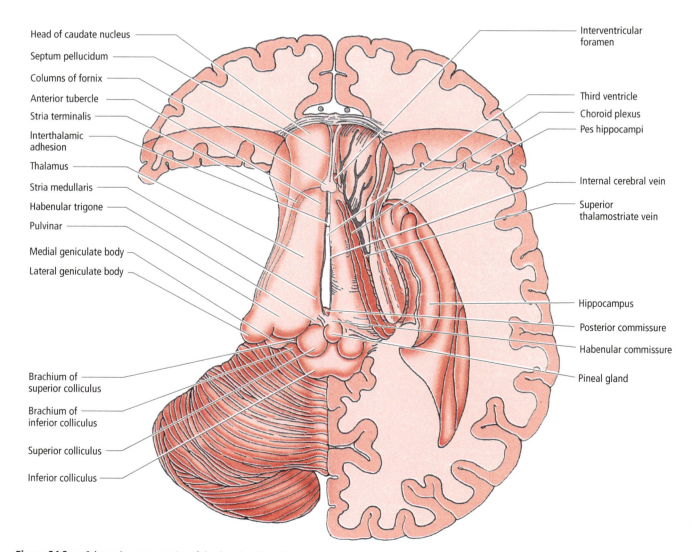

Figure 24.3 ● Schematic representation of the dorsal surface of the thalamus on a horizontally sectioned brain. Part of the cerebral hemisphere has been removed to expose deeper structures such as the thalamus, third ventricle, caudate nucleus, and hippocampus.

The thalamus rests on the mesencephalon; thus, its inferior surface forms a ceiling over the mesencephalic tectum (Fig. 24.1). Numerous tracts such as the medial lemniscus and the trigeminothalamic, rubrothalamic, reticulothalamic, and dentatothalamic (cerebellothalamic) tracts ascend through the brainstem and pass into the thalamus from its inferior surface. In addition, the brainstem reticular formation extends superiorly into the thalamus through this surface.

INTERNAL AND EXTERNAL MEDULLARY LAMINAE

The thalamus is closely associated with two nearly vertical layers of myelinated fibers, the internal and external medullary laminae

The **internal medullary lamina**, a Y-shaped layer of white matter, partitions the thalamus into three main groups of nuclei, the **anterior**, **medial**, and **lateral groups**, based on their relative position (Figs. 24.2 and 24.4). The posterior portion of this lamina, the body of the Y, separates the medial from the lateral thalamic nuclear groups. Its anterior portion, which is divided into two limbs, partially surrounds the anterior nuclear group and separates it from the neighboring medial and lateral nuclear groups. The internal medullary lamina contains afferent and efferent thalamic fibers that enter or leave the thalamic subnuclei.

The **external medullary lamina** is a curved layer of white matter that follows the contour and covers mostly the anterior and lateral surfaces of the thalamus. It is composed of myelinated **thalamocortical fibers** that originate in the thalamus and terminate in the cerebral cortex, and of reciprocal feedback **corticothalamic fibers** that originate in the cerebral cortex and terminate in the thalamus. The external medullary lamina is covered by a slender layer of gray matter, the **thalamic reticular nucleus**, which is the most lateral (and superficial) nucleus of the thalamus (Fig. 24.4B).

The myelinated thalamocortical fibers arising from the thalamus pass through the external medullary lamina, and radiate from the lateral surface of the thalamus, collectively forming the **thalamic radiations**. These fibers become

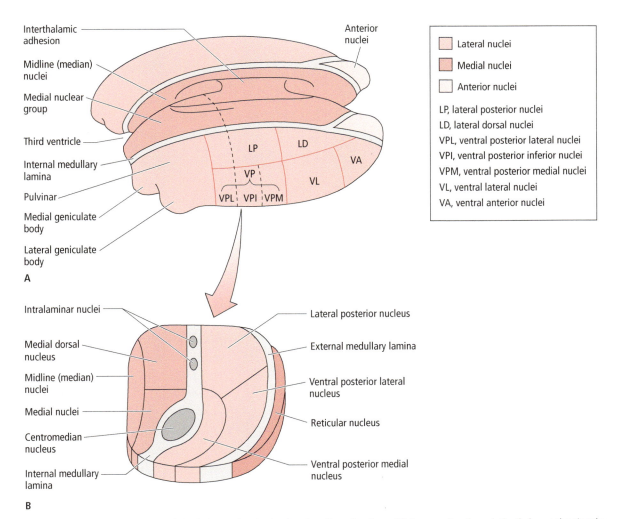

Figure 24.4 ● (A) Schematic representation of the thalamus showing its dorsal and lateral surfaces. (B) Cross-section through the thalamus showing the periventricular, medial, intralaminar, and lateral groups of nuclei.

484 ● ● ● **CHAPTER 24**

incorporated in the various parts of the internal capsule and the corona radiata as they ascend to terminate in the cerebral cortex. Corticothalamic fibers also course through the corona radiata and the internal capsule, and when they approach the thalamus, they converge to pass through the external medullary lamina to terminate in the various thalamic subnuclei.

The thalamocortical and corticothalamic fibers of the thalamic radiations interconnecting the thalamus and the cerebral cortex of the five lobes, are organized into four peduncles, the anterior, superior, posterior, and inferior thalamic peduncles, consisting of white matter. The anterior thalamic peduncle connects the thalamus with the frontal lobe. It traverses the anterior limb of the internal capsule to interconnect the thalamus with the prefrontal cortex and the cingulate gyrus. The superior thalamic peduncle traverses the posterior limb of the internal capsule to interconnect the thalamus to the motor and sensory cortices in the frontal and parietal lobes, respectively. The posterior thalamic peduncle fibers become incorporated into the retrolentiform part of the internal capsule where they appear to form a spray of fibers that interconnect the thalamus with the occipital lobe and surrounding areas of the parietal and temporal lobes. The fibers of the inferior thalamic peduncle cradle the putamen and globus pallidus and connect the thalamus to the anterior temporal cortex.

THALAMIC NUCLEI

Anatomical classification

The thalamus is partitioned by the internal medullary lamina into the anterior, medial, and lateral nuclear groups (subdivisions)

Based on their afferent and efferent connections, each thalamic nuclear group functions as an independent unit and serves a particular function (Figs. 24.2 and 24.4, Table 24.1). The nuclear groups are further divided into **subnuclei**, and the role of each thalamic subnucleus is associated with the source(s) of incoming afferents, and the cortical target areas of its efferents (Table 24.2).

With the exception of the reticular nucleus of prethalamus, all thalamic subnuclei possess thalamic **projection neurons** that relay processed information to the cerebral cortex. In addition, the thalamic subnuclei also have inhibitory gamma-aminobutyric acid (GABA)-ergic **interneurons** whose cell bodies and processes are confined to a single subnucleus.

Although a number of classifications exist, this textbook uses the most common of these and subdivides the subnuclei into anterior, medial, and lateral nuclear groups. The **anterior** and **medial nuclear groups** are phylogenetically the older thalamic nuclei and are collectively referred to as the **paleothalamus**. The subnuclei of the **lateral nuclear group** are newer and are referred to as the **neothalamus**. Additionally, because of their unique characteristics, two of the thalamic nuclei – the **intralaminar** and **reticular**

Table 24.1 ● Anatomical and functional classification of the thalamic nuclei.

Nuclear group	Component subnuclei	Functional group
Anterior	Anterior ventral (AV)	Specific relay
	Anterior medial (AM)	Specific relay
	Anterior dorsal (AD)	Specific relay
Medial	Lateral dorsal (LD)	Association (multimodal)
	Medial dorsal (MD)	Non-specific
	Periventricular nuclei:	Non-specific
	Parataenial nucleus	
	Paraventricular nucleus of thalamus	
	Medial ventral nucleus	
Lateral	Dorsal tier:	
	Dorsal lateral (DL)	Association (multimodal)
	Lateral posterior (LP)	Association (multimodal)
	Pulvinar (P)	Association (multimodal)
	Ventral tier:	
	Ventral anterior (VA)	Motor relay
	Ventral lateral complex (VL)	Motor relay
	Ventral posterior complex (VP)	Motor relay
	Ventroposterior complex:	
	Ventral posterior medial nucleus (VPM)	Sensory relay
	Ventral posterior lateral (VPL) nucleus	Sensory relay
	Ventral posterior inferior nucleus (VPI)	Sensory relay
	Geniculate nuclei:	
	Medial geniculate nucleus (MGN)	Sensory relay
	Dorsal lateral geniculate nucleus (LGN)	Sensory relay
Intralaminar	Anterior group:	
	Central medial	Nonspecific
	Paracentral	Nonspecific
	Central lateral	Nonspecific
	Central group:	
	Centromedian (CM)	Nonspecific
	Parafascicular (PF)	Nonspecific
	Subparafascicular	Nonspecific
	Posterior group:	
	Limitans nucleus	Nonspecific
	Suprageniculate nucleus	Nonspecific
Reticular		Nonspecific

nuclei – are usually not included in any of the three thalamic nuclear groups listed previously, but instead are grouped separately.

Table 24.2 ● Connections of the thalamic subnuclei.

Nucleus/subnucleus	Afferents (input)	Efferents (output)	Function(s)
Anterior (A)	Mammillary body via mammillothalamic fasciculus	Cingulate gyrus	Expression of emotions
	Hippocampal formation via fornix	Parahippocampal gyrus	Learning, memory
Medial dorsal (MD)			
Parvocellular	Prefrontal association cortex	Prefrontal association cortex	Integration of sensory information
Magnocellular	Amygdaloid body	Prefrontal association cortex	Expression of emotions
	Orbitofrontal cortex		
	Hypothalamus		
	Intralaminar, midline, and lateral posterior thalamic nuclei		
Midline			
Dorsal midline	Striatum, limbic cortex	Striatum, limbic cortex	Modulation of cortical excitability
Ventral midline	Hippocampus	Hippocampus	
Lateral dorsal (LD)	Mammillary body via mammillothalamic tract	Cingulate gyrus	Expression of emotions
		Parahippocampal gyrus	
Lateral posterior (LP)	Superior parietal lobule	Superior parietal lobule	Sensory integration
	Precuneus	Precuneus	
	Superior colliculus		
	Pretectal area		
	Occipital lobe		
Pulvinar (P)	Association areas of parietal, temporal, and occipital lobes	Association areas of parietal, temporal, and occipital lobes	Integration of visual, auditory, and somatosensory information
	Retina, superior colliculus, dorsal lateral geniculate nucleus, medial geniculate nucleus, cerebellum		
Ventral anterior (VA)			
Magnocellular	Substantia nigra (reticular part) via nigrothalamic pathway	Frontal eye field	Control of eye, face, and head movements
	Intralaminar and midline nuclei of the thalamus		
	Reticular formation		
	Frontal eye field		
Parvocellular	Globus pallidus (medial segment) via pallidothalamic pathway	Premotor cortex	Control of limb movement
	Intralaminar and midline nuclei of the thalamus	Supplementary motor area	Control of limb movement
	Reticular formation		
Ventral lateral (VL)			Control of movement
VLa			
Magnocellular	Substantia nigra (reticular part)	Supplementary motor area	
Parvocellular	Globus pallidus (medial segment)	Premotor cortex	
VLp	Contralateral cerebellar nuclei via dentatothalamic tract	Premotor cortex	Control of head and limb movement
		Premotor cortex	
Ventral posterior medial (VPM)	Orofacial region via trigeminothalamic tracts	Primary somatosensory cortex in postcentral gyrus	Processes touch, pressure, pain, and temperature sensation, and proprioception
	Nucleus of the solitary tract		Taste
Ventral posterior lateral (VPL) and ventral posterior inferior (VPI)	Anterior and lateral spinothalamic tracts	Primary somatosensory cortex in the postcentral gyrus	Processes pain, temperature, and nondiscriminatory touch sensation from the body
	Medial lemniscus	Primary somatosensory cortex in the postcentral gyrus	Processes discriminatory touch, pressure, joint movement, and vibratory sense from the body

Table 24.2 ● *Continued*

Nucleus/subnucleus	Afferents (input)	Efferents (output)	Function(s)
Ventral posterior inferior (VPI)	Vestibulothalamic fibers	Vestibular area of the somatosensory cortex	Control of motor functions
Medial geniculate nucleus (MGN)	Inferior colliculus, Other auditory nuclei via brachium of inferior colliculus, Lateral lemniscus, Primary auditory cortex	Auditory radiation to the primary auditory cortex	Hearing
Dorsal Lateral geniculate body (LGN)	Retina via optic tract, Primary visual cortex, Pulvinar, Other thalamic nuclei	Optic radiation to the primary visual cortex, Pulvinar, Other thalamic nuclei	Vision, arousal, visual attention, regulation of visual input transmission to primary visual cortex
Intralaminar Centromedian	Globus pallidus Premotor cortex Primary motor cortex Ascending sensory systems Reticular formation, Striatum	Striatum (caudate nucleus and putamen) Other intralaminar nuclei	Sensorimotor integration Maintenance of arousal Pain perception Control of cortical activity
Parafascicular	Supplementary motor area Ascending sensory systems Reticular formation, Striatum	Widespread areas of the cerebral cortex	
Reticular	Thalamocortical, corticothalamic, thalamostriatal, and pallidothalamic collaterals	Other thalamic nuclei Reticular formation	Integrates and controls thalamic activity

Anterior nuclear group

The anterior nuclear group is the smallest and consists of a composite of three subnuclei: the AV, AM, and AD subnuclei

The anterior nuclear group consists of a composite of four subnuclei: the **anterior ventral (AV)**, **anterior medial (AM)**, **anterior dorsal (AD)**, **lateral dorsal subnuclei**. The subnuclei of the anterior group are wedged in the anterior tubercle of the thalamus, which forms an elevation on the anterior part of the lateral wall of the third ventricle. This is the smallest thalamic nuclear group, and due to the comparable connections and roles of its three subnuclei, they are collectively referred to as the **anterior nuclei of the thalamus**.

The anterior nuclear group, a relay nucleus of the medial limbic circuit (eponym: Papez circuit), has a significant number of connections with the limbic system. It is the primary target of the afferent fibers relaying information from the mammillary body of the hypothalamus via the mammillo-thalamic fasciculus, and the hippocampal formation via the fornix. It in turn projects to the cingulate gyrus of the limbic association cortex and the parahippocampal gyrus. Due to its limbic connections, the anterior nuclear group functions in the expression of emotions, and due to its considerable connections with the hippocampal formation, this group of nuclei is also believed to be associated with learning and memory processes. The anterior nuclei are also associated with attention.

The **lateral dorsal (LD) nucleus** is the caudal continuation of the anterior thalamic nuclear group and may play a role in the expression of emotions. It receives inputs from the mammillary body via the mammillothalamic fasciculus, and in turn projects to the cingulate and parahippocampal gyri of the limbic system.

Medial nuclear group

The medial nuclear group resides medial to the internal medullary lamina, which separates it from the anterior and lateral nuclear groups

The **medial nuclear group** includes the more prominent **medial dorsal nucleus** (MD; also referred to as the medi-odorsal, or dorsomedial nucleus), as well as the diffuse periventricular nuclei (also referred to as the median or midline nuclei) that are located on the most medial surface of the thalamus (adjoining the third ventricle) and within the interthalamic adhesion. It should be noted that some authors classify the periventricular nuclei as a separate group of nuclei (and not part of the medial group).

The **MD nucleus** functions in the processing of information related to emotion. It is subdivided into the parvocellular (small cell) and magnocellular (large cell) divisions. The **parvocellular** division has reciprocal connections with the prefrontal association cortex, which functions in the integration of sensory information associated with higher-order cognitive functions. As part of the association loop of the basal nuclei, the globus pallidus and the substantia nigra project to the ventral anterior and the MD nuclei of the thalamus. These nuclei in turn project via thalamocortical fibers to the prefrontal cortex. The **magnocellular** division is associated with limbic system components such as the amygdaloid body and the orbitofrontal cortex. Additionally, the MD nucleus has extensive connections with the olfactory system, the hypothalamus, and the intralaminar, periventricular, and lateral posterior subnuclei of the thalamus. A lesion damaging the MD nucleus produces a diminution of anxiety, aggression, or obsessive thinking.

The **periventricular nuclei** function in the modulation of cortical excitability. They include the **dorsal periventricular nuclei**, which are associated with the striatum and the limbic cortex, and the **ventral periventricular nuclei**, which are associated with the hippocampus.

Lateral nuclear group

The lateral nuclear group is interposed between the internal and external medullary laminae, and consists of numerous subnuclei that collectively form the largest nuclear group of the thalamus

The **lateral nuclear group** is interposed between the internal and external medullary laminae, and consists of numerous subnuclei that collectively form the largest nuclear group of the thalamus. The subnuclei of this nuclear group are arranged into two step-like rows, the **dorsal** and **ventral tiers of nuclei** (subnuclei).

Dorsal tier of the lateral thalamic nuclei

These nuclei are connected with the association areas of the cerebral cortex and function in the integration of sensory input

The **dorsal tier of the lateral thalamic nuclei** is composed of lateral posterior nucleus and the pulvinar nucleus. With the exception of the lateral dorsal nucleus, these nuclei are connected with the *association areas of the cerebral cortex* and function in the integration of sensory input.

The **lateral posterior (LP) nucleus** is the rostral continuation of the pulvinar nucleus and is believed to play a role in complex sensory integration. Since these nuclei have comparable connections, they are also called the **pulvinar–LP complex**. The LP nucleus is interconnected with the superior

parietal lobule (eponym: Brodmann areas 5 and 7) and the precuneus. It also receives afferent terminals from components of the visual system: the superior colliculus, the pretectal area, and the occipital lobe.

The **pulvinar nucleus** (L., "cushion") is the most prominent nucleus of the thalamus and functions in the integration of visual, auditory, and somatosensory information. It has reciprocal projections with the association areas of the parietal, temporal, and occipital lobes. It also receives sensory information from components of the visual system (superior colliculus and dorsal lateral geniculate nucleus), the auditory system (medial geniculate nucleus), and the cerebellum. The pulvinar is involved in the integration of visual and auditory information as well as with eye movements.

Ventral tier of the lateral thalamic nuclei

The **ventral tier of the lateral thalamic nuclei** includes the **ventral anterior (VA) nucleus**, **ventral lateral (VL) complex,** and **ventroposterior (VP) complex**. The latter is further subdivided into the ventral posteromedial (VPM) nucleus, the ventral posterolateral (VPL) nucleus, and the ventral posterior inferior (VPI) nucleus. The ventral tier also includes the geniculate nuclei, including the **medial** and **dorsal lateral geniculate nuclei** (MGN and LGN, respectively).

The nuclei residing in the anterior aspect of the ventral tier of the system serve as relay stations for the somatic motor system and the brainstem reticular formation

The nuclei residing in the *anterior aspect* of the ventral tier of nuclei (VA and VL) serve as relay stations for the somatic motor system including the basal nuclei and the cerebellum. They also process inputs from the brainstem reticular formation.

The nuclei residing in the posterior aspect of the ventral tier of the thalamic nuclei serve as relay stations for all of the sensory systems

The nuclei residing in the *posterior aspect* of the ventral tier of nuclei (VPM, VPL, VPI, MGN, and LGN) serve as relay stations for all the sensory systems (the somatosensory, visual, auditory, and gustatory systems, but not the olfactory system).

Ventral anterior nucleus

The VA nucleus plays a role in the control of the movement of the eyes, face, head, and limbs

The **ventral anterior (VA) nucleus** consists of magnocellular (VAmc) and parvocellular (VApc) components. The **magnocellular** component is the site of termination of the nigrothalamic pathway, which conveys information from the ipsilateral reticular part of the substantia nigra. The **parvocellular** component is the site of termination of the pallidothalamic pathway, which transmits information from the medial segment of the globus pallidus. In addition, fibers from the intralaminar and midline nuclei of the thalamus, as well as from the brainstem reticular formation, terminate in the VA nucleus. This nucleus sends a spray of fibers throughout the frontal lobe. The magnocellular component of this nucleus is interconnected with the frontal eye field (eponym: Brodmann area 8). The parvocellular component of this nucleus is interconnected with the premotor cortex and supplementary motor area (eponym: Brodmann area 6). The VA nucleus plays a role in the control of movement. The information arising from the reticular part of the substantia nigra is involved in movements of the eyes, face, and head, whereas that arising from the medial segment of the globus pallidus is associated with movement of the limbs.

Ventral lateral complex

The VL complex exerts its influence on somatic motor activity via the basal nuclei and the cerebellum

The **ventral lateral (VL) complex** is further subdivided anatomically into an anterior (VLa) portion and a posterior (VLp) portion. This nucleus, as does the VA nucleus, consists of magnocellular and parvocellular components.

The **VLa** receives afferent fibers from the basal nuclei. Its magnocellular component receives projections from the reticular part of the substantia nigra, and its parvocellular component receives projections from the medial segment of the globus pallidus. The VLa in turn sends efferents to the medial surface of the **supplementary motor area** and the lateral surface of the **premotor cortex** (eponym: Brodmann area 6) of the cerebral cortex.

The **VLp** receives prominent projections arising from the deep cerebellar nuclei (mainly from the dentate nucleus) of the opposite side, via the dentatothalamic tract. The VLp then relays the information to the premotor and primary motor cortex (eponym: Brodmann areas 6 and 4, respectively). The medial region of the VLp projects to the head region of the **primary motor cortex**, whereas the lateral region of the VLp projects to the lower limb region of the primary motor cortex.

It is important to note that there is no overlap in the specific target site of termination of the afferent fibers arising from the substantia nigra, the globus pallidus, and the cerebellum. In addition, fibers from other thalamic nuclei and the reticular formation terminate in the VL nucleus. The VL nucleus exerts its influence on somatic motor activity via the basal nuclei and the cerebellum.

Ventroposterior complex

The VP complex relays processed sensory information to the primary somatosensory cortex located in the postcentral gyrus of the parietal lobe (eponym: Brodmann areas 3, 1, and 2)

The **ventral posterior (VP) complex**, also known as the ventrobasal complex, is subdivided into the VPM nucleus, VPL nucleus, and the smaller VPI nucleus, which is interposed

between the latter two nuclei. The VP complex relays processed sensory information to the primary somatosensory cortex located in the postcentral gyrus of the parietal lobe (eponym: Brodmann areas 3, 1, and 2). These cortical areas send corticothalamic fibers back to the VP nucleus.

The **VPM nucleus** receives the terminals of the trigeminothalamic tracts (the ventral and dorsal trigeminal lemnisci) relaying *general somatic afferent information* (touch, pressure, pain and temperature sensation, and proprioception) from the **orofacial region**, and special visceral afferent sensation (taste) from the nucleus of the solitary tract. Proprioceptive sensory input is relayed anteriorly, touch sensation in the middle, and nociceptive input posteriorly.

The **VPL** and **VPI nuclei** receive terminals of the anterior and lateral spinothalamic tracts relaying *general somatic afferent information* (pain, temperature, and nondiscriminatory [crude] touch sensation) from the body, terminals of the medial lemniscus transmitting *general somatic afferent information* (discriminatory [fine] touch, pressure, joint movement, and vibratory sensation) from the **body**. Fibers relaying sensory input from the cervical region terminate medially within the VPL, whereas fibers relaying sensory input from the sacral region terminate laterally. Also, proprioceptive sensory input is relayed anteriorly, touch sensation in the middle, and nociceptive input posteriorly within the VPL.

Note that the clinical case at the beginning of the chapter refers to a patient whose symptoms include headache and complete unilateral loss of sensation on the left side of the face, upper limb, and lower limb. A CT scan was diagnostic of a right thalamic hemorrhage.

1 Which thalamic nuclei were damaged by this cerebral hemorrhage, which caused a complete unilateral sensory loss from the face and upper and lower limbs in this patient?

2 Which ascending sensory tracts terminate in these thalamic nuclei?

Additionally, the VPI nucleus receives fibers from the vestibular nuclei (vestibulothalamic fibers) and projects to the vestibular area of the somatosensory cortex located deep within the central sulcus. These projections may be associated with the control of motor functions.

The VP complex sends projections to the primary somatosensory cortex (postcentral gyrus of the parietal lobe) and the secondary somatosensory cortex (parietal operculum)

Geniculate nuclei (Metathalamic nuclei)

The MGN processes auditory input from both ears, whereas the LGN processes visual input from both eyes

The **geniculate nuclei** include the medial and dorsal lateral geniculate nuclei that form two oval-shaped elevations (located inferior to the pulvinar of the right and left sides), referred to as the **medial** and **dorsal lateral geniculate bodies**, respectively.

The **medial geniculate nucleus** (MGN) is located on the posterior aspect of the thalamus, inferior to the pulvinar, and lateral to the superior colliculus. This nucleus receives afferent fibers relaying **auditory input** (from both ears) from the inferior colliculus, as well as some projections from other nuclei of the auditory system via the brachium of the inferior colliculus, and directly via the lateral lemniscus (the principal pathway of the auditory system). Sensory input of low-pitched sounds is relayed to the lateral aspect of the MGN, whereas high-pitched sounds are relayed to the medial aspect of the MGN. The MGN gives rise to the auditory radiation that terminates in the primary auditory cortex residing in the superior transverse temporal gyrus (eponym: gyrus of Heschl or Brodmann areas 41). The auditory association cortex (eponym: Brodmann area 42) is reciprocally connected with the MGN.

The **dorsal lateral geniculate nucleus** (**LGN**), a thalamic relay nucleus for **vision**, is larger than its medial counterpart. It usually consists of six laminae. Laminae 1 and 2 contain large cells and are referred to as the magnocellular layers. Layers 4–6 contain small cells and are referred to as the parvocellular layers. The majority of fibers of the optic tract, bringing information from both eyes, terminate here. Laminae 1, 4, and 6 receive visual input from the contralateral nasal hemiretina, whereas laminae 2, 3, and 5 receive visual input from the ipsilateral temporal hemiretina. This nucleus projects via the optic radiation to the **striate** (primary visual) cortex (eponym: Brodmann area 17) on the banks of the calcarine fissure of the occipital lobe. The LGN is reciprocally connected with the striate (primary visual) cortex; additionally, it is interconnected with the pulvinar and other subnuclei of the thalamus. The thalamocortical projections may play a role in arousal of the organism, the control of visual attention, and the regulation of visual input transmission to the striate (primary visual) cortex.

The **posterior nuclei of the thalamus** consist of the lateral posterior nucleus and the pulvinar (anterior/inferior/lateral/medial) nuclei. It processes nociceptive information relayed by the ascending sensory systems and in turn projects to the cerebral cortex.

Intralaminar nuclei

The intralaminar nuclei mark the rostral continuation of the ascending reticular activating system (ARAS) into the thalamus

The **intralaminar nuclei** mark the rostral continuation of the ascending reticular activating system (ARAS) into the thalamus. They consist of multiple, diffuse aggregations of nerve cell bodies embedded within the *internal medullary lamina* and the *interthalamic adhesion* of the thalamus. These nuclei receive sensory information from the spinothalamic tract, the ventral trigeminothalamic tract (ventral trigeminal lemniscus), and the multisynaptic ascending pathways of the reticular formation that relay pain sensation, pain perception, and arousal of the organism. Two of these intralaminar nuclei, the centromedian nucleus (CM) and parafascicular nucleus (PF), are readily perceptible in the posterior part of the internal medullary lamina. The centromedian nucleus is the most prominent, and is located medial to the VPM and VPL nuclei. The parafascicular nucleus is located medial to the centromedian nucleus.

The prominent **centromedian nucleus** (**CM**) is the site of termination of the fibers arising from the globus pallidus and the premotor (eponym: part of Brodmann area 6) and primary motor cortical areas (eponym: Brodmann area 4), whereas the **parafascicular nucleus** (**PF**) is the site of termination of fibers arising from the supplementary motor area (eponym: part of Brodmann area 6). Both of these nuclei send prominent projections to the basal nuclei (caudate nucleus and putamen). Their projections to the somatosensory areas of the cerebral cortex suggest that they function in *sensorimotor integration*. Furthermore, because of their widespread cortical connections they are believed to play a role in the control of cortical activity.

The intralaminar nuclei are not interconnected with other thalamic nuclei; they only project to each other.

Reticular nucleus of prethalamus

The reticular nucleus of prethalamus (thalamic reticular nucleus) receives collaterals of the thalamocortical, corticothalamic, thalamostriatal, and pallidothalamic fibers

The **reticular nucleus of prethalamus (thalamic reticular nucleus)** was named for its reticular (mesh-like) appearance and is not considered part of the brainstem reticular formation. It consists of a thin layer of nerve cell bodies that covers the external medullary lamina on the lateral surface of the thalamus. Some cells are embedded within the external medullary lamina. This is the lateral-most nucleus of the thalamus that is in direct contact with the posterior limb of the internal capsule, laterally.

The reticular nucleus of the prethalamus receives collaterals from almost all of the **thalamocortical fibers** destined for the cerebral cortex, and the **corticothalamic fibers** on their way to the thalamic subnuclei. The **thalamostriatal** and reciprocal **pallidothalamic fibers** also give rise to collaterals that synapse in the reticular nucleus of prethalamus.

Unlike the other thalamic nuclei, which project to the cerebral cortex, the reticular nucleus of prethalamus has *no known direct cortical connections*. Instead, it relays sensory input locally within the reticular nucleus of prethalamus, to the other thalamic nuclei, and to the reticular formation.

The reticular nucleus of prethalamus resides in a location where it not only receives a copy of, but can also monitor, thalamic communication with the cortex and the basal nuclei. This nucleus may play a role in the control of thalamic function.

Functional organization of the thalamic nuclei (subnuclei)

Functional organization of the thalamic subnuclei is based on projections between the thalamic nuclei and the cerebral cortex

In addition to the anatomical classification of the thalamic nuclei that was discussed in the previous sections, these nuclei are also categorized according to their function. Functional organization of the thalamic subnuclei is based on projections between the thalamic nuclei and the cerebral cortex (see Table 24.2). The thalamic nuclei have been classified into specific relay (sensory relay and motor relay) nuclei, association (multimodal) nuclei, and nonspecific nuclear functional groups.

The **specific relay nuclei** have reciprocal feedback connections with particular, well-defined, sensory or motor areas of the cerebral cortex (Fig. 24.5). All of the specific relay nuclei rest in the ventral tier of the lateral nuclear group.

The **sensory relay nuclei** of the thalamus include the VPM and VPL nuclei, the MGN, and the LGN. The VPM and VPL nuclei function in the processing and integration of somatosensory information arising from the orofacial region and body, respectively. The MGN and LGN function in the processing of special sensory information related to hearing and vision, respectively. The sensory relay nuclei in turn project specific sensory input from receptors located in peripheral structures to specific regions of the sensory neocortex (e.g., visual input from the retina to the primary visual cortex).

The VPM nucleus gives rise to axons that project to the head region on the lateral surface of the postcentral gyrus. The VPM relays somatosensory and taste sensation. Taste is relayed to the postcentral gyrus adjacent to the lateral sulcus.

Neurons located in the medial aspect of the VPL nucleus give rise to axons that ascend to terminate in the upper limb region located on the lateral surface of the postcentral gyrus, whereas neurons located in the lateral aspect of the VPL give rise to axons that ascend to the lower limb region located on the medial surface of the postcentral gyrus.

The MGN receives auditory input via the brachium of the inferior colliculus and then projects to the primary auditory cortex located on the transverse temporal gyrus (eponym: gyri of Heschl or Brodmann area 41) on the superior surface of the temporal lobe. Other auditory fibers terminate in the auditory association cortex located in the surrounding cortical association area (eponym: Brodmann area 42).

The dorsal lateral geniculate nucleus (LGN) receives fibers relaying visual input from both retinas. The right hemifield is relayed to the left LGN (and vice versa), which projects to the primary visual cortex (eponym: Brodmann area 17). The **motor relay nuclei** of the thalamus include the VA and VL nuclei. These nuclei relay motor information arising from the somatic motor system – including the basal nuclei and the cerebellum – to the motor cortical areas.

The dentate nucleus of the cerebellum gives rise to output fibers, the cerebellothalamic fibers that ascend to terminate in the posterior part of the contralateral VL nucleus of the

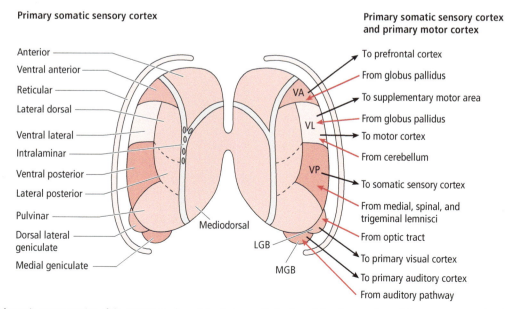

Figure 24.5 • Schematic representation of the superior surface of the thalamus. Thalamic nuclei and connections of the component nuclei of the thalamus. LGB, dorsal lateral geniculate body; MGB, medial geniculate body. Source: Modified from Fitzgerald, MJT & Folan-Curran, J (2002) *Clinical Neuroanatomy and Related Neuroscience*. WB Saunders, New York; fig. 22.1.

thalamus. In contrast, the anterior part of the VL nucleus receives input from the ipsilateral internal segment of the globus pallidus and the substantia nigra, both of which also project to the VA nucleus. The VL nucleus gives rise to fibers that ascend to project to the primary motor cortex (eponym: Brodmann area 4) and premotor cortex (eponym: Brodmann area 6), whereas the VA sends fibers to the premotor cortex and supplementary motor cortex (eponym: Brodmann area 6).

In addition to the nuclei mentioned above, the *anterior nuclear group*, the smallest collection of thalamic nuclei, is also included in the specific relay nuclei group. The anterior nuclear group, as a component of the **medial limbic circuit** (eponym: Papez circuit of the limbic system), receives afferent fibers from the subiculum of the hippocampal formation and from the mammillary bodies via the mammillothalamic fasciculus. The anterior nuclear group then projects to the cingulate gyrus which projects to the hippocampal formation in the temporal lobe. The anterior nuclear group plays a role in the processing of emotions, learning, and memory processes, via its connections to the limbic system.

The **association** (**multimodal**) **nuclei**, unlike the specific relay nuclei, receive sensory and motor information from the sensory and motor systems *indirectly* via a relay in other thalamic nuclei, and from many different parts of the brain.

The thalamic association (multimodal) nuclei include the MD, LD, LP, and pulvinar nuclei. These nuclei are referred to as association nuclei because they have widespread connections with the cortical association areas of the frontal, parietal, and temporal lobes. These nuclei are also referred to as multimodal nuclei because they process various modalities. They are involved in the processing of information related to memory and emotional expression, and also consolidate various types of sensory information.

The **nonspecific nuclei** of the thalamus include the intralaminar nucleus and reticular thalamic nucleus. These nuclei receive terminals arising from the caudate nucleus, putamen, cerebellum, and motor cortex. They are involved in the control of arousal and consciousness. Some authors also include the **periventricular nuclei** in this functional group. The periventricular nuclei are associated with the striatum and limbic structures and are believed to function in the modulation of cortical excitability.

CLINICAL CONSIDERATIONS

A tumor or stroke that results in thalamic damage and affects the VPM or VPL nuclei may at first result in diminished or complete loss of touch, pressure, pain or temperature sensation, or proprioception from the *contralateral* orofacial region or body, respectively. Recall that the VPM nucleus of the thalamus receives projections from the main sensory nucleus of the trigeminal (relaying discriminatory touch and pressure sensation) via the posterior trigeminal lemniscus, and the spinal nucleus of the trigeminal (relaying nondiscriminatory touch and nociceptive and temperature sensation) via the anterior trigeminal lemniscus from the orofacial region. The VPL nucleus receives projections from the posterior horn of the spinal cord relaying nondiscriminatory touch, pressure, and nociceptive and temperature sensation from the contralateral side of the body via the anterolateral system. In addition, the VPL also receives the axon terminals of the medial lemniscus which relays discriminative (fine) touch, vibratory sense, and proprioception from the contralateral side of the body.

Subsequent to the loss of sensation caused by a thalamic lesion, the individual experiences **paresthesias** (abnormal sensations such as tingling,

pinching, or burning, in the absence of an actual stimulus), which may progress to agonizing pain in the parts of the body where loss of sensation has occurred. This condition is referred to as the **thalamic pain syndrome** (eponym: Déjérine–Roussy syndrome). In many cases, even nonpainful, innocuous stimuli such as a light touch or the contact of clothing on the skin may trigger a painful sensation. This hypersensitivity is referred to as **allodynia** (a painful sensation elicited from an innocuous stimulus on normal skin), **hyperpathia** (an unusual and intense response to nociceptive stimuli), or **dysesthesia** (an abnormal sensation triggered by the sense of touch). The pain may originate in the thalamus, but the individual perceives the pain sensation as if it arises in the part(s) of the body where loss of sensation has occurred.

A lesion damaging the ascending sensory pathways to the thalamus (deafferentation), or the absence of sensory input to the thalamus from a certain part of the body (as occurs following amputation of a limb), may result in neural plasticity within the thalamus. Alterations in neural tissue may produce changes in neural circuitry within the thalamus, causing excruciating chronic pain.

SYNONYMS AND EPONYMS OF THE THALAMUS

Name of structure or term	Synonym(s)/eponym(s)
Association nuclei of the thalamus	Multimodal nuclei of the thalamus
Auditory radiations	Thalamocortical projections of the auditory system
Basal nuclei	Basal ganglia
Dentatothalamic tract	Cerebellothalamic tract
Medial dorsal (MD) nucleus of the thalamus	Dorsomedial (DM) or mediodorsal nucleus of the thalamus
Frontal eye field	Brodmann area 8
Optic radiation	Geniculocalcarine tract
	Geniculostriate tract
	Thalamocortical projections of the visual system
Interthalamic adhesion	Massa intermedia
Dorsal Lateral geniculate nucleus (LGN)	Lateral geniculate nucleus or body (LGB)
Medial geniculate nucleus (MGN)	Medial geniculate body (MGB)
Medial segment of the globus pallidus	Internal segment of the globus pallidus
Periventricular nuclei of the thalamus	Median nuclei of the thalamus
	Midline nuclei of the thalamus
Nondiscriminatory touch sensation	Crude touch sensation
Premotor cortex	Part of Brodmann area 6
Primary auditory cortex	Transverse temporal gyrus of Heschl
Primary motor cortex	Precentral gyrus
	Brodmann area 4
Striate cortex	Brodmann area 17
	Primary visual cortex
Somatosensory cortex	Somesthetic cortex
Supplementary motor area (SMA)	Part of Brodmann area 6
Thalamic radiations	Thalamocortical projections of the thalamus
Trigeminothalamic tracts	Dorsal and ventral trigeminal lemnisci
Ventral posterior complex	Ventrobasal complex

FOLLOW-UP TO CLINICAL CASE

A sensory deficit with sudden onset and focal symptoms may be caused by ischemic stroke or hemorrhage of a cerebral vessel affecting one or more of the sensory pathways on the right side of the brain. Headache is a nonspecific symptom and may be seen in either ischemic stroke (non-hemorrhagic) or in cerebral hemorrhage, though it may favor the latter. Even a very small hemorrhage is easily detectible on brain CT, and it becomes evident immediately. The CT in this case showed a moderately sized right thalamic hemorrhage. If this were a non-hemorrhagic stroke the CT would have been normal or may have revealed only very subtle changes. With time the stroke would have become apparent on CT (within 48 hours) as increasing hypodensity in the right thalamus.

The main priority in the treatment of this patient is to stabilize their vital signs. The location of this hemorrhage and the high blood pressure indicate that the probable cause of the hemorrhage was hypertension. However, the patient's blood pressure should not be abruptly lowered to "normal" levels. Intracerebral pressure increases following an ischemic stroke or hemorrhage, and it is essential that brain perfusion is maintained. This is accomplished by a modest reduction of the patient's blood pressure.

A lesion affecting a single side of the thalamus is usually easily diagnosed since it causes a pure or relatively pure contralateral sensory loss or alteration of sensation. Ischemic strokes and hemorrhages are the most common causes of thalamic lesions. Sometimes a thalamic lesion can cause severe, excruciating pain (referred to as thalamic pain) that is difficult to treat.

A bilateral thalamic lesion caused by an infarction of a variant of the posterior cerebral circulation (eponym: artery of Percheron) presents with patients obtunded, comatose, or agitated, with supranuclear vertical gaze palsy and memory deficits.

In contrast to a stroke that affects the thalamus, a stroke that produces a lesion involving a particular cortical area will produce symptoms such as aphasia (language dysfunction) if the language-dominant hemisphere (usually the left) is affected, or contralateral neglect if the nondominant hemisphere (usually the right) is affected. These symptoms are "cortical" symptoms and usually are indicative of a lesion (most often a stroke) affecting a particular area of the cerebral cortex.

Hypertension (systolic blood pressure > 130 mmHg and/or diastolic blood pressure > 80 mmHg) is one of the most common causes of intracerebral hemorrhages. It usually affects the basal nuclei, thalamus, brainstem, or cerebellum. Other causes of intracranial hemorrhage include anticoagulant medication, trauma, aneurysm, or vascular malformation. Blood pressure will resolve spontaneously with time. Depending on the size of the thalamic lesion, symptoms may resolve completely, improve partially, or remain the same over time. A very small thalamic stroke is sometimes clinically "silent" and is only noted incidentally and after the fact on a CT or MRI done for some other reason. Most hemorrhages require a neurosurgical consultation. Large-size hemorrhages sometimes need to be evacuated (removed).

3 Why is it crucial not to lower this patient's blood pressure to normal levels?

QUESTIONS TO PONDER

1. Which group of thalamic nuclei has extensive connections with the limbic system?

2. Which thalamic nuclei are included in the specific relay nuclear group?

3. Which thalamic nuclei receive projections from the cerebellum?

4. Which thalamic nuclei receive sensory information from the orofacial structures and body?

5. Which thalamic nuclei are involved in the control of arousal and consciousness?

CHAPTER 25

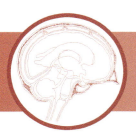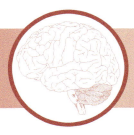

Cerebral Cortex

- CLINICAL CASE
- CELLS OF THE CEREBRAL CORTEX
- TYPES OF CORTEX
- CELL LAYERS OF THE NEOCORTEX
- VERTICAL COLUMNAR ORGANIZATION OF THE CEREBRAL CORTEX
- AFFERENTS (INPUT) TO THE CEREBRAL CORTEX
- EFFERENTS (OUTPUT) FROM THE CEREBRAL CORTEX
- INTERNAL CAPSULE AND CORONA RADIATA
- LOBES OF THE CEREBRAL CORTEX
- FUNCTIONAL ORGANIZATION OF THE CEREBRAL CORTEX
- CEREBRAL DOMINANCE
- CLINICAL CONSIDERATIONS
- SYNONYMS AND EPONYMS OF THE CEREBRAL CORTEX
- FOLLOW-UP TO CLINICAL CASE
- QUESTIONS TO PONDER

CLINICAL CASE

A 45-year-old patient, P.S., who was previously in perfect health, had two seizures in the past week. Both occurred suddenly and each lasted less than a minute. The first seizure happened while in bed, and it started as twitching of the right side of the face, which progressed to twitching of the right hand and then the right leg and foot. P.S. felt some perioral numbness as well. There was no loss of consciousness or other symptoms, and P.S. was aware of what had occurred. The second seizure, as described by P.S.'s partner, started with facial twitching, but P.S. rapidly lost consciousness and then had a generalized convulsion. P.S. bit their tongue and had blood in their mouth. P.S. also had urinary incontinence. P.S. began responding a few minutes after the convulsion but was still confused and somewhat agitated for about 30 minutes after that. P.S. had never had a seizure before. In between these spells P.S. had not experienced symptoms, and examination was normal.

A Textbook of Neuroanatomy, Third Edition. Maria A. Patestas, Amanda J. Meyer, and Leslie P. Gartner.
© 2025 John Wiley & Sons, Inc. Published 2025 by John Wiley & Sons, Inc.
Companion website: www.wiley.com.go/patestas/neuroanatomy3e

The **cerebral cortex** is the most complex component of the human brain, due to its complex and widespread connections. It functions in the planning and initiation of motor activity, perception and conscious awareness of sensory information, learning, cognition, comprehension, memory, conceptual thinking, and awareness of emotions.

The cerebral cortex (L., *cortex*, "bark") consists of 50–100 billion nerve cell bodies arranged into a three- to six-layered sheet that laminates the brain surface covering the paired and prominent cerebral hemispheres of the cerebrum, forming a superficial layer much like the bark that covers a tree. The majority of the cerebral cortex consists of phylogenetically the most highly evolved and complex neural tissue of the human brain.

The cerebral cortex overlies the subcortical white matter of the cerebral hemispheres. In other animals, the brain's cortical surface appears smooth, whereas in humans the brain surface is convoluted displaying prominent, alternating grooves and elevations as a result of the folding of the cerebral cortex during embryonic development. The elevations are referred to as **gyri**, whereas the grooves are referred to as sulci or fissures (Fig. 25.1). **Sulci** are shallow, short grooves, whereas **fissures** are deeper and more constant grooves, with a consistent location on the brain surface. The cortex forming the **gyri** dips down into the pit of the adjacent sulci or fissures to line them. Certain gyri, sulci, and fissures are similar in all typical human brains. Others, however, may vary in different brains and in the two cerebral hemispheres of the same brain. The gyri and sulci greatly increase the total surface area of the cerebral cortex. If the cerebral cortex of a normal human brain were spread out (that is, if the pleats formed by the sulci and fissures were stretched out), the cortex would extend over $0.23\,m^2$ ($2.5\,ft^2$). In the course of evolution, the cerebral cortex became increasingly folded into alternating gyri and sulci or fissures to facilitate its accommodation in the limited space available in the cranial vault.

The cerebral cortex covering the cerebral hemispheres is divided into the **frontal, parietal, temporal, occipital,** and **limbic lobes.** All of these lobes, with the exception of the limbic lobe, are visible on the lateral aspect of the brain. The limbic lobe is only visible on the medial surface of a midsagittally sectioned brain. In addition to the limbic lobe, the medial surface of the other lobes can also be seen on a midsagittal section of the brain.

The cerebral cortex is classified phylogenetically into the allocortex (heterogenetic cortex), mesocortex (juxtallocortex), and isocortex (homogenetic cortex). The **allocortex** (G., *allo*, "other") consists of the most primitive **archicortex (archipallium)** (G., *archaios*, "ancient") of the hippocampal formation and the more recent **paleocortex (paleopallium** or **periallocortex)** (G., *palaios*, "old") of the parahippocampal gyrus (entorhinal cortex), uncus (piriform cortex), and lateral olfactory gyrus of the olfactory system. The **mesocortex (juxtallocortex)** is a transitional cortex between the allocortex and isocortex, located in the cingulate gyrus and the insula. The **isocortex** is the most recent cortex, also referred to as the **neocortex (neopallium)**, constituting the frontal, parietal, temporal, and occipital lobes. The neocortex forms over 90% of the human cerebral cortex. The remainder is allocortex and mesocortex.

The archicortex consists of three cell layers, the paleocortex from three to five cell layers, and the mesocortex from three to six cell layers; the neocortex has six distinct cell layers.

The thickness of the cortex also varies from 1.5 mm in the primary sensory areas (being thinnest in the calcarine sulcus of the striate cortex) to being three times thicker (4.5 mm; in the motor and association cortical areas). In general, the

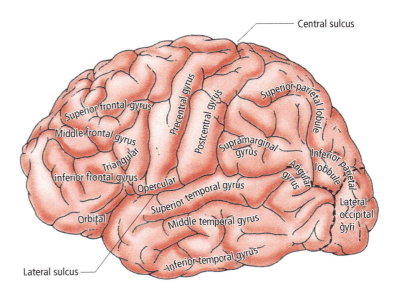

Figure 25.1 • Lateral view of the cerebral hemisphere showing the principal gyri and sulci of the cerebral cortex.

cortex is usually thickest at the peak of each gyrus and thin-nest at the bottom of each sulcus. The thickness reflects the different cytoarchitectonic arrangements of the various corti-cal areas. These include cell density, the types of cells present, fiber density, and the arrangement of the cells and fibers in each of the layers.

The neocortex of the frontal, parietal, temporal, and occip-ital lobes is anatomically divided into the **primary cortical areas** and **association cortical areas**. The primary cortical areas are subdivided into a primary motor area and the pri-mary sensory areas:

1 The **primary motor area** located in the precentral gyrus of the frontal lobe (eponym: Brodmann area 4) contains approximately one-third of the cell bodies of the upper motor neurons whose axons form the corticospinal tract.
2 The **primary sensory areas** receive sensory information from specific **thalamic nuclei**, including the following four nuclei:

- The ventral posterior medial (VPM) nucleus of the thalamus, which relays sensory information (touch, pressure, pain, temperature, and proprioception) from the orofacial region to the corresponding head region mapped in the **primary somatosensory cortex** in the postcentral gyrus of the parietal lobe (eponym: Brodmann areas 3, 1, and 2). It also relays input about head movement and orientation to the **primary ves-tibular area**, the inferior parietal lobule.
- The ventral posterior lateral (VPL) nucleus of the thalamus, which relays sensory information (touch, pressure, pain, temperature, and proprioception) from various parts of the body to the corresponding body part region mapped in the primary somatosen-sory cortex.
- The dorsal lateral geniculate nucleus (LGN) of the thala-mus, which relays a single modality – visual input from the contralateral visual field to the **striate (primary vis-ual) cortex** located in the banks of the calcarine sulcus of the occipital lobe (eponym: Brodmann area 17).
- The medial geniculate nucleus (MGN) of the thala-mus, which relays a single modality – auditory infor-mation from both ears to the **primary auditory cortex** in the transverse temporal gyri (eponym: transverse temporal gyri of Heschl in the lateral fissure of Syl-vius, Brodmann area 41).

The thalamus projects to the cerebral cortex of all five lobes via thalamocortical fibers.

Thalamocortical fibers relay most neural signals that arise from extracortical sources.

Most of the cerebral cortex consists of association cortex, which is classified into **unimodal association cortex** and **multimodal (heteromodal) association cortex**. Unimodal association areas are located next to, near, or around the pri-mary sensory cortices and expand on the functions of the respective primary areas. As an example, the striate (primary

visual) cortex (eponym: Brodmann area 17) projects to the visual association cortex (eponym: Brodmann areas 18, 19, 20, 21, and 37), which processes only a *single modality*, in this case vision, and is thus referred to as unimodal association cortex. In contrast, the multimodal association cortical areas receive inputs from *multiple sensory modalities*; they integrate the information and formulate a composite experience via their higher-order cognitive functions. The multimodal asso-ciation cortex is associated with imagination, judgment, deci-sion making, and formulating long-term plans.

CELLS OF THE CEREBRAL CORTEX

The multilayered cerebral cortex, which contains an esti-mated 86 billion neurons, is surprisingly populated by only three kinds of neurons: **pyramidal cells**, **fusiform cells**, and **stellate (granule) cells** (Fig. 25.2). Cells of the cerebral cor-tex have also been classified to include other cells; however, these other cells are merely variations of the pyramidal, fusi-form, and stellate cells.

Pyramidal cells

Pyramidal cells are the most abundant cortical cells and constitute about 75% of the cells of the cerebral cortex

Pyramidal cells may be small- to giant-sized and con-stitute approximately 75% of the cells of the cerebral cor-tex. They all have a pyramid-shaped cell body, with a dendrite emerging from its apex (Fig. 25.3). This **apical dendrite** is oriented perpendicular to, and courses toward, the cortical surface. Additional den-drites, **basal dendrites** arising from the basal aspect of the cell body, run horizontally and parallel to the cortical surface. A single axon emerges from the base of the pyramidal cell body. Some pyramidal cells have a short axon that courses away from the cortical surface and terminates locally in other cortical layers. The majority of pyramidal cell axons, however, are long, give rise to a recurrent collateral branch prior to leaving the cortex, and then enter the subcortical white matter. These long axons are destined to serve as com-missural or association or projection fibers. **Commissural fibers** form most of the corpus callosum and cross to the opposite cerebral hemisphere to synapse in the cerebral cor-tex. **Association fibers** project to cortical association areas in the ipsilateral hemisphere, whereas **projection fibers** leave the cortex, and project to other regions of the CNS, such as the thalamus, striatum, brainstem, or spinal cord.

Pyramidal cells are the main output neurons of the cerebral cortex

The majority of the pyramidal cells serve in the transmis-sion of information away from the cerebral cortex. Pyramidal cells liberate the neurotransmitters glutamate or aspartate and form excitatory synapses at their termination. Pyrami-dal cells are the main output neurons of the cerebral cortex.

CEREBRAL CORTEX ••• 497

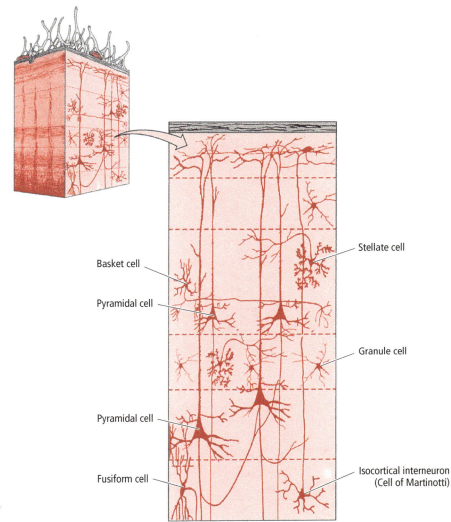

Figure 25.2 • The different types of neurons of the cerebral cortex.

Fusiform cells

The projection fibers arising from fusiform cells project mainly to the thalamus

Fusiform cells reside in the deepest cortical layer, and their dendrites project toward the cortical surface (Fig. 25.2). These cells, like the pyramidal cells, project via association, commissural, or projection fibers. The projection fibers arising from fusiform cells project mainly to the thalamus.

Stellate (granule) cells

Stellate cells have a short axon and project locally in the cerebral cortex

Stellate cells are small cells and are also referred to as **granule cells**. Since the dendrites of these cells radiate from their cell body in all planes, these cells have a star-shaped appearance (Fig. 25.2). A short axon emerges from the cell body, which projects locally, in the cerebral cortex.

There are two types of stellate cells based on their morphological and functional characteristics: one type is referred to as **aspiny stellate cells** (**smooth stellate cells**) since they do not have spines on their dendrites. These neurons are believed to be inhibitory interneurons synapsing with pyramidal cells; they utilize the inhibitory neurotransmitter, gamma-aminobutyric acid (GABA). The other type, located mostly in Layer IV of the cerebral cortex, are referred to as **spiny stellate cells** since they have spines on their dendrites. They are believed to be excitatory interneurons, which release the excitatory neurotransmitter, glutamate. Spiny and aspiny cells are located in all cortical layers with the exception of the most superficial layer.

Isocortical interneurons

The isocortical interneurons (eponym: cells of Martinotti) are located throughout the cerebral cortex and are most populous in the deepest cortical layer

The isocortical interneurons (eponym: cells of Martinotti) are multipolar interneurons displaying short dendrites (Fig. 25.2). Although they are located throughout the cerebral cortex, they are most populous

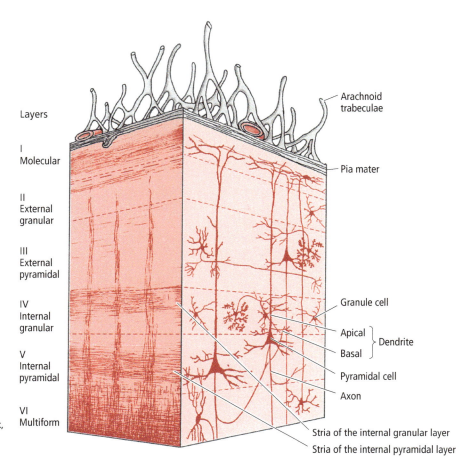

Figure 25.3 ● The histology of the cerebral cortex, illustrating the layers and cell types in each of the layers.

in the deepest cortical layer. Their axons, which give rise to several collateral branches, course toward the surface of the cortex. These cells regulate pyramidal cell activity in Layer I via synapses with apical dendrites of pyramidal cells to modulate dendritic spike generation and frequency-dependent disynaptic inhibition.

TYPES OF CORTEX

Allocortex

The **allocortex** consists of **archicortex** and **paleocortex**.

Archicortex

The archicortex is phylogenetically, the most ancient component of the cerebral cortex

The **archicortex** is phylogenetically, the most ancient component of the cerebral cortex. It consists of only three cell layers: a deep *polymorphic* layer, an *intermediate* pyramidal layer, and a superficial *molecular* layer. The archicortex is located only in the hippocampal formation. This type of cortex is associated with the limbic system, which is involved in memory processes and the expression of emotions.

Paleocortex

Phylogenetically, the paleocortex formed after the archicortex

Phylogenetically, the **paleocortex** formed after the archicortex and consists of three to five cell layers. It is located in the primary olfactory cortex (the piriform cortex, consisting of the lateral olfactory gyrus, anterior half of the uncus, and periamygdaloid cortex) and the secondary olfactory cortex, which includes the entorhinal cortex of the parahippocampal gyrus. It mediates the sense of smell and may also be involved in the processing of emotions.

Mesocortex

The mesocortex is a transitional cortex between the older allocortex and newer isocortex

The **mesocortex** (**juxtallocortex**) is a transitional cortex between the allocortex and isocortex. It consists of three to six layers and is located in the cingulate gyrus and the insula.

Isocortex (neocortex)

In general, the first four layers (I–IV) of the cerebral cortex serve as input stations (receiving corticopetal fibers), whereas the remaining layers (V and VI) are a major source of output (corticofugal) projection fibers

The **neocortex** consists of six cell layers. In addition to being named, its cell layers are numbered sequentially with Roman numerals (I–VI) in the order that they appear from superficial to deep, that is, from immediately deep to the pia mater to the subcortical white matter (Fig. 25.3).

Although the neocortex consists of six layers, each of these layers is not easily distinguishable in all cortical areas. In the **motor cortex (precentral gyrus)**, Layers II–V contain an enormous number of large pyramidal cells whose axons leave the cortex, and in their descent to lower centers course in the corona radiata, internal capsule, cerebral peduncle, anterior pons, and pyramids of the medulla oblongata to terminate in the brainstem and spinal cord. Non-pyramidal (granule) cells, on the other hand, are not as numerous in these layers of the motor cortex. Due to this morphological characteristic, the motor cortex is referred to as **agranular cortex**. Agranular cortex is typical of the cortical motor areas that are heavily populated by pyramidal cells.

Unlike the motor cortex, the primary **sensory cortex (postcentral gyrus)** contains small pyramidal cells and is heavily populated by non-pyramidal (granule) cells. These cells give the cortex of the primary sensory areas a granular appearance, so it is referred to as **granular cortex**.

In general, the first four layers (I–IV) of the cerebral cortex serve as input stations, receiving **corticopetal fibers**, whereas the remaining layers (V and VI) are a major source of output **corticofugal projection fibers.**

CELL LAYERS OF THE NEOCORTEX

Layer I: molecular (plexiform) layer

The molecular (plexiform) layer is the most superficial layer of the cerebral cortex, underlying the pia mater

The **molecular (plexiform) layer** is the most superficial layer of the cerebral cortex, underlying the pia mater (Fig. 25.3). As its name implies, the molecular layer consists mainly of interlacing nerve cell processes. The interlacing fibers present in this layer consist of both dendrites and axons from various cells. The dendrites arise from the apices of pyramidal and fusiform cells whose cell bodies reside in other (deeper) cortical layers. The axons consist of afferent (thalamocortical) axon terminals arising from the non-specific, intralaminar, and midline nuclei of the thalamus. These thalamocortical fibers also send collaterals to Layers V and VI of the cerebral cortex. In addition, other axons arise from stellate cells and isocortical interneurons. This layer contains a myriad of synaptic contacts.

Layer II: external granular layer

The external granular layer consists mainly of small granule (stellate) cells that give this layer a granular, stippled appearance

The **external granular layer** (Fig. 25.3) consists mainly of small granule (stellate) cells that give this layer a granular, stippled appearance; other cells include small pyramidal cells. The dendritic trees emerging from these cells extend superficially to the molecular layer. Their axons course in the opposite direction to either synapse locally in other cortical layers, or to join the axons of the subcortical white matter, and project to other cortical areas as association fibers.

Layer III: external pyramidal layer

The pyramidal cell axons of this layer are the principal source of corticocortical fibers (association or commissural fibers)

The **external pyramidal layer** (Fig. 25.3) consists predominantly of pyramidal cells. Those pyramidal cells that lie more superficially are medium sized, whereas those lying deeper within this layer, are large-sized cells. Some granule and isocortical neurons are also present. The apical dendrites of the pyramidal cells extend through the external granular layer to reach the molecular layer, whereas their axons join the subcortical white matter to project to other parts of the cerebral cortex mostly as association or commissural fibers. The pyramidal cell axons of this layer are the principal source of corticocortical fibers. In addition, some pyramidal cell axons, referred to as projection fibers, leave the cortex to terminate in other parts of the CNS. Some thalamocortical fibers from the specific relay thalamic nuclei terminate in this layer of the primary motor cortex. These thalamocortical fibers also send collaterals to Layers V and VI.

Layer IV: internal granular layer

Layer IV is the main afferent (input) station of the cerebral cortex and is thicker in the primary sensory areas

Layer IV is the main afferent (input) station of the cerebral cortex and is thicker in the primary sensory areas. The **internal granular layer** (Fig. 25.3) consists mostly of stellate (granule) cells. The stellate cells are further classified as **smooth (aspiny) stellate cells** and **spiny stellate cells**. The axons of the stellate cells remain in the cerebral cortex. The axons of pyramidal cells course through the deeper cortical layers either to synapse there, or join the axons of the subcortical white matter. The stellate cells receive synapses from fibers arising in the thalamic nuclei:

- the stellate cells of the **primary somatosensory cortex** receive **thalamocortical fibers** from the VPL and VPM nuclei of the thalamus;

- those in the **striate (primary visual) cortex** receive afferent fibers from the LGN; whereas
- those of the **primary auditory cortex** receive afferents from the MGN.

As the myelinated thalamocortical fibers enter the internal granular layer, they become oriented horizontally to form a layer of white matter, the **stria of the of the internal granular layer** (eponym: outer band of Baillarger). This layer of white matter (consisting of thalamocortical fibers from the LGN) is very prominent grossly in the striate (primary visual) cortex (and is eponymously referred to as the "stripe of Gennari"). Due to the striped appearance of the primary visual cortex, it is called the **striate cortex**. The thalamocortical fibers also send collaterals to Layers V and VI. Layer IV is the main **afferent (input) station** of the cerebral cortex and is especially thicker in the primary sensory areas.

Layer V: internal pyramidal layer

The internal pyramidal layer is the source of the majority of corticofugal (output) fibers of the cerebral cortex

The **internal pyramidal layer** (Fig. 25.3) is thickest in the motor areas of the cerebral cortex since it is a major source of efferent (output) fibers contributing to the corticonuclear and corticospinal tracts. This layer is the source of the majority of corticofugal (output) fibers of the cerebral cortex that descend to terminate in the corpus striatum, brainstem and spinal cord.

The internal pyramidal layer consists primarily of large and medium-sized pyramidal cells, but stellate and isocortical interneurons are also present in this layer. The **primary motor cortex** (eponym: Brodmann area 4) located in the precentral gyrus of the frontal cortex contains a group of **giant pyramidal cells** (eponym: cells of Betz), which are unique to Layer Vb of this cortical area, the source of the majority of **corticofugal (output) fibers** of the cerebral cortex. The pyramidal cells of this layer give rise to axons that form the corticotectal, corticorubral, corticoreticular, corticopontine, corticonuclear, and corticospinal tracts. All of these tracts terminate at various levels of the brainstem, except for the corticospinal tracts that terminate in the spinal cord. Moreover, corticostriate fibers arising from pyramidal cells of this layer descend to terminate in the striatum (caudate nucleus and putamen) of the basal nuclei. Layer V gives rise to fibers that function in *motor* activity.

In addition to the neurons and their processes, this layer also contains a group of horizontally oriented myelinated fibers, collectively forming a prominent stripe of white matter, the **stria of the internal pyramidal layer** (eponym: inner band of Baillarger), located in the deep aspect of Layer V. This stripe of white matter consists of collateral branches arising from the axons of cortical association neurons that are destined for Layers II and III of the cerebral cortex, as well as collateral branches arising from the pyramidal cells residing in this layer. Layer V is thickest in the motor areas

of the cerebral cortex since it is a major source of efferent fibers contributing to the corticonuclear and corticospinal tracts and thinnest in the primary sensory areas.

In addition to the prominent horizontally oriented stria present in Layers IV and V, there are also vertically oriented, but less prominent, groups of afferent and efferent axons that pass through the cortical layers.

Layer VI: multiform (fusiform) layer

The multiform (fusiform) layer is the deepest layer of the cerebral cortex and lies immediately next to the white matter underlying the cortex

The **multiform (fusiform) layer** is the deepest layer of the cerebral cortex and lies immediately next to the white matter underlying the cortex (Fig. 25.3). The multiform layer consists mostly of fusiform cells, although pyramidal cells and interneurons are also present. The axons of the fusiform and pyramidal cells course in the underlying white matter as association and commissural fibers that terminate in the cerebral cortex, or as corticothalamic projection fibers that terminate in the thalamus.

VERTICAL COLUMNAR ORGANIZATION OF THE CEREBRAL CORTEX

The cerebral cortex is not only organized into six cell layers that run parallel to the cortical surface, but it is also organized into cell **columns**, which are aligned perpendicular to the cortical surface. The cortical columns are believed to hold the key to cortical function and represent specific functional units. As an example, within the somatosensory cortex, the cells of certain specific columns are stimulated by impulses related to touch, other columns respond to nociception, whereas others respond to proprioceptive input.

AFFERENTS (INPUT) TO THE CEREBRAL CORTEX

The cerebral cortex receives **afferent (input) fibers**: corticocortical (association and commissural) excitatory fibers (which release glutamate or aspartate) from ipsilateral and contralateral cortical areas, thalamocortical excitatory fibers from thalamic nuclei, cholinergic excitatory fibers from the basal forebrain nuclei, noradrenergic inhibitory fibers from the brainstem locus caeruleus, and serotonergic inhibitory fibers from the brainstem raphe nuclei.

The afferent fibers arising from the VPL, VPM, ventral lateral (VL), and ventral anterior (VA) nuclei and the MGN and LGN of the thalamus establish synaptic contacts with the **stellate cells** residing in **Layer IV** of the cerebral cortex. Fibers arising from other thalamic nuclei and the remaining cortical input sources establish synaptic contacts in Layers I–IV of the cerebral cortex.

The cholinergic and serotonergic afferents to the cerebral cortex function in the control of sleep and arousal.

EFFERENTS (OUTPUT) FROM THE CEREBRAL CORTEX

The cerebral cortex gives rise to efferent fibers that are grouped as association, commissural, or projection fibers

Cortical **efferent fibers** arise from the **output neurons** of the cerebral cortex: the **pyramidal** and **fusiform cells**. Their myelinated axons course deep, pass into the subcortical white matter, and then are distributed to widespread regions of the CNS. Thus, the subcortical white matter consists of these three main types of fibers: **association, commissural,** and **projection fibers**.

Association fibers

The association cortex of the parietal, occipital, and temporal lobes is connected with the frontal lobe via prominent fasciculi composed of association fibers spanning the four lobes

Association fibers consist of axons arising from small pyramidal cells, primarily from cortical layers II and III. These fibers vary in length from short to long, and project to the ipsilateral hemisphere. Association fibers that connect various cortical areas make up most of the subcortical white matter. These fibers gather to form fasciculi that connect different lobes, but, similar to a two-way highway, fibers merge into and exit these fasciculi all along their course.

The **short association fibers** connect adjacent gyri. These fibers leave the cortex of one gyrus, enter the underlying white matter, and then loop around the pit of a sulcus and project into the cortex of an adjacent gyrus. Consequently, these fibers are also referred to as **U-fibers** or **arcuate fibers** (Fig. 25.4A). These fibers bridge the primary sensory areas with the adjacent cortical association areas.

The **long association fibers** connect different lobes of the same hemisphere. These fibers leave the cortex of one lobe, enter the underlying white matter, gather to form a bundle, and then project to the cortex of a different lobe of the same hemisphere.

Bundles of long association fibers form the following fasciculi: the superior and inferior longitudinal fasciculi, the superior and inferior occipitofrontal fasciculi, and the cingulum (Fig. 25.4).

The **superior longitudinal fasciculus** interconnects the frontal lobe with the parietal, temporal, and occipital lobes. As these fibers link these lobes, they form a dome over the insula and radiate within the parietal, temporal, and occipital lobes.

The superior longitudinal fasciculus contains a group of fibers, the **arcuate fasciculus**, which forms a distinct arch posterior to the insula. This fasciculus links the frontal lobe motor area for speech in the **inferior frontal gyrus** (eponym: Broca area) with the temporal lobe area for language comprehension in the posterior part of the **superior temporal gyrus** (eponym: Wernicke area).

The **inferior longitudinal fasciculus** connects the temporal and occipital lobes.

The **superior occipitofrontal fasciculus** interconnects the frontal lobe with the occipital lobe. The fibers of the **inferior occipitofrontal fasciculus**, which like their superior counterpart also interconnect the frontal and occipital lobes, radiate in the frontal lobe, pass through the temporal lobe inferior to the insula, to radiate also in the occipital lobe. Some of its fibers curve around on the deep aspect of the lateral sulcus, interconnecting the frontal and temporal lobes, and as a result of their hook-like path are referred to as the **uncinate fasciculus** (L., *uncus*, "hook"). Deficits of connectivity between the frontal and temporal lobes have been linked to schizophrenia.

The **cingulum**, a part of the medial limbic circuit, is embedded in the cingulate gyrus, and its fibers form a connection between the anterior perforated substance and the parahippocampal gyrus. Deficits of connectivity in the cingulum have been linked to bipolar disorder.

Commissural fibers

Commissural fibers arise in one cerebral hemisphere and cross the midline to terminate in the corresponding cortical area of the contralateral hemisphere

Commissural fibers arise in one cerebral hemisphere and cross the midline to terminate in the corresponding cortical area of the contralateral hemisphere (Fig. 23.5A). These fibers, consisting of the myelinated axons of medium-sized pyramidal neurons residing in various layers of the cerebral cortex, provide an avenue of communication between the corresponding cortical areas of the two hemispheres.

The **corpus callosum** is the largest commissure in the entire nervous system, comprising more than 200 million myelinated neuronal axons linking mainly *corresponding* areas of the neocortex of the two cerebral hemispheres. The corpus callosum, however, also contains commissural fibers that arise in certain cortical areas of one cerebral hemisphere and cross to the opposite side to terminate in *noncorresponding* areas of the contralateral hemisphere. This is true for fibers arising in the **primary visual cortex** of one cerebral hemisphere that cross to the opposite side to terminate in the **visual association cortex** of the contralateral hemisphere. Similarly, parts of the **motor** and **somatosensory cortical areas**, where the hand is represented, do not receive commissural fibers from the corresponding motor and sensory cortical areas of the contralateral hemisphere. However, the somatosensory association areas receive commissural fibers that connect the two cerebral hemispheres; thus, each hemisphere samples information available to the contralateral hemisphere.

The corpus callosum consists of the following component parts: the rostrum and genu anteriorly, the body forming

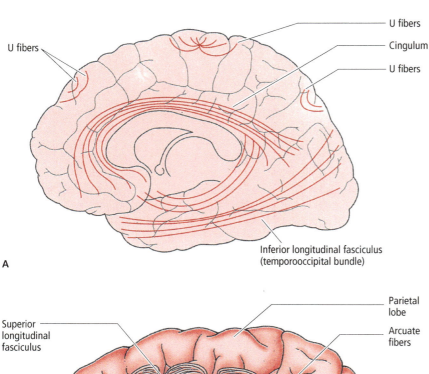

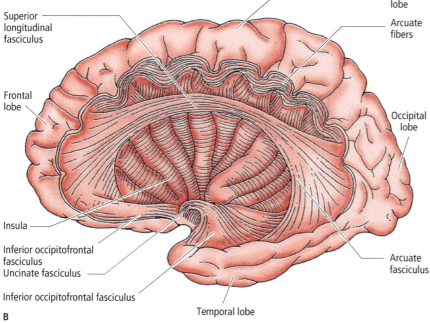

Figure 25.4 ● (A) Schematic representation of the major association fiber system (lateral view). (B) Three-dimensional schematic representation of the major association fiber systems (lateral view).

the roof of the lateral ventricles, and the splenium posteriorly. The **rostrum** and **genu** consist of commissural fibers connecting the anterior cortical areas of the frontal lobe. The **body** consists of most of the commissural fibers that connect the remaining frontal lobe, parietal lobe, and part of the temporal lobe. Posteriorly, the **splenium** consists of commissural fibers that connect the occipital lobes. Cortical commissural fibers emerge mainly from the pyramidal cells residing in the cortical superficial layers and the external granular (layer II) and external pyramidal (layer III) layers.

Due to its size and position, the corpus callosum is the most vulnerable of the white matter tracts and is the most frequently reported site of damage following mild traumatic brain injury. The corpus callosum is also a target of multiple sclerosis, Alzheimer disease, bipolar disorder, and schizophrenia.

Other commissures of the brain include the **anterior** (Fig. 25.5B) and **posterior commissures** that bridge the temporal lobes, and the **hippocampal commissure (commissure of the fornix)** connecting the hippocampal formation of the two hemispheres.

Commissural fibers form an anatomical bridge connecting the two cerebral hemispheres, which is essential in the integration of information in the two sides of the brain.

CEREBRAL CORTEX ••• 503

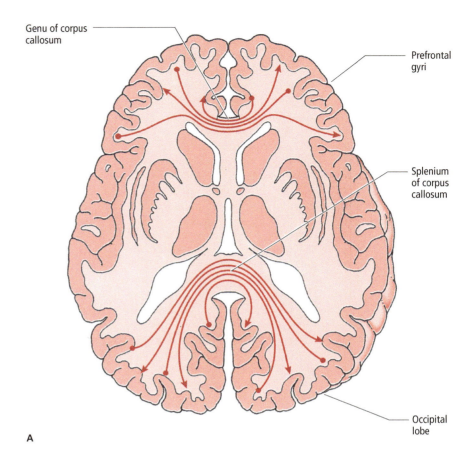

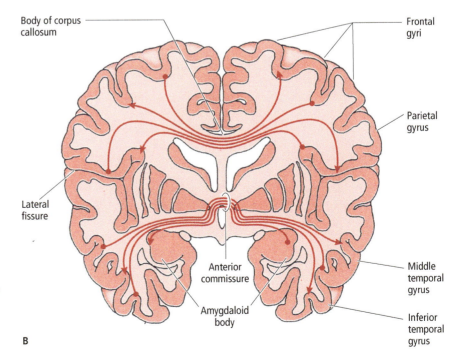

Figure 25.5 • (A) Horizontal schematic representation of the brain showing the fiber connections of the genu and splenium of the corpus callosum. (B) Coronal schematic representation of the brain showing the fibers of the anterior commissure and the body of the corpus callosum.
Source: Adapted from Young, PA & Young, PH (1997) *Basic Clinical Neuroanatomy*/Wolters Kluwer.

Projection fibers

Projection fibers consist of both afferent and efferent fibers to and from the cerebral cortex, that connect it to the thalamus, basal nuclei, brainstem, and spinal cord

The **afferent fibers** to the cerebral cortex are also referred to as **corticopetal fibers**. They include: (1) **thalamocortical fibers** arising from the **VPL and VPM nuclei of the thalamus** relaying sensory information such as touch, pressure, conscious proprioception, two-point tactile sensation, as well as pain and temperature sensation from the body and head to the **sensory cortex**; (2) the **MGN of the thalamus** relaying auditory information via the auditory radiation to the **auditory cortex**; and (3) the **LGN of the thalamus** relaying visual information via the optic radiation to the **visual cortex**. Although these afferent fibers terminate in Layers I–IV of the cerebral cortex, most terminate in Layer IV.

The **efferent fibers** from the cerebral cortex are referred to as **corticofugal fibers**. These fibers consist of axons arising from large pyramidal cells that course in the corona radiata and internal capsule to terminate in the basal nuclei, the brainstem, and the spinal cord. Corticofugal fibers include fibers of the corticostriate, corticoreticular, corticorubral, corticonuclear, corticopontine, corticotegmental, and corticospinal tracts.

INTERNAL CAPSULE AND CORONA RADIATA

The internal capsule is a massive, fan-shaped collection of fibers (white matter) that connect the thalamus to the cerebral cortex

The **internal capsule** is a massive, fan-shaped collection of fibers (white matter) that connect the thalamus to the cerebral cortex. On horizontal section, it appears as a wide letter "L" with its apex directed toward the midline of the cerebrum. It is divided into an **anterior limb**, a **genu** (L., "knee"), and a **posterior limb**. The two limbs converge toward the genu, the angle between the anterior and posterior limbs. The anterior limb is interposed between the head of the caudate nucleus and the lentiform nucleus. The posterior limb is interposed between the thalamus and the globus pallidus. The majority of the internal capsule fibers traversing the cerebrum connect the thalamus to the cerebral cortex. Descending motor fibers arising from the motor cortex destined for the brainstem and the spinal cord also descend in the internal capsule. In addition to the above components, the internal capsule also has a **sublentiform limb** whose fibers run anterior to the lentiform nucleus and a **retrolentiform limb** whose fibers run posterior to the lentiform nucleus.

The **anterior limb** of the internal capsule includes the (1) anterior thalamic radiations (corticothalamic and thalamocortical fibers), which serve as a channel between the anterior and medial dorsal nuclei of the thalamus and the cerebral cortex, the (2) corticostriate fibers to the striatum (caudate nucleus and putamen), and the (3) frontopontine fibers to the pontine nuclei in the basal pons.

The **genu** of the internal capsule transmits corticonuclear fibers arising from the motor cortex in the frontal lobe and the somatosensory cortex of the parietal lobe. The corticonuclear fibers descend to terminate in the cranial nerve motor nuclei (of the trigeminal, facial, glossopharyngeal, vagus, accessory, and hypoglossal nerves) located in the pons, medulla oblongata and spinal cord. As the corticonuclear fibers descend in the genu of the internal capsule, they shift posteriorly to descend in the anterior part of the **posterior** limb of the internal capsule. Other fibers traversing the genu are the thalamocortical fibers from the motor nuclei of the thalamus, the ventral anterior (VA), and ventral lateral (VL) nuclei.

The **posterior limb** accommodates the (1) corticonuclear fibers that shift posteriorly from the genu in their descent in the internal capsule, the (2) parietopontine, thalamoparietal, corticoreticular, and corticorubral fibers immediately posterior to the genu, whereas the (3) corticospinal tract descends in the posterior third or half of the posterior limb. As the fibers of the corticospinal tract descend in the internal capsule, they shift posteriorly. The central thalamic radiations include the thalamocortical projections arising from the ventral posterior (VP) complex of the thalamus that terminate in the primary somatosensory cortex. The **sublentiform limb** carries the auditory radiations arising from the MGN to terminate in the transverse temporal gyri (eponym: of Heschl). The **retrolentiform limb** carries visual information from the LGN to the visual cortex via the optic radiations.

As the fibers of the internal capsule emerge from between the caudate nucleus and the lentiform nucleus, they spread out forming a curved radiating crown, the **corona radiata** (the continuation of the internal capsule), containing afferent and efferent fibers to and from the cerebral cortex.

LOBES OF THE CEREBRAL CORTEX

The cerebral cortex of each cerebral hemisphere is divided anatomically into five lobes: the frontal, parietal, temporal, occipital, and limbic lobes

The cerebral cortex of each cerebral hemisphere is divided anatomically into five lobes: the frontal, parietal, temporal, occipital, and limbic lobes (Fig. 25.6). All of the lobes are visible on the lateral aspect of the brain, with the exception of the limbic lobe, which is visible only from the medial surface of the brain.

The **frontal lobe** is involved in motor control, cognitive functions, the processing of emotions, and the motor aspects of language. The **parietal lobe** functions in sensation; it processes touch, pressure, pain and temperature sensations, and proprioception. The **temporal lobe** contains the cortical areas that process hearing, memory, emotions, and the sensory aspects of speech. The **occipital lobe** has the exclusive function of processing visual information, and the **limbic lobe** functions in the processing of emotions.

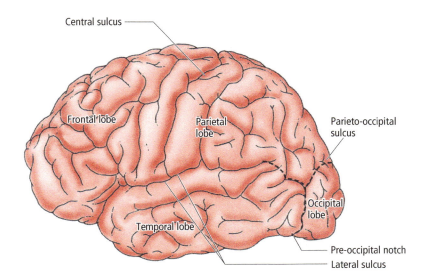

Figure 25.6 • Lateral view of the brain showing four of the five lobes of the cerebrum.

Borders

The **frontal lobe** extends from the anterior pole of the cerebral hemisphere, posteriorly to the central sulcus (eponym: of Rolando) and inferiorly to the lateral fissure (eponym: of Sylvius) (see Fig. 25.5).

The **parietal lobe** extends from the central sulcus anteriorly, to the parieto-occipital sulcus posteriorly and inferiorly, and the lateral fissure laterally and inferiorly.

The **temporal lobe** extends from the lateral fissure superiorly, to its free inferior border, and posteriorly to an imaginary line extending from the posterior extent of the lateral fissure inferior–posteriorly, separating it from the occipital lobe.

The **occipital lobe** extends from the parieto-occipital sulcus to the preoccipital notch, which accommodates the lateral pole of the cerebellar hemisphere. It consists of the cortex covering the posterior inferior pole of the cerebral hemisphere.

The **limbic lobe** is not visible laterally. Unlike the other lobes, which consist of only neocortex, it is composed of archicortex, mesocortex, and some neocortex and surrounds the corpus callosum.

FUNCTIONAL ORGANIZATION OF THE CEREBRAL CORTEX

In the early 20th century the cerebral cortex was classified by Brodmann into 52 different cytoarchitectural areas

The cerebral cortex consists of certain cell types arranged into three to six layers, as described previously. The thickness of these layers, and the number and type of cells within each layer, however, varies in different regions of the cortex.

In the early 20th century the cerebral cortex was classified by Brodmann into 52 different cytoarchitectural areas, but he did not assign a function to each area at that time. He numbered them 1–52, which reflected the order in which he examined and mapped them (Fig. 25.7). Additional studies have provided information correlating a function to many of these areas, referred to as **Brodmann areas**. The anatomical/functional correlation, however, is not as accurate as was thought to be in the past. Some of these areas are so commonly referred to that both name and number should be known by the student. These areas are: the primary somatosensory cortex (Brodmann areas 3, 1, and 2), motor cortex (Brodmann areas 4, 6, and 8), secondary sensory cortex (Brodmann areas 5 and 7), striate (visual) cortex (Brodmann areas 17, 18, and 19), and auditory cortex (Brodmann areas 41 and 42). The functional areas of the cerebral cortex are shown in Fig. 25.8.

More recent studies using multi-modal magnetic resonance images from the Human Connectome Project have delineated 180 areas per hemisphere with unique cortical morphology, function, and connectivity.

Cortical areas controlling motor activity

The motor areas of the cerebral cortex located anterior to the central sulcus control the motor activity of the entire opposite side of the body

The *motor activity* of the entire opposite side of the body is controlled by the **motor areas** of the cerebral cortex that are located anterior to the central sulcus. Fibers arise from the primary motor cortex (MI), the premotor cortex, the supplementary motor area, and the primary somatosensory (somesthetic) cortex (SI) of the frontal and parietal lobes to terminate in the motor nuclei of the brainstem and spinal cord.

Primary motor cortex

The precentral gyrus of the cerebral hemisphere controls movement of the contralateral head and body

The **primary motor cortex** (eponym: Brodmann area 4) resides in the precentral gyrus and in the paracentral lobule of the *frontal lobe*. It plays an important role in the execution of distinct,

506 ●●● CHAPTER 25

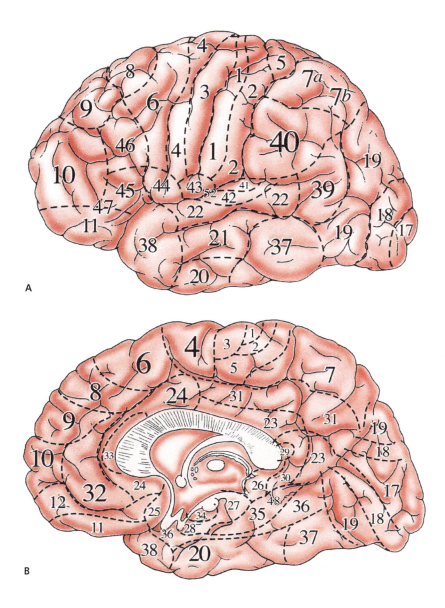

Figure 25.7 ● Views of the cerebral hemisphere showing the cytoarchitectural map of the cerebral cortex: (A) lateral view and (B) medial view. Numbers represent Brodmann's areas. Source: Adapted from Noback, CR et al. (1996) *The Human Nervous System*/Springer Nature.

well-defined, *voluntary* movement (Fig. 25.8). The precentral gyrus of the right cerebral hemisphere controls movement of the left side of the head and body, whereas that of the left cerebral hemisphere controls movement of the right side of the head and body. The body is mapped on the primary motor cortex somatotopically as an upside down homunculus (L., "little person") (Fig. 25.9). The region of the primary motor cortex responsible for toe movement is situated on its superior medial aspect above the corpus callosum, whereas the area of the precentral gyrus governing movement of the tongue, mouth, and larynx lies near its inferior lateral aspect, adjacent to the lateral fissure. Notably, a substantial portion of the primary motor cortex in humans is dedicated to controlling hand, tongue, lip, and larynx movements, underscoring human capacity for manual dexterity and speech production. Furthermore, it plays a crucial role in regulating limb muscle activity, particularly in controlling the fine muscles of the hands and feet. More recently identified are three distinct Somato-Cognitive Action Networks (SCAN) linking the motor regions for the lower limb, upper limb, and head. These SCAN regions exhibit reduced cortical thickness, display strong interconnectivity among themselves and with the cingulo-opercular network – an essential component for coordinating actions, physiological responses, attention, error monitoring, and pain processing. Nerve cells in the primary motor cortex are organized into groups, each group sending its axons to the cranial nerve motor nuclei, or the reticular formation, or the anterior horn of the spinal cord gray matter, where they control the motor activity of a single muscle. The total cortical area that mediates motor activity of a particular body region is proportional to the complexity of the movements produced in that region.

The primary motor cortex receives afferent fibers from the other motor areas and the somatosensory cortex. Prior to the onset of movement, the primary motor cortex receives

CEREBRAL CORTEX • • • 507

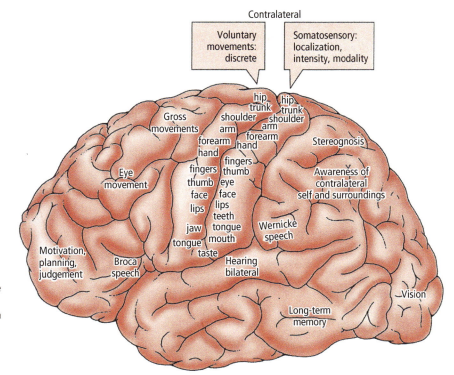

Figure 25.8 • Lateral view of the cerebral hemisphere showing the functional areas of the cerebral cortex. Broca = motor speech area. Wernicke = sensory speech area. Stereognosis is the ability to identify an object by touch without visual or auditory input. Source: Young, PA & Young, PH (1997) Basic Clinical Neuroanatomy/ Wolters Kluwer.

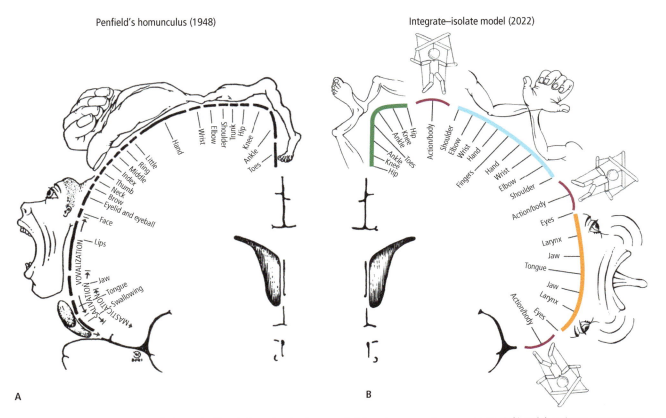

Figure 25.9 • Coronal section through the cerebral hemisphere showing homunculi of the primary somatosensory cortex (left) and the primary motor cortex (right). Source: Gordon et al. (2023)/Springer Nature/CC BY 4.0.

508 ●●● **CHAPTER 25**

instructions and information about the pattern of the intended movement from the other motor areas of the cortex.

Premotor cortical areas

The principal function of the premotor cortical areas is the programing of complex motor activity, which is then relayed to the primary motor cortex, where the execution of motor activity is initiated

The **premotor cortical areas** consist of three regions: the premotor cortex, the supplementary motor area, and the frontal eye field. All of these motor cortical areas reside in the frontal lobe (anterior to the central sulcus). The posterior parietal area located in the parietal lobe is involved with movements that require visual guidance. The principal function of the premotor cortical areas is the programing of complex motor activity, which is then relayed to the primary motor cortex where the execution of motor activity is initiated. The primary motor cortex then translates this information into the execution of movement and relays it mainly to the brainstem or spinal cord. Most of the nerve signals that arise in the premotor cortical areas mediate complex movements produced by groups of muscles performing a particular task, unlike the discrete muscle contractions elicited by stimulation of the primary motor cortex.

The **premotor cortex** (**PMC**) corresponds eponymously to most of Brodmann area 6 on the lateral aspect of the frontal lobe. The PMC receives axon terminals from the cerebellum via a relay in the ventral lateral (VLp) nucleus of the thalamus, and fibers from the cortical association areas. Input from the cerebellum assists the PMC in its role in movement. The main function of this area is the motor control of axial and proximal limb musculature. It also plays a role in orienting the body and upper limbs in the direction of a target. Once a motor task has been initiated, activity in the PMC diminishes, reflecting its key function in the planning phase of motor activity.

The **supplementary motor area** (**SMA**) corresponds to the medial and superolateral aspect of Brodmann area 6. This area receives axon terminals from the basal nuclei (the globus pallidus and the reticular part of the substantia nigra) via a relay in the ventral lateral (VLa) nucleus of the thalamus. Input from the basal nuclei assists the SMA in its role in the programing phase of the patterns and sequences of complex movements, and the coordination of bilateral movements. The SMA mediates muscle contractions of the axial (trunk) and proximal limb (girdle) musculature (i.e., the muscles controlling movements of the arm and thigh). Interestingly, functional magnetic resonance imaging (fMRI) studies have revealed that during execution of a complex movement, both the supplementary motor cortex and the primary motor cortex are involved and become active, whereas when an individual merely mentally practices the movement, only the supplementary motor cortex is active, and the primary motor cortex is quiescent. Stimulation of the SMA prompts a desire to move, or that of an impending movement.

The **frontal eye field** (**FEF**) corresponds eponymously to Brodmann area 8. This area is located anterior to the premotor area, on the posterolateral aspect of the prefrontal cortex. The FEF area contains nerve cell bodies whose axons form the corticotectobulbar tract and descend to terminate in the eye movement control centers located in the midbrain reticular formation and the paramedian pontine reticular formation. They, in turn, transmit information to the motor nuclei of cranial nerves III, IV, and VI, which control voluntary eye movements. The FEF plays a role in the coordination of eye movements, particularly those mediating voluntary visual tracking of a moving object via conjugate deviation of the eyes. Stimulation of the FEF causes conjugate ocular deviation to the opposite side. A lesion damaging the FEF results in deviation of the eyes to the side of the lesion with the inability to look to the opposite side.

The **posterior parietal area** corresponds to the superior parietal lobule (eponym: Brodmann areas 5 and 7). Area 5 is associated with tactile discrimination and stereognosis, as well as statognosis in relation to reaching and guiding movements (Fig. 25.8). Area 7 (the posterior parietal visual area) is involved with movements that require visual guidance.

Primary sensory areas of the cerebral cortex

The **primary sensory areas** are involved in distinguishing and integrating sensory input relayed there primarily by the thalamic nuclei.

Primary somatosensory (somesthetic) cortex (SI)

The primary somatosensory cortical area receives sensory information arising from the contralateral half of the head and body

The **primary somatosensory cortex** is located in the postcentral gyrus on the lateral surface of the parietal lobe (eponym: Brodmann areas 3a, 3b, 1, and 2; see Fig. 25.1A). The primary somatosensory cortex also comprises the dorsal part of the paracentral lobule, which resides and is visible only on the medial surface of the cerebral hemisphere. This cortical region is the site of termination of the thalamocortical fibers arising from the VPL and VPM nuclei of the thalamus. These fibers relay sensory information from the contralateral half of the body and head, respectively. The body is represented on the somatosensory cortex by an upside-down homunculus (Fig. 25.9). The foot is represented on the superior medial surface of the hemisphere, the leg, thigh, trunk, shoulder, arm, and forearm on the superior surface of the postcentral gyrus, whereas the hand, head, teeth, tongue, larynx, and pharynx are on the lateral surface of the postcentral gyrus. The total cortical area representing a body part is proportional to the discriminative sensory capability of the body part.

The primary sensory cortex also includes Brodmann areas 3a and 3b, which are buried in the central sulcus. Area 3a, located anteriorly, adjacent to area 4, receives sensory input from muscle receptors (muscle spindles), whereas area 3b, located in the posterior surface of the central sulcus, receives input from skin receptors (cutaneous receptors).

The primary somatosensory cortex receives commissural fibers from the contralateral corresponding cortical area and short association fibers from the primary motor cortex located next to it.

The primary somatosensory cortex projects to the precentral gyrus and to the superior parietal lobule of the same hemisphere via association fibers, and to the contralateral corresponding sensory cortex via commissural fibers.

Some of the cells residing in the primary somatosensory cortex give rise to projection fibers that descend to terminate in sensory relay nuclei of the brainstem and spinal cord gray matter. There, they influence motor activity, not by synapsing with motor neurons, but instead by modulating the transmission of sensory input from peripheral structures into the posterior column nuclei of the brainstem and the posterior horn of the spinal cord.

Striate area (Primary visual cortex)

The primary visual cortex receives visual input from the LGN, which relays visual information arising from the contralateral visual field

The neurons residing in the visual cortex respond to different visual stimuli relayed by neurons conveying color, motion, three-dimensional vision, or a combination of various visual stimuli.

The **striate area (primary visual cortex) the thinnest cortex of the entire brain,** is located mainly in the medial surface of the occipital lobes on the banks of the **calcarine fissure**, although it extends into their posterolateral surface. However, Layer IV of this striate area is prominent due to a thick layer of myelinated fibers, which is characteristically so thick in this cortex that it gives the striate cortex a grossly visible striped appearance contributing to its name.

The dorsal lateral geniculate nucleus (LGN) projects visual information to the striate area (primary visual cortex; eponym: Brodmann area 17). This cortical area receives visual input arising from the contralateral visual field: the temporal half of the ipsilateral retina and the nasal half of the contralateral retina.

The striate area receives fibers of the optic radiation arising from the LGN of the thalamus. Table 25.1 lists the topographic relationships of fibers from the retina to the LGN, and then to the striate area on banks of the calcarine sulcus.

The posterior third of the striate area receives visual input from the macula lutea of the retina, whereas the anterior two-thirds receive visual input from the paramacular (and more peripheral) fields of the retina. A larger proportion of the cortex of the striate area is dedicated to macular, compared to paramacular or peripheral, representations reflecting the high visual acuity of the macula lutea.

The striate area is arranged into columns oriented perpendicular to the cortical surface, extending from the cortical surface (deep to the pia mater) to the subcortical white matter. Nerve cells in a particular cortical column are functionally similar. Columns in the striate area include the **ocular dominance columns**, the site where fibers relaying information from the ipsilateral or the contralateral eye terminate, and the **orientation columns**, where nerve cells are responsive to visual stimuli that have a comparable spatial orientation. Layers II and III of the striate cortex also contain vertically oriented aggregates of nerve cells that are responsive to color.

Transverse Temporal Gyrus (Primary Auditory Cortex)

The transverse temporal gyrus (primary auditory cortex) resides deep in the lateral fissure in the hidden superior surface of the superior temporal gyrus

The **transverse temporal gyrus** (primary auditory cortex; Brodmann area 41) resides deep in the lateral fissure in the hidden superior surface of the superior temporal gyrus, containing the eponymous gyri of Heschl, and in the floor of the lateral fissure. The primary auditory cortex of each side receives information from both ears by way of the medial geniculate nucleus of the thalamus and then sends projections to the contralateral side via the corpus callosum. The primary auditory cortex plays an important function in the detection of pattern alteration as well as in the location of the sound.

The primary auditory cortex receives the **auditory radiations** from the **medial geniculate nucleus**. This cortical region has a tonotopic representation of frequencies, that is, neurons responding to low frequencies reside in its anterior extent, whereas neurons responding to high frequencies reside in its posterior extent. The primary auditory cortex is arranged into two-dimensional, alternating, vertically oriented columns of neurons. One dimension of the auditory cortex is composed of **frequency columns**. The cells in each frequency column respond to an auditory stimulus of a particular frequency. As indicated above, cells responding to low frequencies reside in the frequency columns located in the anterior extent of the transverse temporal gyri, whereas frequency columns containing cells responding to gradually higher frequencies are lined up in sequence toward the posterior extent of the primary auditory cortex.

The other dimension of the auditory cortex is composed of alternating **binaural columns**. There are two types of binaural columns: summation columns and suppression columns.

Table 25.1 ● The topographic relationships of fibers from the retina to the dorsal lateral geniculate nucleus (LGN) and the banks of the calcarine sulcus.

Visual field quadrants	Retinal field quadrants	LGN	Optic radiations	Calcarine banks
Upper	Lower	Lateral half	Inferior part	Inferior
Lower	Upper	Medial half	Superior part	Superior

The neurons residing in the **summation columns** respond to an auditory stimulus that stimulates both ears simultaneously. In contrast, the neurons in the **suppression columns** respond maximally to an auditory stimulus that stimulates only one ear but respond minimally when the auditory stimulus stimulates both ears (Fig. 25.10).

Primary vestibular area

The primary vestibular area receives input related to the orientation and movement of the head

The cortical area that receives vestibular information is ambiguous. However, it is believed that the vestibular nuclei project primarily crossed fibers to the contralateral VPM nucleus of the thalamus, which in turn relays this input to the **inferior parietal lobule**, the **primary vestibular area**, and the **superior temporal gyrus** next to the primary auditory cortex. This area receives input related to head movement and position.

Taste area

The cortical area where taste information is relayed resides next to the area where general sensory input from the tongue is transmitted

Taste receptors relay gustatory information to the **nucleus of the solitary tract** in the brainstem. The nucleus of the solitary tract then transmits the information via the ipsilateral central tegmental tract to the VPM nucleus of the thalamus. The thalamus then projects this input to the inferior extent of the **postcentral gyrus** (eponym: Brodmann area 43) in the parietal operculum. The cortical area where taste information is relayed resides next to the area where the general sensory input from the tongue is transmitted.

Olfactory area

The **primary olfactory area** is located in the **prepiriform** (Brodmann area 51) and **periamygdaloid regions** of the cerebral cortex.

Sensory association areas of the cerebral cortex

The sensory association areas receive inputs from the primary sensory areas and further process, integrate, and interpret the incoming sensory input

The cerebral cortex of the human brain is classified into **sensory**, **motor**, and **association cortex**. The human brain contains more association cortex than the brain of any other animal, and thus is the principal component of the human cerebral cortex. **Sensory association areas,** also referred to as **secondary sensory areas,** usually surround the primary sensory areas. They

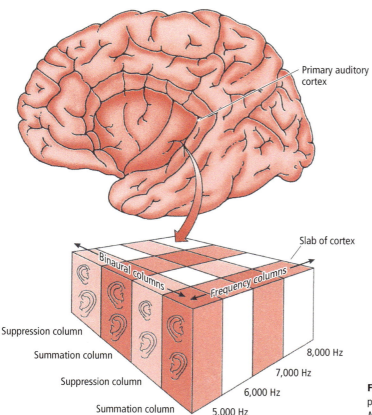

Figure 25.10 • The location and functional organization of the primary auditory cortex. Source: Adapted from Matthews, G (2001) *Neurobiology*/Oxford University Press.

receive input from the primary sensory areas and further process, integrate, and interpret the incoming sensory input. Our ability to learn, reason, and plan our daily activities and lives, and our language skills is an example of association cortical functions.

Somatosensory (somesthetic) association area

The somatosensory association area functions primarily in the integration of sensory input from various sensory systems

The secondary somatosensory cortex is the inferior continuation of the postcentral gyrus forming the parietal operculum (L. "covering"), covering the insula. A somatotopic map of the body where the head and face are represented anteriorly and the sacral region, posteriorly, has been located in the secondary somatosensory cortex.

A **somatosensory association area (parietal association area)** is located in the cortex of the **superior parietal lobule** and the **inferior parietal lobule**. The superior parietal lobule includes eponymous Brodmann areas 5 and 7 which process somatic sensory information. The **inferior parietal lobule**, involved in sensory perception, and language, consists of the **supramarginal gyrus** (area 40) and the **angular gyrus** (area 39). The supramarginal gyrus integrates somatosensory (tactile), auditory, and visual information. The superior parietal lobule receives input mainly from the primary somatosensory cortex (to area 5), but also projections from the visual cortex (to area 7); however, fibers arising from the VPL and VPM nuclei of the thalamus also terminate here. Broadmann area 40 functions primarily in the integration of sensory input from various sensory systems (visual, auditory, and somatosensory). One way of testing the function of the superior parietal lobule of the somatosensory association cortex of a patient is by placing an object (e.g., a spoon) in the patient's hand, and while the patient's eyes are closed ask the patient to identify the object by using other senses (e.g., touch). In order to identify the object, an individual needs to feel the size, shape, texture, and weight of the object, but in addition, to recognize it by associating it with previous sensory experiences linked to it. This should enable the individual to recognize and identify the object without looking at it.

Visual association areas

The secondary visual areas function in identifying an object, determining its location and color, comparing it to prior visual experiences, and determining its significance

The **striate** (primary visual) cortex projects to the surrounding **extrastriate cortex** (visual association cortex, Brodmann area 18), **parastriate cortex** (V2, V3), and **peristriate cortex** (Brodmann area 19, V4). The secondary visual areas (V2–V4) function in identifying an object, determining its location and color, comparing it to prior visual experiences, and determining its significance. Furthermore, the middle temporal area has an important function in detecting moving objects.

Auditory association areas

The secondary auditory cortical areas function in the interpretation of sounds and the comprehension of language

The secondary auditory area corresponds to the anterior part of the superior temporal gyrus (eponym: Brodmann area 42). The auditory association area is also part of the superior temporal gyrus (eponym: Brodmann area 22). The **secondary auditory cortical areas** surround and form reciprocal connections with the primary auditory cortex. The secondary auditory cortical areas function in the interpretation of sounds and, via their connections with the posterior part of the superior temporal gyrus (eponym: Wernicke area), function in the comprehension of language.

Language areas

There are two cortical areas that function in the processing of language: posterior part of the superior temporal gyrus (eponym: Wernicke area) and the triangular and opercular parts of the inferior frontal gyrus (eponym: Broca area)

There are two cortical areas that function in the processing of language: the posterior part of the superior temporal gyrus, the sensory speech area (eponym: Wernicke area) and the triangular and opercular parts of the inferior frontal gyrus, the motor speech area (eponym: Broca area). A review of two decades of PET and fMRI studies has demonstrated that both cerebrums are activated during processing of speech and non-speech: left cerebral responses were reported for the discrimination of fast changing verbal and non-verbal sounds, whereas right lateralized responses were detected with changes of intonation, stress, pitch, and tempo which convey emotion and intention. Therefore, bilateral activation of the auditory areas is crucial not only to hear, but also to comprehend the content, context, and non-verbal cues of speech.

The sensory speech (Wernicke) area (see Fig. 25.8) consists of auditory association cortex (part of Brodmann area 22), as well as parts of the supramarginal and angular gyri (Brodmann areas 37, 39, and 40, usually in the left, dominant hemisphere). This area is also referred to as the receptive (sensory) language area, general interpretative area, or the gnostic area. It plays an important role in the comprehension and formulation of language. In the right cerebral hemisphere this cortical area is an auditory association area that serves to determine the emotional undertones of language. When we hear someone speak, the person's voice often reveals the individual's state of mind: happiness, dismay, fear, surprise, or disgust.

The motor speech (Broca) area (see Fig. 25.8), (Brodmann areas 44 and 45), is located in the opercular and triangular portions of the inferior frontal gyrus. This area not only functions in the initiation of a sequence of complex movements (via its connections with the primary motor cortex), which are essential in the production of speech (a motor function), but also has expressive language capacities (e.g., grammar). The supplementary motor area is also necessary in the production of language.

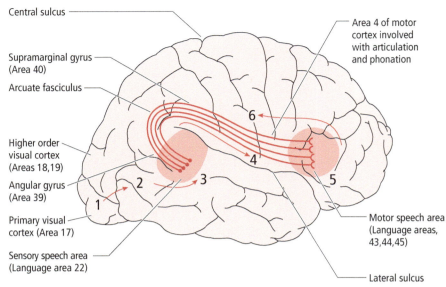

Figure 25.11 • Diagram showing the possible progression of neural transmission from the visual cortex through the cortical areas associated with speech to the primary motor cortex in the production of speech. For example, when an individual sees an object and is asked to name it, information is relayed from (1) the visual cortex (Brodmann areas 17, 18, and 19) to (2) the angular gyrus (Brodmann area 39) to (3) sensory speech (Wernicke) area (Brodmann area 22), by way of (4) the arcuate fasciculus to (5) motor speech (Broca) area of speech (Brodmann areas 43, 44, and 45) and finally to (6) the primary motor cortex (Brodmann area 4), which gives rise to the corticonuclear tract that terminates in the brainstem motor nuclei of the cranial nerves associated with vocalization. Source: Adapted from Noback, CR et al. (1996) The Human Nervous System/Springer Nature.

Sensory speech (Wernicke) and motor speech (Broca) areas are interconnected via the **arcuate fasciculus** (Fig. 25.11), which is necessary for language comprehension and the ability to communicate.

The following is a general description of how a word that is heard is repeated (Fig. 25.12A). Auditory sensory input (a word) reaching the ear is relayed via the auditory pathway to the primary auditory cortex where it is "heard." The primary auditory cortex transmits the input to the auditory association cortex, which functions in the interpretation of sound. This sound is comprehended in the sensory speech (Wernicke) area, which projects the information by way of cortical association fibers (the arcuate fasciculus) to the motor speech (Broca) area. The motor speech area is involved in the planning of the motor activity that is necessary for the word to be produced. This area projects this information to the premotor cortex where motor activity is planned and relayed to the primary motor cortex, which initiates the movements necessary to produce speech. When an individual reads and says the words (Fig. 25.12B), the visual pathway relays the sensory visual input to the striate (primary visual) cortex where the words are "seen." The striate (primary visual) cortex transmits this input to the association visual cortex where the significance of the words is determined. The visual information is then relayed to the sensory speech (Wernicke) area where the words are comprehended, and the creation of the words takes place. This area then projects to the motor speech (Broca) area, which, as described dictates the motor activity necessary to produce the words.

Prefrontal cortex

The prefrontal cortex is associated with the thalamus, limbic system, and basal nuclei

The **prefrontal cortex** is located in the frontal lobe anterior to the motor cortex, as well as on the medial and inferior surfaces of the frontal lobe. The cortex on the inferior surface of the frontal lobe consists of the orbital gyri, whereas the cortex on its medial surface is located rostral to the corpus callosum. The prefrontal cortex consists of multimodal association cortex. It has extensive reciprocal connections with the medial dorsal (MD), ventral anterior (VA), and intralaminar thalamic nuclei via corticothalamic and thalamocortical projection fibers. In addition, its orbital region is associated with the limbic system, whereas its dorsolateral region (located anterior to the motor cortex) is interconnected with the association areas of the parietal, temporal, and occipital lobes via long association bundles. Furthermore, both cortical regions are a source of fibers that terminate in the caudate nucleus and putamen of the basal nuclei. Commissural fibers arising from the prefrontal cortex pass via the corpus callosum to the opposite hemisphere to terminate not only in the cortex of the prefrontal area, but also in that of the other lobes.

The **dorsolateral region** of the prefrontal cortex functions in *working memory*. Observing a laboratory demonstration at school and then being able to repeat the task soon afterward, is an example of working memory. Individuals with lesions to the dorsolateral region have difficulty in problem solving and in paying attention. The main function of the **orbital region** is the expression of an appropriate behavioral response in a particular situation. Individuals with lesions in the orbital region are impulsive and exhibit an inappropriate behavioral response for a given circumstance.

The prefrontal cortex may play a role in the individual's level of intelligence, education, as well as psychological state. It is believed that the prefrontal cortex may correspond to the neocortical area of the limbic system.

CEREBRAL DOMINANCE

In approximately 95% of the human population, the left cerebral hemisphere is usually the dominant hemisphere, whereas the right hemisphere is the nondominant hemisphere

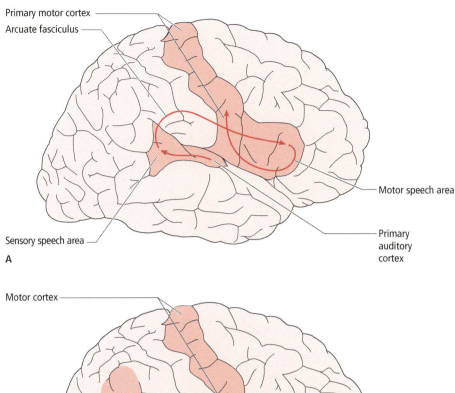

Figure 25.12 • Neural pathways associated with: (A) hearing a word that is then spoken, and (B) reading a word that is then spoken.

(Fig. 25.13). The cerebral cortex is involved in a wide spectrum of functions including the processing of somatosensory, visual, auditory, olfactory, gustatory, and vestibular input, as well as language, learning, memory, emotions, and motor control. Although both sides of the brain have areas that are involved in these functions, for a particular function, one of the cerebral hemispheres plays a more dominant role than the other hemisphere as determined by the number of neurons allocated to that function. The dominant hemisphere mediates processes related to language, speech, problem solving, and mathematical skills. The nondominant hemisphere mediates spatial perception, recognition of faces, expression of emotions, as well as music and poetry skills. Cerebral dominance for speech is demonstrated in individuals who have suffered strokes damaging the speech area of the left cerebral hemisphere. These individuals are more likely to exhibit speech deficits than individuals who had strokes damaging the speech area of the nondominant (right) cerebral hemisphere. Another example that demonstrates cerebral dominance for speech is to ask an individual with a corpus callosotomy "split brain" to *name* an object (without seeing it) placed in their hand. Individuals suffering from intractable seizures may have a neurosurgical procedure to section the commissures that connect the two cerebral hemispheres (to create two separate hemispheres) to stop the spread of seizures. If an object is placed in the right hand of an individual with a corpus callosotomy, the person is able to name the object. But when an object is placed in the left hand, the individual

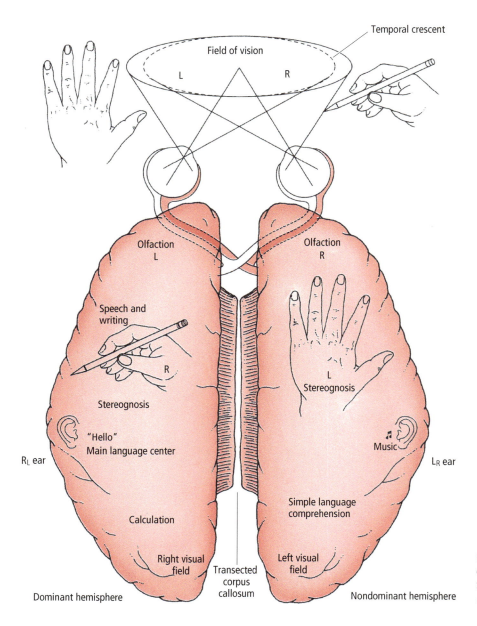

Figure 25.13 ● Functions associated with the dominant and nondominant cerebral hemispheres. Source: Adapted from Noback, CR et al. (1996) *The Human Nervous System*/Springer Nature.

is unable to name the object. When the object is placed in the right hand, tactile sensory information is relayed to the left, dominant hemisphere for speech. However, when the object is placed in the left hand, sensory information is relayed to the right, nondominant cerebral hemisphere, which has a minor function in speech and the individual is unable to *name* the object. Another function of cerebral dominance is **handedness**. Many individuals are right-handed, that is handedness is dominant in their left cerebral hemisphere. Although learning and social factors play a role in handedness, handedness is controlled by the left cerebral hemisphere, which is also dominant for speech. For right-handed individuals, the left cerebral hemisphere is dominant for both speech and handedness. In contrast, for left-handed individuals, the left cerebral hemisphere is dominant for speech, whereas the right hemisphere is dominant for handedness.

CEREBRAL CORTEX ••• 515

CLINICAL CONSIDERATIONS

Although the brain processes pain sensation arising from other injured body tissues, the brain tissue itself has no pain sensation

Over the years, clinical information has been derived from studying patients treated for various cortical lesions resulting from trauma, cerebral vascular accident (stroke), or tumors. In addition, cortical function has also been studied by observations made on conscious patients (under local anesthesia) while undergoing brain surgery. Interestingly, although the brain processes pain sensation arising from other injured body tissues, the brain itself has no pain sensation; thus, recordings and observations can be made during brain surgery while the patient is awake. The cerebral cortex is an extraordinary structure since, following a cerebral lesion, some cortical functional reorganization can result to some degree in restored cortical function.

Sectioning of the corpus callosum
The corpus callosum forms an anatomical bridge connecting the two cerebral hemispheres, which is essential for the integration of information in the two sides of the brain

Historically, surgical sectioning of the corpus callosum (corpus callosotomy) has been performed to alleviate symptoms of epilepsy. This surgical procedure has proved to be effective in preventing seizures from being transferred from one cerebral hemisphere to the other. Although individuals seem normal following this surgery, postoperative studies have revealed some interesting findings. When patients were asked to read words that were presented in their right visual field, they were able to see and to read the words. However, when words were presented in their left visual field, they could not read them as they were not even aware that the words were there. One explanation is as follows: visual information from the right visual field is relayed to the left cerebral hemisphere (via the visual pathway, which was not affected when the corpus callosum was sectioned), which is the dominant hemisphere in the processing of language in most individuals. Thus the patients were able to see and read the words. In contrast, visual input from the left visual field is relayed to the right cerebral hemisphere (again, via the visual pathway, which was not affected when the corpus callosum was sectioned), the nondominant hemisphere for language. As a result of the bisection of the corpus callosum the visual input relayed to the right cerebral hemisphere has no way of *also* reaching the left hemisphere containing the language areas. Consequently, patients could not see or read the words, as if they had a left homonymous hemianopia. These individuals are *not* blind. When they were asked to select with their hand an object that corresponds to the object that was presented in their left visual field, they selected the matching object. This showed that the visual system is functioning properly, and that it is the language function that is not. Although the above studies support the finding that the left cerebral hemisphere is the "dominant" hemisphere in the processing of language, other studies indicate that the right hemisphere is involved to some extent in language comprehension.

Lesions involving the corpus callosum
Lesions involving the corpus callosum, which unites the two cerebral hemispheres, results in disconnection syndromes

The corpus callosum receives its blood supply from the anterior and posterior cerebral arteries. Its genu and body are supplied by the anterior cerebral artery, and the splenium is supplied by both the anterior and posterior cerebral arteries. A stroke involving the **anterior cerebral artery** may affect the **anterior corpus callosum** and result in **apraxia** (impairment of the capacity to execute various movements) of the left upper limb because of the isolation/disconnection of the dominant (left) hemisphere from the right motor cortex. Apraxia may affect the muscles used to produce speech which results in speech deficits.

A stroke involving the **posterior cerebral artery** or a lesion involving the **posterior corpus callosum** on the left side may result in **alexia** (the inability to recognize words or to read). Visual input from the left visual field is relayed to the right, intact visual cortex, which is isolated/disconnected from the language-dominant (left) hemisphere. If the basilar artery is affected, which gives rise to both the posterior cerebral arteries, infarction of the visual cortex will be bilateral and will cause cortical blindness; however, the affected individual is unaware that they cannot see (anosognosia).

Lesions in the motor cortex
The **middle cerebral artery** is buried in the lateral fissure that separates the parietal lobe from the temporal lobe. Its branches that emerge from the lateral fissure, supply structures in the frontal, parietal, and temporal lobes. The lateral surface of the primary motor cortex (corresponding to the precentral gyrus, Brodmann area 4) controlling movement of the head, upper limb and trunk, as well as the lateral surface of the premotor and supplementary motor cortices (corresponding to Brodmann area 6) and the frontal eye field (corresponding to Brodmann area 8) receive their arterial blood supply from branches of the middle cerebral artery. The medial extensions of the motor cortices located in the medial surface of the frontal lobe, that is, the primary motor cortex controlling the movement of the lower limb and the medial extension of the premotor and supplementary motor cortices, are supplied by the **anterior cerebral artery**. A vascular infarct resulting from occlusion of the anterior or middle cerebral artery will result in motor deficits in the contralateral side of the body.

A lesion damaging upper motor neurons that is, either their cell bodies (residing in the motor cortex) or their axons anywhere along their descending path in the white matter, results in **upper motor neuron lesion signs**. It is important to note that the motor cortex not only gives rise to the *corticospinal* and *corticonuclear tracts*, but in addition, it gives rise to other descending fibers of the motor system, including the *corticorubral fibers* terminating in the red nucleus, and the corticoreticular fibers terminating in the pontine and medullary reticular formation, all of which descend in the internal capsule. It is also important to note that *rubrospinal, reticulospinal,* and *vestibulospinal neurons* are also considered to be upper motor neurons and when damaged produce upper motor neuron lesion signs. Since the upper motor neuron axons of the various descending motor tracts (named previously) *intermingle* to some extent as they descend through different levels (in the internal capsule, the brainstem, or the spinal cord), upper motor neuron lesion signs are usually produced following damage of a mix of descending motor axons due to their intermingling or descending in close proximity to one another.

With an upper motor neuron lesion, initially there is a temporary **contralateral weakness** or **flaccid paralysis, hypotonia**, and **hyporeflexia**. About two weeks (or a variable amount of time) following the lesion, the individual will regain function of the proximal limb musculature and eventually **spasticity** will appear in the distal limb muscles. **Hypertonia**, characterized by an increase in muscle tone (muscle tension), and **hyperreflexia** (exaggerated tendon reflexes) will appear in the distal limb muscles. Hypertonia can be demonstrated during the passive manipulation of an extremity. The hypertonic extremity will exhibit an increase in resistance to the passive movement/stretch by the physician. Exaggerated hypertonia is referred to as spasticity. The hypertonia is more severe in the antigravity muscles (the upper limb proximal

CLINICAL CONSIDERATIONS (continued)

flexors and lower limb extensors). With this type of lesion, fine movements of the distal limbs (hand and foot) will be affected most.

A typical response to the increase in resistance to passive stretch is the **clasp-knife response**. As an extremity (e.g., forearm) is extended by the physician against the spastic resistance there is an abrupt loss of resistance. The clasp-knife response is believed to be caused by activation of the tendon organs (TOs) as muscles are stretched during the limb manipulation. This in turn stimulates the TOs Ib sensory afferent fibers that relay excitatory messages to inhibitory spinal cord interneurons. These interneurons inhibit the alpha motor neurons that overstimulated the skeletal muscles to create the hypertonia and increased resistance to passive stretch.

Hypertonia, hyperreflexia, and spasticity are not well understood. There are several hypotheses that have offered an explanation of the cause of **spasticity** and related **hypetonia** and **hyperreflexia**.

One hypothesis suggests that an upper motor neuron (UMN) lesion diminishes or eliminates descending inhibitory influences on the dynamic gamma motor neurons. This in turn causes overactivity of the gamma motor neurons resulting in contraction of the contractile portions of intrafusal muscle fibers, and stretching of the central noncontractile portion of the intrafusal fibers. This overstimulates the type Ia annulospiral afferent fibers from the muscle spindles whose central processes synapse with alpha motor neurons, which are overstimulated, causing an abnormal increase in skeletal muscle tone.

Another hypothesis suggests that an UMN lesion may result in diminution of cortical stimulation of spinal cord inhibitory interneurons, resulting in hypertonia and spasticity.

Cortical UMN axons stimulate a recurrent inhibitory interneuron eponymously known as the Renshaw cell. This interneuron is also stimulated by a collateral branch of a lower motor neuron and in turn inhibits the lower motor neuron. These neural connections inhibit the reflex stimulation of antagonist muscles when the agonists are contracting. An UMN lesion involving cortical neurons would eliminate the inhibitory effect on antagonist muscles. This causes alternating contraction of agonists and antagonist muscles known as **clonus**, which accompanies spasticity and hyperreflexia.

A unilateral lesion in the primary motor cortex will affect mainly the fine movements of the contralateral distal limbs (hand and foot)

A *unilateral* lesion in the **primary motor cortex** (Brodmann area 4) will affect mainly the fine movements of the contralateral distal limbs (hand and foot). Lesions confined to the primary motor cortex are rare.

A *unilateral* lesion in the **premotor cortex** (**PMC**, Brodmann area 6) results in the inability to produce voluntary movements in the contralateral side of the body although there is no apparent weakness, paralysis, or change in muscle tone. The premotor cortex plays a role in the programing of postural adjustments. Recall that the premotor cortex, along with the supplementary motor cortex, has a key function in the *planning* phase of motor activity. Once the plan (sequence of movements) is created, it is relayed to the primary motor cortex for execution.

A *unilateral* lesion in the **supplementary motor cortex** (**SMC**, Brodmann areas 6) results in the inability to coordinate different voluntary movements *bilaterally*. As an example, the affected individual will have no difficulty in coordinating the same movement with both hands concurrently. However, when attempting to perform different movements with each hand concurrently, it will be difficult or impossible to execute. A lesion limited to the supplementary motor cortex does not produce apparent muscle weakness or paralysis and has no effect on muscle tone. Interestingly, damage to the supplementary motor area results in akinesia, a deficit associated with the dysfunction of the basal nuclei.

A *unilateral* lesion of the **primary motor cortex** and *another* motor cortical area will produce more severe motor deficits, typical of upper motor neuron lesion signs: a **contralateral spastic paralysis** with **hypertonia** and **hyperreflexia**.

Following trauma to the head involving the brain, scar tissue formation in the brain often results in epileptic seizures. A lesion of the primary motor cortex, with subsequent scar tissue formation, results in **Jacksonian epileptic seizures**. In this type of seizure convulsions may occur only in a certain part of the body, or additional parts of the body may be involved as well, depending on the site of the primary motor cortex that is affected by the lesion.

Note that the clinical case at the beginning of the chapter refers to a patient who had two seizures. Each started with twitching of the right side of the face, the right hand, and the right leg, and then progressed to loss of consciousness and generalized convulsions.

1 Based on the twitching of the right face, hand, and leg, which part of the cerebral cortex do you suspect was the focus of the seizure?

2 What is the name of the seizures resulting from a lesion (and subsequent scar tissue formation) in the primary motor cortex?

3 This patient did not have any prior history of seizures. What could cause the onset of seizures in an adult?

Lesions in the frontal eye field

Normally, stimulation of the frontal eye field results in deviation of both eyes toward the contralateral side. A unilateral lesion in the frontal eye field (Brodmann area 8) results in both eyes deviating to the side ipsilateral to the lesion. In addition, the affected individual will be unable to turn the eyes contralateral to the lesion. The effects of frontal eye field lesions are not permanent.

Lesions in the motor speech (Broca) area

A unilateral lesion in Broca's area of speech in the dominant cerebral hemisphere results in Broca's aphasia

A unilateral lesion in Broca's area (inferior frontal gyrus, Brodmann areas 44 and 45) (see Fig. 25.11) of speech in the dominant (left) cerebral hemisphere results in **motor** (expressive, nonfluent, Broca) **aphasia** (G., a, "absence"; *phasis*, "speech"). An individual with this condition has difficulty expressing a thought, is nonfluent, and speaks slowly and with difficulty, even though these individuals do not have a lower motor neuron lesion innervating the muscles used in phonation and these muscles are normal. In addition, their use of words is distorted. They only use the main or necessary words of a sentence (this is referred to as **telegraphic speech**). An individual who has a severe form of motor aphasia is incapable of speaking. This is known as **mutism**. Although spoken and written language comprehension in motor aphasia is a relatively preserved skill, it is still affected by the lesion. Interestingly, the individual is aware that what they are saying makes no sense. As indicated above, the motor speech (Broca) area is involved in the planning of motor activity that is necessary for words to be produced. This area projects this information to the premotor cortex where motor activity is planned and relayed to the primary motor cortex, which initiates

CLINICAL CONSIDERATIONS (*continued*)

the movements necessary to produce speech. A lesion in the motor speech (Broca) area will disrupt its input to the premotor cortex. Motor aphasia also impairs an individual's ability to express a thought or concept by writing it down. This is known as **agraphia** (G., a, "absence;" *graphein*, "to write").

Lesions in sensory speech (Wernicke) area
A unilateral lesion in Wernicke's area in the dominant hemisphere results in Wernicke's aphasia

A unilateral lesion in the sensory speech (Wernicke) area in the dominant hemisphere (usually the left hemisphere), involving the posterior region of the superior temporal gyrus (Brodmann area 22), and supramarginal and angular gyri (Brodmann areas 37, 39, and 40) (see Fig. 25.11), results in **sensory (Wernicke)** (receptive, fluent) **aphasia**. Since this area plays a role in the comprehension and formulation of language, individuals with a severe form of this condition have difficulty comprehending the spoken word and are unable to read (**alexia**: G., a, "absence"; *lexis*, "word") or write in understandable language (**agraphia**), even though their visual and auditory systems are normal. Even if speech is fluent, and patients can speak with ease, the combination and order of the words selected is meaningless and makes no sense to them or to others. This is referred to as **fluent paraphasic speech** ("word salad"). Interestingly, these individuals are unaware of their difficulties. In milder forms of sensory (Wernicke) aphasia, individuals may use the wrong word (e.g., "airplane" for "bird") or use a word that sounds like the word they intended to say in a sentence (such as "cook" instead of "book").

Lesions in sensory (Wernicke) and motor (Broca) areas
A lesion affecting both sensory and motor speech areas in the dominant hemisphere results in a severe condition referred to as **global aphasia**. An affected individual is unable to comprehend what he hears or reads, cannot write, and in addition is unable to formulate normal language.

Lesions in the superior longitudinal fasciculus
A lesion in the arcuate fasciculus component of the superior longitudinal fasciculus results in conduction aphasia

A lesion in the arcuate fasciculus component of the superior longitudinal fasciculus will interrupt the connection between motor and sensory speech areas (see Fig. 25.11), the two language areas, and will result in **conduction aphasia**. Individuals with this type of lesion are able to comprehend what they hear (since the sensory speech area is intact) and their language expression is fluent; however, they have difficulty processing what they hear and then formulating a coherent response.

Lesions in other afferent fibers terminating in motor (Broca) or sensory (Wernicke) speech areas
Lesions involving other afferent fibers from various cortical areas terminating in Broca's or Wernicke's areas produce aphasic syndromes referred to as transcortical aphasias

Lesions involving other afferent fibers from various cortical areas terminating in motor or sensory speech areas also produce aphasic syndromes, referred to as **transcortical aphasias**. The lesions are extrinsic and do not involve motor or sensory speech areas. However, affected individuals present symptoms characteristic of motor or sensory aphasia as a consequence of the lack of input to the language areas, which is essential for normal linguistic function and capacity.

Lesions in the primary somatosensory area
A unilateral lesion in the primary somatosensory area results in contralateral loss of two-point discrimination, graphesthesia, stereognosis, and vibratory and position sense

A unilateral lesion in the primary somatosensory area (Brodmann areas 3, 1, and 2) (see Fig. 25.7A) results in **contralateral loss** of **two-point discrimination**, **graphesthesia** (the ability to recognize letters or numbers as they are stroked on the skin), and **stereognosis** (the ability to determine the size, shape, and texture of an object following tactile examination), as well as the loss of **vibratory** and **position sense**. In addition, the significance of the qualitative and quantitative perception of the stimulus is diminished or absent. Stimulation of this area does not evoke pain sensation. Due to the complex projections of the pain pathways to more than one cortical area, pain sensation persists. Usually, pain, temperature, and light touch sensation are only minimally impaired; however, the individual is unable to localize the stimulus.

Lesions in the somatosensory association areas/cortex
A lesion in the somatosensory association areas results in agnosia, a sensory perception disorder

A lesion in the somatosensory association area results in minimal sensory loss (provided that the primary somatosensory cortex is intact). However, since the somatosensory association area plays an important role in the memory of somatosensory information, sensory integration, and interpretation of the sensory information, the individual has a condition referred to as **agnosia** (G., a, "absence"; *gnosis*, "knowing"). Agnosias are classified according to the particular secondary (association) cortex affected.

If there is a lesion in the somatosensory association cortex the deficits resulting from the lesion are referred to as **astereognosis** and **tactile agnosia**. An individual with this type of lesion (which involves the somatosensory association cortex representing the contralateral hand) are unable to identify a familiar object such as a spoon placed in their palm while their eyes are closed. Although they can feel the object in their palm, and information about it (such as its size, shape, weight, and texture) is transmitted to the primary somatosensory cortex, they are unable to identify it as the somatosensory association cortex (which stores tactile sensory information based on prior experiences, with a spoon in this case) is not functioning. Thus, this new information cannot be assembled and compared to stored information and matched, which is necessary for the identification of the object.

Astereognosis does not only affect an individual's ability to recognize objects without looking at them, but also results in them experiencing a loss of awareness of the position of parts of their body contralateral to the lesion. A condition, referred to as **cortical neglect**, is a severe form of astereognosis, in which the affected individuals ignore or deny the presence of one side of their body. This condition most frequently results from an extensive lesion in the superior aspect of the nondominant (right) parietal lobule (the posterior parietal association cortex: Brodmann areas 5, 7, 39, and 40). Since affected individuals do not recognize certain parts of their body as belonging to them, they will not wash the left hand, or moisturize the left side of their face, and so on.

Lesions in the primary visual area
A unilateral lesion in the primary visual area results in blindness of the contralateral visual field

A unilateral lesion in the striate (primary visual) area (Brodmann area 17) (see Fig. 25.7A) results in blindness of the contralateral visual field, referred

CLINICAL CONSIDERATIONS (*continued*)

to as a **contralateral homonymous hemianopsia**. If destruction of the striate (primary visual) cortex is the result of a vascular lesion, **macular sparing** may occur, that is, central vision is unaffected. It is believed that this is due to the presence of collateral anastomotic channels between the middle and posterior cerebral arteries.

A bilateral lesion in the primary visual area results in complete blindness, also referred to as **cortical blindness**.

Lesions in the association visual areas
A lesion in the visual association areas results in visual agnosia

A lesion in the visual association areas (Brodmann areas 18 and 19) (see Fig. 25.7A) (usually bilaterally) results in **visual agnosia**. Since the visual pathways and striate (primary visual) cortex are intact, the individuals are able to "see" an object or person. However, although they may be looking at a familiar object (such as a toaster) or person, they are unable to identify them, or describe the functions of the object. Individuals with visual agnosia may recognize an object one day but not the next. Also, they may recognize certain familiar objects but not others. Interestingly, an object may be identified by other senses, such as touch.

Dyslexia is a type of visual agnosia. Individuals with this disorder are unable to read and write words clearly, although they can see and recognize letters.

Lesions in the primary auditory area
A unilateral lesion in the primary auditory area results in partial deafness

A unilateral lesion in the primary auditory area results in **partial deafness**. This is due to a bilateral deficit in hearing since auditory information arising in each ear is relayed to the cortical primary auditory area *bilaterally*. Usually, individuals with this type of lesion experience a greater loss of hearing in the ear contralateral to the cortical lesion. Recall that a greater number of fibers relaying auditory input to the cortex decussate, whereas a lesser number of fibers project ipsilaterally. An individual with this lesion has difficulty in localizing sounds on the opposite side.

A bilateral lesion of the primary auditory area results in **total deafness**.

Lesions in the auditory association area
A unilateral lesion in the auditory association areas results in different agnosias

A unilateral lesion in the auditory association area of the dominant (left) hemisphere results in **receptive aphasia**. A unilateral lesion in the nondominant (right) hemisphere results in **amusia**, a condition in which the individual is unable to recognize familiar voices or music.

A bilateral lesion of the auditory association area results in **auditory agnosia** in which the individual does not recognize complex sounds.

Lesions in the cortical vestibular area
A lesion in the cortical vestibular area results in vertigo

A unilateral lesion in the superior temporal gyrus (see Fig. 25.1) results in **subjective vertigo**, in which individuals have the strange experience of feeling as though they are spinning around their environment.

In contrast, a unilateral lesion in the parietal lobe in the lower margin of the intraparietal sulcus results in **objective vertigo**, in which individuals feel as though the environment is revolving around them.

Lesions in the prefrontal association cortex
An individual with a lesion in the prefrontal association cortex displays changes in various personality traits as well as in higher-order functions and/or executive functions

An individual with a lesion in the prefrontal association cortex displays changes in various personality traits as well as changes in higher-order functions and/or executive functions such as judgment, creativity, analytic thinking, and responsibility. These are more pronounced if the lesion is *bilateral*. In addition, affected individuals behave in a manner that is socially inappropriate. Their behavior is characterized by a lack of self-respect or concern for others; they lose the social skills necessary to interact respectfully with others, are easily distracted, and are emotionally unstable.

Dorsolateral convexity lesions result in apathetic, lifeless behavior; orbitofrontal lesions result in disinhibited behavior and poor judgment; left frontal lesions result in depressed behavior; and right frontal lesions result in manic behavior.

Lesions in the supramarginal gyrus (Brodmann area 40)
A unilateral lesion in the parietal lobe of the dominant hemisphere results in various conditions collectively known as apraxias

A unilateral lesion in the parietal lobe of the dominant hemisphere results in various conditions collectively known as apraxias (G., a, "absence"; *praxia*, "action"). **Apraxia** is a condition in which an individual's ability to perform certain skilled movements that are normally triggered by a stimulus is impaired, although the individual does not have any sensory or motor deficits.

A unilateral lesion in the supramarginal gyrus (Brodmann area 40) (see Fig. 25.1) of the parietal lobe results in an **ideomotor apraxia**, in which the individual's ability to perform a desired activity is impaired. For example, although affected individuals can perform certain motor activities, their impairment prevent them from doing them on command. **Ideational apraxia** is a condition in which affected individuals can perform certain activities separately, for example picking up a glass, filling it with water, and drinking it; however, when asked to perform these activities in one sequence, they are unable to do them.

Lesions in the supramarginal and angular gyri of the dominant hemisphere
Lesions in the supramarginal and angular gyri of the dominant hemisphere may result in dysgraphia, dyscalculia, or dyslexia

Lesions in the supramarginal and angular gyri (see Fig. 25.1) of the dominant hemisphere result in one or more of the following clinical symptoms of Gerstmann syndrome: (1) **dysgraphia** (G., *dys*, "faulty"; *graphein*, "to write"), a condition in which the affected individual has difficulty writing although the sensory and motor systems are intact; (2) **dyscalculia**, a condition in which the individual has difficulty making calculations, right–left confusion, and **finger agnosia** (in which the individual does not know one finger from another); and (3) **dyslexia**, in which the lesion interferes with the ability to read.

Lesions in the parietal association area of the nondominant (right) hemisphere
Lesions in the parietal association area that involve the inferior parietal lobule result in the following deficits: (1) distorted concepts of body image,

CEREBRAL CORTEX ● ● ● **519**

CLINICAL CONSIDERATIONS (*continued*)

in which individuals are not aware of the left side of their body (e.g., does not shave the left side of the face, does not wash or dress the left side of their body, etc.), referred to as **asomatognosia**; (2) a lack of awareness that they are ill; (3) spatial disorientation, e.g., they do not know where objects are located in their house, or where locations are on a map; and

(4) an inability to draw simple figures, and ignores the left side of the object they are asked to draw (e.g., when asked to draw the clock on the wall, they only draw the right side of the clock showing numbers 1–6); (5) **contralateral neglect**, which is the lack of awareness of the left half of their environment.

SYNONYMS AND EPONYMS OF THE CEREBRAL CORTEX

Name of structure or term	Synonym(s)/eponym(s)	Name of structure or term	Synonym(s)/eponym(s)
Allocortex	Heterogenetic cortex,	Multimodal association areas of the cerebral cortex	Heteromodal association areas of the cerebral cortex
Archicortex	Archipallium		
Astereognosis	Stereoanesthesia	Stria of internal granular layer	Outer line of Baillarger
Auditory radiations	Geniculotemporal radiations		Outer stripe of Baillarger
Broca area	Brodmann areas 44 and 45	Paleocortex	Paleopallium
Central sulcus	Central sulcus of Rolando	Posterior parietal area	Brodmann areas 5 and 7
Corticofugal fibers	Output projection fibers from the cerebral cortex	Posterior limb of the internal capsule	Thalamolenticular part (older term)
		Prefrontal cortex	Brodmann areas 9–12 and 45–47
Corticonuclear fibers	Corticobulbar fibers (older term)	Premotor cortex (PMC)	Premotor area
Corticopetal fibers	Afferent fibers to the cerebral cortex		Part of Brodmann area 6
Frontal eye field (FEF)	Brodmann area 8	Primary auditory cortex	Brodmann areas 41 and 42
Optic radiation	Geniculocalcarine tract		Transverse temporal gyri of Heschl
	Geniculostriate tract	Primary motor cortex	Precentral gyrus
Girdle musculature	Proximal limb musculature		Brodmann area 4
Hippocampal commissure	Commissure of the fornix	Primary somatosensory cortex	Primary somesthetic cortex
Horizontal cells	Horizontal cells of Cajal		Brodmann areas 3, 1, and 2
Stria of internal pyramidal layer	Inner line of Baillarger	Primary visual cortex	Brodmann area 17
	Inner stripe of Baillarger		Striate area
Isocortex	Neocortex,	Retrolentiform part of the internal capsule	Retrolenticular limb of the internal capsule
	Neopallium,		
	Homogenetic cortex		
Jacksonian seizure	Jacksonian march	Secondary somatosensory cortex	Secondary somesthetic cortex
Lateral fissure	Lateral fissure of Sylvius	Short association fibers (of the cerebral cortex)	Arcuate fibers
Layer I of the cerebral cortex	Molecular layer		U-fibers
	Plexiform layer	Somesthetic association area	Superior parietal lobule
Layer II of the cerebral cortex	External granular layer		Brodmann areas 5, 7, and 40
Layer III of the cerebral cortex	External pyramidal layer	Stellate cells of the cerebral cortex	Granule cells of the cerebral cortex
Layer IV of the cerebral cortex	Internal granular layer	Sublentiform limb of the internal capsule	Sublenticular part of the internal capsule
Layer V of the cerebral cortex	Internal pyramidal layer	Superior occipitofrontal fasciculus	Subcallosal bundle
Layer VI of the cerebral cortex	Multiform layer	Supplementary motor area (SMA)	Part of Brodmann area 6
	Fusiform layer	Trunk musculature	Axial musculature
Medial geniculate nucleus (MGN)	Medial geniculate body (MGB)	Visual association cortex	Extrastriate cortex
Mesocortex	Periallocortex		Brodmann areas 18 and 19
	Juxtallocortex	Wernicke area	Brodmann area 22

FOLLOW-UP TO CLINICAL CASE

The onset of seizures in a previously healthy individual without a history of seizure mandates careful evaluation. Brain imaging is required; preferably with MRI (a CT is done in the emergency room setting to look for a cerebral hemorrhage). MRI in the patient discussed revealed a meningioma over the midlateral aspect of the left hemisphere, in the vicinity of the primary motor cortical area.

The first seizure experienced by this patient was a focal motor, or simple partial, seizure. Normal consciousness was maintained and motor manifestations occurred only on the contralateral side to where the brain focus of the seizure was. In this case, the seizure focus began in or around the facial part of the left primary motor cortex and spread progressively to involve the upper extremity and then the lower-extremity part of the primary motor cortex (refer to the somatotopic representation of the face and upper and lower limbs on the motor homunculus; see Fig. 25.9). This is referred to as a Jacksonian march or Jacksonian seizure. It is rare to see the full march of clonic activity, but it is historically and conceptually important to know about it.

The second seizure almost certainly had its onset in the same spot as the first one, except that this one spread rapidly and became generalized, that is, involving both hemispheres. This generalization results in loss of consciousness and often generalized convulsions (generalized tonic-clonic seizures). Tongue biting and/or urinary incontinence is also often associated with generalized tonic–clonic seizures. When there is alteration of consciousness with a seizure, but it is not generalized, it is called complex partial.

Seizures are a cortical phenomenon. A seizure indicates an abnormality of the electrophysiology of the cortical neurons. The full electrophysiology of seizure initiation and progression is beyond the scope of this discussion. The cortical neurons become hyperexcitable and typically produce "positive" symptoms rather than symptoms indicating loss of function. The manifestations of seizures are widely varied and depend on which part of the cerebral cortex is affected. The most "epileptogenic" regions, in descending order, are the temporal lobes, frontal lobes, parietal lobes, and occipital lobes. The electrical activity of the brain is investigated with an electroencephalogram.

Seizures are either primary or secondary. A secondary etiology is more common with new-onset seizures in adults, and can be caused by strokes, tumors, vascular malformations, hemorrhage, injury, and any other lesion causing cortical irritation. Subcortical lesions do not cause seizures. Primary seizures are more common in children, who often "outgrow" them, and there is no identifiable lesion on brain imaging. There is often a hereditary component. However, seizures in children can certainly be secondary and new-onset seizures in adults can be primary (or at least show no identifiable lesion).

4 Which lobe of the cerebral cortex is the most epileptogenic lobe?

QUESTIONS TO PONDER

1. Distinguish between unimodal and multimodal association cortices.

2. Which cells are the most common cell type and the main output neurons of the cerebral cortex?

3. Why is it that patients who have their corpus callosum sectioned to relieve symptoms of epilepsy can see and read words presented in their right visual field, but cannot see or read words presented in their left visual field?

4. Describe what is likely to happen following a stroke involving one or more of the blood vessels that supply the corpus callosum.

5. Name the artery that supplies motor (Broca) and sensory (Wernicke) speech areas, and the symptoms that may occur following the occlusion of the vessel's branches that supply these language areas.

CHAPTER 26

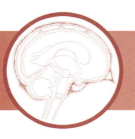

Evolution of the Human Brain

This is the final chapter of our textbook. Over the previous 25 chapters, you have explored the intricate complexities of the human brain. Perhaps you've wondered how this marvel of evolution came to be. How did the brain adapt over millions of years? How did we gain the ability to speak and perform complex actions?

EVOLUTIONARY BIOLOGY FUNDAMENTALS

Evolution is the process by which organisms change over time through genetic variations influenced by environmental pressures, with natural selection favoring traits that enhance reproductive success

Evolution refers to the process of descent with modification where an organism, influenced by environmental pressures, produces offspring with new genetic variants (mutations) which may lead to new traits, and indeed, a new species. Natural selection is the mechanism whereby genetic variants which increase the reproductive fitness (number of offspring surviving and reproducing) of an organism are favored, and those that do not, are eliminated from the genetic pool. These processes lead to adaptation, survival, and evolution of the species.

EARLY HOMININS

Homo sapiens and the Great Apes evolved from a common ancestor and our species underwent significant brain size increases through ancestors like Homo erectus and Homo neanderthalensis, yet our more advanced brains are smaller, suggesting a trade-off for greater cognitive complexity

As members of the species *Homo sapiens*, we share ancestry with the Great Apes (*Hominidae*) and that relationship existed as many as 20 million years ago. More recently, between 1.3 million and 110,000 years ago, our direct ancestors, *Homo erectus,* had an average cranial capacity of 959 mL (based on 15 specimens). Following *Homo erectus, Homo heidelbergensis* emerged 600,000–200,000 years ago with a much larger average cranial capacity of 1,227 mL (based on 6 specimens). This substantial increase in cranial capacity is believed to have corresponded with an increase in brain size and general intelligence, evidenced by the use of hand-crafted tools, spear throwing, and shelter constructions.

The most recent relative, *Homo neanderthalensis*, was walking the earth between 175,000 and 40,000 years ago with an average cranial capacity of 1,415 mL (based on 25 specimens). *Homo neanderthalensis* used more sophisticated hunting and building techniques and exhibited cultural practices such as the intentional burial of the dead within caves or rock shelters. Modern *Homo sapiens* then appeared with a slightly smaller cranial capacity of average volume 1,330 mL (based on more than 500 specimens). Despite this reduction in cranial capacity, and by inference in brain size, we are the most advanced of the hominins. Could it be that a smaller brain is the metabolic trade-off for greater cognitive and behavioral complexity?

NON-HUMAN PRIMATES

Compared to living non-human primates, Homo sapiens have larger brains relative to our body size, with more neurons, a process that has come at a high metabolic cost

Our closest living relatives are non-human primates, specifically the chimpanzee (*Pan troglodytes*) and the bonobo (*Pan paniscus*). Four to six million years ago, the chimpanzees and bonobos diverged from the line of hominins that led to modern humans. Compared to chimpanzees (EQ: 2.6) and bonobos (EQ: 2.8), humans have a much higher encephalization

---

A Textbook of Neuroanatomy, Third Edition. Maria A. Patestas, Amanda J. Meyer, and Leslie P. Gartner.
© 2025 John Wiley & Sons, Inc. Published 2025 by John Wiley & Sons, Inc.
Companion website: www.wiley.com/go/patestas/neuroanatomy3e

quotient (EQ: 7), a measure of relative brain size compared to other species of similar body size. Human brains are not only larger, but have undergone increased gyrification, which increases the number of neurons. The metabolic cost of increasing the number of neurons has been estimated as $6 \times 10^{-9} \mu mol$ glucose/min across a number of mammalian species.

MODERN HUMAN BRAINS

Modern human brains, averaging 1,400 g and containing 86 billion neurons, showcases significant evolutionary advancements in organization and connectivity which have enhanced cognitive and motor capabilities to support complex behaviors essential for survival and adaptation

Modern human brains weigh an average 1,400 g and contain 86 billion neurons and an equal number of neuroglia. The brain cannot make or store glucose; therefore, it must receive a constant supply via the circulation. Although the brain only constitutes about 2% of body weight, it consumes 20% of the cardiac output.

Interestingly, the ratio of cerebral to cerebellar neurons is consistent across mammals, with approximately four cerebellar neurons for every neuron in the cerebral cortex. This suggests that evolutionary pressures affected both structures equally and in tandem. Improved hunting, gathering, and cooking techniques might have provided the necessary caloric energy for human brains to support more neurons.

Early neural evolution likely improved the *reorganization* of neurons to enhance performance in vision, sensation, and dexterity, ultimately improving *Homo sapiens'* reproductive fitness prior to the increase in neuron numbers. Notably, humans exhibit higher densities of neurons in critical brain regions including the striate (primary visual cortex) and extrastriate areas (visual association areas), the postcentral gyrus (primary somatosensory cortex), and the superior temporal gyrus (primary auditory cortex). Furthermore, the human brain contains a significantly greater number of cortical areas, averaging 180 per cerebral hemisphere, compared to approximately 50 in non-human primates. This increased cortical specialization supports multiple processing networks, enhancing connectivity and facilitating rapid multitasking, improved performance, and more complex behaviors. Such specialization is particularly evident in the precentral gyrus, where distinct domains for motor control of the head and neck, upper limbs, and lower limbs are interspaced with somato-cognitive action networks. Additionally, the positioning of upper motor neurons in the precentral gyrus, superior to brainstem, effectively reduces the functional distance of the corticospinal tract from the neocortex to the digits, thus enhancing efficiency. Evidence also indicates that humans possess more corticospinal neurons than chimpanzees and that these neurons exhibit faster conduction speeds. Collectively, these evolutionary advancements in neuronal organization and connectivity have endowed humans with exceptional cognitive and motor

capabilities, enabling complex behaviors that are crucial for survival and adaptation.

PRENATAL AND POSTNATAL DEVELOPMENT

The extended development of the human brain results in an increased number of neurons and synapses, leading to complex neural circuitry, particularly in neocortical areas associated with language processing

Humans have significantly longer gestational period (280 days) compared to bonobos (240 days) and chimpanzees (243 days). This extended period allows for prolonged neural development, facilitating the production of more neurons. However, this also results in a longer window of sensitivity to teratogens (drugs, chemicals, infections, and environmental factors which may cause developmental issues). Postnatally, humans also have extended child-to-juvenile-to-adult phases during which cortical plasticity and synaptic pruning take place, further enhancing neural complexity.

Experimental limitations using post-mortem fetal brain tissue have spurred the use of humanized mouse models, organoid cell cultures, and induced pluripotent stem cells to elucidate the molecular differences between humans and non-human primates. Genomic studies have implicated several genes, such as *SRGAP2, ARHGAP11B, ZEB2, PPP1R17, FZD8, FOXP2, CNTNAP2, NOTCH2NL, mTOR, ITGb8,* and *CROCCP2*, as significant in neural development. In particular, *SRGAP2* prolongs synapse maturation leading to increased synaptic density, *ARHGAP11B* increases basal cell progenitor mitosis, *ZEB2* delays the transition of neuroepithelial cells to radial glial cells which leads to greater cortical expansion, and *PPP1R17* and *FZD8* accelerate cell-cycle progression. The differential expression of these genes in human brains contributes to our increased neuron number and synaptic complexity.

Current estimates suggest the human brain may house between several hundred trillion to over a quadrillion synapses, indicating a vast potential for connectivity. Human brains also exhibit distinct regional localization and varying abundances of excitatory and inhibitory neurons, neuromodulatory neurons, and neuroglial cells, which may contribute to more complex neural connectivity. The neocortex, particularly the prefrontal cortex, has experienced significant changes relative to body size with increased thickness in Layers II and III indicating more local short-range connections (including U-shaped fibers). Another enhanced connection is the arcuate fasciculus, playing a crucial role in language processing by connecting the neocortex of the inferior frontal gyrus and the superior temporal gyrus, which is more active in humans compared to non-human primates.

In summary, the evolution of the human brain is a sophisticated process shaped by genetic, environmental, and cultural factors, making *Homo sapiens* the most cognitively advanced species among the hominins.

Questions to ponder: answers to odd questions

CHAPTER 1

Question 1: Although these two anatomical components of the nervous system are treated as if they were separate entities, they are in fact intimately related to one another. Although the central nervous system is housed in the skull and spinal column, many of its neurons, whose cell bodies are located in the brain and in the spinal cord, have processes that leave the confines of their bony housing and enter other regions of the body. Here they are referred to as nerve fibers of the peripheral nervous system. Therefore, in many instances, the peripheral nervous system is an extension of the central nervous system.

Question 3: Oligodendroglia are one of the types of neuroglia located in the central nervous system. These cells act as "insulators," protecting neurons from coming into contact with extraneous cells and/or elements located in the central nervous system. Since most neurons have numerous processes, many of which may be quite long, each neuron requires a large number of oligodendroglia to form that physical barrier.

Question 5: There is little voluntary control over a two neuron reflex arc. As an example, the patellar reflex, which is a two neuron reflex arc, cannot be suppressed by an individual. However, due to the presence of the interneuron in a three neuron reflex arc, the motor action of the three neuron reflex arc can be suppressed. As an example, the normal reflex in response to a needle prick to the finger tip is to withdraw the finger from harm's way; however, if the finger is being pricked in a physician's office for the purposes of a blood test, the patient, although aware of the pain, will (usually) suppress the withdrawal component of the reflex.

CHAPTER 2

Question 1: The process of transforming the embryoblast to a trilaminar germ disc is known as gastrulation. This process establishes the presence of the three germ layers – the ectoderm, mesoderm, and endoderm – each of which will be responsible for the development of specific structures of the body. As an example, the ectoderm gives rise to the central nervous system, the mesoderm to the development of the vertebral column, and the endoderm to the epithelium of the gut.

Question 3: Genes code for the synthesis of specific proteins, many of which are enzymes that are responsible for the occurrence of specific events. Certain genes have been conserved through evolution and their study in lower organisms, such as the fruitfly, has provided information that is directly applicable to the developmental processes of higher organisms, including humans. A family of genes, known as the homeobox genes, code for the synthesis of transcription factors, proteins that bind to and regulate the expression of other genes.

The time sequence of the events controlled by products of these homeobox genes and by growth factors are essential for normal development because certain genes can be switched on only if some other genes have already been activated and if still other genes have not as yet been switched on. In other words, there are "windows of opportunity" for the occurrence of certain events and if that timeframe is missed then development will not proceed normally. Therefore, these homeobox genes become activated in a specific order, and their sequential expression establishes a pattern of developmental events. In order for development to progress in a

---

A Textbook of Neuroanatomy, Third Edition. Maria A. Patestas, Amanda J. Meyer, and Leslie P. Gartner.
© 2025 John Wiley & Sons, Inc. Published 2025 by John Wiley & Sons, Inc.
Companion website: www.wiley.com.go/patestas/neuroanatomy3e

QUESTIONS TO PONDER: ANSWERS TO ODD QUESTIONS • • • **525**

normal manner, cells must interact with each other. These interactions may involve the physical contact of two cells or the release of a particular substance by one cell, referred to as the signaling cell, to act as a message for the other cell, known as the target cell.

Question 5: Both develop as an outgrowth of the forming brain and an invagination of the ectoderm.

The hypophysis (pituitary): The infundibulum of the diencephalon is a downward evagination of the floor of the third ventricle. As the infundibulum grows down, it meets an ectodermal diverticulum of the oral cavity, known as adenohypophyseal pouch (Rathke's pouch). These two structures fuse with each other to form the hypophysis (pituitary gland). The neurohypophysis (median eminence, stalk, and pars nervosa) arises from the infundibulum whereas the adenohypophysis (pars tuberalis, pars distalis, and pars intermedia) derives from adenohypophyseal pouch). A small colloid-filled cleft frequently remains as a vestige of the lumen of adenohypophyseal pouch, in the pars intermedia.

The eye: During the third week of gestation, the ventral lateral wall on each side of the future diencephalon gives rise to an optic vesicle. These hollow vesicles grow in a lateral direction; they reach and induce the ectoderm to form the lens placode, the precursor of the lens of the eye. The optic vesicle invaginates to form a two-layered structure, known as the optic cup, which gives rise to the nervous and pigmented layers of the retina; the connection between the optic cup and the forming brain becomes known as the optic stalk, the future optic nerve. The rim of the optic cup encircles the prospective lens to form the iris.

CHAPTER 3

Question 1: Although the cell body of a neuron is relatively large, the axon and dendrites of many neurons possess a lot more cytoplasm than does the cell body. Therefore, the nerve cell body is responsible for the formation of much of the protein, macro-molecules, and neurotransmitter material that the entire cell requires. Because of this high biosynthetic demand, the neuron must be equipped with the biosynthetic machinery to manufacture these substances. Since the nerve cell body is the only expanded portion of the neuron, its organelles must be localized at that site.

Question 3: Since the intracellular concentration of K^+ is greater than the extracellular concentration, potassium is driven by a concentration gradient from the cell into the extracellular space. Potassium ions travel through ion channels that are not gated and are known as K^+ leak channels. Because only K^+ ions are permitted through these channels an electrical imbalance is established (more positive ions are present on the extracellular aspect of the membrane than on the intracellular aspect). At the point where the concentration gradient's "desire" to push potassium out of the cell and the electrical imbalance's "desire" to return potassium into the cell equal each other, the net flow of K^+ ions becomes zero. This does not mean that there is no flow of K^+ ions across the membrane, it simply means that the same number of K^+ ions flow into the cell as out of the cell.

The establishment of this equilibrium potential (modified for K^+ ions) is described by the Nernst equation. This equation can be applied for all ions, and is solved for the most common ions. However, the Nernst equation assumes that only the particular ion in question is permitted to traverse the cell membrane. However, this is not a true situation as ion channels of the plasma membrane are open simultaneously, thus permitting several ions to cross the plasmalemma. The Goldman equation describes the membrane potential as a function of the relative permeability of the cell membrane to more than a single species of ions. Since the neuron membrane is influenced by at least three ions – Na^+, K^+, and Cl^- – all three are considered in the Goldman equation. The Goldman equation is very similar to the Nernst equation, but it considers more than just a single ion type. It is interesting to note that the two equations would be identical if the permeability of two of the three ions were zero. Therefore, the Goldman equation describes the complexity of the existing situation better than does the Nernst equation and provides a more realistic prediction of resting membrane potential for neurons.

Question 5: An increase in the number of postsynaptic potentials (e.g., by the rapid release of multiple quanta of chemical messengers from a single axon terminal or from many axon terminals on the same neuron), results in their summation. Hence, hundreds of nearly simultaneous, small postsynaptic potentials may be necessary to achieve the threshold required for the formation of an action potential when those currents arrive at the initial segment of the axon. Generally, inhibitory synapses are closer to the initial segment of the axon than are excitatory synapses. Since inhibitory synapses function to hyperpolarize the postsynaptic membrane, they tend to diminish the flow of currents formed at the excitatory potential making it more difficult to reach threshold levels at the initial segment of the axon. This sequence of events is referred to as the summation of inhibitory postsynaptic potentials and excitatory postsynaptic potentials.

CHAPTER 4

Question 1: The neurotransmitter substances that bind specifically to ligand-gated ion channels cause those channels to open directly, thus the target neuron responds immediately, resulting in an ionotropic effect. Those neurotransmitter substances that bind to G-protein-coupled receptors exert an indirect effect on ion channels. Since the G-proteins act as intermediaries, the effect on the target neuron is slower than in the previous instance. Thus, these neurotransmitter substances are known as neuromodulators, and they are said to exert a metabotropic effect.

Question 3: Biogenic amines are derived from amino acids and exert a metabotropic effect. They bind to G-protein-coupled receptors and activate the second messenger system within the target neuron, which in turn will open ion channels. These substances include the catecholamines dopamine, norepinephrine, and epinephrine as well as histamine and serotonin. Upon release from the presynaptic terminal, biogenic amines are endocytosed by the nerve terminals and glial cells,

526 ● ● ●   **QUESTIONS TO PONDER: ANSWERS TO ODD QUESTIONS**

and are digested by the enzymes cathecol O-methyltranferase and monoamine oxidase. In this fashion, each release of these neurotransmitters is responsible for only a limited number of depolarizations/hyperpolarizations.

Question 5: Acetylcholine is a neurotransmitter substance localized both in the central and peripheral nervous systems. It is synthesized in the presynaptic terminal from acetyl CoA and choline by choline acetyl transferase, the rate-limiting enzyme of its synthesis. Acetylcholine has the ability to bind to both ionotropic receptors and metabotropic receptors. In skeletal muscle it is an excitatory neurotransmitter substance, binding directly to ligand-gated sodium (nicotinic receptors) as well as to potassium ion channels, and causing them to open. In cardiac muscle it is an inhibitory neurotransmitter substance, binding to G-protein-linked receptors (muscarinic receptors) that facilitate the opening of potassium ion channels.

CHAPTER 5

Question 1: The cervical and lumbar enlargements are necessary to provide space for the many neurons that are required for the innervation of the upper and lower extremities, respectively. Because the upper extremity is much more complex in humans, in that the hands and fingers possess a very rich nerve supply, the cervical enlargement is greater than the lumbar enlargement.

Question 3: White matter is composed mostly of myelinated and unmyelinated nerve fibers, which bring information into the central nervous system and transmit it to higher levels, and transmit information from higher levels to the spinal cord and to the muscles and glands. Based on the presence of sulci and fissures, the white matter is subdivided into three major columns, posterior, lateral, and anterior funiculi. Within the funiculi the nerve fibers that have similar destinations are arranged in bundles known as tracts (fasciculi).

Question 5: The spinal cord equals the length of the vertebral canal only until the end of the first trimester of prenatal life. Thereafter the trunk elongates much faster, so that by the time of birth the conus medullaris is only at the level of the third lumbar vertebra and in the adult it extends only as far as the caudal aspect of the first lumbar vertebra. Therefore, lower levels of the spinal cord segments are displaced rostrally in relation to their corresponding vertebral levels. Because of this differential growth, the subarachnoid space caudal to the conus medullaris, known as the lumbar cistern, is devoid of spinal cord. It is this space that may be accessed by physicians to remove cerebrospinal fluid for diagnostic purposes without causing injury to the spinal cord.

CHAPTER 6

Question 1: They are the commissural, projection, and association fibers. Commissural fibers connect the right and left cerebral hemispheres. Projection fibers connect the cerebral cortex to lower levels, namely the corpus striatum, diencephalon, brainstem, and spinal cord; the majority of these fibers are axons of pyramidal cells and fusiform neurons. Association fibers are restricted to a single hemisphere.

Question 3: Peduncles are large fiber tracts that connect parts of the brain to the brainstem. There are two separate groups of peduncles, the cerebral and cerebellar peduncles. The cerebral peduncles are two very large, thick fiber bundles that connect the cerebral hemispheres to the brainstem. The cerebellar peduncles are three pairs of fiber bundles that connect the cerebellum to the midbrain, pons, and medulla of the brainstem.

CHAPTER 7

Question 1:
a. Left oculomotor palsy, the left pupil will not constrict following stimulation of either eye
b. Upon protrusion, the jaw will deviate to the right
c. Weakness of the right lower face
d. Uvula will deviate to the left
e. Upon protrusion, the tongue will deviate to the right
f. Weakness in right upper and lower limbs.
These are the deficits characteristic of Weber syndrome.

Question 3:
a. Right oculomotor nerve palsy, right pupil will not constrict following stimulation of either eye
b. The jaw will deviate to the left
c. Weakness of the left lower face
d. Uvula deviates to the right
e. Tongue deviates to the left
f. Weakness in left upper and lower limbs.
These are the deficits characteristic of Benedict syndrome.

Question 5:
a. Movement of the mandible is not affected since the corticonuclear fibers to the trigeminal motor nucleus have already synapsed at the midpontine level (which is located above the level of the lesion), and thus those corticonuclear fibers have not been damaged.
b. Lower face weakness on the right
c. The uvula deviates to the left
d. The tongue deviates to the right
e. Dysarthria
f. Upper and lower limb weakness on the right
g. Ataxia on right (due to damage of the pontine nuclei and pontocerebellar fibers arising from them).
Deficits are characteristic of Ataxic hemiparesis.

CHAPTER 8

Question 1: The dura mater of the spinal cord differs from that of the cranial dura mater because the spinal dura is composed only of the meningeal layer. The periosteal layer is the true periosteum of the vertebral canal and is separated from the meningeal layer by a loose connective tissue and fat-filled epidural space.

Question 3: Although the brain itself may not feel pain, the dura mater possesses a very rich sensory nerve supply, derived mostly from cranial nerve V (the trigeminal nerve) but also from the first three cervical spinal nerves that serve the dura mater of the posterior cranial fossa. It is these sensory fibers that become stimulated and the patient may experience severe pain.

Question 5: Arachnoid granulations are small, mushroom-shaped evaginations of the arachnoid that protrude into the lumen of the dural sinuses. Most of the arachnoid granulations are associated with the lacunae lateralis – diverticula of the superior sagittal sinus – although some jut into the lumen of the sinus. The core of an arachnoid granulation is continuous with the subarachnoid space and is surrounded by the epithelioid layer of the arachnoid and the dura, forming a membrane that is two cell layers thick. As the arachnoid granulation evaginates into the lacuna lateralis it is invested by some cellular and collagenous elements of the meningeal dura mater, which in turn is surrounded by the endothelial lining of the blood vessel. Cerebrospinal fluid from the subarachnoid space enters into the core of the arachnoid granulation and from there penetrates, probably by osmosis, the epithelioid layers of the arachnoid granulations and the endothelial lining to escape into the lacuna lateralis. Therefore, the function of the arachnoid granulations is transporting cerebrospinal fluid manufactured by the choroid plexuses of the brain ventricles into the vascular system. It is important to note that although the arachnoid granulations protrude into the venous sinuses, they are always separated from the blood by the endothelial lining of the dural sinus/lacuna lateralis.

CHAPTER 9

Question 1: Radicular arteries are extremely important for the vascularization of the spinal cord because, with the exception of the cervical region, the anterior and posterior spinal arteries by themselves are unable to provide an adequate vascular supply to the spinal cord. Therefore, an injury to a spinal nerve damages not only the afferent and efferent fibers of a particular spinal cord level, but may also damage the segmental white and gray matter by producing ischemic conditions due to the reduction in blood supply from the radicular artery serving that region.

Question 3: The posterior communicating artery connects the cerebral portion of the internal carotid artery to the posterior cerebral branch of the basilar artery. The right and left arteries are not identical, in that one is frequently smaller than the other, and, in fact, one may be entirely absent or doubled. The main function of this vessel is to ensure a viable blood supply to the brain in case the internal carotid or vertebral artery becomes occluded. However, it also has central branches that pierce the region of the base of the cerebrum and serve part of the posterior limb of the internal capsule, the medial aspect of the thalamus, and the tissue forming the lateral border of the third ventricle.

Question 5: Four major vessels – the two internal carotid arteries and the two vertebral arteries – bring blood to the brain. Although numerous veins receive blood from the brain, they all deliver their contents into the venous sinuses of the meninges. The venous sinuses, in turn, deliver their blood into the superior bulb of the right and left internal jugular veins.

CHAPTER 10

Question 1: Preganglionic sympathetic nerve cell bodies are located in the lateral horn of the spinal cord only at levels T1 through L2, therefore the preganglionic fibers are located only in the corresponding spinal nerves. They leave the spinal nerves via the white rami communicantes to join the sympathetic chain ganglia. There they may synapse with postganglionic sympathetic nerve cell bodies at various levels along the sympathetic trunk. Postganglionic sympathetic fibers of these nerve cell bodies re- enter the spinal nerve (through gray rami communicantes) at the level that leads to the region of their destination. Since there are 31 pairs of spinal nerves there has to be a corresponding gray rami communicantes for each spinal nerve.

Question 3: The autonomic nervous system is designed in such a fashion that preganglionic fibers can synapse only with their own postganglionic nerve cell bodies, therefore a preganglionic parasympathetic fiber cannot synapse with a post-ganglionic sympathetic nerve cell body. The postganglionic nerve cell body can thus only be stimulated by its own preganglionic neuron. However, a target cell, such as the conducting system of the heart, may be innervated by both postganglionic sympathetic and parasympathetic fibers and the target cell has to know whether it is responding to a sympathetic or a parasympathetic message.

Question 5: As feces accumulate, the sigmoid colon distends and the individual becomes conscious of the need to defecate, but the feces are retained in the colon by the two anal sphincter muscles, the internal smooth muscle sphincter and the external skeletal muscle sphincter. The former is supplied by postganglionic sympathetic fibers from the hypogastric plexus as well as by postganglionic parasympathetic fibers located in the myenteric plexus (Auerbach's plexus) (whose preganglionic fibers arise from the sacral spinal cord). The external sphincter is supplied by the somatic nervous system, namely the inferior rectal nerve. Thus, when the sigmoid colon becomes distended, the internal sphincter muscle relaxes (due to the parasympathetic nerve supply) and at the same time the external sphincter muscle contracts, preventing the feces from exiting the bowel. While the individual is asleep, this contraction of the external sphincter muscle is a reflex response. When the individual is awake, the contraction of the external sphincter is a voluntary response.

CHAPTER 11

Question 1: An interneuron

Question 3: The muscle stretch reflex compensates for the muscle stretch and serves to protect the muscle fibers from overstretching that may cause injury to the muscle.

528 ••• QUESTIONS TO PONDER: ANSWERS TO ODD QUESTIONS

Question 5: A single (noxious) stimulus sets off two reflexes: the flexor (withdrawal) reflex for one lower limb (affected by the noxious stimulus), and the crossed extension reflex for the other (unaffected) lower limb.

Question 7: (a) when the gross muscle is passively stretched, as in tapping its tendon (this stretches the extrafusal fibers and in addition, the muscle spindles containing intrafusal fibers). The central (non-contractile) portion of each intrafusal fiber stretches, which disturbs and stimulates the Ia annulospiral endings, or (b) the central portion stretches when the polar (contractile) ends of the intrafusal fibers contract, in response to gamma motoneuron stimulation, thereby disturbing and stimulating the Ia annulospiral endings.

CHAPTER 12

Question 1: Some receptors respond quickly and maximally at the onset of the stimulus, but stop responding even if the stimulus continues; these are known as rapidly adapting (phasic) receptors. They are essential in responding to changes but they ignore ongoing processes, such as when one wears a wristwatch and ignores its presence and continuous pressure on the skin of the wrist. However, there are other receptors, slowly adapting (tonic) receptors, that continue to respond as long as the stimulus is present.

Question 3: It is the degree of stretching that is proportional to the load placed on the muscle. The larger the load, the more strongly the spindles are depolarized and the more extrafusal muscle fibers are in turn activated.

Question 5: Pain perception is not only processed in the primary somatosensory cortex, but also in the anterior cingulate and anterior insular cortices. Recent studies support the finding that pain sensation persists following a lesion to the primary somatosensory cortex as a result of these additional cortical representations of pain.

CHAPTER 13

Question 1: By alpha and gamma motor neuron coactivation. When the gross muscle contracts (following stimulation by alpha motor neurons), the contractile portions of the intrafusal fibers undergo a simultaneous corresponding contraction (following stimulation by gamma motor neurons), stretching their central noncontractile portion. The muscle spindles are then ready to detect muscle stretch when the muscle becomes stretched from the contracted state or from any length. Therefore, when the gross muscle begins to stretch from the contracted state, the intrafusal fibers will be immediately stretched and their afferent (sensory) fibers will fire, informing the central nervous system that muscle stretching is occurring.

Question 3: Although the upper motor neurons arising from the primary motor cortex may be damaged, the tectospinal, reticulospinal, and vestibulospinal tracts and the upper motor neurons contributed by the secondary motor cortex provide a motor input to the interneurons and lower motoneurons innervating the skeletal muscles.

Question 5: The cortical group, which includes the lateral corticospinal tract.

CHAPTER 14

Question 1: The limbic system consists of a group of cortical and subcortical structures that are involved in the processing of emotions. Via its connections to the basal nuclei, the limbic system can influence the underlying motivational aspects of behavior or movement. Here is an example of the role played by the limbic system in movement. The pleasant aroma of freshly brewed coffee and baked cinnamon rolls are enough to motivate anyone of us to walk toward the coffee and cinnamon rolls and taste them by lifting the coffee cup and roll to take a sip and a few bites. Tasting the cinnamon roll may elicit certain facial expressions that may indicate to those looking at us that we are enjoying the delicious tasting cinnamon roll. Furthermore, the warm cinnamon aroma may evoke fond memories of the first time we tasted a cinnamon roll. In this example, the olfactory system (mediating the sense of smell), the limbic system (processing emotions and memory), and the basal nuclei involved in movement are all interconnected.

Question 3: Normally, the subthalamic nucleus projects glutaminergic (excitatory) fibers to the two output nuclei of the basal nuclei, the medial segment of the globus pallidus and the reticular part of the substantia nigra, which in turn send inhibitory projections to the thalamus. Thalamocortical fibers that are excitatory project to the motor cortex. Projections up to this point are ipsilateral; however, the upper motor neurons arising from the cortex decussate in the pyramids, thus affecting the opposite side of the body. A lesion in the subthalamic nucleus would result in disinhibition (excitation) of the thalamus, which in turn becomes overactive, overstimulating the motor cortex, which in turn results in hyperkinetic symptoms in the opposite side of the body.

Question 5: If excessive l-DOPA is administered to an individual with Parkinson disease, the individual will manifest choreic movements. l-DOPA administered to a patient with Huntington disease will intensify the choreic movements.

CHAPTER 15

Question 1: The cerebellar efferents and the descending motor pathways (i.e., the corticospinal and rubrospinal tracts) are crossed. Thus a lesion in the right cerebellar hemisphere via its efferents (which decussate) affects the left side of the brain. As the left corticospinal tract descends to the pyramids, most of its fibers cross to the opposite (right) side as the lateral corticospinal tract, terminating in the right spinal cord, and synapsing with lower motor neurons that innervate the muscles on the right side of the body. The corticorubral fibers arising from the left side of the brain descend to the left red nucleus, which in turn gives rise to the rubrospinal tract whose fibers cross to the right side in

QUESTIONS TO PONDER: ANSWERS TO ODD QUESTIONS ●●● 529

the midbrain. The right rubrospinal tract will descend to terminate on lower motor neurons in the right spinal cord that innervate muscles on the right side of the body.

Question 3: The flocculonodular lobe of the cerebellum. It is interesting to note that although both motor and cognitive functions are affected by the consumption of alcohol, the cells of the cerebellum are very susceptible to the effects of alcohol. This becomes apparent by the awkward, uncoordinated execution of movements by the intoxicated individual, especially when the eyes are closed.

Question 5: A decrease in resistance to passive manipulation resulting in excess movement of the limbs (referred to as "rag-doll appearance") is caused by the absence of cerebellar influence on the stretch reflex.

CHAPTER 16

Question 1: Nociceptive input from the body is transmitted to the spinal cord posterior horn by first order pseudounipolar neurons whose cell bodies are located in the posterior root ganglia. Second order neurons, whose cell bodies are located in the dorsal horn of the spinal cord, give rise to axons that ascend as the spinothalamic or spinoreticular tracts. The spinothalamic tract terminates in the ventral posterior lateral nucleus of the thalamus, but also sends collaterals to the brainstem reticular formation. The spinoreticular tract terminates in the reticular formation, and also sends collaterals to the thalamus.

Nociceptive input from the orofacial region is transmitted to the pons by the pseudounipolar neurons whose cell bodies are located in the trigeminal ganglion. Fibers conveying nociceptive input descend in the spinal tract of the trigeminal nerve to terminate in the caudal subnucleus of the spinal nucleus of the trigeminal. This nucleus also receives the central processes of sensory neurons of other cranial nerves relaying nociceptive input from structures in the head. The axons of second order neurons housed in the caudal subnucleus cross to the opposite side of the pons, and join the anterior trigeminal lemniscus to terminate in the ventral posterior medial nucleus of the thalamus with collaterals to the reticular formation.

Question 3: Through their ascending fibers the neurons of the medial zone influence the autonomic nervous system, and the level of arousal. Via their descending fibers, which join the reticulospinal tracts, the neurons of the medial zone influence the motor control of the axial and proximal limb musculature.

Question 5: Lesions damaging the midbrain reticular formation can cause hypersomnia associated with slow respiration. When the midbrain is compressed with consequent extensive damage of the ARAS pathways, a comatose state ensues.

CHAPTER 17

Question 1: Interruption of the parasympathetic innervation of the sphincter pupillae muscle causes the pupil ipsilateral to the lesion to remain dilated (mydriasis) and to not respond (constrict) to a flash of light. This may be the first clinical sign of intracranial pressure on the general visceral efferent fibers of the oculomotor nerve. The ciliary muscle is also nonfunctional due to interruption of its parasympathetic innervation, and cannot accommodate the lens for near vision (that is, cannot focus on near objects).

Question 3: The abducens nucleus innervates a single extraocular muscle, the lateral rectus. A lesion in the abducens nucleus damages not only the general somatic efferent (GSE) motor fibers innervating the lateral rectus muscle, but also the internuclear fibers projecting from the abducens nucleus to the contralateral oculomotor nucleus, that synapse with the motor neurons that innervate the medial rectus muscle.

A lesion in the abducens nucleus that damages the GSE motor fibers innervating the lateral rectus results in paralysis of the lateral rectus muscle, which normally abducts the eye. Since the individual will be unable to move the ipsilateral eye laterally, it deviates medially as a result of the unopposed action of the medial rectus. The individual can turn the ipsilateral eye from its medial position to the center (looking straight ahead), but not beyond it. This paralysis results in medial strabismus (convergent, internal strabismus, esotropia). Since the eyes become misaligned, the individual experiences horizontal diplopia (double vision). The diplopia is greatest when looking toward the side of the lesion and is reduced by looking towards the unaffected side since the visual axes become parallel. The individual realizes that the diplopia is minimized by turning his head slightly so that the chin is pointing toward the side of the lesion. Bilateral abducent nerve lesion results in the individual becoming "cross-eyed."

A lesion involving the abducens nucleus damaging the internuclear neurons results in the inability to turn the opposite eye medially as the individual attempts to gaze toward the side of the lesion. This condition is referred to as lateral gaze paralysis.

Question 5: A lesion in the vicinity of the abducens nucleus that involves the entire ipsilateral abducens nucleus as well as the decussating medial longitudinal fasciculus fibers arising from the contralateral abducens nucleus, results in a rare condition referred to as "one-and-a-half."

If, for example, a lesion is present in the vicinity of the left abducens nucleus the following will occur.

1. The GSE motoneurons whose axons form the left abducens nerve innervating the left lateral rectus muscle are damaged. Therefore that lateral rectus muscle is paralyzed.
2. The internuclear neurons housed in the left abducens nucleus are also damaged. Therefore their crossing fibers do not form excitatory synapses with the motor neurons of the contralateral (right) oculomotor nucleus that innervate the right medial rectus muscle.
3. The crossing fibers of the internuclear neurons arising from the contralateral (right) abducens nucleus are also damaged. Thus they do not form excitatory synapses with the motoneurons of the left oculomotor nucleus that innervate the left medial rectus muscle.

530 ● ● ● QUESTIONS TO PONDER: ANSWERS TO ODD QUESTIONS

Therefore, when attempting to gaze to the left, the left eye will not abduct and the right eye will not adduct during conjugate horizontal gaze to the left. When attempting to gaze to the right, the right eye abducts with nystagmus, whereas the left eye will not be able to adduct during conjugate horizontal gaze to the right.

It is important to note that the innervation to all the extra-ocular muscles of both eyes is intact, except to one – the left lateral rectus. If you ask this individual to look at a near object placed directly in front of his face, both eyes will converge since both medial recti and their innervation (branches of the oculomotor nerve) are intact. This type of lesion (causing "one-and-a-half") becomes apparent only during conjugate horizontal eye movement.

Question 7: The first cranial nerve that is likely to be damaged from a growing pituitary tumor is the optic nerve. The decussating axons at the optic chiasma will be compressed, followed by the nondecussating fibers. If the tumor becomes very large, the cranial nerves located in the cavernous sinus (namely the oculomotor, trochlear, and ophthalmic division of the trigeminal and the abducens nerves) will likely be compressed and cause deficits in the structures that these nerves innervate.

CHAPTER 18

Question 1: Although light travels from the inner to the outer layers of the retina (internal limiting membrane to pigmented epithelium), visual electrical signals are transmitted in the opposite direction from the outer to the inner layers of the retina (rod and cone layer to layer of nerve fibers of the retina (optic nerve fiber layer).

Question 3: When the right eye was illuminated, only the consensual (left) papillary light reflex was elicited. This indicates that a lesion may be present in the efferent limb (oculomotor nerve) of the pupillary light reflex arc of the right side. When the left eye was illuminated, again, only the left eye responded, which indicates that the lesion may be present in the efferent limb (oculomotor nerve) of the pupillary light reflex arc on the right side.

Question 5: The patient had a lesion affecting the ganglion cell axons arising from the nasal half of each retina, which cross at the optic chiasma over the body of the sphenoid bone. The crossing fibers may be damaged due to compression by a tumor of the hypophysis (pituitary gland).

CHAPTER 19

Queston 1: The middle ear functions in the transmission and transduction of tympanic membrane vibrations to the inner ear (from an air to a fluid medium). The vestibular (oval) window and the cochlear (round) window both serve necessary functions in these processes.

Question 3: The basilar membrane is an elastic structure exhibiting a gradual increase in width, and a decrease in stiffness from the vestibular (oval) window (at the cochlear

base) to the helicotrema (at the cochlear apex). The gradual decrease in stiffness permits this membrane to be sensitive to high frequency vibrations near the cochlear base, and low frequency vibrations near its apex, at the helicotrema.

Question 5: The olivocochlear bundle has an inhibitory effect on cochlear nerve activity, modulating and sharpening auditory transmission.

CHAPTER 20

Question 1: Sensory input from the visual, vestibular, and proprioceptive systems, is integrated by the nervous system, especially the cerebellum, to generate motor responses. These motor responses maintain equilibrium, posture, and muscle tone, as well as reflex movements of the eyes, all of which are carried out at the subconscious level.

Question 3: The neuroepithelia of the ampullary crest, macula of utricle, and macula of saccule are covered with a gelatinous glycoprotein membrane. The stereocilia and kinocilia are embedded in this membrane. In the utricle and saccule, the gelatinous glycoprotein membranes overlying their macula have, on their free surface, statoconia also called otoconia (G., "ear dust") or otoliths (G., "ear stones") – which are crystals consisting of calcium carbonate. Since the gelatinous glycoprotein membrane of the utricle and saccule has statoconia, it is referred to as the statoconial membrane. The statoconia have a specific gravity that is greater than that of the endolymph which surrounds them, and they are consequently pulled by gravity. Gravity applies a continuous linear acceleration on the head, and thus deflects the gelatinous membrane, which in turn stimulates the hair cells. The glycoprotein membrane of the ampullary crests is dome-shaped and is called the cupula; it is devoid of crystals. The cupula increases resistance to the flow of endolymph and, unlike the otolithic membrane, is not influenced by gravitational forces. Instead, it responds to endolymphatic flow. The cupula is a mechanical structure that narrows the lumen of the canal and so increases resistance to flow. The three ampullary crests are enclosed in the semicircular duct ampullae (one crista in each ampulla) and detect angular acceleration or deceleration (rotational movement) of the head (as in turning or tilting the head). The receptors in the utricle and saccule (the macula of utricle and macula of saccule within the utricle and saccule, respectively) detect spatial orientation of the head in space relative to gravity and linear acceleration or deceleration forces.

Question 5: The vestibular nerve is unique since it is the only cranial nerve that sends some of its first order afferent fibers relaying sensory input directly from the peripheral receptors to the cerebellar cortex. The vestibular nuclei and the cerebellum serve as the first relay centers of the vestibular pathway.

Question 7: When an individual fixes his gaze on ("stares at") an object and then begins to turn his head, for example to the right, the eyes will reflexly (via the medial longitudinal fasciculus vestibular connections to the extraocular nuclei) "turn" to the left (as the head is moving), which is opposite to the direction of head movement. This turning of the eyes

QUESTIONS TO PONDER: ANSWERS TO ODD QUESTIONS ••• 531

(dictated by vestibular reflex connections) compensates for shifting of the head position in order to maintain visual fixation on an object. Without this vestibular reflex mechanism coupling head and eye movements, as the erect head moves from a certain position (to the right or to the left or is flexed or extended) the eyes would remain still, perceiving a changing visual field during the head movement.

CHAPTER 21

Question 1: Olfactory receptor cells, the first order bipolar neurons of the olfactory pathway, reside in a peripheral structure, the olfactory epithelium (instead of being housed in a sensory ganglion). They are the only neurons that are in direct contact with the environment and they give rise to unmyelinated axons, the slowest impulse-conducting axons of the central nervous system. They serve the functions of: (i) sensory receptors (chemoreceptors that become activated by chemical stimuli in their surroundings); (ii) transducers (generate graded potentials); and (iii) first order neurons (relaying olfactory sensation centrally to the olfactory bulbs).
Question 3: Odorant-receptor molecules (transmembrane G-protein-coupled receptors) are embedded in the plasmalemma of the ciliary membrane of the olfactory receptor cells.
Question 5: The binding of an odorous substance to the odorant-receptor protein molecules (on the ciliary membrane of the olfactory receptor cell), stimulates the initiation of a second messenger pathway associated with an olfactory-specific G-protein, which subsequently stimulates adenyl cyclase, which generates cAMP. The cAMP elevation unlocks a cyclic nucleotide-gated cation channel embedded in the ciliary membrane, permitting sodium and calcium cations to pass into the cell. This stimulates the chemosensitive cilia of the olfactory receptor cells, and a wave of depolarization (generator potential) is propagated from the dendrites to the cell body of the olfactory receptor cell. A high enough depolarization triggers a nerve impulse that is propagated along the axon of the olfactory receptor cell, centrally to the olfactory bulb.

CHAPTER 22

Question 1: The limbic lobe consists of the subcallosal, cingulate, and parahippocampal gyri, as well as the hippocampal formation, all of which collectively form a cortical perimeter around the corpus callosum. The limbic lobe is considered to be a component of the more extensive limbic system.

The limbic system includes all the structures of the limbic lobe as well as the amygdaloid body, septal area, and some nuclei of the thalamus and hypothalamus. The two main structures of the limbic system are the hippocampal formation (which includes the hippocampus proper, dentate gyrus, and the subiculum) and the amygdaloid body.
Question 3: The dentate gyrus gets its name from its toothlike configuration, created by numerous blood vessels that pierce the ventricular surface of the hippocampus and then the dentate gyrus, imparting a notched appearance.

Question 5: The entorhinal cortex. Widespread areas of the association cortex relay input to the entorhinal cortex and it is via the entorhinal cortex that these cortical association areas have access to the hippocampus and play a role in memory consolidation.
Question 7: The hypothalamus mediates autonomic nervous system (visceral) responses that accompany the expression of emotions.
Question 9: The perforant pathway from the entorhinal cortex.

CHAPTER 23

Question 1: The nuclei of this zone include the periventricular, suprachiasmatic, and arcuate nuclei. The periventricular and arcuate nuclei control the functions of the adenohypophysis, (anterior pituitary gland) which synthesizes and releases various endocrine hormones into the bloodstream. The suprachiasmatic nucleus functions in the control of circadian rhythms, and is referred to as the "master clock" of the body.
Question 3: The lateral zone of the hypothalamus includes the following nuclei: the lateral preoptic nucleus anteriorly, the lateral tuberal nuclei in the tuberal region, and the lateral hypothalamic nucleus, which extends throughout the anteroposterior extent of the lateral zone. The lateral zone receives input from the limbic system and plays an important role in the behavioral expression of emotions.
Question 5: The supraoptic region contains nuclei that function in the control of circadian rhythms, temperature regulation, and the control of water balance.
Question 7: The mammillary region includes the prominent mammillary nuclei and the posterior hypothalamic nucleus. The mammillary nuclei are associated with emotions, whereas the posterior hypothalamic nucleus serves as a "thermostat" and regulates body temperature.
Question 9: In addition to the neural input, the hypothalamus also receives non-neural input from the vascular system. Hypothalamic neurons serve as receptors that are sensitive to rapid changes in the temperature, osmotic pressure, and hormone levels of the circulating blood. These receptors are referred to as circumventricular organs (CVOs). They consist of hypothalamic neurons that are embedded in the wall of the third ventricle, and are sensitive to alterations in the chemical composition of the cerebrospinal fluid. One of these structures is associated with the lamina terminalis, and is referred to as the vascular organ of the stria terminalis. The vascular organ of the stria terminalis is a chemosensory structure that detects and responds to peptides and macromolecules present in the bloodstream. The other CVO is the subfornical organ located inferior to the fornix, near the interventricular foramen. The subfornical organ serves in the regulation of body fluids. The fenestrated capillaries supplying these receptors do not have a blood-brain barrier, thus they permit substances present in the bloodstream to pass through and enter the extracellular space surrounding the receptor cells.

532 ● ● ● QUESTIONS TO PONDER: ANSWERS TO ODD QUESTIONS

Question 11: The fornix.

Question 13: The ascending reticular activating system relays nociceptive input to the hypothalamus, which mediates the autonomic and reflex responses to nociception.

Question 15: The tuberohypophyseal tract carries fibers from the arcuate and periventricular nuclei to the infundibular (pituitary) stalk (of the posterior pituitary). The axon terminals of the tubero hypophyseal tract release "releasing hormones" or "release-inhibiting hormones" in the hypophyseal portal system, which carries them to the anterior lobe of the pituitary gland, where they regulate the synthesis and release of anterior pituitary hormones. The tuberohypophyseal tract and the hypophyseal portal system serve as a connection between the hypothalamus and anterior pituitary gland.

Question 17: There are two main problems that might cause this condition: (i) a blockage of the superior hypophyseal artery, thus depriving the anterior pituitary of nutrition; or (ii) a lesion in the periventricular zone of the hypothalamus.

CHAPTER 24

Question 1: The anterior nuclear group, a relay nucleus of the Papez circuit of emotion, has extensive connections with the limbic system. It receives afferent fibers relaying information from the mammillary body and the hippocampal formation. It in turn projects to the cingulate gyrus of the limbic association cortex. Due to its limbic connections, the anterior nuclear group functions in the expression of emotions and, due to its considerable connections with the hippocampal formation, this nuclear group is also believed to be associated with learning and memory processes.

The lateral dorsal nucleus (of the lateral group of thalamic nuclei) is the caudal continuation of the anterior thalamic nuclear group and may play a role in the expression of emotions. Like the anterior thalamic nuclear group, it receives inputs from the mammillary body and in turn projects to the cingulate and parahippocampal gyri of the limbic system. The dorsomedial nucleus (of the medial group of thalamic nuclei) also functions in the processing of information related to emotion via its connections to the limbic system components such as the amygdala and the orbitofrontal cortex.

Question 3: The posterior portion of the ventral lateral (VLp) nucleus receives prominent projections arising from the deep cerebellar nuclei (mainly from the dentate nucleus) of the opposite side, via the dentatothalamic tract. The VLp then relays the information to the premotor and primary motor cortex (Brodmann areas 6 and 4, respectively).

Question 5: The nonspecific nuclei of the thalamus, which includes the intralaminar and reticular thalamic nuclei.

CHAPTER 25

Question 1: Most of the cerebral cortex consists of association cortex, which is classified into unimodal association cortex and multimodal (heteromodal) association cortex. Unimodal association areas are located next to, near, or around the primary sensory cortices that expand on the functions of the respective primary areas. As an example, the primary visual cortex (Brodmann area 17) projects to the visual association cortex (Brodmann areas 18, 19, 20, 21, and 37), and only processes a single modality, in this case vision, and is thus referred to as a unimodal association cortex. In contrast, the multimodal association cortical areas receive inputs from multiple sensory modalities; they then integrate the information and, via their higher order cognitive functions, formulate a composite experience. The multimodal association cortex is associated with imagination, judgement, decision making, and making long-range plans.

Question 3: Studies of patients who had their corpus callosum sectioned showed the following: When patients were asked to read words that were presented in their right visual field, they were able see and to read the words. However, when words were presented in their left visual field, they could not read them, since they were not even aware that the words were there. One explanation is as follows. Visual information in the right visual field is relayed to the left cerebral hemisphere (via the visual pathway, which was not affected when the corpus callosum was sectioned), which is the dominant hemisphere in the processing of language in most individuals. Thus the patients were able to see and read the words. In contrast, visual input from the left visual field is relayed to the right cerebral hemisphere (again, via the visual pathway, which was not affected when the corpus callosum was sectioned), the nondominant hemisphere for language. As a result of the bisection of the corpus callosum, the visual input relayed to the right cerebral hemisphere has no way of also reaching the left hemisphere containing the language areas. Consequently, patients could not see or read the words – as if they had a left homonymous hemianopia. These individuals are not blind. When they were asked to select with their hand an object that corresponds to the object that was presented in their left visual field, they selected the matching object. This showed that the visual system is functioning properly, and that it is the language function that is not.

Question 5: Broca and Wernicke areas are supplied by the middle cerebral artery. This vessel has an extensive distribution and supplies parts of the frontal, parietal, temporal, and occipital lobes. If only the cortical branches of the middle cerebral artery supplying Broca area are occluded, the cerebral cortex of Broca area becomes infarcted, resulting in Broca aphasia. If the proximal segment of the middle cerebral artery is occluded, the cerebral cortex of the insula and subcortical white matter will also become infarcted, and a severe form of Broca aphasia results, which is accompanied by motor deficits in the contralateral lower face, tongue, and upper limb. If only the cortical branches of the middle cerebral artery supplying Wernicke area are occluded, the cerebral cortex of Wernicke area becomes infarcted, resulting in Wernicke aphasia. If the subcortical white matter and geniculocalcarine tract (optic radiation) are also involved, visual deficits will also be present.

Questions to ponder: answers to even questions

CHAPTER 1

Question 2: The "fight or flight or freeze" response is a function of a component of the autonomic nervous system. The autonomic nervous system has three components – sympathetic, parasympathetic, and enteric – each with well defined, discrete functions whose interplay maintains homeostasis. One of the major responsibilities of the sympathetic nervous system is to ensure that the body is ready to act in an "emergency" situation that may be life-threatening, so that the individual has the ability to either run away from the threatening situation or to face and oppose it.

Question 4: Although all of the neuroglia of the central nervous system have different functions and different morphological characteristics, they are all related to each other because they all have a common embryologic origin within the central nervous system. The only exception to this rule are the microglia, since they are macrophages derived from the cells of the bone marrow and they migrate into the central nervous system using blood vessels as their mode of entry.

CHAPTER 2

Question 2: During the transformation of the bilaminar germ disc to the trilaminar germ disc, the small, anterior region of the primitive streak, known as primitive node (Hensen's node) gives rise to a group of cells that migrate into the primitive pit, pierce its floor, and travel *en masse* between the epiblast and hypoblast until they reach the prochordal plate. They form the notochordal process, whose cells are in close contact with the cells of the overlying ectoderm. These notochordal cells possess inductive properties and cause the transformation of the ectodermal cells lying above them to form the neural plate, the beginning of the nervous system. If the primitive node does not give rise to the notochordal process the embryo will be unable to form a nervous system.

Question 4: As the neuroepithelial cells continue to proliferate, they establish a thicker wall composed of a pseudostratified epithelium, each of whose cells initially extend the entire thickness of the developing neural tube. As development progresses, some of these neuroepithelial cells remain adjacent to the lumen, whereas most cells will migrate away from it, resulting in a three-layered neural tube. The layer adjacent to the lumen is known as the ventricular zone (ependymal layer), some of whose cells will develop into ependymal cells that will line the central canal of the spinal cord. The middle region of the developing neural tube is the mantle layer (intermediate zone). The third layer of the developing neural tube, located the farthest from the lumen is the marginal layer (marginal zone).

CHAPTER 3

Question 2: Oligodendroglia, in the central nervous, and neurolemmocytes (Schwann cells), in the peripheral nervous system, are responsible for the formation of myelin sheaths around axons. Since axons may be very long, numerous oligodendroglia or neurolemmocytes are required to line up along the axon to perform the myelination. The length of an axon covered by the myelin sheath of a single oligodendroglion or neurolemmocyte is referred to as an internode, whereas the part of the axon left uncovered at the junction of two neurolemmocytes or oligodendroglia is referred to as the myelin sheath gap (node of Ranvier). It is at the myelin sheath gaps that numerous ion channels are located and thus, as the nerve impulse jumps from one node to the other, nerve conduction occurs very rapidly; in this fashion the myelin sheath gaps are responsible for the process of saltatory conduction.

A Textbook of Neuroanatomy, Third Edition. Maria A. Patestas, Amanda J. Meyer, and Leslie P. Gartner.
© 2025 John Wiley & Sons, Inc. Published 2025 by John Wiley & Sons, Inc.
Companion website: www.wiley.com.go/patestas/neuroanatomy3e

QUESTIONS TO PONDER: ANSWERS TO EVEN QUESTIONS

Question 4: When a section of an axon is removed from the body and is stimulated in vitro at the midpoint of its length, the action potential travels in both directions, starting at the point of stimulation. However, when the nerve is stimulated under normal circumstances *in vivo*, the reason why the action potential travels only in an anterograde direction is twofold. First, the initial depolarization usually begins at a proximal point rather than at the midpoint of the axon. Second, after Na^+ ions rush in through a voltage-gated sodium channel, that channel becomes refractory and does not permit the passage of Na^+ ions to pass through it for 1–2 ms, and the Na^+ ion channel cannot respond to alterations in the resting potential. Thus, the sodium channel is inactivated; it cannot be opened, and the axon is said to be in the absolute refractory period. Those Na^+ ions that enter at the Na^+ ion channel and disperse toward the nerve cell body encounter Na^+ ion channels that are still inactivated and whose inactivation gate can not be opened. Since the inactivation gate remained closed, Na^+ ions could not pass through the Na^+ ion channel and the membrane could not be depolarized at the old point. What this means then is that the action potential in a normal *in vivo* situation can travel only in a single (anterograde) direction, away from the nerve cell body toward the axon terminal and not in a retrograde direction.

CHAPTER 4

Question 2: Molecules must meet the following five criteria to be considered neurotransmitters (or neuromodulators): (1) they must be synthesized by presynaptic neurons; (2) they have to reside within the synaptic terminals (enclosed in synaptic vesicles); (3) they have to be released from the presynaptic terminals by way of a calcium-dependent mechanism; (4) they have to bind to specific receptors on the postsynaptic membrane; and (5) they must be inactivated in the synaptic cleft. Each neuron has at least one, but usually a number of neurotransmitter substances, cotransmitters, in its presynaptic terminal. These cotransmitters may be released individually or together, and they are usually housed in different synaptic vesicles.

Question 4: Neuropeptides are small polypeptides, cleaved post-translationally before they reach the Golgi apparatus, from larger propeptide molecules synthesized on the rough endoplasmic reticulum in the nerve cell body. The cleaved neuropeptides are packaged into secretory vesicles in the *trans*-Golgi network and transferred, via rapid anterograde transport, to the presynaptic terminal. Often, a presynaptic terminal may possess and release a number of different neuropeptides simultaneously, but once released, instead of being recycled, they are destroyed by peptidases. Thus, a single release of a quantum of neuropeptides cannot elicit multiple responses from the postsynaptic cell. Depending on the specific neuropeptide and/or its receptor, these small peptides may be excitatory or inhibitory neurotransmitters or neuromodulators. The most common neuropeptides are somatostatin, substance P, and the opioid neuropeptides – endorphins, enkephalins, and dynorphins.

CHAPTER 5

Question 2: The spinal nerves are extremely important structures because they (along with the cranial nerves) act as the pathways for communication from the body to the central nervous system (CNS) and from the CNS to the body. The length of a spinal nerve is immaterial since each spinal nerve divides into a dorsal primary ramus and a ventral primary ramus, and these two components are merely continuations of the spinal nerve.

Question 4: Although the radicular arteries are quite small, they are extremely important for the vascularization of the spinal cord. The reason is that, with the exception of the cervical region, the anterior and posterior spinal arteries by themselves are unable to provide an adequate vascular supply to the spinal cord. Therefore, an injury to the spinal nerve damages not only the afferent and efferent fibers of a particular spinal cord level, but may also damage the segmental white and gray matter by producing ischemic conditions due to the reduction in blood supply from the radicular artery serving that region.

CHAPTER 6

Question 2: The basal ganglia are a collection of nerve cell bodies located in the white matter of the cerebrum. Since, by definition, a group of nerve cell bodies located outside the central nervous system (CNS) are known as ganglia, and those located within the CNS are known as nuclei, it is not truly correct to call these nuclei ganglia (basal ganglia).

Question 4: Unlike other cranial nerves (originating from the anterior aspect of the brainstem), the trochlear nerve originates from the posterior aspect of the brainstem.

CHAPTER 7

Question 2:
a. Ipsilateral (right) oculomotor nerve palsy, and the right pupil will not constrict following stimulation of either eye.
b. Contralateral ataxia and intension tremor (on left).
These are the deficits characteristic of Claude syndrome.

Question 4:
Deficits resulting from a lesion to the corticonuclear (corticobulbar) tract (upper motor neurons, include:
a. contralateral lower face weakness.
b. Dysarthria (motor speech disorder due to weakness/paralysis of the mouth (lips, tongue), and lower face.
Deficits resulting from a lesion to the corticospinal tract (upper motor neurons), include:
a. contralateral upper and lower limb weakness (motor hemiparesis).
Deficits are characteristic of dysarthria hemiparesis (pure motor hemiparesis).

CHAPTER 8

Question 2: The periosteal layer of the dura mater is richly supplied by blood vessels, whereas the meningeal layer has no vascular supply. The blood vessels of the periosteal layer include the middle meningeal and accessory meningeal arteries of the middle cranial fossa, as well as the meningeal branches of the anterior and posterior ethmoidal arteries and the meningeal branches of the internal carotid artery of the anterior cranial fossa. Meningeal branches of the vertebral, occipital, middle meningeal, and ascending pharyngeal arteries serve the periosteal dura of the posterior cranial fossa. Most of these vessels enter the cranial fossa via foramina and canals of the skull, such as the jugular and mastoid foramina and the hypoglossal canal. Blood is drained from the dura by meningeal veins that empty their blood (indirectly through venous lacunae) into several of the sinuses as well as into nearby emissary veins and diploic veins.

Question 4: The cisterna magna (cisterna cerebellomedullaris), the largest of the subarachnoid cisterns, is formed between the caudal aspect of the cerebellum and the posterior surface of the medulla oblongata as the arachnoid stretches across these two structures rather than following the contour of the brain. The cisterna magna communicates with the fourth ventricle of the brain via the medial apperture (foramen of Magendie) through which the cerebrospinal fluid formed by the choroid plexuses of the ventricles of the brain enters the subarachnoid space. The interpeduncular cistern, another expanded portion of the subarachnoid space, located between the right and left cerebral peduncles, receives cerebrospinal fluid from the two lateral appertures (foramina of Luschka). It is via these three foramina of the fourth ventricle that cerebrospinal fluid leaves the ventricular system of the brain and enters the subarachnoid space.

CHAPTER 9

Question 2: The basal lamina-lined endothelial cells of the capillaries of the brain and spinal cord form a blood–brain barrier. These endothelial cells not only form fasciae occludentes with each other, but they have only a limited ability to form caveolae, relying instead mostly on receptor-mediated transport to transfer material between the capillary lumen and the neural tissue of the central nervous system. Therefore, most macromolecules injected into the capillary lumen are unable to gain access into the intercellular spaces of the brain and spinal cord. Similarly, most macromolecules injected into the intercellular compartment of the central nervous system are unable to enter the capillary lumen, unless the endothelial cells possess specific receptors for them. Small, noncharged molecules (e.g., O_2, H_2O, CO_2) and numerous small lipid-soluble substances, including some drugs, can dissolve through the lipid membrane of the endothelial cells and thus are able to penetrate the blood–brain barrier. Molecules, such as some of the vitamins, amino acids, glucose, and nucleosides, can penetrate the blood–brain barrier via carrier proteins, with or without passive diffusion. Ion channels are also present in the endothelial cell membranes and they are responsible for the transport of ions across the blood–brain barrier. Although astrocyte end-feet surround the basal lamina of the capillaries and assist in the formation of the blood–brain barrier, it is the endothelial cells that bear the brunt of the responsibility in the establishment and maintenance of this barrier.

Question 4: The cerebral arterial circle (of Willis) connects branches of the two internal carotid arteries to branches of the two vertebral arteries, forming a method of shunting blood from one side to the other in case of a blockage in any one of the vessels. The anterior communicating artery forms the anterior connection and the posterior communicating artery forms the posterior connection between the right and left sides. The arteries involved in the circle are: the right internal carotid, right anterior cerebral, anterior communicating, left anterior cerebral, left internal carotid, left posterior communicating, left posterior cerebral, right posterior cerebral, right posterior communicating, and back to the right internal carotid artery.

CHAPTER 10

Question 2: The location of the postganglionic sympathetic nerve cell body is usually close to the spinal cord (in the sympathetic chain ganglia, near the vertebral column), whereas the location of the postganglionic parasympathetic nerve cell body (usually within the target organ) is far from the central nervous system. Therefore the preganglionic fiber has to be the appropriate length.

Question 4: Short ciliary nerves arise from the ciliary ganglion, whereas the long ciliary nerves bypass the ciliary ganglion without contacting it. Postganglionic sympathetic fibers have already synapsed and may take any convenient route to their destination (that is either the short or long ciliary nerves). Preganglionic parasympathetic fibers must synapse in the ciliary ganglion. The postganglionic fibers also take the most convenient route to their destination and since the short ciliary nerves arise from the ciliary ganglion, they have to travel within the short ciliary nerves and cannot conveniently reach the long ciliary nerves.

CHAPTER 11

Question 2: The tendon organs (Golgi tendon organs) do not have contractile portions as the intrafusal fibers of the muscle spindles do.

Question 4: In autogenic inhibition, the tendon organs (Golgi tendon organs) play a role in the fine-tuning of skeletal muscle contractile force during movement.

Question 6: Gamma motor neurons are always under supraspinal control (that is, controlled by higher levels in

QUESTIONS TO PONDER: ANSWERS TO EVEN QUESTIONS

the brain) and they do not receive input from sensory (either Type Ia or Type II) fibers innervating intrafusal fibers of muscle spindles and, therefore, there are no messages relayed from the muscle spindles back to the gamma motor neurons.
Question 8: Alpha-gamma coactivation is a mechanism via which the muscle spindles sensitivity to stretch is maintained (during voluntary movement) regardless of muscle length at the onset of muscle stretch. During voluntary movement, it is the alpha-gamma coactivation that modulates the sensitivity of the intrafusal fibers so that they can retain their ability to serve as muscle stretch receptors by detecting even a slight change in muscle length (resulting from stretch of the gross muscle) regardless of muscle length at the onset of muscle stretch. The gamma motor neurons stimulate and cause contraction of the polar, muscular ends of the intrafusal fibers, exerting a pull (tension) on the central (non-contractile) portions of the intrafusal fibers, keeping the central part *taut* so even the slightest additional stress placed on it will stimulate the sensory fibers.

CHAPTER 12

Question 2: The gamma motor neurons (innervating the skeletal, contractile portion of the intrafusal fibers) cause corresponding contraction of the contractile portions of the intrafusal fibers that stretch the central, noncontractile region of the intrafusal fiber. Thus the sensitivity of the intrafusal fibers is constantly maintained by continuously readapting to the most current status of muscle length, and they are therefore always ready to detect any change in muscle length.
Question 4: During slight stretching of a relaxed muscle, the muscle spindles are stimulated, whereas the tendon organs remain undisturbed and quiescent. During further stretching of the muscle, which produces tension on the tendons, both the muscle spindles and the tendon organs are stimulated. Thus tendon organs monitor and check the amount of tension exerted on the muscle (regardless of whether it is tension-generated by muscle stretch or contraction), whereas muscle spindles check muscle fiber length and rate of change of muscle length (during muscle stretch or contraction).
Question 6: There are two theories that may explain this phenomenon of "referred pain." (1) The convergence–projection theory of visceral pain suggests that the central processes of sensory neurons (general visceral afferent (GVA) neurons) innervating a visceral structure such as the heart terminate at the same spinal cord level as the central processes of sensory neurons (general somatic afferent (GSA) neurons) innervating a somatic structure such as the upper limb. There they synapse with interneurons and/or second order neurons of the anterolateral system to higher brain centers. (2) The concept of referred pain suggests that second order GSA projection neurons of the anterolateral system (relaying nociceptive input from the upper limb in this case) are continuously being activated by GVA neurons (relaying nociceptive input from the heart). GVA nociceptive input (from the heart) is eventually relayed to the patch of somesthetic cortex that normally receives somatic information from the upper limb.

The brain in turn perceives the nociceptive input as coming from the upper limb, and therefore "feels" the painful sensation in the upper limb.

CHAPTER 13

Question 2: The noise enters both ears, and is relayed via the auditory pathways to the inferior colliculi (reflex nuclei of the auditory system) in the midbrain. The inferior colliculi project to the superior colliculi (reflex nuclei of the visual system) in the midbrain (connecting the special senses of vision and hearing). The visual association areas send corticotectal fibers to the oculomotor accessory nuclei and the deep layers of the superior colliculus. The superior colliculus gives rise to the fibers of the tectospinal tract, which descends to terminate in the cervical spinal cord levels. The **tectospinal tract** is involved in the coordination of head movements with eye movements elicited by visual, auditory, and vestibular stimuli.
Question 4: The upper motor neuron fibers that are damaged in this lesion are those that descend to synapse with the interneurons and lower motor neurons that innervate the muscles of the upper limb, trunk, and lower limb on the left side of the body. The individual will have a hemiplegia of the upper and lower limbs contralateral to the side of the lesion, since most of the corticospinal tract fibers cross at the pyramidal decussation in the medulla.

CHAPTER 14

Question 2: One of the normal functions of the basal nuclei (basal ganglia) when stimulated, is to relay signals to the motor cortex and brainstem, which ultimately inhibit excess muscle tone.

Following a lesion to the basal nuclei, this inhibitory influence is diminished or lost, and symptoms of hypertonicity are manifested contralateral to the side of the lesion.
Question 4: Hyperkinetic disorders are caused by a disturbance of the excitatory subthalamopallidal neurons of the indirect loop. Hypokinetic disorders result from neuronal degeneration in the neostriatum, disrupting the direct loop of the basal nuclei.

CHAPTER 15

Question 2: Tremor following a lesion in the basal nuclei (basal ganglia) is characterized as "tremor at rest." It is apparent at rest and disappears when a voluntary movement is initiated. On the contrary, a lesion involving the cerebellum is characterized as "intention tremor," which is manifested during a voluntary movement and is especially apparent toward the end of a movement, such as when approaching the bridge of the nose when putting on a pair of glasses.
Question 4: The output of the cerebellum is excitatory.

CHAPTER 16

Question 2: The neurons of the reticular formation possess elaborate dendritic trees whose branches radiate in all directions. Some neurons give rise to an ascending or descending axon, with numerous collateral branches. Other neurons give rise to a primary axon, which bifurcates forming an ascending and a descending branch, both of which then give rise to collateral branches. The dendrites are oriented perpendicular to the long axis of the brainstem. The elaborate radiating dendritic trees, and the axons and their collaterals, facilitate the ability of neurons of the reticular formation to collect information from ascending and descending fibers as they traverse the dendritic trees of neurons in the reticular formation.

Question 4: The brainstem reticular formation contains interneurons that are located near the cranial nerve nuclei and participate in the coordination of reflex activity associated with the cranial nerves. Interneurons of the lateral reticular formation are associated with the trigeminal, facial, and hypoglossal nerve nuclei that innervate structures of the mouth and face. The motor control exerted by these cranial nerves on the structures they innervate in the orofacial region is coordinated by the reticular formation. Although motor control of the muscles of mastication and those of the tongue, lips, soft palate, and upper pharynx – is under voluntary control, during chewing the action of all of these structures is automated and is executed without our being aware of it. The branches of the trigeminal nerve transmit general proprioceptive information from the temporomandibular joint, teeth, and muscles of mastication, and general sensation information from the mouth (thermal sensation and information about food consistency) to the sensory nuclei of the trigeminal nerve. These nuclei in turn project to the motor nuclei of the trigeminal and facial nerves and to the reticular formation, which in turn coordinates the movements of the orofacial structures involved in chewing.

CHAPTER 17

Question 2: The motor fibers arising from the trochlear nucleus decussate posteriorly in the midbrain, and emerge from the brainstem at the junction of the pons and midbrain, just below the inferior colliculus. These fibers then course anteriorly to the orbit to innervate a single extraocular muscle, the superior oblique. Thus, a lesion to the trochlear nucleus results in paralysis or paresis of the contralateral superior oblique muscle, whereas a lesion in the trochlear nerve (after it decussates and emerges from the brainstem) results in the same functional deficits but in the ipsilateral muscle.

When the superior oblique muscle is paralyzed, the ipsilateral eye will extort (rotate around an anteroposterior axis) accompanied by a simultaneous upward and outward (lateral) movement of the eye. This is caused by the unopposed inferior oblique muscle, and results in external strabismus. Since the eyes become misaligned following such a lesion, the individual with trochlear nerve palsy experiences vertical diplopia (double vision) accompanied by a weakness of downward movement of the eye, most notably in an effort to adduct the eye (turn medially). The diplopia is most apparent

to the individual when descending stairs (looking down). To counteract the diplopia and to restore proper eye alignment, the individual realizes that the diplopia is reduced as they tilt their head with their chin pointing towards the side of the unaffected eye. Normally, tilting of the head to one side elicits a reflex rotation about the anteroposterior axis of the eyes in the opposite direction so that an object will remain fixed on the retina. Tilting of the head towards the unaffected (normal) side causes the unaffected eye to intort and become aligned with the affected eye, which is extorted.

Chin tuck (moving chin downward) causes the elevation of the normal eye aligning it with the hypertopic, affected eye.

Question 4: A lesion to one medial longitudinal fasciculus (MLF) interrupts the interneurons connecting the abducens nucleus and the contralateral oculomotor nucleus. This is known as internuclear ophthalmoplegia. In this condition horizontal ocular movements will not occur in a conjugate fashion.

When there is a lesion of the right MLF, and the individual attempts to gaze to the right, the lesion is not apparent since both eyes can move simultaneously to the right. However, when attempting to gaze to the left, the right eye cannot move inward (medially beyond the midline) but the left eye, which should move outward (laterally) in this lateral gaze, does so since it is not affected.

If you ask this same individual to look at a near object placed directly in front of them, which necessitates that both eyes adduct (converge), they are able to do so. This indicates that: (1) both oculomotor nerves (which innervate the medial recti) are intact; and (2) the upper motor neurons arising from the motor cortex, which stimulate the motor neurons of the oculomotor nuclei, are also intact. Therefore, a unilateral lesion of the MLF becomes apparent only during conjugate horizontal eye movement away from the side of the lesion.

Question 6: A unilateral lesion of the facial nerve proximal to the geniculate ganglion causes a loss of salivation by the ipsilateral submandibular gland. A condition referred to as the "crocodile tear syndrome" (tearing/lacrimation while eating) may result as follows. As the preganglionic parasympathetic ("salivation") fibers originating from the superior salivatory nucleus are regenerating, they may be unsuccessful at finding their way to their intended destination, the submandibular ganglion. Instead they may take a wrong route to terminate in the pterygopalatine ganglion, establishing inappropriate synaptic contacts there with ("lacrimation") postganglionic neurons whose fibers project to the lacrimal gland stimulating it to produce tears during eating.

Question 8: A growing tumor in the vicinity of the cerebellopontine angle known as an acoustic neuroma (arising from the neurolemmocytes wrapped around the axons of the vestibular nerve), may compress the cranial nerves that emerge from the brainstem at the cerebellopontine angle. The most susceptible nerves are the facial and vestibulocochlear.

CHAPTER 18

Question 2: When the right eye was illuminated, both the direct (right) and consensual (left) pupillary light reflexes were elicited (both pupils constricted). This is an indication that the

538 ●●● QUESTIONS TO PONDER: ANSWERS TO EVEN QUESTIONS

right afferent limb (optic nerve) and both the right and left efferent limbs (oculomotor nerves) of the pupillary light reflex arc and both pretectal areas were intact. When the left eye was illuminated, neither pupil responded, which indicates that the left afferent limb (optic nerve) of the pupillary light reflex arc may be damaged. The information was not relayed to the central nervous system (to the parasympathetic Edinger–Westphal nucleus of the oculomotor nerve) to elicit a response. Since both pupils constricted when the right eye was illuminated, the pretectal areas and the parasympathetic nuclei and parasympathetic fibers of the oculomotor nerves were intact.

Question 4: The afferent limb (ophthalmic division of the trigeminal nerve) of the corneal blink reflex is intact on both sides. However, the efferent limb (branch of the facial nerve to the orbicularis oculi muscle) on the left side may be damaged.

Question 6: A lesion that damages the sympathetic (preganglionic or postganglionic) fibers of the pupillary dilation pathway, results in oculosympathetic paresis (Horner's syndrome) ipsilateral to the lesion. One of the characteristics of this syndrome is a persistent miosis (pupillary constriction) ipsilateral to the lesion due to the unopposed action of the parasympathetic nervous system. Although an individual with oculosympathetic paresis exhibits loss of the pupillary dilation reflex, the pupillary constriction reflex is intact, both ipsilateral and contralateral to the lesion. The anatomical components of the pupillary dilation reflex pathway (sympathetic) are different from those of the pupillary constriction pathway (parasympathetic).

CHAPTER 19

Question 2: The auditory (pharyngeal, Eustachian) tube serves to equalize the atmospheric pressure between the middle and external ear. At higher elevations, the atmospheric pressure in the external auditory meatus is lower than that of the middle ear cavity and causes the tympanic membrane to curve outward (towards the external auditory meatus), stimulating pain receptors of the tympanic membrane.

In higher atmospheric pressure (such as when flying in an airplane), the atmospheric pressure in the external auditory meatus is higher than that of the middle ear cavity and causes the tympanic membrane to curve inward (toward the middle ear cavity), stimulating pain receptors of the tympanic membrane.

Since normally the wall of the auditory tube is collapsed, pressure differences can be relieved by the action of swallowing, chewing, yawning, or coughing which open the Eustachian tube, allowing equalization of pressure on the two sides of the tympanic membrane.

Question 4: The tympanic membrane and the ossicles. The auricle and external auditory meatus channel sound waves to the tympanic membrane, which causes it to vibrate. Oscillations of the tympanic membrane cause the three ossicles to vibrate in sequence, conducting the vibration to the vestibular (oval) window. The articulating ossicles form a lever

system, and not only mediate vibrations from the tympanic membrane to the membrane of the vestibular window, but – a fact of paramount importance – amplify these vibrations as they are converted from air waves of the outer ear into fluid waves (perilymph) of the inner ear. Since liquids are more resistant to the conduction of sound waves, the incoming sound waves hitting the tympanic membrane have to be amplified by the ossicles in order to ensure adequate vibration of the perilymph. Since the tympanic membrane is much larger than the vestibular window membrane, and the sound waves reaching the vestibular window membrane are amplified by the ossicles, the compound effect is that the vibrations reaching the oval window have been amplified approximately 20-fold from the original vibrations at the tympanic membrane.

Question 6: The cochlear, trigeminal, and facial nerves. The sound waves entering one ear follow the normal auditory pathway via the cochlear nerve to the ipsilateral anterior cochlear nucleus. Each anterior cochlear nucleus projects bilaterally to the superior olivary nuclei. They in turn project bilaterally to the facial nuclei and the motor nuclei of the trigeminal nerve. Each facial nucleus projects via the nerve to the stapedius (a branch of the facial nerve) to the ipsilateral stapedius muscle. Each motor nucleus of the trigeminal nerve projects via the branch to the tensor tympani muscle (a branch of the trigeminal nerve) to the ipsilateral tensor tympani muscle. Both stapedius muscles and both tensor tympani muscles are stimulated concurrently to contract, dampening the auditory input.

CHAPTER 20

Question 2: Each semicircular duct has a single dilated segment, an ampulla (L., "dilation"), near one of its ends. The ampullae of the three semicircular ducts, as well as the utricle and saccule, contain in their interior the sensory receptors of the vestibular system. The receptors in the ampullae of the semicircular ducts are the ampullary crests, whereas the receptors in the utricle and the saccule are the macula of utricle and the macula of the saccule, respectively. Thus there are a total of five receptors (three crests and two maculae) in each vestibular sensory apparatus. The ampullary crests of the semicircular canals rest in a patch of neuroepithelium at the base of each ampulla. The macula of the utricle rests in a strip of neuroepithelium on the base of the utricle, whereas the macula of the saccule is oriented vertically on the medial wall of the saccule.

Question 4: The flow of endolymph in the direction of the ampulla of the horizontal canal results in deflection of the stereocilia toward the kinocilium, which stimulates the hair cells. In the anterior and posterior semicircular canals, flow of endolymph in the direction of the ampulla has the opposite effect, that is the hair cells are inhibited. If rotation is sustained, the endolymph of the semicircular canals eventually moves at the same rate as the cupula. When this occurs, the cupula is no longer pushed and bent by the endolymph, and the stereocilia and kinocilia are no longer deflected. Since the hair cells are no longer depolarized, the semicircular canal receptors are not activated by sustained rotation.

Question 6: The cerebellum.

CHAPTER 21

Question 2: Odorant-binding proteins are immersed in the mucus layer covering the olfactory epithelium, and bind to the odorous substances that enter through the nasal cavities.

Question 4: Each olfactory receptor cell contains about a dozen identical receptors (all of them being identical structurally), all of which are coded by one gene. Each olfactory receptor cell contains only one type of receptor; however this same type of receptor may also be exhibited by thousands of other olfactory receptor cells distributed throughout the olfactory epithelium.

CHAPTER 22

Question 2: The hippocampal formation consists of three regions: the hippocampus proper, the dentate gyrus, and the subiculum.

Question 4: The granule cell layer of the dentate gyrus archicortex contains the cell bodies of granule cells, whose axons form the output of the dentate gyrus.

Question 6: Efferent fibers of the hippocampus proper arise mostly from the pyramidal cells of the hippocampal pyramidal cell layer.

Question 8: The fornix, which relays information to the hypothalamus and the septal nuclei.

Question 10: Alzheimer disease is a degenerative disorder of the brain and is the most common form of progressive dementia in the elderly. Individuals with this disease are unable to form new memories. This disease is caused by pathologic alterations including neurofibrillary tangles, neuritic plaques, and neuronal degeneration, which appear initially in the subiculum (of the hippocampal formation) and the pyramidal cell islands (of layer II) of the entorhinal cortex. From there, degeneration spreads to the CA1 zone of the hippocampus proper, and then back to the deeper layers of the entorhinal cortex. Consequently, this neuronal degeneration hinders the normal flow of information through the hippocampal formation. Confusion and deficits in executive function occur following further spread to the temporal pole and prefrontal cortex.

CHAPTER 23

Question 2: The medial zone of the hypothalamus includes the following nuclei: the medial preoptic, anterior hypothalamic, paraventricular, supraoptic, dorsomedial, ventromedial, premammillary, tuberomammillary, supramammillary, mammillary, and posterior hypothalamic nuclei. The nuclei in this zone control the functions of the autonomic nervous system, and the secretory activity of the posterior pituitary gland (neurohypophysis).

Question 4: The preoptic region controls the release of gonadotropic hormones from the adenohypophysis.

Question 6: The tuberal region contains nuclei that are involved in the control of anterior pituitary gland (adenohypophyseal) hormone release, caloric intake, and appetite.

Question 8: Most neural input to the hypothalamus arises from the limbic system. **Question 10**: In addition to the neural output, the hypothalamus also provides output to the adenohypophysis and neurohypophysis.

Question 12: The spinothalamic fibers (a component of the anterolateral system of the ascending sensory pathways) relay nociceptive input from the spinal cord to the autonomic control centers of the hypothalamus. There they synapse with neurons that give rise to the hypothalamospinal tract, which is associated with the autonomic and reflex responses (i.e., endocrine and cardiovascular responses) to nociception.

Question 14: The supraopticohypophyseal tract is a short tract consisting of axons that carry oxytocin or vasopressin synthesized in the cell bodies of hypothalamic neurons residing in the supraoptic and paraventricular nuclei. Oxytocin and vasopressin are synthesized in different populations of hypothalamic neurosecretory neurons. The secretory substances are transported via the axons to the posterior lobe of the pituitary gland (neurohypophysis). Upon stimulation, the hypothalamic neurosecretory cells transmit impulses to their terminals, which results in the release of the hormones into the general circulation.

Question 16:

1 The hypothalamus serves as a "thermostat" and plays a central and vital role in body temperature regulation.
2 The hypothalamus contains a "feeding center" and a "satiety center" and has a very important function in food intake regulation.
3 The hypothalamus contains neurons that play a role in water intake regulation.
4 The hypothalamus is the master controller of autonomic functions.
5 Through its extensive interconnections with the limbic system, which processes emotions, the hypothalamus plays an important role in emotional behavior.
6 Via its connections with the hippocampal formation, the hypothalamus plays a role in memory.
7 The hypothalamus contains the cell bodies of the neurosecretory cells that produce the gonadotropic hormones, antidiuretic hormone and oxytocin.
8 The hypothalamus regulates circadian rhythms.

CHAPTER 24

Question 2: In addition to the anatomical classification of the thalamic nuclei, the thalamic nuclei are also categorized according to their function. Functional organization of the

540 ● ● ●   **QUESTIONS TO PONDER: ANSWERS TO EVEN QUESTIONS**

thalamic subnuclei is based on projections between the thalamic subnuclei and the cerebral cortex. The thalamic nuclei have been classified into specific relay (sensory relay and motor relay) nuclei, association (multimodal) nuclei, and nonspecific nuclear functional groups. The sensory relay nuclei of the thalamus include the ventral posterior medial (VPM), ventral posterior lateral (VPL), medial geniculate nucleus (MGN), and lateral geniculate nucleus (LGN). The motor relay nuclei of the thalamus include the ventral anterior (VA) and ventral lateral (VL) nuclei. In addition to these nuclei, the anterior nuclear group – the smallest group of thalamic nuclei – is included in the specific relay nuclear group.

Question 4: The ventral posterior medial (VPM) nucleus receives the terminals of the trigeminothalamic tracts (anterior and posterior trigeminal lemnisci) relaying general somatic afferent (GSA) information (light touch, pressure, pain and temperature sensation, and proprioception) from the orofacial region, and special visceral afferent (SVA) sensation (taste), from the solitary nucleus.

The ventral posterior lateral (VPL) and ventral posterior inferior (VPI) nuclei receive terminals of the lateral and anterior spinothalamic tracts (relaying pain and temperature sensation, and light touch sensation from the body), as well as terminals of the medial lemniscus (transmitting discriminatory (light) touch, pressure, joint movement, and vibratory sensation from the body).

CHAPTER 25

Question 2: Pyramidal cells.

Question 4: Lesions involving the corpus callosum, which unites the two cerebral hemispheres, result in disconnection syndromes. The corpus callosum receives its blood supply from the anterior and posterior cerebral arteries. Its genu and body are supplied by the anterior cerebral artery, and the splenium is supplied by both the anterior and posterior cerebral arteries. A stroke involving the anterior cerebral artery may affect the anterior corpus callosum and result in apraxia (impairment of the capacity to execute various movements) of the left arm as a consequence of isolation/disconnection of the language-dominant (left) hemisphere from the right motor cortex.

A stroke involving the posterior cerebral artery or a lesion involving the posterior corpus callosum on the left side, may result in alexia (the inability to recognize words or read). Visual input from the left visual field is relayed to the right, intact visual cortex, which is isolated/disconnected from the language-dominant (left) hemisphere.

Answers to clinical case margin questions

CHAPTER 2

Question 1: The upper extremities receive their innervation from C4 to T1 spinal cord levels and the lesion is at the lumbosacral region. Therefore the upper extremities should not be affected.
Question 2: The legs receive their motor innervation from the spinal nerves L2 to S3, regions that are directly involved in the lesion.
Question 3: The legs receive their sensory innervation from the spinal nerves L2 to S3, regions that are directly involved in the lesion.

CHAPTER 3

Question 1: No, because tumors are usually unilateral and do not involve the cranial and/or spinal nuclei of both sides.
Question 2: No, usually a single lesion causes a particular event. However, in some cases, these events may begin to resolve and then recur if the conditions alter.
Question 3: No, usually the lesion results in either motor or sensory loss. However, if the lesion involves a peripheral nerve then both motor and sensory function may become impaired.

CHAPTER 4

Question 1: The basal nuclei (basal ganglia) are suspect when movement disorders occur.
Question 2: The immediate conclusion should be Parkinson disease because it is very common and the patient displays typical signs of the disease.
Question 3: The immediate conclusion should be Parkinson disease because it is very common and the patient displays typical signs of the disease.

CHAPTER 5

Question 1: Acute onset and rapid progression indicate that the disease is a rapidly progressing one and should be diagnosed as soon as possible. If it is caused by a tumor, the lesion should be removed at the earliest possible time.
Question 2: The upper extremity is innervated by the brachial plexus: spinal nerves C4 to T1.
Question 3: Because lesions in the brain that affect the legs usually affect the arm and face of the same side, but almost never bilaterally.

CHAPTER 6

Question 1: Because of the possibility of a subdural hematoma, which may be life threatening.
Question 2: As his skull accelerated forward, it pushed the occipital lobe of the brain. As the skull suddenly stopped, the frontal and parietal lobes "crashed" into the skull. In addition, shearing forces at the level of the upper brainstem region can also cause damage.
Question 3: No, the brain does not swell in the same sense as most tissues. Since the brain is encased in a rigid skull the brain has limited confines, which it cannot overcome. However, portions of the brain can swell at the expense of other portions that may consequently become compressed.

CHAPTER 8

Question 1: Light sensitivity alone should lead the physician to think about migraine headache, but when it is accompanied by fever the thinking should switch to meningitis or encephalitis.

A Textbook of Neuroanatomy, Third Edition. Maria A. Patestas, Amanda J. Meyer, and Leslie P. Gartner.
© 2025 John Wiley & Sons, Inc. Published 2025 by John Wiley & Sons, Inc.
Companion website: www.wiley.com.go/patestas/neuroanatomy3e

542 ● ● ● ANSWERS TO CLINICAL CASE MARGIN QUESTIONS

Question 2: Stiff neck accompanied by the patient's other symptoms is indicative of meningitis and possibly encephalitis. Question 3: A resounding "yes"! At this point meningitis is suspected but it is not known whether it is viral or bacterial. Viral meningitis is usually, but not always, self-limiting, whereas bacterial meningitis can be lethal unless treated. Therefore it is always wise to assume bacterial meningitis and treat it with the appropriate antibiotics.

CHAPTER 9

Question 1: Sudden cases of speech disorder, blindness (not involving eye damage), partial paralysis, etc., should trigger the possibility of a stroke in the physician's mind.
Question 2: Sudden cases of weakness on one side should be regarded as indicative of a possible stroke or hemorrhage.
Question 3: Most strokes involve the middle cerebral artery and the symptoms of middle cerebral artery stroke include hemiparesis, hemisensory loss, aphasia, and hemineglect. The regions of the brain responsible for being alert are not affected.

CHAPTER 10

Question 1: Salivation and the release of tears are regulated by the autonomic nervous system.
Question 2: Orthostatic hypotension is the term that describes a sudden decrease of blood pressure when an individual makes a sudden move, such as standing up quickly.
Question 3: Acetylcholine is the neurotransmitter at motor end-plates (skeletal muscle neuromuscular junctions) and postganglionic parasympathetic fiber terminations. Both lacrimal and salivary glands receive their secretomotor supply from the parasympathetic nervous system.

CHAPTER 12

Question 1: The gracile fasciculus and cuneate fasciculus, bilaterally.
Question 2: In the posterior root ganglia.
Question 3: In the posterior funiculus (posterior column) of the spinal cord.
Question 4: Both the gracile and cuneate fasciculi have been affected, thus damage is extensive.
Question 5: The brisk reflexes in the lower limbs indicate that peripheral neuropathy is unlikely in this patient.
Question 6: This patient has a vitamin B_{12} deficiency that has led to the degeneration of white matter in general. The motor fibers of the corticospinal tract (which synapse with lower motor neurons in the spinal cord that directly innervate the muscles of the body) in the spinal cord will also be affected. Thus if this patient's condition remains untreated this will lead to motor problems as well. Dementia from degeneration of the white matter in the brain and visual disturbance

from optic nerve involvement may also occur if this condition remains untreated.

CHAPTER 13

Question 1: The corticospinal tracts.
Question 2: Both the upper motor neurons (of the corticospinal tracts) and the lower motor neurons located in the ventral horn of the spinal cord (that innervate the skeletal muscles of the body) have been affected.
Question 3: A contralateral hemiplegia (paralysis) of the upper and lower limbs, pathologically brisk reflexes, extensor plantar response (Babinski's reflex), and spasticity.
Question 4: Flacid paralysis, atrophy, and fasciculations in the affected skeletal muscles.
Question 5: Neurologic symptoms following a stroke are of sudden onset, and head imaging in these patients will reveal changes in the brain tissue. In contrast, the patient in the clinical case described started noticing symptoms of the motor system 5 months prior to neurologic evaluation. Head imaging in patients affected by amyotrophic lateral sclerosis is usually normal.

CHAPTER 14

Question 1: Huntington disease.
Question 2: The caudate nucleus and putamen.
Question 3: Genetic testing will confirm that this patient has Huntington disease.
Question 4: Psychiatric disturbances and dementia. This disease causes progressive and relentless deterioration of mental function, which leads to incapacity and death 10–20 years after onset. Chorea and psychosis can be treated by antidopaminergic medication; however, there is no effective treatment for the progressive dementia.

CHAPTER 15

Question 1: The left side since symptoms appear ipsilateral to the side of cerebellar lesion.
Question 2: The vermal, inferior midline parts, and lateral hemisphere of the cerebellum. The nystagmus, vertigo, and loss of equilibrium are indicative of a lesion in the flocculonodular lobe. The dysarthria, dysmetria, and ataxia of the left arm indicate lateral hemisphere involvement.
Question 3: The posterior inferior cerebellar artery (PICA).

CHAPTER 16

Question 1: Damage of the right upper motor neurons of the corticospinal tracts caused the complete unilateral paralysis and extensor plantar response (Babinski reflex), on the left side of the body.

Damage of the right upper motor neurons of the corticonuclear tract caused the slurred speech.

Question 2: Normal function of the ascending reticular activating system (ARAS) is critical for the maintenance of consciousness. Lesions involving the ARAS result in coma. Respiratory difficulties and loss of all brainstem reflexes are caused by lesions to the brainstem reticular formation. This patient had a massive stroke. The above scenario is a typical progression indicative of brain herniation secondary to progressive edema from the stroke and subsequent brainstem damage.

Question 3: Stretching and compression of the ipsilateral oculomotor nerve (CN III) over the tentorial notch leads to ipsilateral pupillary dilation and fixed pupil.

CHAPTER 17

Question 1: The trigeminal nerve.

Question 2: The trigeminal ganglion.

Question 3: The caudal subnucleus of the spinal nucleus of the trigeminal nerve.

Question 4: The ventral posterior medial (VPM) nucleus of the thalamus.

Question 5: This disorder may be treated pharmacologically or surgically. Sectioning of the descending spinal trigeminal tract proximal to its termination in the caudal subnucleus selectively obliterates the afferents relaying nociception, but spares the fibers relaying tactile sensation from the orofacial region. Microvascular decompression surgery has produced excellent results for refractory cases and is widely accepted.

CHAPTER 18

Question 1: The converging axons of the ganglion cells of the retina.

Question 2: When the ganglion cells emerge from the back of the bulb of the eye to form a bundle, the optic nerve, they become myelinated.

Question 3: The afferent limb (formed by the right optic nerve).

CHAPTER 19

Question 1: In conduction deafness, hearing loss is caused by a defect in the mechanical transmission of sound, which interferes with sound conduction. Conduction deafness may result from wax buildup in the external auditory canal, perforation of the tympanic membrane, otitis media, or otosclerosis. Sensorineural (nerve) deafness results following a lesion involving the cochlea, the cochlear nerve, or the central auditory pathways.

Question 2: The acoustic neuroma usually appears in the vicinity of the cerebellopontine angle, which compresses the vestibulocochlear nerve (associated with hearing and balance) as it emerges from the brainstem. The ataxia can arise from cerebellar dysfunction secondary to compression.

Question 3: The facial nerve (CN VII). Compression of the facial nerve may cause facial muscle weakness.

CHAPTER 20

Question 1: The three semicircular ducts, the utricle, and the saccule.

Question 2: The macula of the utricle and the macula of the saccule, respectively.

Question 3: The macula has a gelatinous glycoprotein membrane the statoconial membrane (the otolithic membrane) that has on its free surface statoconia (otoliths), crystals consisting of calcium carbonate. These crystals may become dislodged and float in the endolymph, causing dizziness which most often resolves on its own.

CHAPTER 21

Question 1: The olfactory receptor neurons. Their axons are susceptible to shearing as they pass into the cranial vault via the cribriform plate of the ethmoid bone, severing their connections with the overlying olfactory bulbs.

Question 2: Other causes that may lead to the loss of the sense of smell include upper respiratory infections, sinus disease, or any other nasal obstruction or congestion. Medications can induce loss of or alteration of the sense of smell. Neurodegenerative diseases, such as Alzheimer disease, Parkinson disease, or Huntington disease, are also associated with loss of olfactory sensation.

Question 3: Recovery of normal olfactory function following head trauma is rare, and there is no effective treatment.

CHAPTER 22

Question 1: The hippocampus, located in the temporal lobe, has an exceptionally low threshold for epileptic seizures, which may be transmitted to other parts of the limbic lobe and to the neocortex.

Question 2: Psychological or emotional symptoms are associated with seizures generated in the frontal or temporal lobes. The most epileptogenic regions of the brain are the temporal and frontal lobes, especially the limbic regions, which are associated with the regulation of emotional expression.

Question 3: The temporal lobes exhibit medial temporal sclerosis. This appears as subtle scarring and loss of volume of the parahippocampal gyri. A loss of neurons and gliosis in the hippocampus is also evident.

544 ●●● ANSWERS TO CLINICAL CASE MARGIN QUESTIONS

CHAPTER 23

Question 1: Orexin (hypocretin).

Question 2: Fasting causes an increase in orexin synthesis by the hypothalamus. At night, hypothalamic orexin production decreases, causing an individual to fall asleep.

Question 3: Individuals with narcolepsy lack the normal number of hypothalamic neurons that synthesize orexins (hypocretins), which are essential for normal sleep–wake cycle function.

CHAPTER 24

Question 1: The ventral posterior medial (VPM) and ventral posterior lateral (VPL) nuclei of the thalamus.

Question 2: The anterior and posterior trigeminal lemnisci terminate in the VPM nucleus of the thalamus, whereas the spinothalamic tract and the medial lemniscus terminate in the VPL nucleus of the thalamus.

Question 3: The patient's blood pressure should not be abruptly lowered to "normal" levels due to the increased intracerebral pressure caused by edematous swelling of the brain following an ischemic stroke or hemorrhage. It is essential that brain perfusion is maintained, and that is accomplished by a modest reduction of the patient's blood pressure.

CHAPTER 25

Question 1: The left primary motor cortex strip.

Question 2: Jacksonian epileptic seizures.

Question 3: New-onset seizures can be caused by strokes, tumors, vascular malformations, hemorrhage, or any other lesion resulting in scar tissue formation and cortical irritation.

Question 4: The temporal lobe. The hippocampus of the temporal lobe has an exceptionally low threshold for epileptic seizures, which may be transmitted to other parts of the limbic lobe and to the neocortex.

Index

abducens nucleus 102, 341–342
abducens nerve 103, 127, 341–346
accessory cuneate nucleus 98
accessory nerve 354–357
accessory nucleus 244, 355
accessory oculomotor nucleus 164, 330, 381, 469
acetylcholine 56
acetylcholine receptor 58
acetylcholine-releasing neurons 275
acetylcholinesterase 56
acoustic neuroma 403, 404
action potentials 35, 45–47, 187
acute meningitis 132
adenohypophysis 28, 457, 471–475
adenosine triphosphate (ATP) 58
ADH *see* antidiuretic hormone
adrenergic receptors 169
afferent fibers 66, 110
    basal nuclei 263–268, 273–274
    cerebellum 291, 297, 299–304
    cerebral cortex 500–501, 504, 517
    hypothalamus 463, 466
    limbic system 441, 444, 446
    reticular formation 314, 318
    spinal cord 22, 66–67
    spinal reflexes 177
    thalamus 485–486
    vestibular system 414
    visual system 378, 386
aggression 469, 476
AICA *see* anterior inferior cerebellar artery
alar plates 21–22, 24, 94
allocortex 495, 498
alpha efferents 69
alpha-gamma coactivation 180–181, 242
alpha latrotoxin (black widow spider venom) 58
alpha motor neurons 175, 178–179, 181
ALS *see* amyotrophic lateral sclerosis; anterolateral system
alveus 447–448
Alzheimer disease 432–433, 450
amnesia 450
amnestic confabulatory syndrome 450

ampullary crests 408–409, 411
amygdaloid body 434, 444–446
    basal nuclei 257, 261, 263
    limbic system 434, 444–446
amyotrophic lateral sclerosis (ALS) 254
analgesia 227–228
aneurysms 152
angular gyrus 79, 511
anopsia 384
anosmia 432
ANS *see* autonomic nervous system
ansa lenticularis 269
anterior bundle of optic radiation (Meyer loop) 372, 374, 375
anterior cerebellar lobe 295
anterior cerebral artery 139, 249, 515
anterior commissure of spinal cord 29, 179
anterior commissure of cerebrum 449
anterior corticospinal tract 238, 241
anterior gray funiculus 67, 69
anterior hypothalamic nucleus 462, 476
anterior inferior cerebellar artery (AICA) 114, 341
anterior perforated substance 77, 137, 139, 150, 430–432, 501
anterior spinal arteries 70, 134
anterior spinal artery syndrome 224–225
anterior spinocerebellar tract 216
anterior stream of the visual pathway 213
anterograde amnesia 450
anterolateral system (ALS)
    ascending sensory pathways 186, 197–207
    brainstem 102
    cingulate and insular cortices 206
    first order neurons 206, 208–212, 215
    neuron sequence 197–198
    pain pathways from the body 198–206
    second order neurons 197, 201–204, 212
    somatosensory cortex 204–206
    spinothalamic and spinoreticular pathways 197–207
    temperature pathways from the body 206–207
    third order neurons 198, 204–206, 213
    visceral pain 206, 207
antidiuretic hormone (ADH) 54, 157, 462, 470–472 *see also* "vasopressin"

A Textbook of Neuroanatomy, Third Edition. Maria A. Patestas, Amanda J. Meyer, and Leslie P. Gartner.
© 2025 John Wiley & Sons, Inc. Published 2025 by John Wiley & Sons, Inc.
Companion website: www.wiley.com.go/patestas/neuroanatomy3e

546 ● ● ● **INDEX**

aphagia 476
apoptosis 12
appetite 462
arachnoid mater
    cranial 122–125
    spinal 126
    cisternae 122, 124
    granulations 118, 124–125
ARAS *see* ascending reticular activating system
arbor vitae 86, 288, 289
archicerebellum 293
archicortex 82, 495, 498
arcuate fasciculus 501, 512
arcuate fibers 314, 462, 472
ascending reticular activating system (ARAS)
    hypothalamus 464
    reticular formation 318
    thalamus 489
ascending sensory pathways 185–230
    anatomy and function 186–187
    anterolateral system 186, 197–207
    classification of receptors by stimulus modality 188–194
    classification of receptors by stimulus source 188
    clinical case 185, 230
    clinical considerations 218–226
    descending analgesia-producing pathways 227–228
    gate control theory of pain 227
    modulation of nociception 226–228
    muscle spindles and tendon organs 194–197
    neuroplasticity 229
    nociception 190, 203, 211, 220, 226–228
    pain pathways from the body 198–206
    sensory pathways to the cerebellum 210, 214–217
    sensory receptors 187–201
    spinothalamic and spinoreticular tract 186–187, 199–203
    synonyms and eponyms 229
    tactile sensation and proprioception 207–214
    temperature pathways from the body 206–207
    visceral pain 206, 207, 211
aspiny projection neurons 257
aspiny stellate cells 497, 499
association fibers 84, 288, 496, 501, 502
association loop 272
association nuclei 491
astrocytes 20, 36, 40–41
ataxia 323
ataxic hemiparesis syndrome 113
ATP *see* adenosine triphosphate
auditory association areas 511, 518
auditory system 388–404
    auditory transmission 393–395
    central auditory pathways 394–402
    clinical case 388, 404
    clinical considerations 403
    descending auditory projections 401, 402
    ear anatomy and physiology 388–393
    external ear 389
    inferior colliculus 396, 398
    internal ear 391, 392
    lateral lemniscus 396
    medial geniculate nucleus 398–399
    middle ear 389–391
    primary auditory cortex 399–401
    sound attenuation reflex 401, 402
    superior olivary nuclei 396, 401
    synonyms and eponyms 404
Auerbach (myenteric) plexus 167, 172, 527
autogenic inhibition 177, 178

autonomic motor system 449–450
autonomic nervous system (ANS) 3, 9–10, 154–172
    central connections and functions 157
    cholinergic and adrenergic receptors 169
    clinical case 154, 172
    clinical considerations 171
    enteric nervous system 155, 167–168
    functional components 154–155
    hypothalamus 465, 466
    limbic system 445–446
    neurotransmitter substances 157
    parasympathetic nervous system 155, 164–167
    pelvic functions 169–171
    reticular formation 315, 318–320
    sympathetic nervous system 155–164
    synonyms and eponyms 172
axons 20, 33–35
axon terminal 35–38

Babinski sign (Extensor plantar response) 179, 251, 252
balance maintenance 405–406, 412
basal cells 425
basal nuclei 77, 84, 85, 255–284
    associated nuclei 260–261
    caudate nucleus 256, 258–259, 261, 263–265
    cerebellum 256
    circuits 270–274, 277
    classification of 256, 257
    clinical case 255, 284
    clinical considerations 278–282
    components of 256–260
    connections 263–270
    direct, indirect, and hyperdirect loops/pathways 275–277
    globus pallidus 256–260, 267–270
    input, intrinsic, and output nuclei 261–264
    lentiform nucleus 29, 143, 256–260, 504
    modulatory circuits 277
    motor cortex 232
    motor function 256
    neurotransmitter substances 274–275
    nucleus accumbens 256, 269
    putamen 256, 258–265
    striatum 263–267
    substantia nigra 257, 258, 260, 262
    subthalamic nucleus 258, 260
    synonyms and eponyms 283
basal plates 21, 94
basilar artery 143–146
basilar artery thrombosis 152
basilar membrane 393, 394
basilar plexus 128
basket cells 290, 293, 305, 306
basolateral (ventrolateral) nucleus 445
Bell palsy (Idiopathic facial nerve paralysis) 403
Benedikt syndrome 115
benign position vertigo 423
Betz (Giant pyramidal) cells 240
bifurcating fibers 208
bilaminar germ disc 12, 13
biogenic amines 52–54
bipolar neurons
    histophysiology 39–40
    olfactory system 425
    retina 21, 329
    spinal cord 21
    vestibular system 411
black widow spider venom (alpha latrotoxin) 58
blastocyst 12, 13

INDEX ● ● ● 547

blood–brain barrier (BBB) 122, 131, 135–136
body temperature regulation 467, 468
Bowman (olfactory) glands 427
brachium pontis 104, 309
bradykinesia 280, 282
brain 6
    anatomy 75–90
    arterial supply 135–148
    basal nuclei 77, 84, 85
    basilar artery 144–146
    blood–brain barrier and neurovascular unit 135–136
    Brodmann classification 81, 83
    cerebellar veins 149, 150
    cerebellum 26, 27, 85–87
    cerebral arterial circle (of Willis) 133, 139, 140
    cerebral veins 149–150
    cerebrum 76–84
    circumventricular organs 136
    clinical case 75, 90
    clinical considerations 30
    cranial nerves and arterial supply 81–82
    derivatives of primary divisions 23
    diencephalon 84–85
    early hominins 522
    evolutionary biology fundamentals 522
    histology of the cerebral cortex 81–83
    internal carotid artery 136–143
    lateral ventricles 77, 78
    lateral view 76, 87
    limbic lobe 70–81
    mesencephalon 6, 20, 26, 29–30
    metencephalon 6, 20, 24–25
    middle cerebral artery 140, 141
    modern human brains 523
    myelencephalon 6, 20, 24
    neurodevelopment 19–21, 23–30
    non-human primates 522–523
    prenatal and postnatal development 523
    prosencephalon 20, 25, 27–29
    rhombomeres 25–26
    sagittal section 80
    synonyms and eponyms 90
    telencephalon 6, 20, 29
    vascular supply 135–148
    venous drainage 149–151
    ventricles of the brain 130–131
    vertebral artery 143–144
    white matter of the cerebral hemispheres 83–84
brainstem 6, 91–116
    alar plate derivatives 94
    auditory system 394
    basal plate derivatives 94
    brain anatomy 87–88
    cerebellum 297
    clinical case 91, 116
    clinical considerations 111–115
    cranial nerve nuclei 94–96
    internal organization 91–96
    limbic system 446
    medulla oblongata 96–102
    metencephalon 24–25, 87
    mesencephalon 87, 107–111
    neurodevelopment 94–96
    pons 94, 102–107, 114
    pontomedullary border 102
    posterior surface 92
    synonyms and eponyms 116

tegmentum 115
    vascular supply 118–120
    ventral surface 93
branchiomotor (special visceral efferent) neurons 243, 334, 339, 349
Broca (motor speech) area 77, 511–512, 516–517
Brodmann classification 81, 83
Brown–Sêquard syndrome 73, 218, 220, 221
bulbous corpuscles 193–194
bulbous corpuscles (Ruffini endings) 193

calcarine sulcus 146, 369, 373, 495, 496, 509
caloric nystagmus 421–422
capillary plexus 463, 475
carbamazepine 360
cardiogenic cells 14
cataplexy 478
catecholamines 52–54
cauda equina 23, 62, 126
caudal (inferior) medulla oblongata 96–98, 113, 115, 208, 225, 236, 238, 241, 243, 358
caudal (inferior) midbrain 107–108, 110–113, 115, 296, 331, 344–346, 398, 465
caudal (inferior) pons 94, 95, 102–104, 106, 112–114, 344, 346, 414
caudal subnucleus 337
caudate nucleus 258–259, 263–265
cavernous sinuses 127, 129
cavum septum pellucidum 458–460, 476
cavum trigeminale 120, 122
cell body 34
cells of Martinotti (multipolar neurons) 497
central canal 6, 62, 67
central nervous system (CNS) 3, 6
    ascending sensory pathways 185, 236
    autonomic nervous system 155
    basal nuclei 261
    brain 6, 7
    cells of the CNS 6
    common terms in neuroanatomy 4
    gray matter and white matter 8–9
    histophysiology 32, 40–44
    meninges 122
    olfactory system 428
    reticular formation 311
    spinal cord 6, 7
    vascular supply 133–153
    visual system 352
central pain 226
central sulcus 77
centromedian (CM) nucleus 261, 489
cerebellothalamic fibers 111, 113, 115
cerebellum 285–309
    afferent fibers 288, 291, 296–304
    ascending sensory pathways 214–217
    basal nuclei (basal ganglia) 286
    brain anatomy 85–87
    clinical case 285, 309
    clinical considerations 307–308
    deep cerebellar nuclei 287, 288, 293, 298–300
    efferent fibers 281, 298, 299, 304–305
    folding of the cerebellar cortex 288
    functional organization 305–307
    gray matter and white matter 287–288
    intrinsic circuitry 305–307
    layers of the cerebellar cortex 288–291
    lobes 293–296
    morphology 287–296
    motor cortex 232, 285–287

548 ● ● ●　**INDEX**

cerebellum (*continued*)
　movement production 285–287
　neurodevelopment 26, 27
　neurons of the cerebellar cortex 291–293
　peduncles 296–298
　synonyms and eponyms 309
　vascular supply 143–145
　vestibular system 413–417
　zones 293, 294
cerebral anterior commissure 449
cerebral arterial circle (of Willis) 140, 146–148
cerebral cortex 494–521
　afferent fibers 500–501
　allocortex 495, 498
　association fibers 496, 501, 502
　basal nuclei 270–273
　borders 505
　brain anatomy 76
　Brodmann classification 81, 83
　cell layers of the neocortex 498–500
　cerebellum 295–296
　cerebral dominance 512–514
　clinical case 494, 520
　clinical considerations 515–519
　commissural fibers 496, 499
　descending motor pathways 235–236
　efferent fibers 501–504
　functional organization 505–512
　fusiform cells 497, 501
　histology 81–84
　internal capsule and corona radiata 504
　lobes 504–505
　mesocortex 498
　motor activity 505–508
　neocortex 495, 499–500
　neurons 496–498
　phylogenetic classification 495, 498
　primary sensory areas 508–510
　projection fibers 504
　pyramidal cells 496
　sensory association areas 510–512
　stellate cells 497
　structure and function 495–496
　synonyms and eponyms 519
　vertical columnar organization 500
cerebral falx 120–121
cerebral peduncle 27, 87, 107
cerebrocerebellum 295, 298, 301
cerebrospinal fluid (CSF)
　brainstem 107
　composition 130
　histophysiology 40
　meninges 117, 122, 130
　spinal cord 60, 67
cerebrovascular accident *see* stroke
cerebrum 76–84
　frontal lobe 76, 77, 81, 82
　insula 77, 80
　lobes of the cerebral hemispheres 77–81
　occipital lobe 79–80
　parietal lobe 77–79
　temporal lobe 79
Charcot–Bouchard aneurysm 152
chemosensitive cilia 427, 428
cholinergic receptors 169
choroid layer of the eye 362, 363, 364
choroid plexus 5, 22, 27, 28, 30, 41, 77, 117, 125, 130, 131, 138, 139, 144, 145, 146, 150, 362, 476, 527, 535

ciliary ganglion 330, 378
ciliary muscles 329–330
cingulate cortex 205–206
cingulate gyrus 80, 435, 436, 443
cingulate motor cortex 235
cingulate sulcus 77
cingulum 443, 448, 501
circumventricular organs (CVO) 136, 463
cisterna magna 123
Clarke column (posterior thoracic nucleus) 68
clasp-knife response 249, 516
clathrin-coated pits 37
Claude syndrome 115
claustrum 261
climbing fibers 100, 290, 301–303
CNS *see* central nervous system
cochlea 391–393
cochlear duct 391–394, 403, 404, 408
cochlear nerve 104, 348, 349, 388, 391, 393
cochlear nuclei 101, 304, 349, 395, 396
cognitive loop 272
commissural fibers 83, 496, 498, 501–502
commissural interneurons 179
computed tomography (CT) 492
conduction deafness 403
cones 365–367, 376
confluence of sinuses 121, 130
congenital megacolon 171
consciousness 320–321
contact receptors 188
contralateral homonymous hemianopsia 384
contralateral muscle 332–333
convergence accommodation reflex 379–381
convergence–projection theory 226
cornea 362
corneal blink reflex 381–382, 386
corona radiata 504
corpus callosum
　brain anatomy 76, 83
　cerebral cortex 501–503, 515
　limbic system 436
corpus striatum 29, 256, 258–260
corticomedial (dorsomedial) nucleus 431
corticonuclear (corticobulbar) tract 108, 236, 237, 242–244
corticonuclear fibers 114–115
cortico-olivocerebellar pathway 302–303
corticopontocerebellar pathway 296, 298, 301
corticotectal tract 244–245
corticoreticular fibers 245–247, 301, 315–316
corticoreticular tract 286
corticoreticulocerebellar pathway 301
corticorubral tract 245
corticospinal tract
　anterior corticospinal tract 241
　cerebellum 295
　clinical considerations 115
　lateral corticospinal tract 240
　motor cortex 236–242
　muscle spindles 242
　somatosensory fibers 236, 242
corticostriatal fibers 264
corticothalamic fibers 483–484
corticotropin-releasing hormone (CRH) 472
co-transmitters 52, 157
cranial nerves 323–360
　abducens nerve 341–346
　accessory nerve 6, 26, 69, 88, 244, 325, 326, 329, 353, 354–537
　brainstem 94–96

clinical case 323, 360
clinical considerations 349–350, 353, 354, 357, 358
components, distribution, and function 326–328
facial nerve 346–348
functional components 324–325
glossopharyngeal nerve 349–353
hypoglossal nerve 357–358
internal organization 323–325
oculomotor nerve 329–331
olfactory nerve 329
optic nerve 329
parasympathetic ganglia 325
reticular formation 320
sensory and motor nuclei 325
sensory ganglia 324
synonyms and eponyms 359
trigeminal nerve 334–340
trochlear nerve 331–333
vagus nerve 353–355
vestibulocochlear nerve 348–349
CRH *see* corticotropin-releasing hormone
crocodile tear syndrome 350
crossed extension reflex 179
CSF *see* cerebrospinal fluid
CT *see* computed tomography
cuneate fasciculus 70, 88, 97, 98, 186, 207, 208, 212, 216, 218, 221
cuneate nucleus 70, 88, 97, 98, 186, 207, 208, 212, 216, 218, 221
cuneocerebellar tract 99, 187, 216, 217, 295
CVO *see* circumventricular organs

deep cerebellar nuclei 26, 86, 287, 288, 293, 298–299
defecation 170–171
Déjérine–Roussy syndrome 491
dementia 450
demyelination 360
dendrites 5, 21, 33–35
dentate gyrus 435, 439
dentate nucleus 293, 298–299
denticulate ligaments 61, 127
depolarization 36, 38, 45, 46, 51, 52, 56–58, 305, 306, 394, 411, 428, 444, 526, 531, 534
dermatomes 9, 64, 65, 219
descending analgesia-producing pathways 227–228
descending auditory projections 401, 402
descending motor pathways 231–254
cerebral cortex 235–236
clinical considerations 249–252
corticonuclear (corticobulbar) tract 242–244
corticoreticular fibers and reticulospinal tracts 245–247
corticorubral, rubrobulbar, and rubrospinal tracts 245
corticospinal tract 236–242
corticotectal and tectospinal tracts 244–245, 247
functional classification 247–249
muscle spindles 242
somatosensory fibers 236, 242
spinal cord 70
vestibulospinal tracts 246
diabetes insipidus 476
diagonal band (of Broca) 449
diaphragma sella 122
diencephalon 6, 20, 27–29, 84
diffuse neuroendocrine system (DNES) 54, 167, 168
dilator pupillae 161, 346, 347
diplopia 114, 332–333, 342–346
dizziness 423
DNES *see* diffuse neuroendocrine system
dominant hemisphere 492, 512–514
dopamine 52

dopamine-releasing neurons 274–275, 281–282
dorsal (posterior) acoustic striae 396, 398
dorsal (posterior) basal nuclei 241, 256
dorsal (posterior) longitudinal fasciculus 465, 466
dorsal (posterior) raphe 313, 446, 462
dorsal (posterior) spinocerebellar tract 68, 97, 98, 99, 215, 217, 297, 300, 301, 307
dorsal (posterior) trigeminothalamic tract 105, 107, 335, 336, 338
dorsolateral pontine reticular formation (DPRF) 228
dorsomedial (DM) nucleus 337, 492
DPRF *see* dorsolateral pontine reticular formation
dura mater
cranial 118–122, 126
reflections of the meningeal layer 120–122
spinal 126–127
vascular and nerve supply 118–121
venous sinuses 127–130
dynorphin 55
dysarthria 308
dysarthria hemiparesis syndrome 113
dysdiadochokinesia 308
dyskinesias 278, 279
dysmetria 308
dysrhythmokinesis 308

ear *see* auditory system
Eaton–Lambert syndrome 58
E-cadherins 14
ectoderm 13
ectomesenchymal cells 26
ectomesenchymal derivatives 15, 16
Edinger–Westphal (accessory oculomotor) nucleus 164, 330, 381, 469
EEG *see* electroencephalogram
effector 174
efferent fibers
auditory system 394
basal nuclei 267, 268, 270
cerebellum 288, 291, 299
cerebral cortex 501–504
hypothalamus 463–466
limbic system 442, 445, 446
reticular formation 315
spinal reflexes 174
vestibular system 414–416
visual system 378–379, 386
ejaculation 171
electroencephalogram (EEG) 451
elliptic nucleus (nucleus of Darkschewitsch) 111, 330, 415
emboliform nucleus 298, 299
embryogenesis 11–12
emotion 469
empty spiracles-like 1 (EMX1) gene 27
encapsulated mechanoreceptors 190, 192, 193, 208
endocrine system 320, 449–450
endoderm 13
endolymph 393, 408
endorphins 55
enkephalins 54, 55, 163, 164, 277
enteric nervous system 10, 167–168
entorhinal cortex 440, 442
ependymal cells 5, 20, 41
ependymomas 73
epidural venous network 73
epilepsy 233, 451, 515
epinephrine 53
epithalamus 27, 84
epithelial tactile discs (Merkel tactile discs) 186, 191, 192

EPSP *see* excitatory postsynaptic potential
erection 171
Eustachian (auditory) tube 390
excitatory postsynaptic potential (EPSP) 36
executive loop 272
extensor plantar response 179, 251, 252, 310
external capsule 152, 261, 264
external carotid nerves 161–162
external granular layer 499
external pyramidal layer 499
exteroceptors 188
extrafusal fibers 180, 194, 195
extrastriate areas 373
eye *see* visual system

facial nerve
   autonomic nervous system 166
   cranial nerves 346–348
   motor cortex 243
   neurodevelopment 18
   visual system 382
facial nucleus 103
falx cerebelli 120
falx cerebri 120
fasciculi 8
fasciola cinerea 435
fastigial nucleus 293, 298, 299
FEF *see* frontal eye field
fertilization 11, 12
fibroblast growth factor (FGF) 13–14, 29–30
fibrous astrocytes 6
fimbria 447–448
first order neurons 198–201, 206, 208–212, 428
flaccid paralysis 181, 340, 349
flexor reflex 178–179
flocculonodular lobe 293, 294
flocculonodular syndrome 307
fluid intake 466, 469
follicle-stimulating hormone (FSH) 457, 472
fMRI *see* functional magnetic resonance imaging
food intake 468, 476
fornix
   hypothalamus 463, 469
   limbic system 448
   thalamus 482
Foville syndrome 113
free nerve endings 190, 406
Friedreich ataxia 223
frontal eye field (FEF) 235, 240, 244, 271, 508
frontal lobe 77, 81, 83, 504–505
frontal stream 214
functional magnetic resonance imaging (fMRI) 206, 233, 234
funiculi 8, 70
fusiform cells 497, 501

GABA-releasing neurons 274, 277
gait disorders 307
gamma aminobutyric acid (GABA) 57
gamma efferents 69
gamma loop 180–181
ganglion cells 15, 54, 166, 329, 336, 362, 363, 365, 367, 368, 369–371, 373, 374, 376
gate control theory of pain 227
general somatic afferent (GSA)
   ascending sensory pathways 186
   brainstem 95
   cranial nerves 325–326, 340, 346, 351
   neurodevelopment 23

   spinal cord 66
   visual system 381
general somatic efferent (GSE)
   brainstem 95
   cranial nerves 326, 356–357
   neurodevelopment 23
   spinal cord 63
general visceral afferent (GVA)
   ascending sensory pathways 206
   brainstem 95
   cranial nerves 326, 346, 349
   neurodevelopment 23
   reticular formation 315
   spinal cord 63
general visceral efferent (GVE)
   autonomic nervous system 155
   brainstem 95
   cranial nerves 315, 326, 346
   neurodevelopment 23
   reticular formation 315
   spinal cord 63
geniculate ganglion 346–348
geniculocalcarine tract (optic radiation) 369, 372–373, 384
GFAP *see* glial fibrillary acidic protein
giant pyramidal neurons 240, 500
gigantocellular nucleus 313–315
glia *see* neuroglia
glial fibrillary acidic protein (GFAP) 20, 40
glioblastoma 47
glioblasts 20
gliotransmitter substances 6
globose nucleus 293, 298
globus pallidus 256–263
glossopharyngeal nerve 166, 349–353
glutamate 56–58
glutamate-releasing neurons 274
glutamine–glutamate cycle 56–58
glycine 56, 57
glymphatic system 125
GnRH *see* gonadotropin-releasing hormone
Goldman equation 44–45
Golgi cells 290, 292, 293, 305, 306
gonadotropin-releasing hormone (GnRH) 472
G-proteins 37
gracile nucleus 88
gracile fasciculus 70, 88, 97, 98, 186, 206, 207, 208, 211, 212, 215, 218, 221
gracile tubercle 88, 92, 98
granule cells 290–292, 305, 439, 496, 497
gray commissure 67, 68, 70
gray matter 8–9
   brain anatomy 76, 77, 84
   cerebellum 287–288
   cerebral cortex 506
   internal morphology of the spinal cord 67–70
   motor cortex 241
   neurodevelopment 21
   neuronal groups of the posterior gray funiculus 68–69
   neuronal groups of the gray commissure 70
   neuronal groups of the lateral gray funiculus 69
   neuronal groups of the anterior gray funiculus 69
   reticular formation 311, 314
   reticular formation and central canal 67–68
   thalamus 482
gray rami communicantes 163, 164
growth factors 12
GSA *see* general somatic afferent
GSE *see* general somatic efferent

gustatory pathway 346–349
GVA *see* general visceral afferent
GVE *see* general visceral efferent

habenulointerpeduncular tract 449
habenular nuclei 446
hair cells 392, 393, 408, 409
handedness 514
helicotrema 393, 394
hemianopsia 384, 385
hemiparalysis 358
hemispheric zone 301, 305
hemorrhage 492
hemorrhagic stroke 152, 492
hepatolenticular degeneration (Wilson disease) 282, 283
hippocampal formation
    afferent and efferent fibers 441–442
    associated structures 440–442
    dentate gyrus 437, 439–440
    function in memory 443–444
    hippocampus proper 436–439, 441–442
    intrinsic circuitry 442–443
    layers of the hippocampal archicortex 438–439
    limbic system 428, 436–441
    subiculum 440
histamine 53–54
histophysiology 32–48
    clinical case 32, 48
    clinical considerations 47
    generation and conduction of nerve impulses 44–47
    neuroglia 32, 40–44
    neurons 33–40
homeobox genes (HOX) 13, 25
homeostasis 3, 10, 154, 454
horizontal nystagmus 421
hormones
    autonomic nervous system 167–168
    hypothalamus 457, 462–463, 466–470
Horner syndrome (oculosympathetic paresis) 112, 114, 171, 221, 386, 538
HOX *see* homeobox
Huntington disease 58, 278, 279
hydrocephalus 30, 131
hyoid arch 17
hyperacusis 340, 349, 403
hyperkinetic disorders 278
hyperphagia 476
hyperpolarization 38, 45, 52, 57, 366, 411
hyperreflexia 249, 250, 515–516
hypertension 492
hyperthermia 475
hypertonia 249, 515–516
hypertonicity 278
hypertropia 332–333
hypnagogic hallucinations 478
hypoglossal nerve 357–358
hypoglossal nucleus 99, 112, 244
hypokinetic disorders 280, 281
hypophysis 28
hyporeflexia 181, 249, 515
hyposmia 432
hypothalamohypophyseal connections 469–475
hypothalamospinal tract 466
hypothalamothalamic fibers 465
hypothalamus 453–478
    adenohypophysis 457, 469, 471–475
    anterior and posterior mammillary regions 460, 462
    autonomic nervous system 466, 469, 476

    basal nuclei 262
    bidirectional pathways 466, 467
    body temperature regulation 468
    borders 454, 455
    brain anatomy 84–85
    cerebellum 303
    clinical case 453, 478
    clinical considerations 475–476
    connections 462–463
    emotion, memory, and aggression 469
    fluid/water intake 469
    food intake 468, 476
    functions 453–454, 466–469
    hypophysis (pituitary gland) 469–475
    hypothalamohypophyseal connections 469–475
    internal organization 453–454
    lateral zone 457
    limbic system 446
    mammillothalamic tract 448
    medial zone 457–461
    neurohypophysis 469–471
    pathways 463–467
    periventricular zone 455–459
    regions and component nuclei 457–462
    reticular formation 315
    sleep–wake cycle 465, 468
    synonyms and eponyms 477
    tuberal region 462
    visual system 368, 379
    zones and component nuclei 441–446
hypotonia 181, 249, 307–308

induction 12
indusium griseum 435, 440–441
infarction 152
inferior cerebellar peduncle 287, 297–298
inferior cervical ganglion 162–163
inferior colliculus 107, 108, 398
inferior hypophyseal artery 475
inferior longitudinal fasciculus 501
inferior occipitofrontal fasciculus 501
inferior olivary nucleus 100
infundibulum 27
inhibitory postsynaptic potential (IPSP) 36
INO *see* internuclear ophthalmoplegia
insula 77
insular cortex 205–207
intermediate acoustic striae 396
internal capsule 29, 83, 236, 238, 504
internal carotid artery 136–143
    anterior cerebral artery 139, 140
    anterior choroidal artery 138
    anterior communicating artery 139
    cavernous portion 138
    central branches 139–140
    cortical branches 140
    internal portion 138
    posterior communicating artery 138–139
internal carotid nerves 161
internal granular layer 499–500
internal pyramidal layer 500
interneurons 8
    ascending sensory pathways 202, 207
    autonomic nervous system 174
    brainstem 103
    cranial nerves 329
    histophysiology 33

552 ● ● ● **INDEX**

interneurons (*continued*)
  motor cortex 236, 240
  reticular formation 311
  spinal reflexes 174, 179
  thalamus 484
  visual system 363, 381
internuclear neurons 343–344
internuclear ophthalmoplegia (INO) 343–344
interoceptors 188
interoculomotor nucleus (of Perlia) 380, 381
interpolar subnucleus 337
interposed nuclei 288, 293, 294, 299
intersegmental reflexes 174
interstitial nucleus (of Cajal) 111, 244, 330, 415
interthalamic adhesion 27, 84, 489
interventricular foramina 20, 131
intervertebral foramen 63–64
intrafusal fibers 174–177, 181, 194
intralaminar nuclei 205, 206, 277, 489
intrasegmental reflexes 174
intrinsic fibers 288, 305
inverse myotatic reflex 177, 178
iodopsin 366
ionotropic effect 49, 50
ionotropic receptors 39, 56
ipsilateral muscle 332–33, 340
IPSP *see* inhibitory postsynaptic potential
ischemia 152
ischemic stroke 152, 492
isocortical interneurons 497–498
itch sensation 335–340

Jacksonian epileptic seizures 252, 516
Jervell–Lange–Nielsen syndrome 403
juxtarestiform body 101, 104, 297–300, 413

K+ (potassium) channels 44–46, 56, 57, 526
Klüver–Bucy syndrome 450
Korsakoff syndrome 443, 450

labyrinthine artery 144
language processing 400, 504–505
lamellar corpuscle (Pacinian corpuscle) 186, 190–193, 206, 208
lateral corticospinal tract 238, 240
lateral dorsal (LD) nucleus 446, 487
lateral geniculate nucleus (LGN)
  cerebral cortex 509
  thalamus 490–496
  visual system 370–371, 373, 376
lateral gray funiculus 67, 69
lateral hypothalamic nucleus 457, 462
lateral lemniscus 104, 107, 396
lateral medullary syndrome 112–113, 152
lateral posterior (LP) nucleus 487
lateral strabismus 114, 331
lateral sulcus 88
lateral ventricles 20, 77, 130
lateral zone 293, 314
LD *see* lateral dorsal
Leigh syndrome 47
lens (of the eye) 15, 26, 28, 58, 166, 362, 365–367
lenticular fasciculus 269
lentiform nucleus 29, 143, 256–206, 504
LGN *see* lateral geniculate nucleus
limbic lobe 77, 435–446
limbic loop 273
limbic system 434–452
  afferent and efferent fibers 441–442, 444, 445
  amygdaloid 443–446

associated structures 440–441
brainstem centers associated with function of 446
clinical case 434, 451
clinical considerations 450
dentate gyrus 434, 435, 439
endocrine, autonomic, and somatic motor systems 449–450
habenular nuclei 446
hippocampal formation 427, 434–444
hippocampus proper 434–444
hypothalamus 446, 454
intrinsic circuitry 442–443
limbic association cortex 449
limbic lobe 435, 436
locus caeruleus and dorsal raphe 446
memory 435, 443–444, 450
pathways 446–449
reticular formation 446, 450
structure and function 434–435
subcallosal, cingulate, and parahippocampal gyri 435
subiculum 440
synonyms and eponyms 451
thalamus 446
ventral tegmental area 446
lipid droplets 34
lipofuscin 34
locus caeruleus 303, 430, 446
long-term potentiation 444
Lou Gehrig disease (amyotrophic lateral sclerosis) 254, 542
lower motor neurons 174–175, 240, 241
lumbar cistern 23, 61–62, 126

macroglia 5
maculae (of the utricle and saccule) 349, 408–411
macula lutea (of the eye) 363, 365, 509
macular sparing 384–386, 518
magnetic resonance imaging (MRI) 451
magnocellular pathway 373
mammillary bodies 28, 85, 146, 442–444, 447–449, 457, 462, 463, 465, 469, 486, 491, 532
mammillary nuclei 462
mammillary peduncle 462, 463
mammillointerpeduncular tract 448, 465
mammillotegmental tract 449
mammillothalamic tract 435, 436, 448
mandibular arch 17–18
MAO *see* monoamine oxidase
masseteric "jaw jerk" reflex 339, 340
mechanoreceptors 190–194
mechanonociceptors 188
medial aperture (foramen of Magendie) 124, 131
medial forebrain bundle (MFB) 467
medial geniculate nucleus (MGN)
  auditory system 398–399
  cerebral cortex 484
  thalamus 489
medial lemniscus (ML) 92, 96–98, 102, 106, 108–109, 111
medial longitudinal fasciculus (MLF)
  brainstem 92, 96, 101–102, 104, 106, 108
  cerebellum 295
  cranial nerves 341–342
  vestibular system 406, 414–419
medial medullary syndrome 112–113
medial strabismus 342
medial reticulospinal tract 316–317
medial zone of reticular formation 312–318, 320
medulla oblongata
  brain anatomy 87
  brainstem anatomy 94, 96–102

INDEX ● ● ● **553**

caudal (inferior) medulla at the motor decussation 95, 97–98
caudal (inferior) medulla at the sensory decussation 95, 98–100
clinical considerations 111–115
mid-olivary level 100–101
neurodevelopment 23
rostral (superior) medulla 101–102, 413, 414
structures present at all levels of 96–102
vascular supply 134
medullary posterior horn 337
medullary reticulospinal tract 96, 246
medullary white matter 86
melanin pigments 34
melanocyte-stimulating hormone inhibitory factor (MSHIF) 473
melanocyte-stimulating hormone-releasing hormone (MSH) 473
melanopsin-containing ganglion cell 367
membranous labyrinth 391, 408–410
memory
  hypothalamus 469
  limbic system 435, 443–444
meninges 117–132
  cerebrospinal fluid 122, 130
  clinical case 117, 132
  clinical considerations 131
  cranial arachnoid mater 122–126
  cranial dura mater 118, 119, 127–130
  cranial meninges 118–126
  cranial pia mater 125–126
  reflections of the meningeal layer 120–122
  spinal arachnoid 126
  spinal dura mater 126
  spinal meninges 126–127
  spinal pia mater 126–127
  synonyms and eponyms 132
  vascular and nerve supply 118–120
  venous sinuses of the cranial dura mater 127–130
  ventricles of the brain 130–131
meningiomas 131
meningitis 127, 131, 132, 432
mesencephalic nucleus 104, 105, 108, 335–340
mesencephalic tract 107, 336, 338
mesencephalon 6, 20, 25–27, 87
mesocortex 82, 435, 495, 498
mesoderm 13
metabotropic effect 49, 51
metabotropic receptors 39, 56
metathalamic nuclei 488–489
metencephalon 6
  brain anatomy 87
  neurodevelopment 20, 24–25
MFB see medial forebrain bundle
MGN see medial geniculate nucleus
microfilaments 34
microglia 5, 41
microtubules 34
midbrain
  basal nuclei 262
  brain anatomy 87
  brainstem anatomy 91, 107–111
  caudal (inferior) midbrain 95, 107–110
  clinical considerations 114–115
  rostral (superior) midbrain 109–111, 115, 325
middle cerebellar peduncle 114, 296
middle cerebral artery (MCA) 141, 142, 153, 156, 249, 310, 384
middle cervical ganglion 160, 162
midget ganglion cells 367, 376
midline nuclei 487
midpons 104–106
Millard–Gubler syndrome 114

mitral cells 429
ML see medial lemniscus
MLF see medial longitudinal fasciculus
molecular layer 439
molecular (plexiform) layer 499
monoamine oxidase (MAO) 52
monoaminergic fibers 302, 303
morula 12
mossy fibers
  cerebellum 290, 300–302, 305–306
  limbic system 439
motivational loop 272–273
motor cortex 231–254
  basal nuclei 232
  cerebellum 232
  cerebral cortex 232
  clinical case 247, 254
  clinical considerations 249–252
  cortical areas controlling motor activity 232–235
  corticonuclear (corticobulbar) tract 236, 237, 242–245
  corticotectal and tectospinal tracts 236, 244–245, 253, 380
  corticoreticular fibers and reticulospinal tracts 245–247
  corticorubral, rubrobulbar, and rubrospinal tracts 236, 245, 247
  corticospinal tract 231, 236–242
  descending motor pathways 235–249
  functional classification of descending motor pathways 247–249
  internal capsule 236, 238, 243
  muscle spindles 242
  premotor cortical areas 231–235
  primary motor cortex 231–233, 251–252
  reticular formation 243, 244
  somatosensory fibers 236, 242
  synonyms and eponyms 253
  vestibulospinal tracts 236, 246, 248
motor decussation 97–98
motor innervation 176, 194
motor neurons
  ascending sensory pathways 208
  autonomic nervous system 169
  brainstem 103, 108
  cranial nerves 337–340
  descending motor pathways 231, 236, 240–249
  spinal reflexes 174–175, 177, 180–181
motor nucleus
  brainstem 99–100
  cranial nerves 325, 337, 339
  vestibular system 418–420
motor relay nuclei 490
MRI see magnetic resonance imaging
MSH see melanocyte-stimulating hormone
MSHIF see melanocyte-stimulating hormone inhibitory factor
multiform (fusiform) layer 500
multimodal association cortex 496
multimodal nuclei 491
multiple sclerosis (MS) 48, 387
multiple sleep latency test (MSLT) 478
multipolar neurons 5, 21, 39–40
muscarinic receptors 169
muscle spindles
  ascending sensory pathways 194–197
  extrafusal and intrafusal fibers 194, 195
  motor cortex 242
  simple stretch reflex 195
  spinal reflexes 175, 176
  structure and function 195
  vestibular system 406
muscle stretch reflex 176–177
muscle tone 180–181

554 ● ● ● **INDEX**

myasthenia gravis 58
myelencephalon *see* medulla oblongata
myelin 35, 42–44
myelin-associated glycoprotein (MAG) 43–44
myelin sheath gaps (nodes of Ranvier) 35, 42, 43, 46, 47
myenteric plexus (Auerbach plexus) 167, 172
myotatic mandibular reflex (jaw jerk reflex) 336, 339, 340
myotome 17, 65

Na^+/K^+ (sodium/potassium) pump 41, 44, 45
narcolepsy 462, 476, 478
N-cadherins 14
N-CAM *see* neural cell adhesion molecules
neocerebellum 295–296
neocortex 82–83, 495, 499–500
Nernst equation 44
nerve fibers
    action potentials 35, 45–47
    Goldman equation 44–45
    histophysiology 44–47
    Nernst equation 44
    postsynaptic potentials 46–47
    resting potential 44
neural cell adhesion molecules (N-CAM) 14
neural crest 15
neural stem cells 32
neural tube 14–15
neuroblasts 20, 22
neurodevelopment 11–31
    brain 19–21, 23–30, 90
    brainstem 91–96
    clinical case 11, 31
    clinical considerations 30
    embryogenesis and early development 11–14
    neurulation 14–19
    spinal cord 19–23, 30
    synonyms and eponyms of nervous system 31
neuroepithelial cells 19–21
neurofilaments 34
neuroglia 5–6
    histophysiology 32, 40–44
    myelin 42–44
    types and properties of 40–42
neuroglial tumors 47
neurohypophysis 470–472
neurolemmocytes (Schwann cells) 6, 41, 42
neuromeres 25
neuromodulators 49, 51
neurons 5
    anterior gray funiculus 67, 69
    ascending sensory pathways 185–207, 226–227
    auditory system 394
    autonomic nervous system 154–155, 168–169
    axons 33, 35
    axon terminal and its presynaptic membrane 36–38
    basal nuclei 257, 274–282
    cell body 34
    cerebellum 291–293
    cerebral cortex 496–498, 515–516
    classification of 39–40
    cranial nerves 329, 339, 343–346, 353–355
    dendrites 34–36
    descending motor pathways 235–238, 247–249
    gray commissure 67
    histophysiology 32–40
    lateral gray funiculus 67, 69
    neurites 5, 33, 34
    neurodevelopment 21

olfactory system 425
    reticular formation 311
    reticular formation and central canal 67–68
    soma (nerve cell body) 5
    spinal reflexes 173–181
    synapses 33, 35–36
    synaptic cleft and postsynaptic membrane 38–39
    thalamus 484
    vestibular system 411
    visual system 381–382
neuropeptide-releasing neurons 275
neuropeptides 52, 54–55
neurophysin 471
neuroplasticity 229
neurotransmitter substances 5, 49–59
    adenosine triphosphate and nitric oxide 57
    autonomic nervous system 156, 168–169
    basal nuclei 274–275
    biogenic amines 52–54
    classification 52–58
    clinical case 49, 59
    clinical considerations 58
    histophysiology 35, 39
    hypothalamus 469
    ionotropic effect 49, 51
    metabotropic effect 49, 51
    neuropeptides 52, 54–55
    properties of 50
    small molecule neurotransmitters 52, 55–57
neurovascular unit (NVU) 136, 152
neurulation 14–19
    neural crest and ectomesenchymal derivatives 15, 16
    neural tube 14–15
    neuroepithelial cells 19–21
    paraxial mesoderm 16–17
    pharyngeal (branchial) arches and derivatives 12, 17–19, 24, 26, 95, 105, 325, 326, 346, 355, 357
    pharyngeal (branchial) groove derivatives 13, 17, 19
    pharyngeal (branchial) pouch derivatives 17, 19
nicotinic receptors 169
nigrostriate fibers 265
nitric oxide (NO) 58
nociception
    ascending sensory pathways 188–190, 211, 220, 226–229
    cranial nerves 334–339
    spinal reflexes 178–179
nonencapsulated mechanoreceptors 190
nonspecific nuclei of thalamus 490, 491
norepinephrine 53, 163
norepinephrine-releasing neurons 228
notochordal process 14
nuclear bag fibers 176, 194
nuclear chain fibers 176, 194
nucleus accumbens 256, 258, 260
nucleus ambiguus 99, 113, 166, 350, 352–353
nucleus of Luys (centromedian nucleus) 260, 489
nucleus of Perlia (interoculomotor nucleus) 380, 381
nucleus of the posterior commissure 111, 330
nucleus proprius 68
nystagmus 307, 421–422

occipital lobe 79–80, 504–505
occipital sinuses 122, 130
ocular movement
    caloric nystagmus 421–422
    conjugate eye movements 417, 419–420
    cranial nerves 329–333, 342–346
    disconjugate eye movements 417

horizontal movements 417, 419
vertical movements 419–420, 422
vestibular control of eye movements 417
vestibular nystagmus 421, 422
vestibular system 417–420
oculomotor loop 271, 272
oculomotor nerve
autonomic nervous system 164–166
brainstem 114
cranial nerves 329–330
visual system 381
oculosympathetic paresis 171, 221, 386
olfactory-ensheathing cells 425
olfactory glands 427
olfactory nerve 329, 428
olfactory receptor neurons 425
olfactory sulcus 77
olfactory system 424–433
blood supply and drainage 431, 432
central connections 418, 429–431
cerebral cortex 498
clinical case 424, 433
clinical considerations 432
olfactory bulb 425, 426, 428, 430
olfactory nerve 428
olfactory tract 426, 430
primary olfactory cortex 430–431, 498
receptor neurons 425, 426
reticular formation 310
structure and function 424–425
synonyms and eponyms 433
transduction 427–428
oligodendrocytes 35, 41
oligodendroglia 6, 20
olivary nucleus 100
one-and-a-half syndrome 343–344
opioids 55, 228
optic chiasm 28, 29, 81, 124, 139, 329, 368, 369, 374, 378, 384, 385, 454, 461, 530
optic nerve 329, 363, 367, 368
optic neuritis 387
optic radiations 372–373
optic tract 368, 378, 384
optokinetic nystagmus 421
oral subnucleus 337
orexins 468, 476
osseous labyrinth 391
ossicles 390
otoliths 409, 530 *see also* statoconia
oxytocin 462, 465, 470–471

pain
ascending sensory pathways 198–207, 225–228
cingulate and insular cortices 206
clinical considerations 226
descending analgesia-producing pathways 227–228
first order neurons 198–201
gate control theory of pain 227
reticular formation 318, 319
second order neurons 201–204
somatosensory cortex 205–206
third order neurons 198, 204–205
visceral pain 206, 207, 211, 226
paleocerebellum 293–294
paleocortex 495, 498
pallidohabenular fibers 270
pallidonigral fibers 269
pallidosubthalamic fibers 268

pallidotegmental fibers 270
pallidothalamic fibers 269–270
Papez circuit 443
papilledema 368
parahippocampal gyrus 435, 436
parallel fibers 291–292, 306
paramedian pontine reticular formation (PPRF) 113, 244, 318
paramedian zone (of reticular formation) 312–313
parasol cell 367
parasympathetic nervous system 3, 164–167
cranial portion 164–167
parasympathetic fibers 64
parasympathetic neurons 346–348
sacral portion 167
topographic distribution 164
paravermal zone of the cerebellum 214, 295, 296, 301, 304
paravertebral ganglia 158–160
paraxial mesoderm 16–17
paresthesias 491
parietal lobe 77, 79, 504–505
Parkinson disease 58, 280–282
parosmia 432
parvocellular nucleus 314, 320
PCMLS *see* posterior column–medial lemniscal system
pedunculopontine nucleus 273
pedunculopontine tegmental nucleus 266, 274, 313
pelvic functions
autonomic nervous system 169–171
defecation 170–171
erection and ejaculation 171
urine retention and urination 169, 170
perforant pathway 448
periaqueductal gray (PAG) matter
ascending sensory pathways 203
brainstem 105–111
cranial nerves 329–331
hypothalamus 465
reticular formation 315
perilymph 393, 408
peripheral nerve lesions 218
peripheral nervous system (PNS) 3, 9–10
ascending sensory pathways 230
autonomic nervous system 3, 9–10
histophysiology 32, 40–44
somatic nervous system 3, 9, 10
peritrichial nerve endings 190, 191
periventricular bundle 466
petrosal sinuses (superior and inferior) 122, 127–128, 130, 135, 149–151
petrosquamous sinus 130
phantom limb pain 226
phantosmia 432
pharyngeal (branchial) arches and derivatives 12, 17–19
pharyngeal (branchial) groove derivatives 19, 31
pharyngeal (branchial) pouch derivatives 19, 31
photopic vision 365
pia mater 26, 125
Pick bundle (of the uncinate fasciculus) 84, 244, 501
pineal gland 92, 136, 259, 466, 482
pituitary gland (hypophysis)
adenohypophysis 472–475
hypothalamus 469–475
neural control 470
neurodevelopment 28
neurohypophysis 470–472
non-neural control 470, 471
plasticity 444
PNS *see* peripheral nervous system

556 ● ● ●   **INDEX**

polymodal nociceptors 188
polymorphic layer 439, 442
pons
   brain anatomy 87
   brainstem anatomy 93, 102–107
   caudal pons 102–104
   clinical considerations 113–114
   mid pons 96, 105–106, 112, 336, 337
   rostral pons 106–107
   vascular supply of 144
pontine paramedian reticular formation (PPRF) 113, 244, 318, 417
pontine reticulospinal tract 246
pontocerebellum 295–296
pontomesencephalic level of the autonomic nervous system 149
postcentral gyrus 77, 79, 140, 143, 204, 205, 206, 213, 226, 242, 334,
     335, 337, 338, 488, 490, 496, 499, 508, 510, 511, 523
postcentral sulcus 77, 169
postcommissural fibers 448
posterior cerebellar lobe 295–296
posterior column-medial lemniscal system (PCML) 186, 187, 194,
     207, 213, 221
posterior commissure
  of cerebrum 83
  of spinal cord 111
posterior hypothalamic nuclei 457, 468, 475
posterior inferior cerebellar artery (PICA) 143, 152, 309, 341
posterior parietal area 244, 508
posterior parietal cortex 213, 233, 235, 271
posterior perforated substance 93, 455
posterior root lesions 218
posterior spinal artery 70–71, 225
posterior trigeminal lemniscus 338
posterolateral prefrontal cortex 400
posterolateral tract (of Lissauer) 67, 68, 178, 338
postganglionic fibers 9
   parasympathetic nervous system 164–167, 469
   sympathetic nervous system 160–161, 171
postsynaptic membrane 38–39
postsynaptic potentials (PSP) 46–47
precentral gyrus 77, 233, 499
precentral sulcus 77
precerebellar reticular nuclei 314
prefrontal cortex 205, 512, 518
preganglionic fibers 9
   cranial nerves 330
   parasympathetic nervous system 164–167, 330
   sympathetic nervous system 158–160
premotor cortex (PMC) 232, 233, 508
pressure sensation 208–212, 338
presynaptic inhibition 227, 228
presynaptic membrane 36–38
primary auditory cortex 399–401, 496, 500, 510, 512
primary motor cortex 231–233, 505–508
primary olfactory cortex 430–431, 498
primary somatosensory cortex 204, 508–509
primary visual cortex 373, 509
   cerebral cortex 496, 501, 517–518
   columnar organization 376–377
   posterior and anterior stream 373, 375
   visual association cortical areas 373, 375
   visual field representation 372, 374
   "where" and "what" pathways 373
principal mammillary fasciculus, 465
principal nucleus 336
projection fibers 83–84, 504
projection neurons 484
prolactin-inhibitory factor (PIF) 473
prolactin-releasing hormone (PRH) 472

proprioception
   ascending sensory pathways 186, 188, 207–214
     cranial nerves 334–339
prosencephalon 20, 27–30
proteolipid protein (PLP) 43
protoplasmic astrocytes 5
pruriceptors 190
PSP *see* postsynaptic potentials
pterygopalatine ganglion 347
pulvinar nucleus 487
pupil of the eye 362
pupillary constriction 380
pupillary dilation reflex 379, 386
pupillary light reflex 377–379, 386
pure motor hemiparesis syndrome 113
Purkinje cells (of the cerebellum) 27, 57, 86, 288, 290, 292, 293, 294,
     298, 299, 300, 303, 304–305, 306, 307
putamen 256–258, 263–265
pyramid of the brainstem
   brain anatomy 88
   cerebral cortex 496–497
   motor cortex 238, 242
pyramidal decussation 98, 238
pyramidal layer of the cerebral cortex 235, 439

radial glial cells (Müller cells) 363, 365
radicular arteries 71–72
RAS *see* reticular-activating system
receptor molecules 12
reciprocal inhibition 177–178
red nucleus 26, 110, 111, 115, 245, 298, 299, 305, 314
referred pain 10, 226
reflex arcs 32–33
RER *see* rough endoplasmic reticulum
restiform body 99, 101, 297
resting potential 44
reticular-activating system (RAS) 203
reticular formation 67, 96, 310–322
   afferent fibers 314–315
   associated nuclei 314
   autonomic nervous system 315, 318
   clinical case 310
   clinical considerations 321
   consciousness 320–321
   cranial nerve function 320
   efferent fibers 315
   endocrine system 320
   eye movements 318
   functions 315–321
   hypothalamus 464
   morphology 311–312
   motor activity 316–318
   motor cortex 232
   neurons 311
   nuclei 312
   pain sensation 318, 319
   synonyms and eponyms 322
   vestibular system 414
   visual system 368
   zones 312–314
reticulobulbar tract 314
reticulocerebellar fibers 301
reticulospinal tract
   cerebellum 287
   motor cortex 245, 247
   reticular formation 313
retina 362–364
retinal ganglion cells, 367, 370

retinoic acid 14
retinosuprachiasmatic tract 464
retrograde amnesia 450
Rexed (spinal) laminae 68, 69, 70, 197, 198, 313
rhinitis 432
rhodopsin 366
rhombencephalon 20, 23, 25, 26
rods and cones (of the retina) 361–366
rostral medulla oblongata 101–102
rostral pons 106–107
rostral spinocerebellar tract 216, 217
rotatory nystagmus 421
rough endoplasmic reticulum (RER) 34, 38–39
rubrospinal tract 96, 111, 245, 299

saccule 408
sagittal sinuses 124, 128
St. Vitus' dance 280
salivatory nuclei (superior and inferior) 101, 103, 430, 469
SARS-Cov-2 136, 425
scala media (cochlear duct) 391–394, 404
scala tympani (tympanic duct) 391, 392, 393, 394, 404
schwannoma 404
sclera 362
sclerotome 17, 65
scotopic vision 365–366
sea sickness 411
second order neurons
    ascending sensory pathways 197, 201–204, 206, 212
    auditory system 394
seizures 252, 450, 451, 513
semicircular canals 391, 407, 411
semicircular ducts 408–409
sensorineural deafness 403
sensory association areas 510–512
sensory cortex 3, 315, 499, 504
sensory decussation 98–101
sensory ganglia 157
sensory innervation 176, 194
sensory-motor loop 271
sensory neurons
    ascending sensory pathways 195
    autonomic nervous system 154–155, 168
    spinal reflexes 174, 178
sensory nucleus 105, 336, 338, 339
sensory receptors
    ascending sensory pathways 187–197
    classification by stimulus modality 188–194
    classification by stimulus source 188
    muscle spindles and tendon organs 194–197
    pain pathways from the body 198–201
    tactile sensation and proprioception 207–212
septal area 435
septohippocampal tract 448
septum pellucidum 77, 130, 476
SER see smooth endoplasmic reticulum
serotonin 53, 54
serotonin-releasing neurons 227–228
Shy–Drager syndrome 171
sigmoid sinus 130
simple stretch reflex 195
sinusitis 432
skeletal muscle 176
skin 219
sleep–wake cycle 468, 478
SMA see supplementary motor area
small molecule neurotransmitters 52, 55–57
smooth endoplasmic reticulum (SER) 34, 38–39
solitary tract 98, 104

soma 5
somatic motor system 450
somatic nervous system 3, 9, 10, 155
somatosensory cortex 205–206, 208, 209, 213–214
somatosensory fibers 236, 242
somatostatin 54, 473
somatotropin-releasing hormone (SRH) 472
somatosensory association area 511, 517
somites 23
sonic hedgehog (SHH) 14, 30
sound attenuation reflex 401, 402
space sickness 411
spasticity 181, 249, 515–516
special somatic afferent fibers (SSA) 95, 326
special visceral afferent fibers (SVA) 95, 326, 352, 428
special visceral efferent fibers (SVE) 95, 325, 326, 352, 353
specific relay nuclei 490
sphenoparietal sinus 127
sphincter pupillae muscle 329–331, 363, 377–378
spina bifida 15, 31, 73
spinal cord 6–8, 60–74
    ascending sensory pathways 200, 218, 223–225
    basal plates, alar plates, and posterior root ganglia 8, 9, 21
    cerebellum 295
    classification of fiber components 66–67
    clinical case 60, 74
    clinical considerations 30, 73
    development of modalities 23
    external morphology 62–63
    further differentiation of basal and alar plates 21–22
    general structure 61–62
    gray matter 67–70
    histodifferentiation of neuroblasts 21, 22
    internal morphology 67–70
    lumbar cistern 61–62
    modalities of spinal nerves 66
    molecular basis of spinal cord differentiation 22–23
    morphology 61–67
    motor cortex 241–242
    neurodevelopment 19–23, 30
    segmentation and roots of spinal nerves 63–65
    synonyms and eponyms 74
    vascular supply 70–73, 134–135
    vestibular system 413–417
    white matter 67, 70
spinal ganglion 63, 173, 198
spinal lamina I-VII, 68
spinal levels 157
spinal nerves
    classification of fiber components 66–67
    clinical considerations 218
    dermatomes 64–65
    modalities of 66
segmentation and roots of 63–65
spinal nucleus 337
spinal reflexes 173–182
    alpha-gamma coactivation 180–181
    autogenic inhibition 177, 178
    clinical case 173, 182
    components and classification reflexes 173–174
    crossed extension reflex 179
    flexor reflex 178–179
    lower motor neurons 174–175
    muscle stretch reflex 176, 177
    muscle tone and the gamma loop 180
    reciprocal inhibition 177–178
    skeletal muscle innervation 175
    skeletal muscle receptors 175–177
    synonyms and eponyms 182

558 ● ● ● **INDEX**

spinal shock 73
spinobulbar fibers 204
spinocerebellar tract 97, 99, 214, 215, 217
spinocerebellum 295, 296, 298
spinohypothalamic fibers 203, 464
spinomesencephalic fibers 203
spinoreticular tract 197, 199, 202, 203
spinotectal fibers 109, 203
spinothalamic tract
    ascending sensory pathways 197, 198, 201–202
    brainstem 97, 104, 106, 109
spiny stellate cells 497, 499
spiral ganglion 349
spiral organ (of Corti) 388, 392, 402
splanchnic nerves 163–164
splenium 80, 83, 502
spontaneous nystagmus 421
SSA see special somatic afferent
stapedius muscles 390
statoconia (otoconia, otoliths) 408–410, 409
statoconial membrane 409
stellate cells 289, 293, 307, 497, 499
straight sinus 129–130
stretch reflex 308
stria medullaris of prethalamus 450, 466
stria medullaris thalami 466
stria terminalis 445, 449, 463, 482
striate area 509
striatopallidal fibers 267
striatum 263–267
stripe of Gennari (occipital line) 373, 500
stroke 152, 491
stylomastoid foramen 346
subacute combined degeneration 223
subcallosal gyrus 80, 435, 440
subiculum 440, 443
submandibular ganglion 166, 347
submucosal (Meissner) plexus 167
substance P 54–55, 228, 338
substantia gelatinosa 68, 106, 227, 228
substantia nigra
    basal nuclei 260, 261, 273
    brainstem 107
subthalamic nucleus 258, 260, 266, 273–274
subthalamopallidal fibers 268
subthalamus 85, 262
superior cerebellar artery (SCA) syndrome 114
superior cerebellar peduncle 104, 107, 110, 296
superior cervical ganglion 162
superior colliculus 368, 371–372
superior hypophyseal artery 474
superior longitudinal fasciculus 501, 517
superior occipitofrontal fasciculus 501
superior olivary complex 396, 398
superior sagittal sinus 121, 124, 125, 128
supplementary motor area (SMA) 233, 235, 508, 516
suprachiasmatic nucleus 461
supramarginal gyrus 77, 511
supraopticohypophyseal tract 462, 463, 465–466, 470, 471
suprasegmental reflexes 174
SVA see special visceral afferent
SVE see special visceral efferent
Sydenham chorea 280
sympathetic fibers 64
sympathetic nervous system 3, 157–164
    abdominolumbar region 164
    cephalic region 161
    cervical region 162

functions and organization 154–156
    hypothalamus 469
    postganglionic fibers 160–161, 171
    preganglionic fibers and paravertebral ganglia 158–160
    right and left chain ganglia 158
    thoracic region 163
    topographic distribution 161–164
synapses 5, 33, 35–36
synaptic cleft 38–39
synaptic vesicles 37
syphilis 223
syringomyelia 73, 223, 224

tabes dorsalis 73, 223
tactile corpuscles 192–193
tactile sensation
    ascending sensory pathways 192, 207–214
    cranial nerves 334–339
tardive dyskinesia 280
taste sensation 346–347, 353
Tay–Sachs disease 47
tectospinal tract 96, 244–246
tectum 107
tegmentum
    brainstem anatomy 96, 102, 108
    clinical considerations 113–115
telencephalon 6, 20, 29–30, 157
teloreceptors 188
temperature pathways 206–207
temperature regulation 462, 468
temperature sensation 106, 218, 226, 491
temporal lobe 77, 79, 451, 490–491
tendon organs (TOs) 176, 196
    ascending sensory pathways 196
    cerebral cortex 516
    spinal reflexes 176, 177
tensor tympani muscle 18, 105, 334, 337, 339, 340, 389, 390, 401, 402, 538
tentorium cerebelli 120
thalamic pain 225–226
thalamic reticular nucleus 483
thalamocortical fibers 443, 483–484
thalamocortical projections 370
thalamohypothalamic tract 464
thalamus 480–492
    anatomical classification of thalamic nuclei 484–490
    anatomy 482–483
    anterior nuclear group 486
    basal nuclei 257, 260, 261, 272, 273
    borders 480, 481
    cerebral cortex 495
    clinical case 480, 492
    clinical considerations 491
    connections 485–486
    functional classification of thalamic nuclei 490–491
    functions 480
    intralaminar nuclei 489
    lateral nuclear group 487–489
    limbic system 446
    medial nuclear group 486–487
    nuclei 484–491
    synonyms and eponyms 492
    thalamic reticular nucleus 489
    vestibular system 416
thermonociceptors 188
thermoreceptors 190
third order neurons 204–206, 212, 213, 370, 396
thyroid-stimulating hormone-releasing hormone (TRH) 473

tic douloureux 340, 360
tinnitus 403
titubation 307
transcription factors 12, 13, 15, 30
transverse sinuses 122, 130
transverse temporal gyrus 509–510
trigeminal (Meckel) cave 120, 122, 334
trigeminal nerve 334–340
    brainstem anatomy 96, 104–106
    branchiomotor innervation 334, 339
    cerebellum 304
    clinical considerations 340, 360
    masseteric ("jaw jerk") reflex 336, 339
    motor cortex 243
    neurodevelopment 18
    nuclei 335–337
    tracts 337–338
    trigeminal system/pathways 334–335, 338–339
    visual system 379, 381–382
trigeminal neuralgia 340, 360
trilaminar germ disc 12–14
trochlear nerve palsy 331–333, 344–345
trochlear nucleus 108
tryptophan 54
tuber cinereum 85, 146, 454, 457, 462
tuberculum cuneatus 88
tuberculum gracilis (gracile tubercle) 88, 98
tuberohypophyseal tract 462, 472–475
tufted cells 429, 430
tympanic duct 391, 392, 393, 394
tympanic membrane 389, 391
tyrosine 53

uncinate fasciculus 501
uncus 80, 146, 261, 430, 434, 435, 439, 444, 495, 498
unilateral lesion of primary motor cortex 250
unimodal association cortex 496
unipolar brush cells 292–293
unipolar neurons
    cell bodies 353
    histophysiology 39–40
    neurodevelopment 21
    visual system 381–382
upper motor neurons (UMN)
    cerebral cortex 515–516
    descending motor pathways 231–238, 240
urine retention and urination 169, 170
utricle 408–409, 411
uvea 362

vagus nerve 166–167, 353–355
varicosities 160
vascular supply
    arterial supply of the brain 135–148
    basilar artery 144–146
    blood–brain barrier and neurovascular unit 135–136
    brain 80–81, 135–148
    brainstem 111–112, 148, 149, 151
    central nervous system 133–153
    cerebellar veins 150
    cerebral arterial circle (of Willis) 146–148
    cerebral veins 150
    circumventricular organs 136
    clinical case 133, 153
    clinical considerations 111–115, 152
    internal carotid artery 136–143
    meninges 118–120, 123, 143
    middle cerebral artery 140–143

    spinal cord 70–73, 134–135
    synonyms and eponyms 153
    venous drainage of the brain 149–151
    vertebral artery 143–144
vasopressin (antidiuretic hormone, ADH) 465, 470–472
venous sinuses
    anterior inferior group 127–128
    cranial dura mater 127–130
    posterior superior group 128–130
ventral acoustic striae 394
ventral amygdalofugal (amygdaloid) pathway 445
ventral anterior (VA) nucleus 487, 488
ventral basal nuclei 256
ventral lateral (VL) nucleus 271, 487, 508
ventral medial (VMpo) nucleus 202, 205
ventral posterior inferior (VPI) nucleus
    ascending sensory pathways 202, 204, 212
    thalamus 488
    vestibular system 417
ventral posterior lateral (VPL) nucleus
    ascending sensory pathways 187, 202, 206, 209, 212
    cerebral cortex 496, 504
    reticular formation 315
    thalamus 487–488, 491
    vestibular system 416
ventral posterior medial (VPM) nucleus
    cerebral cortex 496, 504
    cranial nerves 335–338
    thalamus 487–488, 491
ventral spinocerebellar tract 214
ventral tegmental area 260, 266, 446
ventral trigeminal lemniscus 489
ventral trigeminothalamic tract 336
ventricles (of the brain) 6
ventromedial nucleus 462
vermal zone 295, 304
vertebral artery 143–144
vertical nystagmus 421
vestibular (Reissner) membrane 392, 393, 394, 408
vestibular nerve 104
vestibular nucleus 101, 104
vestibular nystagmus 421, 422
vestibular system 405–423
    afferent projections to vestibular nuclei 415
    balance maintenance 408–412
    caloric nystagmus 421–422
    central pathways 413–417
    cerebellum 293–296, 304
    cerebral cortex 510, 518
    clinical case 405, 423
    control of ocular movements 417–420
    ear anatomy and physiology 405–406
    efferent projections from vestibular nuclei 415–416
    functions and organization 405–406
    hair cells 410, 411
    mechanisms of action of vestibular apparatus 411, 412
    membranous labyrinth 408–410
    osseous labyrinth 407–408
    receptors 408
    sensory apparatus 406–411
    statoconia 408–410
    synonyms and eponyms 423
    termination of first order afferent fibers 414
    thalamic and cortical projections 416–417
    vestibular nerve 411, 413, 415
    vestibular nuclear complex 414
    vestibular nystagmus 421, 422
vestibule 408

560 ● ● ● **INDEX**

vestibulocerebellum 294, 298, 299
vestibulocochlear apparatus 406–411
vestibulocochlear nerve 348–349, 393
vestibulo-ocular reflex 417–419
vestibulospinal tract 246, 248
vibratory sensation 208–212
visceral branches (of the sympathetic nervous system) 163
visceral pain 206, 207, 211, 226
visual association areas 511, 512, 518
visual system 361–387
    central visual pathways 367–370
    cerebral cortex 496, 501, 508–509, 511, 517–518
    clinical case 361, 387
    clinical considerations 384–386
    columnar organization of the cortex 376–377
    convergence accommodation reflex 379–381
    corneal blink reflex 381–382
    external layer of the eye 362
    eye anatomy and physiology 361–367
    eye movement 318, 329–332, 342–345, 417–423
    geniculocalcarine tract 372
    internal layer of the eye 363, 364
    intermediate layer of the eye 362–363
    lateral geniculate nucleus 365, 367, 369
    lens 366–367
    molecular biology of rods and cones 365, 366
    neurodevelopment 28, 29
    optic radiation 372–373
    posterior and anterior stream 373, 375
    primary visual cortex 373, 376–377, 496, 501, 517–518
    pupillary dilation reflex 379, 386
    pupillary light reflex 377–379
    striate area, 373
    superior colliculus 368, 369
    synonyms and eponyms 387
    vestibular system 417–420
    visual fields 382, 383
    visual reflexes 377–383
    "where" and "what" pathways 373
visuomotor loop 271, 272
vitamin B_{12} deficiency 223
voltage-gated calcium channels 37–38
voltage-gated K^+ (potassium) channels 44, 46
voltage-gated Na^+ (sodium) channels 35, 45–46, 366, 534

Wallenberg (lateral medullary) syndrome 112–113, 152, 259, 319
water intake 469, 476
Weber syndrome 114–115
Wernicke (sensory speech) area 79, 388, 400, 507, 511–512, 517
white matter 8–9
    brain anatomy 77, 83–84, 86
    cerebellum 287–288
    cerebral cortex 495
    funiculi 70, 71
    internal morphology of the spinal cord 67, 70
    neurodevelopment 21
    thalamus 482
Wilson disease (hepatolenticular degeneration) 282
wingless type 1 (WNT1) gene 30
withdrawal reflex 178–179
wrong-way eyes syndrome 113–114

zona incerta 469, 476
zygote 11–12